AA002461

6th International Symposium on Immersion Lithography Extensions 2009

Prague, Czech Republic
22-23 October 2009

ISBN: 978-1-61782-107-3

Printed from e-media with permission by:

Curran Associates, Inc.
57 Morehouse Lane
Red Hook, NY 12571

Some format issues inherent in the e-media version may also appear in this print version.

Copyright© (2009) by SEMATECH
All rights reserved.

Printed by Curran Associates, Inc. (2011)

For permission requests, please contact SEMATECH
at the address below.

SEMATECH
2706 Montopolis Drive
Austin, Texas 78741

Phone: (512) 356-3500
Fax: (512) 356-7848

www.sematech.org

Additional copies of this publication are available from:

Curran Associates, Inc.
57 Morehouse Lane
Red Hook, NY 12571 USA
Phone: 845-758-0400
Fax: 845-758-2634
Email: curran@proceedings.com
Web: www.proceedings.com

TABLE OF CONTENTS

KEYNOTE

Optical Lithography Below 20nm 1
    G. Willson

Line-Edge Roughness and the Ultimate Limits of Lithography 64
    C. Mack

SESSION: DOUBLE PATTERNING I

Density Limits in Logic Metal1Using Double Patterning 89
    V. Wiaux, S. Verhaegen, G. Fenger, P. Wong

Double Patterning Lithography Overlay Components 112
    V. Nagaswami, J. Sinha, S. Veeraraghavan, F. Laske, A. Golotsvan, D. Tien, P. Izikson, J. Robinson, C-S. Koay,
    M. Colburn, S. Holmes, K. Petrillo

SESSION: DOUBLE PATTERNING MATERIALS AND PROCESSING

Development of Freezing Free LLE Process 139
    T. Nakamura, R. Takasu, M. Takeshita, J. Yokoya, Y. Yoshii, H. Saito, K. Ohmori

Exploration of New Resist Chemistries and Process Methods for Enabling Dual-Tone Development 159
    C. Fonseca, M. Somervell, S. Scheer, Y. Kuwahara, K. Nafus, R. Gronheid, S. Tarutani

**Litho-Freeze-Litho-Etch (LFLE) Enabling Coat/Develop Track Process and Freeze CD Tuning Bake
for >200wph Double Patterning** 181
    C. Pieczulewski, C. Rosslee

**Printing the Contact and Metal Layers for the 32 and 22 nm Node: Comparing Positive and Negative
Tone Development Process** 203
    J. Bekaert, L. Van Look, V. Wiaux, V. Truffert, M. Maenhoudt, G. Vandenberghe, M. Reybrouck, S. Tarutani

SESSION: PATTERN SPLIT TECHNIQUES

Double Patterning OPC and Design for 22 nm to 16 nm Device Nodes 234
    K. Lucas, C. Cork, A. Miloslavsky, G. Luk-Pat, X. Li, L. Barnes, W. Gao, V. Wiaux

**Wafer and Simulation Study Comparing 5 LELE Decomposition Algorithms for Both Compliant and
Non-Compliant Layouts** 274
    G. Fenger, P. LaCour, A. Tritchkov, S. Komirenko, V. Wiaux

A New Approach Towards Pitch Division for 193 nm Immersion Lithography 294
    X. Gu, C. Bates, Y. Cho, E. Costner, T. Nagai, T. Ogata, G. Willson, A. Sundaresan, N. Turro, R. Bristol, P.
    Zimmerman

SESSION: IMMERSION LITHOGRAPHY OPTIMIZATION

Double Pattering: Coater/Developer Evolution or Revolution 310
    A. Viswanathan, M. Nakano, R. Crowell, S. Scheer

Challenges Building a 22nm Node 6T-SRAM Cell Using Immersion Lithography 331
    M. Ercken, E. Altamirano-Sanchez, C. Baerts, S. Brus, J. De Backer, M. Demand, C. Delvaux, N. Horiguchi, S.
    Locorotondo, T. Vandeweyer, A. Veloso, S. Verhaegen

**Freeform Illumination Sources: An Experimental Study of Source-Mask Optimization for 22nm
SRAM Cells** 351
    J. Bekaert, B. Laenens, S. Verhaegen, L. Van Look, D. Trivkovic, F. Lazzarino, G. Vandenberghe, P. van
    Adrichem, R. Socha, M. Mulder, S. Baron, M.-C. Tsai, K. Ning, S. Hsu, A. Bouma, E. van der Heijden, K. Schreel,
    R. Carpaij, M. Dusa, J. Zimmermann, P. Graupner, C. Hennerkes

Advances in Modeling and Optical Performance for DOEs used for OAI in Immersion Lithography 388
    J. Carriere, J. Stack, A. Kathman, M. Himel

SESSION: DOUBLE PATTERNING II

Technical and Manufacturing Challenges and the Prospects for HVM using ArF Pitch Division 426
S. Sivakumar

Lithography on the Edge 468
D. Medeiros

Towards 26nm HP: Advances in Litho-Process-Litho 500
P. Wong, D. Vangoidsenhoven, G. Murdoch, M. Maenhoudt, S. Verhaegen, V. Wiaux

Spacer Deposition on Hardened Resists Patterns for Cost Effective SADP 526
S. Gaugiran, I. Guerin, C. Ratin, S. Barnola, K. Menguelti, E. Pargon, A. Pikon, T. Cardolaccia, Y. Bae, Y. Liu, P. Trefonas

SESSION: LITHOGRAPHY TOOLS AND INTEGRATION

Immersion Scanner Nikon NSR-S620 for Double Patterning 547
Y. Shibazaki, M. Hamatani, K. Hirano, J. Ishikawa, Y. Iriuchijima, S. Owa

Improvements in Exposure Systems and Applications to Enable the Ultimate Extension with Immersion Lithography 573
F. de Jong, B. Vleeming, M. Mulder, A. Engelen, J. Mulkens

Reliability Report of High Power Injection Lock Laser Light Source for Double Patterning and Double Patterning ArF Immersion Lithography 596
H. Tanaka, H. Tsushima, M. Yoshino, T. Kumazaki, H. Watanabe, S. Matsumoto, H. Umeda, Y. Kawasuji, T. Suzuki, S. Tanaka, A. Kurosu, T. Matsunaga, J. Fujimoto, H. Mizoguchi

Characterization of Direct Alignment for a LFLE Double Patterning Process 618
D. Laidler, P. Leray, S. Cheng, M. Doytcheva, M. Tenner, R. van Haren

SESSION: ALTERNATIVE LITHOGRAPHY OPTIONS

Can NanoImprint and ML2 Thrive in a 193i World? 639
L. Litt

Readiness of Multiple-E-Beam Maskless Lithography 665
J. Chen, T. Fang, S. Lin, S. Chang, F. Krecinic, W. Wang, B. Lin, B-J. Kampherbeek, G. de Boer, B. Schipper, P. Scheffers, C. van den Berg, M. Wieland

Full-Field Liquid Immersion Interference Lithography 684
J. Jacob, J. Hoffnagle, J. Burnett, E. Benck, D. Armstrong, A. Smith

Interference Assisted Lithography (IAL): A Way to Contain the Lithography Costs for the 32nm and 22nm Half-Pitch Device Generations 703
R. Hendel, D. Markle, J. Petersen, A. Barada, Z. Rao

SESSION: ADVANCES IN PHOTORESISTS AND MATERIALS

Resist Material for Negative Tone Development Process 732
S. Tarutani

Polysulfone Based Non-CA Resists for 193 nm Immersion Lithography: Effect of Increasing Polymer Absorbance on Sensitivity 758
I. Blakey, L. Chen, Y-K. Goh, K. Lawrie, A. Whittaker, E. Piscani, P. Zimmerman

Inorganic Photoresists Based on Hafnium Oxide 782
M. Trikeriotis, W-J. Bae, E. Schwartz, M. Krysak, C. Ober, E. Giannelis, N. Lafferty, P. Xie, B. Smith, P. Zimmerman

POSTERS

A PECVD Bi-layer ARC Solution for Immersion Lithography 801
M. Lin, B. Tang, M. Seamons, R. Ding, L. Miao, H. Dai

Development of Reverse Materials and BARCs for Double Patterning Process 817
Y. Sakaida, H. Yaguchi, R. Sakamoto, B-C. Ho

Bottom-Anti-Reflective Coatings (BARC) for LFLE Double Patterning Process 829
R. Sakamoto, T. Endo, B-C. Ho, S. Kimura, T. Ishida, M. Kato, N. Fujitani, R. Onishi, Y. Hiroi, D. Maruyama

Understanding the Relationship Between the Evaporation Behavior of Water Droplets and the Formation of Stains on Polymer Surfaces 841
Jung-Hoon Kim, Jae Hyun Kim, Wang-Cheol Zin

Process Parameter Influence to Negative Tone Development Process for Double Patterning 842
S. Tarutani, S. Kamimura, J. Yokoyama

Defectivity Investigation with Point-of-Use Filtration Parameter Changes 855
J. Braggin, W. Schollaert, X. Buch, K. Hoshiko

Robust Material and Processes for Double Patterning 856
H. Yaegashi

Tethered Naphthalene Derivatives as Sensitizers for Sequential Two Photon Photoacid Generators for Double Exposure Photolithography 875
P. Zimmerman, A. Sundaresan, Y. Li, N. O'Connor, S. Jockusch, N. Turo, T. Nagai, T. Ogata, Y. Cho, S. Lee, A. Berro, X. Gu, G. Willson

Alternatives to Chemical Amplification for Sub-32nm Photoresists 894
B. Smith, T. Smith, B. Baylav, M. Zhao, R. Yin, P. Xie, C. Scholz, P. Zimmerman

High Fluence Testing of Optical Materials for 193nm Lithography Extensions Applications 909
P. Zimmerman, V. Liberman, S. Palmacci, G. Geurtsen, M. Rothschild

Author Index

Optical Lithography Below 20nm ?

Grant Willson

Department of Chemical Engineering
Department of Chemistry

The University of Texas
Austin, Texas 78712

http://willson.cm.utexas.edu

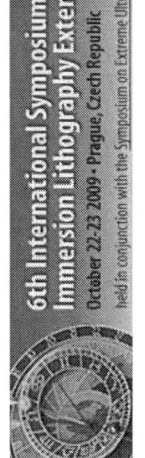

6th International Symposium on
Immersion Lithography Extension:
October 22-23 2009 · Prague, Czech Republic
held in conjunction with the Symposium on Extreme Ultraviolet Lith

Our Job... Make this smaller!

Moore's Law

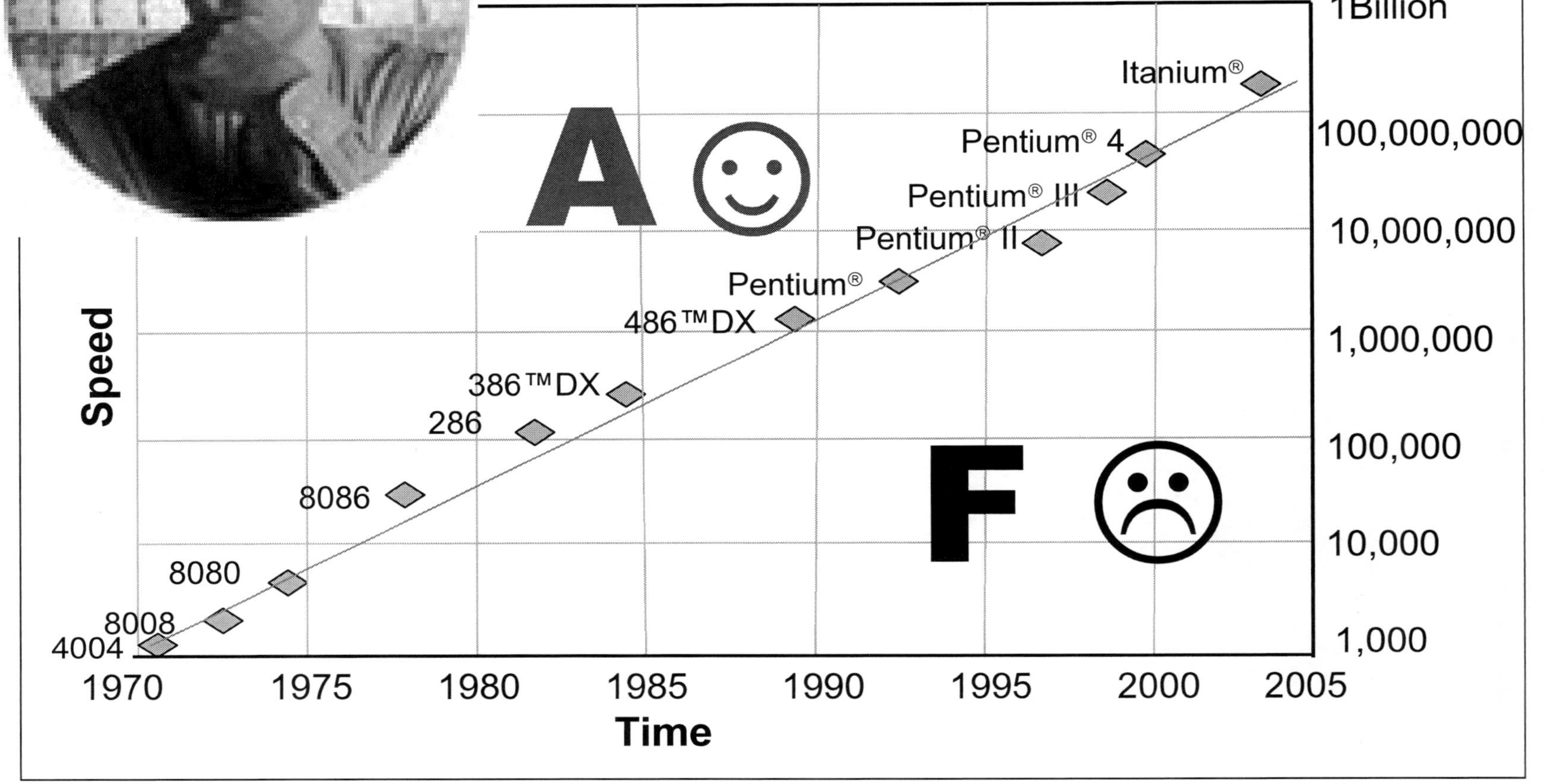

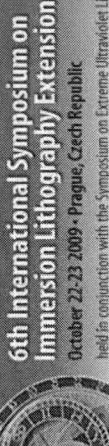

Life of the Chemist in Microelectronics

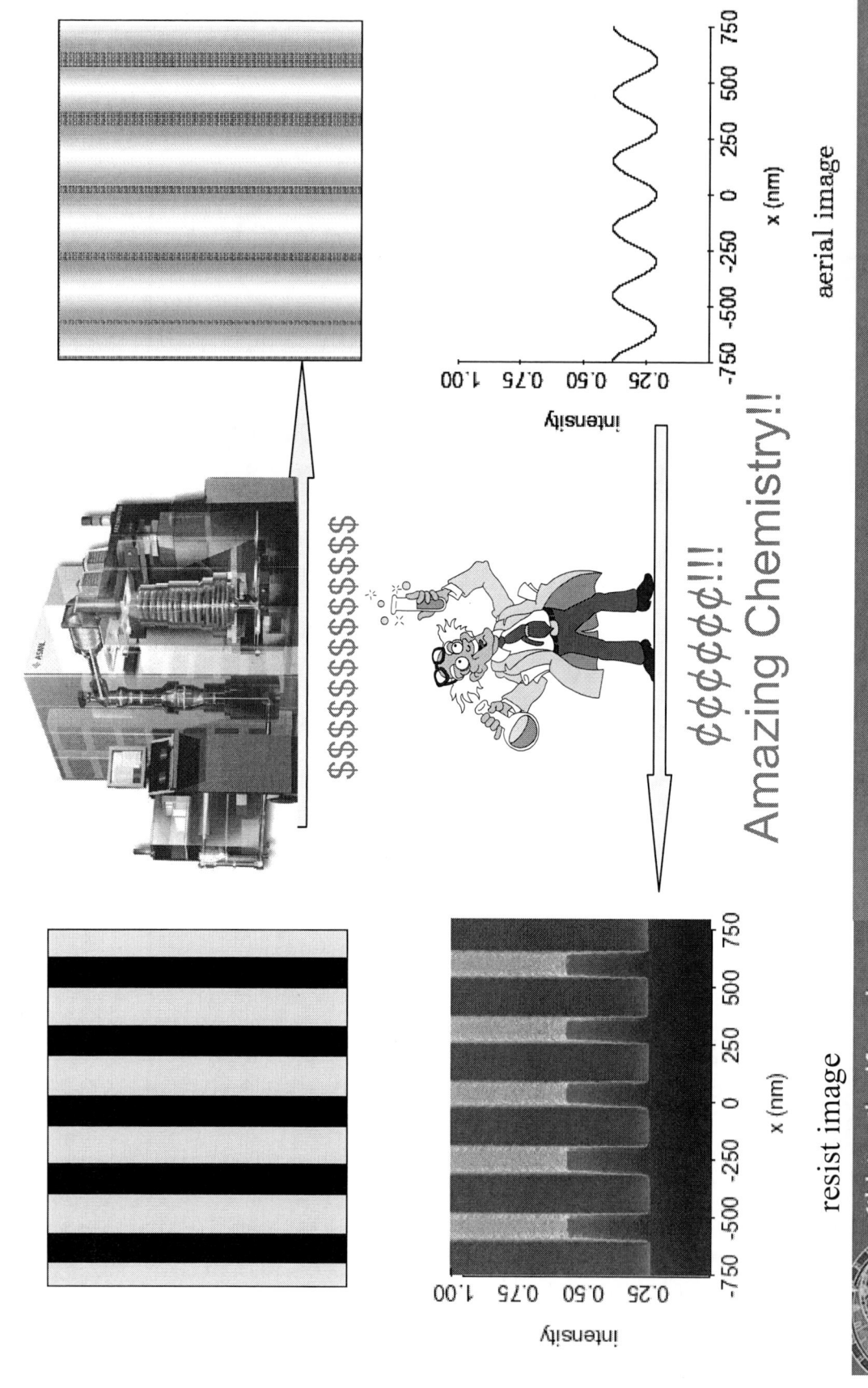

$$$$$$$$$$$$

¢¢¢¢¢!!!
Amazing Chemistry!!

aerial image

resist image

6th International Symposium on
Immersion Lithography Extension:
October 22-23 2009 • Prague, Czech Republic
held in conjunction with the Symposium on Extreme Ultraviolet Lit

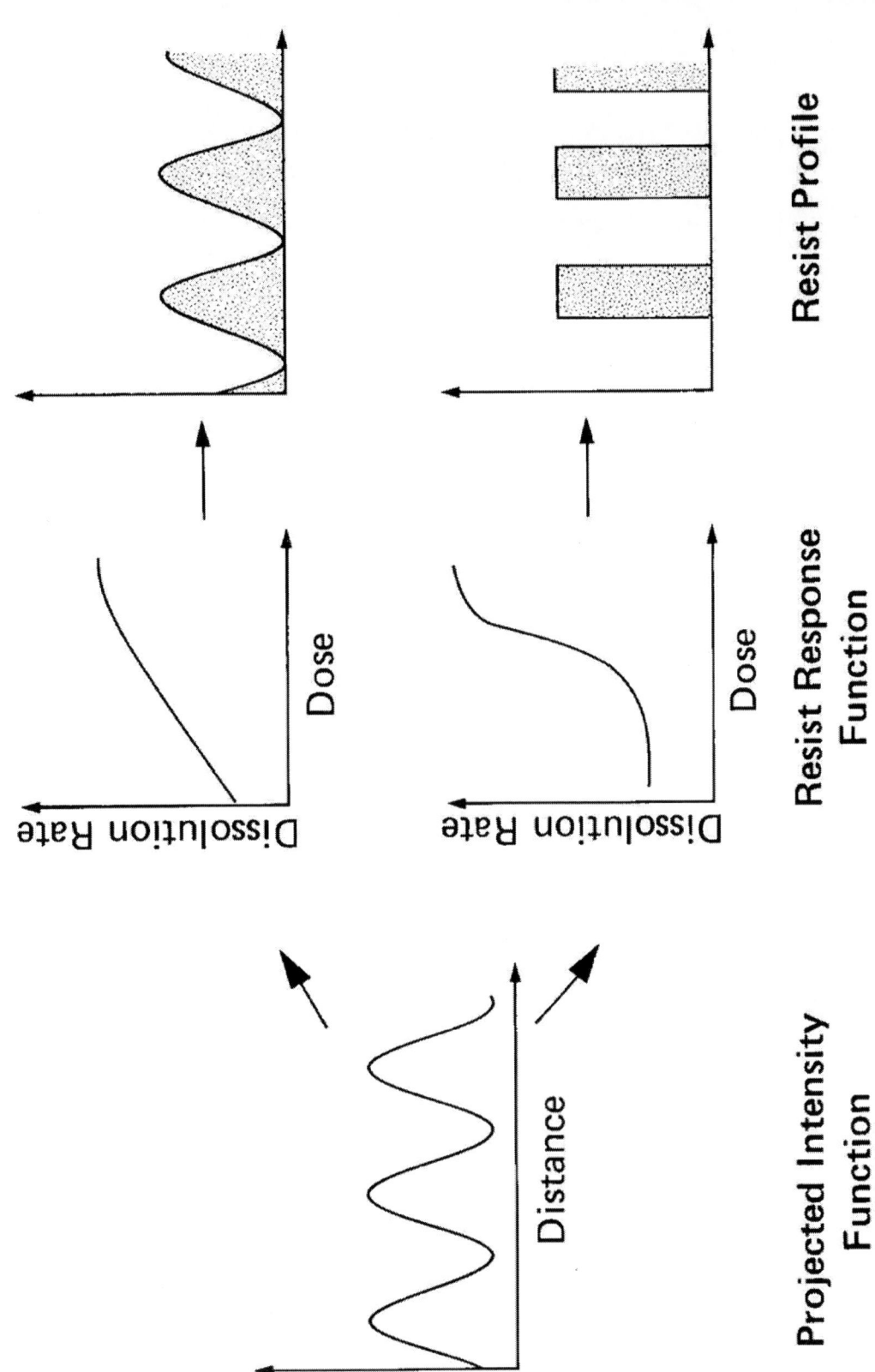

193nm Resist Dissolution Rate Response

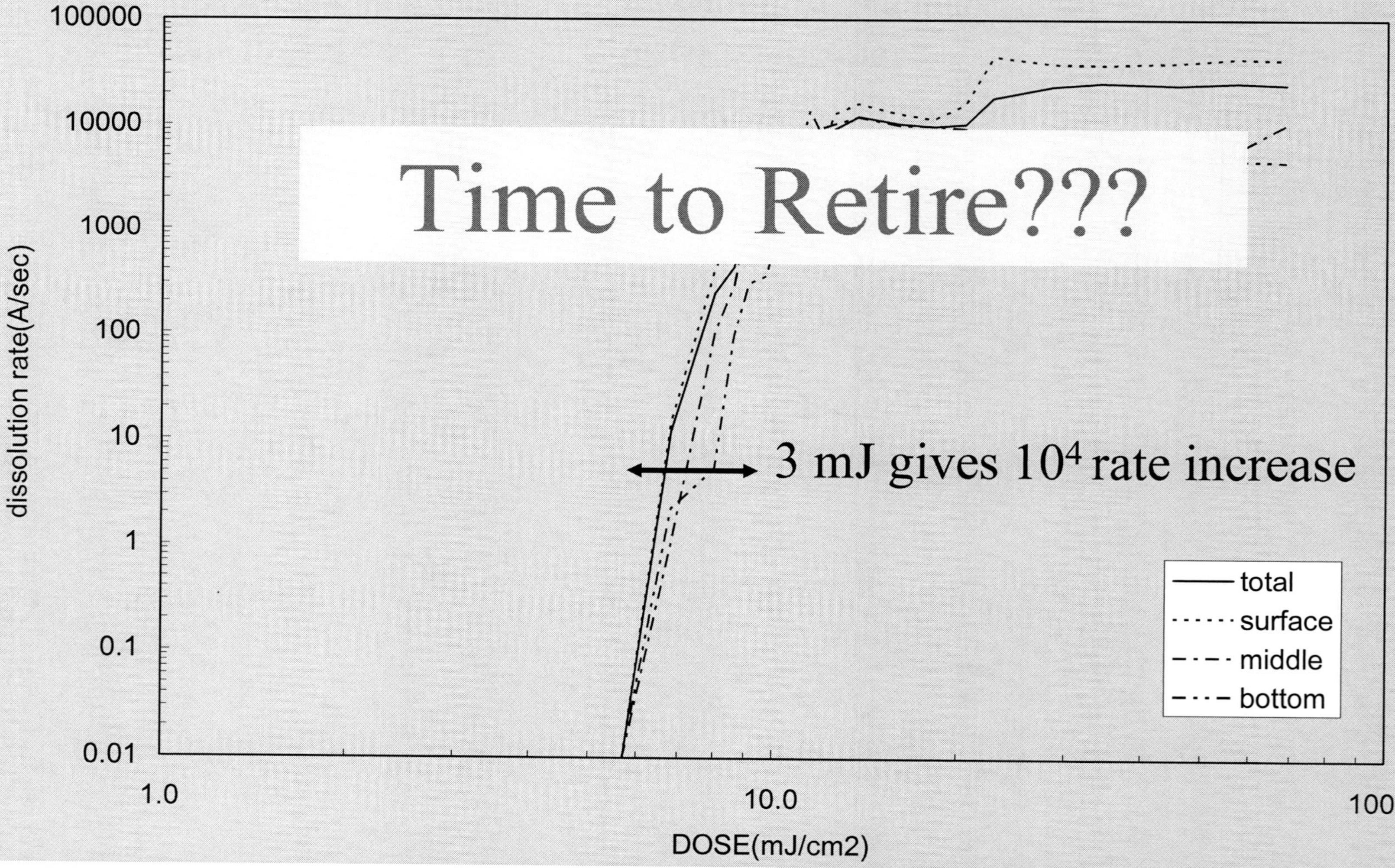

Time to Retire???

3 mJ gives 10^4 rate increase

— total
...... surface
–·– middle
–··– bottom

dissolution rate(A/sec)

DOSE(mJ/cm2)

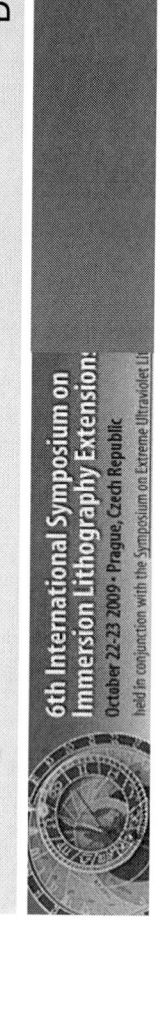

"It" all started here

**Nicéphore Niépce
(1765-1833)**
Chalon-sur-Saône, France

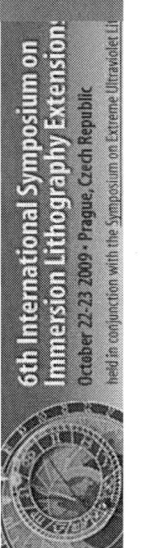

**6th International Symposium on
Immersion Lithography Extension:**
October 22-23 2009 · Prague, Czech Republic
held in conjunction with the Symposium on Extreme Ultraviolet Lit

A Historical Perspective

Printed with Bitumen of Judea 1826

"Point de vue du Gras" by Nicephore Niepce

Dichromated Gelatin

$$\text{Gelatin, } Cr^{6+} \xrightarrow{h\nu} Cr^{3+}$$

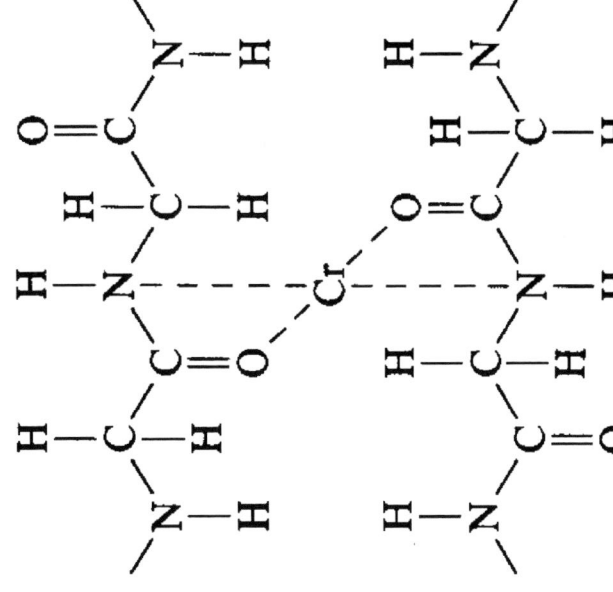

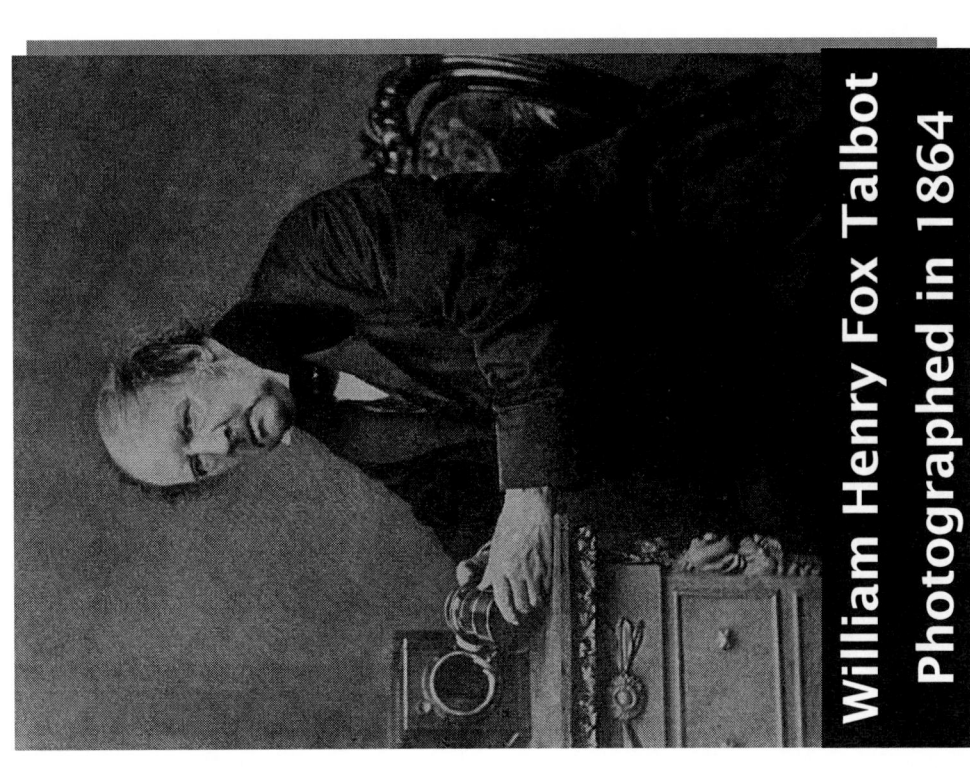

William Henry Fox Talbot
Photographed in 1864

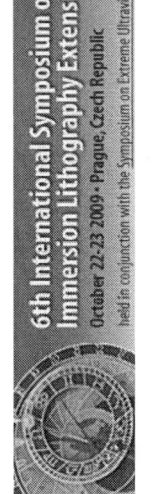

6th International Symposium on
Immersion Lithography Extension:
October 22-23 2009 • Prague, Czech Republic
held in conjunction with the Symposium on Extreme Ultraviolet Lit

The First Synthetic Photopolymer

Poly(vinyl cinnamate)

Adhesion Problems

**Louis Minsk
Eastman Kodak**

Kodak's KTFR

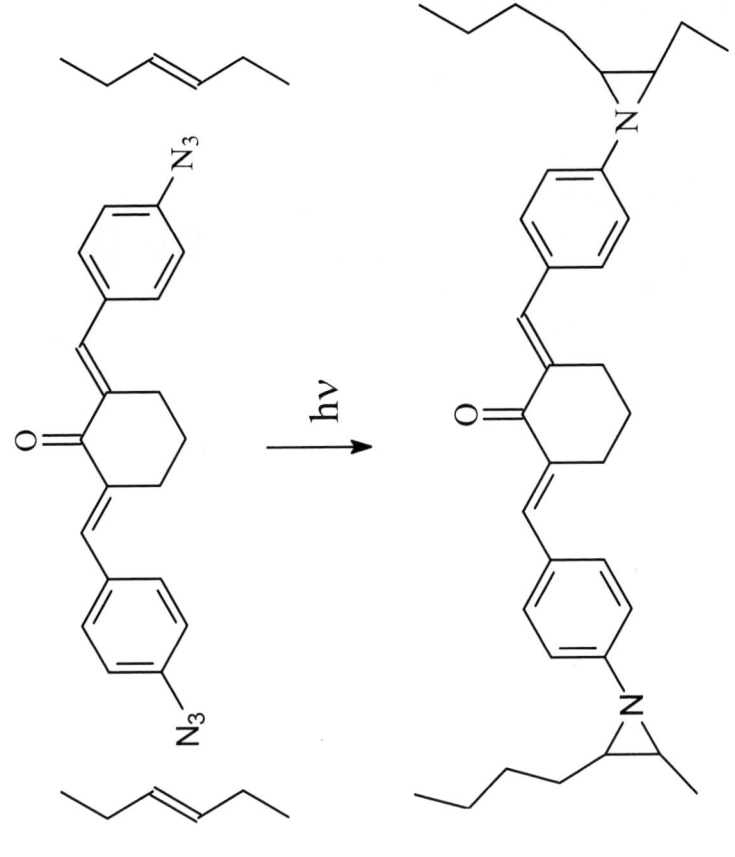

hν

Swelling Problems

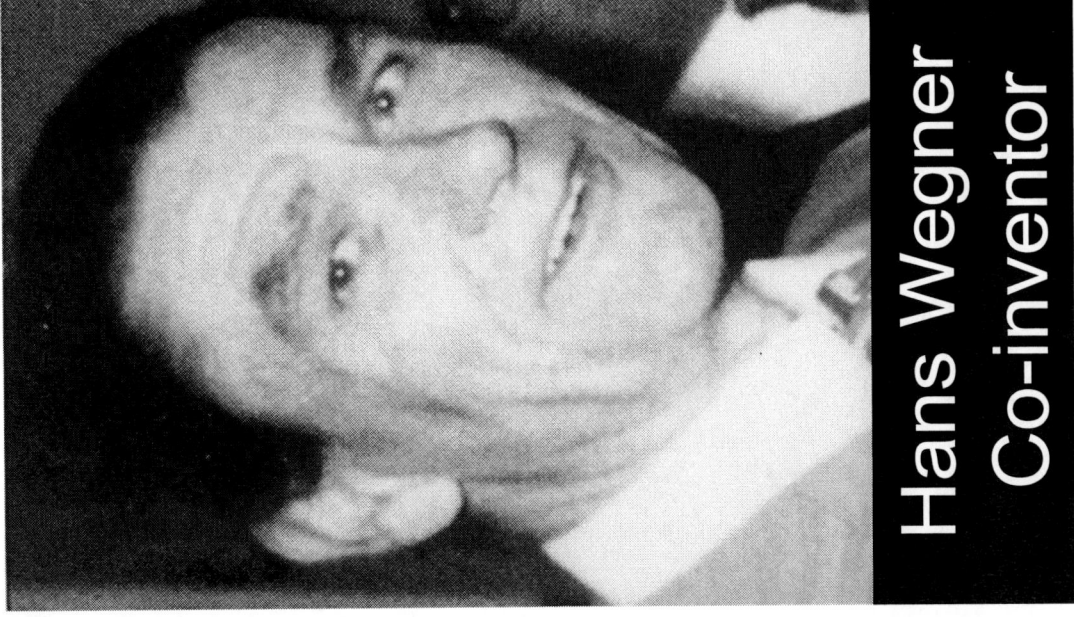

Hans Wegner
Co-inventor

6th International Symposium on
Immersion Lithography Extension:
October 22-23 2009 • Prague, Czech Republic
held in conjunction with the Symposium on Extreme Ultraviolet Lit

KTFR

As aspect ratio approaches 1
Swelling, Snakes, Ugh!!

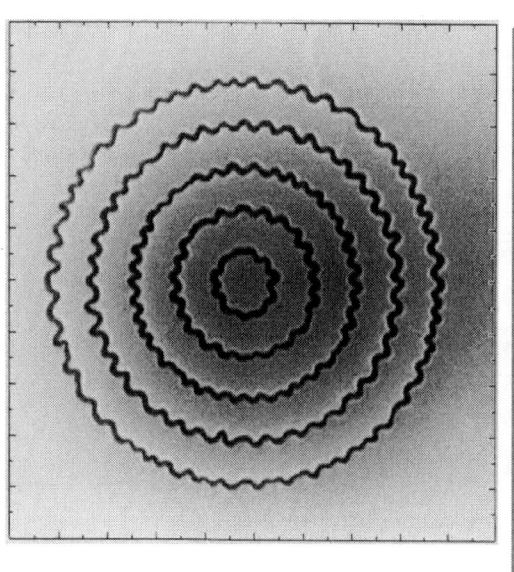

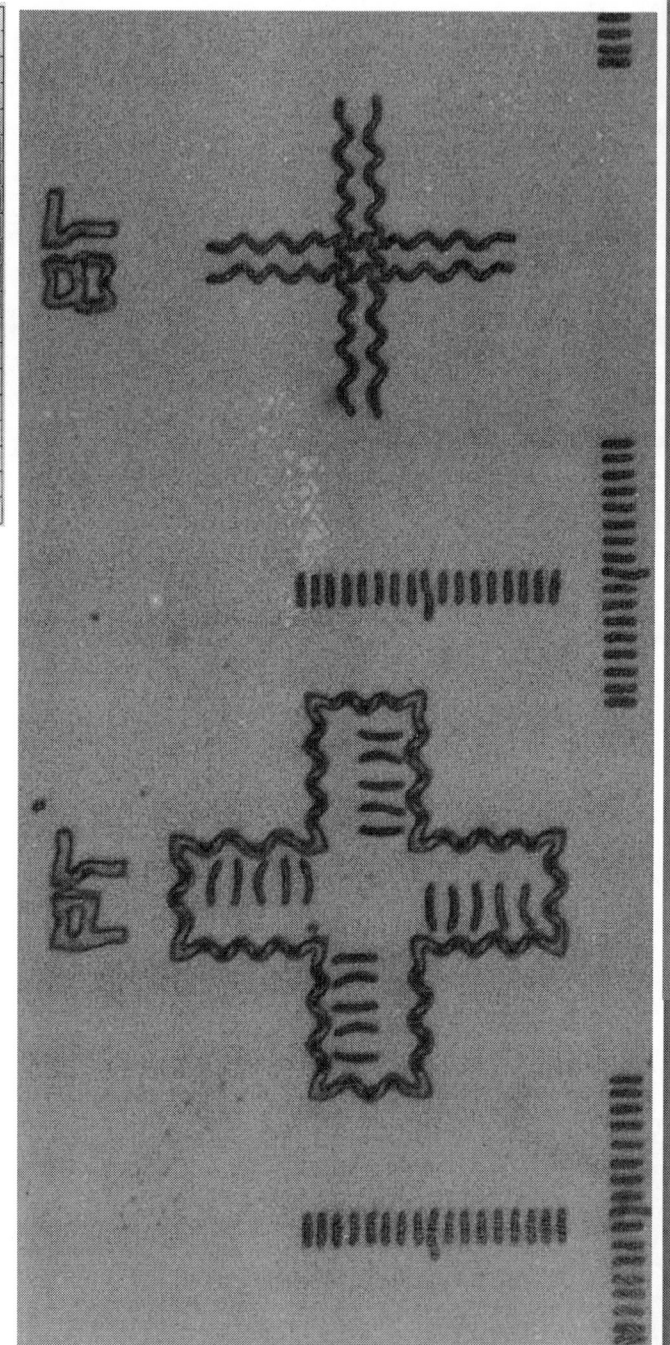

6th International Symposium on
Immersion Lithography Extension:
October 22-23 2009 • Prague, Czech Republic
held in conjunction with the Symposium on Extreme Ultraviolet Lit

The History of Novolac

C.H. Meyer and/or
L.H. Baekeland
Discovered Novolac

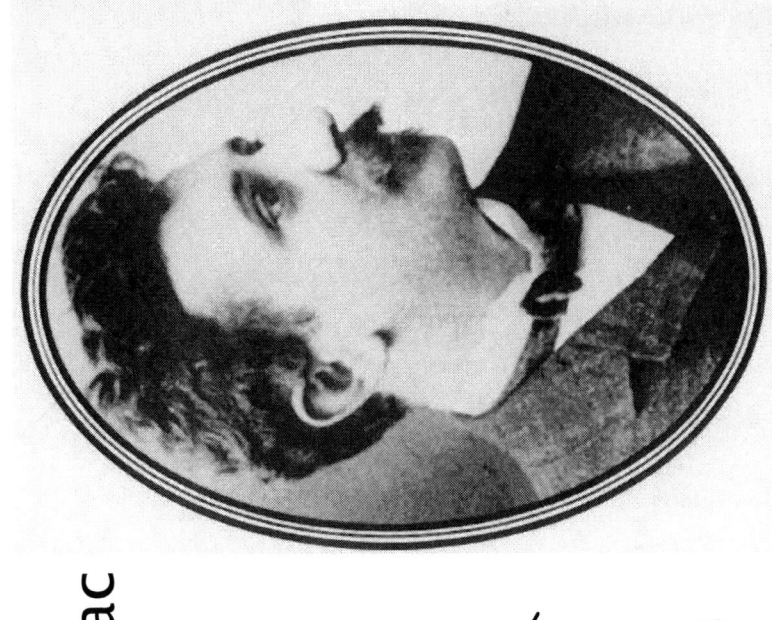

Meyer

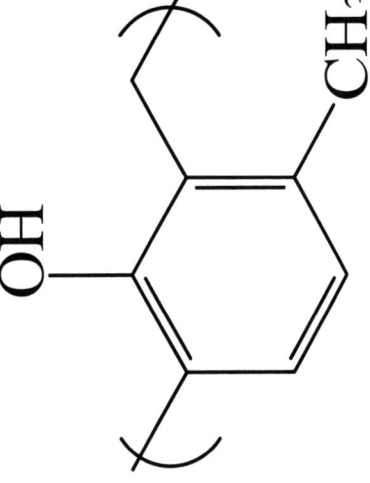

Baekeland

Novolac Resin Production!!

Workers at Albert Co. in Wiesbaden, Germany
"zerkleinert das Harz von Hand"

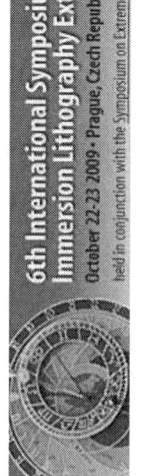

6th International Symposium on
Immersion Lithography Extension:
October 22-23 2009 · Prague, Czech Republic
held in conjunction with the Symposium on Extreme Ultraviolet Li

Diazoquinone Chemisty

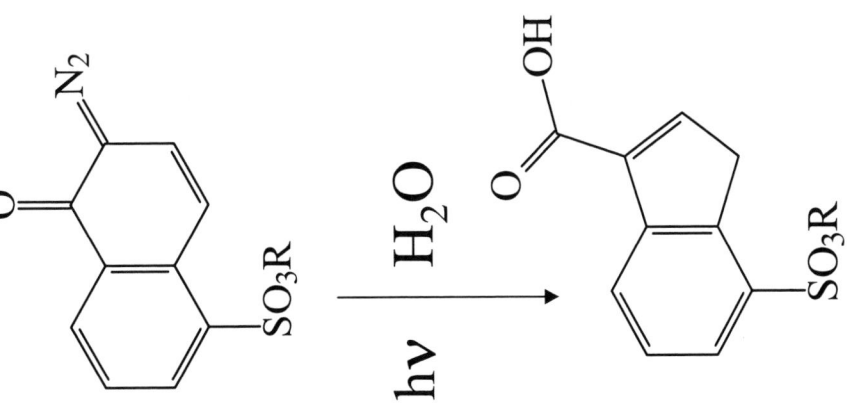

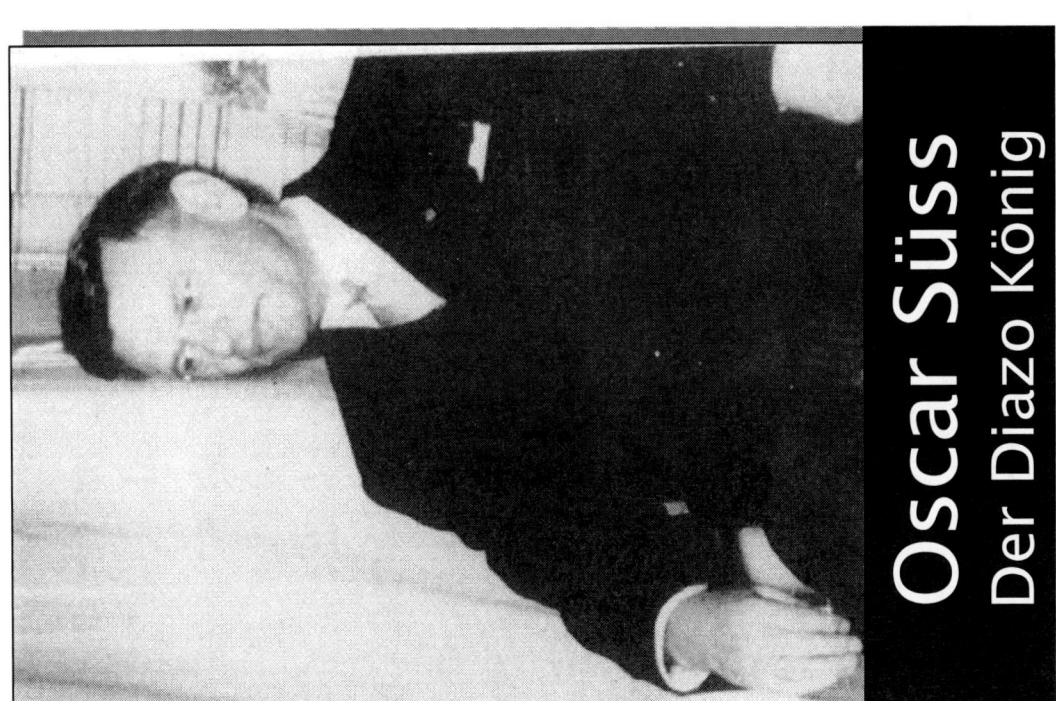

Oscar Süss
Der Diazo König

IBM I line resist development rate curve.

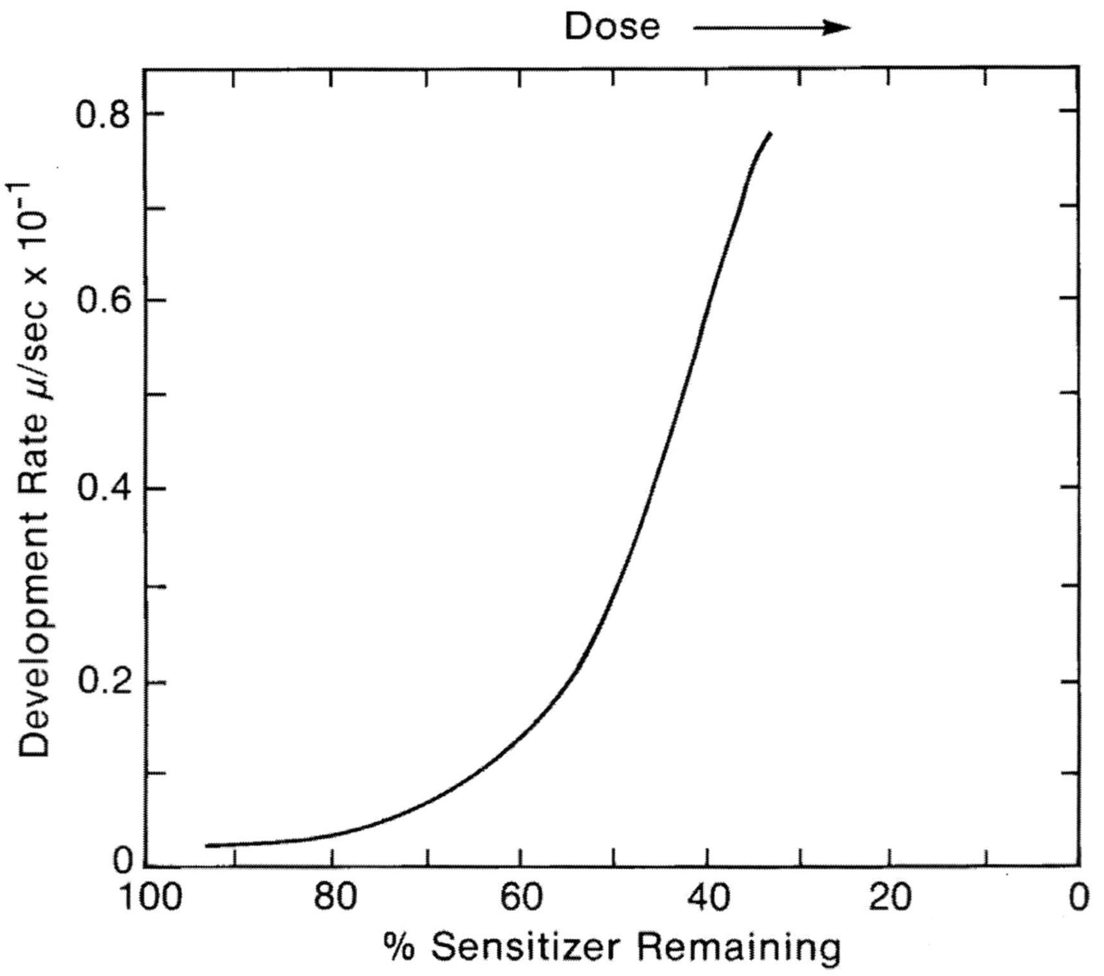

Polyphotolysis

$I \rightarrow S$

$I\text{-}I \rightarrow S\text{-}I \rightarrow S\text{-}S$

$I\text{-}I\text{-}I \rightarrow S\text{-}I\text{-}I \rightarrow S\text{-}S\text{-}I \rightarrow S\text{-}S\text{-}S$

This contributes to a threshold like dissolution rate response to exposure

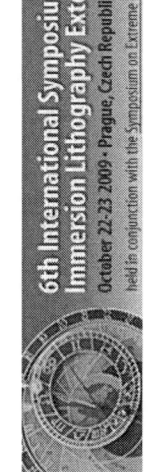

6th International Symposium on
Immersion Lithography Extension:
October 22-23 2009 · Prague, Czech Republic
held in conjunction with the Symposium on Extreme Ultraviolet Lith

IBM I-line Resist

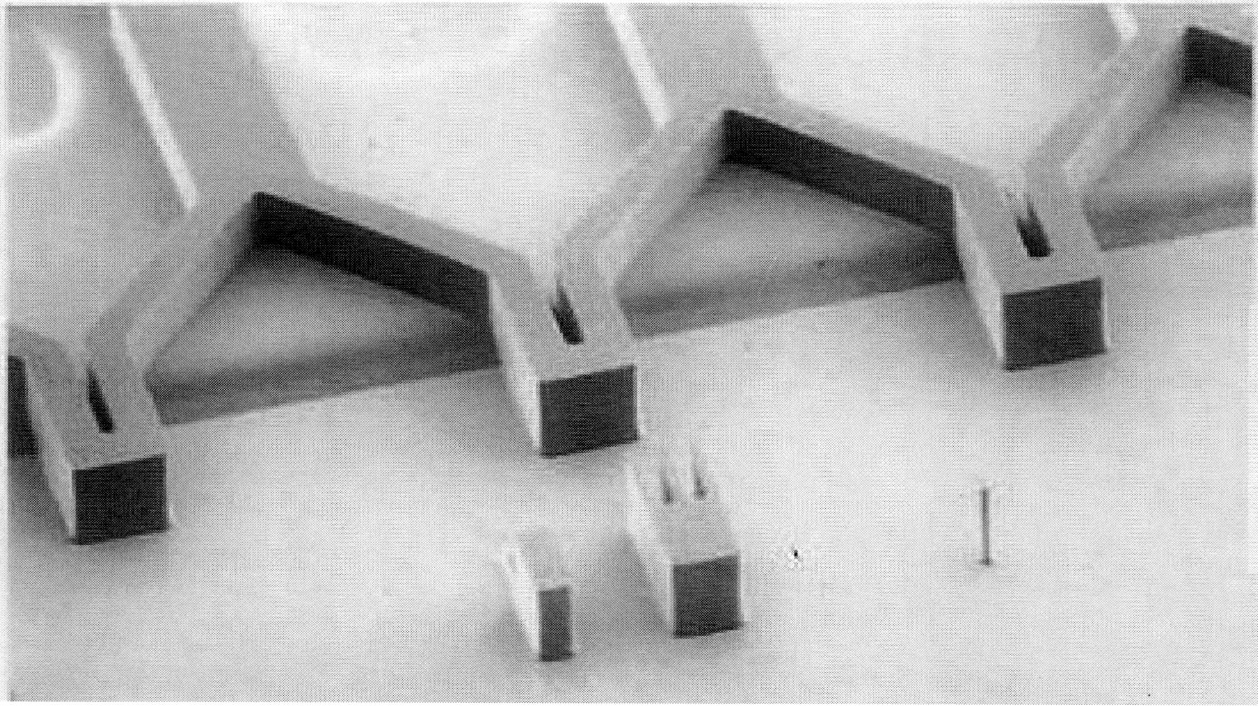

25 μ nominal coating
16 μ Topography
PE-210 exposure
N.A. = 0.167
DOF = ±6.5 μ

Resist and Process Development

Performance (vertical axis)

Time (horizontal axis)

Basic Chemistry — Invention

Formulation and Process Development — Competition

Optimization — Mature products

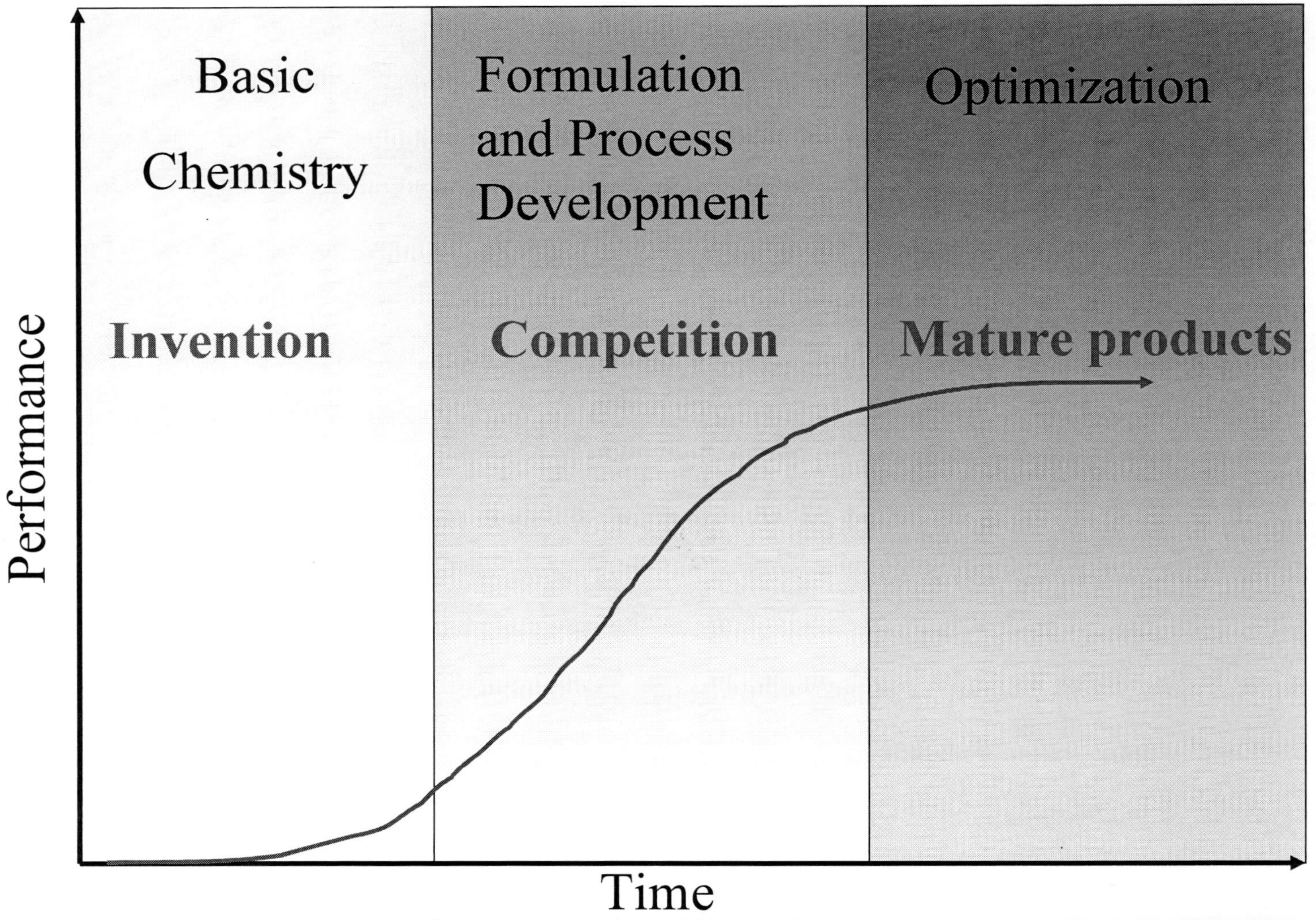

6th International Symposium on
Immersion Lithography Extension:
October 22-23 2009 · Prague, Czech Republic
held in conjunction with the Symposium on Extreme Ultraviolet Lithography

Resist Evolution

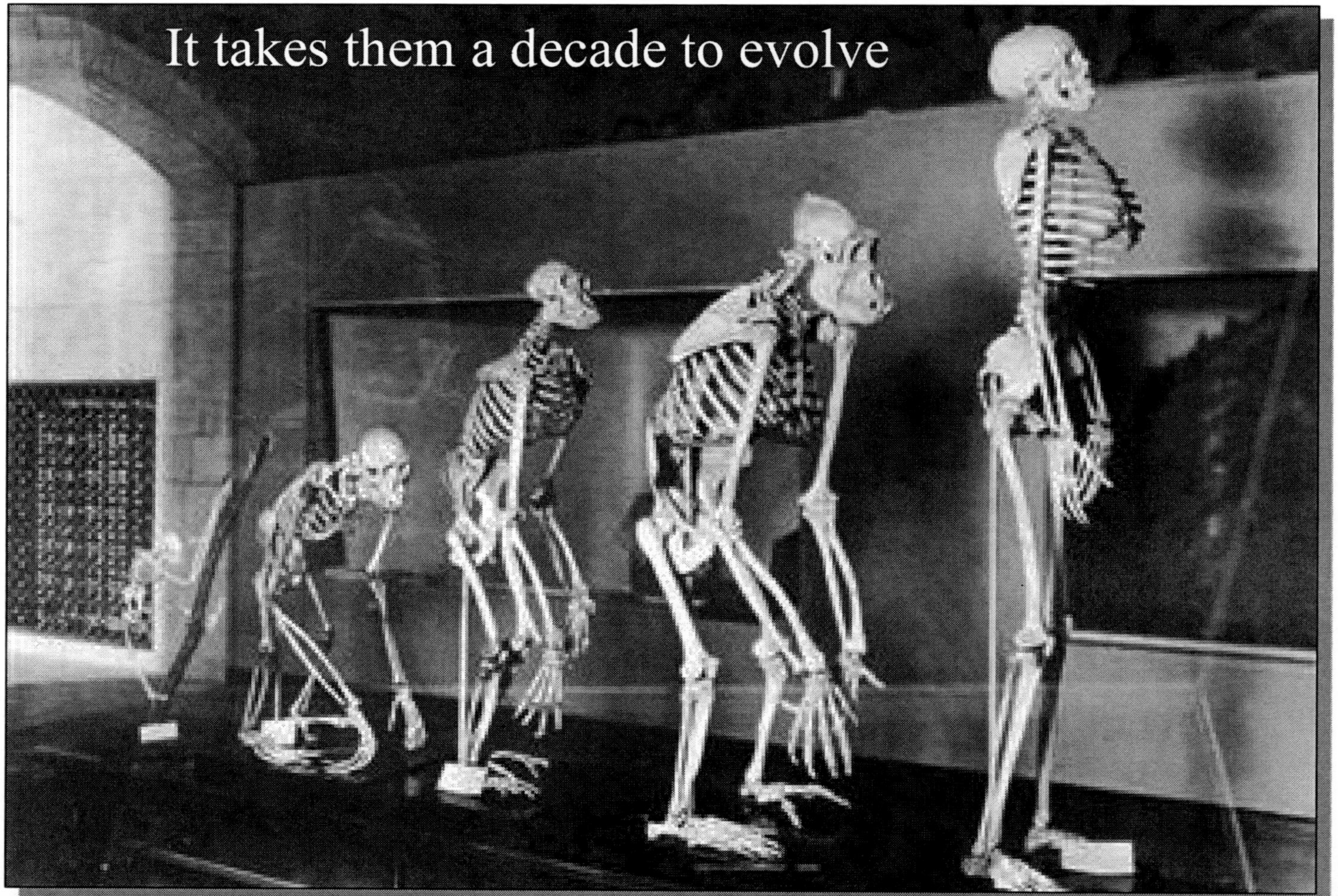

It takes them a decade to evolve

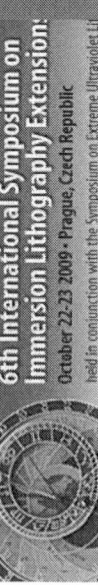

I-line resist Improvement with time

introduced in:	1984	1990	1996
resist name:	AZ®1500	AZ®6200	AZ®7900

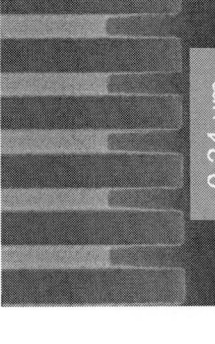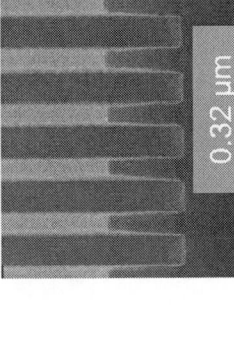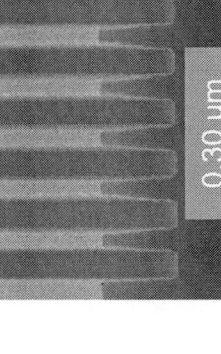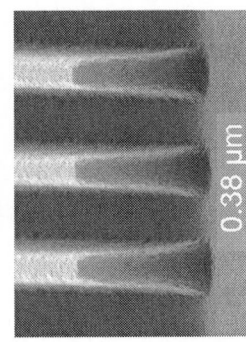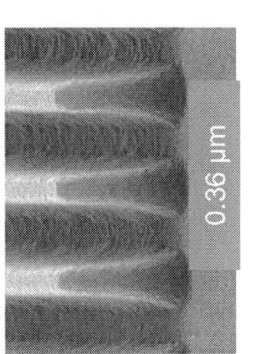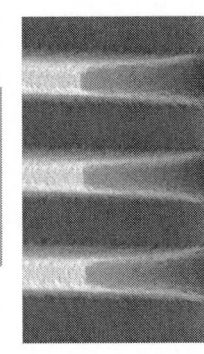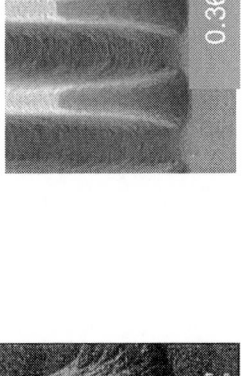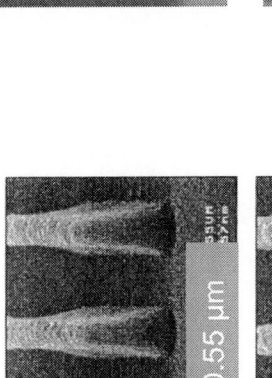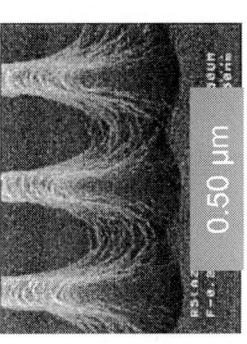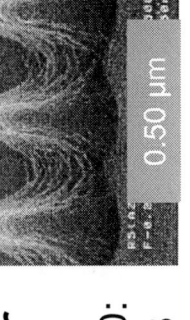

0.34 µm

0.32 µm

0.30 µm

0.40 µm

0.38 µm

0.36 µm

0.60 µm

0.55 µm

0.50 µm

all
resists
on
Nikon®
NA=0.54
i-line
stepper

AZ1500:
no PEB

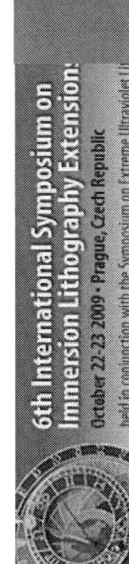

6th International Symposium on
Immersion Lithography Extension:
October 22-23 2009 · Prague, Czech Republic
held in conjunction with the Symposium on Extreme Ultraviolet Lit

Chemical Amplification

Jean Frechet

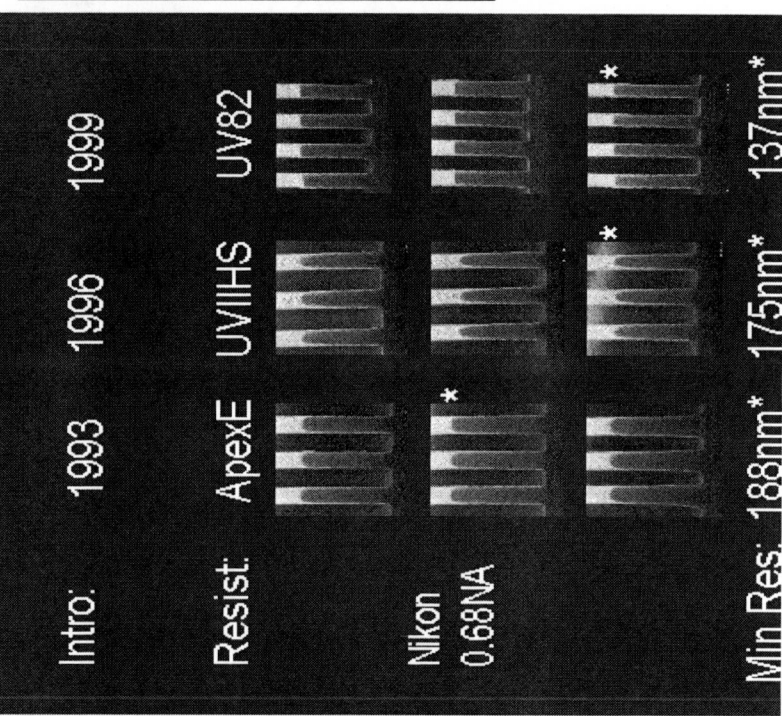

Intro:	1993	1996	1999
Resist:	ApexE	UVIIHS	UV82

Nikon
0.68NA

Min Res: 188nm* 175nm* 137nm*

Hiroshi Ito

Chemically Amplified Resists enabled shorter wavelength exposure and provided much higher sensitivity to light

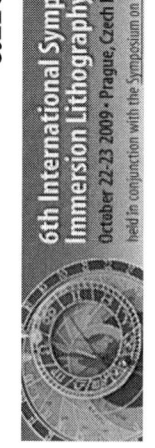

6th International Symposium on Immersion Lithography Extension:
October 22-23 2009 • Prague, Czech Republic
held in conjunction with the Symposium on Extreme Ultraviolet Lit

Chemical Amplification

$$\mathrm{I}nsoluble \quad \xrightarrow[\Phi < 1]{h\nu} \quad \mathrm{S}oluble$$

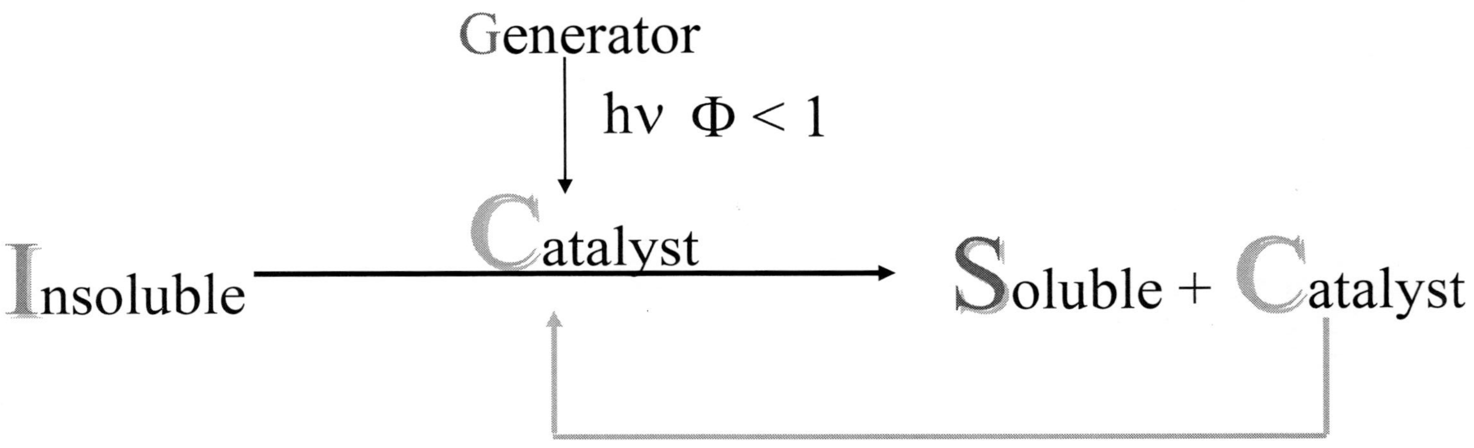

$$\mathrm{I}nsoluble \quad \xrightarrow{} \quad \mathrm{S}oluble + \mathrm{C}atalyst$$

$$\mathrm{G}enerator \quad \xrightarrow{h\nu\ \Phi < 1} \quad \mathrm{C}atalyst$$

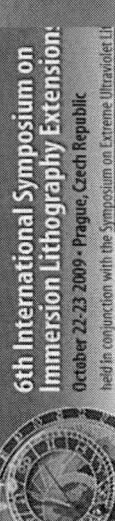

6th International Symposium on
Immersion Lithography Extension
October 22-23 2009 • Prague, Czech Republic
held in conjunction with the Symposium on Extreme Ultraviolet Li

1985: Chemically Amplified Deep UV Resists

Photoacid Generation

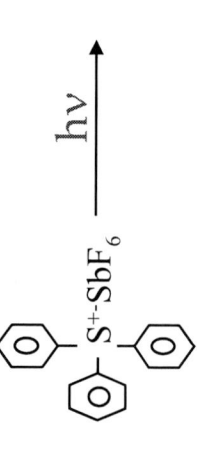

$$\xrightarrow{h\nu} \quad H^+ \cdot SbF_6$$

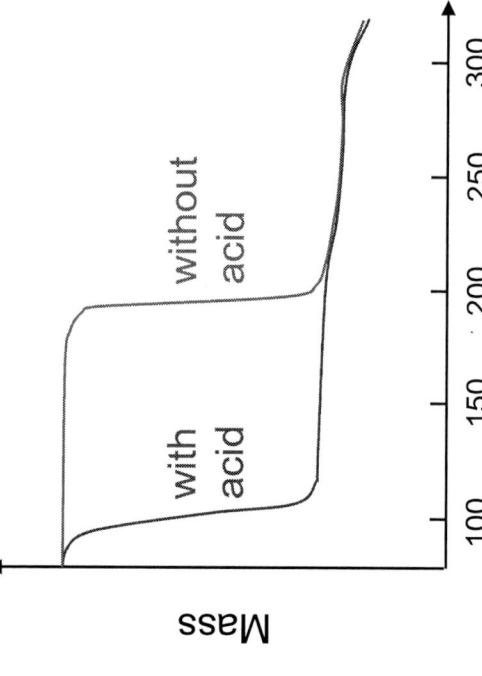

Acid-Catalyzed Deprotection

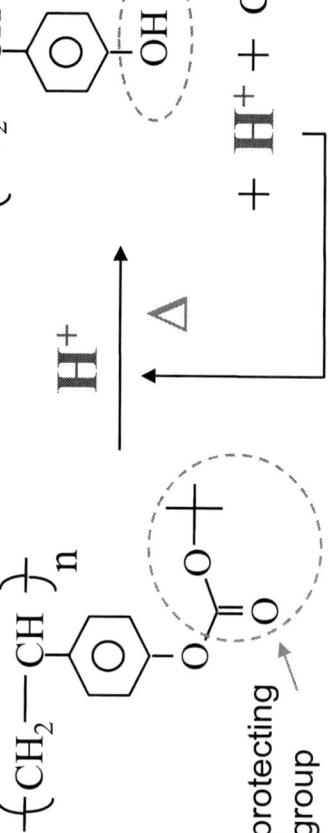

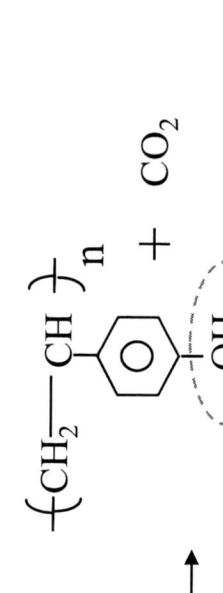

6th International Symposium on
Immersion Lithography Extension
October 22-23 2009 · Prague, Czech Republic
held in conjunction with the Symposium on Extreme Ultraviolet Li

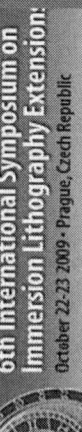

Hiroshi Ito and Nicholas Clecak

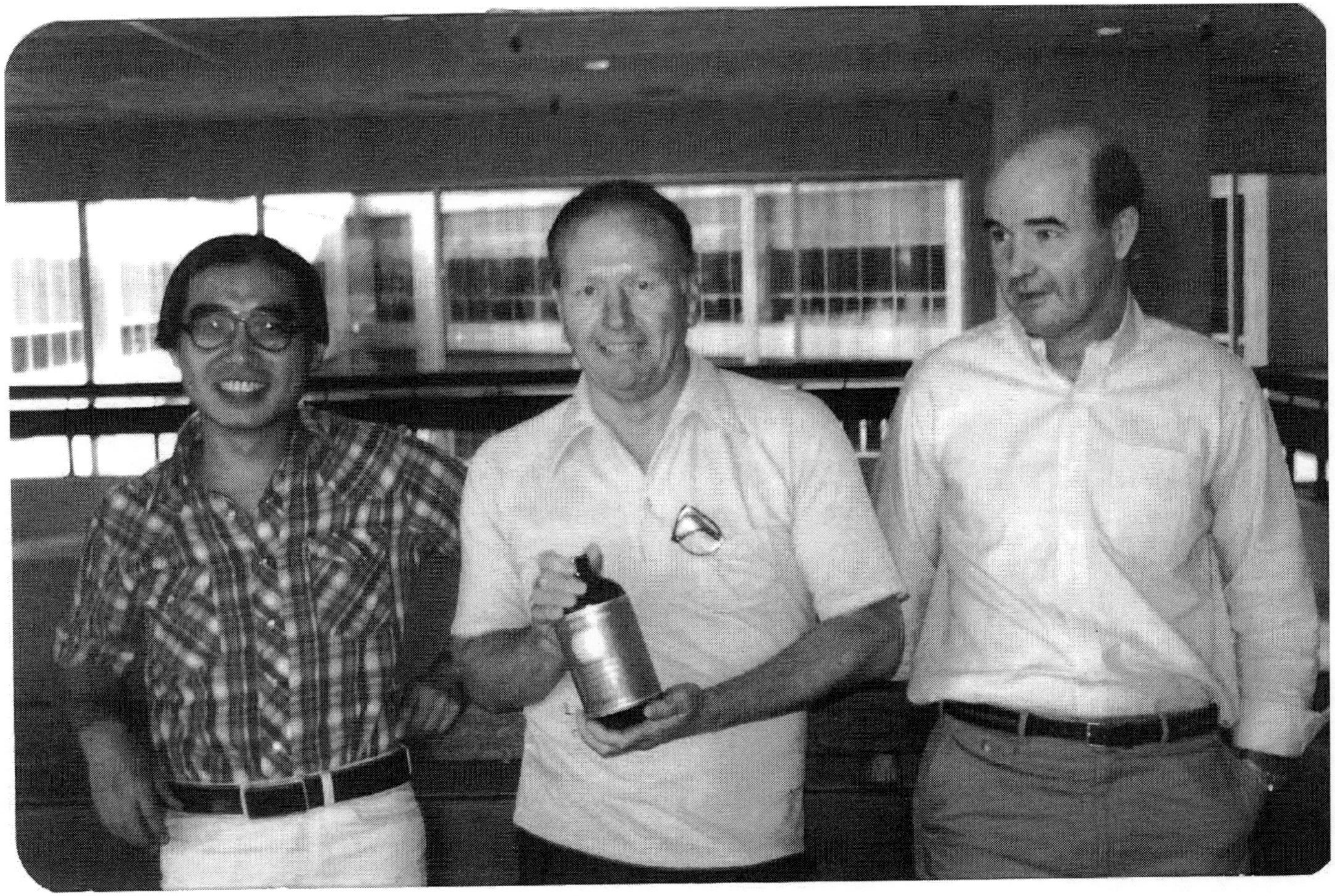

IBM San Jose Research Laboratory

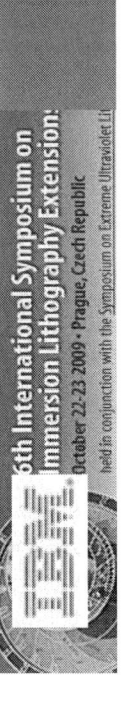

First Commercial Implementation of Deep UV Photolithography

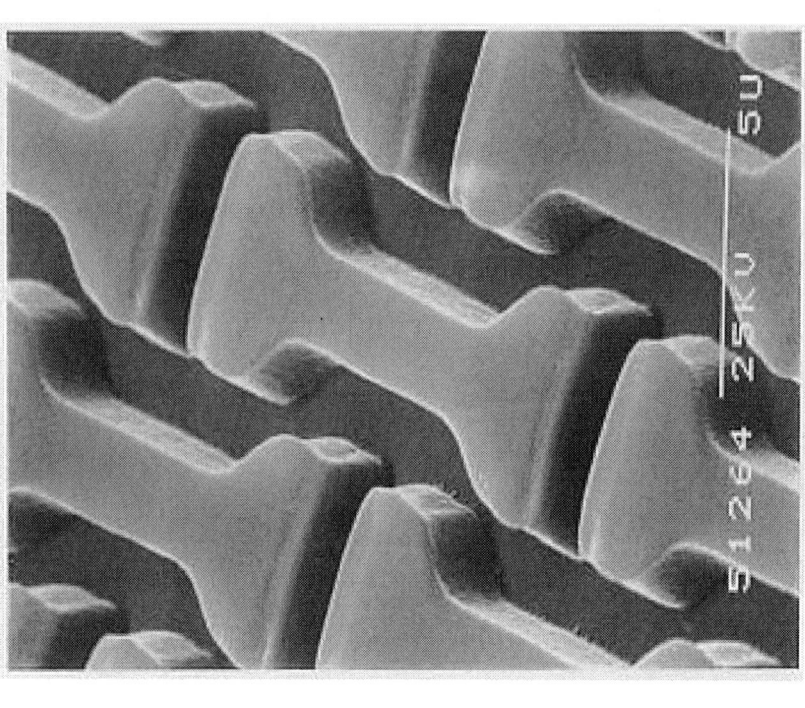

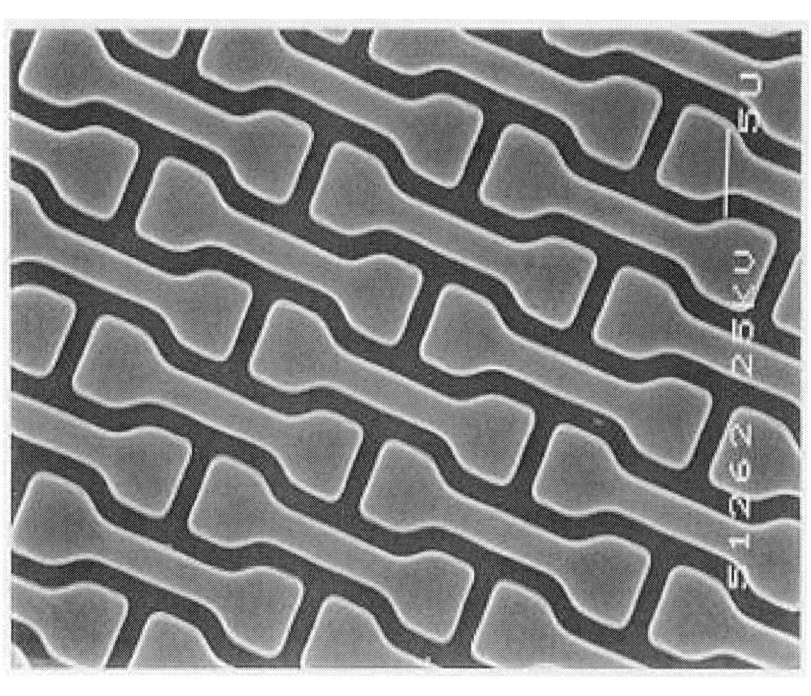

ROX level of 1M DRAM IBM

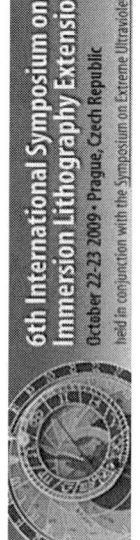

6th International Symposium on Immersion Lithography Extension:
October 22-23 2009 · Prague, Czech Republic
held in conjunction with the Symposium on Extreme Ultraviolet Lit

High And Low Activation Energy Protecting Groups

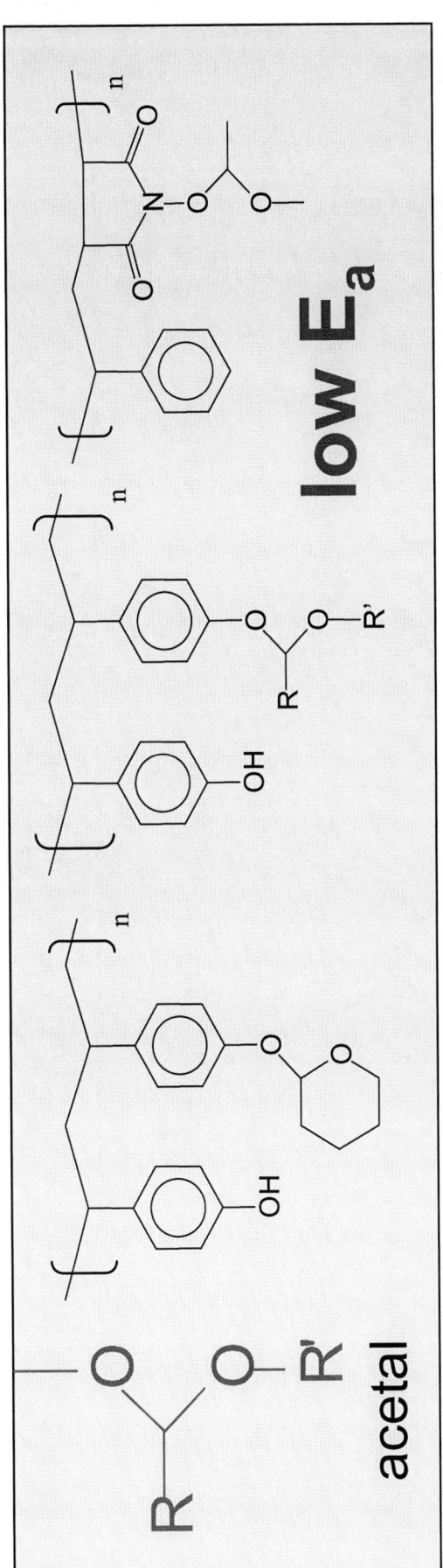

low E_a

acetal

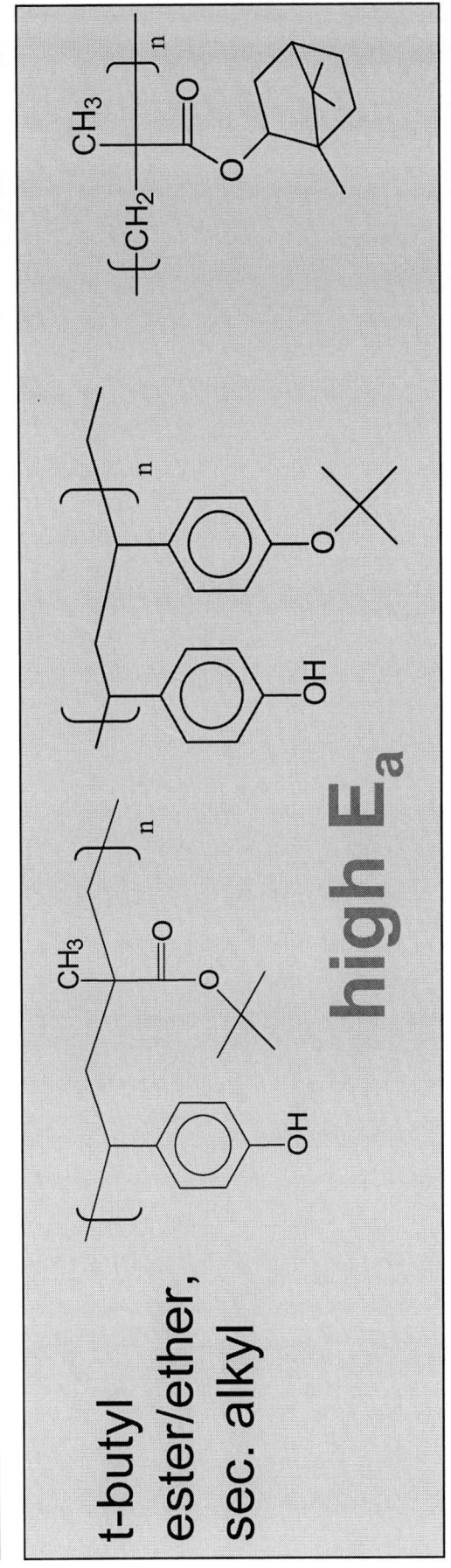

high E_a

t-butyl
ester/ether,
sec. alkyl

KrF Resist Dissolution Rate Response

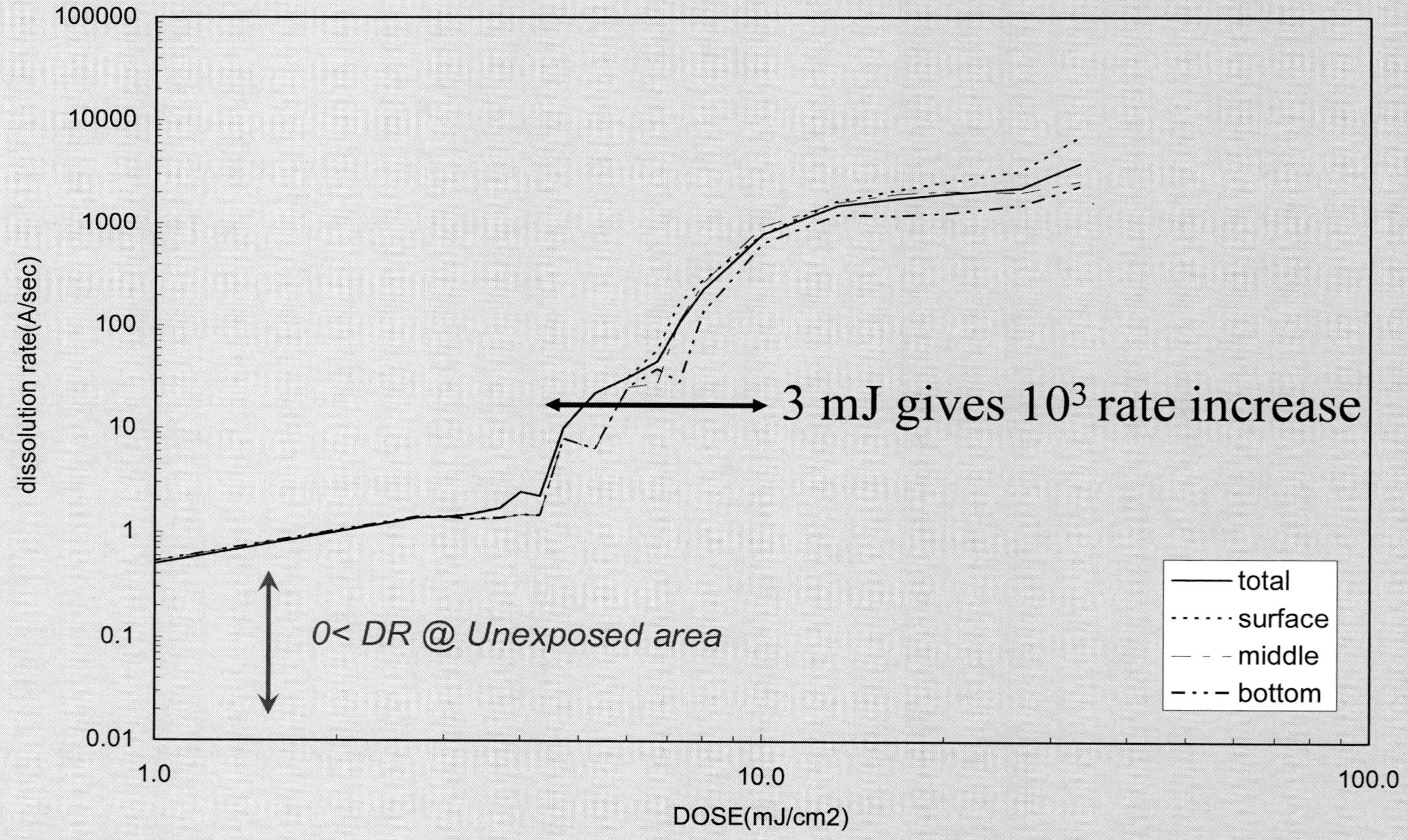

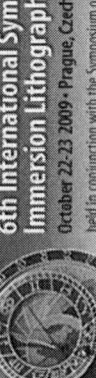

193 nm Resist Materials

Challenge:

- 248 and 365 nm resists are unsuitable for 193 nm imaging because they are opaque at this wavelenth

- Etch resistance requires high carbon/hydrogen ratio but aromatics are precluded because of their absorption

- How do you achieve both 193nm optical transparency and etch resistance?

6th International Symposium on
Immersion Lithography Extension:
October 22-23 2009 • Prague, Czech Republic
held in conjunction with the Symposium on Extreme Ultraviolet Lith

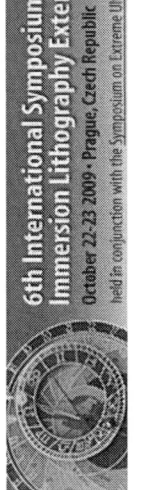

Relative Etch Rate of Polymers

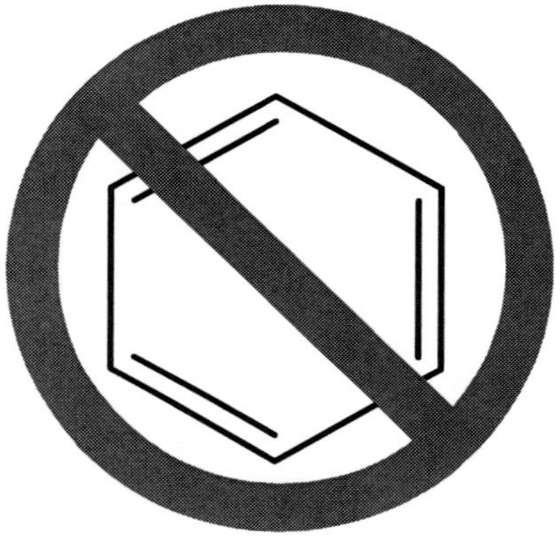

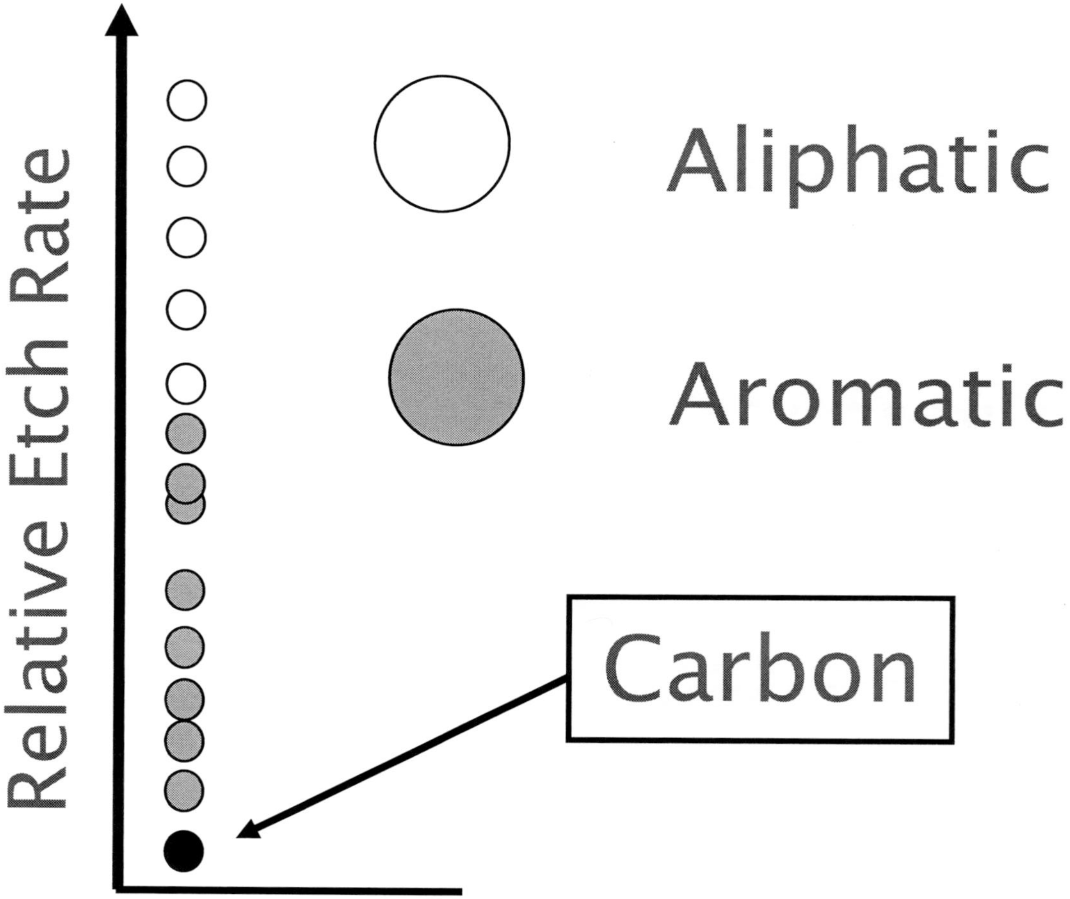

Aliphatic

Aromatic

Carbon

Relative Etch Rate

"Ohnishi factor" also suggests aromaticity

$$R \propto \frac{N_c}{N_t - N_o}$$

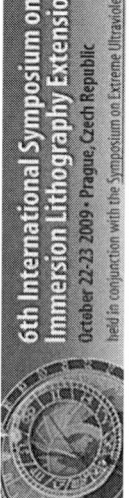

High C:H Ratio of Alicyclic Hydrocarbons

The key!

Structure:	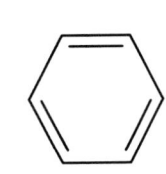	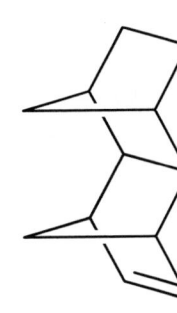	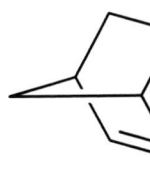 $CH_3(CH_2)_nCH_3$
Formula:	C_6H_6	$C_{12}H_{16}$	C_7H_{10} C_nH_{2n+2}
Double Bond Equivalents:	4	5	3 0

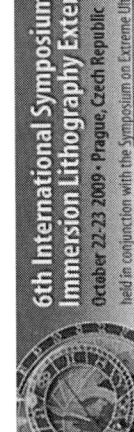

6th International Symposium on Immersion Lithography Extension:

October 22-23 2009 · Prague, Czech Republic

held in conjunction with the Symposium on Extreme Ultraviolet Lit

IBM Version 2 Tetrapolymer

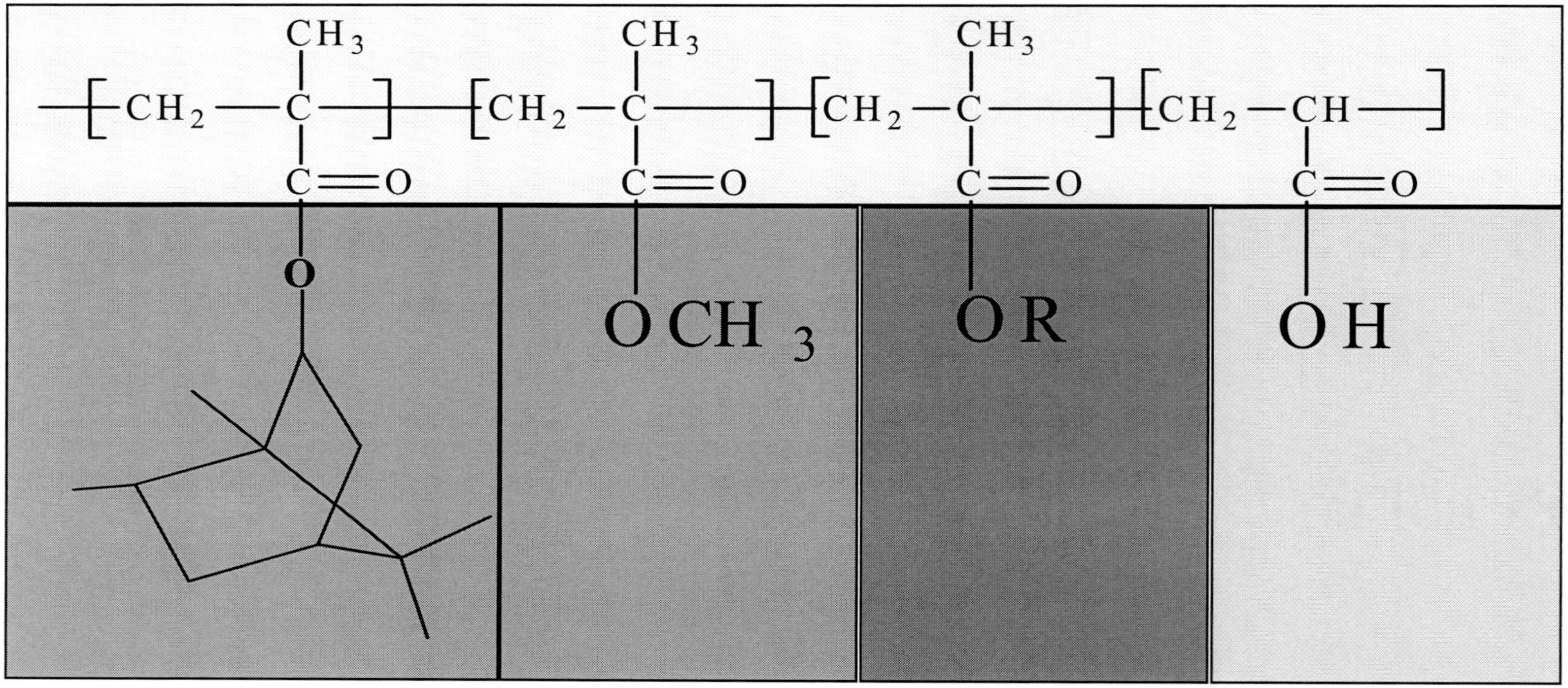

☐ Tethering Function ☐ Acid Lability ☐ Base Solubility

▨ Etch Resistance ▨ Mechanical Properties

Allen, R. D., Wallraff, G. M., DiPietro, R. A., Hofer, D. C., Kunz, R. R. *Proc. SPIE. 2438, 1995, 474*

6th International Symposium on Immersion Lithography Extension:
October 22-23 2009 • Prague, Czech Republic
held in conjunction with the Symposium on Extreme Ultraviolet Lit

High NA Lithography with IBM 193nm Version 1b

Early Images

ISI 193nm, 0.6 NA

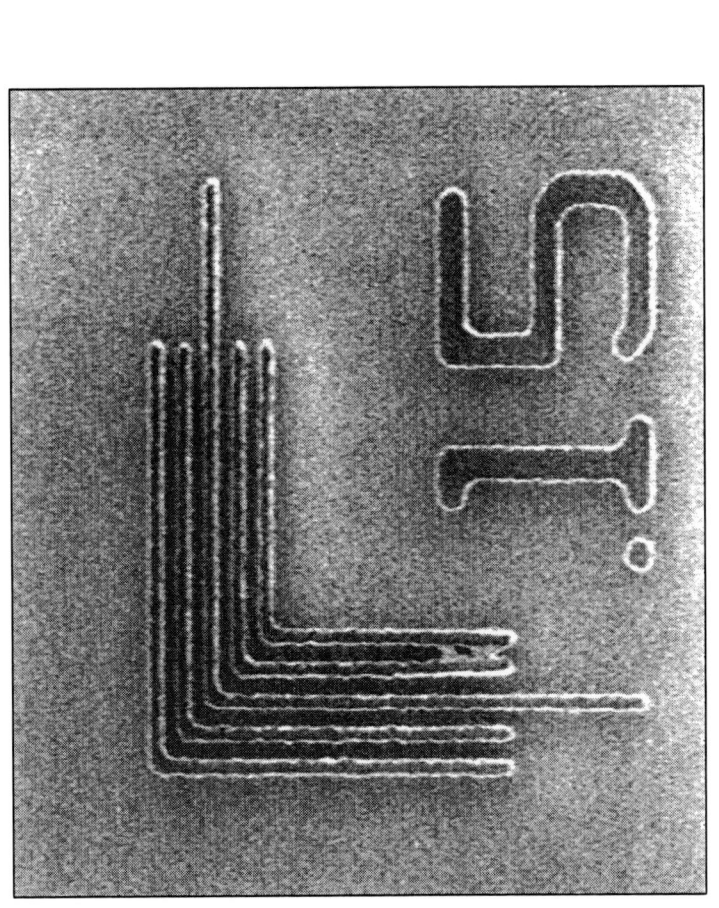

photo courtesy of Andy Grenville, Sid Das & Joe Langston of INTEL

Fujitsu's Acrylic Platform

Free Radical Polymerization (no metals)

Acrylic Polymer Platform

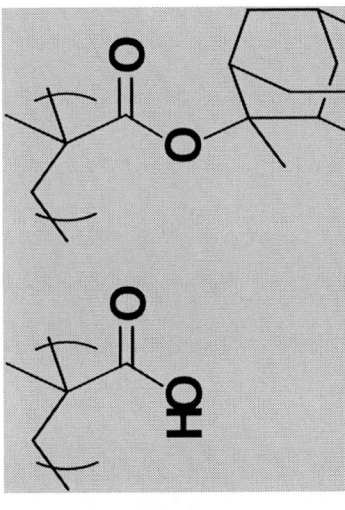

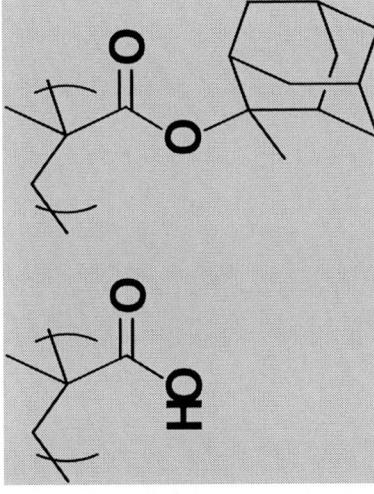

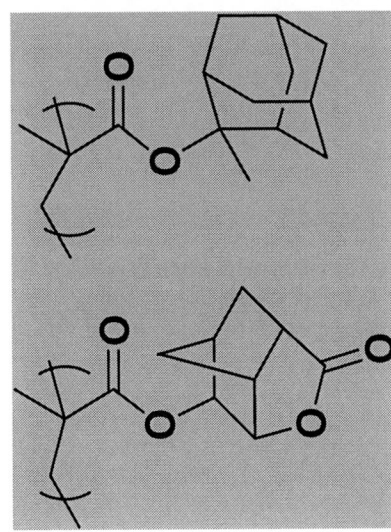

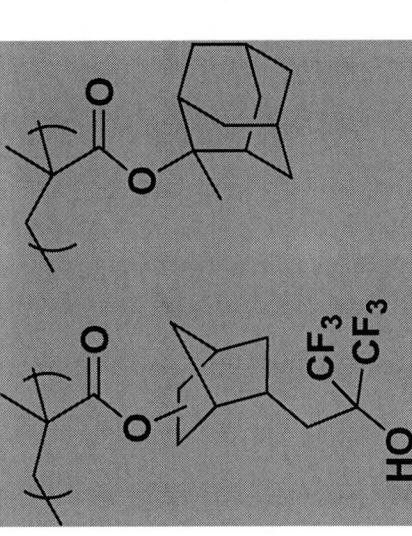

IBM, JSR, TOK etc.

Fujitsu

193nm Resist Dissolution Rate Response

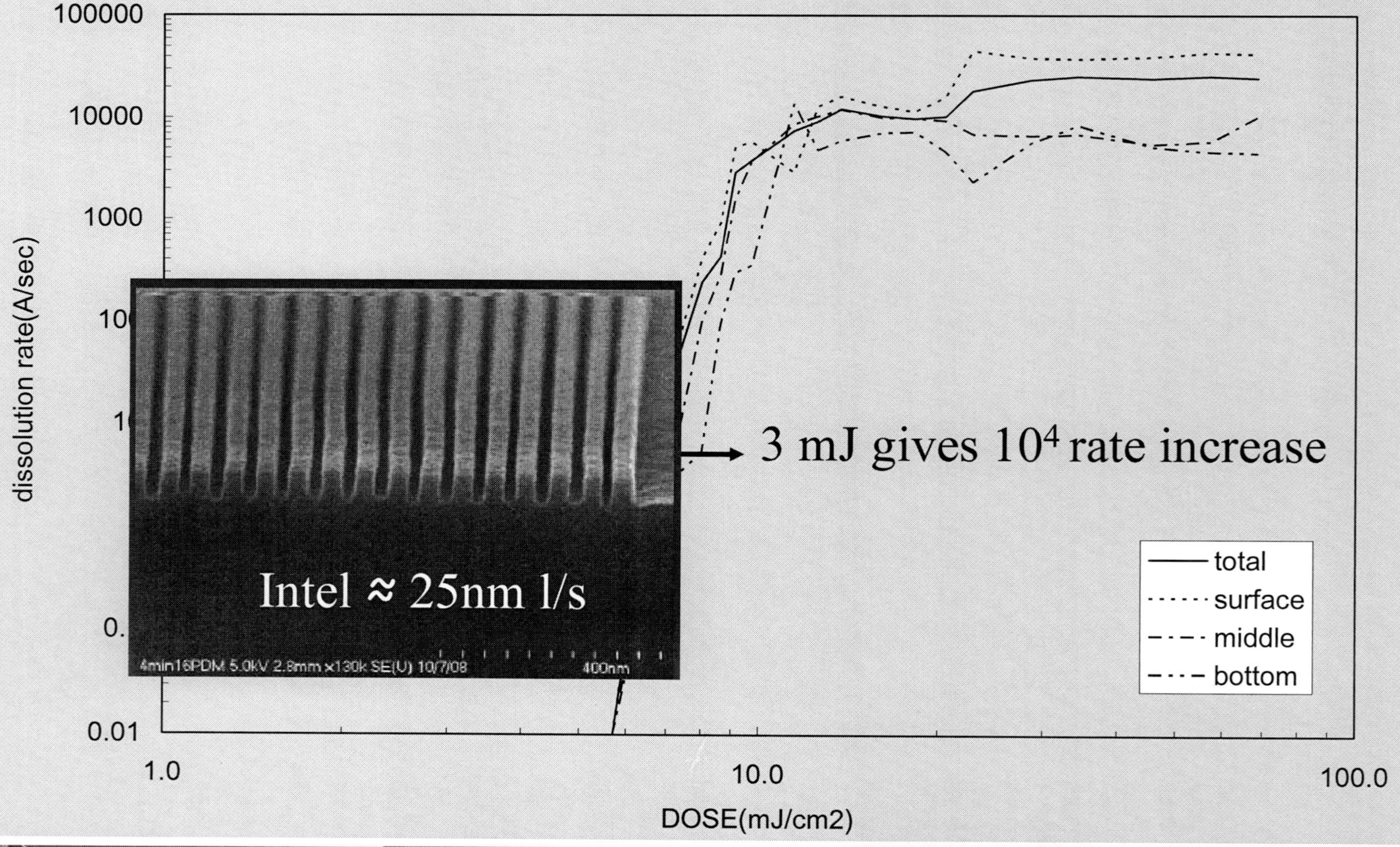

dissolution rate(A/sec)

DOSE(mJ/cm2)

legend:
— total
····· surface
—·—· middle
— — bottom

3 mJ gives 10^4 rate increase

Intel ≈ 25nm l/s

400nm

4min18sPDM 5.0kV 2.8mm x130k SE(U) 10/7/08

6th International Symposium on
Immersion Lithography Extension:
October 22-23 2009 · Prague, Czech Republic
held in conjunction with the Symposium on Extreme Ultraviolet Lit

Intrinsic limitation?

mask

hv

Exposure

Generation of chemically stable catalyst

PEB

Solubility-switching plus transport

Diffusion Bias

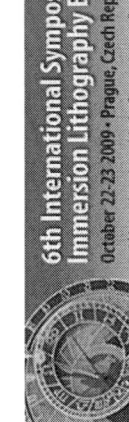

6th International Symposium on Immersion Lithography Extension:
October 22-23 2009 • Prague, Czech Republic
held in conjunction with the Symposium on Extreme Ultraviolet Lith

Simulation of a PE Bake

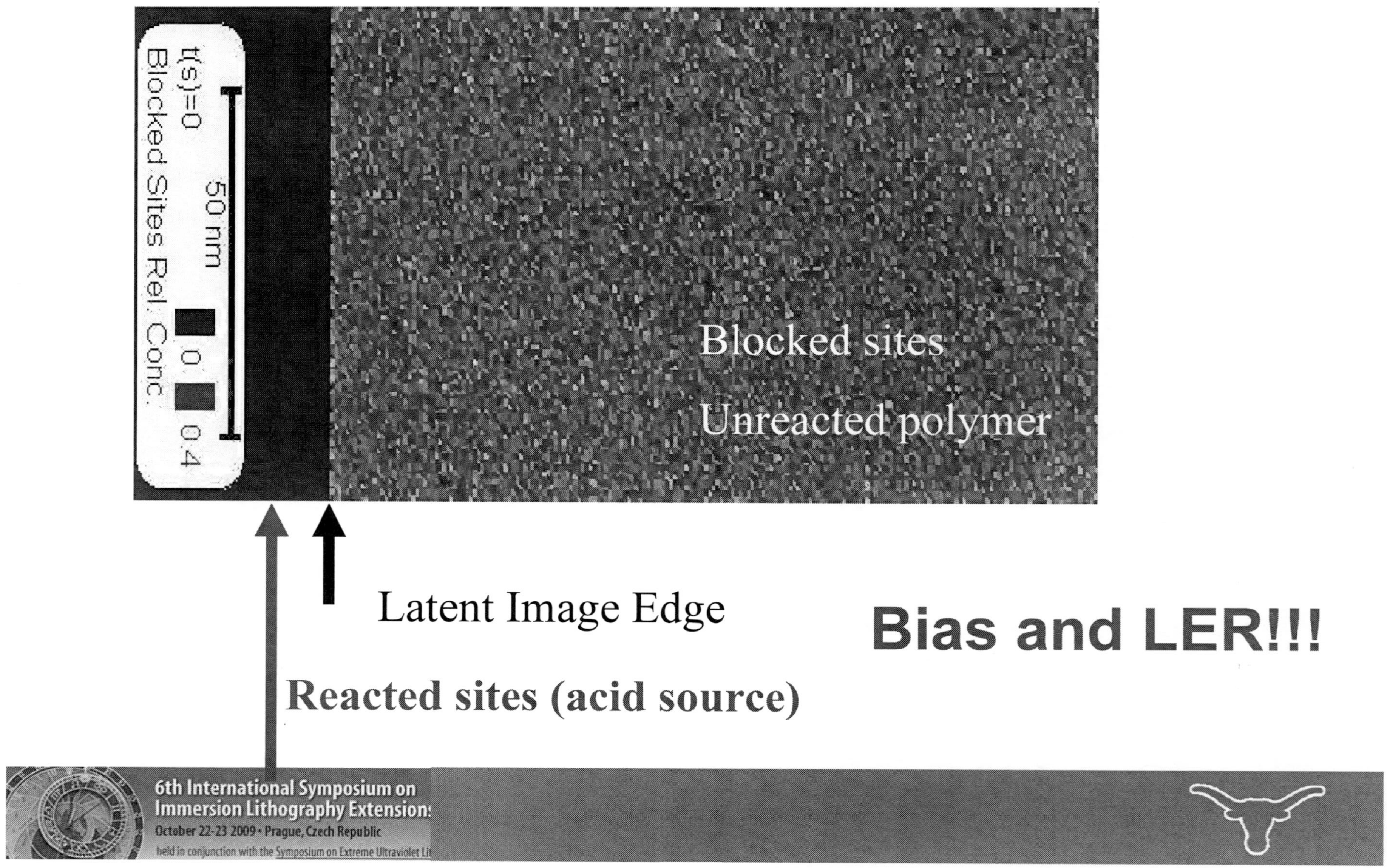

Blocked sites

Unreacted polymer

50 nm

t(s)=0

0   0.4

Blocked Sites Rel. Conc.

Latent Image Edge

Reacted sites (acid source)

Bias and LER!!!

6th International Symposium on
Immersion Lithography Extension:
October 22-23 2009 • Prague, Czech Republic
held in conjunction with the Symposium on Extreme Ultraviolet Lith

Lord Sir Alec and the Queen!

held in conjunction with the Symposium on Extreme Ultraviolet Li

Resolution Limits

With Sir Alec Broers

IBM Yorktown

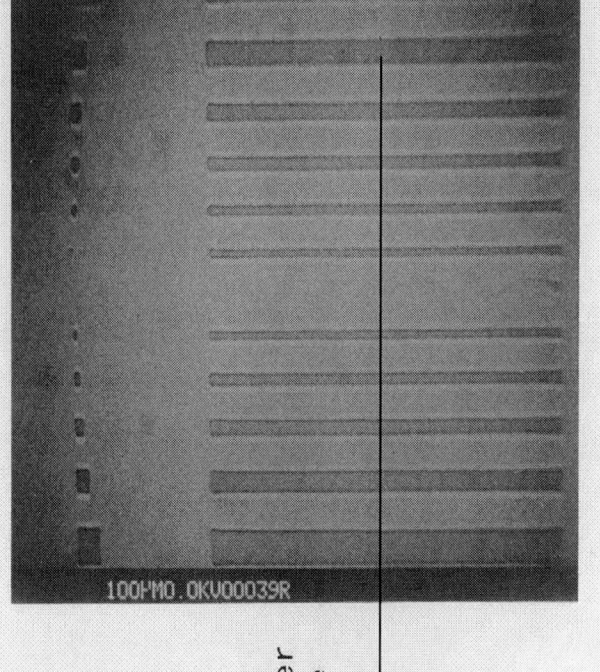

Lower Dose

Higher Dose

6th International Symposium on Immersion Lithography Extension

October 22-23 2009 • Prague, Czech Republic
held in conjunction with the Symposium on Extreme Ultraviolet Lit

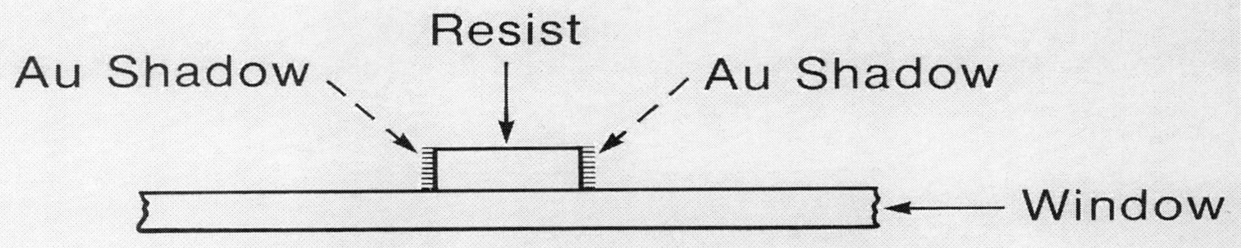
Resist
Au Shadow
Au Shadow
Window

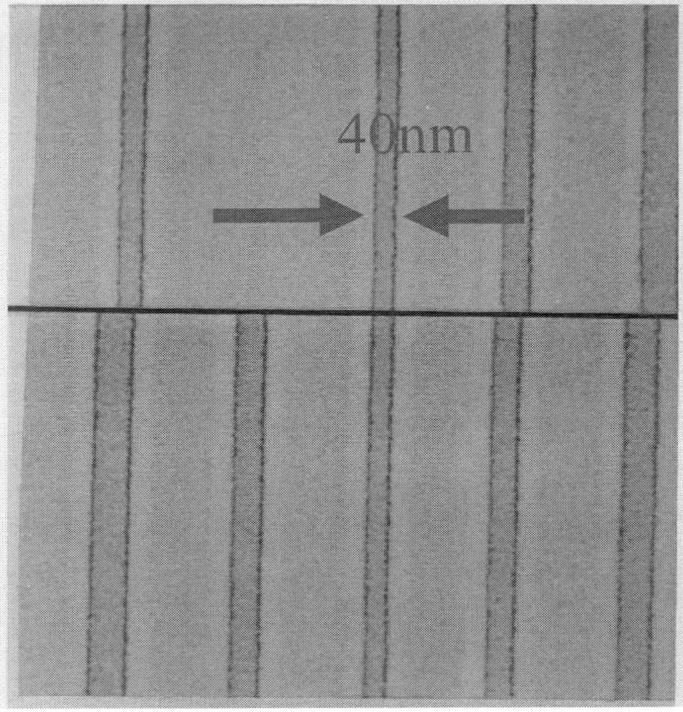
40nm

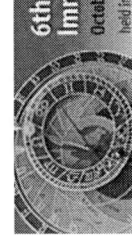

There is a fundamental problem…Bias

The t-BOC Bilayer Experiment

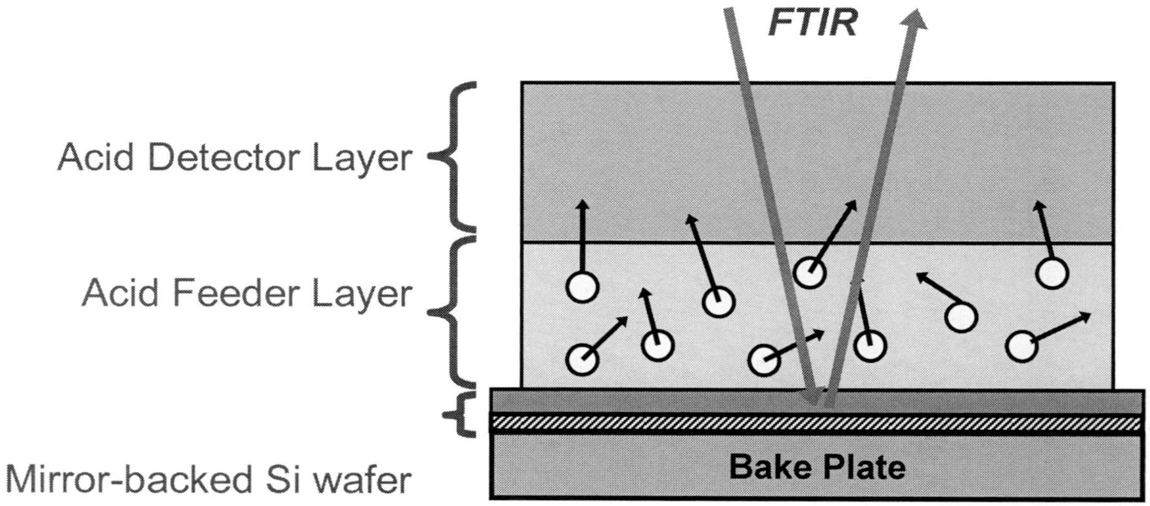

Normalized FTIR Peak at 1760 cm^{-1}

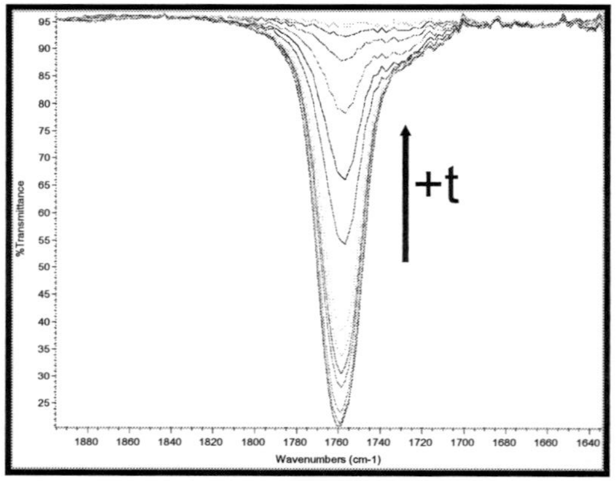

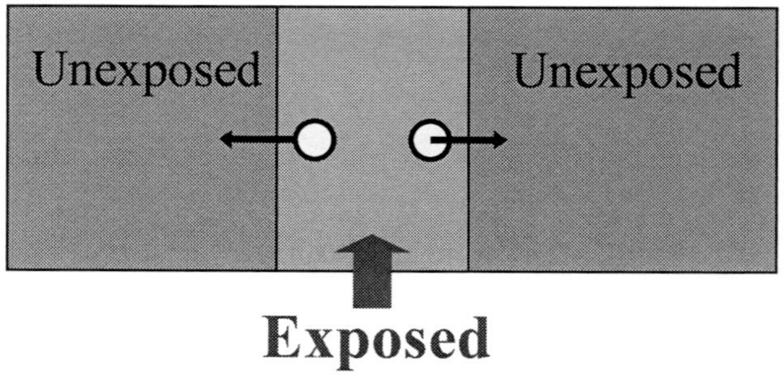

+t

Acid Detector Layer

Acid Feeder Layer

Mirror-backed Si wafer

Bake Plate

FTIR

Unexposed

Unexposed

Exposed

6th International Symposium on Immersion Lithography Extension:
October 22-23 2009 • Prague, Czech Republic
held in conjunction with the Symposium on Extreme Ultraviolet Lit

Bilayer Experimental Results

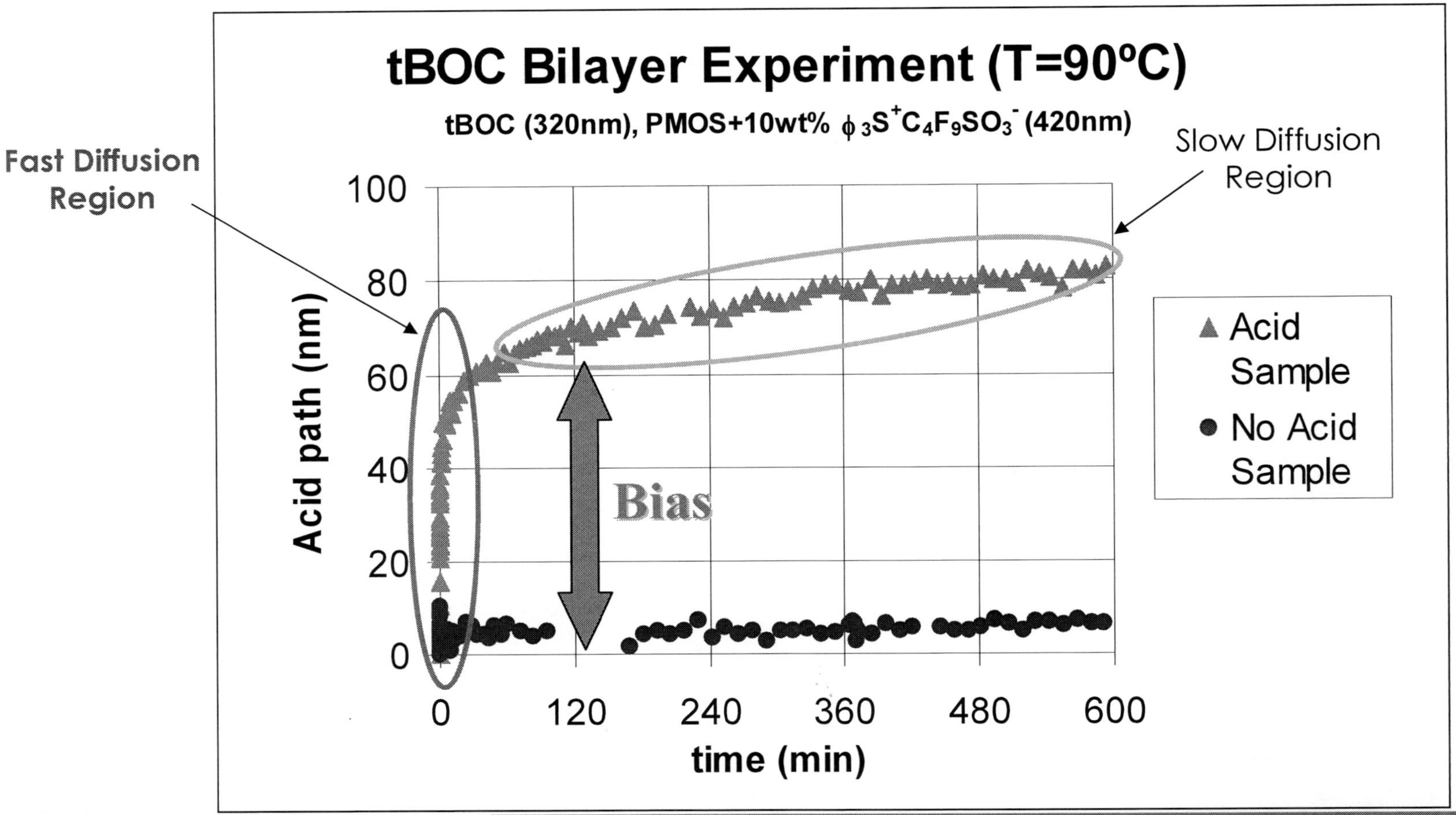

tBOC Bilayer Experiment (T=90°C)
tBOC (320nm), PMOS+10wt% $\phi_3S^+C_4F_9SO_3^-$ (420nm)

Slow Diffusion Region

Fast Diffusion Region

Bias

▲ Acid Sample
● No Acid Sample

Acid path (nm)

time (min)

6th International Symposium on
Immersion Lithography Extension:
October 22-23 2009 · Prague, Czech Republic
held in conjunction with the Symposium on Extreme Ultraviolet Lith

Bilayer Experiments

Acid Transport at Various Temperatures

TBOC (320 nm) Floated onto pMOS + 10wt% PAG($\phi_3SC_4F_9SO_3$)

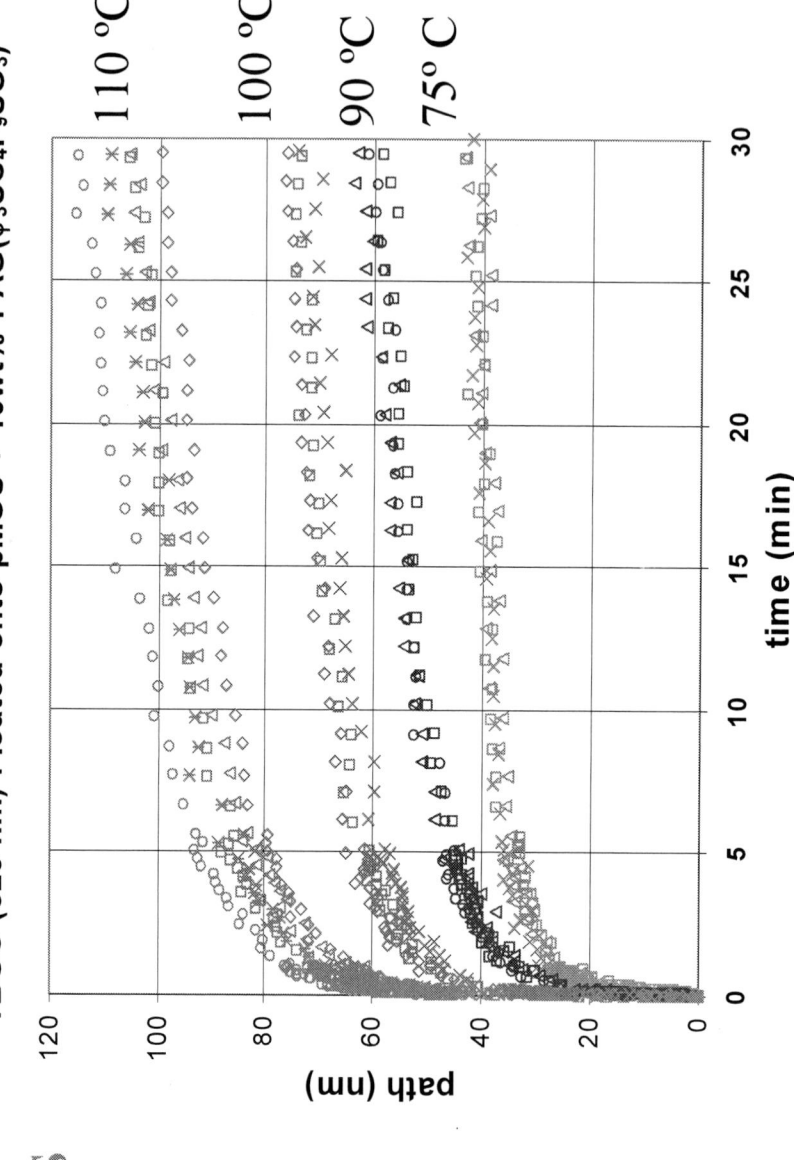

Bias

Effect of Counterion Size on Bias

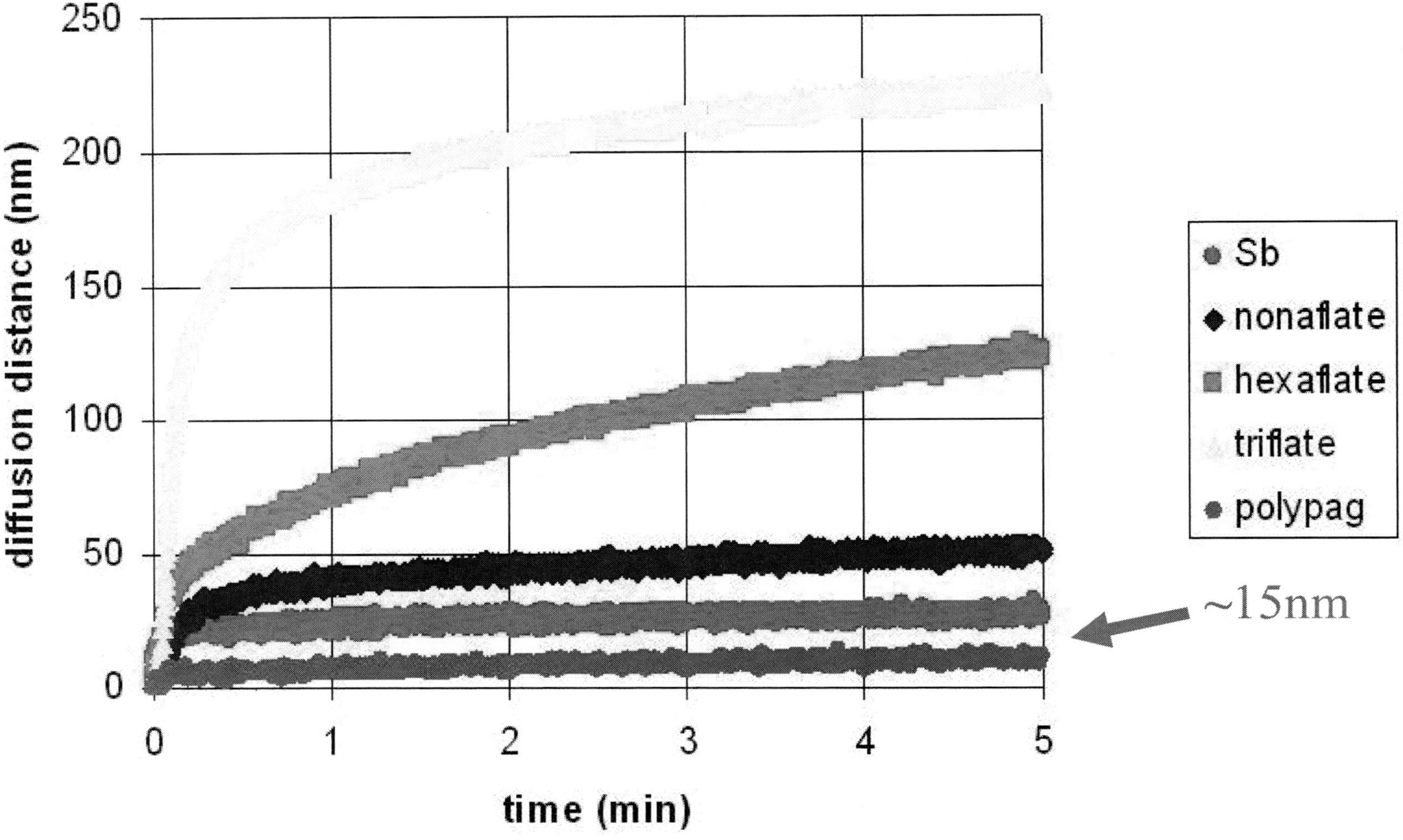

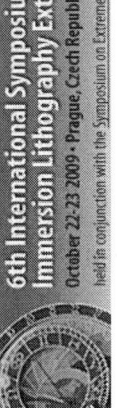

6th International Symposium on
Immersion Lithography Extension:
October 22-23 2009 · Prague, Czech Republic
Held in conjunction with the Symposium on Extreme Ultraviolet Lith

Polymer Bound PAGs

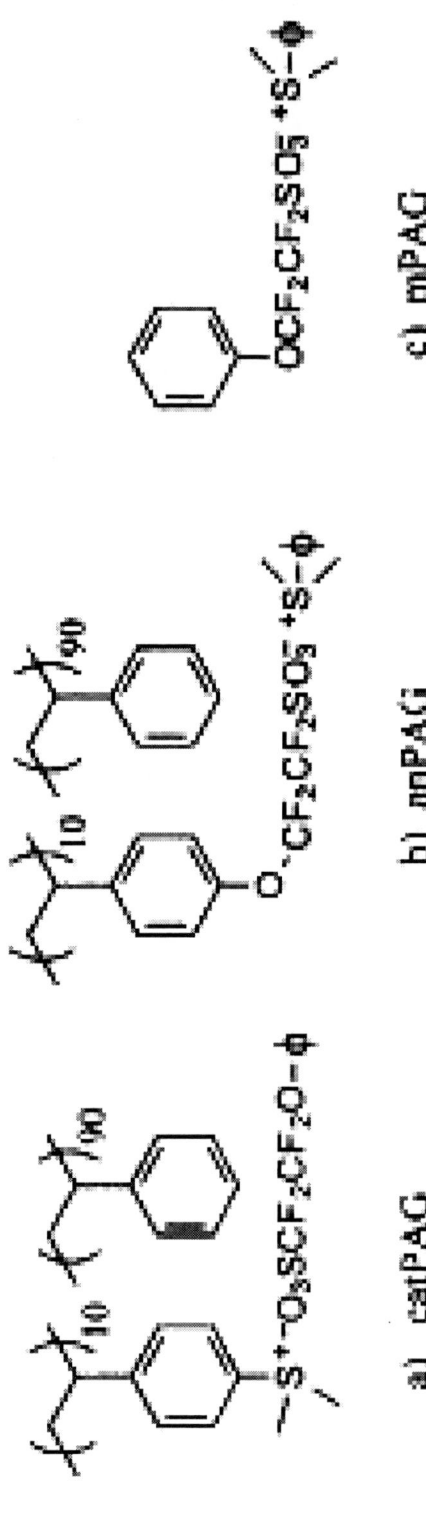

a) catPAG

b) anPAG

c) mPAG

M.D. Stewart and H.V. Tran JVST B . 20(6) 2946-2952 (2003)
R.D. Allen, et al, J. J. Photopolymer Science and Technol. 22, 25 (2009)

Line Spread Function Measurement
F. Houle et al., IBM Almaden

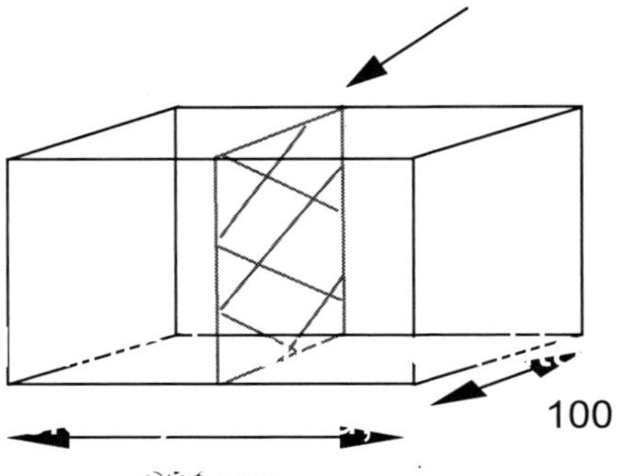

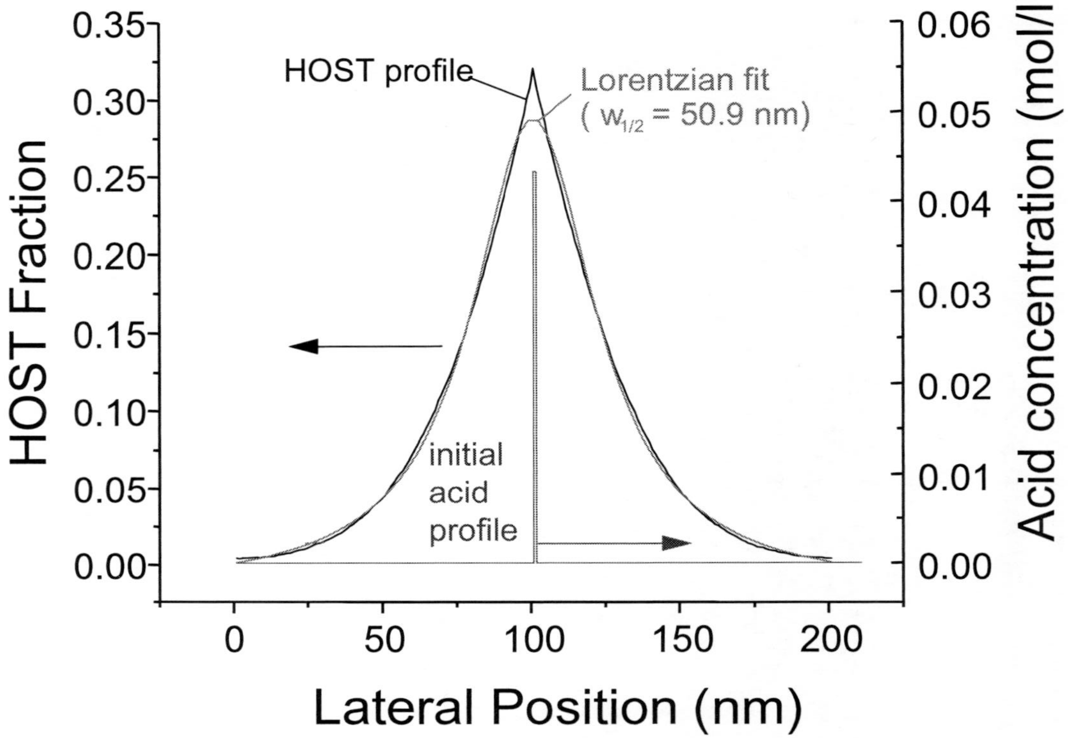

Acid concentration (mol/l)

HOST profile

Lorentzian fit ($w_{1/2}$ = 50.9 nm)

initial acid profile

HOST Fraction

Lateral Position (nm)

100 nm

1 nm width

100 nm

201 nm

Consistent with experimentally measured line spread arising from PEB reaction-diffusion kinetics

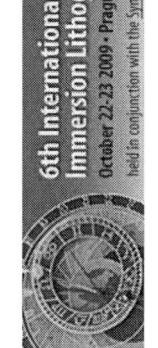

6th International Symposium on
Immersion Lithography Extension:
October 22-23 2009 · Prague, Czech Republic
held in conjunction with the Symposium on Extreme Ultraviolet Lit...

Line edge Roughness

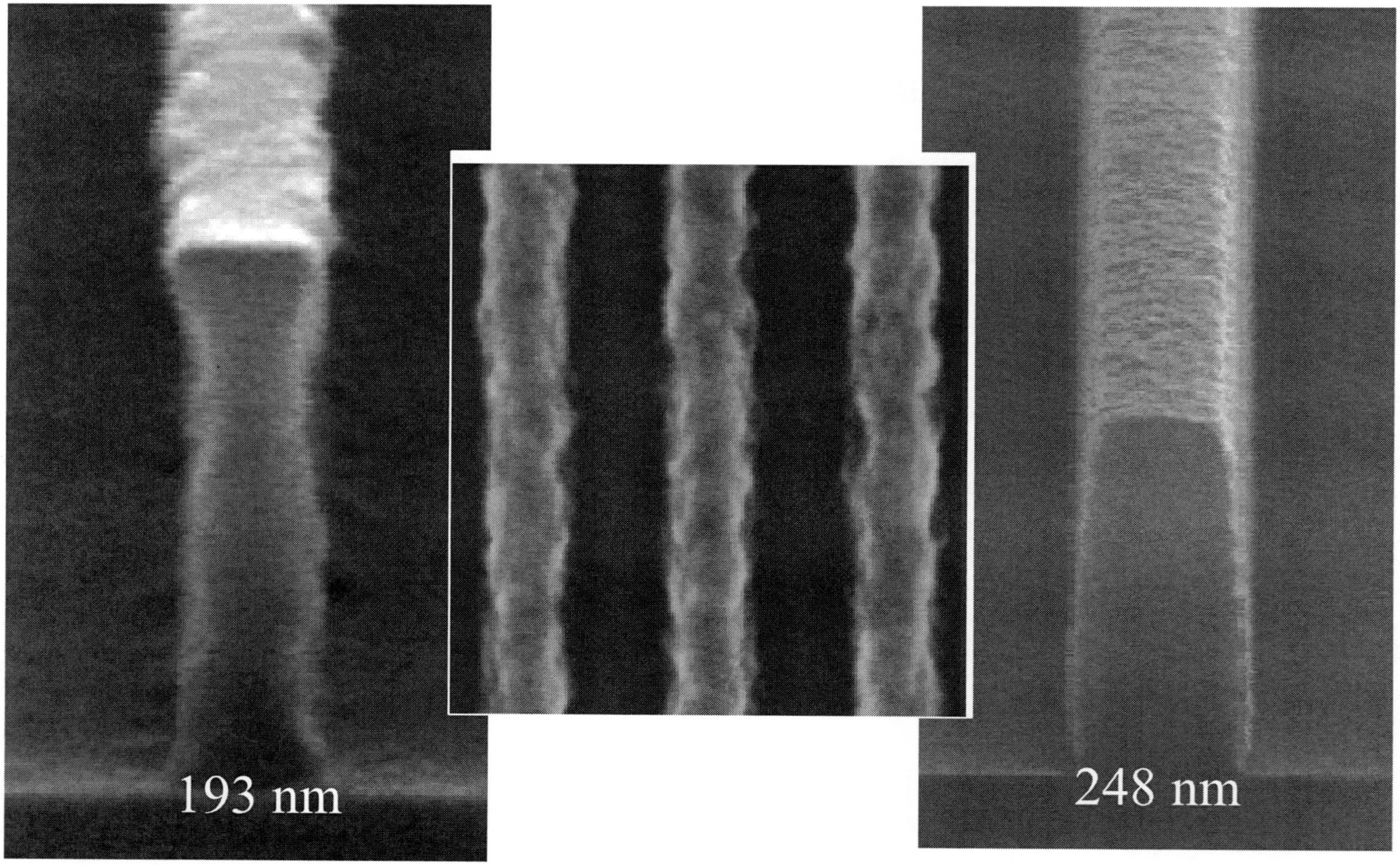

248 nm

193 nm

6th International Symposium on
Immersion Lithography Extension:
October 22-23 2009 · Prague, Czech Republic
held in conjunction with the Symposium on Extreme Ultraviolet Lit

Influence of Base on LER

- Base quencher can decrease the acid sphere of influence in low contrast regions, thereby reducing LER.

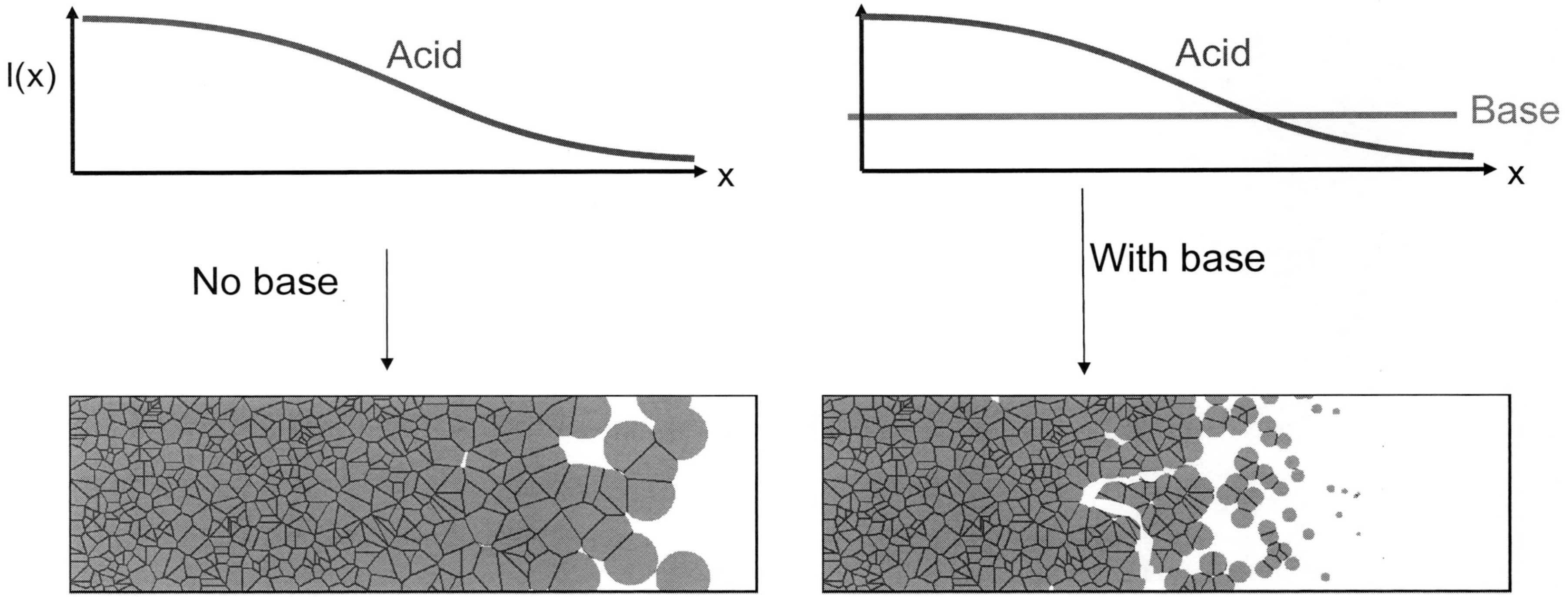

No base

With base

J. E. Meiring, T. B. Michaelson, G. M. Schmid and C. G. Willson, *Proc. SPIE*, **5753** (2005)

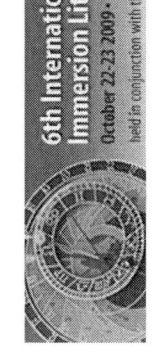

6th International Symposium on
Immersion Lithography Extension:
October 22-23 2009 · Prague, Czech Republic
held in conjunction with the Symposium on Extreme Ultraviolet Li

Exploring Base Effects

- To add base quencher seems to make the contrast higher, thereby LER reducing.

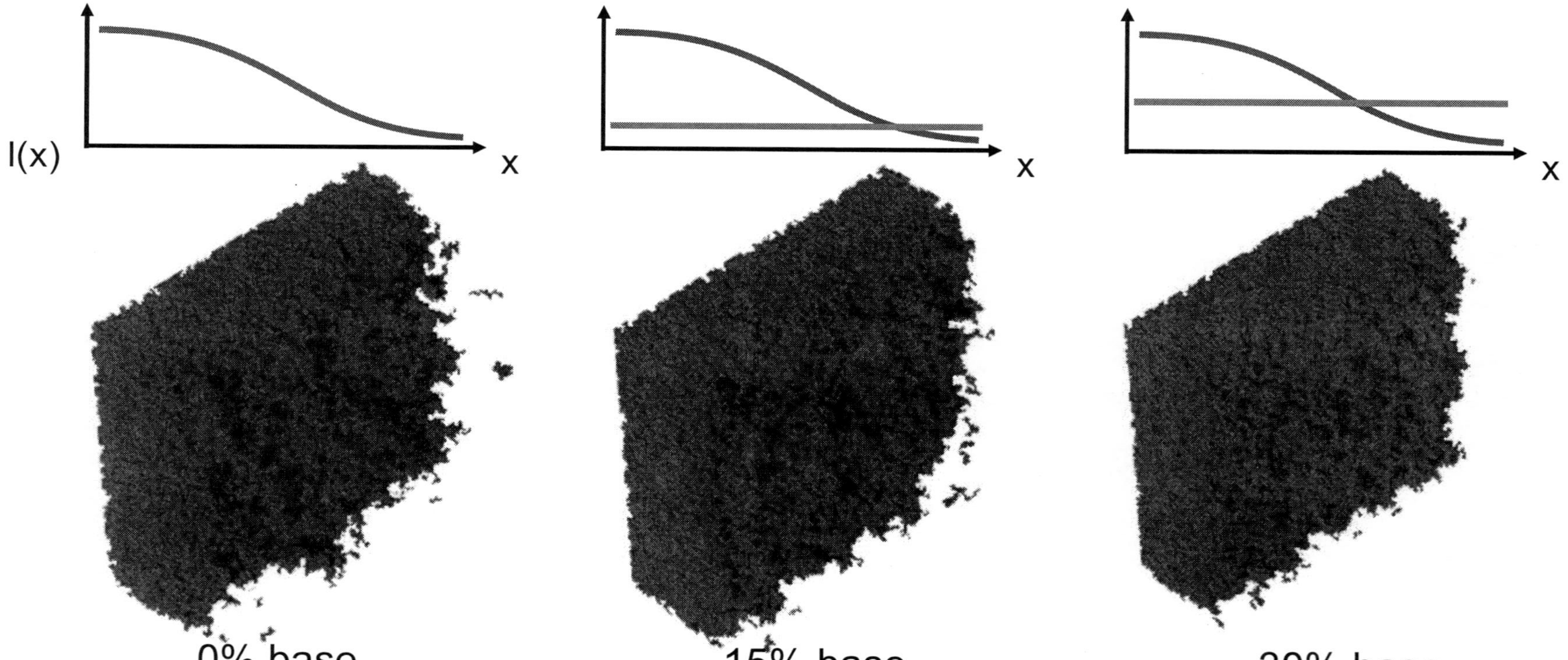

$I(x)$

x

x

x

0% base
6.61 nm RMS

15% base
5.47 nm RMS

30% base
3.89 nm RMS

J. E. Meiring, T. B. Michaelson, G. M. Schmid and C. G. Willson, *Proc. SPIE*, **5753**(2005),

**6th International Symposium on
Immersion Lithography Extension:**
October 22-23 2009 - Prague, Czech Republic
held in conjunction with the Symposium on Extreme Ultraviolet Lit

Influence of ILS on LER

- The most significant contributor to LER is the quality of the aerial image
  → Pitch, illumination, line width, flare, fading, etc.

- LER is a minimum as the image-log-slope (ILS) of the aerial image is a maximum

- As the ILS approaches infinity, the magnitude of LER reaches a non-zero constant

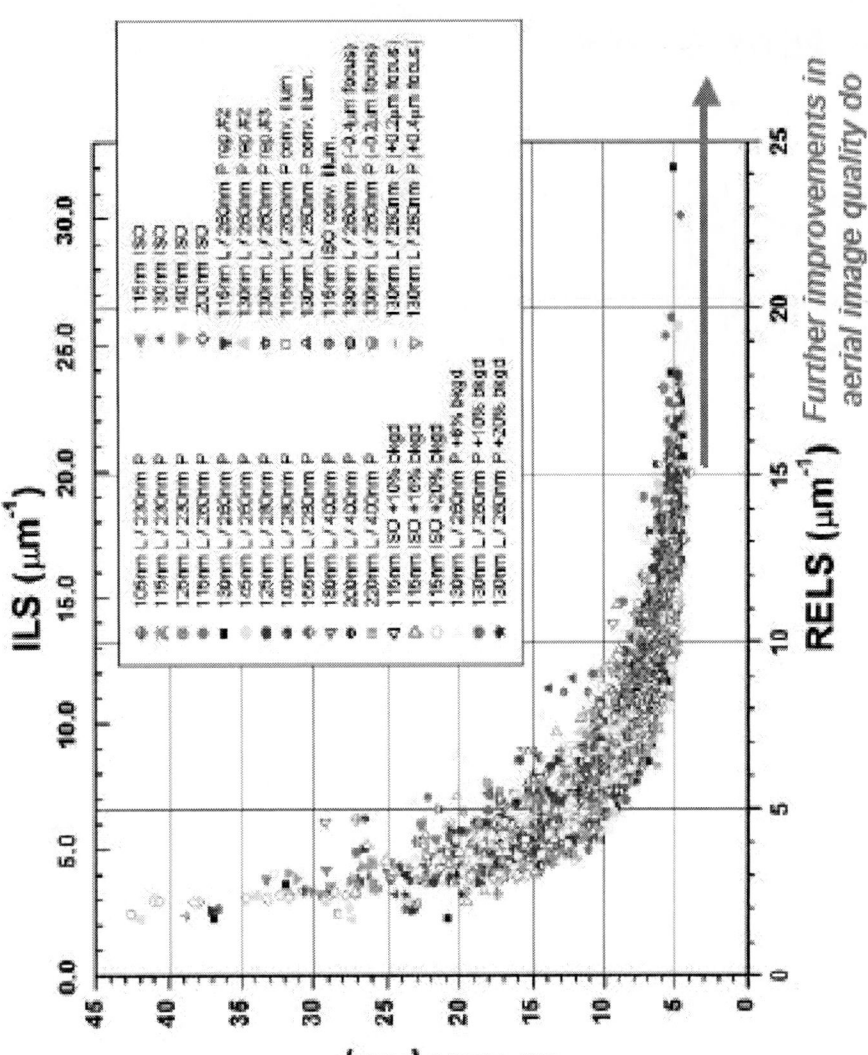

Further improvements in aerial image quality do not reduce LER further

*A.R Pawloski et al., Proc. SPIE, vol. 5378, 2004

Chemical contrast from kinetics

Polymer Protection Gradient

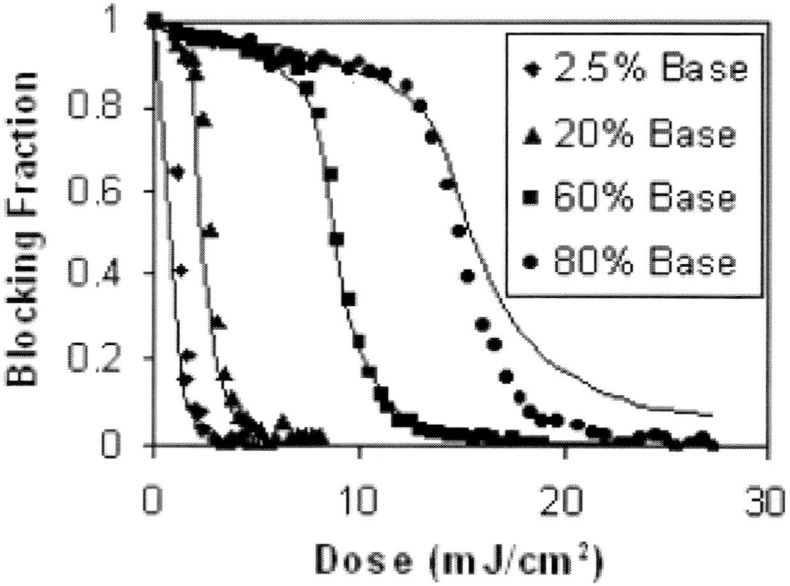

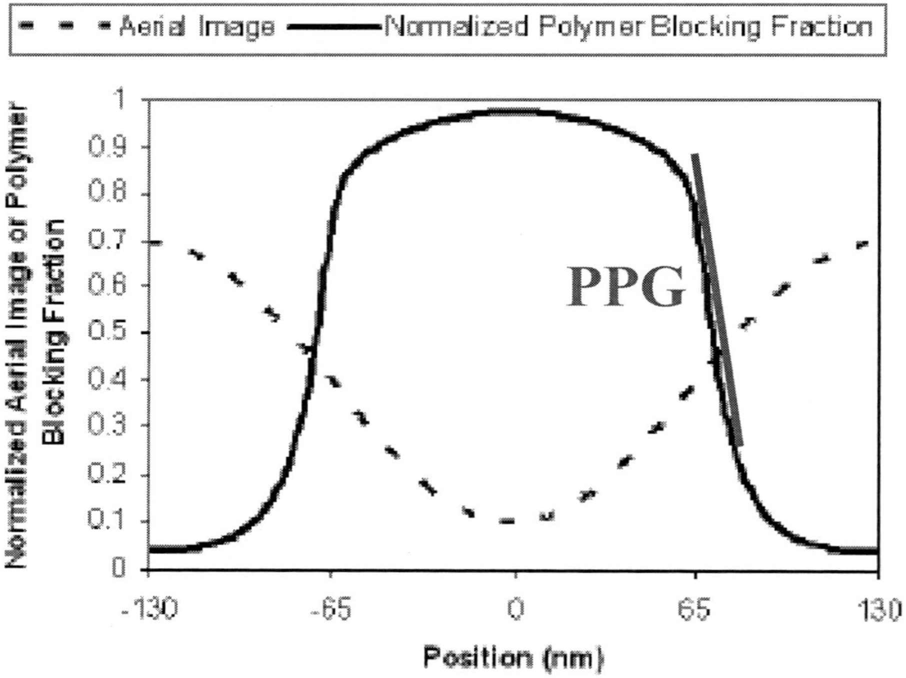

Timothy B Michaelson, et al, Proc SPIE 5753, 368 (2003)

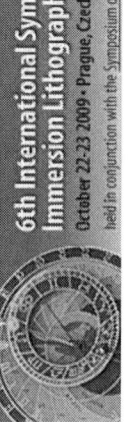

LER vs PPB

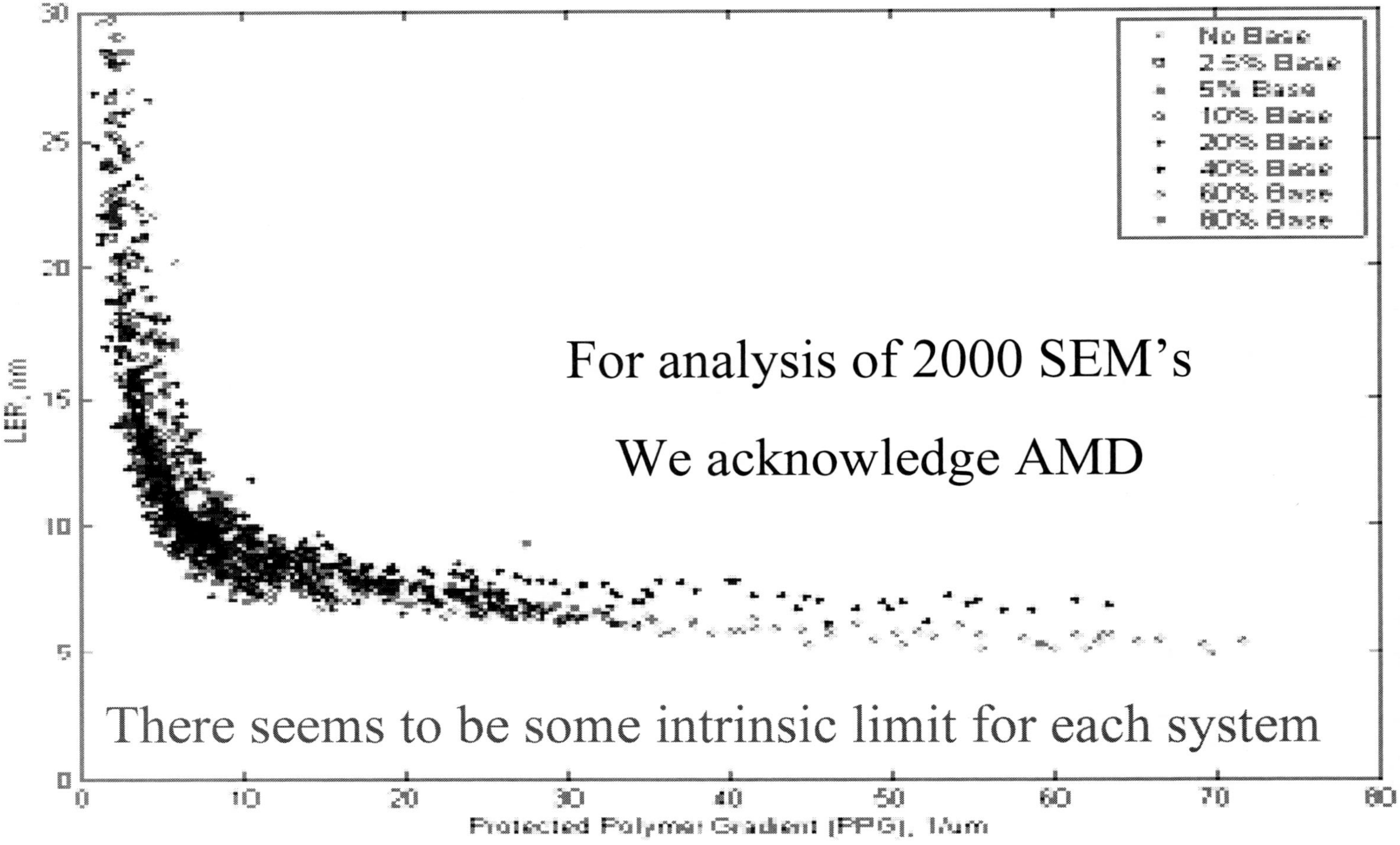

For analysis of 2000 SEM's

We acknowledge AMD

There seems to be some intrinsic limit for each system

Timothy B Michaelson, et al, Proc SPIE 5753, 368 (2003)

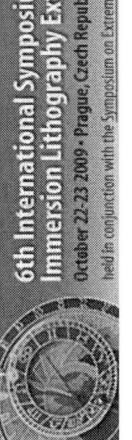

The Triangle of death

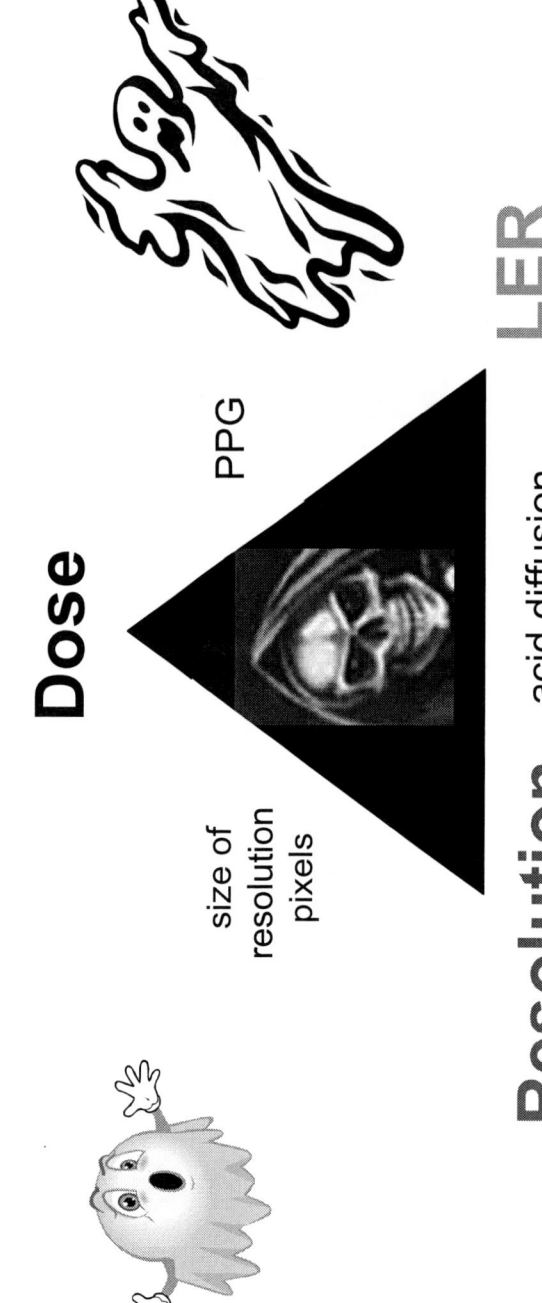

Dose

PPG

size of
resolution
pixels

LER

acid diffusion

Resolution

Pick goodness in any two…

From IMEC, IBM, Intel etc.

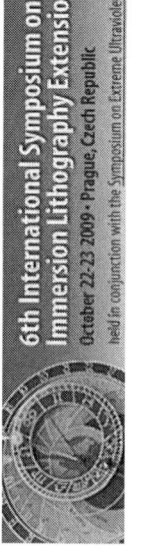

6th International Symposium on
Immersion Lithography Extension:

October 22-23 2009 • Prague, Czech Republic

held in conjunction with the Symposium on Extreme Ultraviolet Li

EUV study from from a chip manufacturer:

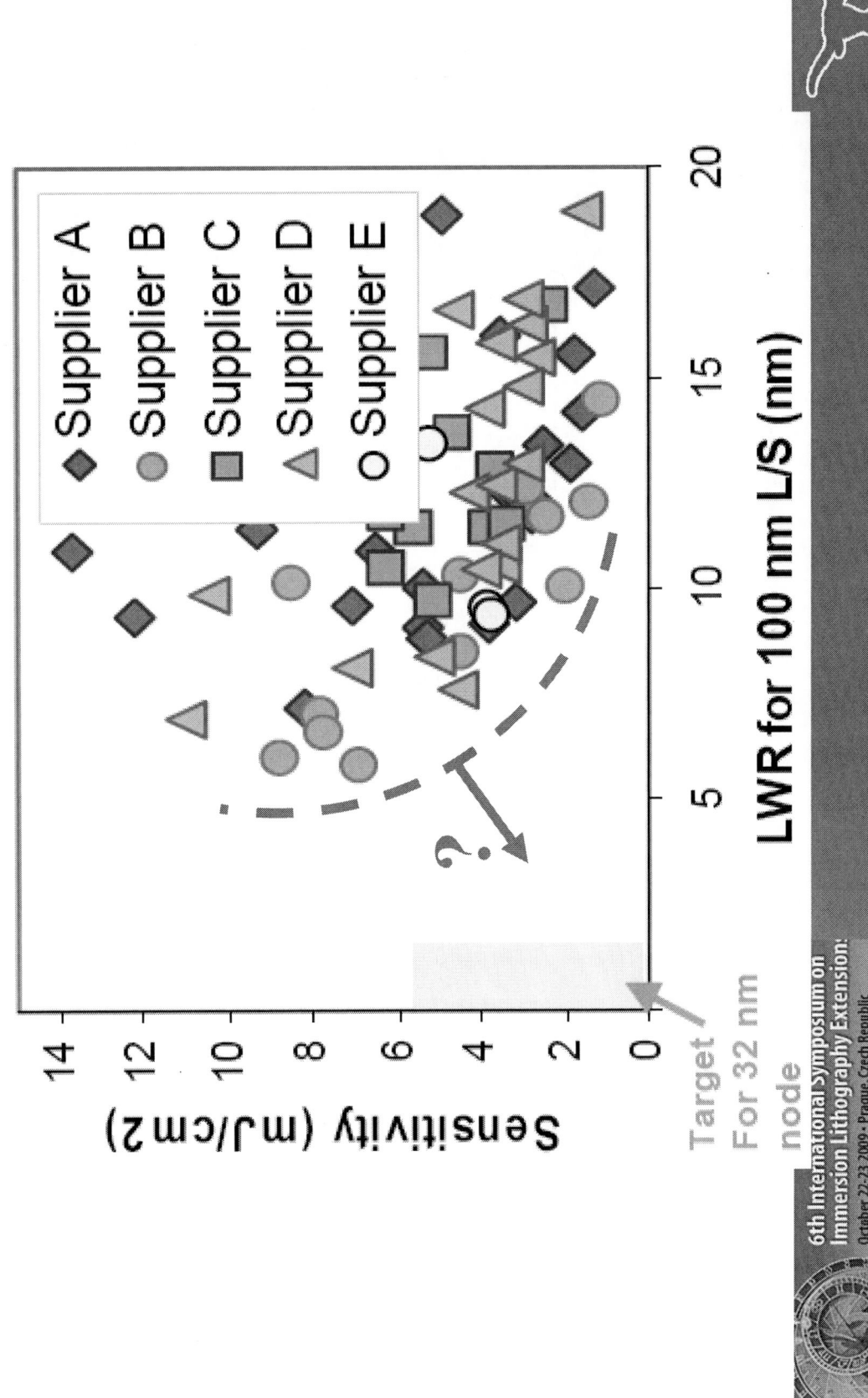

The Big Trade Off

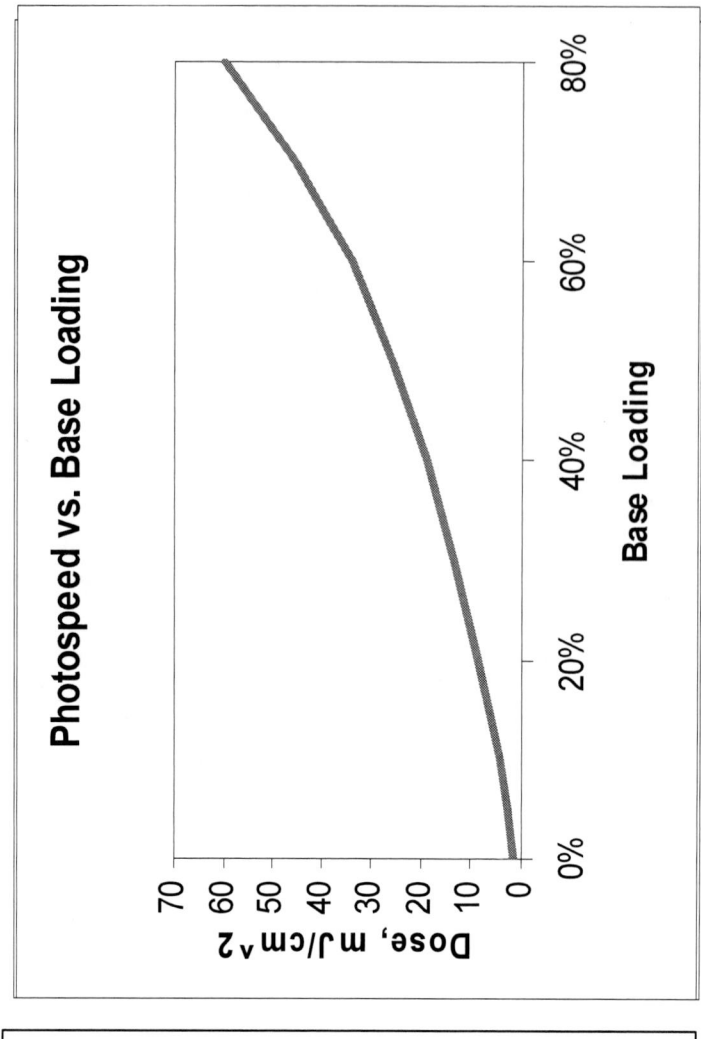

Can we escape the Triangle of Death

- How can we move this limit…change the trade off.

  - Quantified by Z factor (1) and Klup (2)
  - One implication is that higher PAG loading helps (3). There is some experimental evidence to support this prediction (3)
  - EUV people are pushing this HARD!!!

1) D. Van Steenwinckel, et al, Proc SPIE 6519, 65190V (2008)

2) E. Steven Putna, et al Proc SPIE 7273,72731L-9 (2009)

3) Gregg M. Gallatin, et al Proc SPIE 6921, 69211E-3 (2008)

http://www.spectrum.ieee.org/images/nov08/images/doub03.pdf

SIDEWALL SPACER

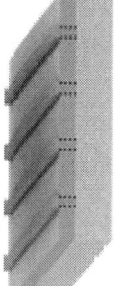
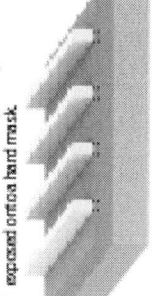

1 Litho + Etch 1 (dummy pattern). A dummy pattern (yellow) is created on silicon (blue).

2 Grow sidewalls. A film (brown) is grown around the dummy lines.

3 Etchback. All of the film is removed except the sidewalls.

4 Strip dummy pattern. The dummy pattern (yellow) is removed, leaving the sidewalls.

5 Etch 2. The remaining double-density sidewall pattern is etched into the silicon (blue).

LITHO-FREEZE-LITHO-ETCH (LFLE)

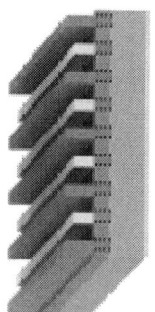
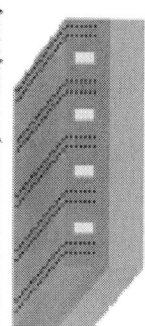

1 Litho 1. The first pattern (yellow) is exposed onto silicon (blue).

2 Freeze, coat with new resist. The already developed layer (yellow) is chemically frozen and coated with a second layer of resist (brown).

3 Litho 2. A second pattern (brown) is exposed, doubling pattern density.

4 Etch. The unprotected silicon is engraved with the final, double-density pattern in a single etching operation.

LITHO-ETCH-LITHO-ETCH (LELE)

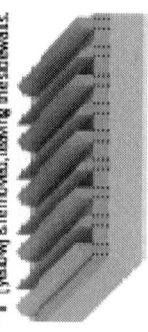

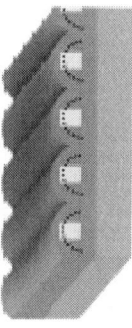

1 Litho 1. The first pattern (yellow) is exposed onto a hard mask.

2 Etch 1. The first pattern is etched into the hard mask (brown).

3 Litho 2. A second pattern (yellow) is exposed onto silicon (blue), doubling pattern density.

4 Etch 2. The final, double-density pattern is engraved into the silicon.

5 Wash. The remaining marks is washed away.

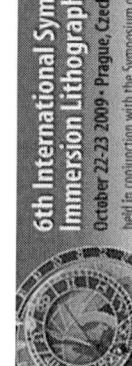

6th International Symposium on
Immersion Lithography Extension:
October 22-23 2009 - Prague, Czech Republic
held in conjunction with the Symposium on Extreme Ultraviolet Lit

Double Exposure

- Please see poster by Paul Zimmerman
  - Progress on double patterning with _no_ freezing and _no_ side wall deposition!

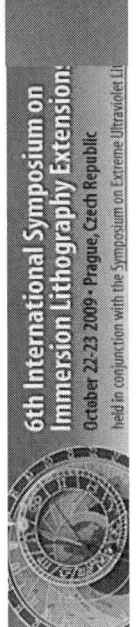

6th International Symposium on
Immersion Lithography Extension:
October 22-23 2009 • Prague, Czech Republic
held in conjunction with the Symposium on Extreme Ultraviolet Li

Pitch Doubling

- Double patterning with a single exposure

  – Please see paper by _Carlos Fonseca_ and Paper by _Xinyu Gu_ for examples.

  – Add magic powders or use magic developers; print a mask with X nm grating and get a resist image with 0.5 X nm grating.

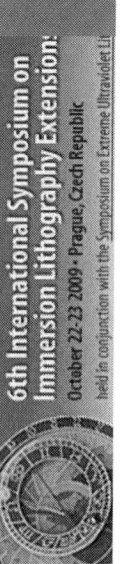

6th International Symposium on Immersion Lithography Extension:
October 22-23 2009 · Prague, Czech Republic
held in conjunction with the Symposium on Extreme Ultraviolet Lit

Can this be done.......Yes!!

Mask

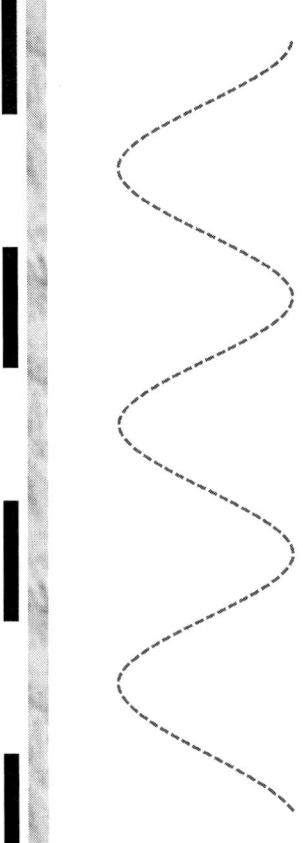

Aerial Image

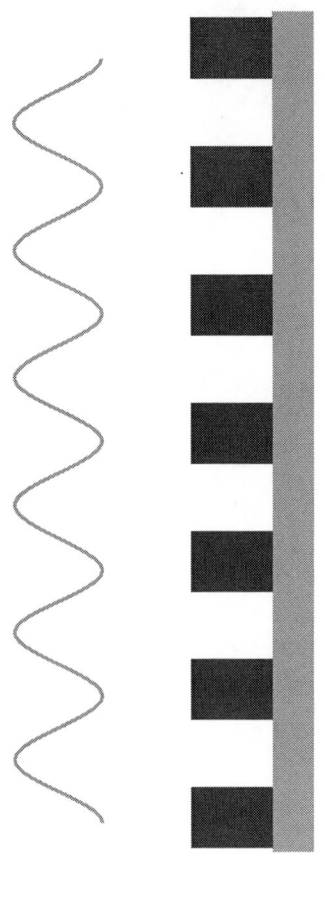

Resist Response
[Acid]

After
Development

Maybe!!!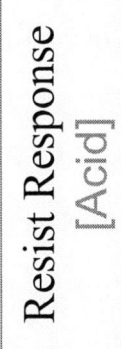

6th International Symposium on Immersion Lithography Extension:
October 22-23 2009 • Prague, Czech Republic
held in conjunction with the Symposium on Extreme Ultraviolet Lit

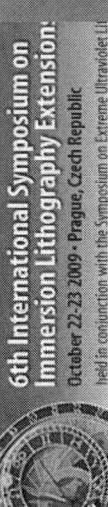

What fun!!!

- Way too early to retire

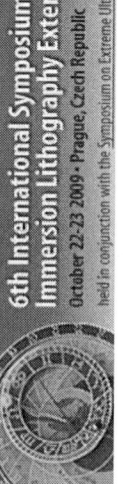

6th International Symposium on Immersion Lithography Extension:
October 22-23 2009 • Prague, Czech Republic
held in conjunction with the Symposium on Extreme Ultraviolet Lithography

Closing Thoughts

As you   you
to think   ntial
are reall   we
can eith   ng
today. I   al
years to

Thank you!

Gordon Moore, Lithography and the Future of Moore's Law
[Proc. SPIE Vol. 2437, May 1995]

Line-Edge Roughness and the Ultimate Limits of Lithography

Chris Mack

www.lithoguru.com

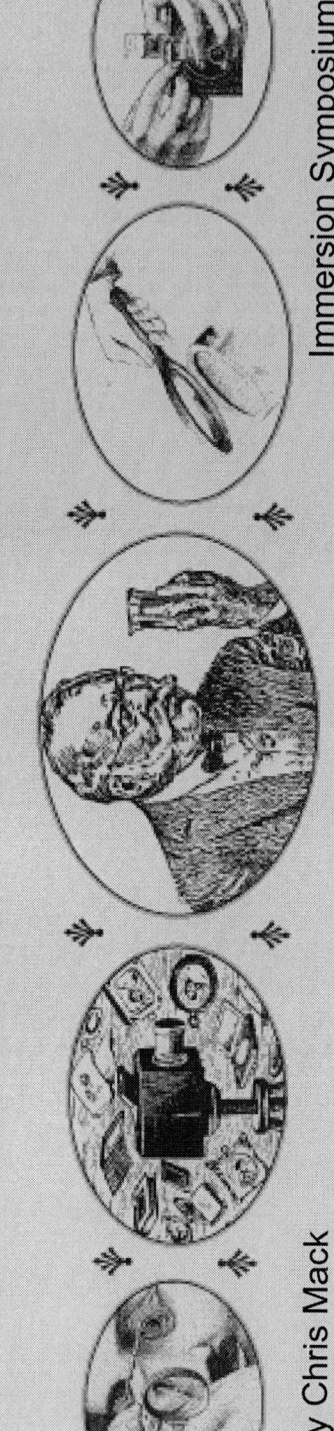

Immersion Symposium - Prague

© 2009 by Chris Mack

Outline:
There's a lot going on in LER

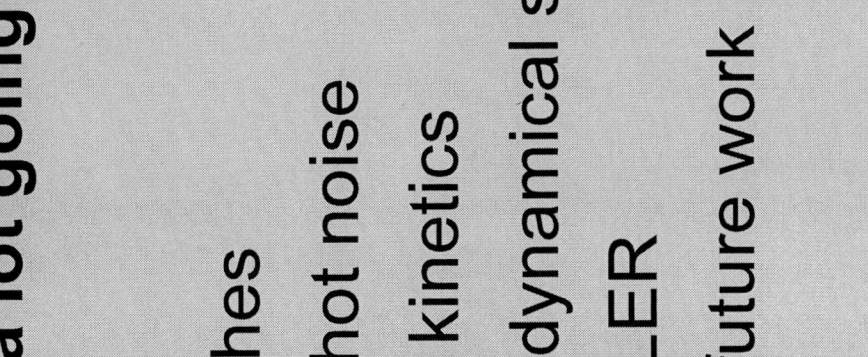

- Modeling Approaches
- Photon and acid shot noise
- Reaction-diffusion kinetics
- Development and dynamical scaling
- Overall model for LER
- What's missing – future work
- Conclusions

(c) 2009

Continuum Approximation

- The real world is discrete (photons, atoms, etc.), but most macroscopic models (e.g., litho simulation) make the *continuum approximation*

  – Matter and energy are described with continuous mathematical functions

  – Ex: aerial image intensity, acid concentration after exposure, resist dissolution

- What are the implications of making the continuum approximation?

  – Line-edge roughness cannot be predicted

(c) 2009

 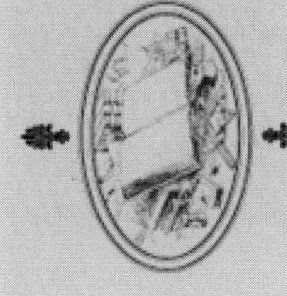

Continuum Approximation Example:
Chemical Concentration

- Concentration: The number of atoms or molecules of a certain type per unit volume
  - By necessity, an average over a volume
- What is the meaning of $H(x,y,z)$ – the concentration of acid at a specific point in space?

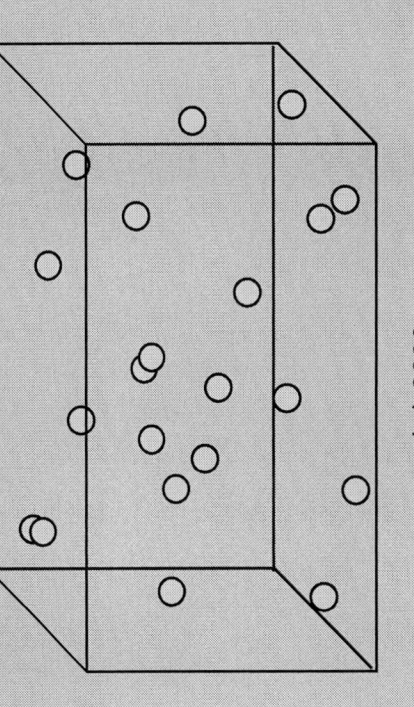

(c) 2009

Modeling LER

- Describe every event probabilistically
- Law of Large Numbers – stochastic models become continuum models when the number of events becomes very large (mean field theory)
- First Stochastic Method: Monte Carlo
  - Can be very rigorous, but very slow
  - Can be difficult to optimize materials and processes (slow, hard to gain intuition)
- Second Stochastic Method: Approximate analytic solution
  - True analytic solution is not possible (how good is the approximation?)
  - Goal: Predict standard deviation, correlation length, and roughness exponent (i.e., predict the full PSD)

(c) 2009

Stochastic View of Chemical Concentration

- Model atom/molecule as a point located at its center of mass
- Consider a volume V – is the molecule in the volume or not?
  - This is a binary proposition, governed by the binomial distribution: $P(n)$ = probability of finding n molecules in V
  - The binomial probability distribution can be well approximated by a Poisson distribution with average concentration C

$$P(n) = \frac{(CV)^n}{n!} e^{-CV} \qquad \langle n \rangle = CV \qquad \sigma_n^2 = CV$$

$$\frac{\sigma_n}{\langle n \rangle} = \frac{1}{\sqrt{\langle n \rangle}} = \frac{1}{\sqrt{CV}}$$

(c) 2009

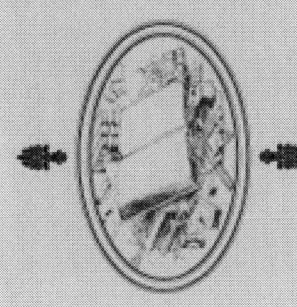

Stochastic Chemical Concentration (Poisson Distribution)

- Example: for a typical 193nm resist formulation, $G_0 N_A = 0.042 /nm^3$ (G_0 = the initial concentration of PAG, N_A = Avogadro's number)

For V = (3 nm)^3 $\langle n \rangle = 1.13$ $\sigma_n / \langle n \rangle = 94\%$

For V = (6 nm)^3 $\langle n \rangle = 9$ $\sigma_n / \langle n \rangle = 33\%$

For V = (10 nm)^3 $\langle n \rangle = 42$ $\sigma_n / \langle n \rangle = 15\%$

(c) 2009

Photon Shot Noise
(Also a Poisson Distribution)

- Example: for a typical 193nm resist with 10 mJ/cm^2 dose to clear,

  For A = (1 nm)^2 $\langle n \rangle = 97$ $\sigma_n / \langle n \rangle = 10\%$

  For A = (10 nm)^2 $\langle n \rangle = 9700$ $\sigma_n / \langle n \rangle = 1\%$

- Example: for an EUV resist with 5 mJ/cm^2 dose to clear,

  For A = (1 nm)^2 $\langle n \rangle = 3$ $\sigma_n / \langle n \rangle = 58\%$

  For A = (10 nm)^2 $\langle n \rangle = 300$ $\sigma_n / \langle n \rangle = 6\%$

(c) 2009

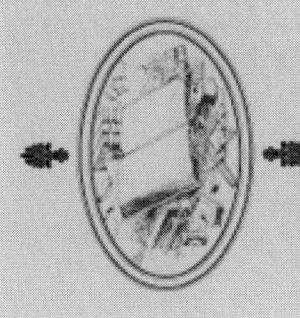

Stochastic View of Exposure Reaction

- Including photon shot noise, acid uncertainty is

$$\sigma_h^2 = \frac{\langle h \rangle}{\langle n_{0-PAG} \rangle} + \frac{\left[(1-\langle h \rangle)\ln(1-\langle h \rangle)\right]^2}{\langle n_{photon} \rangle}$$

- Max value of $\left[(1-\langle h \rangle)\ln(1-\langle h \rangle)\right]^2 = 0.135$

- The pure photon shot noise contribution is usually small, even for EUV

(c) 2009

Stochastic View of Reaction-Diffusion

- Reaction is catalyzed by the diffusing species:

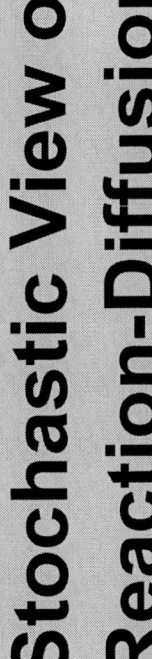

von Smoluchowski Trap:

Reaction can occur once acid approaches the blocking group within its capture radius, a.

$$Rate \propto a$$

—CH$_2$–CH—

M

H+

(c) 2009

Reaction-Diffusion Kinetics

$$h_{eff}(x) = \frac{1}{t_{PEB}} \int_0^{t_{PEB}} h(x, t=0) \otimes DPSF \, dt$$

Relative concentration of blocked polymer sites $\longrightarrow$ $m(x) = e^{-K_{amp} t_{PEB} h_{eff}(x)}$

Deblocking responds to the time-average of the acid latent image

(c) 2009

Stochastic View of Reaction-Diffusion

- Deriving the statistics of reaction-diffusion is hard! For the details, please see: Chris A. Mack, Fundamental Principles of Optical Lithography: The Science of Microfabrication, John Wiley & Sons, (London: 2007).

$$\langle h_{eff} \rangle = \langle h \rangle \otimes RDPSF \qquad RDPSF = \frac{1}{t_{PEB}} \int_0^{t_{PEB}} DPSF \, dt$$

$$\sigma_{h_{eff}} \approx \left(\frac{2a}{\sigma_D} \right) \sigma_h$$

Derivation of this is approximate – more work is needed

(c) 2009

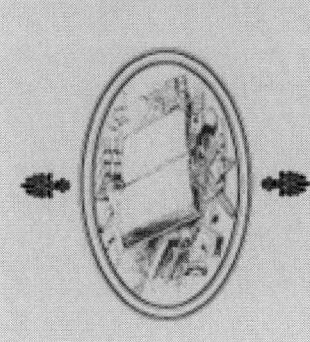

Stochastic View of Deprotection

- Statistical uncertainty in the blocked polymer site concentration comes from the Poisson distribution of the initial blocked sites, plus the stochastics of deblocking

$$\sigma_m^2 \approx \frac{\langle m \rangle}{\langle n_{0-blocked} \rangle} + \langle m \rangle^2 \left(K_{amp} t_{PEB} \sigma_{h_{eff}} \right)^2 = \frac{\langle m \rangle}{\langle n_{0-blocked} \rangle} + \left(\langle m \rangle \ln \langle m \rangle \right)^2 \left(\frac{\sigma_{h_{eff}}}{\langle h_{eff} \rangle} \right)^2$$

- Combining with our previous expressions for $\sigma_{h_{eff}}$ and σ_h gives the final result

(c) 2009

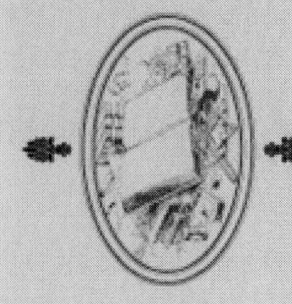

Stochastic View of Exposure + Reaction-Diffusion

- Final expression for the uncertainty in deblocked polymer concentration:

$$\left(\frac{\sigma_m}{\langle m \rangle}\right)^2 = \frac{1}{\langle n_{0-blocked} \rangle \langle m \rangle} + \left(K_{amp} t_{PEB}\right)^2 \left(\frac{2a}{\sigma_D}\right)^2 \left(\frac{\langle h \rangle}{\langle n_{0-PAG} \rangle} + \frac{\left[(1-\langle h \rangle)\ln(1-\langle h \rangle)\right]^2}{\langle n \rangle}\right)$$

Deblocking reaction

Reaction-diffusion

PAG concentration, exposure

Photon shot noise

(c) 2009

Correlation and Acid-Catalyzed Reaction-Diffusion

- As one acid diffuses and catalyzes several deprotection reactions, those deprotections are correlated

$$\text{Correlation Function} \longrightarrow R(\tau) = \frac{RDPSF \otimes RDPSF}{\iiint\limits_{\infty}(RDPSF)^2}$$

- Perform integrations numerically, examine the results
- Results can be almost perfectly fit with the standard exponential correlation function:

$$R(\tau) = e^{-(|\tau|/\xi)^{2\alpha}}$$

(c) 2009

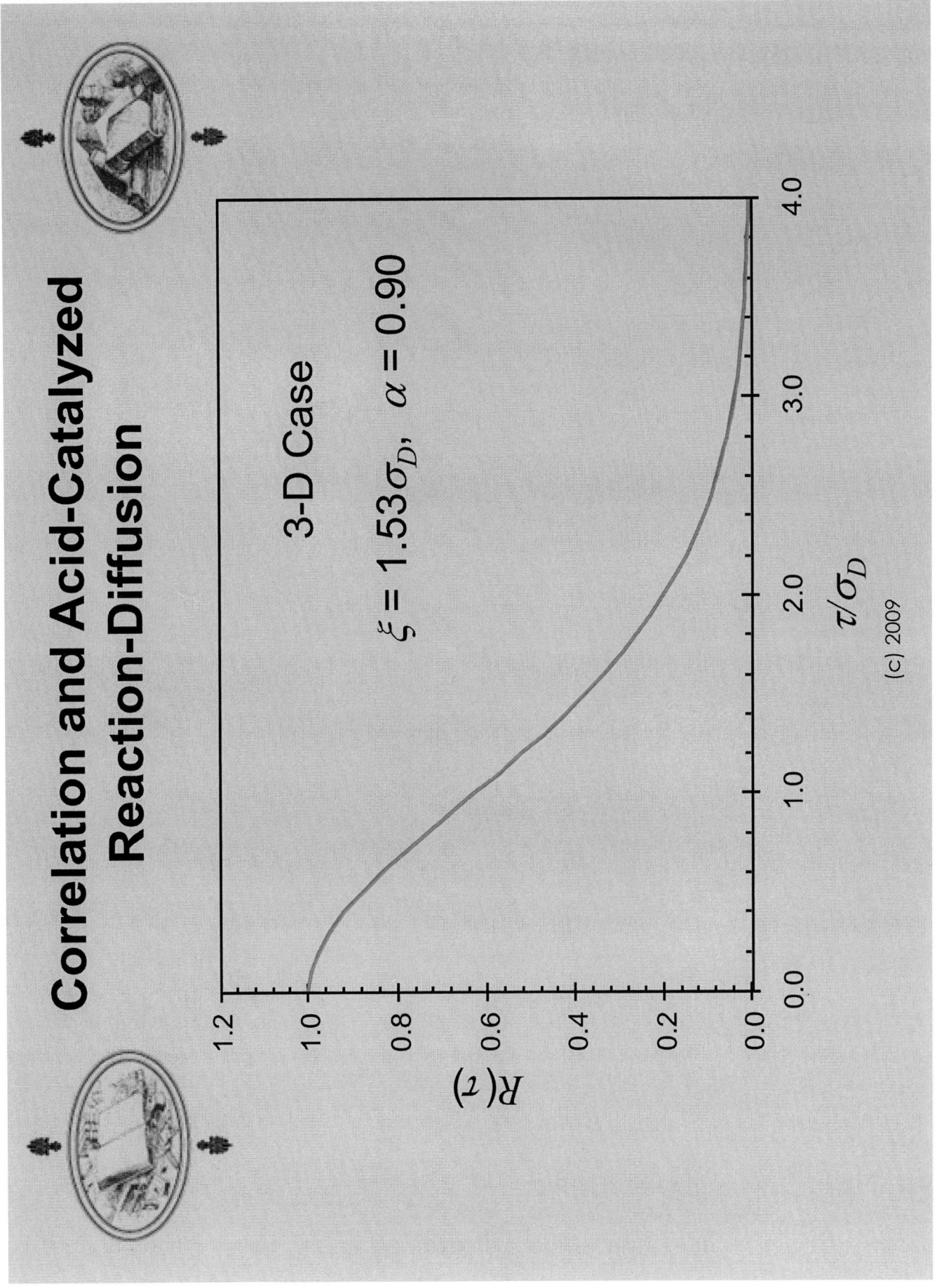

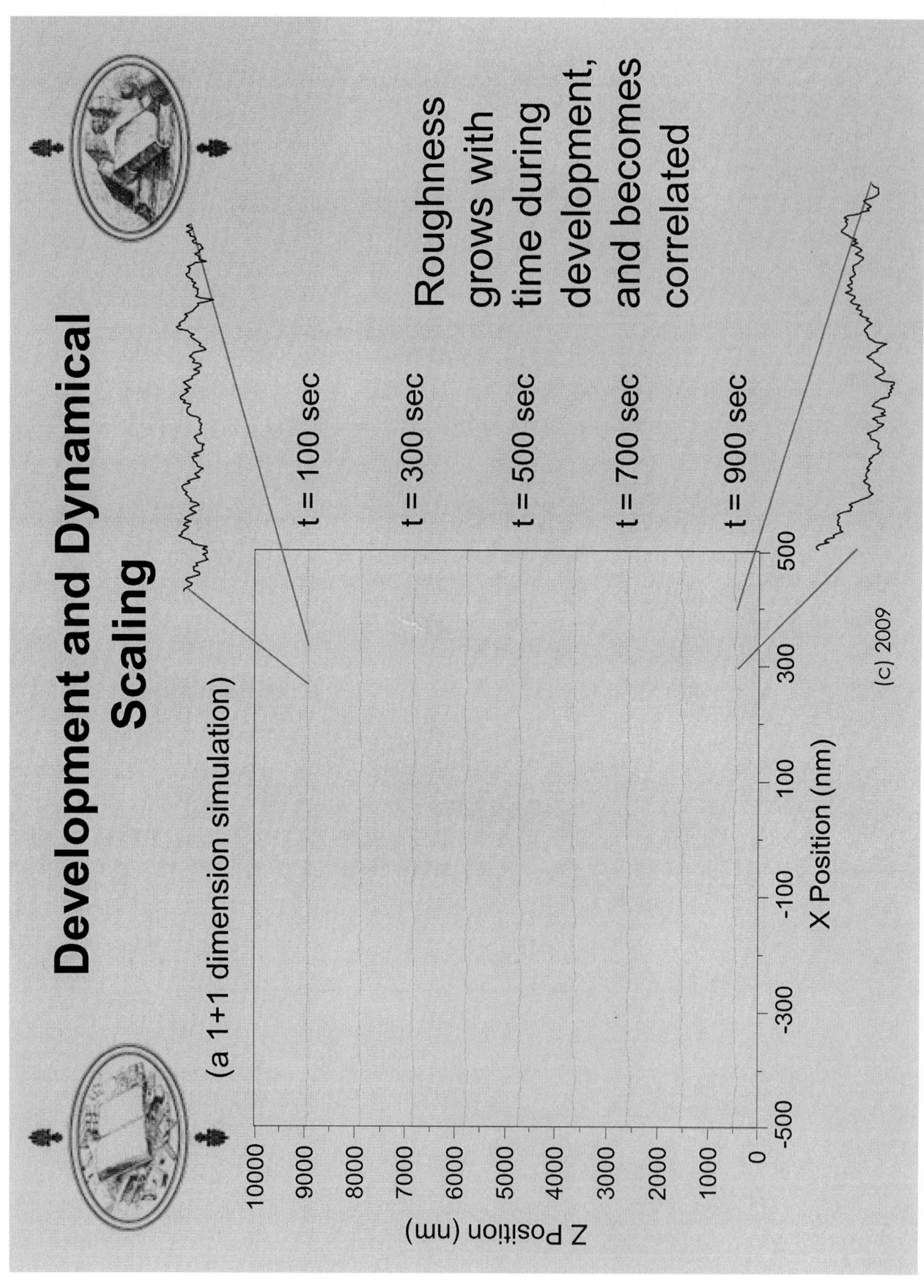

Dynamical Scaling (1+1)

$$\langle r \rangle = 10\,nm/s, \quad \sigma_r = 2\,nm/s$$

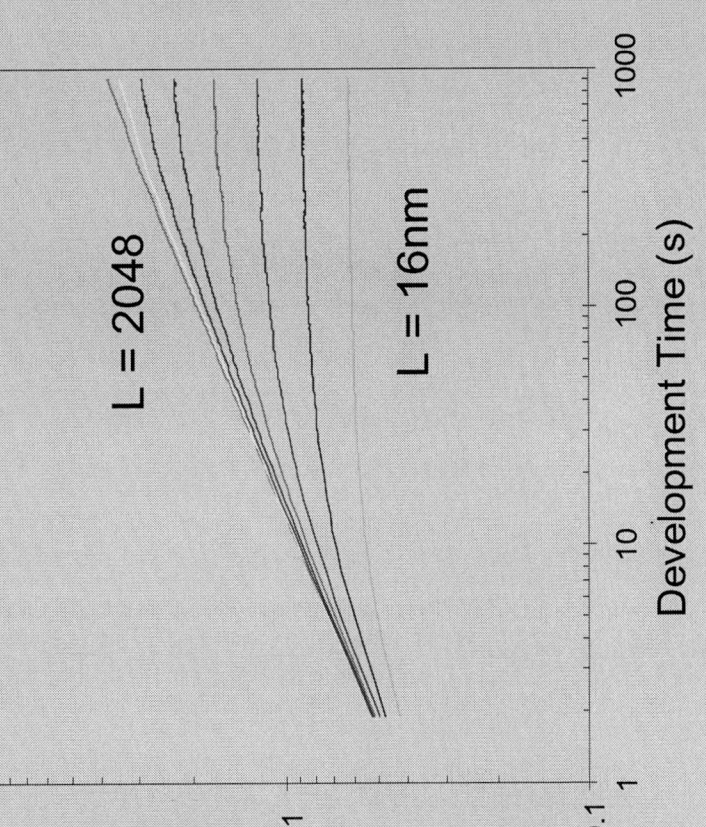

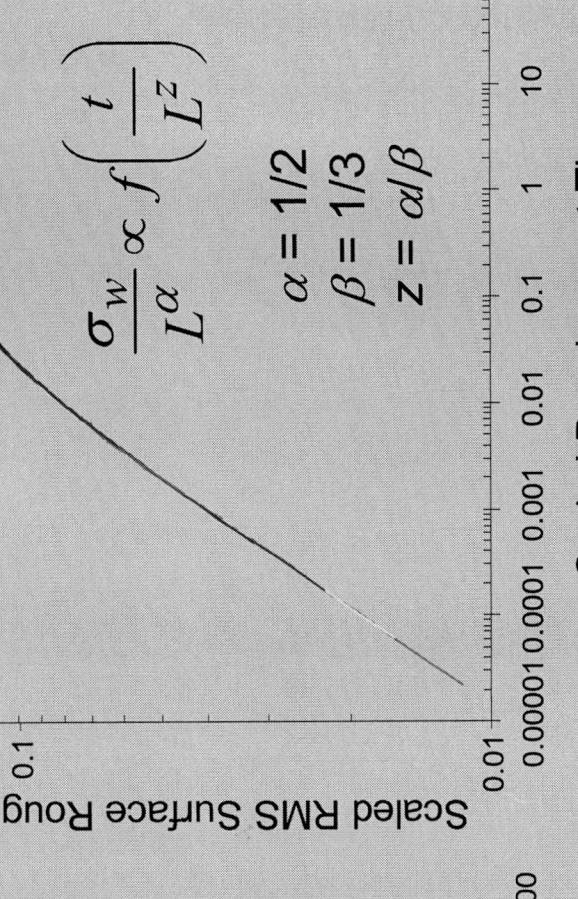

$$\frac{\sigma_w}{L^\alpha} \propto f\left(\frac{t}{L^z}\right)$$

$\alpha = 1/2$
$\beta = 1/3$
$z = \alpha/\beta$

Data collapses to a single curve for the right values of the scaling exponents

(c) 2009

Line-Edge Roughness (Tying it all Together)

• Consider a small deviation in resist development rate. The resulting change in resist edge position will be approximately

$$\Delta x = \frac{dx}{dR}\Delta R$$

• For some variation in development rate σ_R,

$$\sigma_{LER} = \frac{\sigma_R}{dR/dx} = \left(\frac{\sigma_R}{R}\right)\left(\frac{d\ln R}{dx}\right)^{-1}$$

(c) 2009

Line-Edge Roughness
(Tying it all Together)

- The Lithographic Imaging Equation

$$\frac{d\ln R}{dx} = \gamma \frac{d\ln I}{dx} = \gamma(ILS)$$

- Thus,

$$\sigma_{LER} = \left(\frac{\sigma_R}{R}\right)\left(\frac{1}{\gamma\,ILS}\right) \quad \text{or} \quad \frac{\sigma_{LER}}{CD} = \left(\frac{\sigma_R}{R}\right)\left(\frac{1}{\gamma\,NILS}\right)$$

Note: γ is not a bulk resist property, but the value at the line edge (see Chapter 9 of Fundamental Principles of Optical Lithography)

(c) 2009

Line-Edge Roughness
(Tying it all Together)

- How to improve LER:
  - Increase *ILS*
  - Increase γ
  - Decrease σ_R/R

- These terms sometimes work against each other

- The product γ *NILS* controls exposure latitude for a given feature, and thus lithographers already work to maximize this term

(c) 2009

Line-Edge Roughness and Acid Diffusion

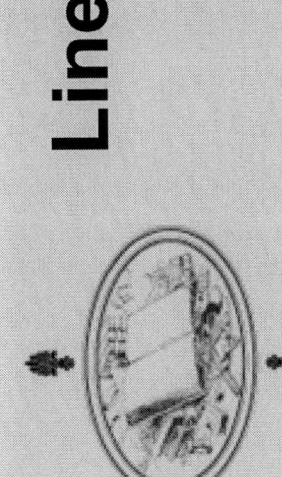

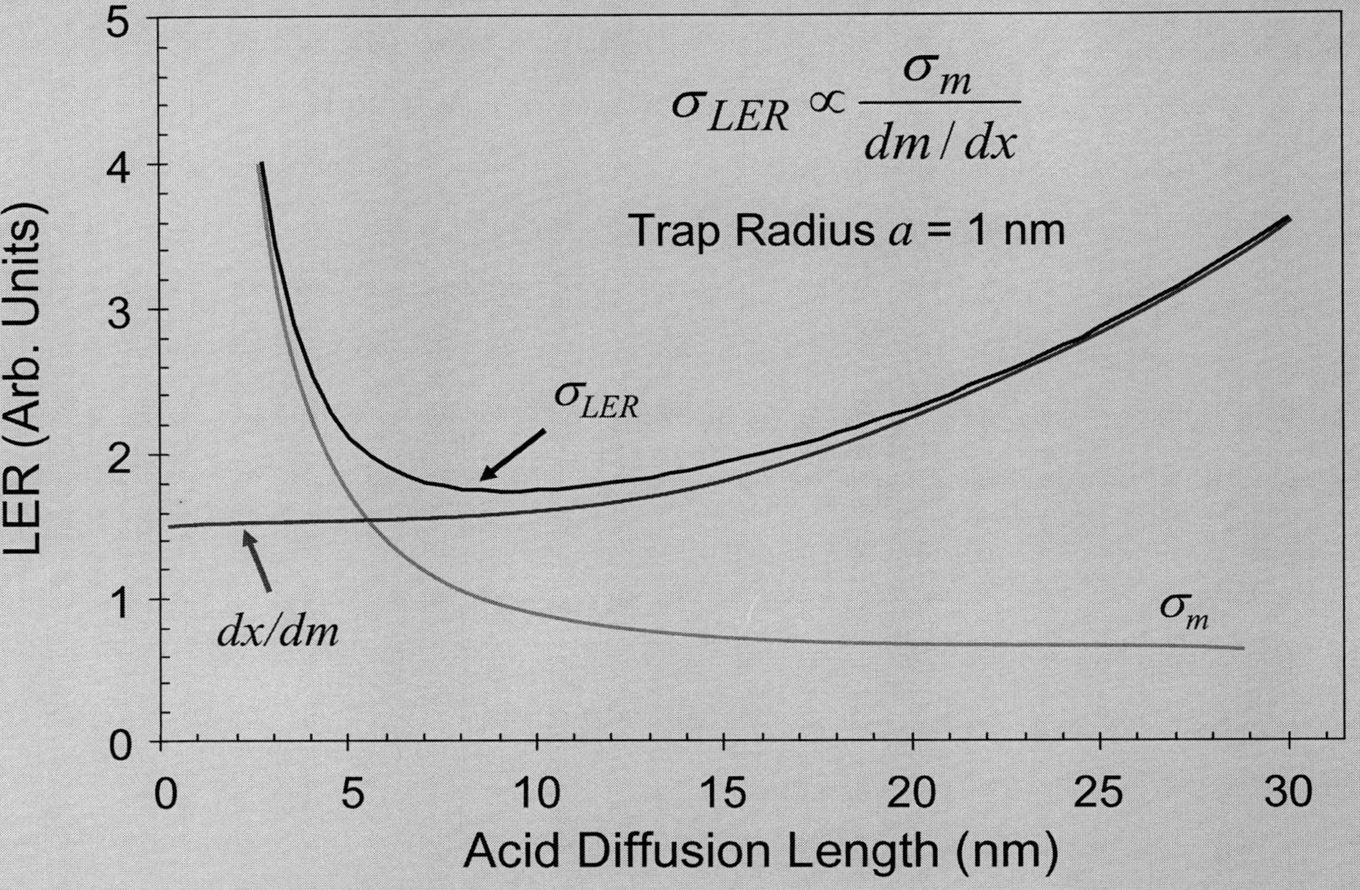

$\sigma_{LER} \propto \frac{\sigma_m}{dm/dx}$
Trap Radius a = 1 nm
LER (Arb. Units)
0
1
2
3
4
5
Acid Diffusion Length (nm)
0
5
10
15
20
25
30
σ_{LER}
dx/dm
σ_m

(c) 2009

Line-Edge Roughness and Acid Diffusion

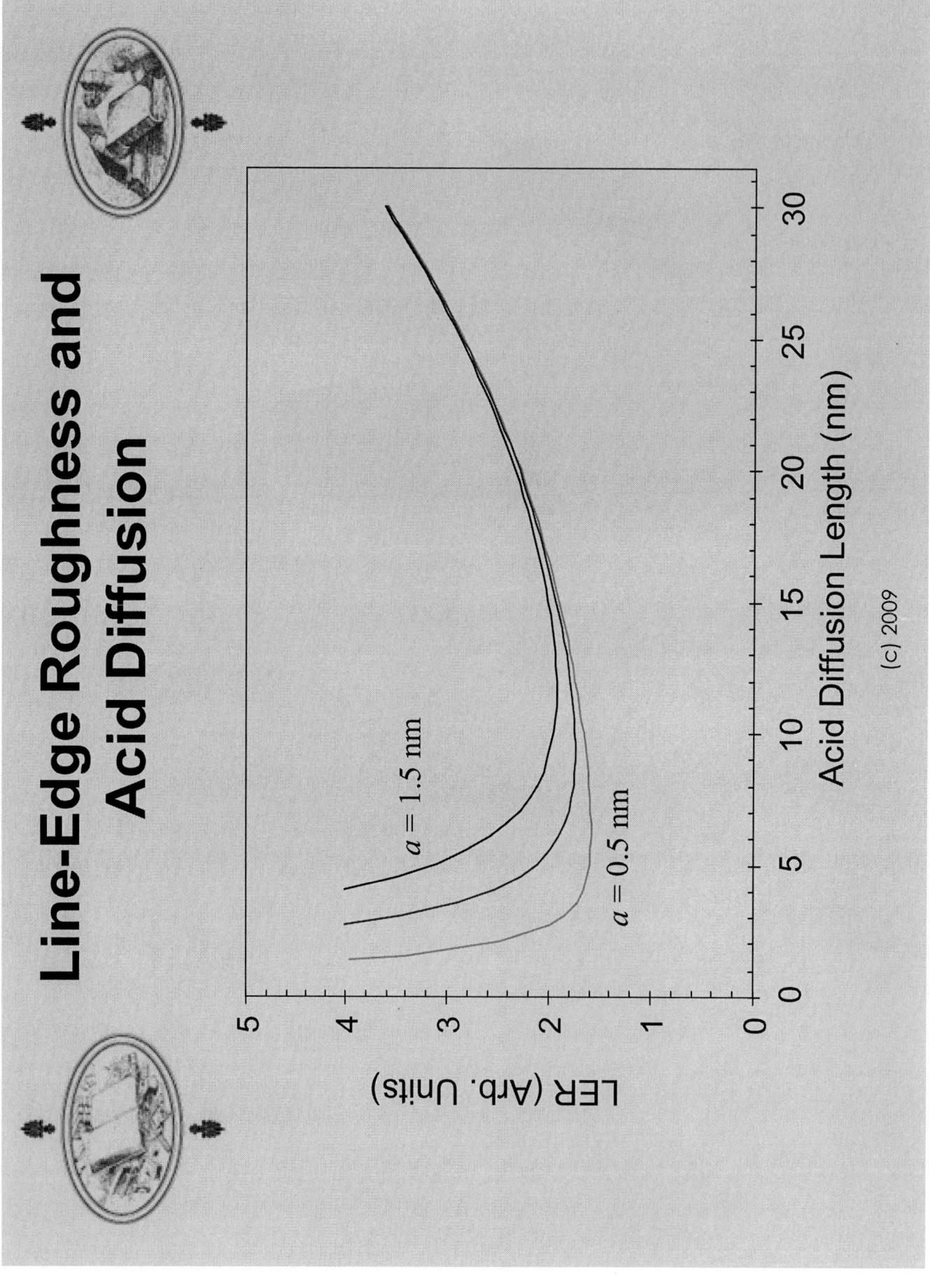

(c) 2009

Future Work (What's Missing)

- Base quencher has been ignored (by me) to date
  - Quencher will always be at lower concentrations than acid, adding an extra term to the final uncertainty in blocked polymer that could be significant
  - Quencher can dramatically improve the latent image gradient, thus quencher concentration and diffusion will be important levers for optimizing LER (there has to be an optimum quencher concentration)
- Development rate uncertainty
  - Examine impact of correlations of development rate noise
  - How does a development rate gradient affect things?
  - What happens as the dissolution rate becomes very slow – will we move into the directed percolation depinning (DPD) universality class?
- Other things
  - Metrology vs. reality, calibrating the model
  - Device impact – how good must the model be?

(c) 2009

Conclusions

- LER is the ultimate limiter to resolution in optical lithography (for both EUV and 193i)

- A good LER model is needed to optimize resist process and material properties and to find the minimum possible LER

  – Progress is being made, but a predictive LER model does not yet exist

  – How low can LER go? What is the ultimate resolution limit?

(c) 2009

Density Limits in Logic Metal1 using Double Patterning

Vincent Wiaux,
Staf Verhaegen, Germain Fenger
Patrick Wong (IMEC)

October 22nd, 2009, Prague

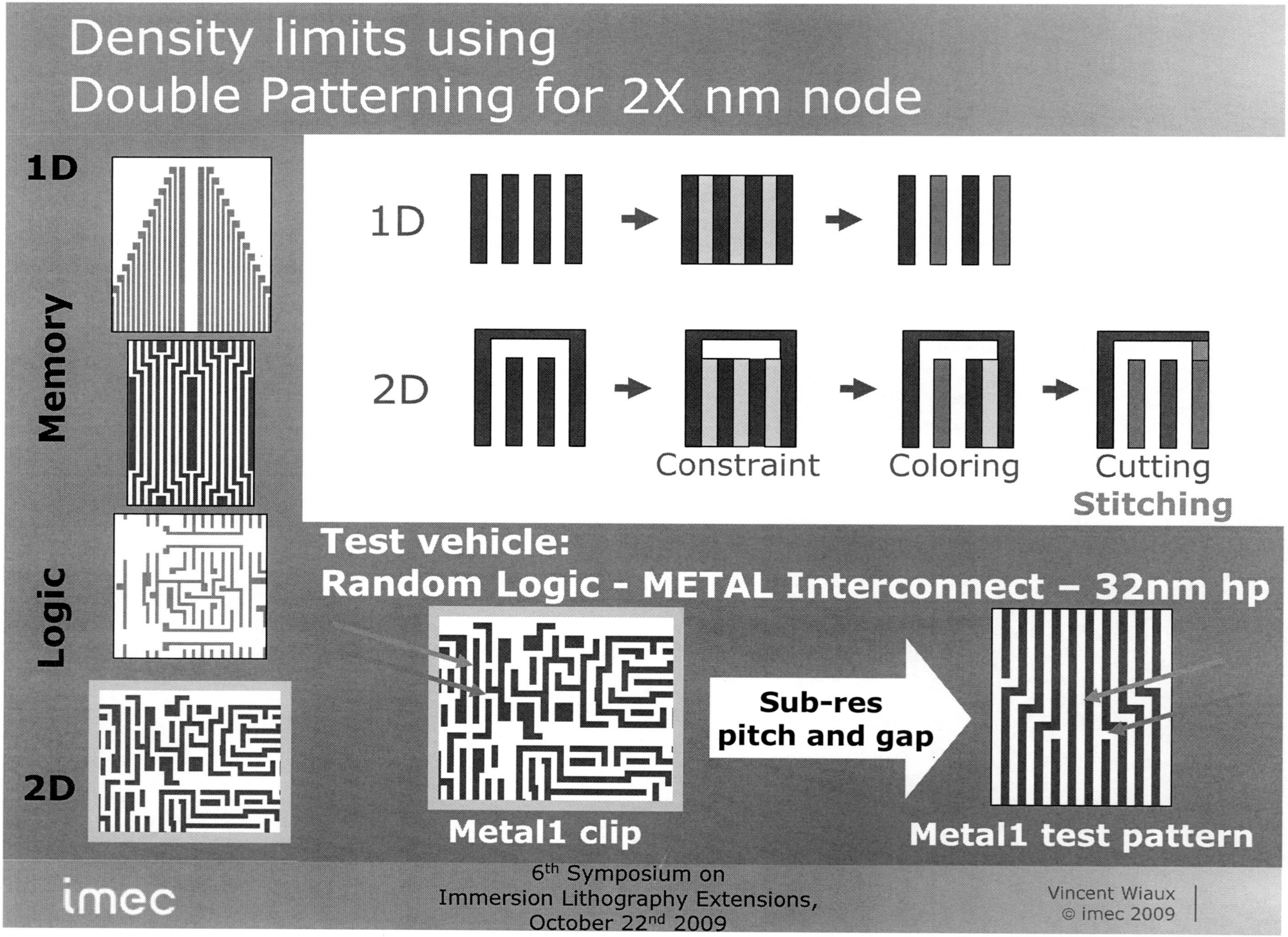

Density limits using Double Patterning for 2X nm node

1D

1D → 1D

2D

Constraint → Coloring

Cutting

Stitching

Test vehicle:
Random Logic – METAL Interconnect – 32nm hp

Metal1 clip

Sub-res pitch and gap

Metal1 test pattern

Memory

Logic

2D

imec

Vincent Wiaux
© imec 2009

6th Symposium on
Immersion Lithography Extensions,
October 22nd 2009

Outline

- Outline

- Trench patterning with Lines in resist

- Trench patterning with Trenches in resist

  - with Positive-tone development

  - with Negative-tone development

- Conclusions

6th Symposium on
Immersion Lithography Extensions,
October 22nd 2009

Vincent Wiaux
© imec 2009

imec

Trench Patterning with lines in resist

- **Trench patterning: Lines in resist – Edge-based split**
  - (locally) *light field mask*
  - Litho-Etch-Litho-Etch or *Litho-Process-Litho-Etch* (cost)

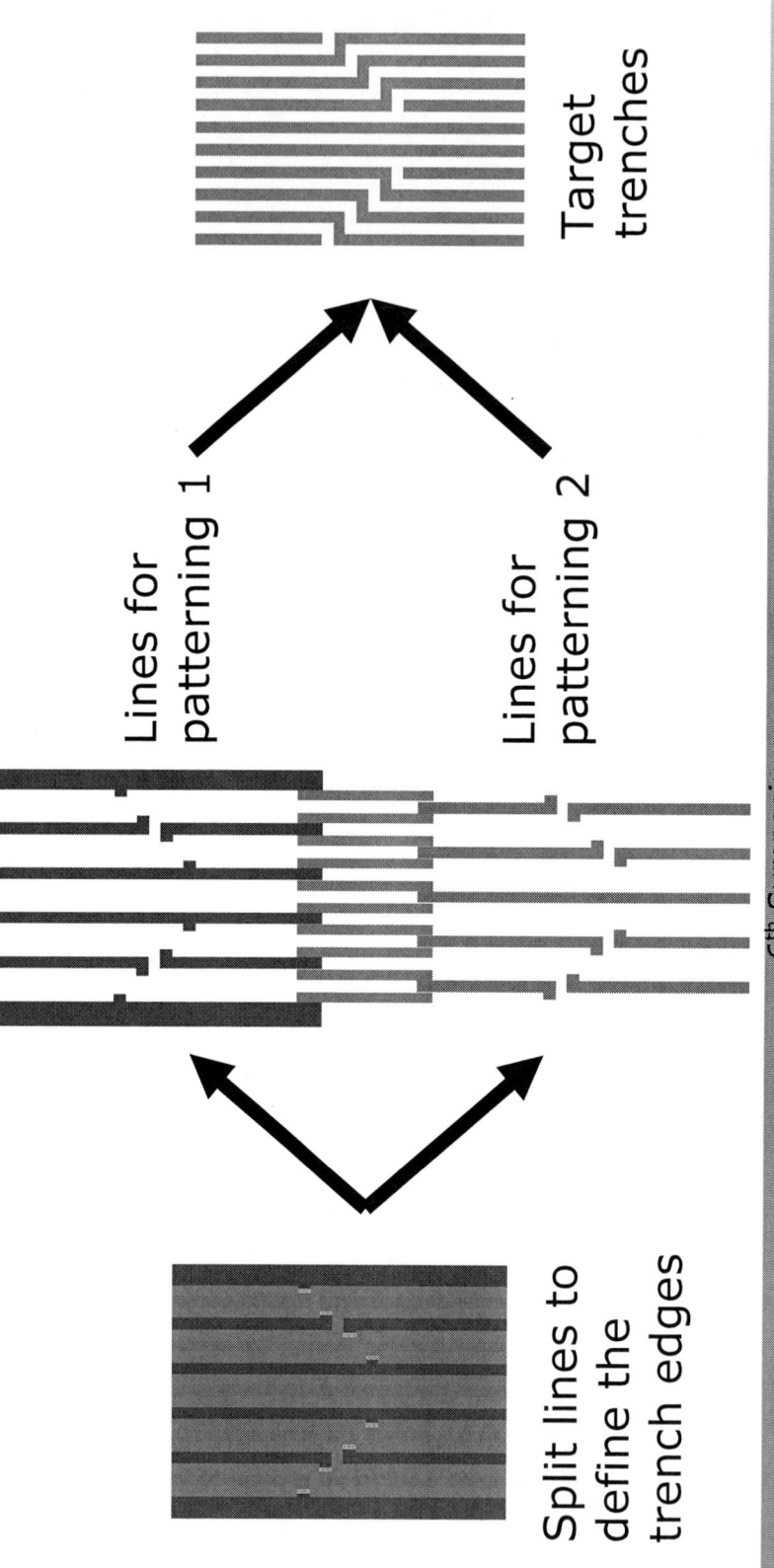

Split lines to define the trench edges

Lines for patterning 1

Lines for patterning 2

Target trenches

6th Symposium on
Immersion Lithography Extensions,
October 22nd 2009

Vincent Wiaux
© imec 2009

Trench Patterning with lines in resist: feasibility

NA1.35 Annular
0.85-0.65 X/Ypol
TOK Posi-Posi

- 76nm pitch in resist using LPLE (freezing process)

Split complexity
CD sensitive to overlay

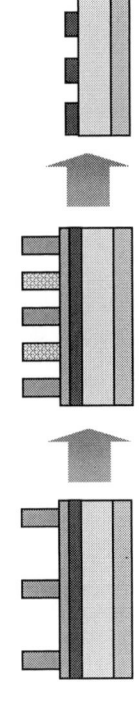

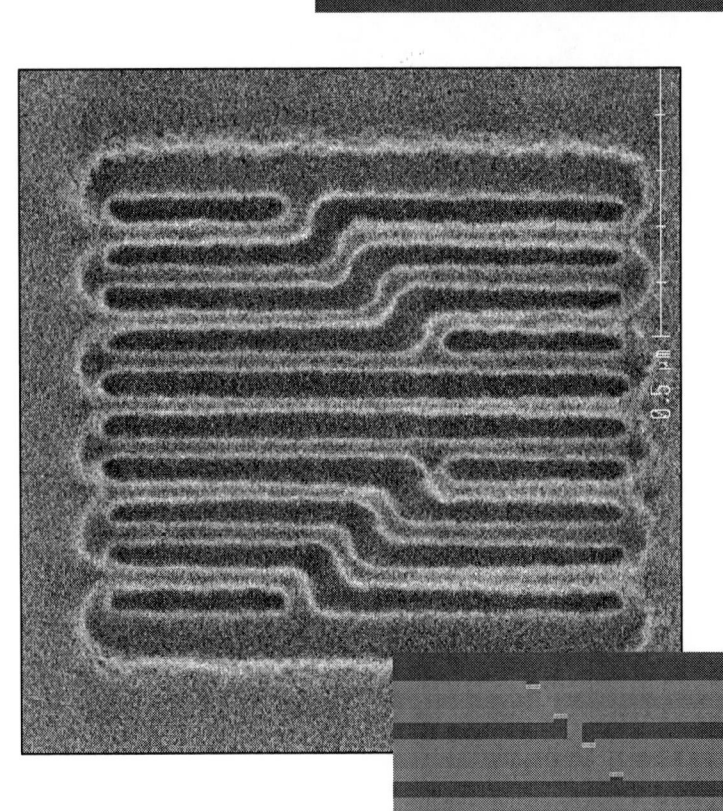

Vincent Wiaux
© imec 2009

6th Symposium on
Immersion Lithography Extensions,
October 22nd 2009

imec

Trench Patterning with trenches in resist

- **Trench patterning: Trenches in resist – Polygon-based split**
  - ***Dark-Field mask with Positive-tone development***
  - Light-Field mask with Negative-tone development

  Litho-Etch-Litho-Etch

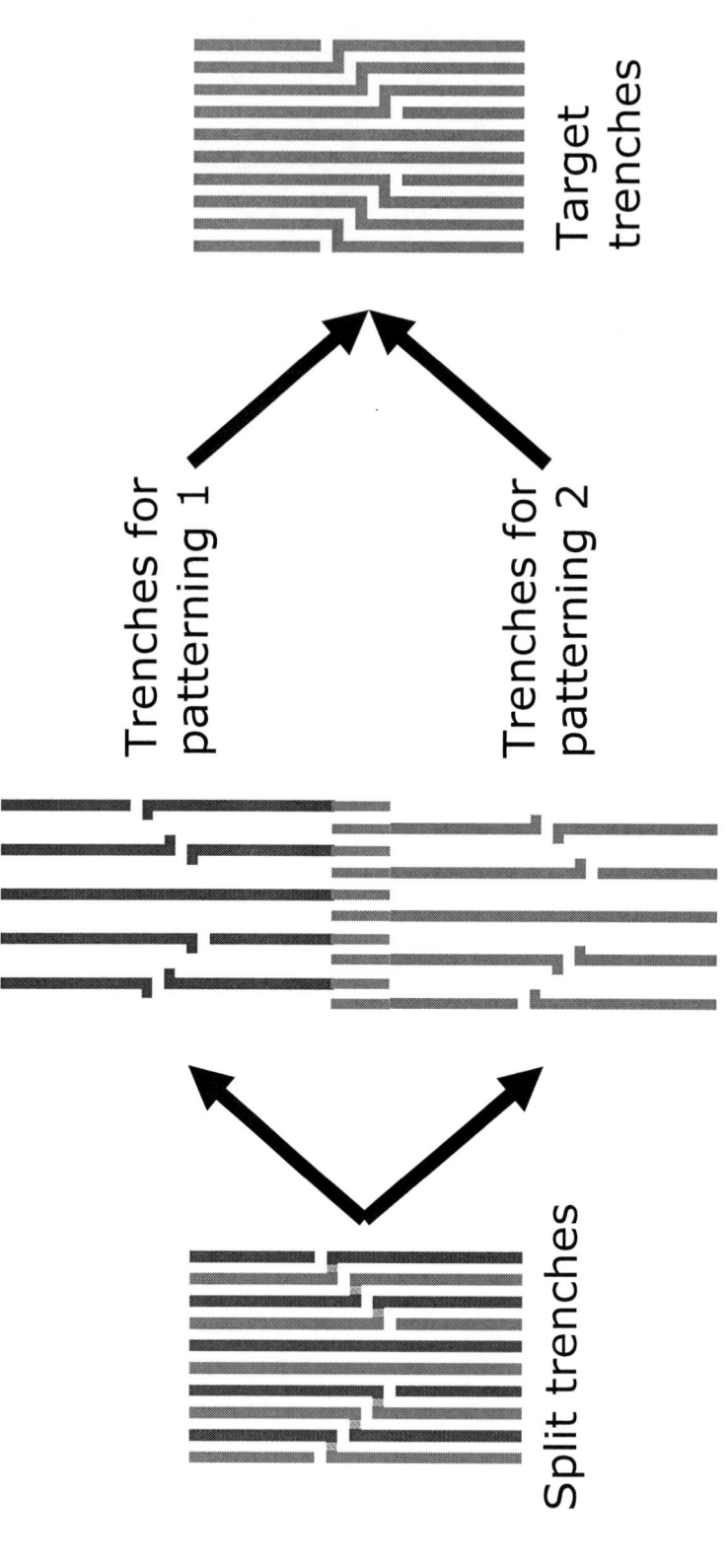

Target trenches

Trenches for patterning 1

Trenches for patterning 2

Split trenches

6th Symposium on
Immersion Lithography Extensions,
October 22nd 2009

Vincent Wiaux
© imec 2009

Trench Patterning with trenches in resist: LELE process

M1 DT
LELE

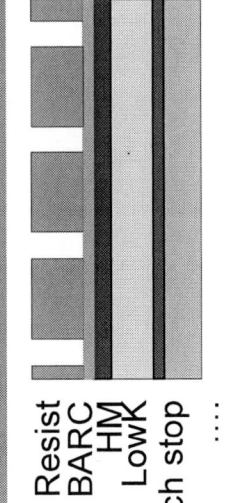

Resist
BARC
HM
LowK
Etch stop
·····

1st trench imaging

CD=60nm

Hard Mask etch
Resist/BARC strip

CD=32nm

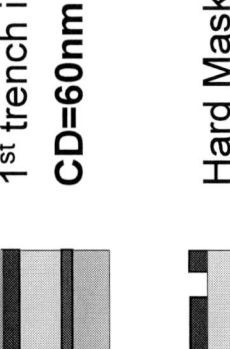

2nd trench imaging

CD=60nm

Hard Mask etch
Resist/BARC strip

CD=32nm

transfer
to dielectric

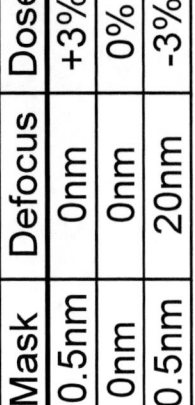

No process failure under various process conditions

Mask	Defocus	Dose
+0.5nm	0nm	+3%
0nm	0nm	0%
-0.5nm	20nm	-3%

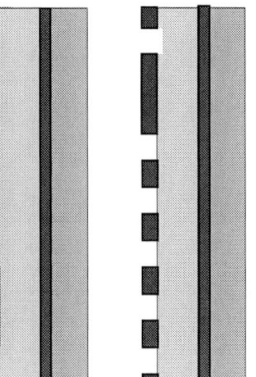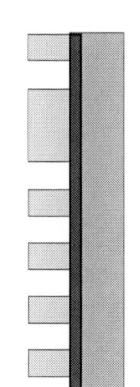

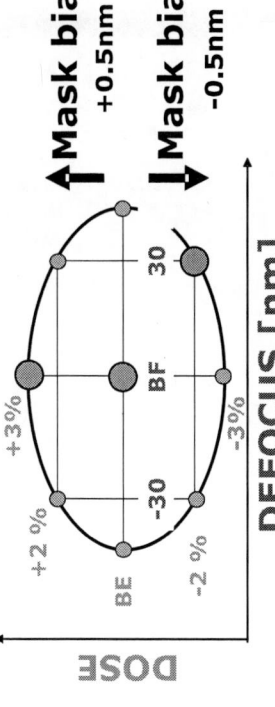

Mask bias +0.5nm

Mask bias -0.5nm

DOSE

DEFOCUS [nm]

+3 %

+2 %

BE

BF

30

-30

-3%

-2 %

imec

Vincent Wiaux
© imec 2009

6th Symposium on
Immersion Lithography Extensions,
October 22nd 2009

M1 DT
LELE

Mask stitching overlap to compensate for...

Annular X/Ypol

- Trench-end rounding

45hp

36nm

Mask overlap 0nm

- Trench-end pullback

32hp

Mask Overlap 20nm

- Overlay error

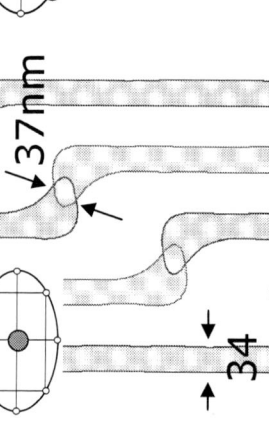

24nm

+6nm

27

27nm

27

37nm

34

Pitch 90nm
Overlap 40nm

Vincent Wiaux
© imec 2009

6th Symposium on
Immersion Lithography Extensions,
October 22nd 2009

imec

Design and split constraints

M1 DT
LELE

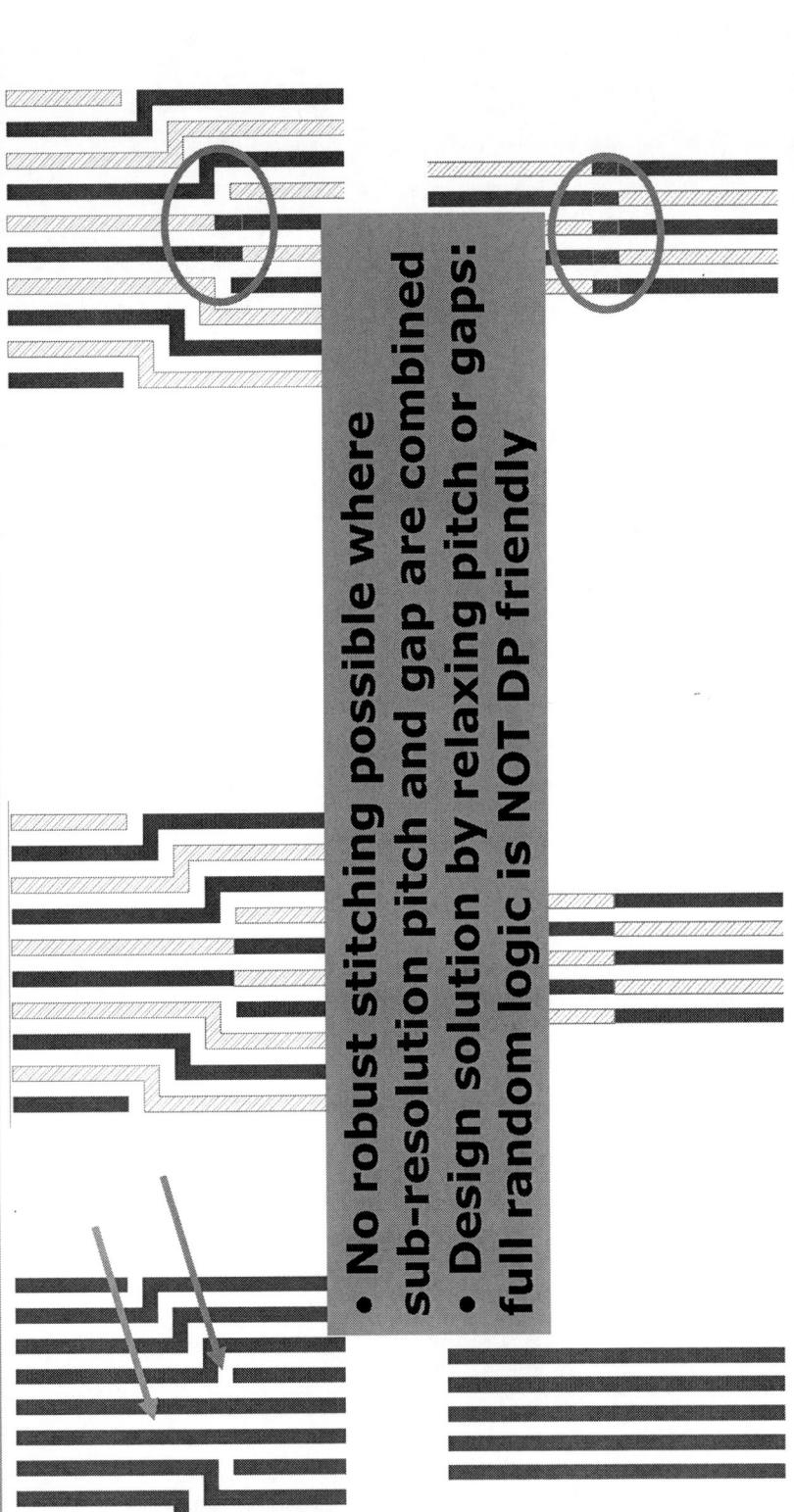

- **No robust stitching possible where sub-resolution pitch and gap are combined**
- **Design solution by relaxing pitch or gaps: full random logic is NOT DP friendly**

Sub-resolution PITCH-GAP solved by split & cutting

Sub-resolution PITCH-GAP re-introduced by the stitching overlap

6th Symposium on
Immersion Lithography Extensions,
October 22nd 2009

Vincent Wiaux
© imec 2009

imec

Design and split constraints: example at 45nm hp

M1 DT
LELE

NA1.20 Annular
0.92-0.72 X/Ypol

E0/M0

E+/M+

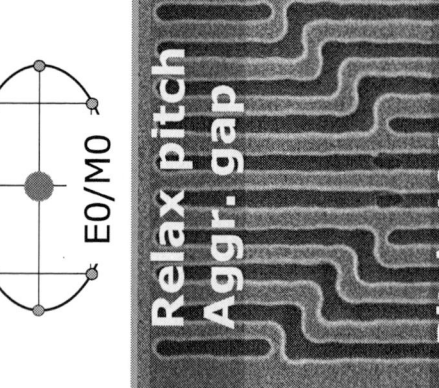

Relax pitch
Aggr. gap

Pitch 104nm
Gap 29/44nm

Bridging

Pitch 100nm
Gap 46nm

Relax gap
Aggr. pitch

Pitch 90nm
Gap 71/68nm

Bridging

Pitch 88nm
Gap 50nm

Target:
pitch 90nm
Gap 45nm

Vincent Wiaux
© imec 2009

6th Symposium on
Immersion Lithography Extensions,
October 22nd 2009

imec

Design and split constraints: down to 32nm hp

M1 DT
LELE

DP P64nm

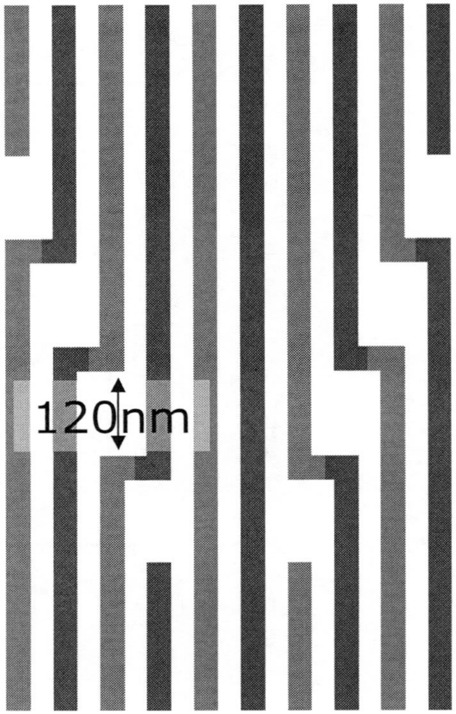

120nm

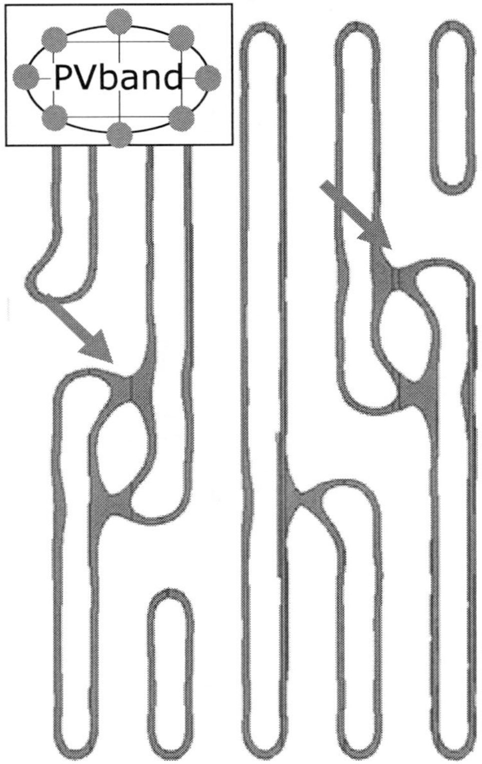

PVband

LithoEtch bias

80nm

Litho gap

Stitching overlap

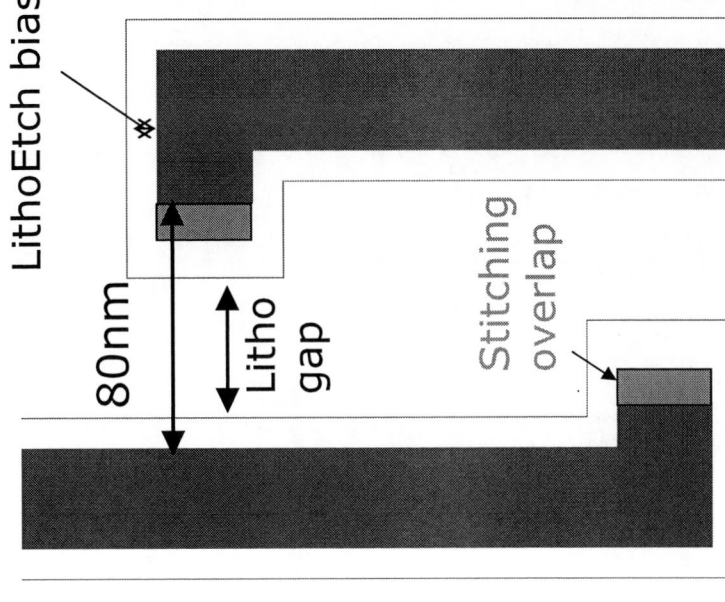

- LITHO GAP = 80nm – Stitching overlap/2 – LithoEtch bias
  - Eg: LITHO GAP = 80nm – 40nm/2 – 28nm = 32nm

- 2D and minimum gap imaging (litho step) limit the density of logic clips using DP Split and Stitching: stitching vs. bridging/pinching

6th Symposium on
Immersion Lithography Extensions,
October 22nd 2009

Vincent Wiaux
© imec 2009

imec

Optimum litho target at 128nm pitch: trench & gap process range

M1 DT
LELE

NA1.35 Annular
0.85-0.65 X/Ypol

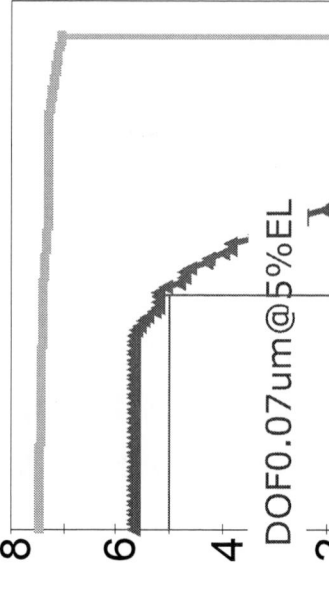

Trench = 60±3.2nm
Exposure Latitude (%)

DOF0.07um@5%EL

Depth of Focus

Litho1 - Trench in Resist
Dark Field Mask / Positive Tone Development

	D0F0M0	D-3%F+20nm M-0.5nm	D+3%F0 M+0.5nm
Target 56nm ✗		pinching	
Target 60nm ○			
Target 64nm ✗			bridging

6th Symposium on
Immersion Lithography Extensions,
October 22nd 2009

Vincent Wiaux
© imec 2009

imec

Design options: diagonal turn

M1 DT
LELE

Right turn

diagonal turn

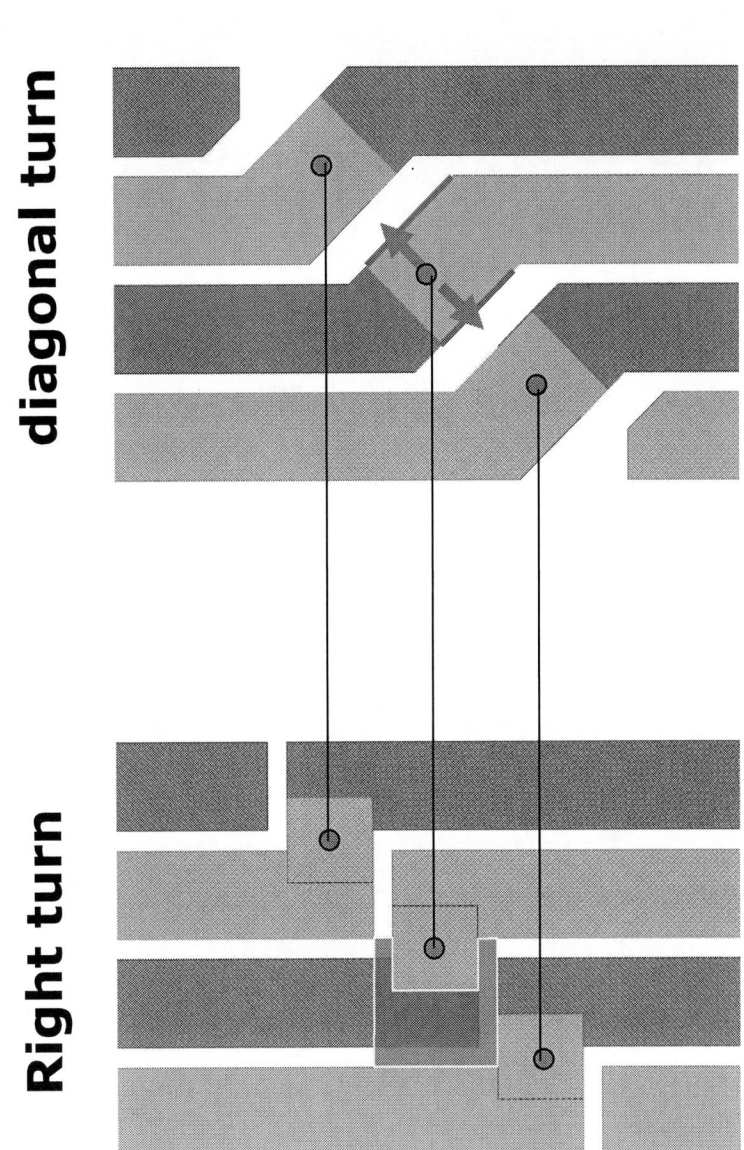

- Diagonal turn can have a larger width.
- Larger jog have better PW; smaller stitching overlap is needed.

imec

6th Symposium on
Immersion Lithography Extensions,
October 22nd 2009

Vincent Wiaux
© imec 2009

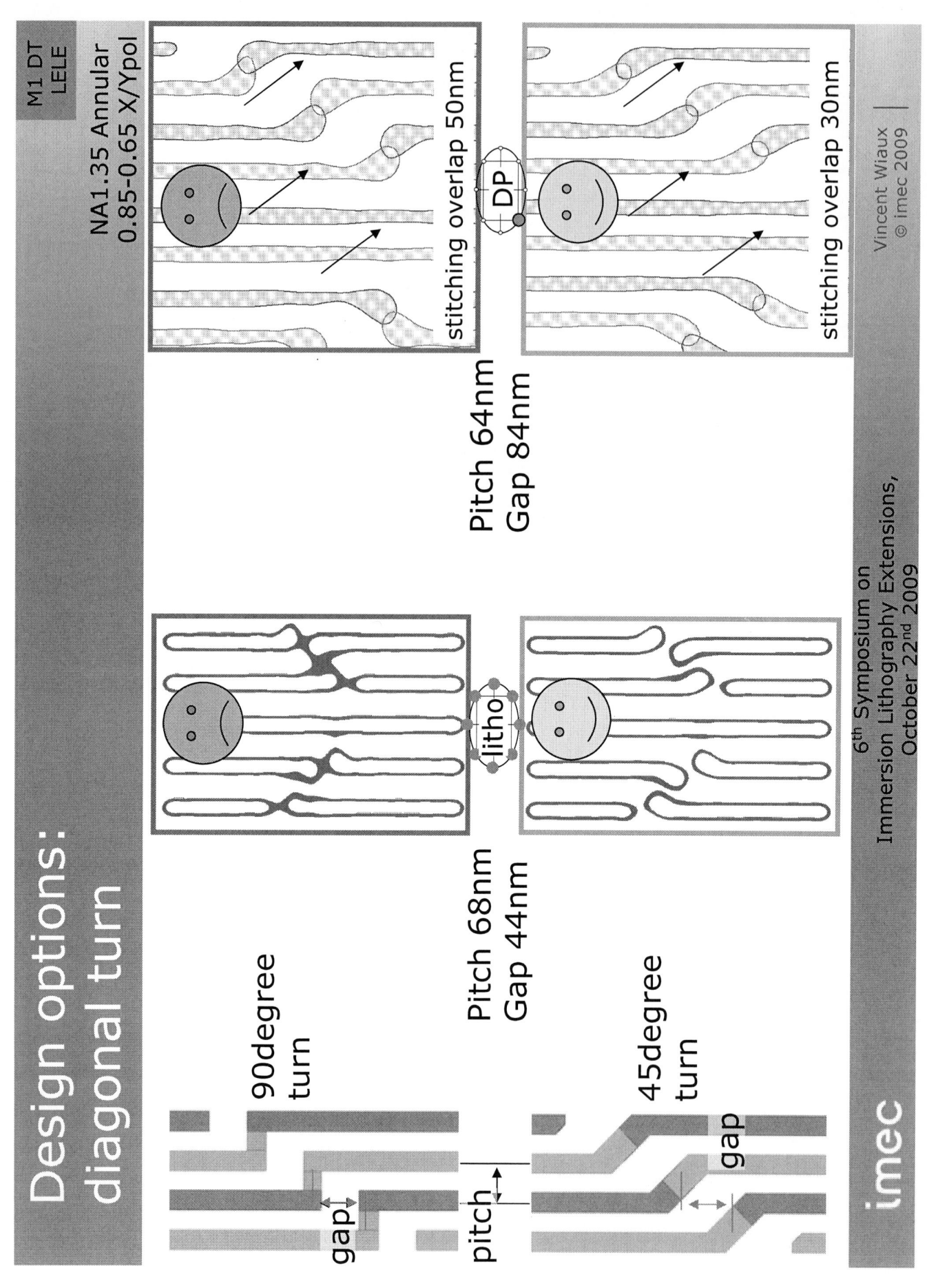

Design options: 64P at fixed stitching overlap

M1 DT LELE

NA1.35 Annular 0.85-0.65 X/Ypol

Double patterning in TiN

Fixed Y-density

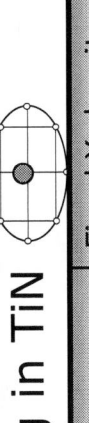
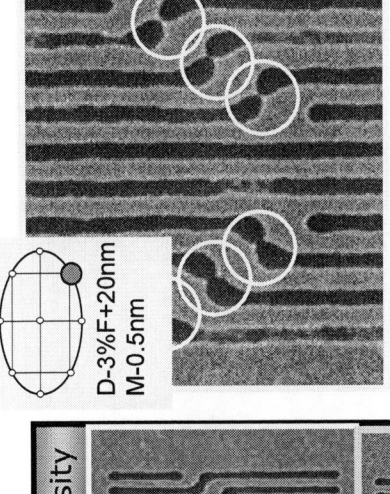
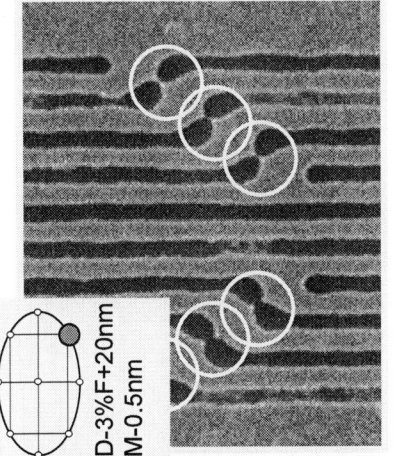
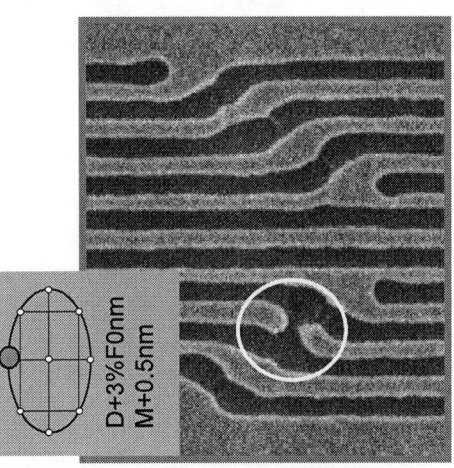
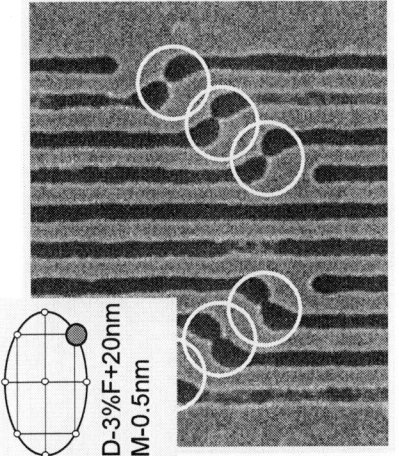
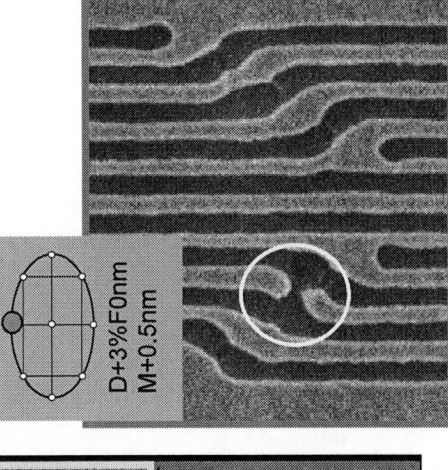

D0F0M0

D-3%F+20nm M-0.5nm

D+3%F0nm M+0.5nm

90 degree turn 32nm Stitch.20nm			
45 degree turn 32nm Stitch.20nm			
45 degree turn 45nm Stitch.20nm			

Vincent Wiaux © imec 2009

6th Symposium on Immersion Lithography Extensions, October 22nd 2009

imec

M1 DT
LELE

Design options:
90 vs 45degree turn at 64P

NA1.35 Annular
0.85-0.65 X/Ypol

- BE/BF: 45degree turn has a better pattern fidelity
- E-/F-/M-: 45degree turn has more robust stitching and less pinching

45 degree turn at 38nm
Stitching overlap 20nm

90 degree turn
Stitching overlap 60nm

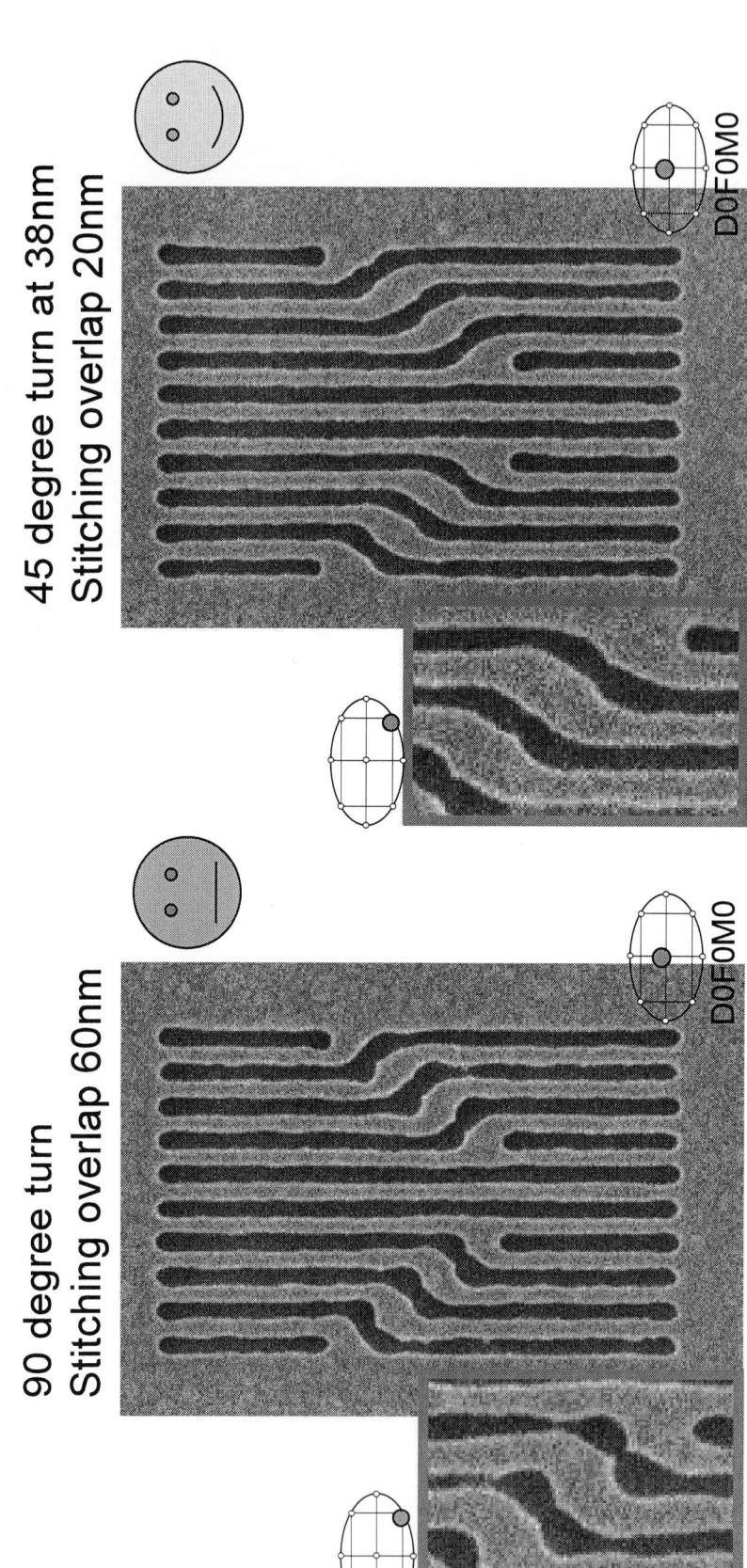

D0F0M0

D0F0M0

Vincent Wiaux
© imec 2009

6th Symposium on
Immersion Lithography Extensions,
October 22nd 2009

imec

M1 DT
LELE

OPC for DP

- PW OPC helps pinching and bridging error, increased EPE in BE/BF.
  PW-OPC: Process Window OPC taking variations of process conditions into account

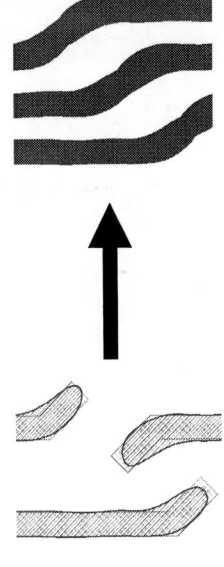

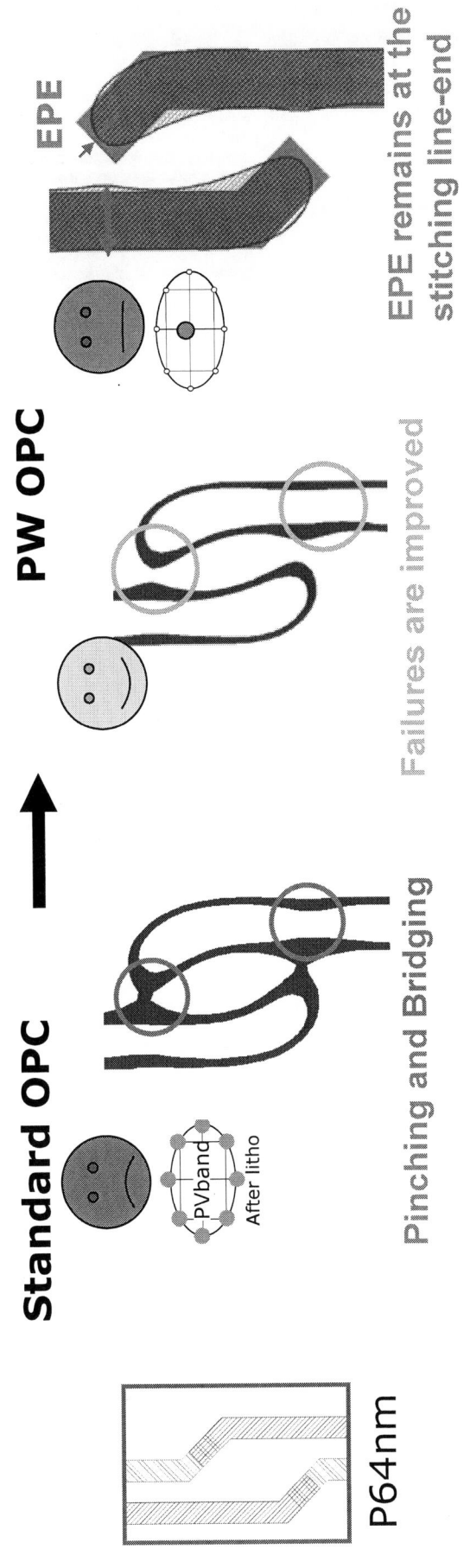

Standard OPC

PW OPC

After litho

PVband

Pinching and Bridging

Failures are improved

P64nm

EPE

EPE remains at the stitching line-end

- A stitching metric is needed for OPC: DP-OPC should ensure a smooth DP contour, the split-line-end is not the final target.

Vincent Wiaux
© imec 2009

6th Symposium on
Immersion Lithography Extensions,
October 22nd 2009

imec

Trench Patterning with trenches in resist

- **Trench patterning: Trenches in resist – Polygon-based split**
  - Dark-Field mask with Positive-tone development
  - **Light-Field mask with Negative-tone development**

  Litho-Etch-Litho-Etch

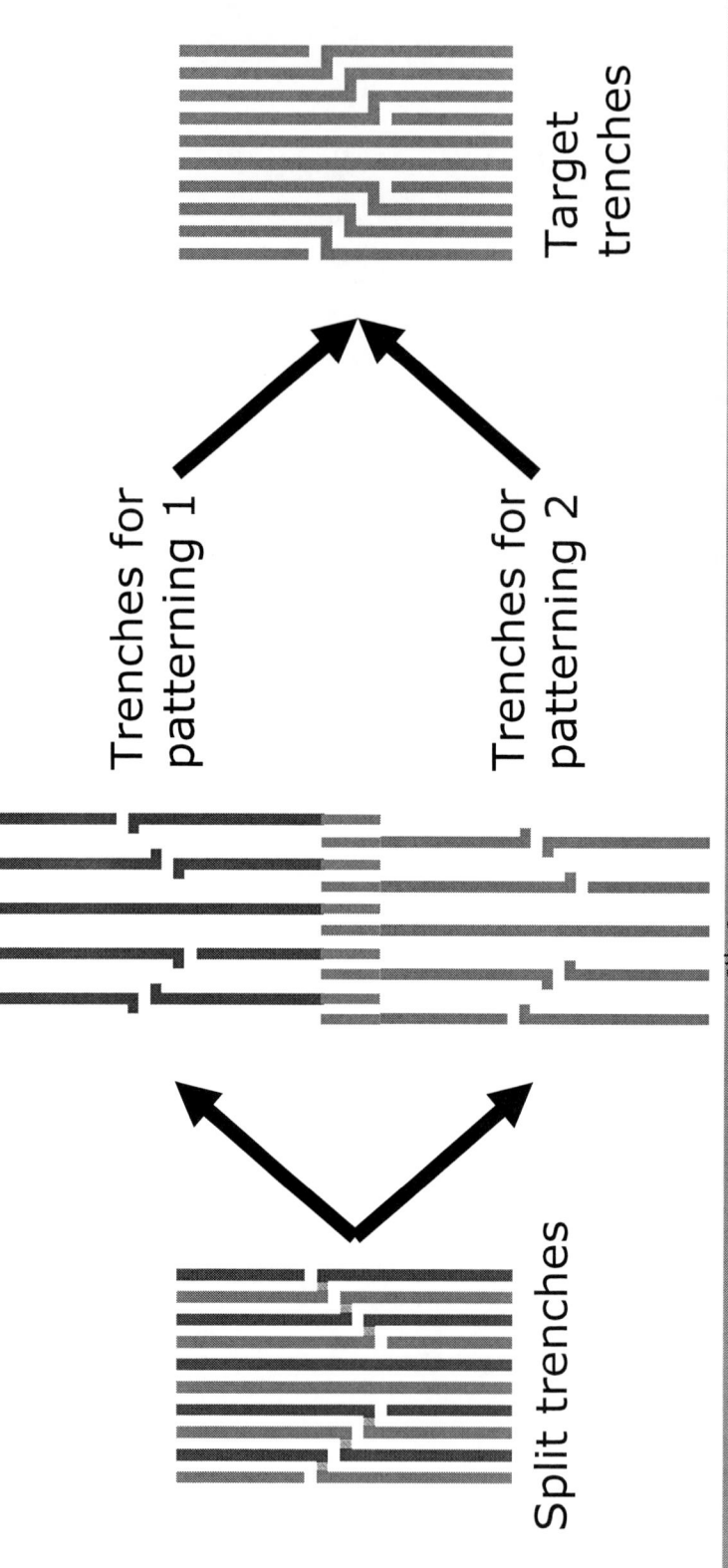

Target trenches

Trenches for patterning 1

Trenches for patterning 2

Split trenches

Vincent Wiaux
© imec 2009

6th Symposium on
Immersion Lithography Extensions,
October 22nd 2009

Comparison Light field imaging with Dark field imaging 'after L1'

M1 DT LELE

NA1.35 Annular 0.85-0.65 X/Ypol

Dark field imaging with PTD

	D0F0M0	D-3%F+20nm M-0.5nm	D+3%F0 M+0.5nm
Target 56nm	pass	pinching	pass
Target 60nm	pass	pass	pass

OPC DF target 52nm

Light field imaging with NTD

	D0F0M0	D+3%F+20nm M-0.5nm	D-3%F0 M+0.5nm
Target 52nm	pass	pass	pass
Target 54nm	pass	pass	pass

OPC LF target 52nm

Light field imaging in NTD enables smaller target in resist with sufficient process.

See J. Bekaert 11:45am

6th Symposium on Immersion Lithography Extensions, October 22nd 2009

Vincent Wiaux
© imec 2009

imec

Outline

- Outline

- Trench patterning with Lines in resist

- Trench patterning with Trenches in resist

  - with Positive-tone development

  - with Negative-tone development

- Conclusions

6th Symposium on
Immersion Lithography Extensions,
October 22nd 2009

Vincent Wiaux
© imec 2009

imec

Conclusions

Density enablers of METAL TRENCH interconnect at 32nm hp using Double Patterning (double trench with LELE)

- **Limits from the litho:**
  - limited O-PW of 2D clips at small CD, semi-dense pitch
  - minimum gap imaging: stitching vs. bridging/pinching

- **Improvement by**
  - **Illumination** setting
  - **Optimum target CD** for the trenches AND small gaps
  - Split and **Design**: need for **relaxed** gaps or pitch to allow robust stitching, **diagonal turns** for increasing the density, design regularity
  - OPC: AF, **PW-OPC**, need for a **stitching-metric**
  - Improved process: e.g. Light Field Imaging with **NTD**

Vincent Wiaux
© imec 2009

6th Symposium on
Immersion Lithography Extensions,
October 22nd 2009

imec

Acknowledgements

Special thanks to
Takashi Matsuda (Panasonic @ IMEC)
Yoichi Nomura (Elpida @ IMEC)

A.Lin (PSC), K.Iwase (Sony), T.Sasaki (Elpida),
Sergei Postnikov (Infineon), Shigeo Irie (Panasonic),
D.Trivkovic, R.DeRuyter, J.Bekaert, L.VanLook,
D.Vangoidsenhoven, G. Vandenberghe

TOPPAN   *M.Lamantia, Toppan's team.*

Hitachi High-Tech T.Ishimoto, N.Yasui, D.Hibino, K.Sekiguchi, J.Kakuta

tok   R.Takasu

JSR   H.Tanaka

FUJiFILM   M.Reybrouck

Mentor Graphics   Pat LaCour, Le Hong, Alex Tritchkov

SYNOPSYS   Weimin Gao, Kevin Lucas

6th Symposium on
Immersion Lithography Extensions,
October 22nd 2009

Vincent Wiaux
© imec 2009

DPL Overlay Components

Venkat R. Nagaswami, Jaydeep Sinha, Sathish Veeraraghavan,
Frank Laske, Anna Golotsvan, David Tien, Pavel Izikson,
John C. Robinson
(KLA-Tencor Corporation)

Chiew-seng Koay, Matthew E. Colburn, Steven Holmes, Karen Petrillo
(IBM Corporation @ Albany NanoTech)

6th International Symposium on Immersion Lithography Extensions
Prague, October 18-23, 2009

Agenda

- Introduction
- Mask Design
- Experiment
  - Mask Contribution to DPL Overlay
  - Wafer Shape Contribution to DPL Overlay
  - Residual correctable errors
  - Remaining residual errors
- Conclusions

10/30/2009

Introduction

- Overlay is a critical component in DPL since design data of a single mask is split into two masks. This could result in doubling the typical errors from mask making and stepper/scanner alignment. The wafer shape can also contribute to overlay budget due to in-plane distortion.

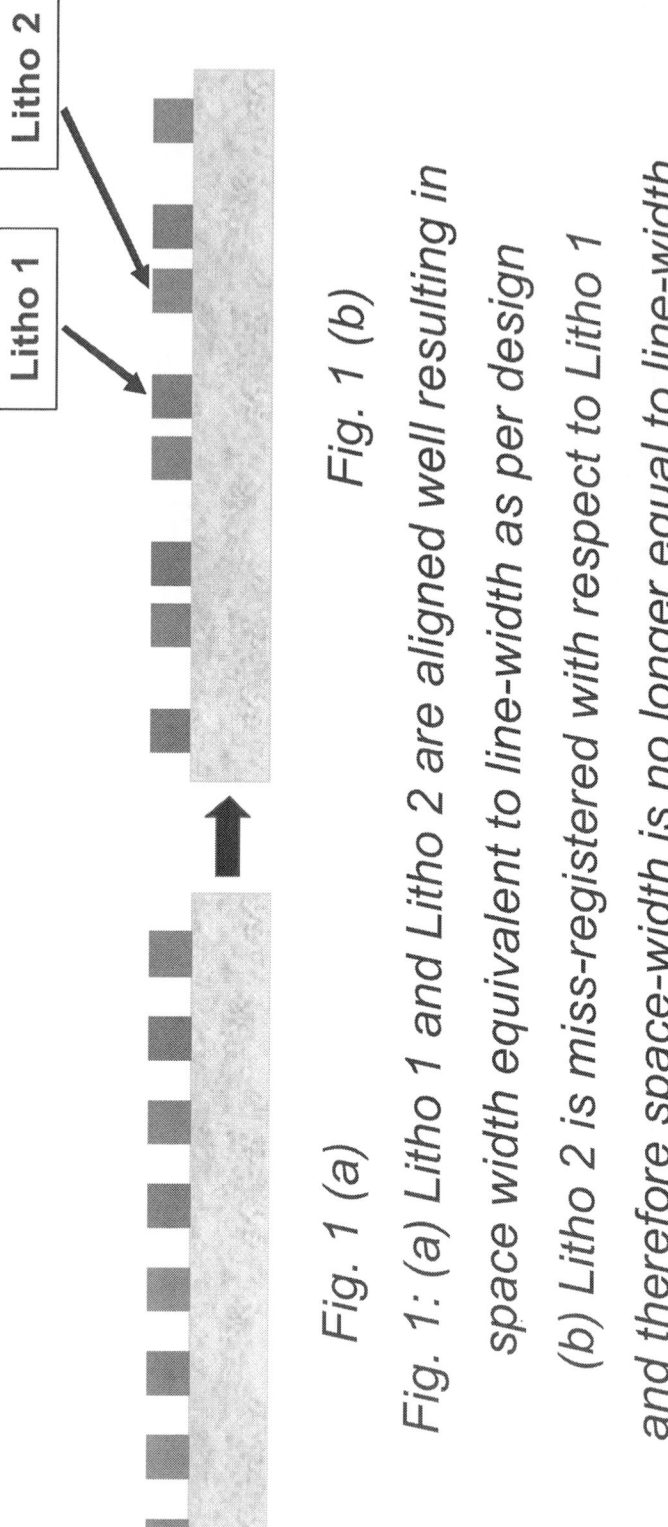

Fig. 1 (a)

Fig. 1 (b)

Fig. 1: (a) Litho 1 and Litho 2 are aligned well resulting in space width equivalent to line-width as per design

(b) Litho 2 is miss-registered with respect to Litho 1 and therefore space-width is no longer equal to line-width

10/30/2009

Introduction (Contd.)

- The components of Overlay errors investigated in this study are:

  - Pattern Placement Error (measured with KLA-Tencor IPRO4 system)

  - Wafer Shape Contribution (measured with KLA-Tencor Geometry Tool)

  - Overlay errors after Stepper/Scanner alignment (measured with KLA-Tencor Archer 200)

- The wafer shape contribution is estimated using a proprietary software based on finite element model developed by KLA-Tencor ADE Division. The residual analysis was performed using KTAnalyzer™, an analytical software available from KLA-Tencor.

10/30/2009

Mask Design

- A Double Patterning mask-set was generated by KLA-Tencor with programmed defects and several types of Overlay Registration marks The zero layer mask was generated using a Pattern Generator and L2 and L3 were generated using an e-beam system at a third party mask-shop.

- The mask-set consisted of a Zero Layer (L1), and a test pattern split in to 2 separate masks (L2 and L3). The IPRO mark was integrated within the KLA-Tencor Archer alignment mark.

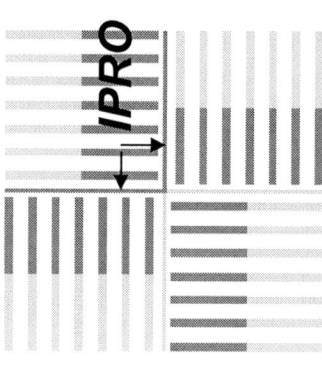

Sample Size

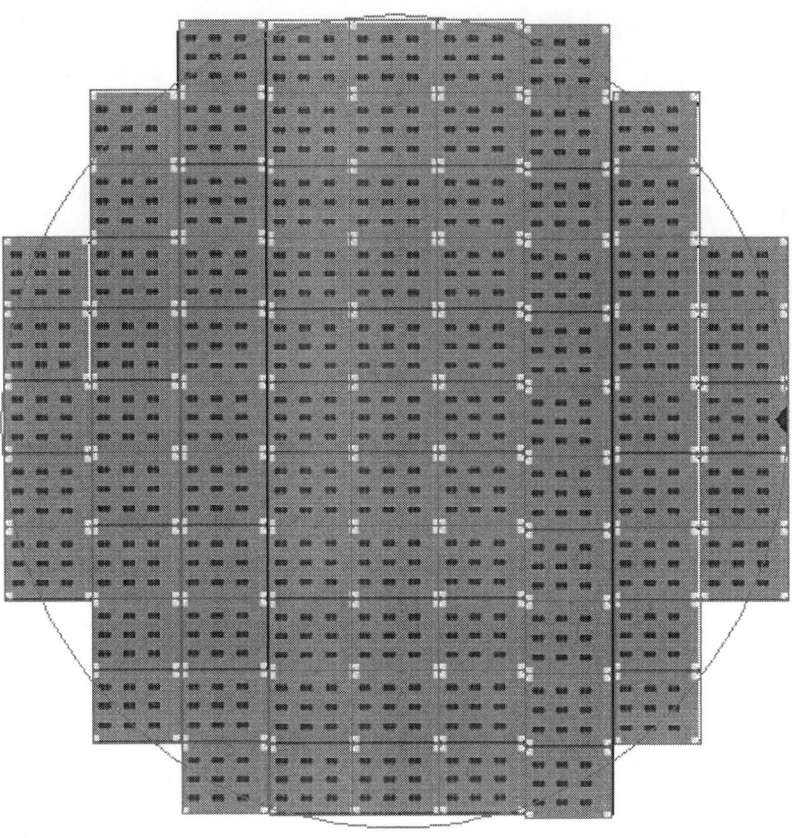

- Field size = 24 x 30 mm

- 49 locations per field

- 71 exposure fields

- Approximately 3500 data points per target per wafer

- 5 wafers

After a detailed analysis of the performance of the targets KLA-Tencor AIM target was selected for the experiment.

10/30/2009

Experimental Setup

- The wafers were patterned using a DPL resist process, using the Stepper/Scanner available at IBM's facility.

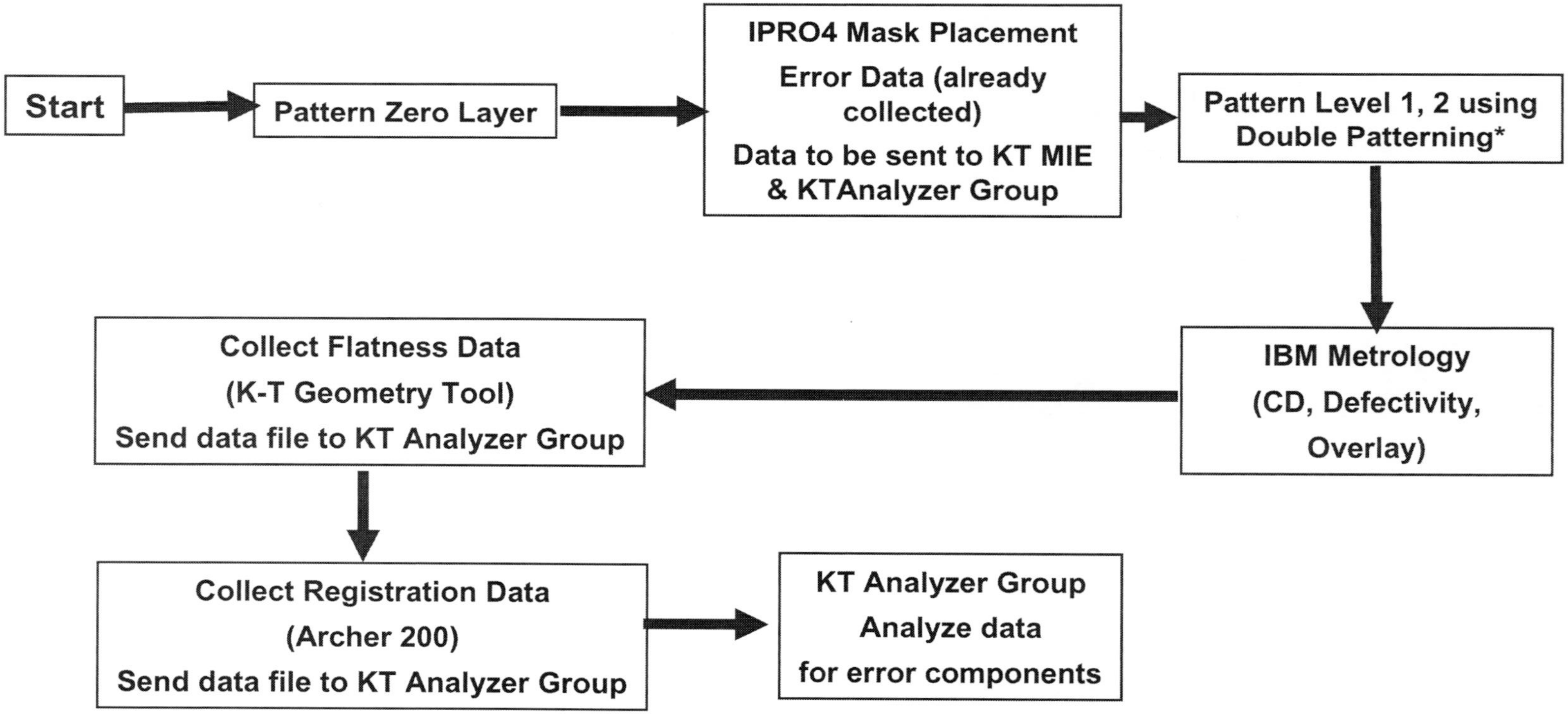

```
Start
  ↓
Pattern Zero Layer
  ↓
IPRO4 Mask Placement
Error Data (already collected)
Data to be sent to KT MIE
& KTAnalyzer Group
  ↓
Pattern Level 1, 2 using
Double Patterning*
  ↓
IBM Metrology
(CD, Defectivity, Overlay)
  ↓
Collect Flatness Data
(K-T Geometry Tool)
Send data file to KT Analyzer Group
  ↓
Collect Registration Data
(Archer 200)
Send data file to KT Analyzer Group
  ↓
KT Analyzer Group
Analyze data
for error components
```

10/30/2009

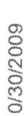

Measured DPL Overlay error (L3 to L2)

Target	Mean X (nm)	Sigma (nm)	3 Sigma (nm)	Mean Y (nm)	Sigma (nm)	3 Sigma (nm)
AIM	-0.131	1.465	4.4	-0.722	1.371	4.1

DPL Overlay Error as measured by Archer® 200

10/30/2009

Mask Contribution

KLA-Tencor
Accelerating Yield

IBM

10/30/2009

Pattern Placement Error on mask (KLA-Tencor IPRO4)

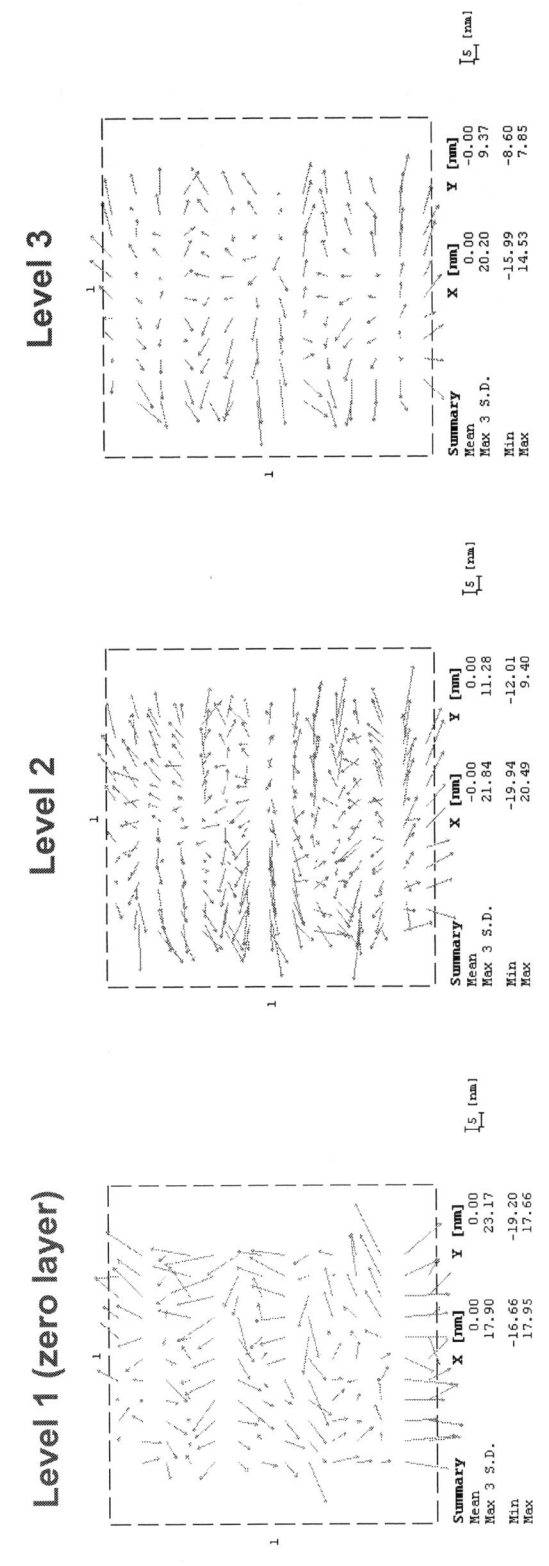

Alignment sites excluded, translation compensation only

10/30/2009

L2/L3 direct comparison (using DEVA software)

Translation compensation only

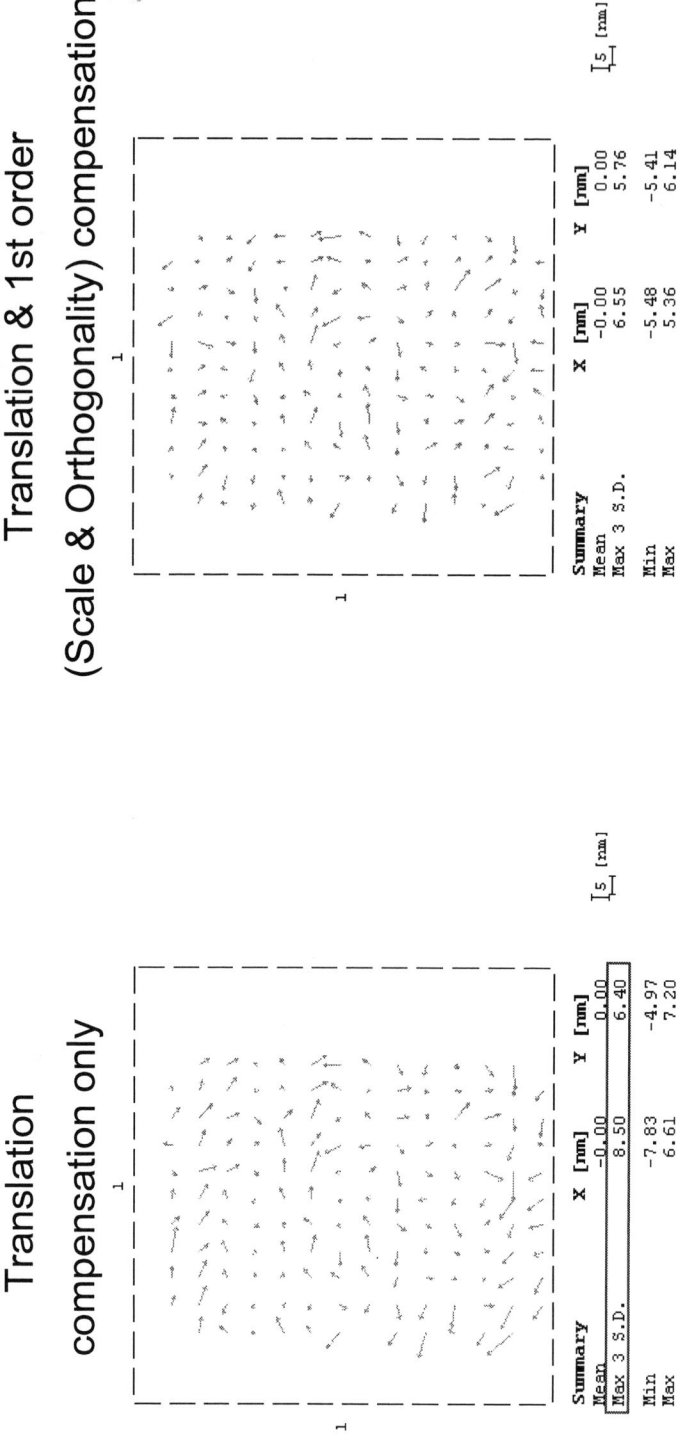

Summary	X [nm]	Y [nm]
Mean	-0.00	0.00
Max 3 S.D.	8.50	6.40
Min	-7.83	-4.97
Max	6.61	7.20

Translation & 1st order (Scale & Orthogonality) compensation

Summary	X [nm]	Y [nm]
Mean	-0.00	0.00
Max 3 S.D.	6.55	5.76
Min	-5.48	-5.41
Max	5.36	6.14

Performing common sites matching with 500µm catch radius

10/30/2009

Placement Error contribution to Overlay

Wafer Level Vector plot

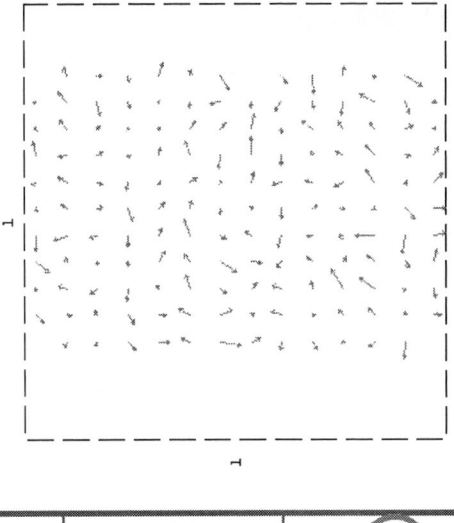

	Mask level	Wafer level	Mask level	Wafer level
	X-axis	X-axis	Y-axis	Y-axis
With Translation Compensation	8.5 nm	2.1 nm	6.4 nm	1.6 nm
With Translation, Mag and Ortho compensation	6.6 nm	1.7 nm	5.6 nm	1.4 nm

Pattern Placement Error on the mask contributes to a very high percentage of the Overlay Error.

10/30/2009

Wafer Shape Contribution

KLA Tencor
Accelerating Yield

IBM

10/30/2009

Wafer Shape and Overlay Errors in Lithography

$$\Delta x \propto \frac{\delta h x}{D^2}$$

δ is the shape

h is the thickness and

D is diameter of wafer

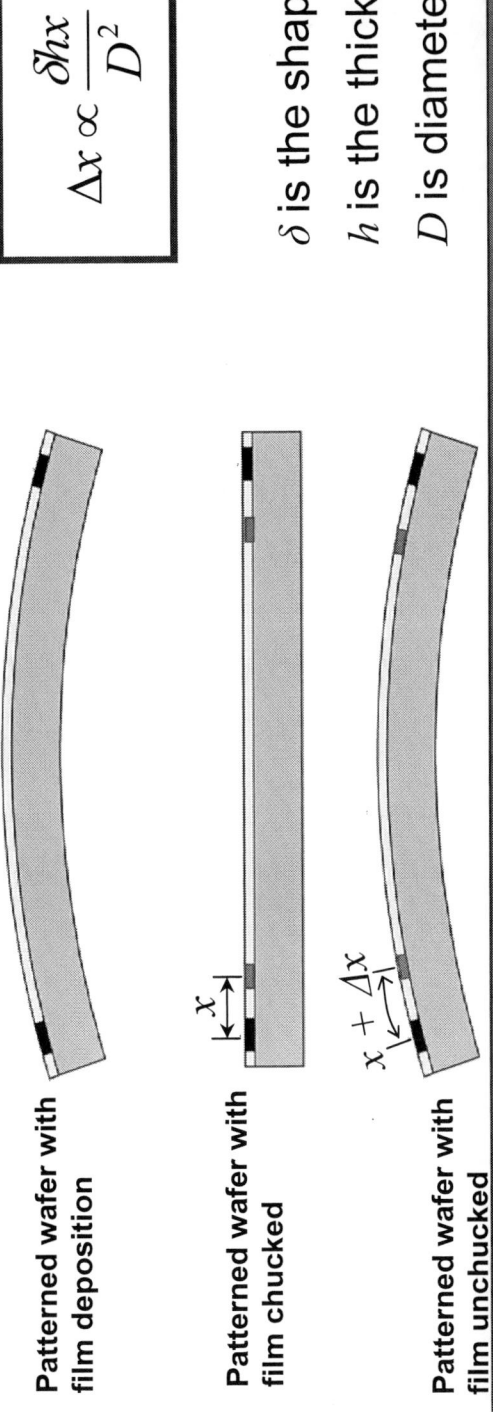

Patterned wafer with film deposition

Patterned wafer with film chucked

Patterned wafer with film unchucked

x

$x + \Delta x$

When the wafer is chucked on the wafer stage of the Stepper/Scanner, the linear component of the pattern shift can be compensated. However, if there is non-linear component, pattern shift cannot be compensated by conventional means.

10/30/2009

Predicted Overlay residuals from Higher Order Shape

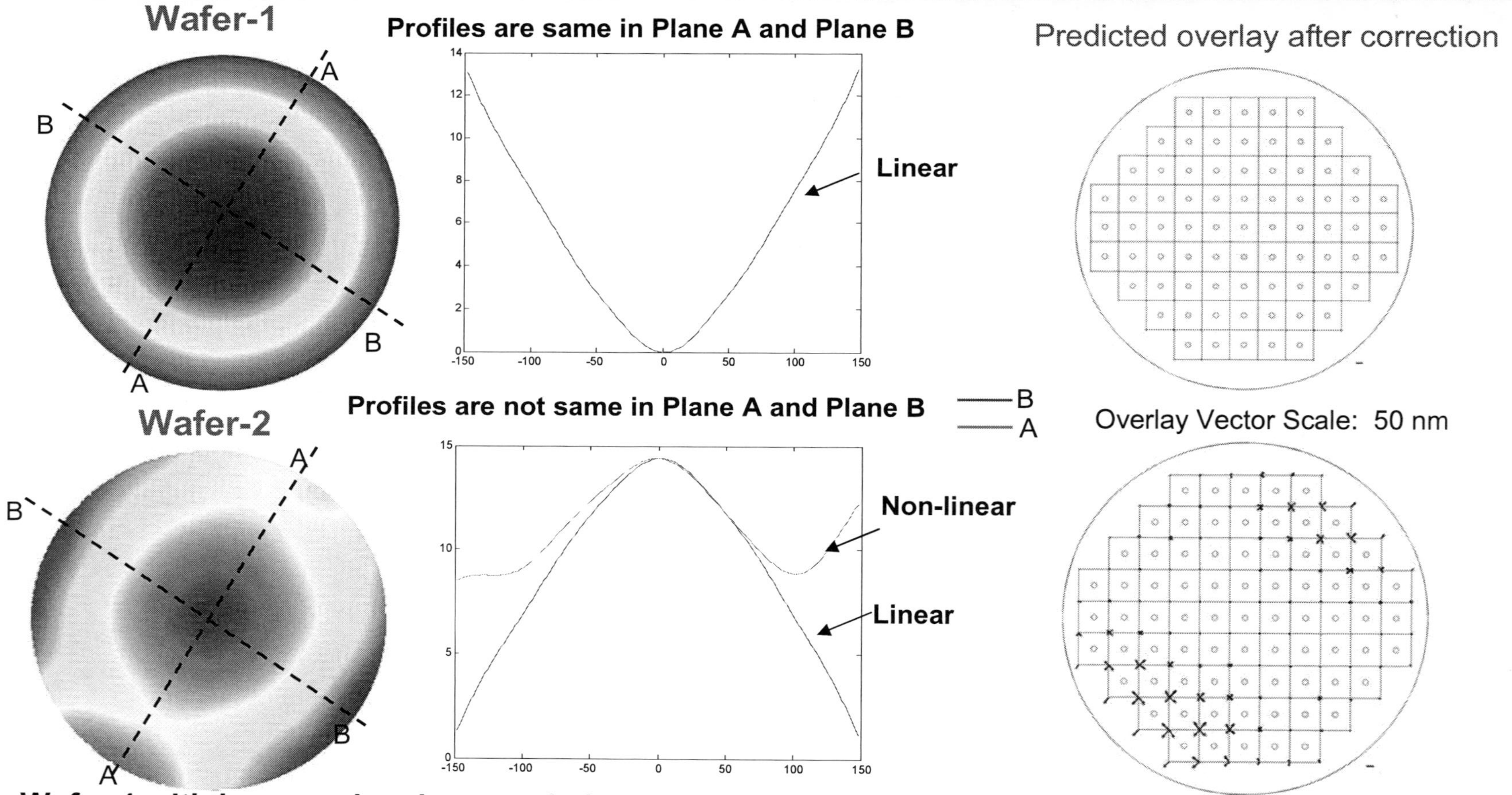

Predicted overlay after correction

Overlay Vector Scale: 50 nm

Profiles are same in Plane A and Plane B

Linear

Profiles are not same in Plane A and Plane B

—— B
—— A

Non-linear

Linear

Wafer-1

Wafer-2

- Wafer-1 with lower order shape variation can be effectively corrected using linear wafer and field level corrections.

- Wafer-2 with higher order shape variations results in overlay residual after linear wafer and field level corrections.

10/30/2009

Wafer Geometry – Shape of Experimental Wafers

- Wafer Shape is defined as the deviation of the wafer surface from a plane when it is unclamped.

- Bow and warp are the metrics used to characterize wafer shape.

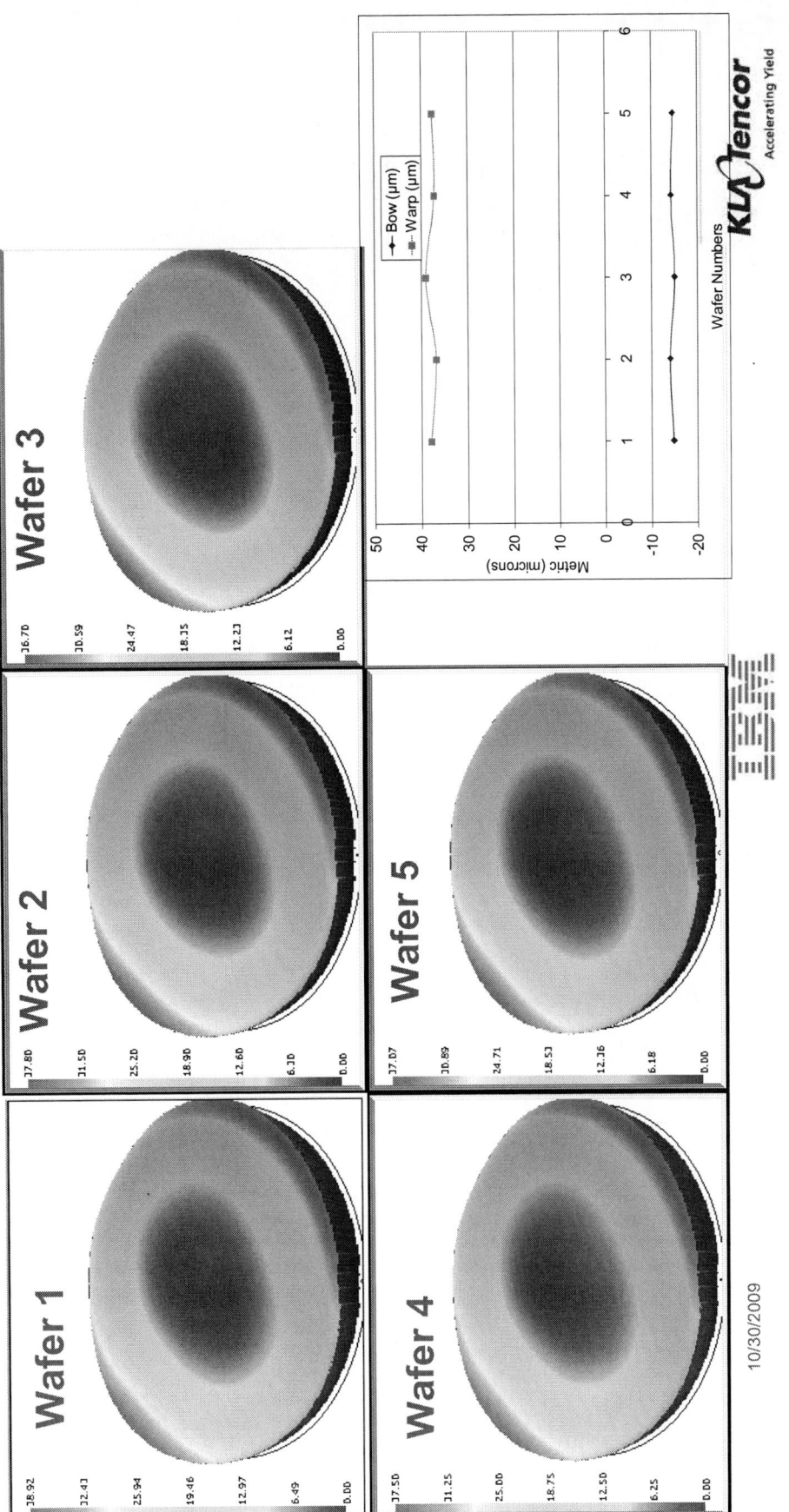

10/30/2009

Modeled Pattern Shift when clamped on scanner chuck (finite element model)

Shape induced pattern shift

Before correction on (26 mm x 33 mm) grid

There is higher pattern shift near the edge compared to the center of the wafer

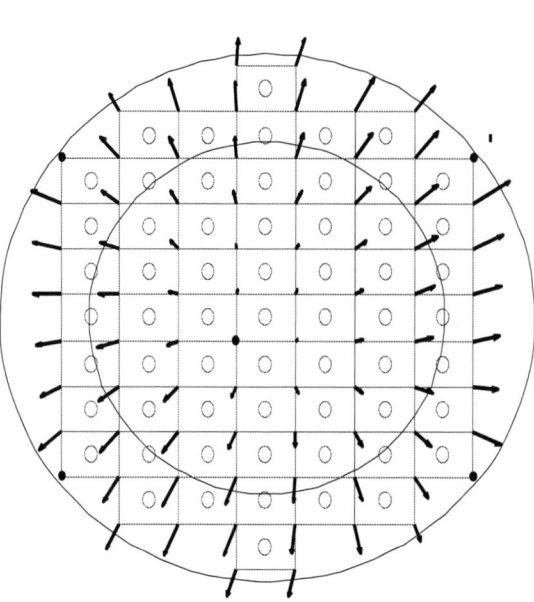

Sqrt(WIS_X^2+WIS_Y^2)

Interpretation:

In a ring sector between radius 30-60 mm has an average WIS_R of 40 nm.

In a ring sector between radius 120-150 mm has an average WIS_R of 170 nm.

Shows near the edge WIS_R is more.

10/30/2009

Correlation of Overlay Range to In-Plane Distortion (IPDX and IPDY)

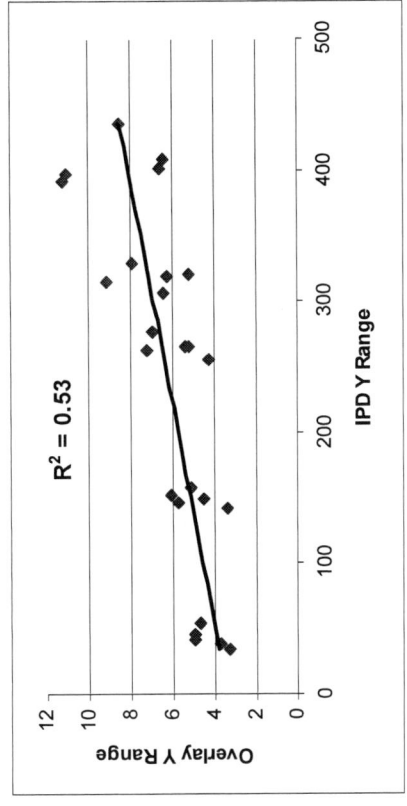

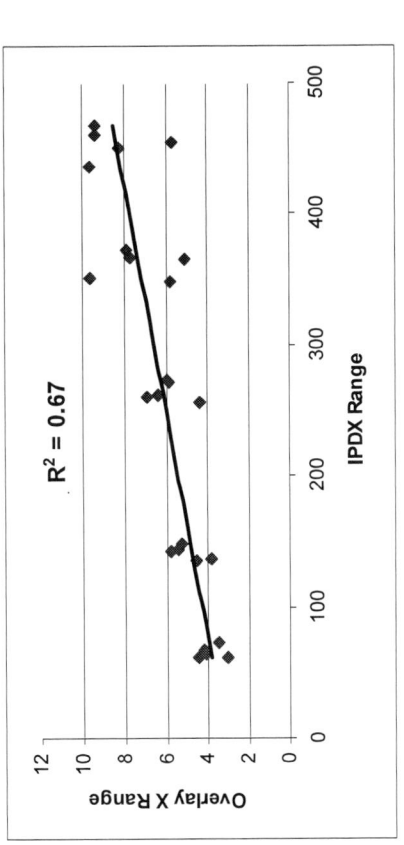

Overlay range in different annular regions correlate to IPD range

10/30/2009

Correctable Overlay residuals

10/30/2009

Overlay residuals

- The Patter Placement Error can be addressed by the e-beam writer

- Wafer shape issues can be addressed through Incoming Quality specifications (or possibly feed forward to the scanner)

- The rest of the overlay residuals can be attributed to the alignment, and process. We investigated how much of the remaining overlay residuals are correctable:

  - By applying suitable corrections at field and wafer level through a feed forward mechanism to the Stepper/Scanner, one could further reduce the overlay errors.

  - The 1st order and 3rd models were applied through simulation to the dataset, to determine how much further the residuals can be reduced.

10/30/2009

Model equations

Wafer 1st Order Model:

$$OverlayX = OffsetX + X1*X + X3*Y$$

$$OverlayY = OffsetY + X2*X + X4*Y$$

Wafer 3rd Order Model:

$$OverlayX = OffsetX + X1*X + X3*Y + X5*X^2 + X7*Y^2 + X9*X*Y + X11*X^3 + X13*Y^3 + X15*X^2*Y + X17*X*Y^2$$

$$OverlayY = OffsetY + X2*X + X4*Y + X6*X^2 + X8*Y^2 + X10*X*Y + X12*X^3 + X14*Y^3 + X16*X^2*Y + X18*X*Y^2$$

Field 1st Order Model:

$$OverlayX = Offsetx + b1*x + b3*y$$

$$OverlayY = Offsety + b2*y + b4*y$$

Field 3rd Order Model:

$$OverlayX = Offsetx + b1*x + b3*y + b5*x^2 + b7*y^2 + b9*x*y + b11*x^3 + b13*y^3 + b15*x^2*y + b17*x*y^2$$

$$OverlayY = Offsety + b2*x + b4*y + b6*x^2 + b8*y^2 + b10*x*y + b12*x^3 + b14*y^3 + b16*x^2*y + b18*x*y^2$$

Where: X,Y Grid coordinates, x,y – Field coordinates

10/30/2009

Modeling scenarios

	Model	Description
1	W1F1	First order wafer model and first order field model
2	W1F3	First order wafer model and third order field model
3	W3F1	Third order wafer model and first order field model
4	W3F3	Third order wafer model and third order field model
5	F2F-F1	First order Field by Field model
6	F2F-F3	Third order Field by Field model

10/30/2009

Overlay Residuals

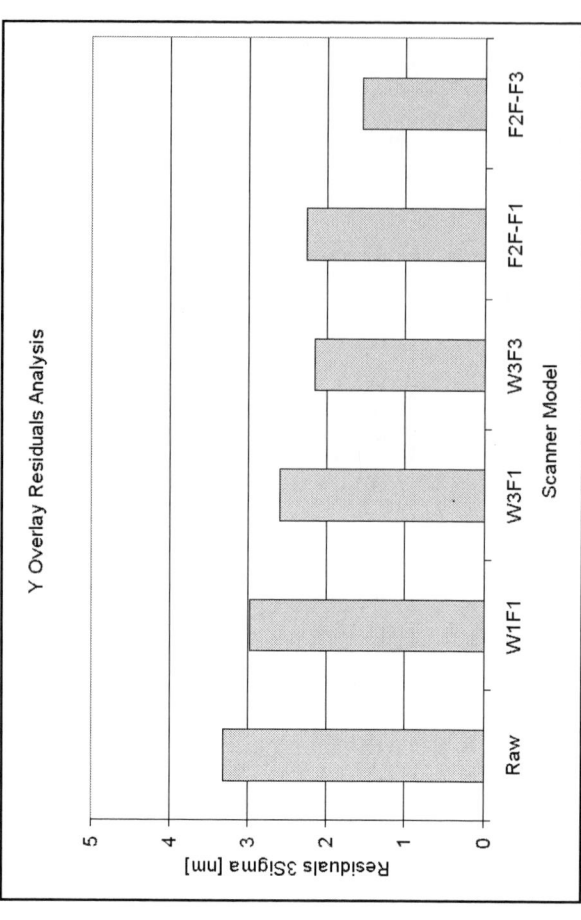

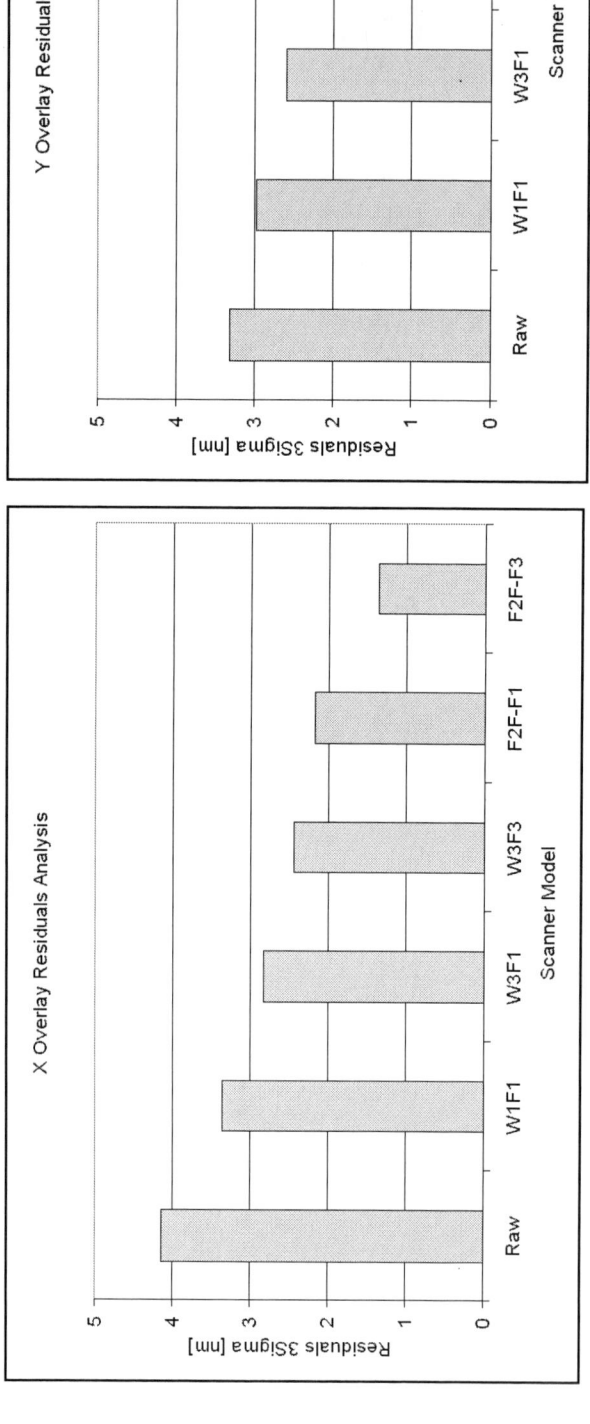

- 3rd Order model reduces overlay error by ~ 40% on the average
- Field by Field High order correction can reduce the residuals to 1.5nm.

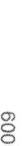

10/30/2009

DPL Overlay Components

10/30/2009

Estimated DPL Overlay Components

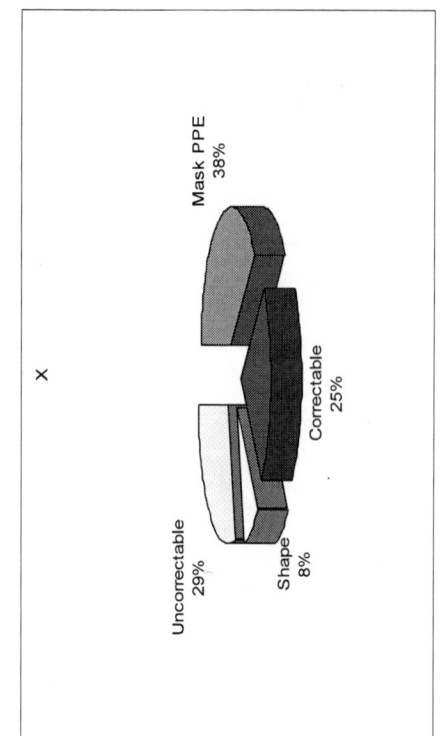

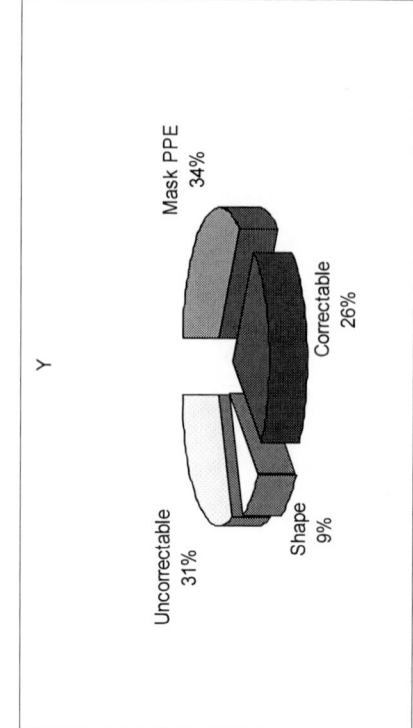

DPL Overlay Component	X	Y	X	Y
	nm	nm	%	%
Mask PPE	1.7	1.4	38.6	34.1
Shape	0.36	0.36	8.2	8.8
Correctable Residuals	1.08	1.08	24.5	26.3
Remaining Uncorrectable Residuals	1.26	1.26	28.6	30.7
Total	4.4	4.1	100	100

10/30/2009

Summary

- Double Patterning Lithography overlay components are studied using a specially designed DPL reticle-set. The reticle contained an array of intrafield alignment marks over the field of 24x30 mm.

- The components of overlay error investigated are;

  - Pattern Placement Error (measured with KLA-Tencor IPRO4 system)

  - Wafer Shape Contribution (measured with KLA-Tencor Geometry Tool)

  - Overlay errors from Stepper/Scanner (measured with KLA-Tencor Archer 200)

- In this investigation, the Pattern Placement Error contributed to 36% and the wafer shape contributed to roughly 9% of the DPL Overlay error. 26% of the overlay error is correctable with a feed forward mechanism.

- The specification for Pattern Placement Error on the mask needs to be tightened to reduce the overlay error further. The contribution from the shape component can be minimized with a tighter specification on the flatness and shape specification.

10/30/2009

Acknowledgement

This work was performed by the Research Alliance Teams at various IBM Research and Development Facilities

We would like to thank MIE, OMD and ADE Divisions at KLA-Tencor as well as

Becky Pinto, Eric Bouche and Brian Trafas for the support.

10/30/2009

Development of Freezing Free LLE Process

Tsuyoshi Nakamura, Ryoichi Takasu, Masaru Takeshita, Jiro Yokoya, Yasuhiro Yoshii, Hirokuni Saito, Katsumi Ohmori

TOKYO OHKA KOGYO CO.,LTD.

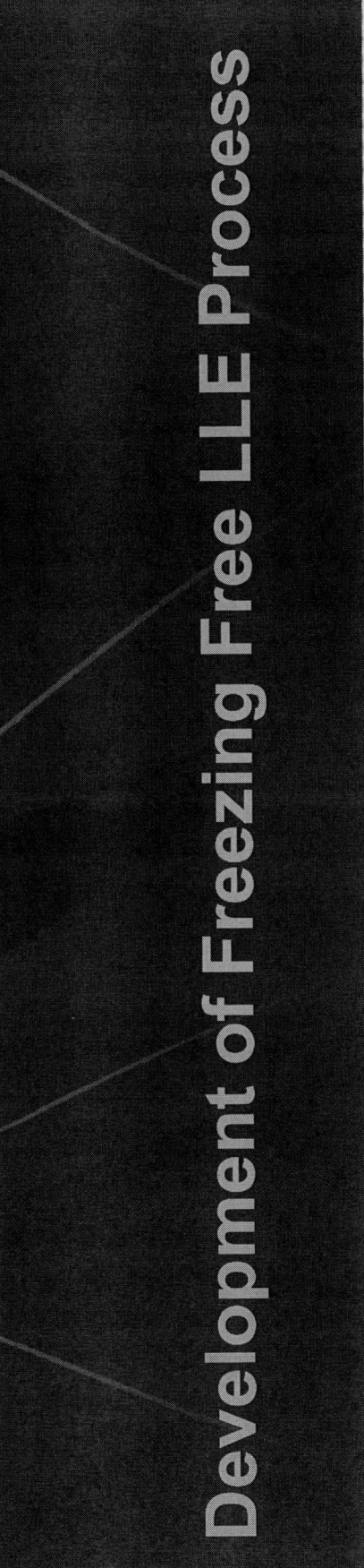

6th International Symposium on Immersion Lithography Extensions

Contents

- **Introduction**
  - √ Introduction of Freezing free LLE process

- **Dual line application**
  - √ Issues
  - √ New 2nd resist
    - ● LWR at single lithography
  - √ Approach for collapse
  - √ Double patterning results
    - ● 32nm DHP
    - ● 26nm DHP process window

- *Cross line application*
  - √ Issue
  - √ Cross-line CH result with new 2nd resist

- **Process feasibility**
  - ● Defectivity
  - ● Etching results

- **Conclusion**

6th International Symposium on Immersion Lithography Extensions

Motivation of Freezing Free LLE process

LELE Process

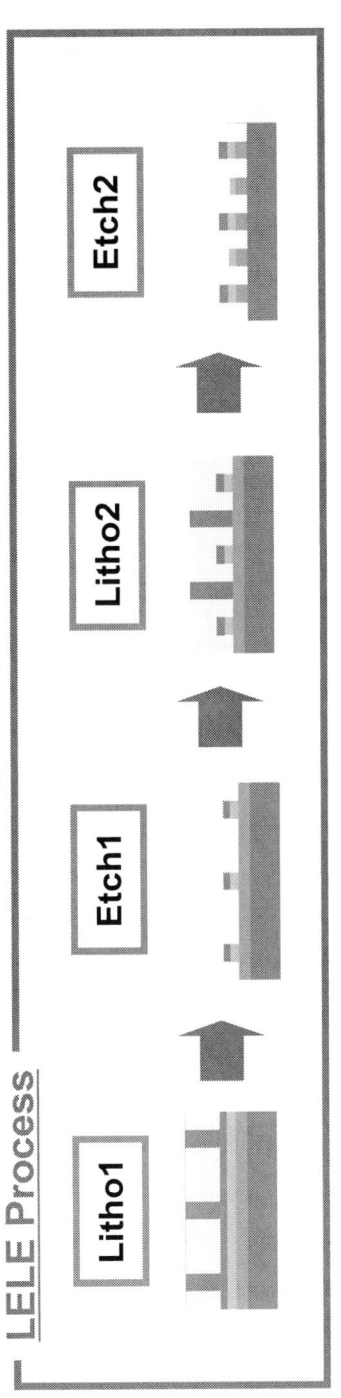

Litho1 → Etch1 → Litho2 → Etch2

Pros.
-Possible to establish a process with current technology

Cons.
-CoO, Process complexity

As alternative process, a simple LLE process is required to reduce CoO

The most simple process of LLE

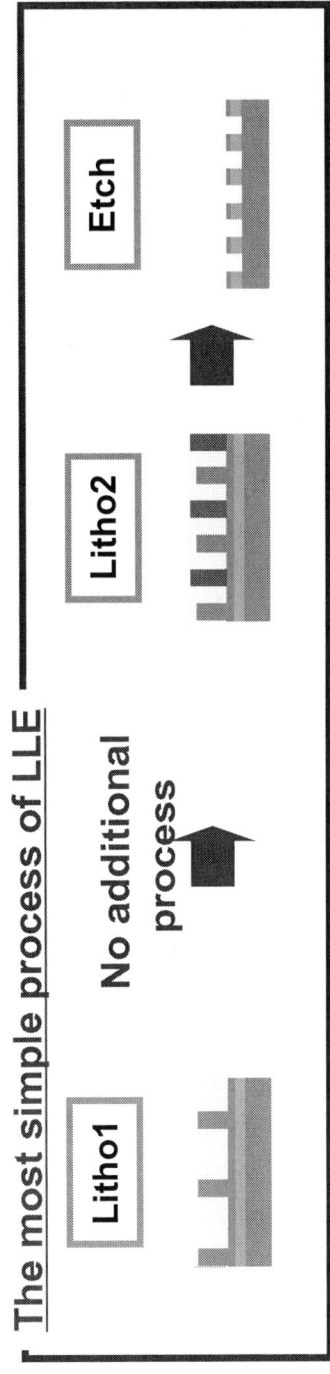

Litho1 → Litho2 → Etch

No additional process

The authors would like to suggest a freezing free LLE process for low CoO

6th International Symposium on Immersion Lithography Extensions

Freezing Free Posi/Posi Process Flow

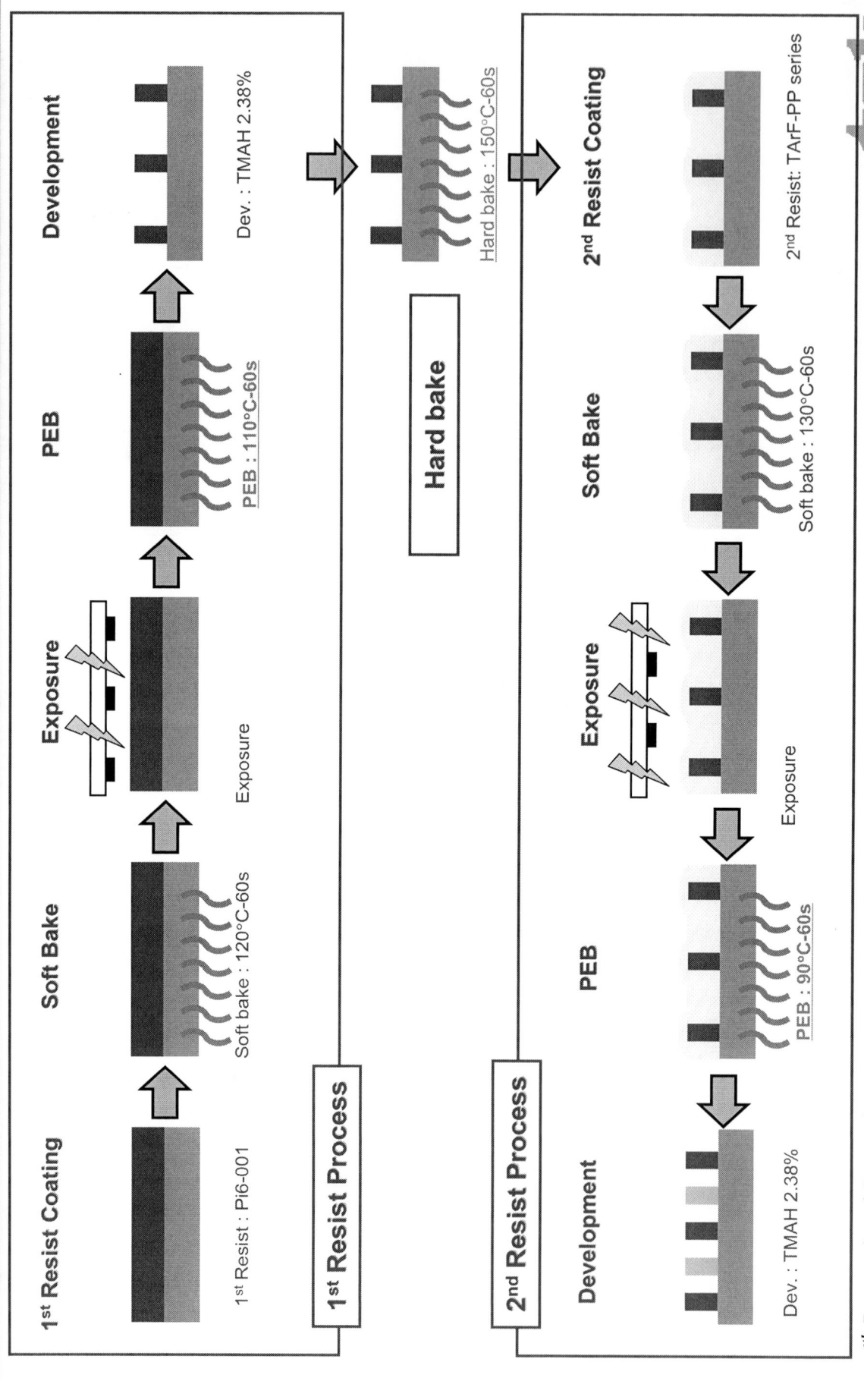

6th International Symposium on Immersion Lithography Extensions

Criteria for Posi/Posi Process

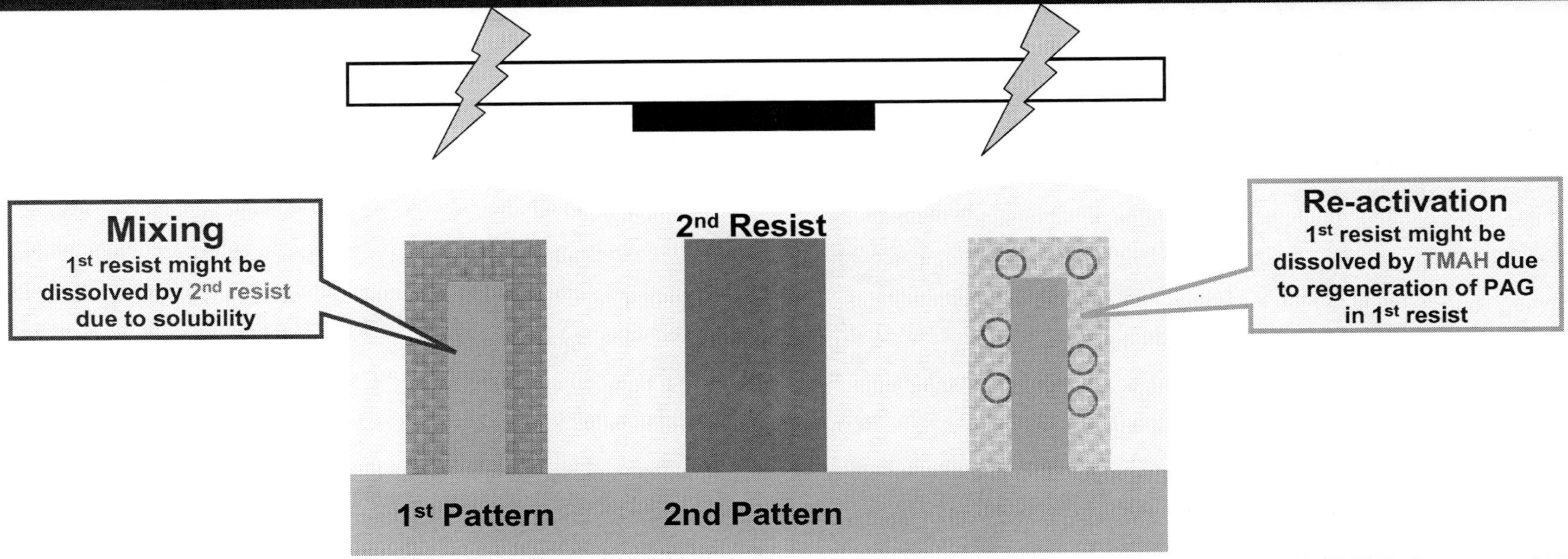

Re-activation
1st resist might be dissolved by TMAH due to regeneration of PAG in 1st resist

Mixing
1st resist might be dissolved by 2nd resist due to solubility

2nd Resist

1st Pattern   2nd Pattern

Object	Countermeasure	Criterion
Mixing	Less solubility	Specific solvent of 2nd resist
Mixing	Less solubility	Specific design of 2nd resist polymer
Mixing	hardening	Hard bake process (150-160C)
Re-activation	No reaction	Different activation energy functional group (1st:High Eact/2nd:Low Eact)- PEB gap 1st and 2nd

Nakamura T. et al., "Newly developed positive tone resist for posi/posi double patterning" SPIE 7273-3 (2009)

6th International Symposium on Immersion Lithography Extensions

Resist Concepts

Special solvent for 2nd resists (alcohol type)

QCM of 1st resist film in solvent

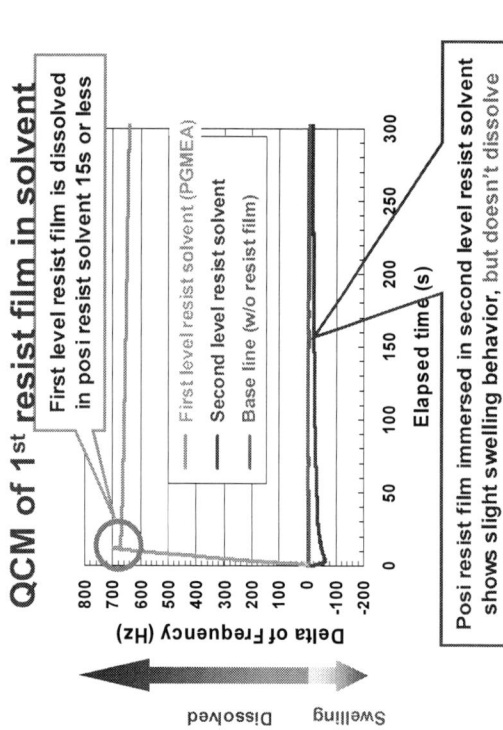

First level resist film is dissolved in posi resist solvent 15s or less

- First level resist solvent (PGMEA)
- Second level resist solvent
- Base line (w/o resist film)

Posi resist film immersed in second level resist solvent shows slight swelling behavior, but doesn't dissolve

PEB Gapping
(High Ea 1st resist and Low Ea 2nd resist)

DRM curve of 1st resist

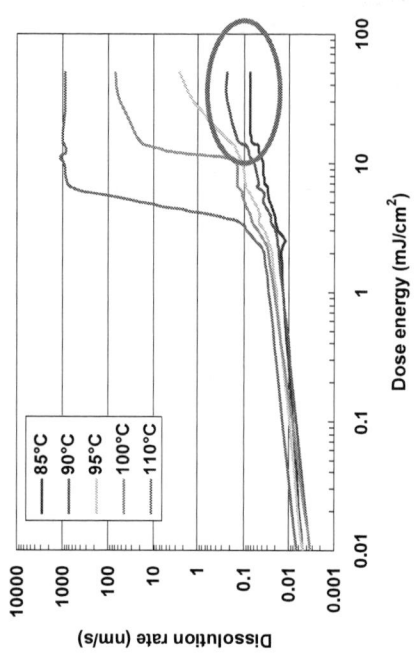

- 85°C
- 90°C
- 95°C
- 100°C
- 110°C

Ando T. et al., Proc. SPIE 7140 (2008)

Hard bake effect – not freezing

Resist film FT-IR analysis w/ and w/o hard bake

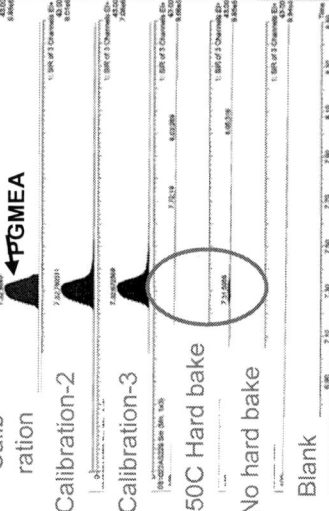

Small difference around 1240cm-1

Resist film GC-MS analysis w/ and w/o hard bake

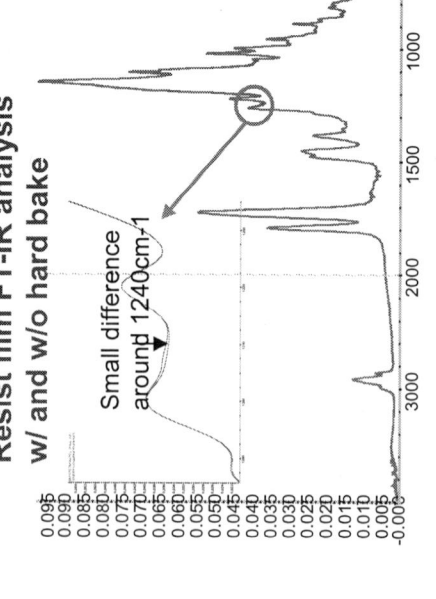

PGMEA

- Calibration
- Calibration-2
- Calibration-3
- 150C Hard bake
- No hard bake
- Blank

There was not cross-linking nor chemical reaction occurring by hard bake.
It occurred just only removing of solvent

Nakamura T. et al., SPIE 7273-3 (2009)

6th International Symposium on Immersion Lithography Extensions

Current Achievement

Dual line patterning process

Data courtesy of IMEC

22nm DHP
1st : 21.4nm(LWR:3.3)
2nd : 25.1nm(LWR:4.1)

22DHP(44MCD)

NA:1.35 Dipole

Nakamura T. et al., J. Photopolymer science and Technology 2009

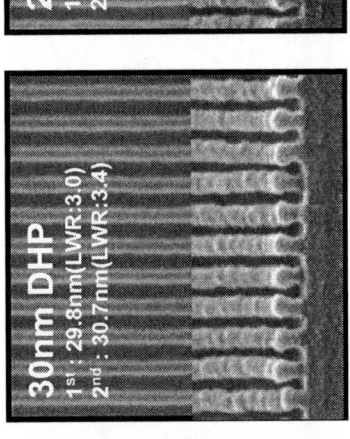

28nm DHP
1st : 28.4nm(LWR:3.1)
2nd : 32.1nm(LWR:4.2)

30nm DHP
1st : 29.8nm(LWR:3.0)
2nd : 30.7nm(LWR:3.4)

32nm DHP
1st : 31.4nm(LWR:2.7)
2nd : 31.9nm(LWR:3.1)

Resolution NA:1.00 Dipole

Cross line patterning process

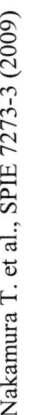

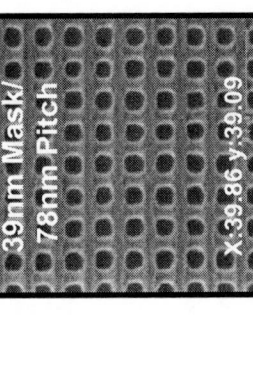

39nm Mask/ 78nm Pitch

X:39.86 y:39.09

Nakamura T. et al., SPIE 7273-3 (2009)

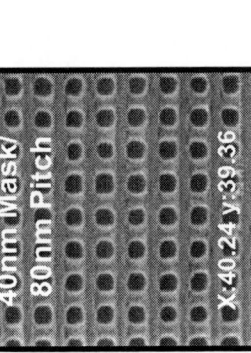

40nm Mask/ 80nm Pitch

X:40.24 y:39.36

45nm Mask/ 90nm Pitch

X:46.43 y:44.59

Resolution NA:1.35 Dipole

6th International Symposium on Immersion Lithography Extensions

Issues For Dual Lines

■ Poor LWR (2nd litho)

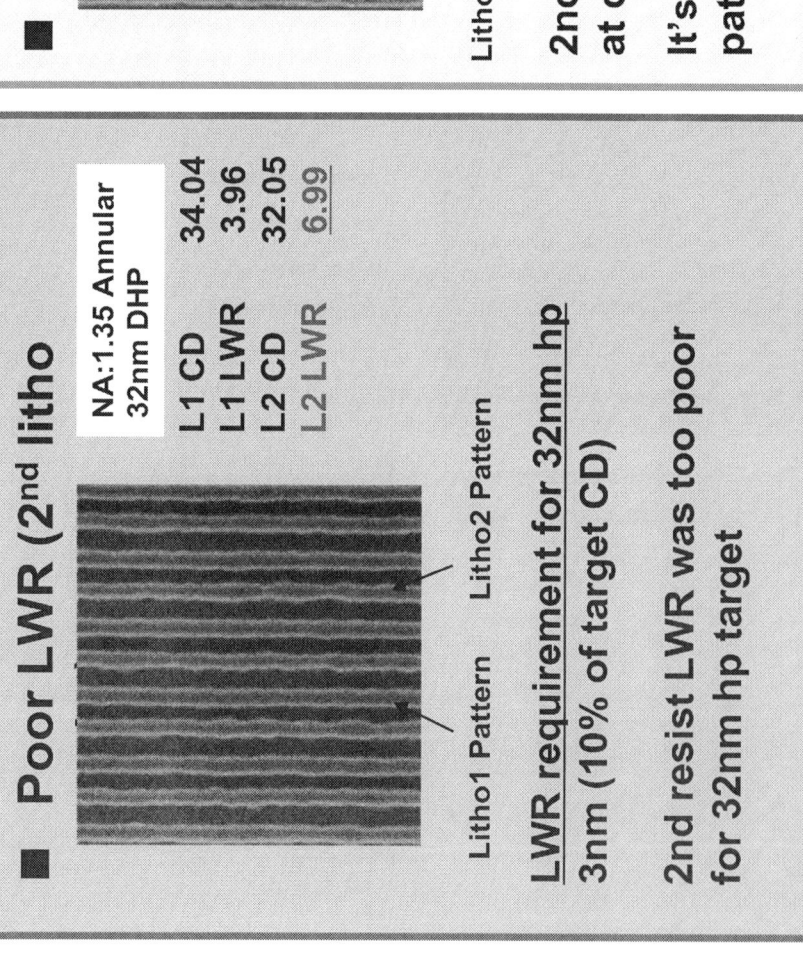

NA:1.35 Annular
32nm DHP

L1 CD	34.04
L1 LWR	3.96
L2 CD	32.05
L2 LWR	6.99

Litho1 Pattern   Litho2 Pattern

<u>LWR requirement for 32nm hp</u>
3nm (10% of target CD)

2nd resist LWR was too poor
for 32nm hp target

■ Pattern Collapse (2nd litho resist)

<u>Litho2 Pattern
Collapse</u>

Litho1 Pattern   Litho2 Pattern

2nd resist pattern tends to collapse
at dual line process

It's necessary to prevent
pattern collapse

2nd resist lithography improvement is necessary
✓ <u>Better contrast for better LWR</u>
✓ <u>Collapse improvement</u>

6th International Symposium on Immersion Lithography Extensions

Better Contrast

Factor	Determiner	Approach
Acid contrast	Acid diffusion Acid density	Soluble PAG with bulky structure anion Acid contrast enhancer
Dissolution contrast	De-protection ratio	Easy cleavable unit

New PAG structure (anion)

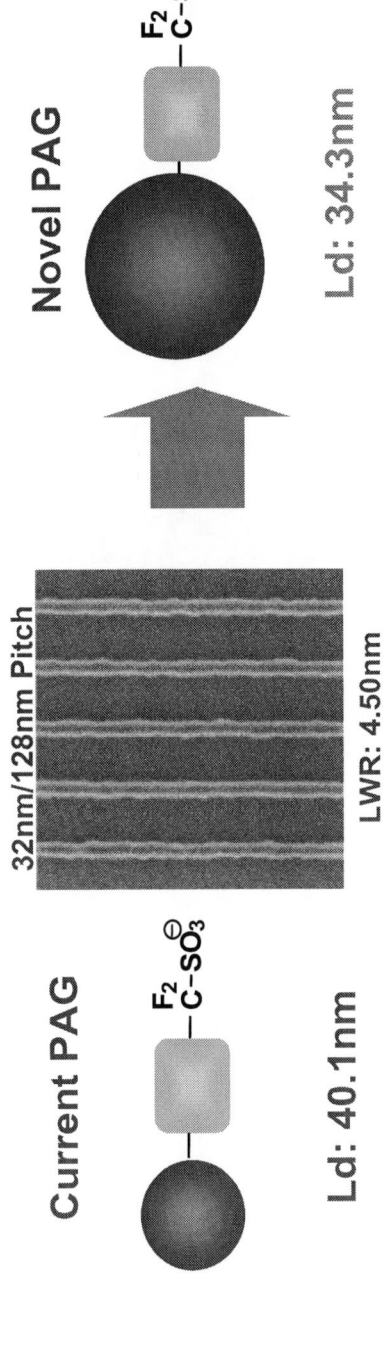

Current PAG

F_2C–$SO_3^{\ominus}$

Ld: 40.1nm

Novel PAG

F_2C–$SO_3^{\ominus}$

Ld: 34.3nm

32nm/128nm Pitch LWR: 4.50nm

32nm/128nm Pitch LWR: 3.64nm

Brunner T. et al., SPIE 5377 (2004)
Van Steenwinckel D. et al., SPIE 5753 (2005)

New PAG improves LWR by control of acid diffusion

2nd Litho Resist Development For LWR

Data courtesy of IMEC   NA:1.35 Annular

LWR (nm) vs **Focus (um)**
- Previous 2nd resist
- New 2nd resist

EL (%) vs **DOF (um)**
- New 2nd Resist

32nm/128nm Pitch

	Previous 33.4mJ	New 2nd resist 22.6mJ
Eop	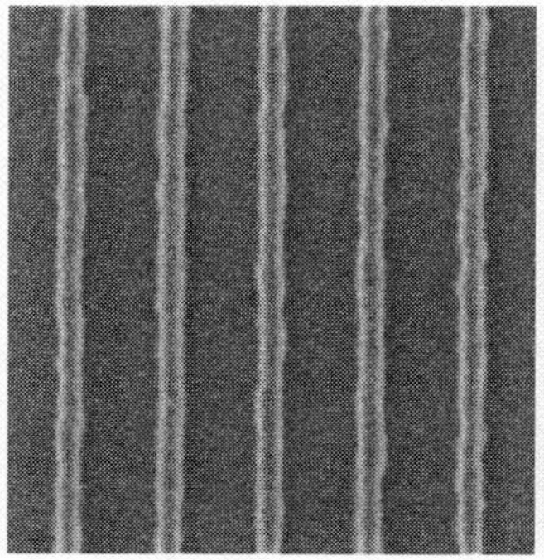	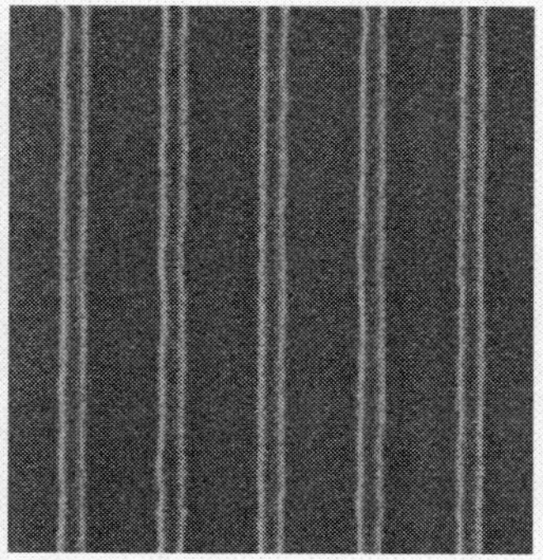
CD / LWR (nm)	32.44 / 4.50	33.19 / 3.33
	41mJ	28mJ
Before collapse	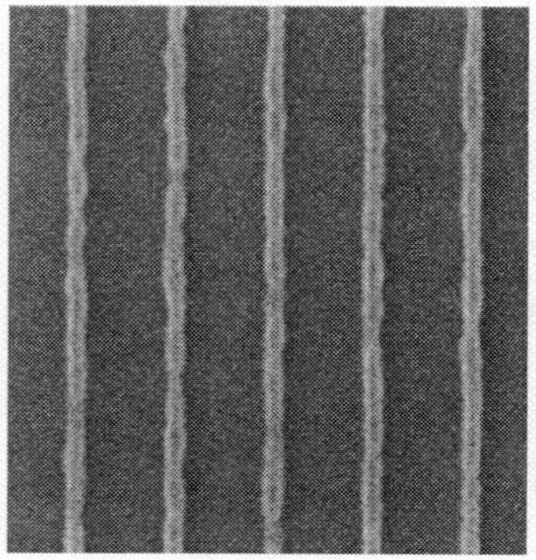	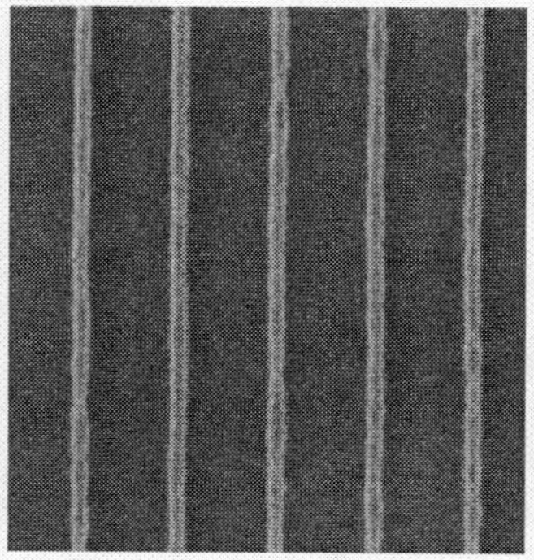
CD / LWR (nm)	22.52 / 5.79	21.33 / 4.03
Ecol / Esize	1.23	1.24

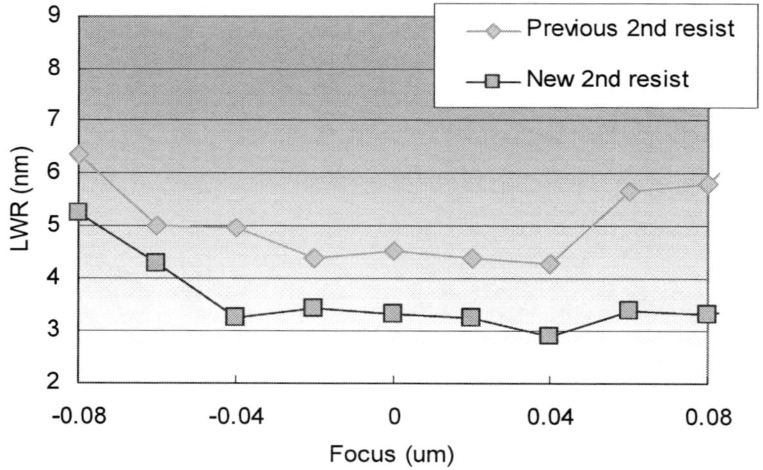

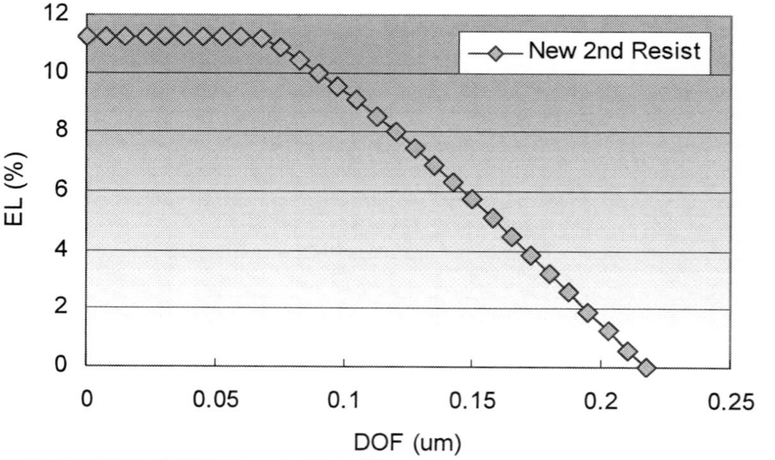

Much better LWR was shown on new 2nd resist

6th International Symposium on Immersion Lithography Extensions

2nd Litho Resist Development For LWR

32nm DHP comparison

NA : 1.35 Dipole40 w/ polarization

Data courtesy of IMEC

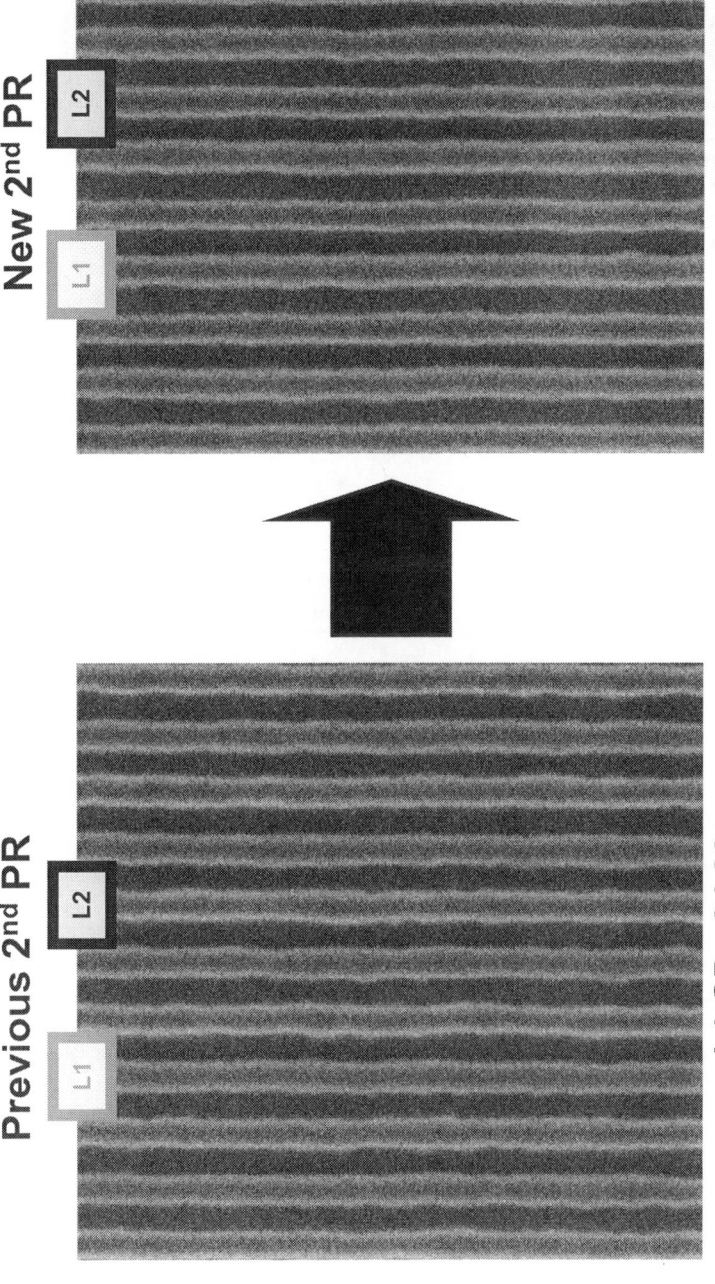

Previous 2nd PR

L1   L2

L1 CD 34.88
L1 LWR3.65
L2 CD 34.83
L2 LWR4.71

New 2nd PR

L1   L2

L1 CD 35.58
L1 LWR3.87
L2 CD 34.76
L2 LWR3.71

LWR of 2nd resist after double patterning was much improved

6th International Symposium on Immersion Lithography Extensions

tok

Process Approach For Pattern Collapse

NA : 1.35 Dipole40X w/ polarization
Target : 26nm DHP

26nm DHP

DIW Rinse

Surfactant Rinse

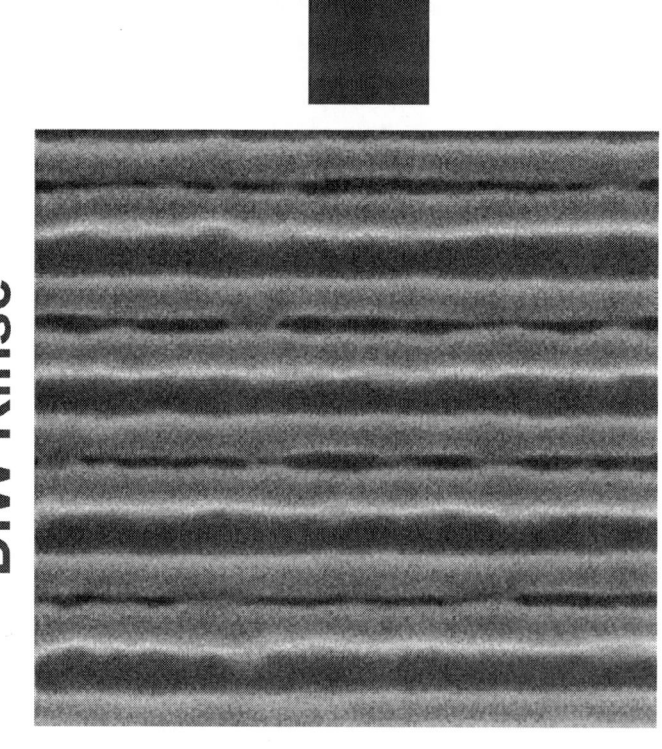

Surfactant rinse is quite effective to prevent pattern collapse at below 26nm hp

6th International Symposium on Immersion Lithography Extensions

26DHP Process Window

NA : 1.35 Dipole40X w/ polarization
Target : 26nm DHP

Data courtesy of IMEC

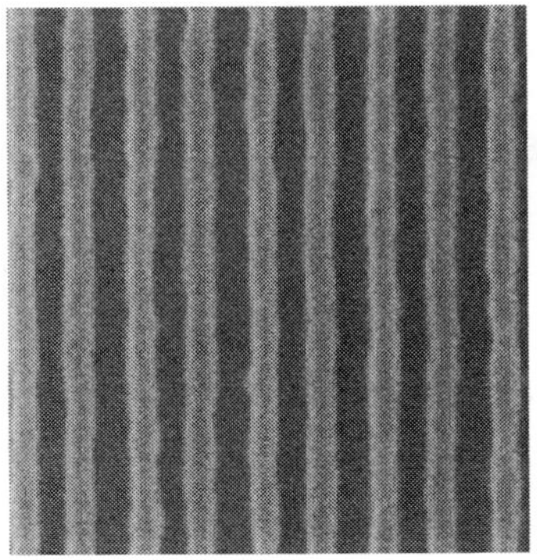

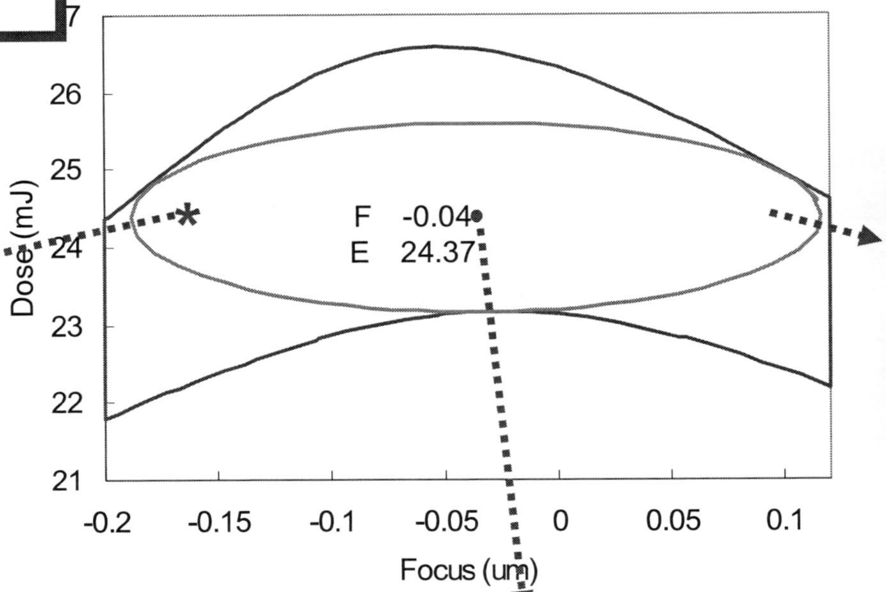

Dose (mJ)
21
22
23
24
25
26
27
Focus (um)
-0.2
-0.15
-0.1
-0.05
0
0.05
0.1
F -0.04
E 24.37

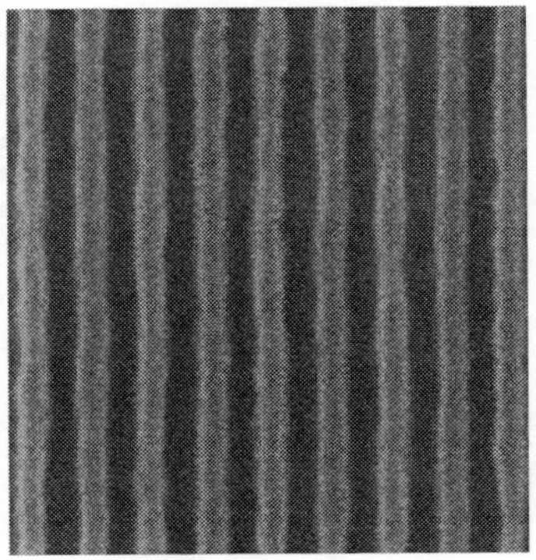

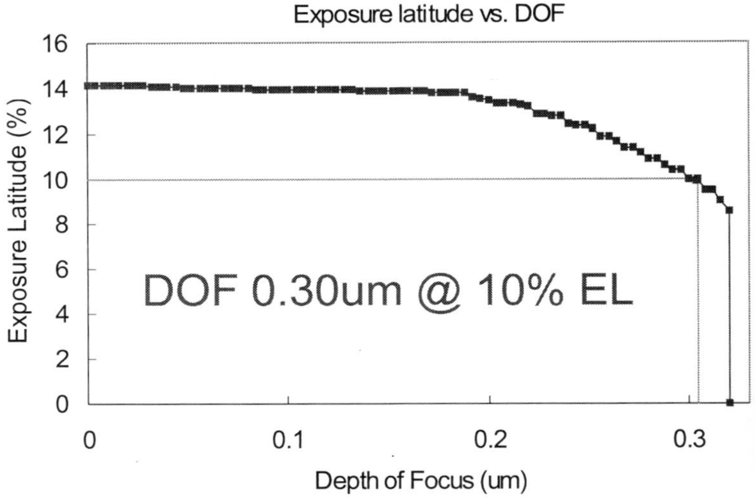

Exposure latitude vs. DOF
Exposure Latitude (%)
0
2
4
6
8
10
12
14
16
Depth of Focus (um)
0
0.1
0.2
0.3
DOF 0.30um @ 10% EL

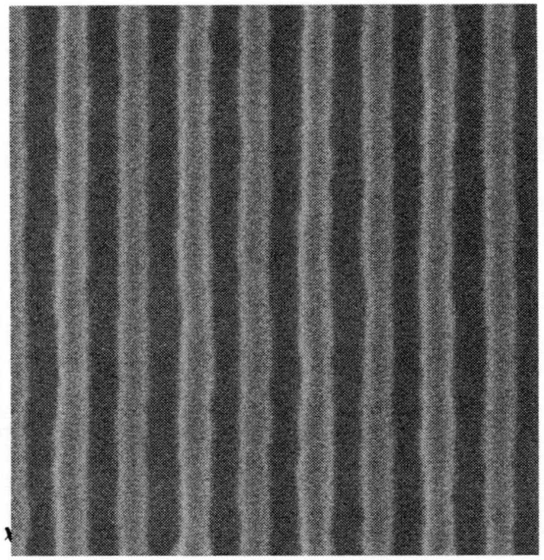

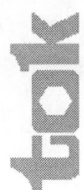

L1 CD 24.76
L1 LWR3.58
L2 CD 26.58
L2 LWR4.14

tok

Issue For Cross-Line CH

■ Local CDU

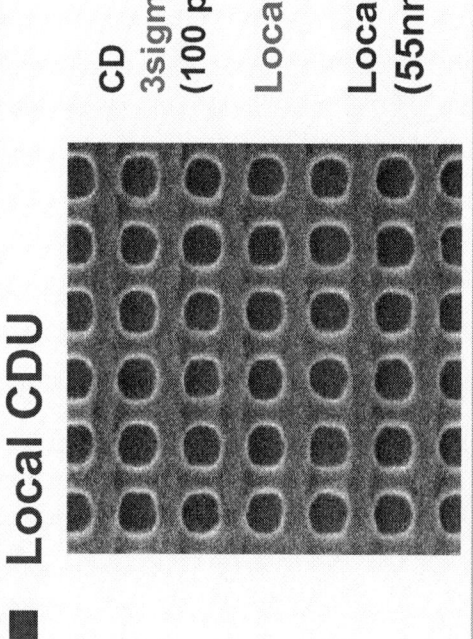

CD : 56.5nm
3sigma : 9.67 nm
(100 points meas.)

Local CDU wasn't good enough

Local CDU should be lower than 10% of CD target at least
(55nm CH ➔ 5.5nm)

2nd resist contrast might be relative to local CDU because it was supposed that 2nd resist lithography was dominant for CH resolution

Better contrast 2nd resist is needed as same as dual line case

6th International Symposium on Immersion Lithography Extensions

Cross Line Patterning

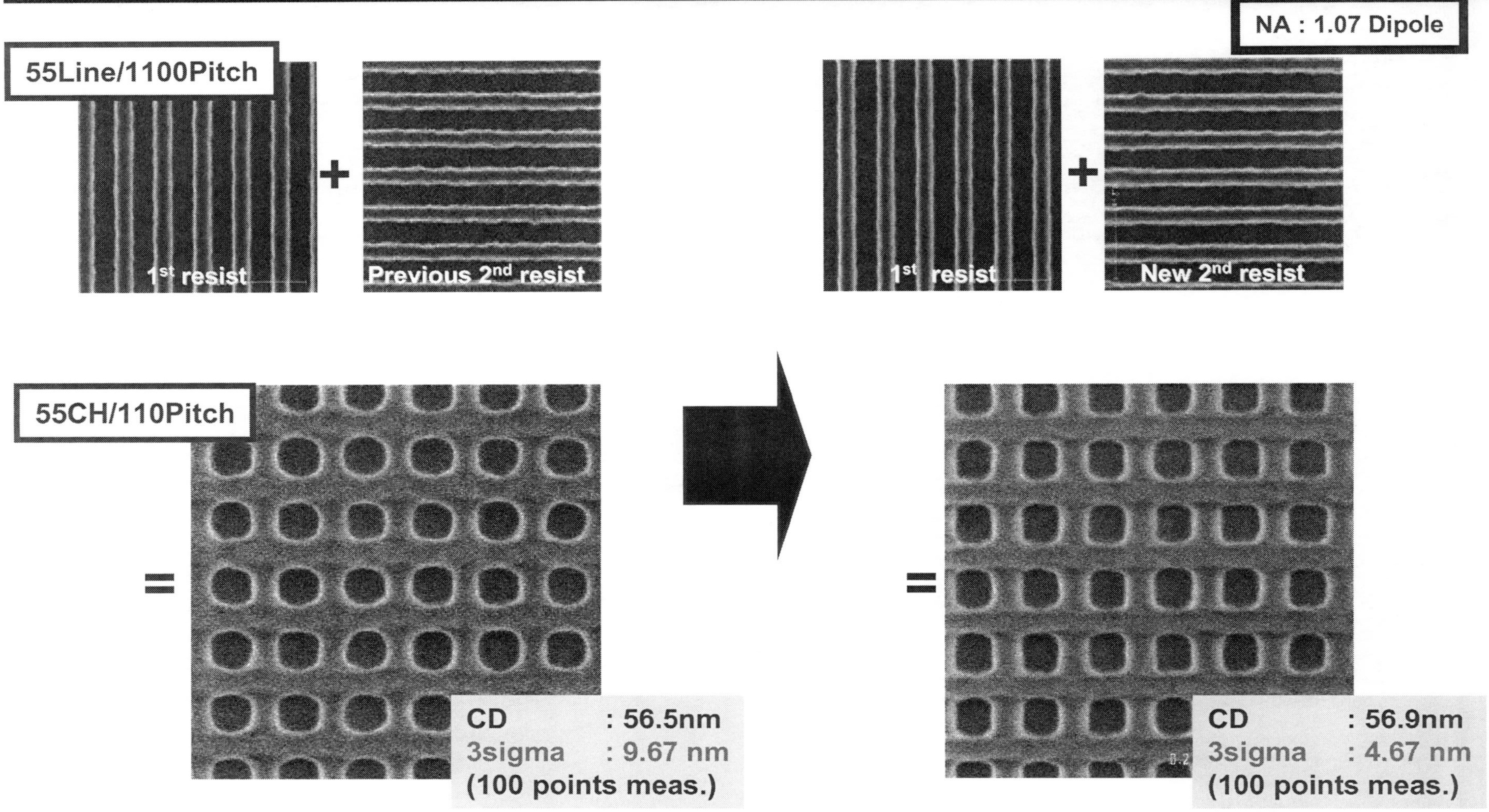

NA : 1.07 Dipole

55Line/1100Pitch

1st resist

+

Previous 2nd resist

=

55CH/110Pitch

CD   : 56.5nm
3sigma : 9.67 nm
(100 points meas.)

1st resist

+

New 2nd resist

=

CD   : 56.9nm
3sigma : 4.67 nm
(100 points meas.)

The local CD uniformity was improved by using new 2nd litho resist

6th International Symposium on Immersion Lithography Extensions

Defectivity 55nm Cross-Line CH

1.07NA dipole
55nm CH/ 110nm Pitch
KLA 2371

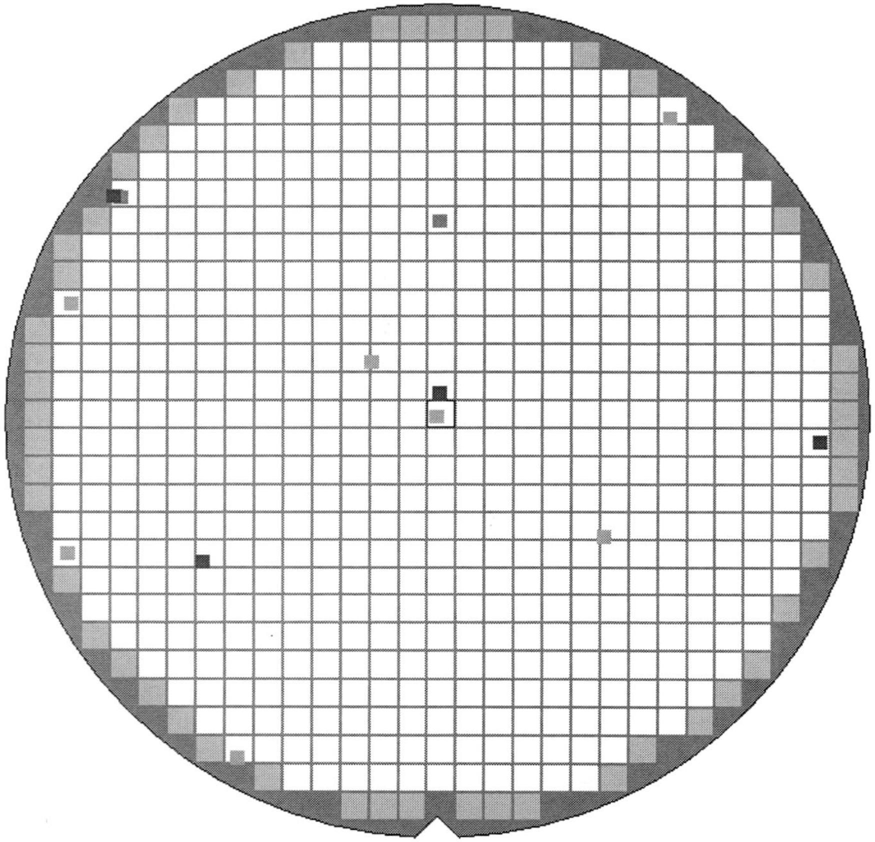

Legend:
- 1st Line Deformation
- 2nd Layer Blob
- 1st Layer Blob

1st Line Deformation

2nd Layer Blob

1st Layer Blob

Defect density 0.06D/cm2

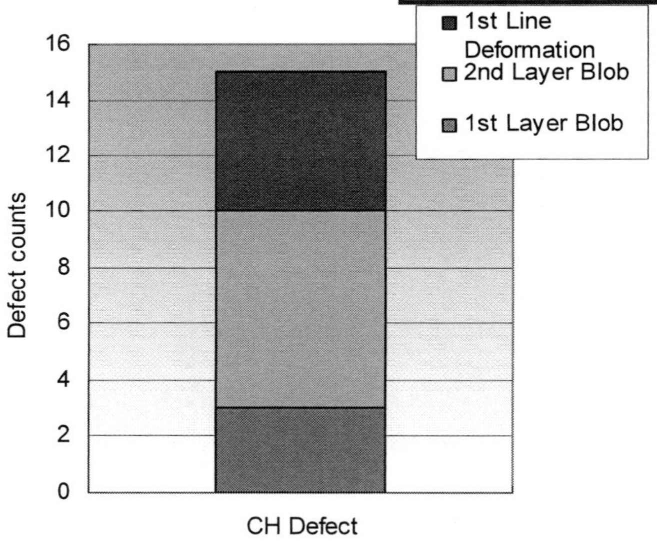

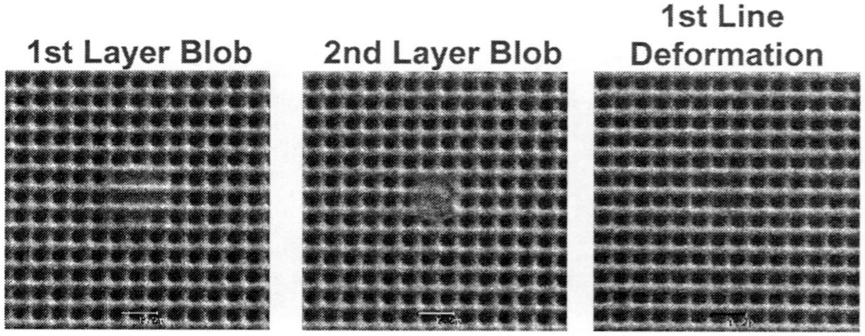

The defectivity of cross-line CH was very low

6th International Symposium on Immersion Lithography Extensions

Etching Test

Data courtesy of Hitachi High-Technologies

60nm CH imaging by NA 1.07

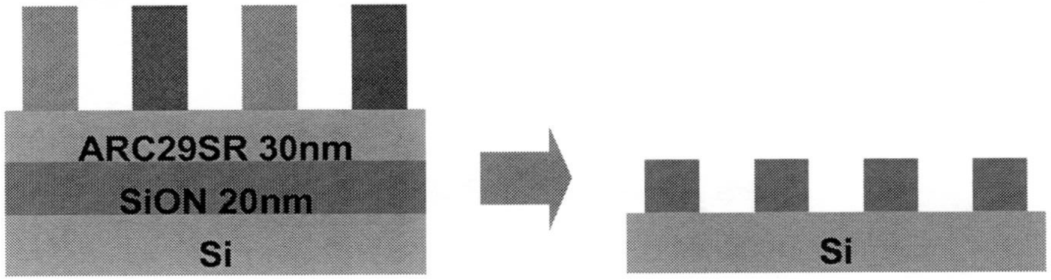

ARC29SR 30nm
SiON 20nm
Si

Tool: Hitachi U-8250
Gas: CF4/CHF3

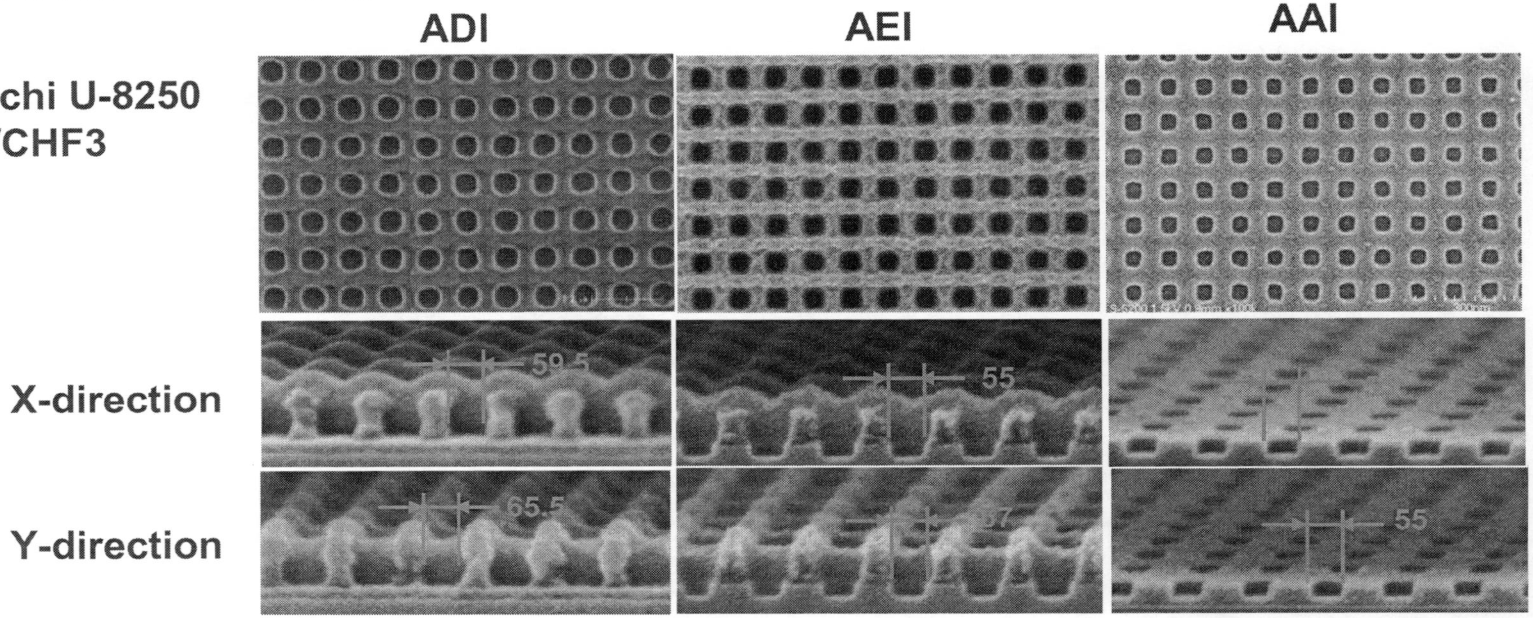

Si

ADI   AEI   AAI

X-direction

Y-direction

CDU-X: 6.77nm
CDU-Y: 5.66nm

Height erosion 1st resist: 32nm
Height erosion 2nd resist: 36nm

CDU-X: 5.92nm
CDU-Y: 5.78nm

CD Bias 1st resist: 6.5nm
CD Bias 2nd resist: 10.5nm

Etching resistance of 1st and 2nd are almost same

6th International Symposium on Immersion Lithography Extensions

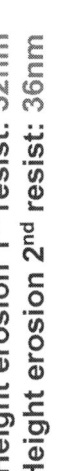

Conclusion

- Freezing free LLE process has been developed as most simple LLE process that doesn't need freezing process

- Freezing free LLE process has already showed 22nm resolution by dual line and 40nm CH by cross-line CH

- New 2nd litho resist can improve following items
  - ✓ 2nd resist LWR from 4.5nm to 3.3nm
  - ✓ Local CDU of CH from 9.6nm to 4.7nm

- In addition, process feasibility have been confirmed
  - ✓ The defectivity of cross-line CH was 0.06D/cm2
  - ✓ Etching tolerance was also confirmed and 2nd resist has same etch tolerance as 1st resist

6th International Symposium on Immersion Lithography Extensions

Acknowledge

We would like to thank IMEC for Imaging data collection and thank Hitachi High-Technologies for Etching support. And we would like to thank engineers in TOK R&D for useful discussion

Hitachi High-Technologies

6th International Symposium on Immersion Lithography Extensions

Thank you for your kind attention !

6th International Symposium on Immersion Lithography Extensions

Exploration of New Resist Chemistries and Process Methods for Enabling Dual-Tone Development

C. Fonseca, M. Somervell, S. Scheer
Y. Kuwahara, K. Nafus
Tokyo Electron Ltd.

R. Gronheid
IMEC

S. Tarutani
FUJIFILM

6th International Symposium on Immersion Lithography Extensions
Prague, Czech Republic

TOKYO ELECTRON

6th International Symposium on Immersion Lithography Extensions, October 22, 2009

Outline

- Background/Goals

- Previous demonstrations and challenges

- Latest results with new resist chemistries for dual-tone imaging

- Outlook/Summary

C. Fonseca et al.

6th International Symposium on Immersion Lithography Extensions, October 22, 2009

Dual-Tone Development Concept

Potential patterning solution for frequency doubling

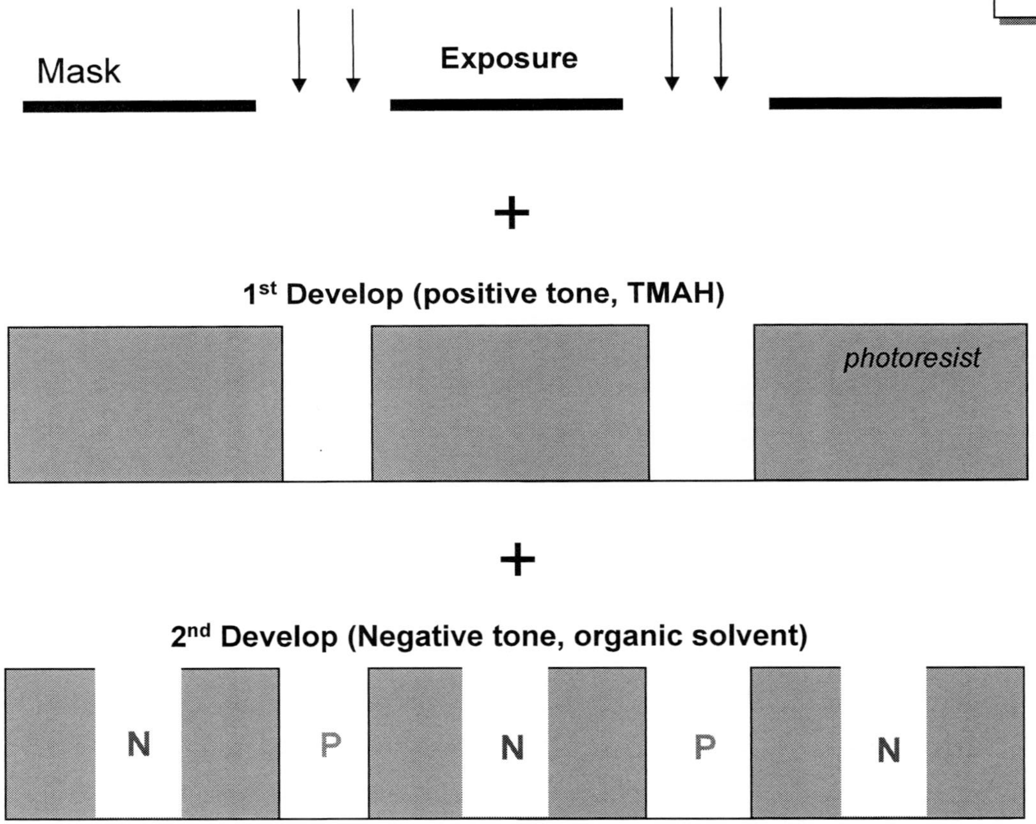

Imaging pitch is 2X of printed pitch (mask features are trenches)

Aqueous base development (TMAH)

Areas with high levels of de-protection are developed away

Organic solvent development

Areas with low levels of de-protection are developed away in solvent.

Mask

Exposure

1st Develop (positive tone, TMAH)

photoresist

+

2nd Develop (Negative tone, organic solvent)

N   P   N

N   P   N

+

C. Fonseca et al.

6th International Symposium on Immersion Lithography Extensions, October 22, 2009

Dual-Tone Double Patterning (DTDP) Goals

Ultimate Goal: Dual-Tone + Sidewall Spacer-defined Process

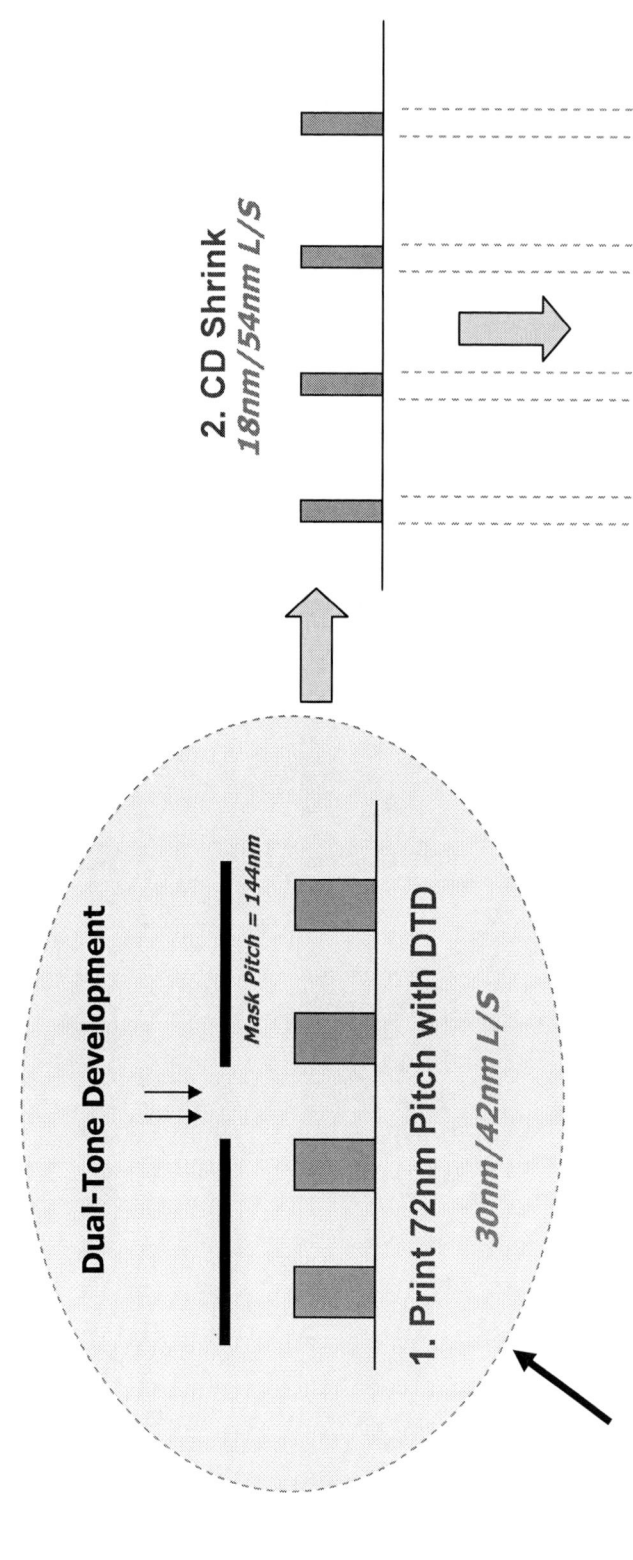

Dual-Tone Development

Mask Pitch = 144nm

1. Print 72nm Pitch with DTD
30nm/42nm L/S

2. CD Shrink
18nm/54nm L/S

3. SWS (Sidewall Spacer)
18nm/18nm L/S

Dual-Tone Patterning Goals:
1) *72nm pitch demonstration*
2) *Ultimate resolution (64nm and below)*

C. Fonseca et al.

6th International Symposium on Immersion Lithography Extensions, October 22, 2009

DTDP Timeline of Events

September 2008

- **Demonstration of 100nm pitch with 0.80NA**
- **PEB2 introduced into process flow**
- **Use of diluted TMAH**

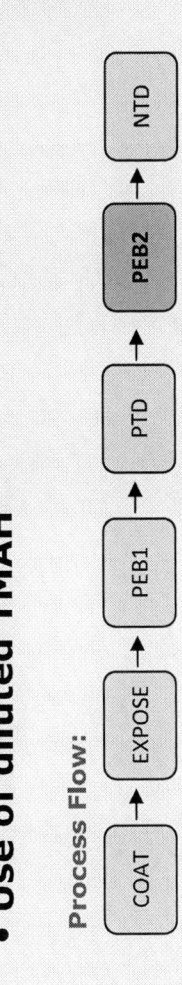

Immersion Symposium 2008

100nm Pitch

Process Flow:

COAT → EXPOSE → PEB1 → PTD → PEB2 → NTD

April 2009

- **Improved resist positive tone resolution (~30nm trenches)**
- **FLOOD EXPOSURE introduced into process flow**

SPIE 2009

Next slide

Process Flow:

COAT → EXPOSE → PEB1 → PTD → FLOOD → PEB2 → NTD

C. Fonseca et al.

6th International Symposium on Immersion Lithography Extensions, October 22, 2009

Highlights from April 2009

Introduced FLOOD EXPOSURE as a "control knob" for sizing negative tone trench

New Process Flow:

COAT → EXPOSE → PEB1 → PTD → FLOOD → PEB2 → NTD

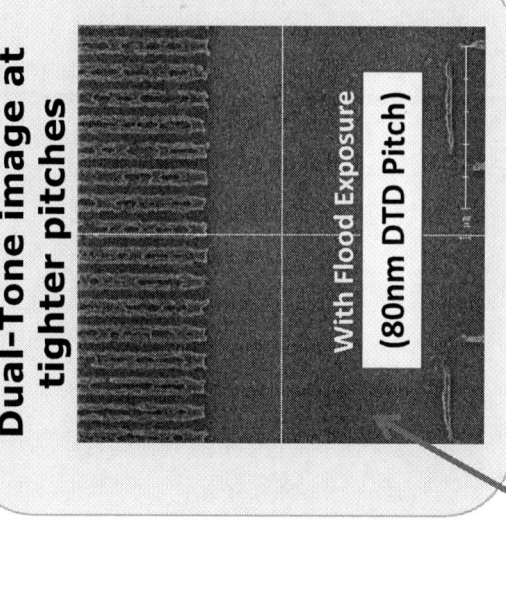

Dual-Tone image at tighter pitches

With Flood Exposure
(80nm DTD Pitch)

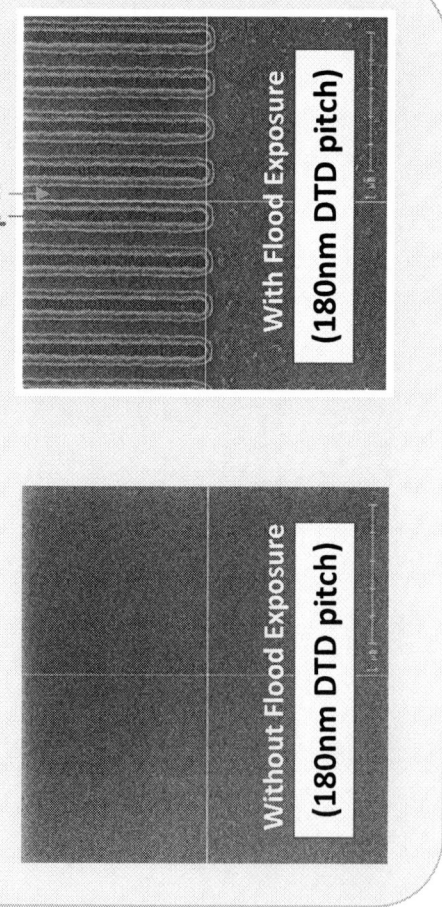

Flood Exposure concept validated

Without Flood Exposure
(180nm DTD pitch)

With Flood Exposure
(180nm DTD pitch)

Hypothesis: Poor acid contrast after positive-tone development

C. Fonseca et al.

6th International Symposium on Immersion Lithography Extensions, October 22, 2009

Resist Chemistries for DTDP

Critical events since April 2009:

1) New resist chemistries have been identified (FUJIFILM)

2) Identification of optimum resist properties (TEL / simulation)

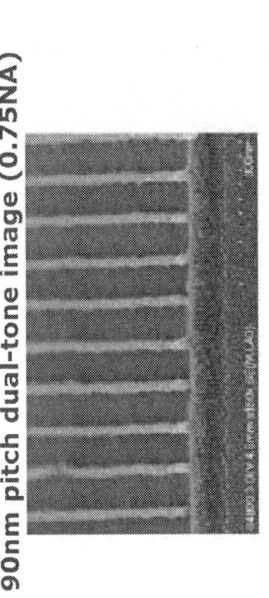

90nm pitch dual-tone image (0.75NA)

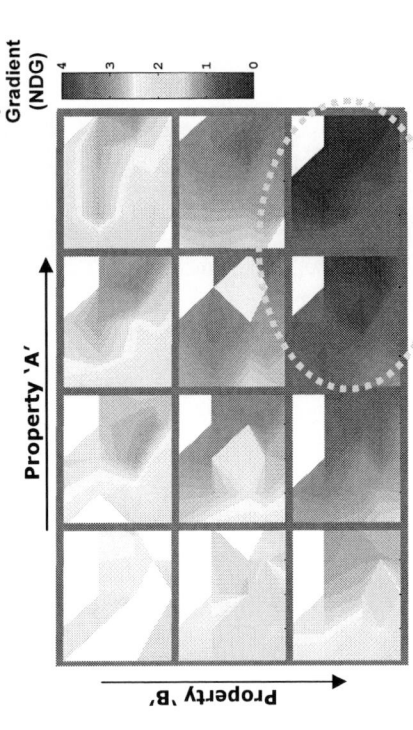

New resist chemistries from FUJIFILM aim to improve inherent dual-tone imaging

Use simulations to find the "optimum" resist properties

C. Fonseca et al.

Experimental Data
(current resist materials)

6th International Symposium on Immersion Lithography Extensions, October 22, 2009

C. Fonseca et al.

6th International Symposium on Immersion Lithography Extensions, October 22, 2009

Process Conditions (baseline)

- **Scanner: 1.26NA, Dipole 40X, polarized illumination**
- **Reticle: Attenuated PSM (6%T)**
  - Mask Feature: 60nm trench at 144nm pitch (72nm after pitch division)
- **Stack: FAiRS-9541EN02 50nm on 85nm ACR29SR**
- **PEB1: 90°C/60sec**
- **PEB2: -varied- (baseline 130°C/60sec)**
- **Flood Exposure (FLOOD): -varied-**
- **PTD: undiluted and diluted TMAH**
- **NTD: 30sec (standard)**
- **Metrology: Hitachi CG4000**

Dual-tone Process Flow:

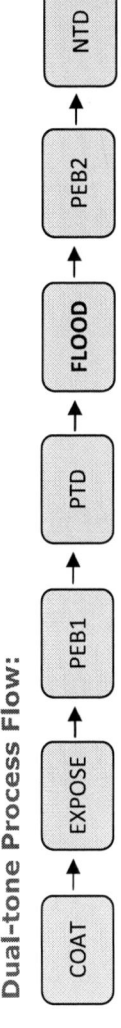

C. Fonseca et al.

PTD Resolution (undiluted TMAH)

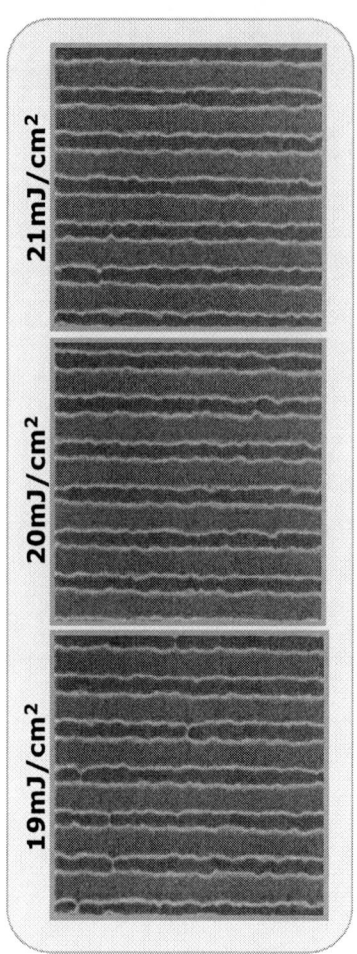

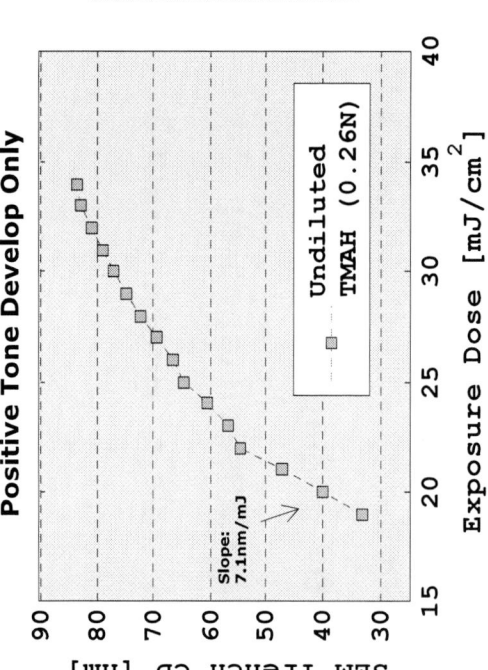

- **PTD trench imaging OK, but needs improvement**

- **Exposure latitude could be improved**
  - Previous materials: slope: ~5nm/mJ

- **This is the "starting image" before negative tone development**

C. Fonseca et al.

6th International Symposium on Immersion Lithography Extensions, October 22, 2009 | Mask Pitch: 144nm | 9541EN02

Dual-Tone Imaging (undiluted TMAH)

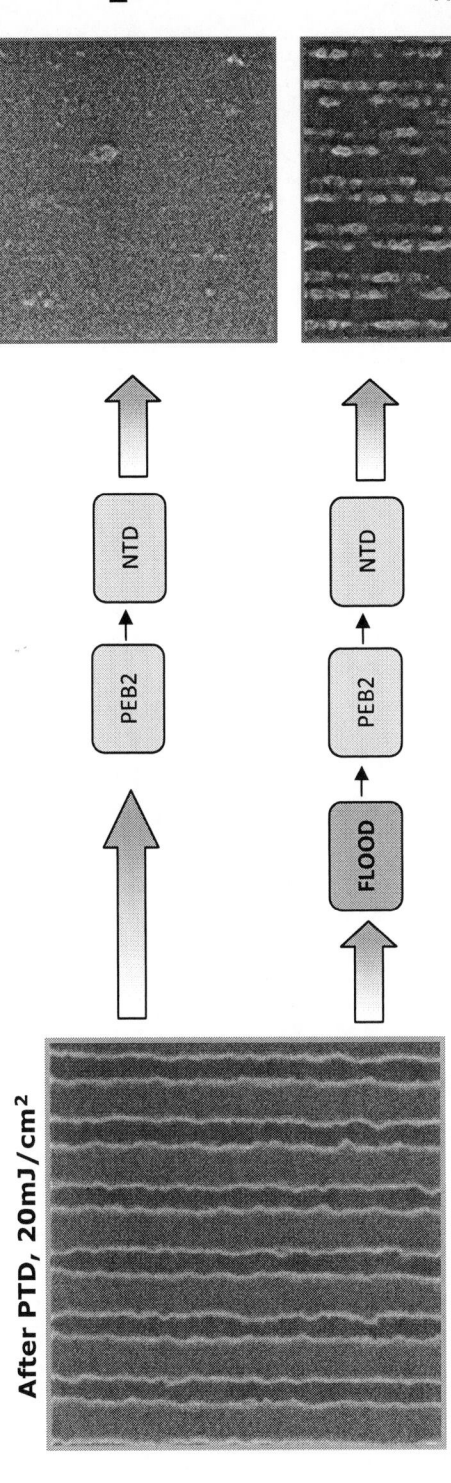

NO FLOOD

1.6mJ FLOOD

1.7mJ FLOOD

- Insufficient overlap between PTD and NTD with <u>undiluted TMAH</u>

- Flood exposure alone is not enough to improve overlap between the two develop steps

C. Fonseca et al.

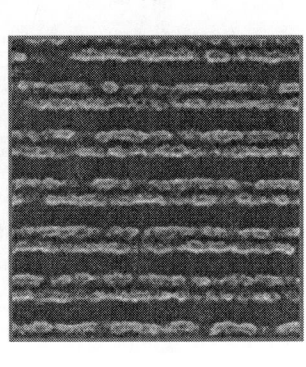

PTD Resolution (diluted TMAH)

Comparable performance across diluted TMAH cases relative to undiluted TMAH

Exposure latitude maintained up to 100X dilution case

What about dual-tone imaging with diluted TMAH?

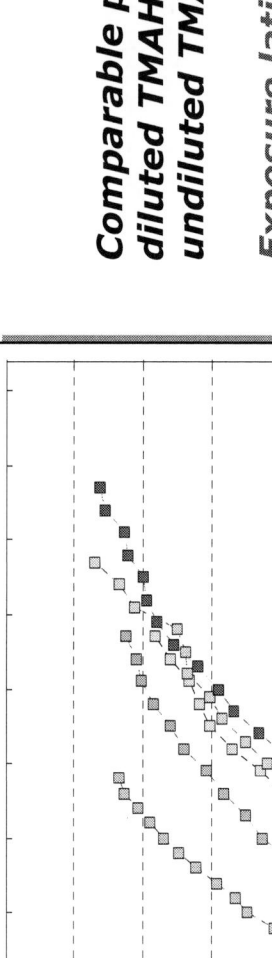

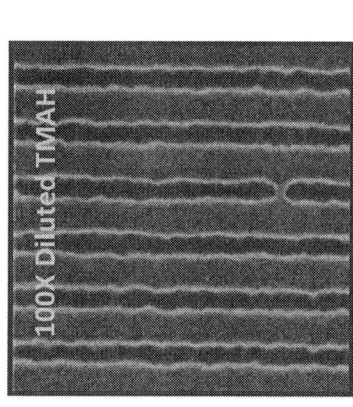

100X Diluted TMAH

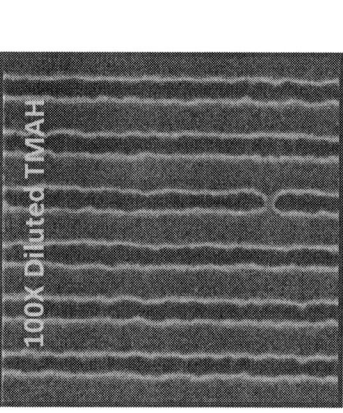

Undiluted TMAH

C. Fonseca et al.

6th International Symposium on Immersion Lithography Extensions, October 22, 2009 **Mask Pitch: 144nm** **9541EN02**

Dual-Tone Imaging (25X diluted TMAH)

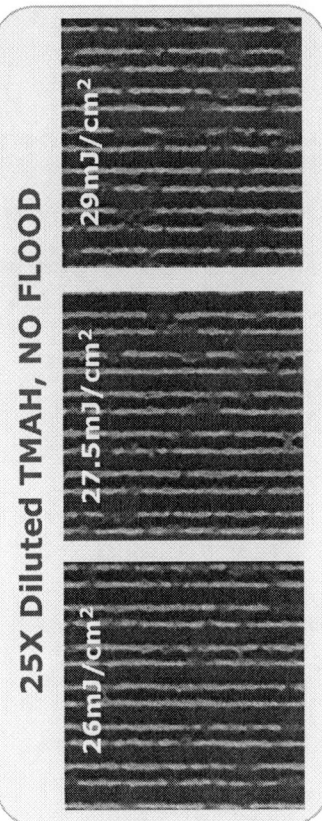

25X Diluted TMAH, NO FLOOD

26mJ/cm² 27.5mJ/cm² 29mJ/cm²

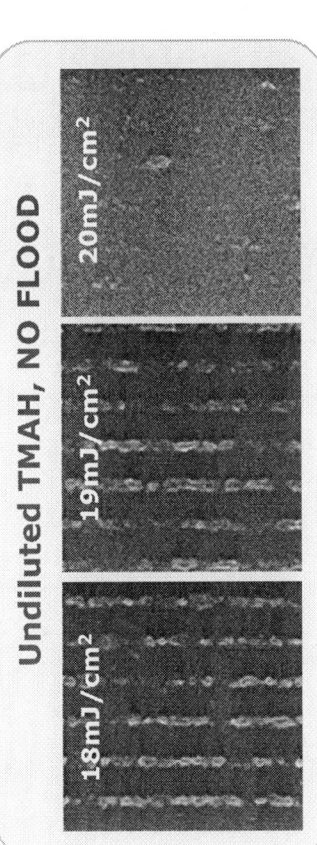

Undiluted TMAH, NO FLOOD

18mJ/cm² 19mJ/cm² 20mJ/cm²

Diluted TMAH improves inherent overlap between PTD and NTD

C. Fonseca et al.

6th International Symposium on Immersion Lithography Extensions, October 22, 2009

Dual-Tone Dissolution Contrast Curves

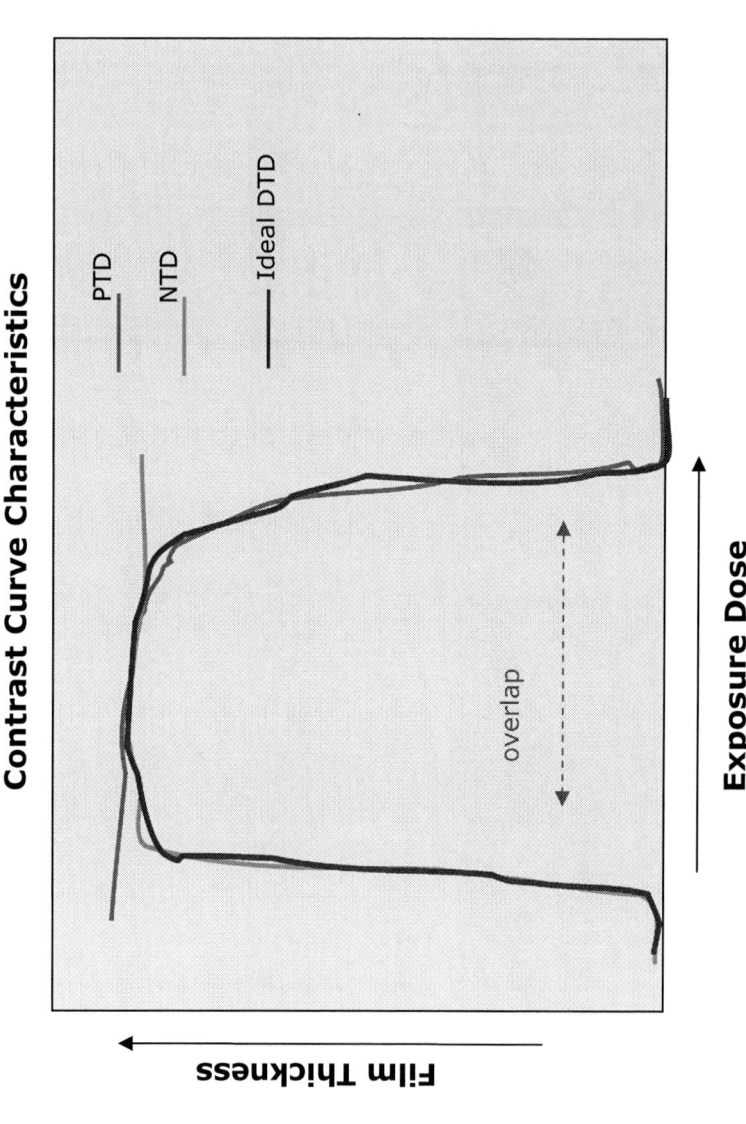

C. Fonseca et al.

6th International Symposium on Immersion Lithography Extensions, October 22, 2009

9541EN02

Dual-Tone Contrast Curves

Without FLOOD

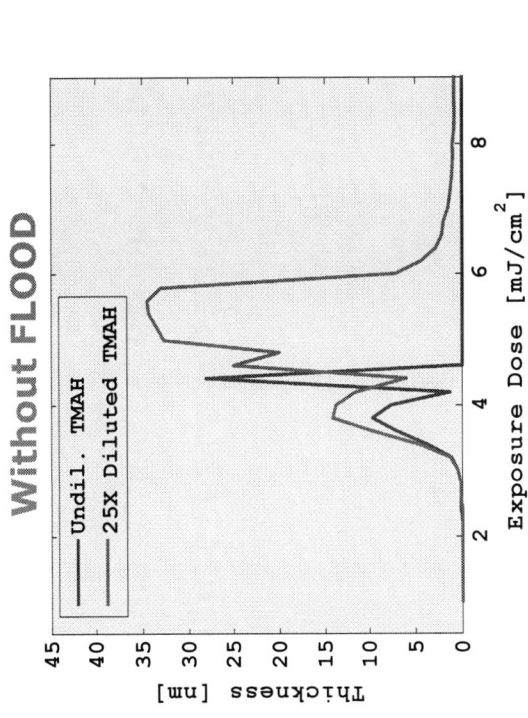

With FLOOD

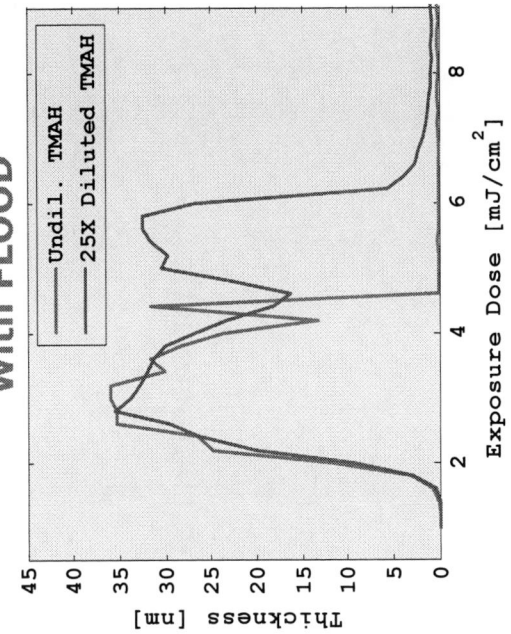

- **"Double-hump" observed in dual-tone contrast curve**
  - Hypothesis: Base from TMAH incorporates into resist film

- **Flood exposure improves overlap, but "double-hump" not completely eliminated**

C. Fonseca et al.

Effect of PEB2 Temperature

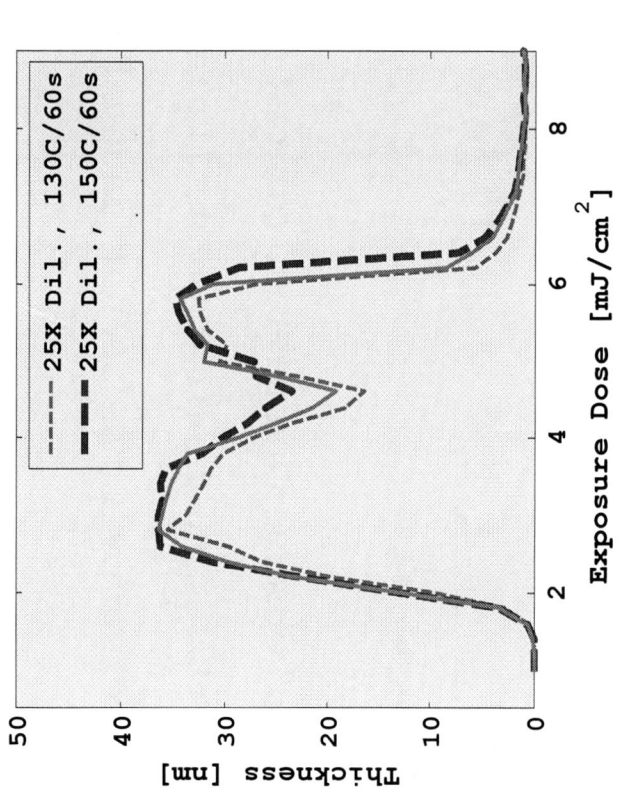

- **Higher PEB2 temperature improves dual-tone characteristic (suppression of "double hump")**

- **Higher dilution is beneficial (resist dependent)**

C. Fonseca et al.

6th International Symposium on Immersion Lithography Extensions, October 22, 2009 | Mask Pitch: 144nm | 9541EN02

Dual-Tone Imaging (100X diluted TMAH, 150°C/60s)

Undiluted TMAH, NO FLOOD, PEB2: 130°C
18mJ/cm² | 19mJ/cm² | 20mJ/cm²

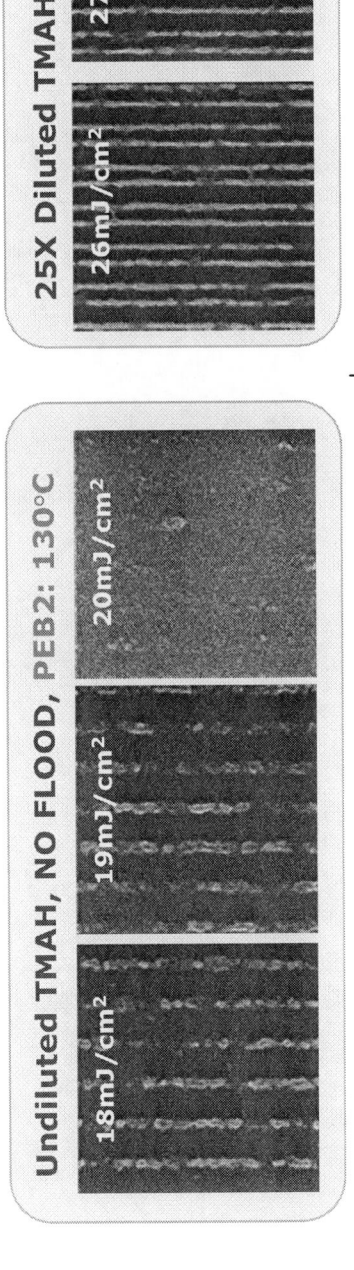

25X Diluted TMAH, NO FLOOD, PEB2: 130°C
26mJ/cm² | 27.5mJ/cm² | 29mJ/cm²

Higher Dilution + Higher PEB2 temperature

100X Diluted TMAH, NO FLOOD, PEB2: 150°C
30.5mJ/cm² | 32mJ/cm² | 33.5mJ/cm² | 35mJ/cm² | 36.5mJ/cm² | 38mJ/cm²

72nm pitch ($k_1 = 0.23$)

C. Fonseca et al.

72nm Pitch Imaging ($k_1 = 0.23$)

6th International Symposium on Immersion Lithography Extensions, October 22, 2009 | Mask Pitch: 144nm | 9541EN02

1.26NA, Dipole
PEB1: 90°C/60s
PEB2: 150°C/60s
Mask: 60nm trench / 144nm pitch
100X diluted TMAH

Some process latitude achieved

Process latitude needs to improve

Future resists should improve process latitude

COAT → EXPOSE → PEB1 → PTD → FLOOD → PEB2 → NTD

C. Fonseca et al.

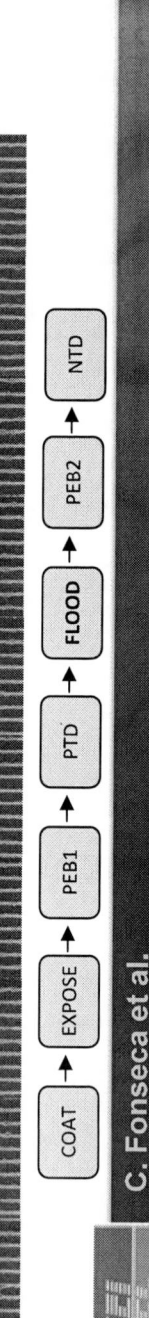

64nm Pitch Imaging

- Single exposure
- Resolution below 1.35NA scanner capability (single exposure and develop)
- Demonstrated image modulation
- Material and process refinement should improve process margin

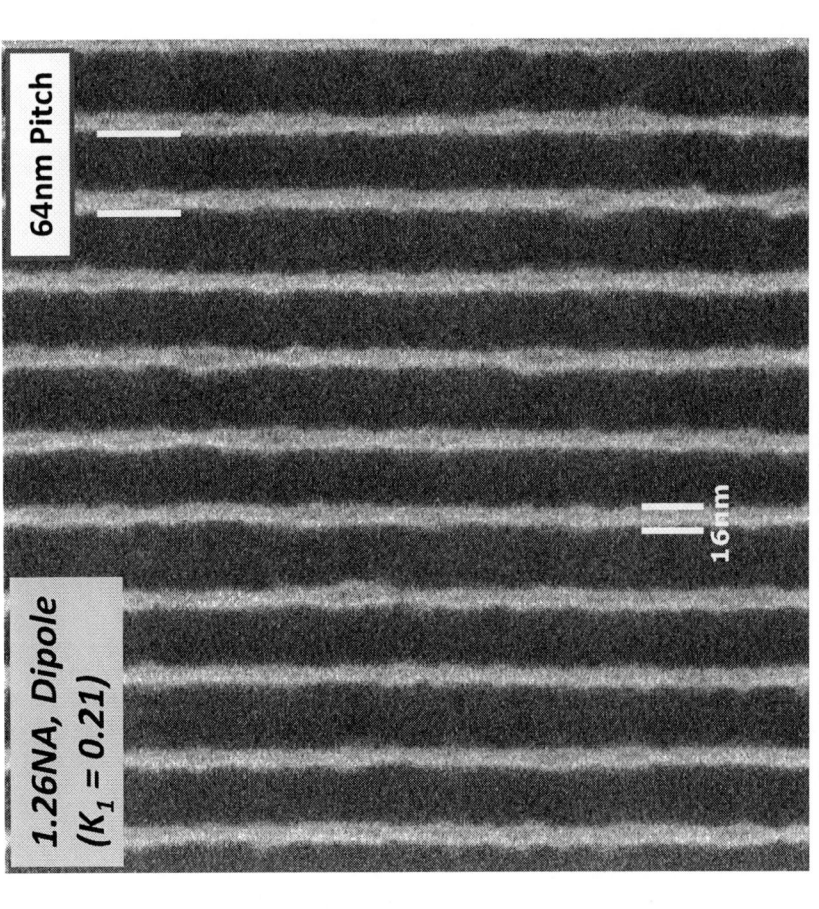

1.26NA, Dipole ($K_1 = 0.21$)

64nm Pitch

16nm

100X diluted TMAH, PEB2: 150C/60s

COAT → EXPOSE → PEB1 → PTD → FLOOD → PEB2 → NTD

C. Fonseca et al.

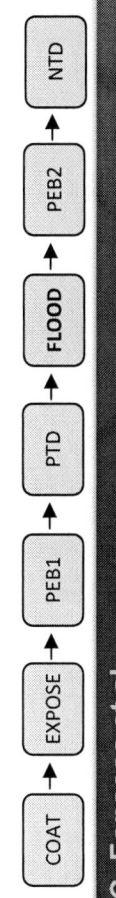

6th International Symposium on Immersion Lithography Extensions, October 22, 2009

Resist Outlook

Use simulations to find the "optimum" resist properties

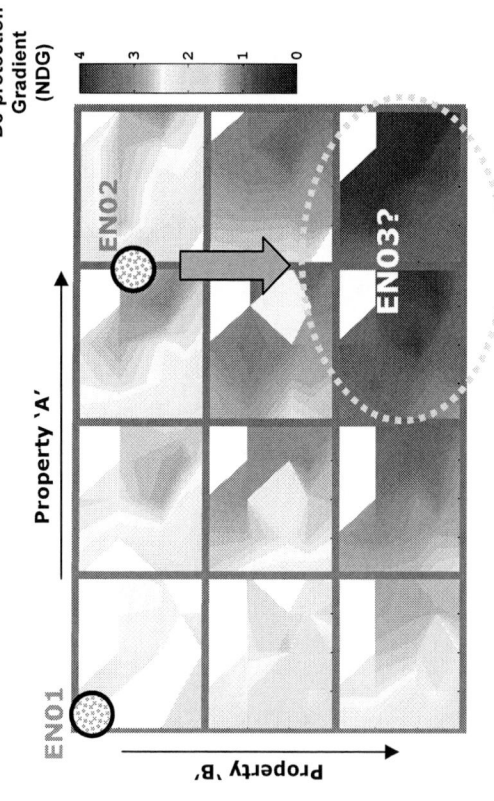

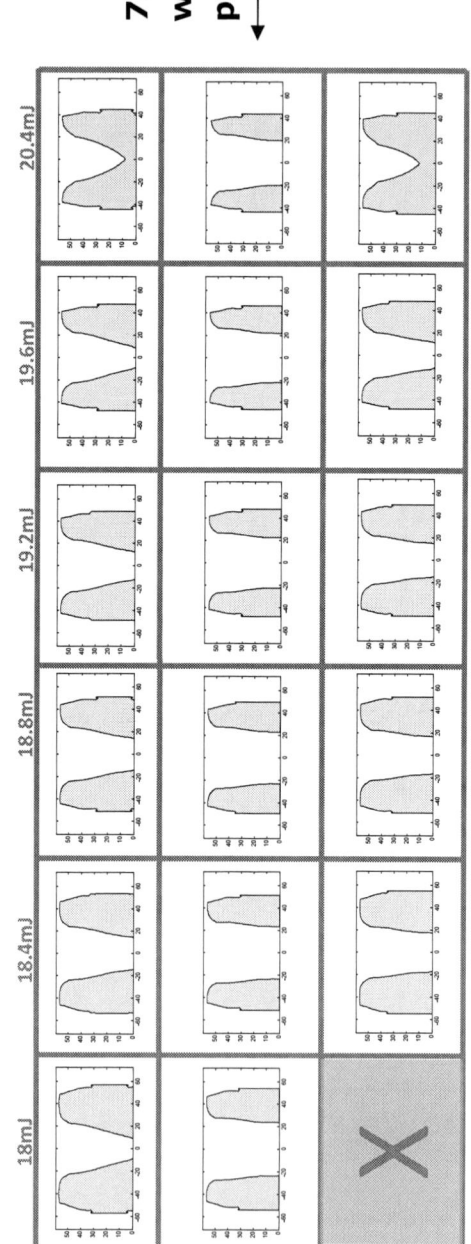

72nm pitch prediction with improved resist properties

C. Fonseca et al.

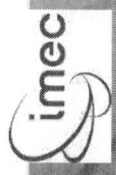

6th International Symposium on Immersion Lithography Extensions, October 22, 2009

Summary

- **Progress made on resist materials for dual-tone development**
  - New resist chemistries
  - Optimum resist properties identified

- **Improved dual-tone resolution demonstrated (version 2)**
  - Achieved 64nm pitch resolution (previous demonstration: 100nm pitch)

- **Co-optimization of developer strength (TMAH), PEB temperature/time, and flood exposure condition is key**

- **Improved resist formulation testing underway (version 3)**
  - Driven by outcome of simulation work

- **Future work: Process performance evaluation**
  - 72nm pitch (after pitch division)
  - Quantify lithography metrics (process latitude, MEEF, etc.)

C. Fonseca et al.

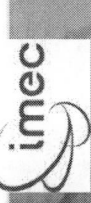

6th International Symposium on Immersion Lithography Extensions, October 22, 2009

Acknowledgements

- **TEL**
  - Wallace Printz (ATG)
  - Junichi Kitano (TKL)

- **FUJIFILM**
  - Yuuichirou Enomoto

- **FFEM**
  - Grozdan Grozev
  - Mario Reybrouck

C. Fonseca et al.

Litho-Freeze-Litho-Etch (LFLE) enabling coat/develop track process and freeze CD tuning bake for >200wph double patterning

Charles Pieczulewski* and Craig Rosslee**

* SOKUDO Co., Ltd., Kyoto, Japan
charles@sokudospeed.com

** SOKUDO assignee at IMEC, Leuven, Belgium
craig.rosslee@usa.sokudospeed.com

6th Int'l Symp. Immersion Lithography Extensions
Double Patterning Materials and Processing
2009 October 22 — Oral Presentation 11:25

LFLE enabling coat/develop track process for >200wph DP

Introduction & Motivation

1. LFL(E) Freeze Coat process CD tuning

- Litho 1 CD, Litho 2 CD matching and CD Uniformity baseline
- Biased Hot Plate (BHP) CD Uniformity tuning advantages

2. >200wph coat/develop track for Double Patterning

- Confirm SOKUDO DUO enables 22nm DP >200wph
  including other LPL (Litho-Process-Litho) and Negative Develop flows
- Change perceptions → LFL is still volume production option

2009 International Symposium on Extreme Ultraviolet Lithography
followed by
6th International Symposium on Immersion Lithography Extensions

October 18-23, 2009
Prague, Czech Republic

LFLE enabling coat/develop track process for >200wph DP

LFL(E) Freeze Coat process & CD tuning References to Previous Work

JSR, IMEC Corp. on Freeze Coat process

- Double-patterning process with freezing technique, Gouji Wakamatsu, et.al. (JSR) [SPIE 7273-10] 2009

- Comparison of LFLE and LELE Manufacturability, Andy Miller, et.al. (IMEC); Takeo Shioya, et.al. (JSR) [5th Int'l Symp. on Imm. Litho. DP-06] 2008

- Double Patterning Process with Resist Freezing Technique, Kenji Hoshiko, et.al. (JSR); Mireille Maenhoudt, et.al. (IMEC) [5th Int'l Symp. on Imm. Litho. DM-08] 2008

- Sub-40-nm half-pitch double patterning with resist freezing process, Masafumi Hori, et.al., (JSR) [SPIE 6923-17] 2008

SOKUDO, IMEC on CD tuning for Double Pattering

- Track Optimization and Control for 32nm Node Double Patterning and Beyond, David Laidler (IMEC), Craig Rosslee (SOKUDO), et.al. [SPIE 7272-128] 2009

- Cluster Optimization to Improve Total CD Control as an Enabler for Double Patterning, David Laidler (IMEC), Craig Rosslee (SOKUDO), et.al. [5th Int'l Symp. on Imm. Litho. P-DP-05] 2008

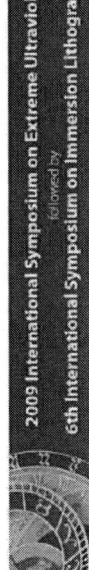

2009 International Symposium on Extreme Ultraviolet Lithography
followed by
6th International Symposium on Immersion Lithography Extensions

October 18-23, 2009
Prague, Czech Republic

LFLE enabling coat/develop track process for >200wph DP

Experimental Conditions

FINAL TARGET CD 32nm 1:1 LS

PROCESS MATERIALS

- **Litho 1**
  - 95nm ARC29A (Brewer Sci.) BARC
    - 205C/60s PAB (Post-Apply Bake)
  - 90nm AR-- (JSR) Conventional Resist
    - 120C/60s PAB, 115C/60s PEB
  - 90nm TCX-- (JSR) Top-Coat
    - 90C/60s PAB

- **Freeze**
  - 150nm FZX-- (JSR) Freeze Coat
    - 130C/90s FZ Bake
    - 165C/90s post develop hard-bake (HB)

- **Litho 2**
  - 90nm AIM-- (JSR) TC-less Resist
    - 100C/60s PAB, 95C/60s PEB

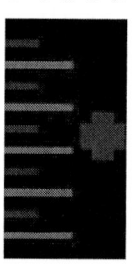

CD-SEM can do separate measurement CD L1, CD L2 for analysis

METROLOGY
- Hitachi CG4000 CD-SEM
- K-T Archer-AIM (Overlay)

Photo Courtesy of IMEC, Belgium

ASML XT:1900Gi + SOKUDO RF3S immersion coat/develop track

- **Exposure illumination mode**
  - Dipole 40X, y-polarized
  - NA=1.0, σ_o=0.85, σ_i=0.65
  - L56P128 structure on reticle

- **Track process elements**
  - Post-Soak (post-exposure rinse)
  - BHP (Biased Hot Plate) with cdTune™
  - ECO develop (dynamic dispense)
  - NOVA 3090 OCD int. metrology

SOKUDO DUO

2009 International Symposium on Extreme Ultraviolet Lithography
followed by
6th International Symposium on Immersion Lithography Extensions

October 18-23, 2009
Prague, Czech Republic

LFLE enabling coat/develop
track process for >200wph DP

CD Tuning Tools for LFL(E)

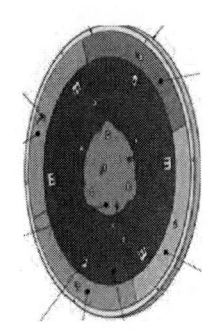

- **Biased Hot Plate (BHP) temp. tuning**

  – Traditional Post-Exposure Bake (PEB) CD bias temp. tuning

  – Freeze Coat (FZX) Post-Apply Bake (PAB) temp. tuning

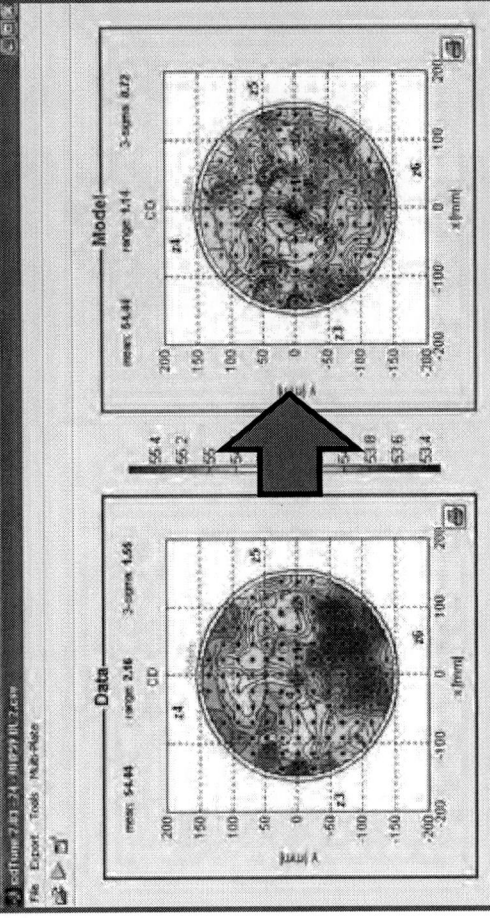

- **cdTune ™ software**

  – Run a CD wafer to one or more bake plates

  – Load the CD data from run into cdTune ™

  – Software generates optimized plate offsets

SOKUDO DUO

2009 International Symposium on Extreme Ultraviolet Lithography
followed by
6th International Symposium on Immersion Lithography Extensions

October 18-23, 2009
Prague, Czech Republic

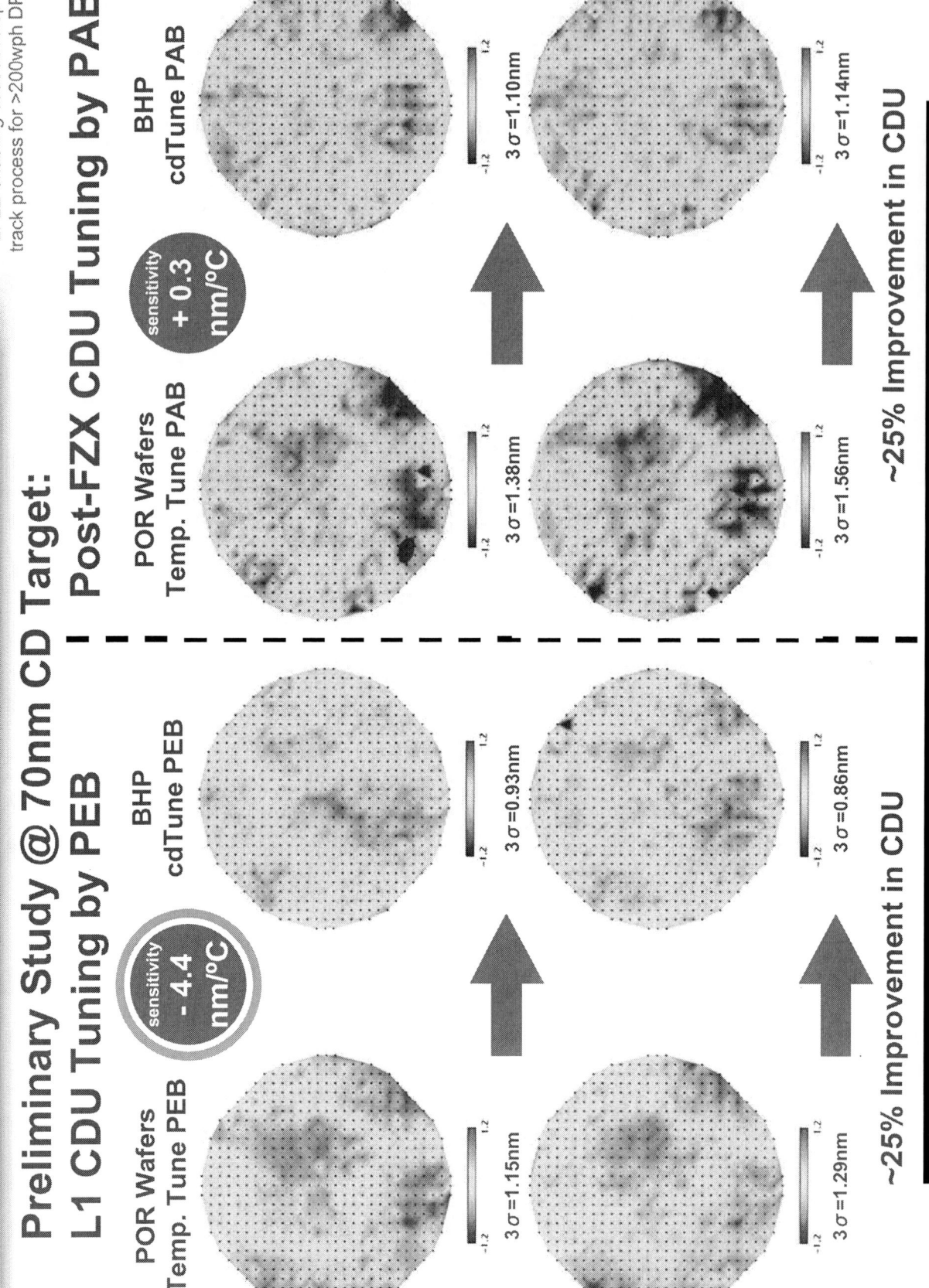

LFLE enabling coat/develop
track process for >200wph DP

Preliminary Study @ 45nm CD Target:
Litho 2 CDU optimization by BHP cdTune PEB

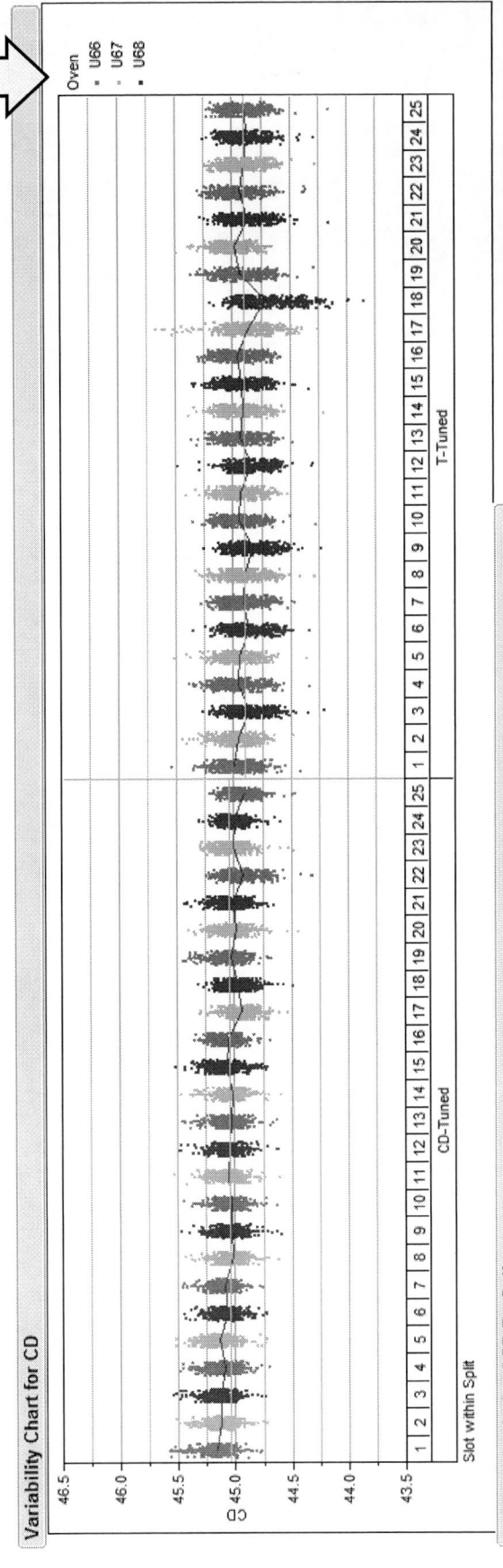

*NOTE: NOVA 3090 OCD Int. Metrology

● **All wafer CDU 3σ:**
  ▸ T-Tuned = 0.5nm
  ▸ CD-Tuned = 0.4nm
  ▸ 25% improvement

● **Mean wiw CDU 3σ:**
  ▸ T-Tuned = 0.4nm.
  ▸ CD-Tuned = 0.3nm.
  ▸ 30% improvement

October 18-23, 2009
Prague, Czech Republic

followed by
6th International Symposium on Immersion Lithography Extensions

LFLE enabling coat/develop track process for >200wph DP

Experiment Set-Up Conditions: 32nm CD Target
Litho-Freeze-Litho Test Patterns & Observations

Litho 1

Freeze Litho 1

Litho 2 (L1+L2 Result)*

~28.5nm CD L1   ~31.0nm CD L1   32nm CD L1:L2

L1 CD grows by ~2.5nm during FZX process so we have to base our targeting on post FZX L1 CD.

L1 also shows a ~1.0nm CD increase post L2 process.

* NOTE: ASML scanner overlay corrections also were subsequently made to improve L1 - L2 gap spacing; details not included herein as it is outside scope of this study

* See also David Laidler, et.al. IMEC, ASML "Characterization of Direct Alignment for LFLE Double Patterning Process" [6th Int'l Symp. on Immersion Lithography Extensions] 2009

Establish target CD L1, L2 dose from FEM (Focus Exposure Matrix)

SOKUDO DUO

October 18-23, 2009
Prague, Czech Republic

2009 International Symposium on Extreme Ultraviolet Lithography
followed by
6th International Symposium on Immersion Lithography Extensions

LFLE enabling coat/develop
track process for >200wph DP

Experiment Set-Up Conditions: 32nm CD Target
Litho 1, Litho 2 CD sensitivity to PEB temperature

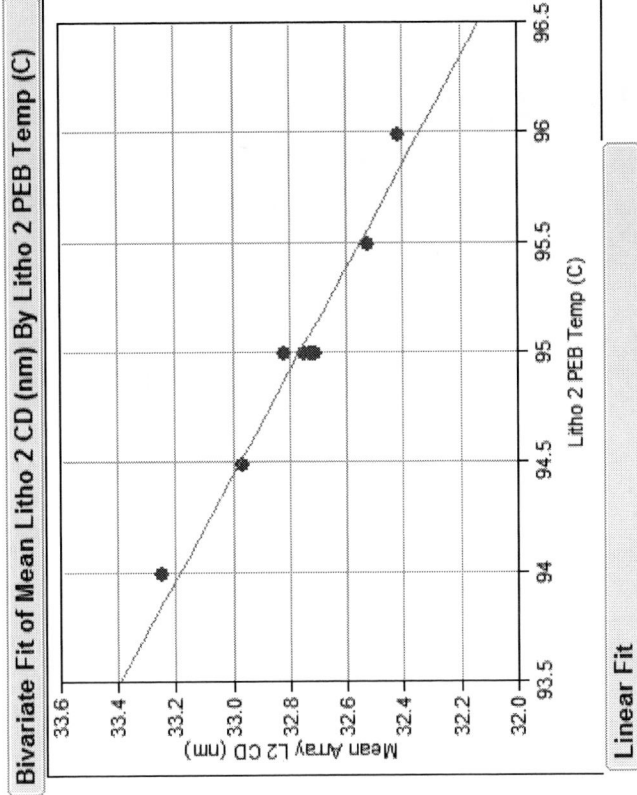

Bivariate Fit of Mean Litho 2 CD (nm) By Litho 2 PEB Temp (C)

Linear Fit

Mean Litho 2 CD (nm) = 72.85347 - 0.4219518 Litho 2 PEB Temp (C)

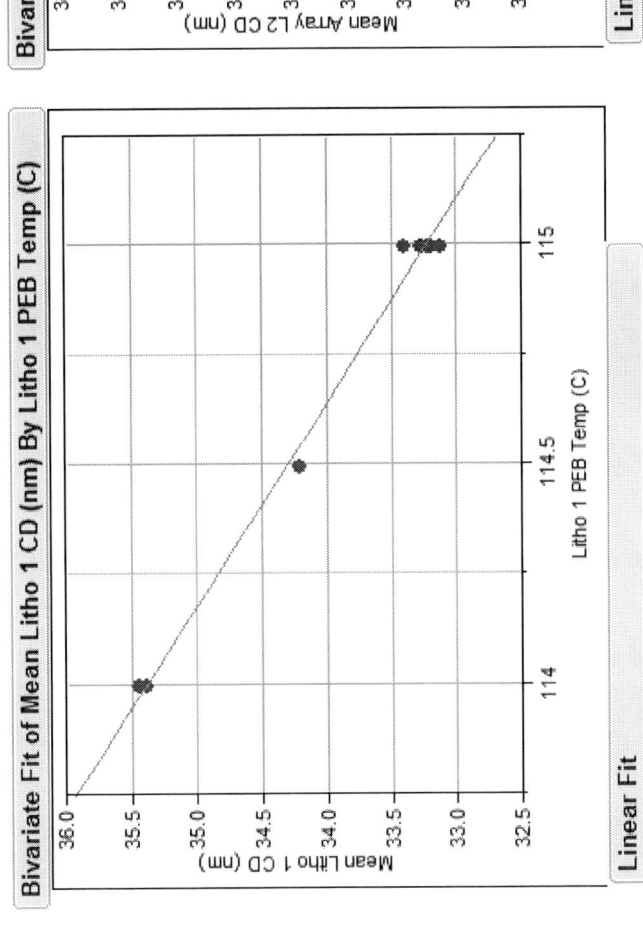

Bivariate Fit of Mean Litho 1 CD (nm) By Litho 1 PEB Temp (C)

Linear Fit

Mean Litho 1 CD (nm) = 281.38172 - 2.1578128 Litho 1 PEB Temp (C)

- Multiple wafers exposed at a range of uniform PEB temperatures for Litho 1 and Litho 2 to determine CD sensitivity
- Litho 1 resist CD sensitivity -2.1nm/°C
- Litho 2 resist CD sensitivity -0.4nm/°C

2009 International Symposium on Extreme Ultraviolet Lithography
followed by
6th International Symposium on Immersion Lithography Extensions

October 18-23, 2009
Prague, Czech Republic

LFLE enabling coat/develop track process for >200wph DP

Calculating Initial cdTune Offsets from POR (Temperature Tuned BHP) PEB Process @ 32nm CD Target

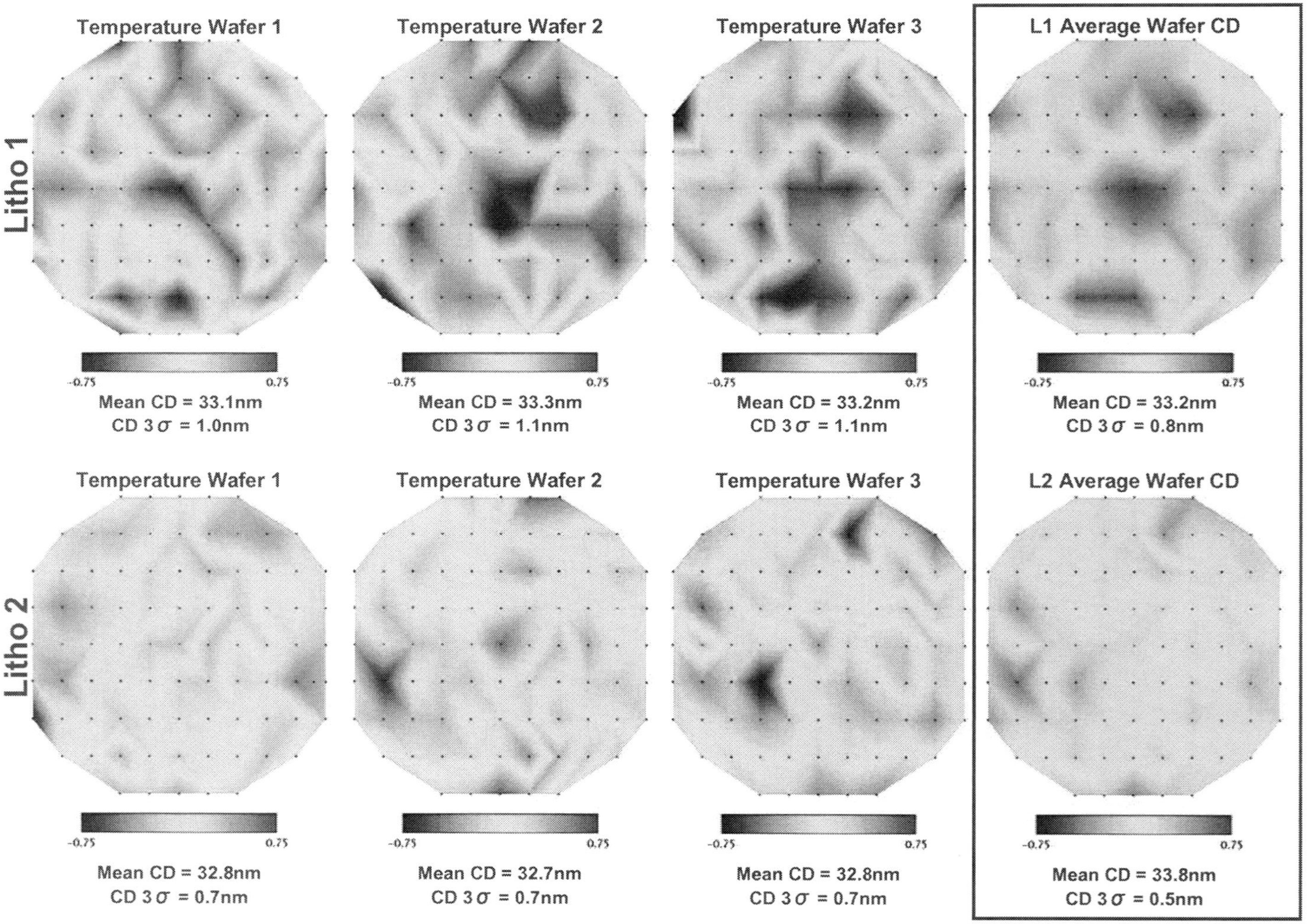

- **Average Wafer CDU isolates CD trend; removes CD-SEM and other random error wafer-to-wafer**

- **Average CD "Temperature" wafer maps were derived and used as input for BHP cdTune PEB bias offsets**

Litho 1

Temperature Wafer 1
Mean CD = 33.1nm
CD 3σ = 1.0nm

Temperature Wafer 2
Mean CD = 33.3nm
CD 3σ = 1.1nm

Temperature Wafer 3
Mean CD = 33.2nm
CD 3σ = 1.1nm

L1 Average Wafer CD
Mean CD = 33.2nm
CD 3σ = 0.8nm

Litho 2

Temperature Wafer 1
Mean CD = 32.8nm
CD 3σ = 0.7nm

Temperature Wafer 2
Mean CD = 32.7nm
CD 3σ = 0.7nm

Temperature Wafer 3
Mean CD = 32.8nm
CD 3σ = 0.7nm

L2 Average Wafer CD
Mean CD = 33.8nm
CD 3σ = 0.5nm

2009 International Symposium on Extreme Ultraviolet Lithography
followed by
6th International Symposium on Immersion Lithography Extensions

October 18-23, 2009
Prague, Czech Republic

BHP cdTune PEB Wafer Maps (Individual Wafers)

LFLE enabling coat/develop track process for >200wph DP

Litho 1

Wafer 1 — Mean CD = 33.3nm, CD 3σ = 0.9nm
Wafer 2 — Mean CD = 33.2nm, CD 3σ = 1.0nm
Wafer 3 — Mean CD = 33.1nm, CD 3σ = 1.1nm
Wafer 4 — Mean CD = 33.2nm, CD 3σ = 1.0nm
Wafer 5 — Mean CD = 33.3nm, CD 3σ = 1.0nm

Litho 2

Wafer 1 — Mean CD = 33.3nm, CD 3σ = 0.6nm
Wafer 2 — Mean CD = 33.2nm, CD 3σ = 0.7nm
Wafer 3 — Mean CD = 33.2nm, CD 3σ = 0.7nm
Wafer 4 — Mean CD = 33.3nm, CD 3σ = 0.6nm
Wafer 5 — Mean CD = 33.3nm, CD 3σ = 0.6nm

SOKUDO DUO

2009 International Symposium on Extreme Ultraviolet Lithography
followed by
6th International Symposium on Immersion Lithography Extensions

October 18-23, 2009
Prague, Czech Republic

LFLE enabling coat/develop
track process for >200wph DP

Individual Wafers CD Summary:
"CD" versus "Temperature" tuned PEB (BHP)

Variability Chart for 3-Sigma (nm)

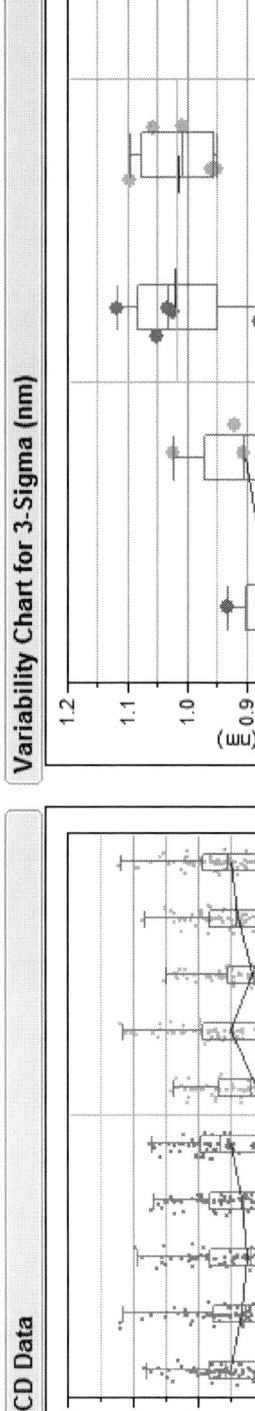

Litho 1 CD Data

Litho 2 CD Data

"CD" tuned PEB Improvements in CDU

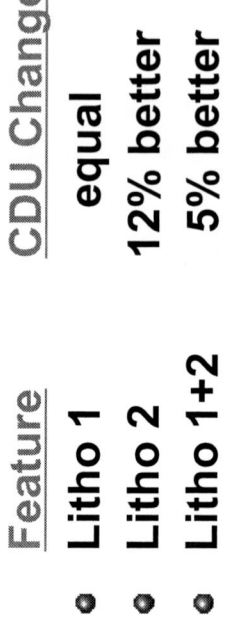

Feature	CDU Change	CDU Result
Litho 1	equal	1.02nm 3σ
Litho 2	12% better	0.64nm 3σ
Litho 1+2	5% better	0.86nm 3σ

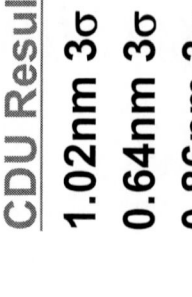

SOKUDO DUO

2009 International Symposium on Extreme Ultraviolet Lithography
followed by
6th International Symposium on Immersion Lithography Extensions

October 18-23, 2009
Prague, Czech Republic

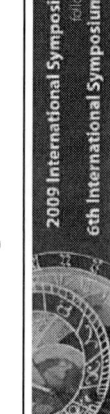

Average Wafer Maps Compared

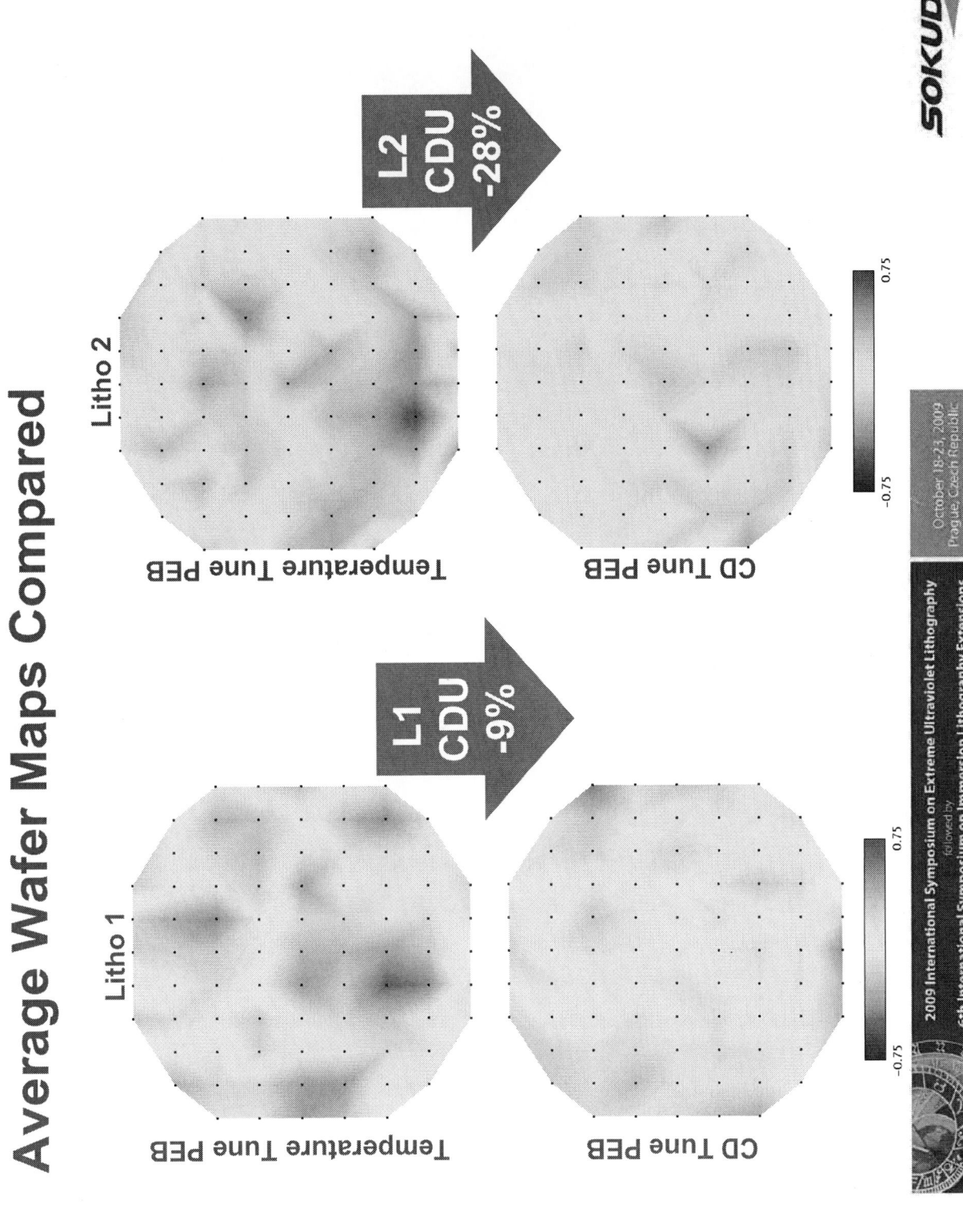

LFLE enabling coat/develop track process for >200wph DP

LFLE enabling coat/develop
track process for >200wph DP

Average Wafer CD Summary:
"CD" versus "Temperature" tuned PEB (BHP)

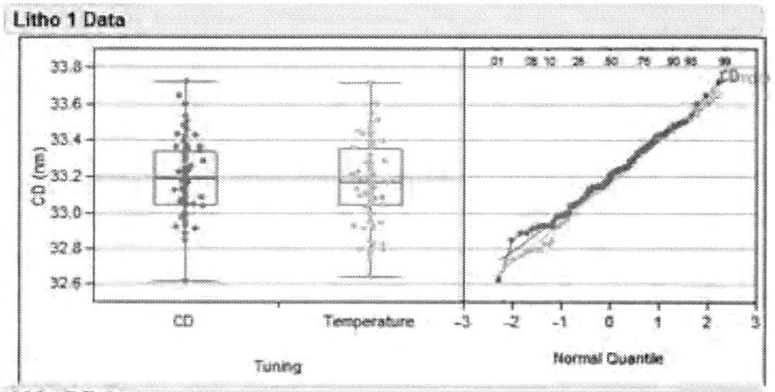

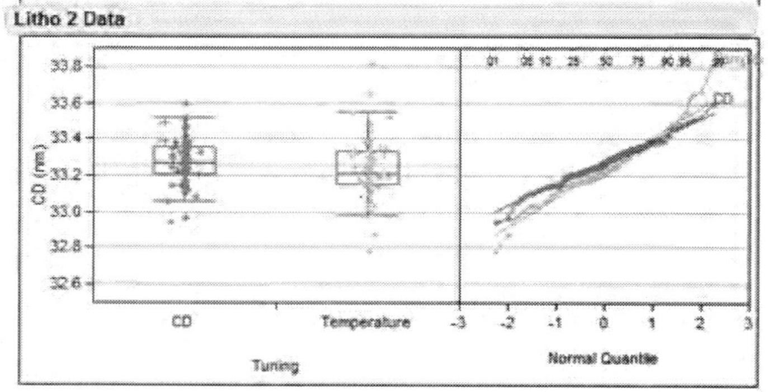

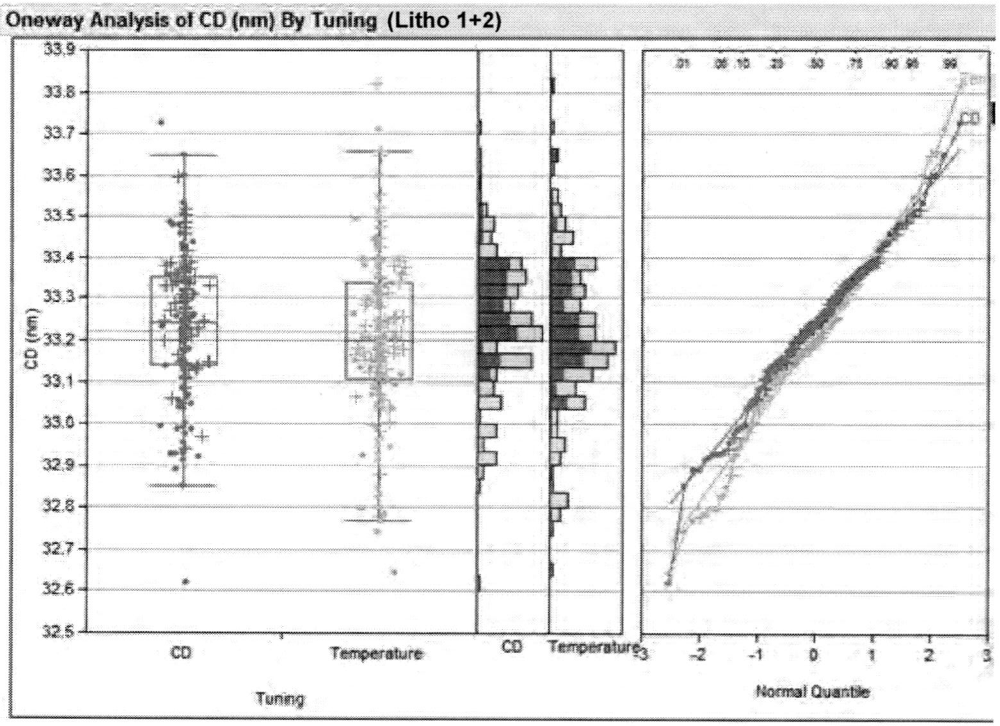

Oneway Analysis of CD (nm) By Tuning (Litho 1+2)

Litho 1 Data

Litho 2 Data

"CD" tune PEB shows lower CDU 3σ for Litho 1, Litho 2 and
combined Litho 1+2 data than Temperature tuned PEB

Feature	CDU Change	CDU Result
Litho 1	9% better	0.6nm
Litho 2	28% better	0.4nm
Litho 1+2	14% better	0.5nm

SOKUDO DUO

2009 International Symposium on Extreme Ultraviolet Lithography
followed by
6th International Symposium on Immersion Lithography Extensions

October 18-23, 2009
Prague, Czech Republic

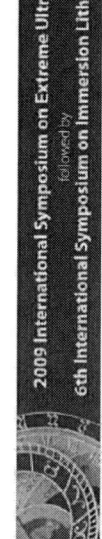

LFLE enabling coat/develop track process for >200wph DP

"Photo" Double Patterning Resist Process Options

	Published Work with Partners	Litho-Freeze chemical b/w 1st & 2nd Resist (LFLE)	Self-Freeze by Bake or 2nd Resist Coating (LPLE)	UV Cure Freeze	Negative Tone Dev. Resist (Nega-Dev)
JSR	IMEC, SUNY Albany	☺	O	△	--
TOK	IMEC, Nikon	--	☺	--	--
FujiFilm	IMEC	--	--	--	☺
Rohm & Haas (Dow Chem.)	CEA-LETI, AMD (GF), SOKUDO	O	O	△	--
Shin-Etsu	AMD (GF), Ushio, SOKUDO	--	O	△	--

☺ Only notes the example used for this study

Claimed to be Complex Process yet Good Results !!

Simpler but Weaker Results (will improve)

Needs More R&D Effort but Attractive if it can work!

High Potential for Trench, CH

See also "Printing the Contact and Metal Layers for the 32 and 22 nm Node: Comparing Positive and Negative Tone Development Process" J. Bekaert, et.al. IMEC, FujiFilm [6th Int'l Symp. ILE] 2009

2009 International Symposium on Extreme Ultraviolet Lithography
followed by
6th International Symposium on Immersion Lithography Extensions

October 18-23, 2009
Prague, Czech Republic

"Photo" Double Patterning Process Flows on Track

LFLE enabling coat/develop track process for >200wph DP

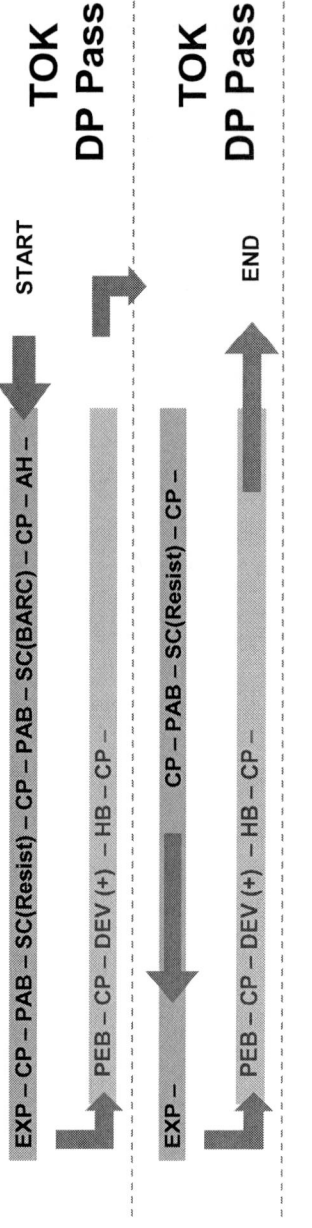

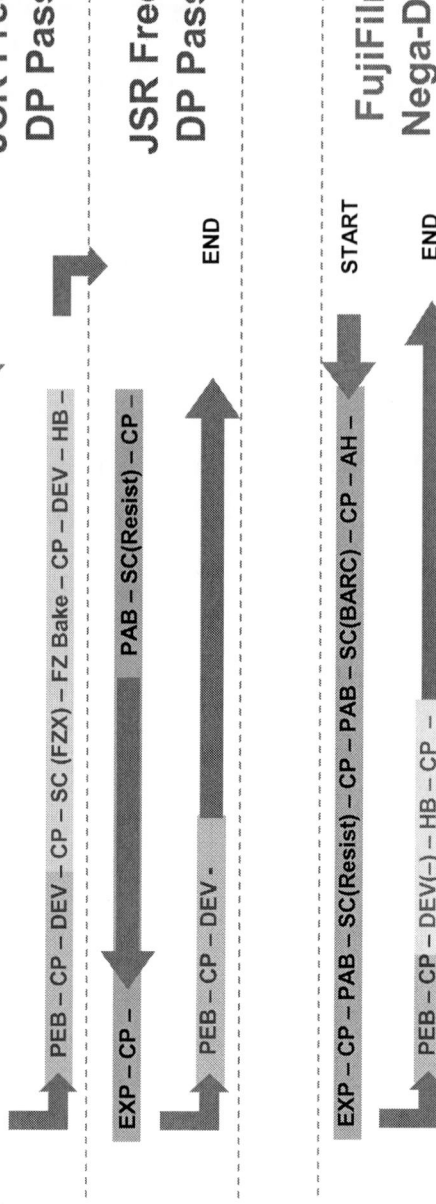

TOK posi-posi (Self-Freeze)

TOK DP Pass 1

COAT — START — EXP – CP – PAB – SC(Resist) – CP – PAB – SC(BARC) – CP – AH –

DEVELOP — PEB – CP – DEV (+) – HB – CP –

COAT — EXP – CP – PAB – SC(Resist) – CP –

DEVELOP — END — PEB – CP – DEV (+) – HB – CP –

TOK DP Pass 2

JSR Freeze-Coat

JSR Freeze DP Pass 1

COAT — START — EXP – CP – PAB – SC(Resist) – CP – PAB – SC(BARC) – CP – AH –

DEVELOP – COAT – DEVELOP — PEB – CP – DEV – CP – SC (FZX) – FZ Bake – CP – DEV – HB –

COAT — EXP – CP – PAB – SC(Resist) – CP –

DEVELOP — END — PEB – CP – DEV –

JSR Freeze DP Pass 2

FujiFilm Nega-Dev

FujiFilm Nega-Dev

COAT — START — EXP – CP – PAB – SC(Resist) – CP – PAB – SC(BARC) – CP – AH –

DEVELOP — END — PEB – CP – DEV(–) – HB – CP –

SOKUDO DUO

2009 International Symposium on Extreme Ultraviolet Lithography
followed by
6th International Symposium on Immersion Lithography Extensions

October 18-23, 2009
Prague, Czech Republic

LFLE enabling coat/develop track process for >200wph DP

Common Myth:
Need one Double Patterning Solution that works? NO

- Reality: One Double Patterning Solution will not cover all IC process pattern scenarios, may need multiple "photo" Double Patterning solutions:

  - Dense lines (straight)
  - Dense lines (elbow)
  - Contact holes, dense / iso
  - Trench patterns
  - Etc.

- Therefore... Coat/Develop Track needs flexibility to run whatever "Double Patterning Solution" that fits the IC process pattern layer

LFLE enabling coat/develop track process for >200wph DP

>200wph Double Patterning Requirements for Coat/Develop Track

- **Flexibility to run whatever "Double Patterning Solution" that fits the IC process pattern layer**
  - Coat/develop track should not limit options of lithography

- **Maintain >200wph throughput even with extra process steps demanded by each Double Patterning solution**
  - Expandable to support adding coat, develop and bake units
  - Enable multi-pass flows (i.e. two-step develop in one flow)

- **Maintain continuous supply of wafers inside track: FOUP / Wafer Buffering**

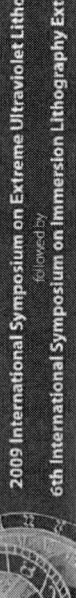

2009 International Symposium on Extreme Ultraviolet Lithography
followed by
6th International Symposium on Immersion Lithography Extensions

October 18-23, 2009
Prague, Czech Republic

LFLE enabling coat/develop track process for >200wph DP

SOKUDO DUO Coat/Develop Track

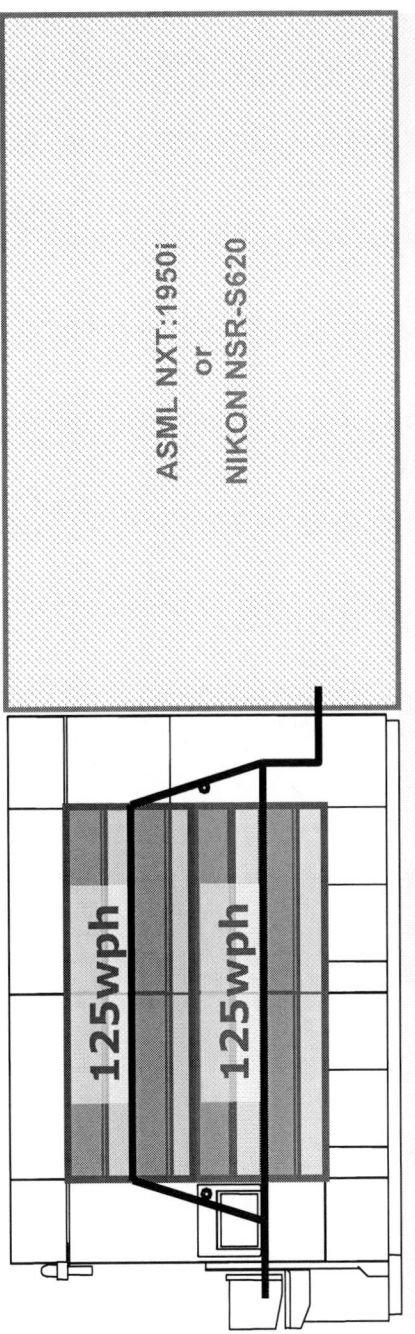

ASML NXT:1950i
or
NIKON NSR-S620

125wph + 125wph = 250wph Throughput

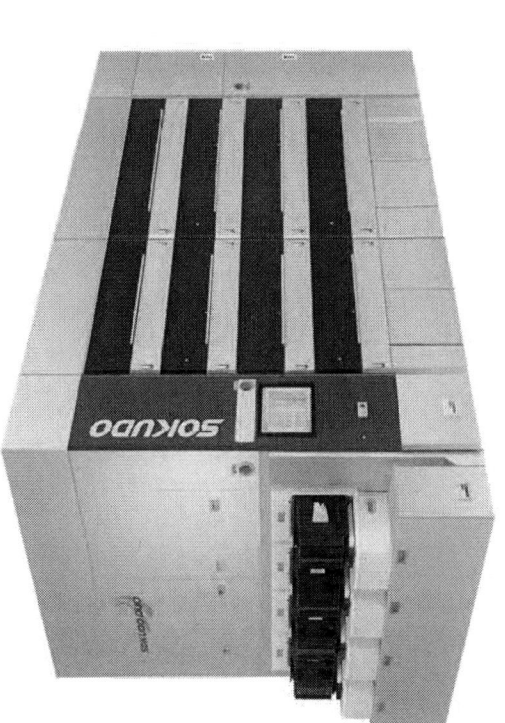

- DUO Top/Bottom Parallel Wafer Flow
- High Throughput 250-300wph Ready
- Flexible Configurations for Immersion Litho. DP

2009 International Symposium on Extreme Ultraviolet Lithography
followed by
6th International Symposium on Immersion Lithography Extensions

October 18-23, 2009
Prague, Czech Republic

wph = wafers per hour

Conclusions

1. **Freeze Coat process CD tuning for LFL(E)**

   - **Discovered L1 post-FZX CDU Tuning by PAB possible with BHP cdTune bake**
     - However, in this study choose higher CD sensitive PEB temp. tuning
   - **LFL (L1+L2) CD Uniformity within wafer can be optimized by BHP cdTune bake**
     - Initial "Temperature Tuned" condition already very good CDU < 1.1nm
     - "CD Tuned" average wafer resulting CDU <0.5nm (individual wafer <0.9nm)

2. **>200wph coat/develop track for Double Patterning (DP)**

   - **SOKUDO DUO configuration flexible to support multiple DP options**
     - Also enables Negative-Tone Develop Process for future CH, Trench patterns
   - **Semiconductor manufacturers no longer need to be limited by track for maximum productivity to implement "photo" DP alternatives**

2009 International Symposium on Extreme Ultraviolet Lithography
followed by
6th International Symposium on Immersion Lithography Extensions

October 18-23, 2009
Prague, Czech Republic

LFLE enabling coat/develop
track process for >200wph DP

Acknowledgements of Support & Advice

- **IMEC**
  - David Laider, Patrick Wong, Philippe Foubert and Andy Miller
- **SOKUDO**
  - Samir Bouchema (AMAT) and Yan Thouroude (SCREEN)
- **JSR**
  - Hiromitsu Tanaka

2009 International Symposium on Extreme Ultraviolet Lithography
followed by
6th International Symposium on Immersion Lithography Extensions

October 18-23, 2009
Prague, Czech Republic

imec FUJIFILM

Printing the contact and metal layers for the 32 and 22 nm node:

Comparing positive and negative development process

Joost Bekaert

L. Van Look, V. Wiaux, V. Truffert, M. Maenhoudt, G. Vandenberghe (IMEC)

M. Reybrouck, S. Tarutani (FujiFilm)

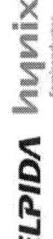

Outline

Negative development process update

- Contact printing: dual line exposure

- Contact printing: single exposure

- Trench printing

- Conclusions

6th International Symposium on
Immersion Lithography Extensions
Prague, 2009

Joost Bekaert
© imec 2009

Negative development process update

• What is negative tone development (FujiFilm – NTD):
Image reversal technique for positive tone resist: an organic solvent dissolves the unexposed areas, while the exposed resist remains.

• Several resists have been screened for their use with negative tone development. In the presented work, we use 100 nm FAIRS9101-A19C resist on 37 nm ARC 160 barc.

• Development scheme further improved for CDU and process time

- 30 seconds FN-DP001 development
- 10 seconds FN-RP002 rinse
- 15 seconds spin dry time

} Total develop time below 1 minute

• Delay effects have been studied   (for dual line exposure)
Limited effects: no particular sensitivity to delay for NTD.
   => Please inquire when interested in results

imec   **FUJiFILM**

Joost Bekaert
© imec 2009

6th International Symposium on
Immersion Lithography Extensions
Prague, 2009

Outline

- Negative development process update

- Contact printing: dual line exposure
  - Principle, and example for dense contact array
  - Application for random logic

- Contact printing: single exposure

- Trench imaging

- Conclusions

imec   **FUJiFILM**

Joost Bekaert
© imec 2009

6th International Symposium on
Immersion Lithography Extensions
Prague, 2009

CH printing using dual line exposure Principle

CH printing with resolution of L/S:

Double exposure of crossing lines produces contacts at the cross-points

Cfr. Presentation at 5th Symp. on Imm. Litho (The Hague, 2008)

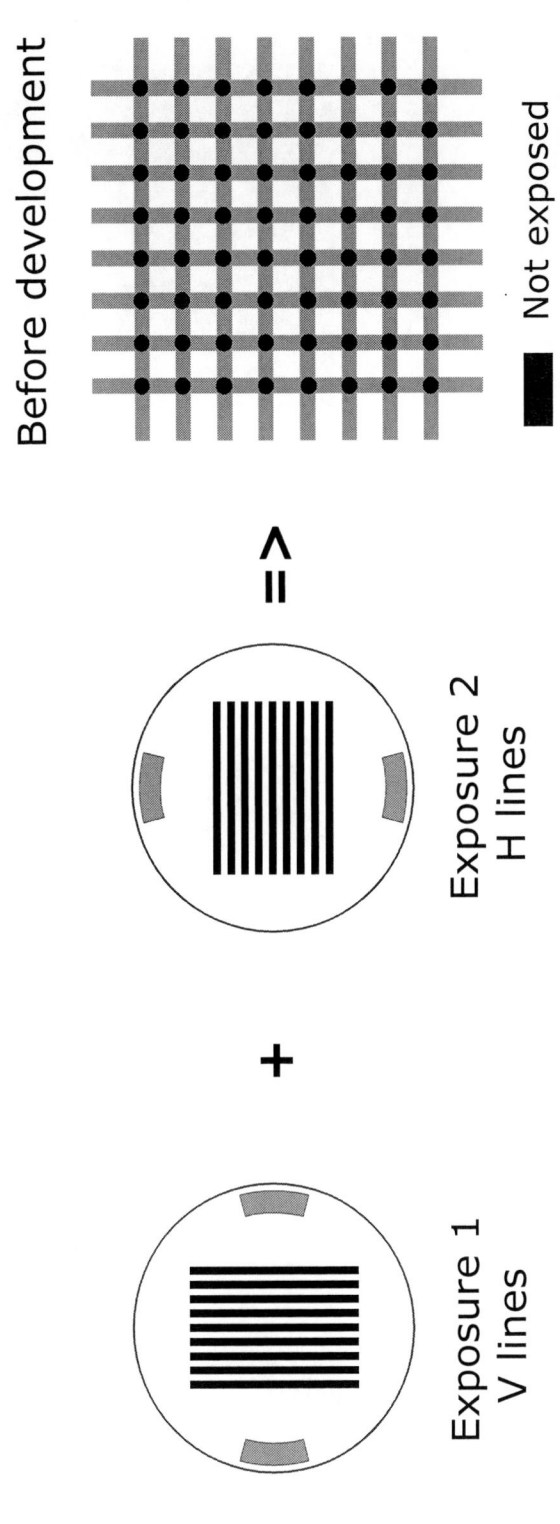

Exposure 1
V lines

+

Exposure 2
H lines

=

Before development

- Not exposed
- Exposed once
- Exposed twice

imec FUJiFILM

6th International Symposium on
Immersion Lithography Extensions
Prague, 2009

Joost Bekaert
© imec 2009

CH printing using dual line exposure Principle

CH printing with resolution of L/S:

Double exposure of crossing lines produces contacts at the cross-points

Cfr. Presentation at 5th Symp. on Imm. Litho (The Hague, 2008)

Exposure 1
V lines

+

Exposure 2
H lines

PTD

=>

Positive development: posts

P80
1.35NA
$k_1$0.28

imec   FUJiFILM

Joost Bekaert
© imec 2009

6th International Symposium on
Immersion Lithography Extensions
Prague, 2009

CH printing using dual line exposure Principle

CH printing with resolution of L/S:

Double exposure of crossing lines produces contacts at the cross-points

Cfr. Presentation at 5th Symp. on Imm. Litho (The Hague, 2008)

Negative development: **holes**

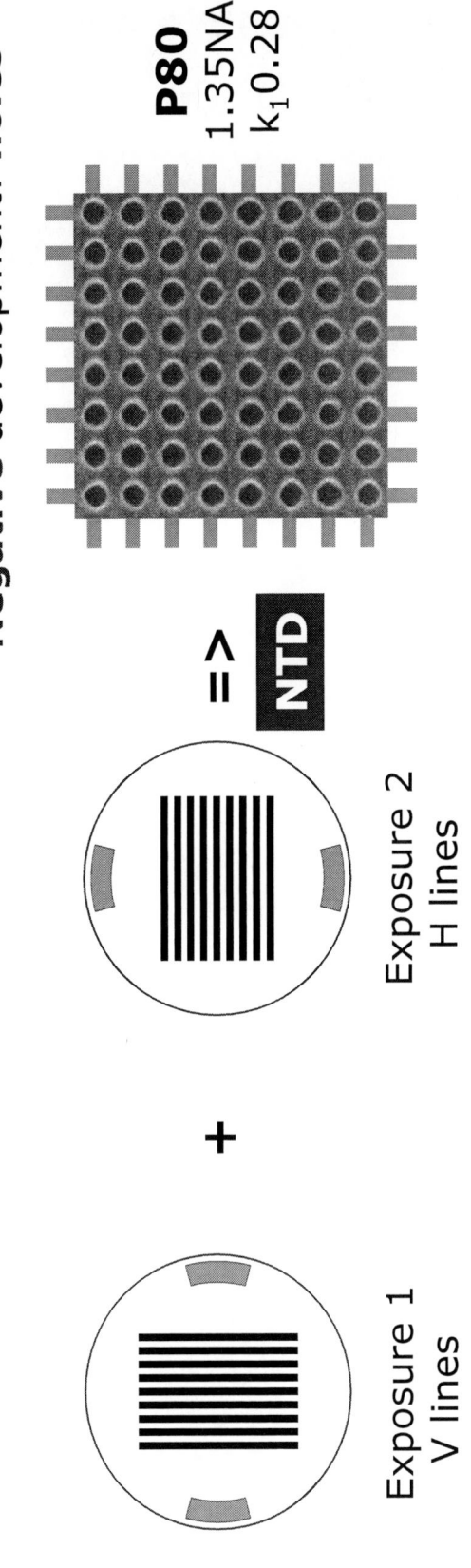

P80
1.35NA
$k_1$0.28

Exposure 1
V lines

+

Exposure 2
H lines

=>

NTD

imec   FUJiFILM

6th International Symposium on
Immersion Lithography Extensions
Prague, 2009

Joost Bekaert
© imec 2009

CH printing using dual line exposure
Performance for dense contact arrays

CH printing with resolution of L/S:

Double exposure of crossing lines produces contacts at the cross-points

Cfr. Presentation at 5th Symp. on Imm. Litho (The Hague, 2008)

Negative development: **holes**

P80
1.35NA
$k_1$0.28

Target 40 nm   (Spec ±10%)

18 % EL – 200 nm DoF

The dual line exposure + NTD shows very good performance for the patterning of dense regular contact patterns.

imec   **FUJiFILM**

Joost Bekaert
© imec 2009

6th International Symposium on
Immersion Lithography Extenstions
Prague, 2009

CH printing using dual line exposure
Random logic application ?

Question:

Can the dual line technique be applied for random logic CH patterns ?

Expo 1
Expo 2

1.35NA
0.95-0.68

Minimum **pitch 100 nm**

- no assists on mask
- aerial image OPC

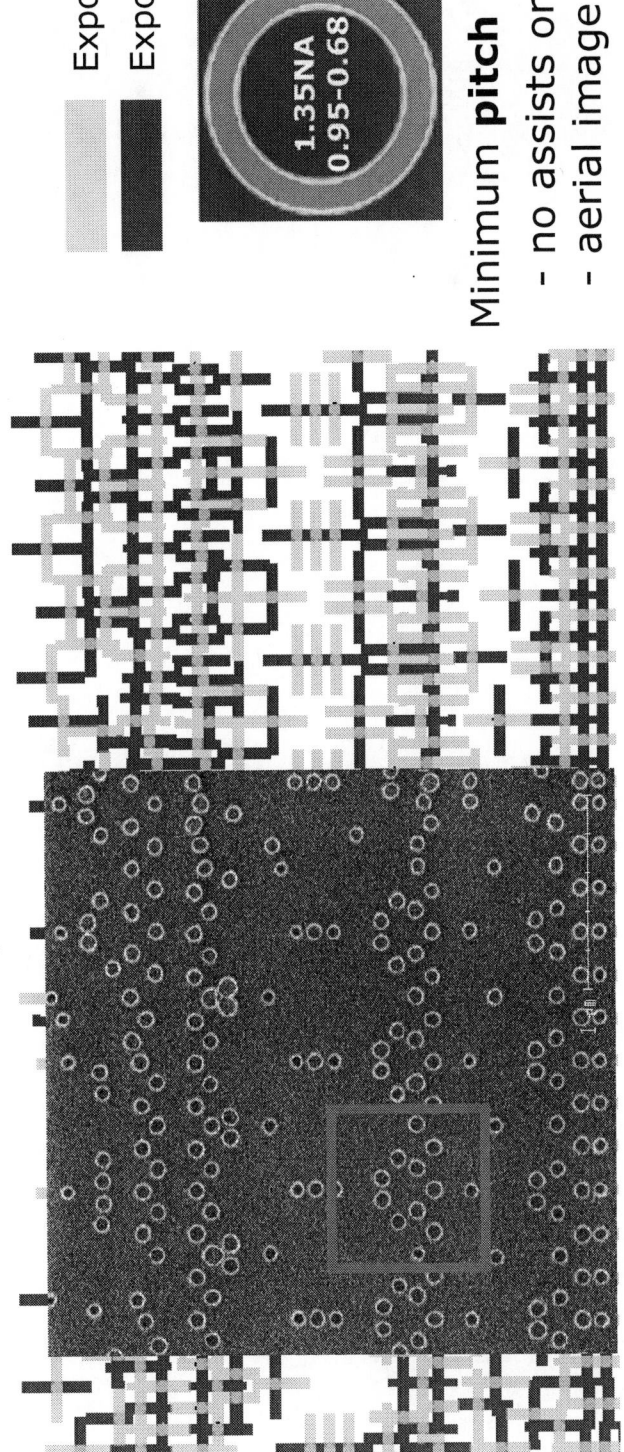

→ Dual line principle works for random patterns
→ Automated split routine and double exposure OPC was achieved

6% attPSM

6th International Symposium on
Immersion Lithography Extensions
Prague, 2009

Joost Bekaert
© imec 2009

CH printing using dual line exposure
Random logic application ?

1.35NA
0.95-0.68

Minimum **pitch 100 nm**

Target 60 nm

7~11 % EL
50~60 nm DoF
Spec of ±10% on CDs

=> Limited process window

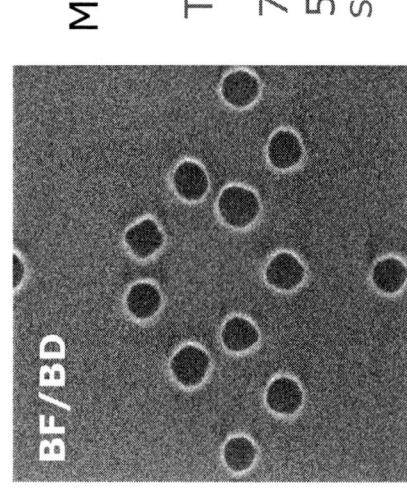

BF/BD

Not straightforward:
- Best split routine
- Dual exposure OPC
- Assists placement/printing

=> Very challenging routine

- Contacts formed near line ends and corners   => CD control is difficult

Although the principle and split automation were demonstrated, this technique has a too small gain/pain ratio for fully random patterns. Moreover, a single exposure solution with NTD will be proposed further on.

imec   **FUJiFILM**

6th International Symposium on
Immersion Lithography Extensions
Prague, 2009

Joost Bekaert
© imec 2009

CH printing using dual line exposure
Memory type application ?

The gain of dual line exposure for CH printing is to be found in applications with regular and simple patterns:

- straightforward split
- very limited variety of pitches and orientations
- illumination can be chosen to optimize process window in both exposures

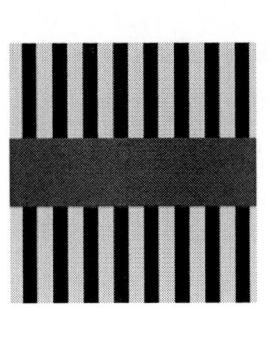

Expo 1
Expo 2

Pitch 80 nm
Contacts are 40 nm by 80 nm

17% EL
160 nm DoF @ 8% EL
Spec of ±10% on CDx and CDy

Example: Chain of CH at P80

BF −80 nm   BF   BF +80 nm

P80

6% attPSM

imec   FUJiFILM

6th International Symposium on
Immersion Lithography Extensions
Prague, 2009

Joost Bekaert
© imec 2009

Outline

- Negative development process update

- Contact printing: dual line exposure

- **Contact printing: single exposure**
  - Generic comparison LF+NTD versus DF+PTD
  - Application of LF+NTD: Random logic with pixbar assists

- Trench imaging

- Conclusions

Joost Bekaert
© imec 2009

6th International Symposium on
Immersion Lithography Extensions
Prague, 2009

imec   FUJiFILM

Contact patterning in single exposure NTD vs PTD for dense contacts (P90)

Dark-field (DF) mask: contacts -> PTD
Light-field (LF) mask: dots -> NTD

Target **dense pitch only:** is NTD better ?

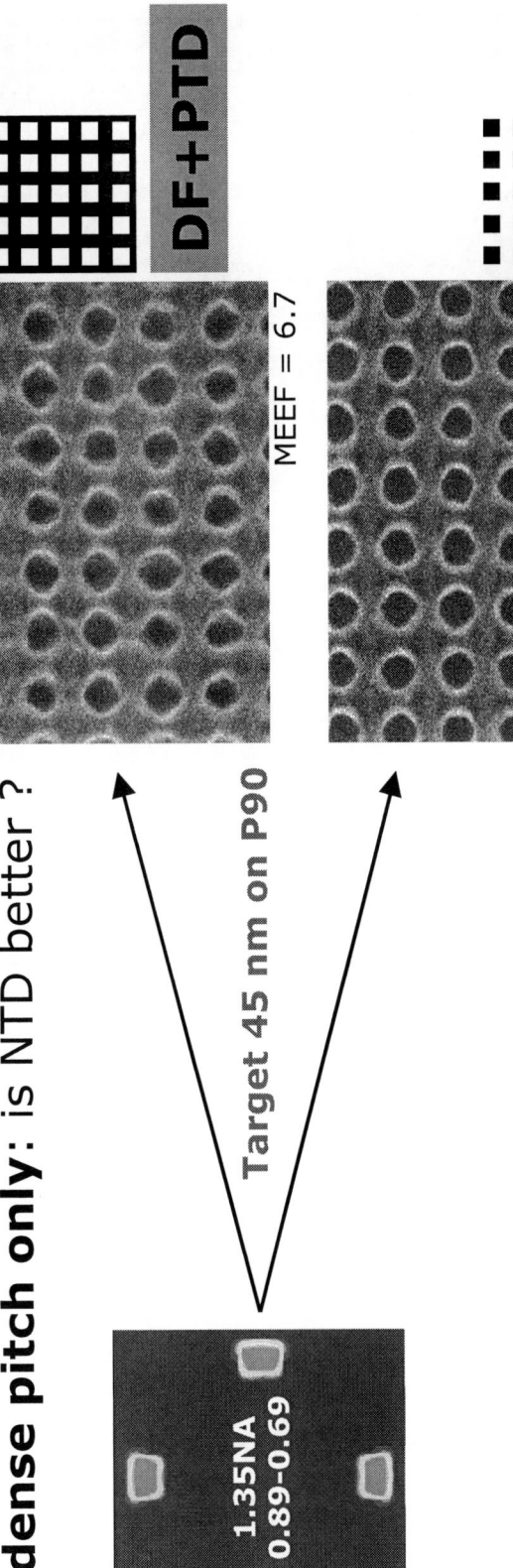

DF+PTD

MEEF = 6.7

LF+NTD

MEEF = 5.4

Target 45 nm on P90

1.35NA
0.89-0.69

LF mask + NTD leads to a better image contrast
=> better EL, MEEF and LWR.

6% attPSM

Joost Bekaert
© imec 2009

6th International Symposium on
Immersion Lithography Extensions
Prague, 2009

imec   FUJiFILM

Contact patterning in single exposure
NTD vs PTD through pitch

Similar comparison, but with setting for through pitch.

Target **50 nm contacts through pitch.**

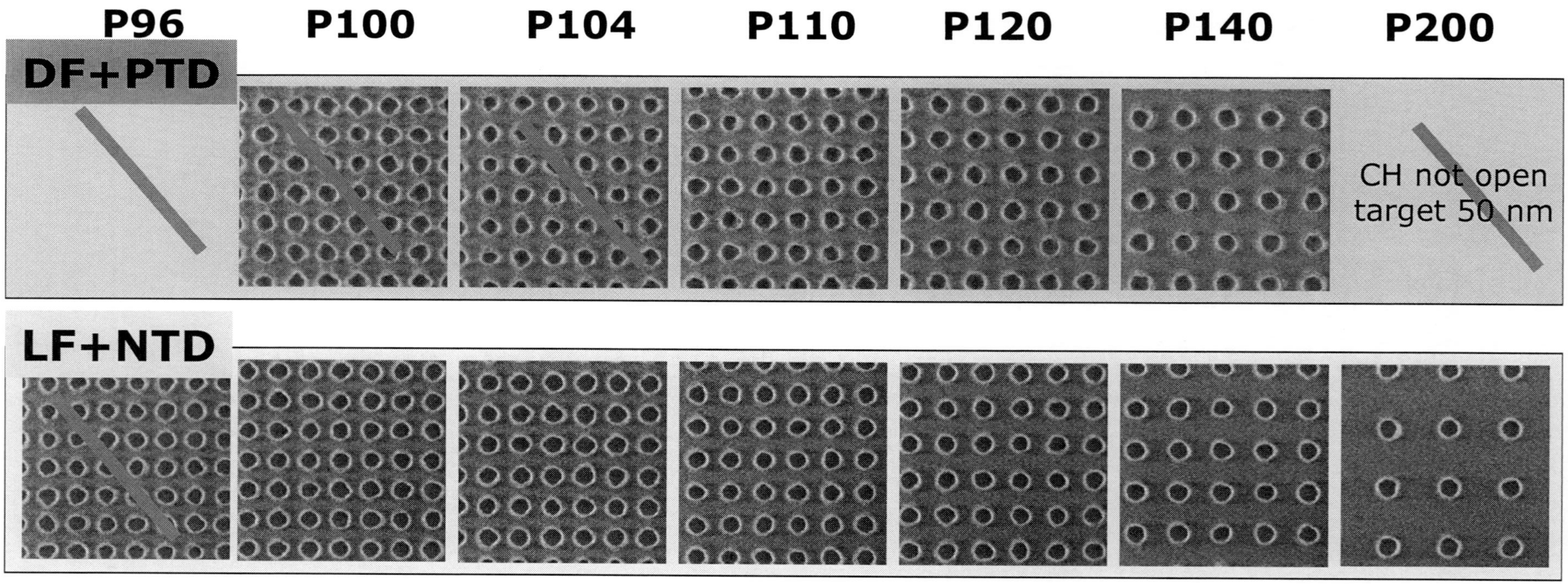

1.35NA
0.93-0.69

| P96 | P100 | P104 | P110 | P120 | P140 | P200 |

DF+PTD

CH not open
target 50 nm

LF+NTD

6% attPSM

imec FUJiFILM

6th International Symposium on
Immersion Lithography Extensions
Prague, 2009

Joost Bekaert
© imec 2009

Contact patterning in single exposure NTD vs PTD through pitch

1.35NA
0.93-0.69

Similar comparison, but with setting for through pitch.

Target **50 nm contacts through pitch**.

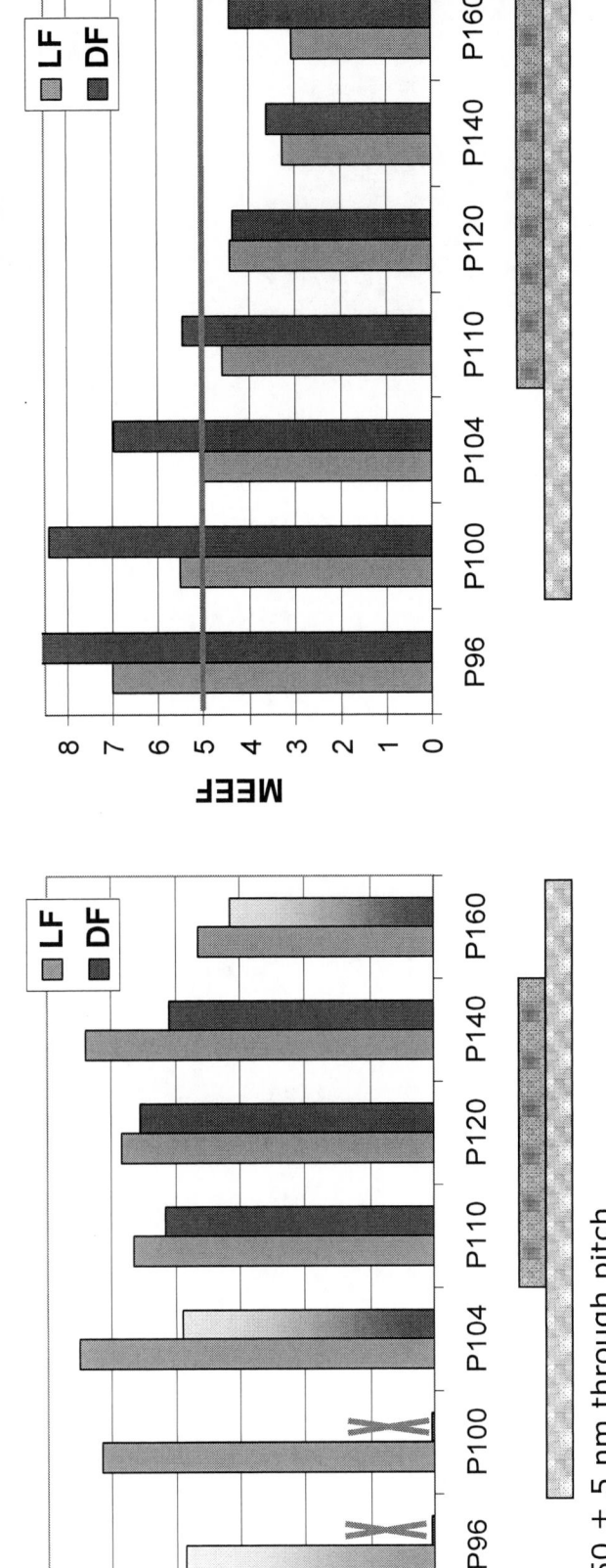

CD spec 50 ± 5 nm through pitch

LF imaging with NTD improves EL and MEEF for smaller contact targets.
Under equal conditions as DF, the LF scheme allows denser CH printing.

Joost Bekaert
© imec 2009

6th International Symposium on
Immersion Lithography Extensions
Prague, 2009

imec   FUJiFILM

Contact patterning in single exposure
LF+NTD: random logic with pixbar assists

A.k.a 'inverse litho', the pixbar technique optimizes the mask for given illumination condition through 'freeform' assist feature placement.

Application:
Random logic clip, treated with pixbar for a light-field mask and NTD.
Basic aerial image OPC.

Ref for DF+PTD:
E. Hendrickx
SPIE 6924-20 (2008)

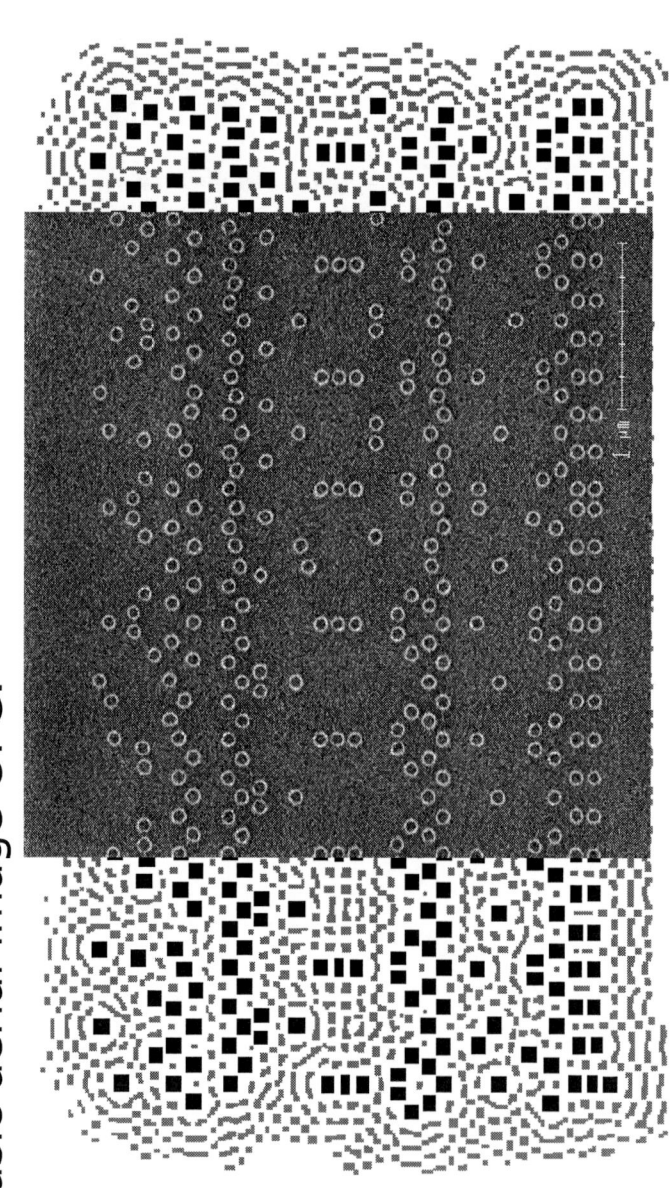

6% attPSM

6th International Symposium on
Immersion Lithography Extensions
Prague, 2009

Joost Bekaert
© imec 2009

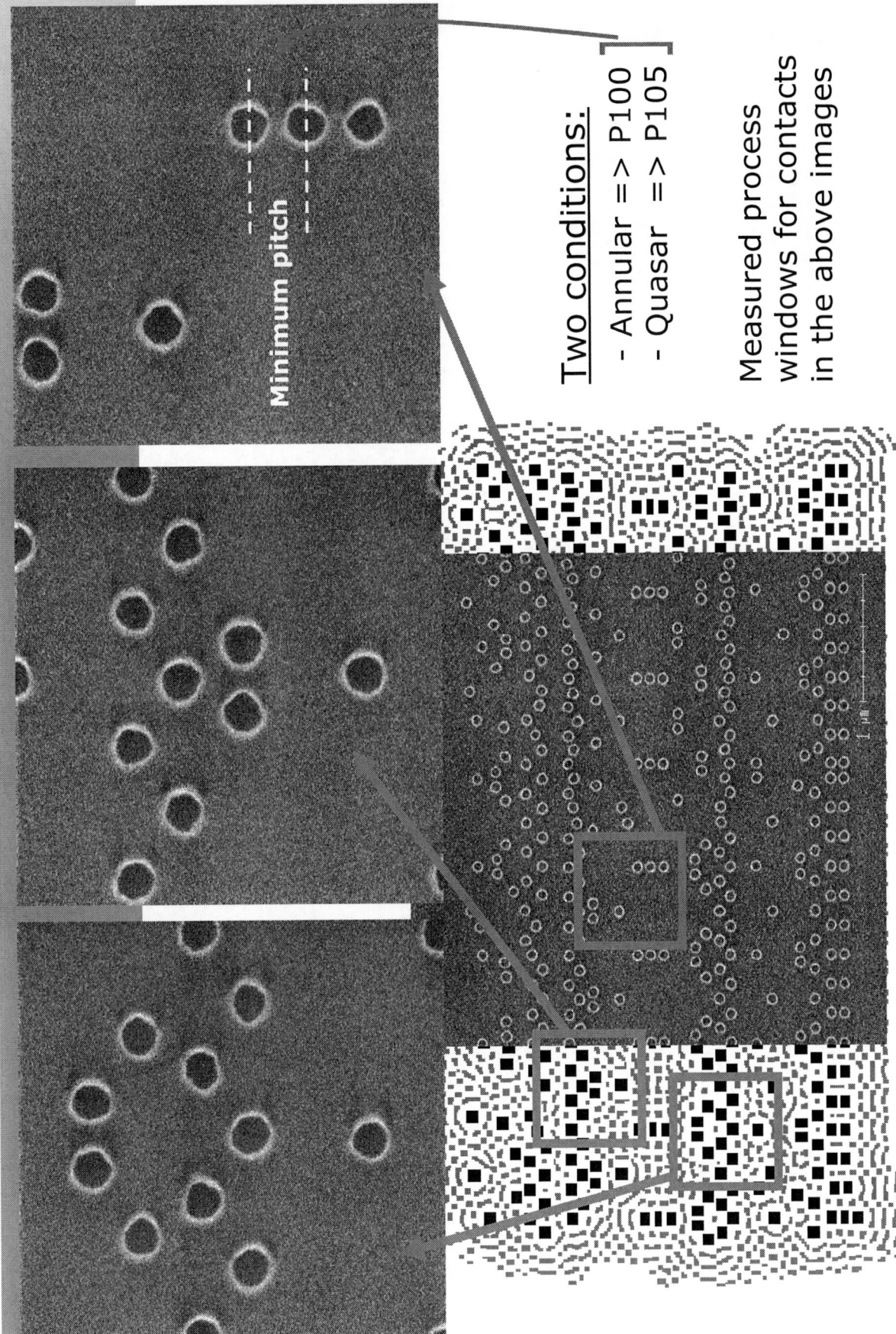

Minimum pitch

Two conditions:
- Annular => P100
- Quasar => P105

Measured process windows for contacts in the above images

6% attPSM

Joost Bekaert
© imec 2009

6th International Symposium on
Immersion Lithography Extensions
Prague, 2009

imec   FUJiFILM

Contact patterning in single exposure
LF+NTD: random logic with pixbar assists

Annular condition minimum pitch of 100 nm

Quasar condition minimum pitch of 105 nm

CD 60 nm ±10% spec

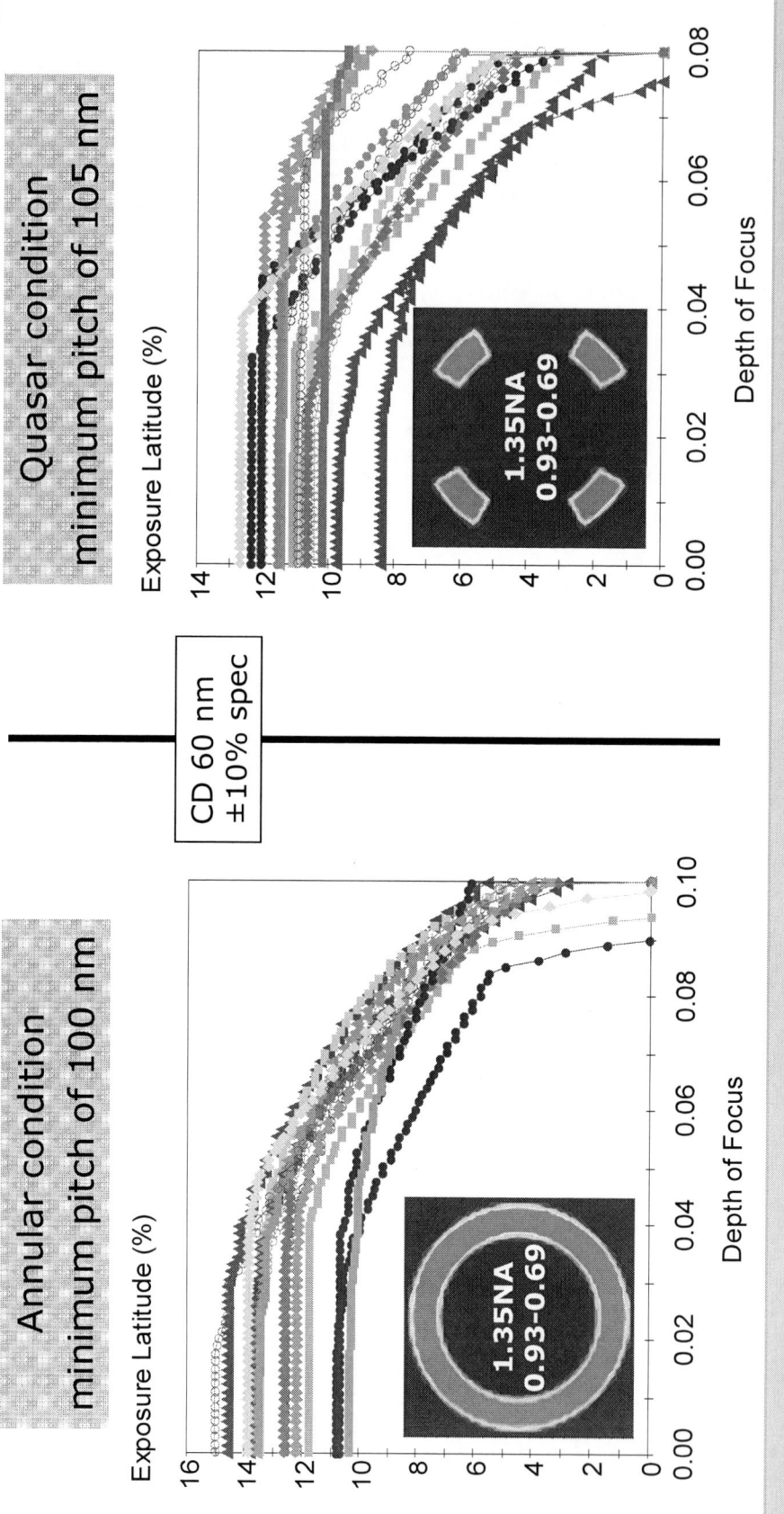

This single exposure technique with NTD has potential for extending the patterning of (e.g. logic) CH patterns to denser pitches.

6th International Symposium on
Immersion Lithography Extensions
Prague, 2009

Joost Bekaert
© imec 2009

imec   FUJiFILM

Outline

- Negative development process update

- Contact printing: dual line exposure

- Contact printing: single exposure

- Trench printing

- Conclusions

6th International Symposium on
Immersion Lithography Extensions
Prague, 2009

Joost Bekaert
© imec 2009

imec   FUJiFILM

Trenches
Target of the study

Goal:
32 nm node metal layer (double patterning):
Compare the imaging performance for LF+NTD versus DF+PTD

Case Study:
NA 1.35, Annular 0.85-0.65, XY Polarized
Trenches on P128 to isolated
6% att PSM

See presentation earlier today by V. Wiaux:
"Density limits in logic metal1 using DP"

Criteria:
- 1D trenches of 36 nm on wafer
- Gaps on these trenches *through pitch* and *through gap width*

(CD spec: ±10% for 1D, and +-5 nm for the gaps)

imec   FUJ:FILM

6th International Symposium on
Immersion Lithography Extensions
Prague, 2009

Joost Bekaert
© imec 2009

Trenches
1D trenches of CD 36 nm through pitch

Trench CD 36 nm

PTD
Dark Field Imaging

NTD
Light Field Imaging

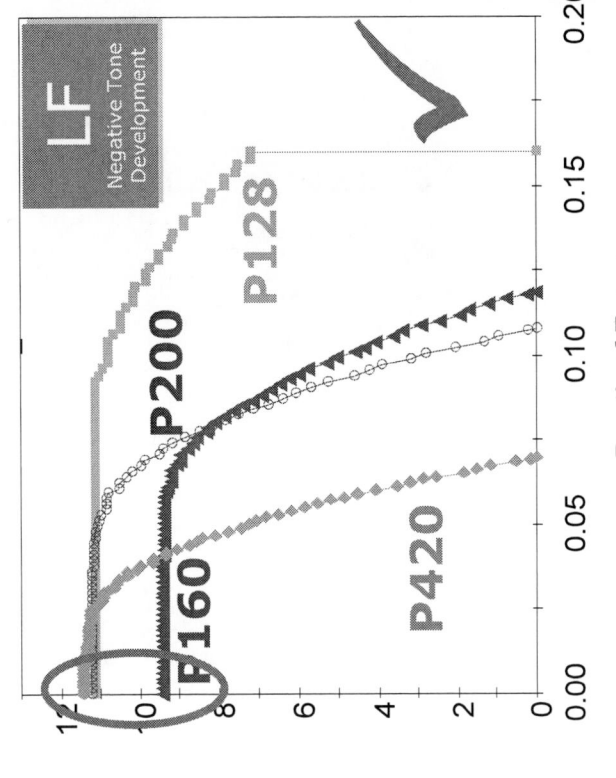

Exposure Latitude (%)

Depth of Focus

NTD has huge advantage over PTD process for Exposure Latitude for imaging small trenches through pitch

Criterion for EL: +/- 10% of target CD

6th International Symposium on
Immersion Lithography Extensions
Prague, 2009

Joost Bekaert
© imec 2009

imec   FUJiFILM

Trenches
1D trenches of CD 36 nm through pitch

Trench CD 36 nm

Exposure Latitude

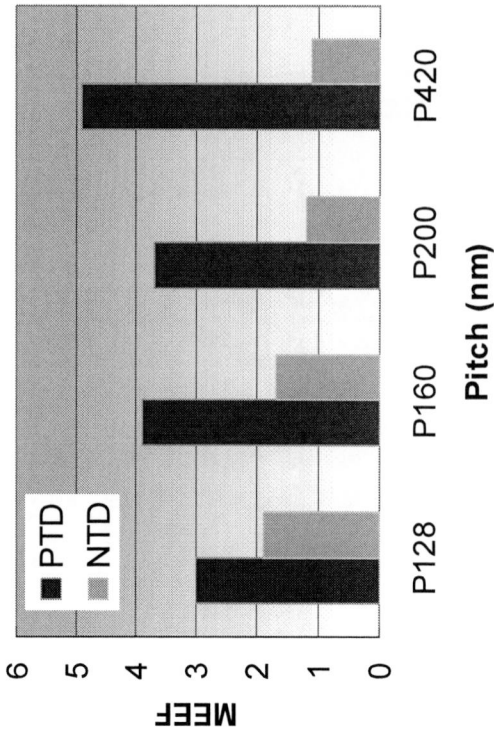

MEEF

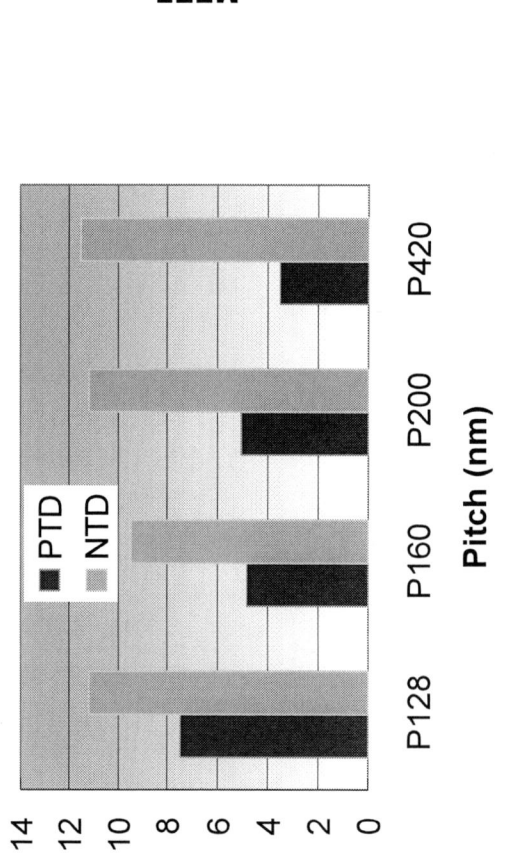

NTD has huge advantage over PTD process for Exposure Latitude for imaging small trenches through pitch

Criterion for EL: +/- 10% of target CD

6th International Symposium on
Immersion Lithography Extensions
Prague, 2009

Joost Bekaert
© imec 2009

imec   FUJiFILM

Trenches
Line End behavior through line pitch

Trench CD 36 nm

Gaps 80 nm

Fixed gap size on wafer – Varying trench pitch

Pictures @ BF, BD

P128   P160   P200

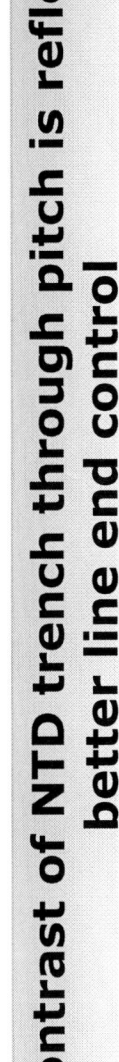

DF Positive Tone Development

LF Negative Tone Development

Better contrast of NTD trench through pitch is reflected in better line end control

6th International Symposium on
Immersion Lithography Extensions
Prague, 2009

Joost Bekaert
© imec 2009

imec   FUJiFILM

Trench CD 36 nm

Gaps 80 nm

Trenches
Line End behavior through line pitch

Fixed gap size on wafer – Varying trench pitch

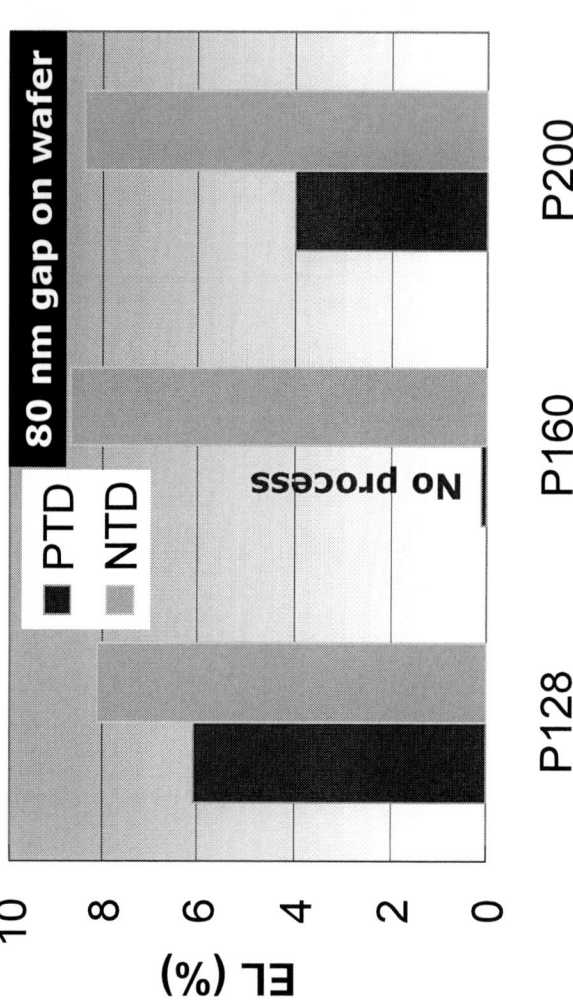

Better contrast of NTD trench through pitch is reflected in better line end control

Criterion for gap EL: +/- 5 nm, for all gap sizes

6th International Symposium on
Immersion Lithography Extensions
Prague, 2009

Joost Bekaert
© imec 2009

imec   FUJIFILM

Trenches
Line End behavior through gap width

Trench CD 36 nm

Trenches P128

Fixed trench pitch – Varying gap size on wafer

Pictures @ BF, BD

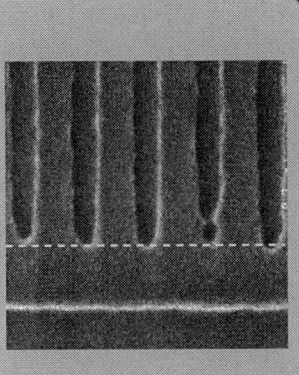 **Gap 73 nm**

 **Gap 105 nm**

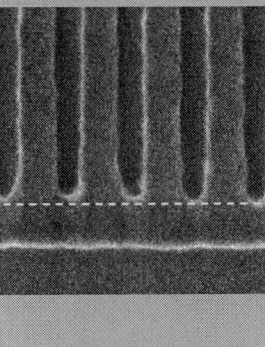 **Gap 160 nm**

DF
Positive Tone Development

LF
Negative Tone Development

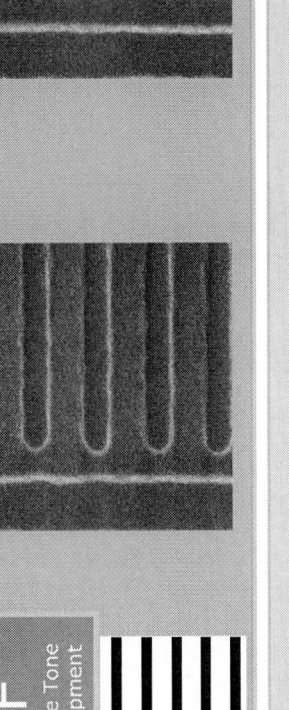
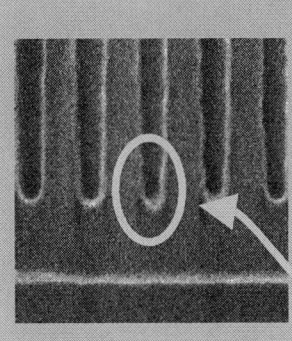
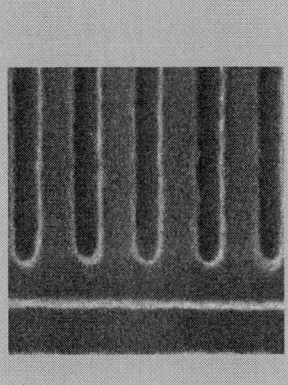
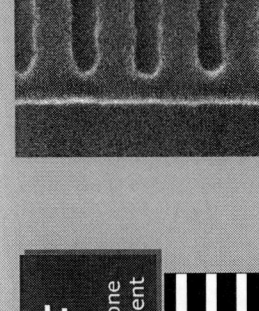

NTD shows better line end control for all gap sizes but currently has somewhat worse resist profiles

imec FUJIFILM

Joost Bekaert
© imec 2009

6th International Symposium on
Immersion Lithography Extensions
Prague, 2009

Trench CD 36 nm

Trenches
Line End behavior through gap width

Trenches P128

Fixed trench pitch – Varying gap size on wafer

The benefit of NTD increases with gap width

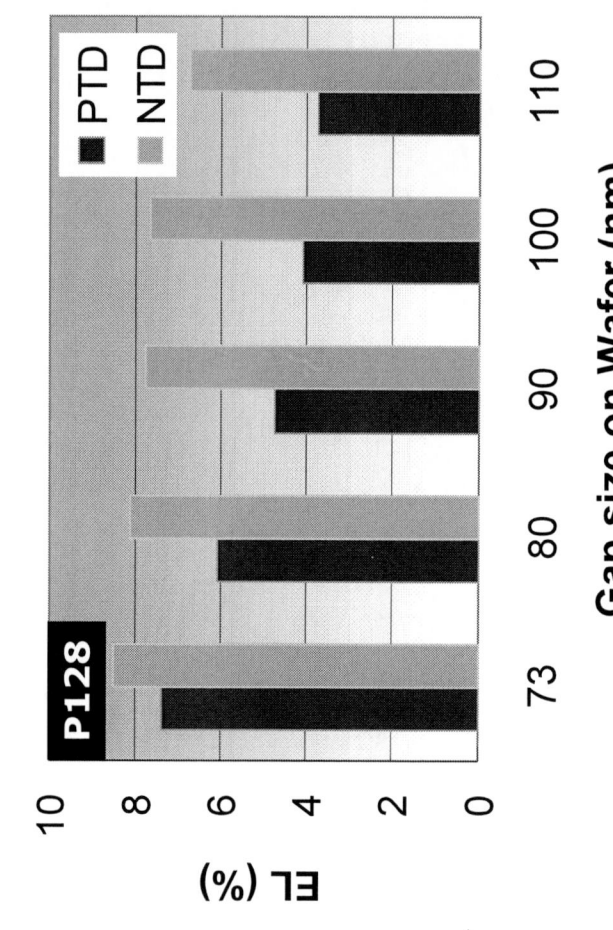

PTD: Relaxing gap makes line end control more difficult
NTD: Substantially better line end control for all gap sizes

Criterion for gap EL: +/- 5 nm, for all gap sizes

6th International Symposium on
Immersion Lithography Extensions
Prague, 2009

Joost Bekaert
© imec 2009

imec   FUJiFILM

Trenches
2D pattern fidelity for NTD compared to PTD

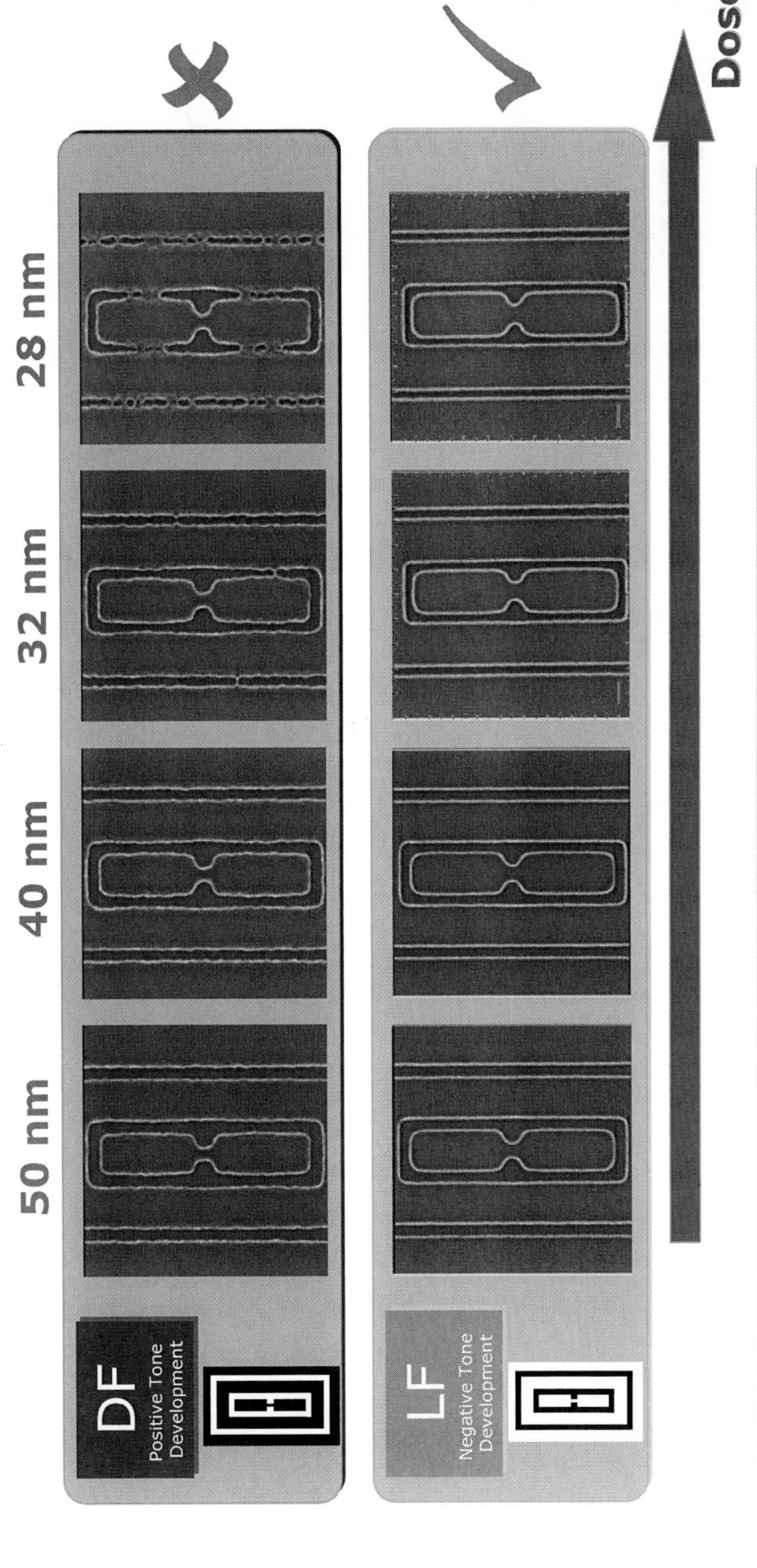

NTD offers a better 2D pattern fidelity than PTD

6th International Symposium on
Immersion Lithography Extensions
Prague, 2009

Joost Bekaert
© imec 2009

Outline

- Negative development process update

- Contact printing: dual line exposure

- Contact printing: single exposure

- Trench printing

Conclusions

6th International Symposium on
Immersion Lithography Extensions
Prague, 2009

Joost Bekaert
© imec 2009

imec   FUJiFILM

Conclusions

- **Dual line exposure for CH imaging** seems best appropriate for simple and very regular (dense) patterns, where design split is straightforward and process latitude can be supported by choice of source. It is not the best path for random logic.

- **Contact patterning in single exposure** clearly benefits from the use of light-field masks combined with NTD, both for dense and through pitch applications. Pixbar was demonstrated as a viable option for random logic. See also: "Source-Mask optimization for SRAM", today at 15:55.

- Also for **trench patterning**, the use of light-field masks in combination with NTD provides better imaging performance for trenches through pitch as compared to traditional dark-field masks with PTD. In particular, better line end control was demonstrated through pitch and through gap width. Also, 2D pattern fidelity is better for NTD.

Joost Bekaert
© imec 2009

6th International Symposium on
Immersion Lithography Extensions
Prague, 2009

imec   FUJiFILM

Tomorrow:

Tomorrow, please see presentation:

Friday at 15:15

Shinji Tarutani – FujiFilm

"Resist materials for the negative development process"

Thank you !

Joost Bekaert
© imec 2009

6th International Symposium on
Immersion Lithography Extensions
Prague, 2009

imec FUJiFILM

Double Patterning OPC and Design for 22nm to 16nm Device Nodes

Kevin Lucas, Chris Cork, Alex Miloslavsky, Gerry Luk-Pat, Xiaohai Li, Levi Barnes, Weimin Gao

Synopsys Inc.

Vincent Wiaux

IMEC

Outline

- Introduction & DPT goals
- Decomposition and coloring
- Design
- OPC & Verification
- Wafer results
- Conclusions

Outline

- ## Introduction & DPT goals

- Decomposition and coloring

- Design

- OPC & Verification

- Wafer results

- Conclusions

DPT Styles

- ## Line-end cutting DPT

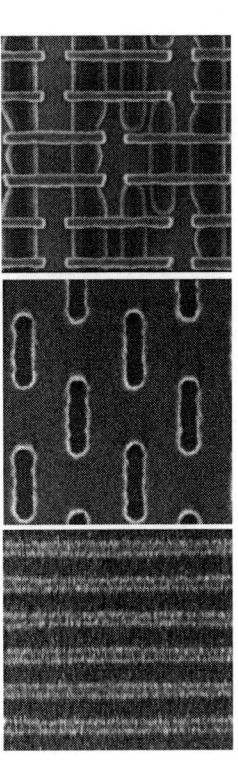

H. Haffner, 2008 Sematech Litho Forum

- ## Litho-etch-litho-etch

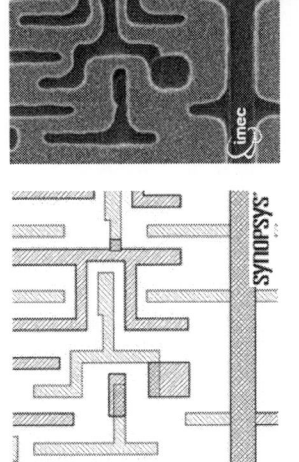

K. Lucas, SPIE 2008

- ## Litho-freeze-litho-etch

M. Maenhoudt, 2008 Sematech Litho Forum

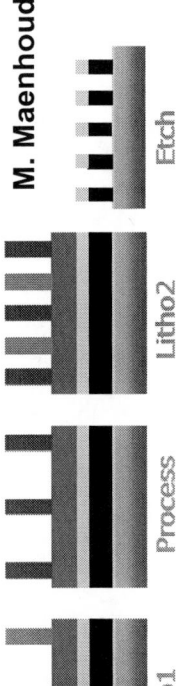

Litho1   Process   Litho2   Etch

- ## Spacer DPT (self-aligned DPT)

Inoue-san, 2008 Sematech Litho Forum

©Synopsys 2009

Product goals for DPT

- High density, performance and yield; reasonable cost and speed; low risk.
  - Design rule values reduced 30%
  - Low design effort overhead
    - automated layout creation, low rework
  - Fast to develop and deploy
  - Fully verifiable before patterning silicon
  - Integrates with other process/device modules
    - Large process margin
    - Good overlay margin
    - Symmetric patterns
    - uniform mask density

Outline

- Introduction & DPT goals
- **Decomposition and coloring**
- Design
- OPC & Verification
- Wafer results
- Conclusions

Benefits of cost-model decomposition
- with user defined constraints

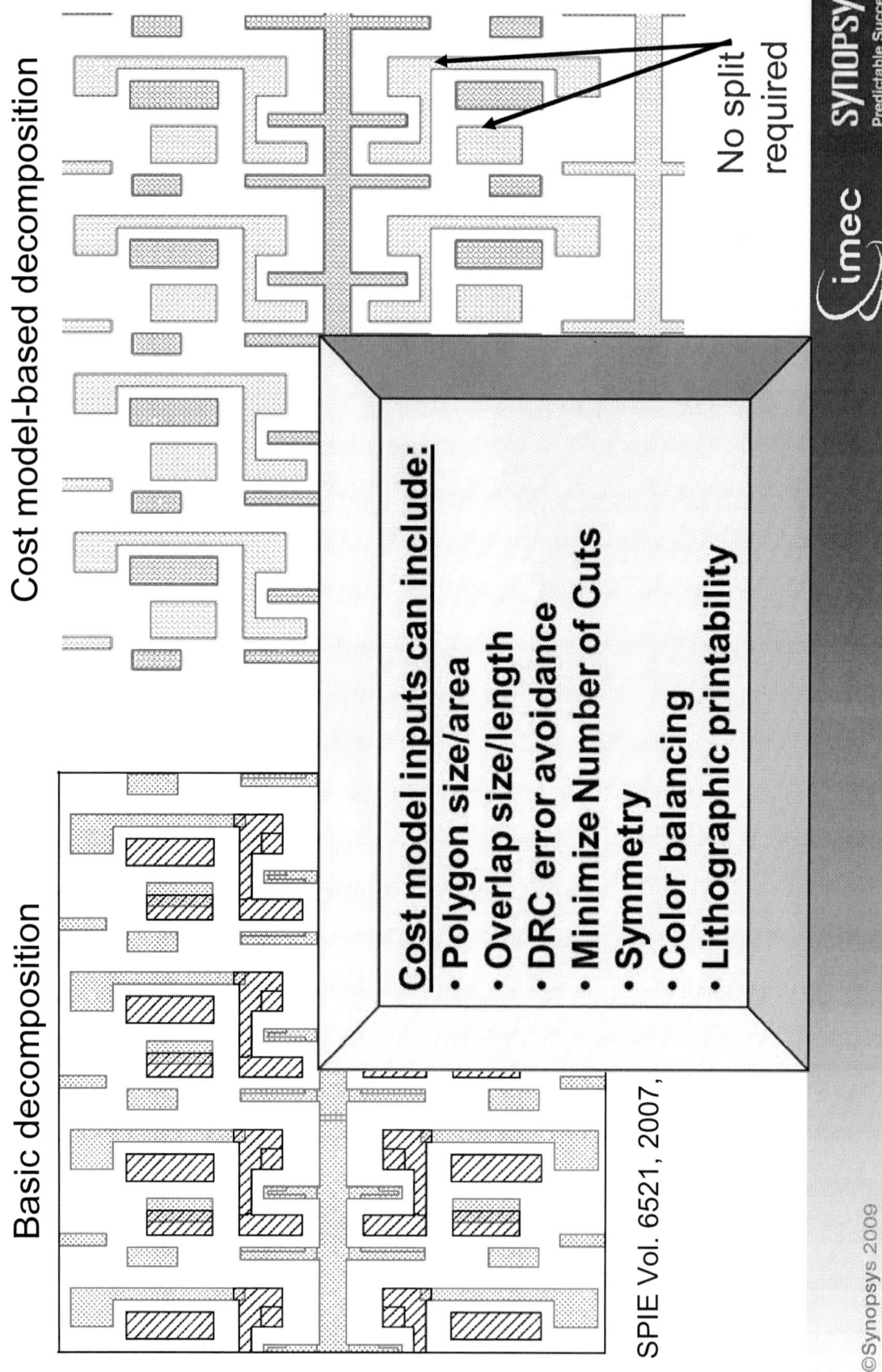

Cost model-based decomposition

Basic decomposition

Cost model inputs can include:
- Polygon size/area
- Overlap size/length
- DRC error avoidance
- Minimize Number of Cuts
- Symmetry
- Color balancing
- Lithographic printability

No split required

SPIE Vol. 6521, 2007,

©Synopsys 2009

Coloring is a global problem

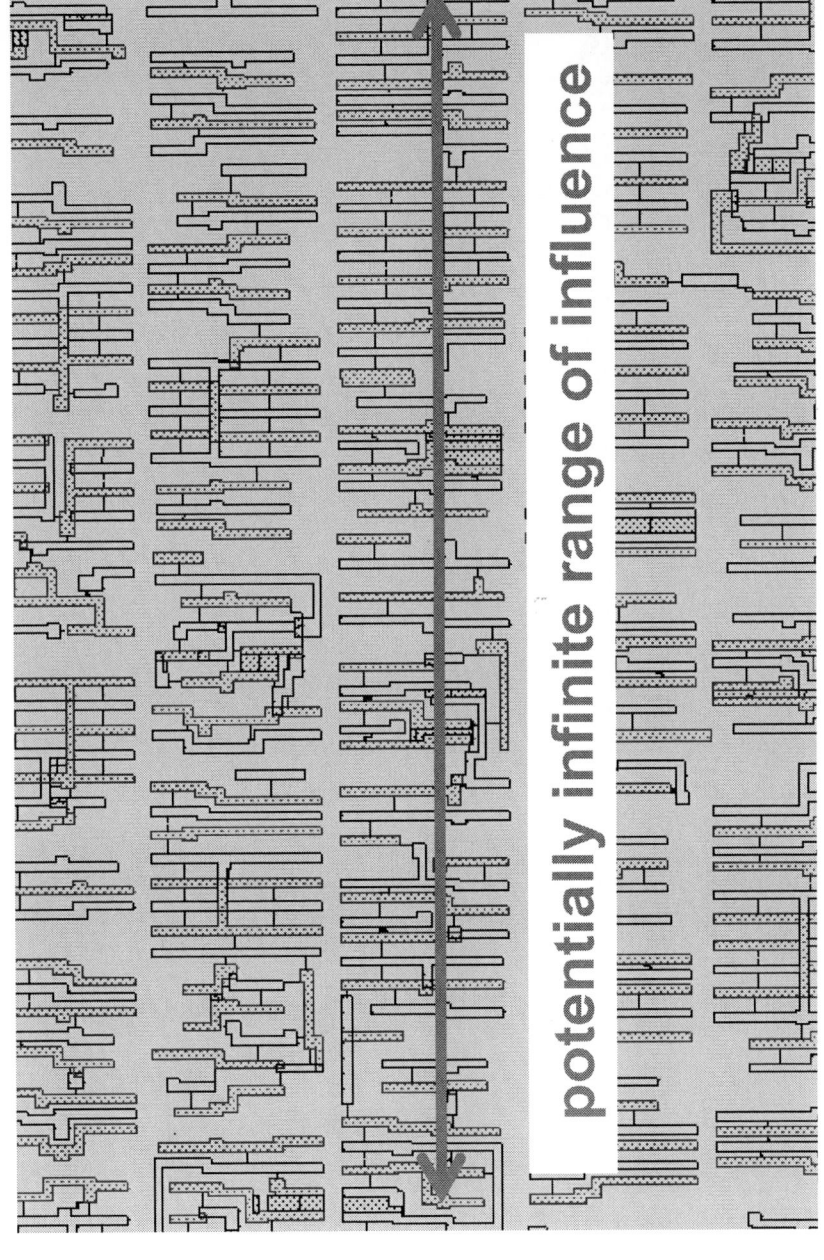

potentially infinite range of influence

L. Barnes, et. al. Bacus 2009

Convert polygon coloring to graph coloring

- A DPT coloring tool looks at polygons as nodes to be colored in a network.
- Converts a layout coloring problem into a graph coloring problem to simplify computational solution.
- Connected nodes need to be assigned opposite colors (if a solution exists, e.g., no DPT coloring compliance error).

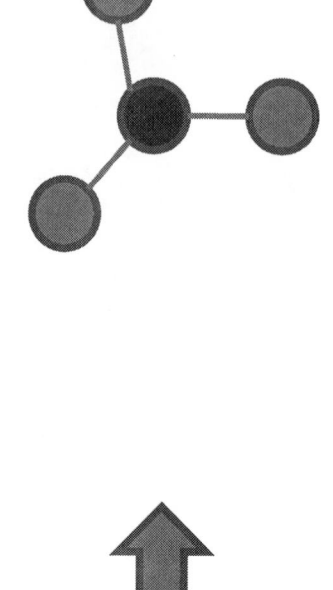

Graph representation

Layout polygons

©Synopsys 2009

Distributing by network

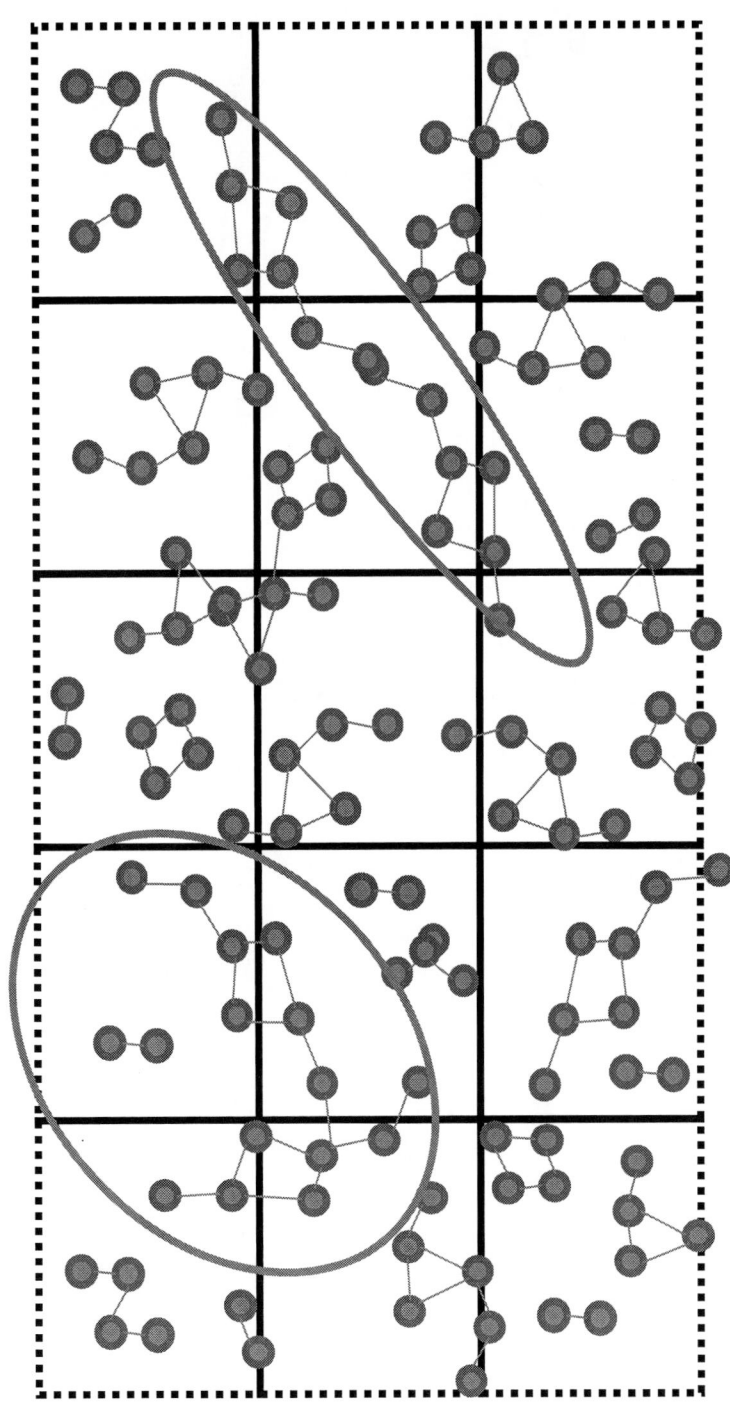

- Straightforward to distribute if many small networks
- But, what about huge networks?

L. Barnes, et. al. Bacus 2009

Network of connected nodes to color

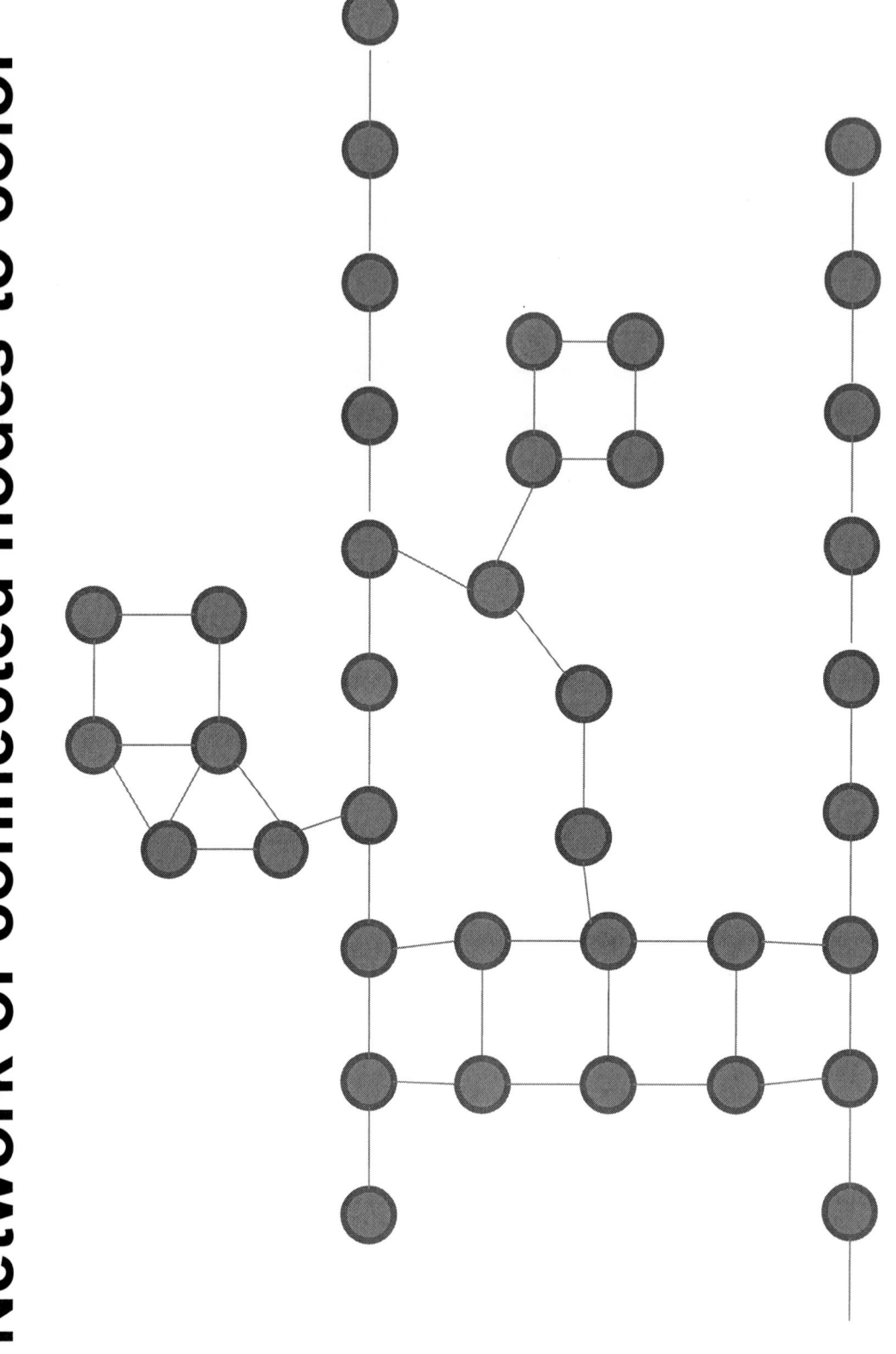

L. Barnes, et. al. Bacus 2009

Pruning

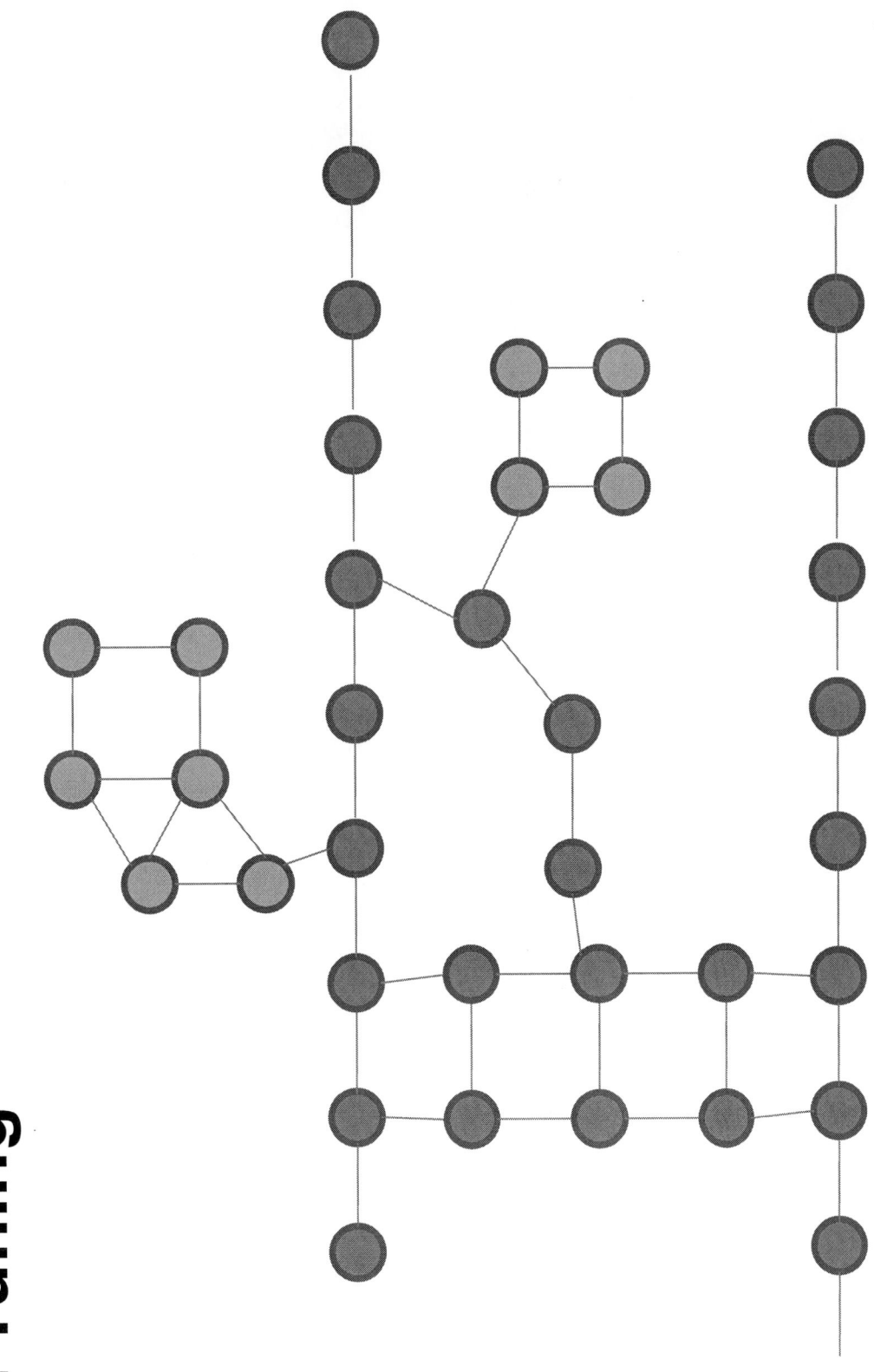

L. Barnes, et. al. Bacus 2009

Color global networks

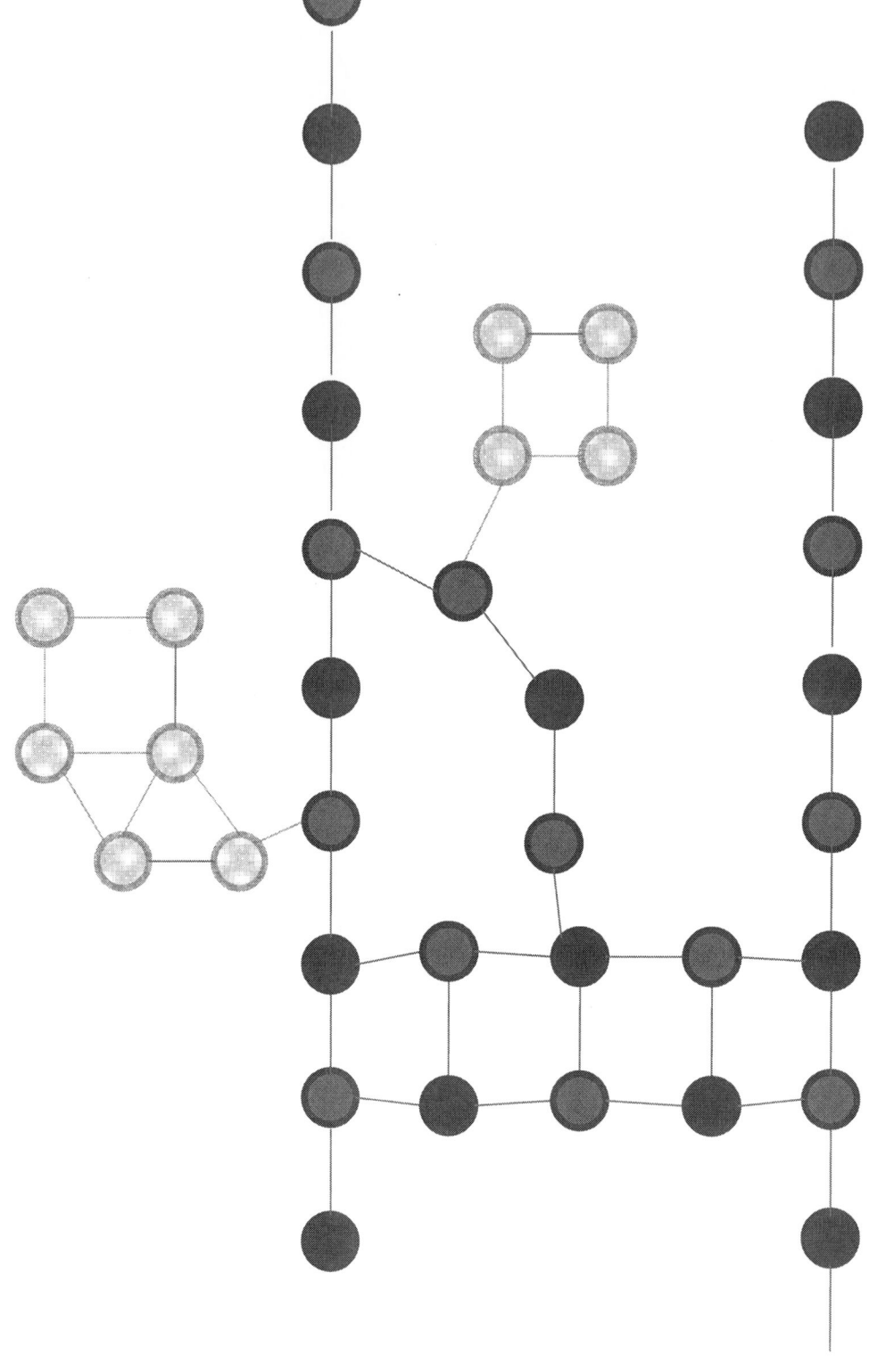

L. Barnes, et. al. Bacus 2009

Color pruned nodes

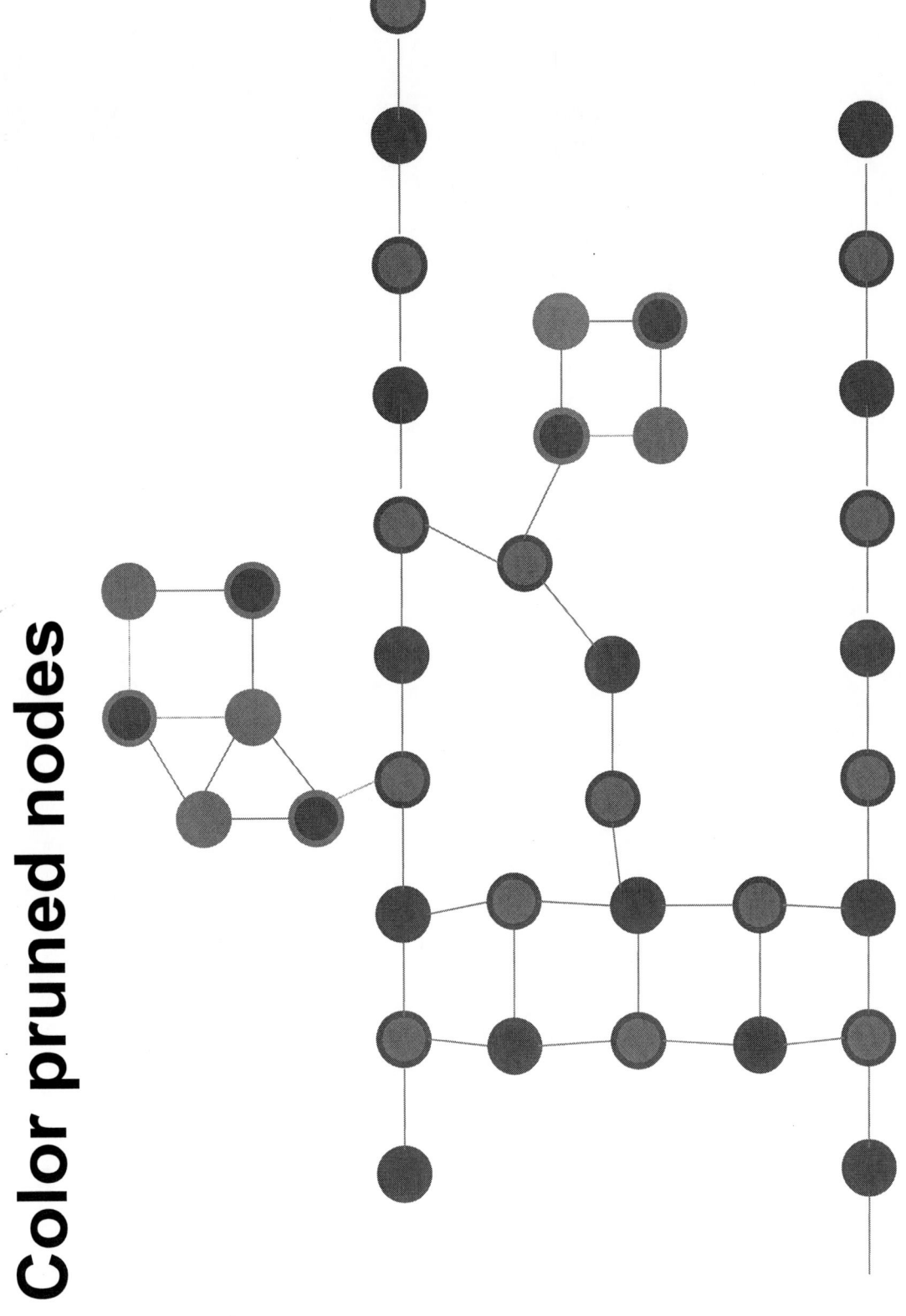

L. Barnes, et. al. Bacus 2009

Pruning w/ grafting

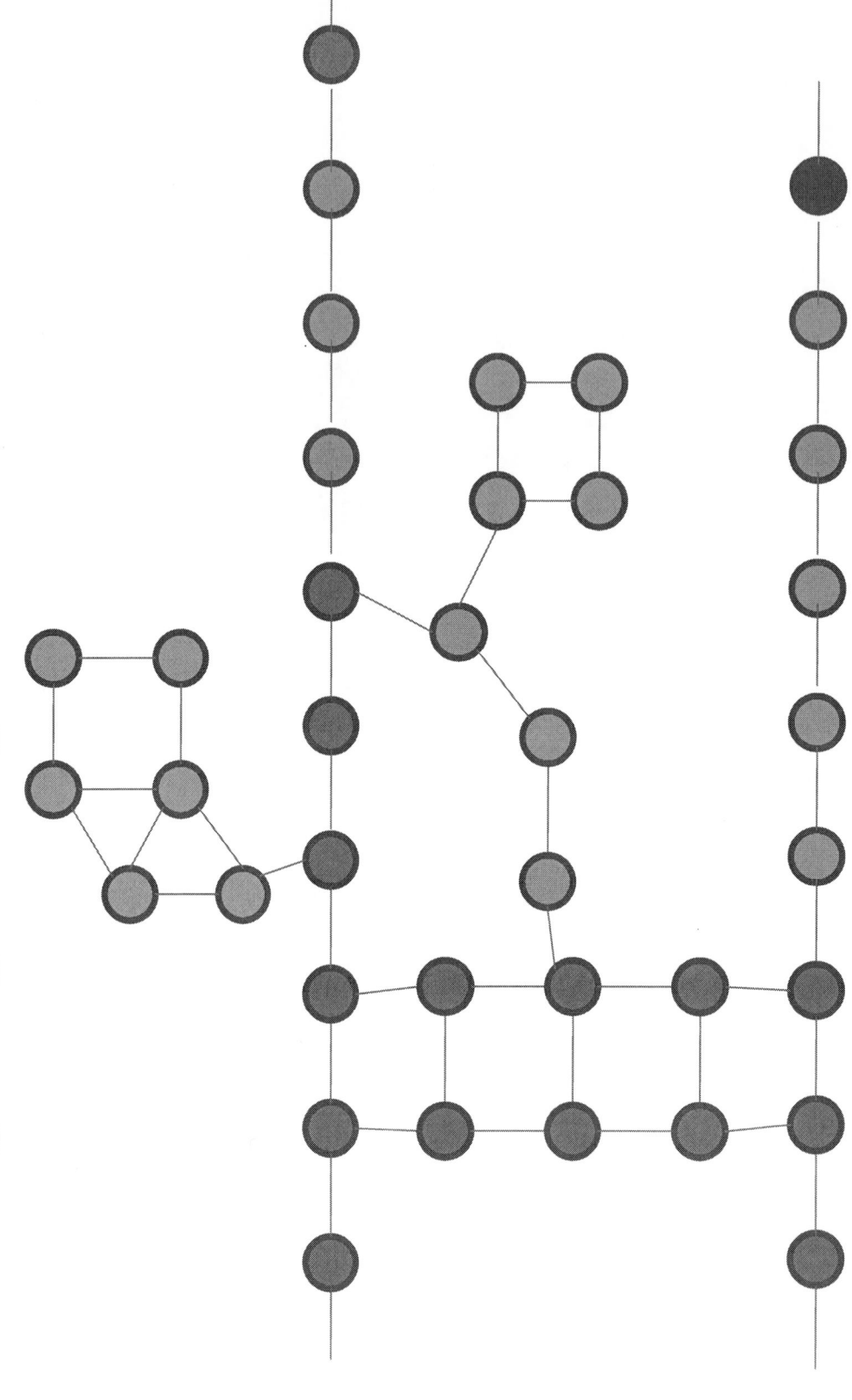

L. Barnes, et. al. Bacus 2009

Color global networks and grafts

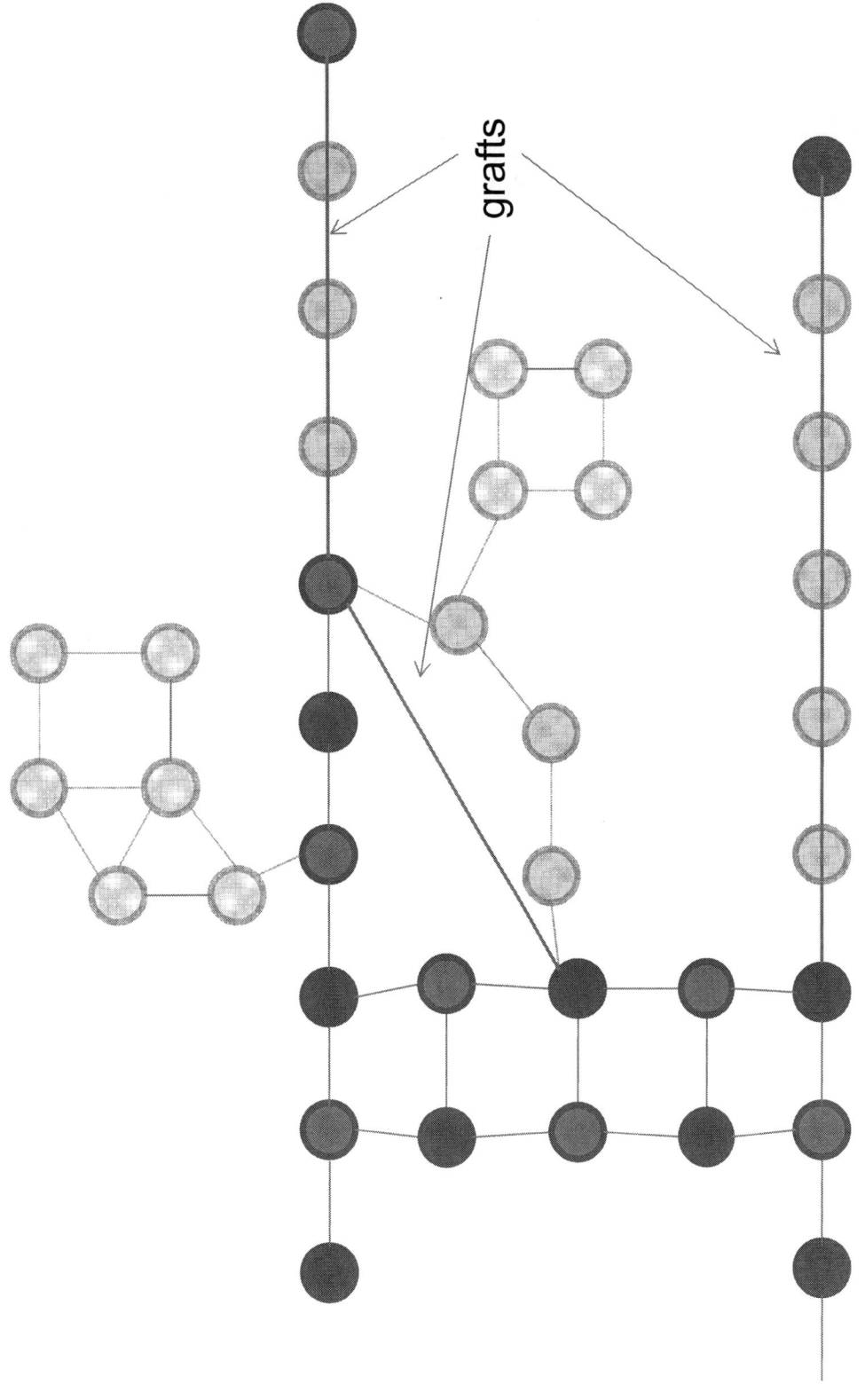

grafts

L. Barnes, et. al. Bacus 2009

Color pruned and grafted nodes

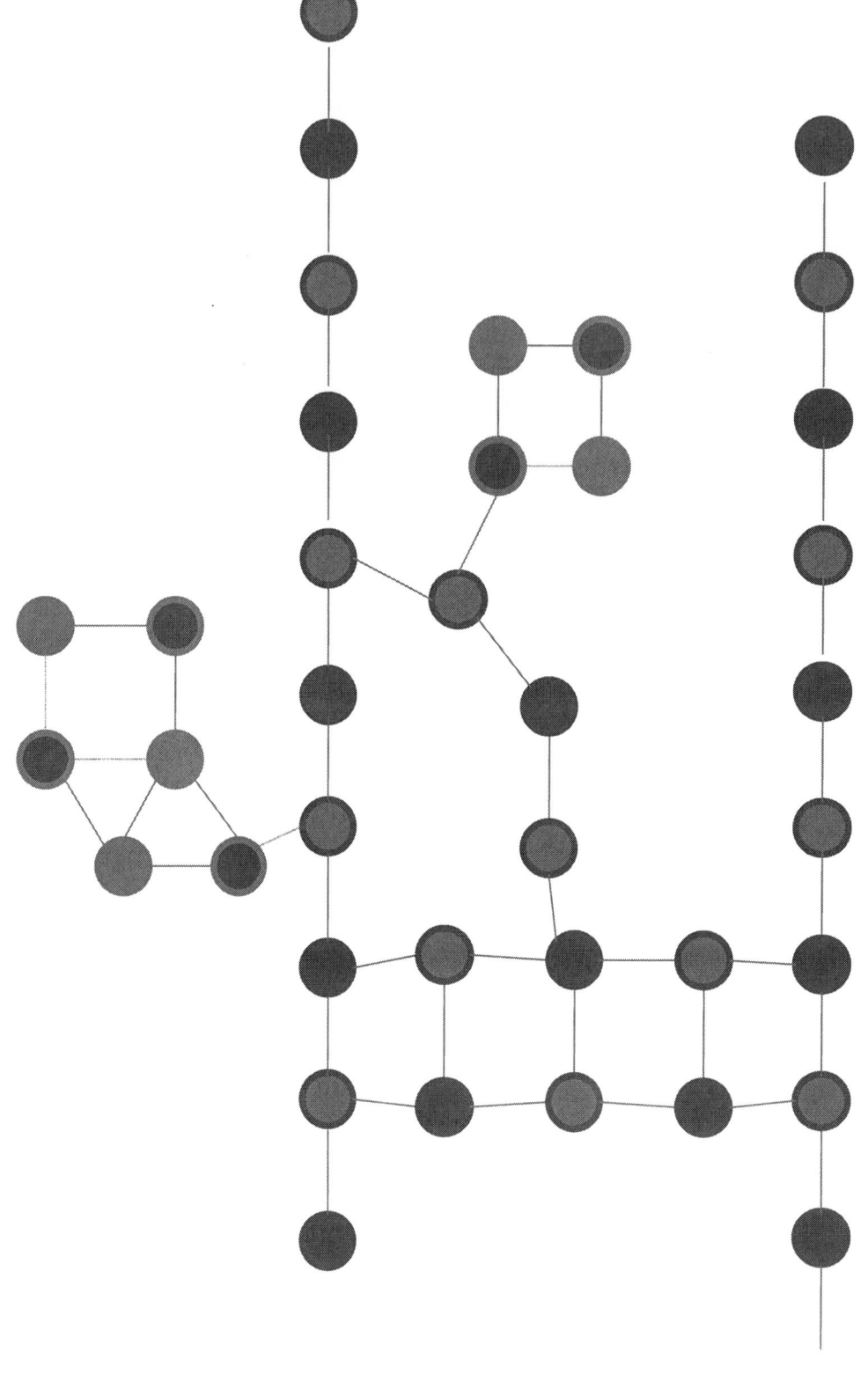

L. Barnes, et. al. Bacus 2009

Pre-coloring of memory arrays

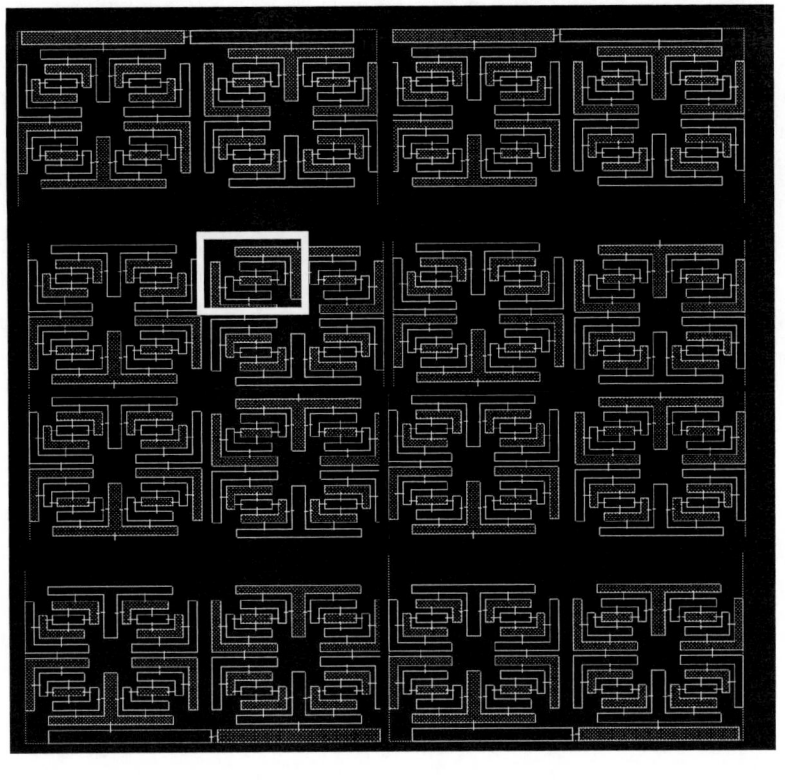

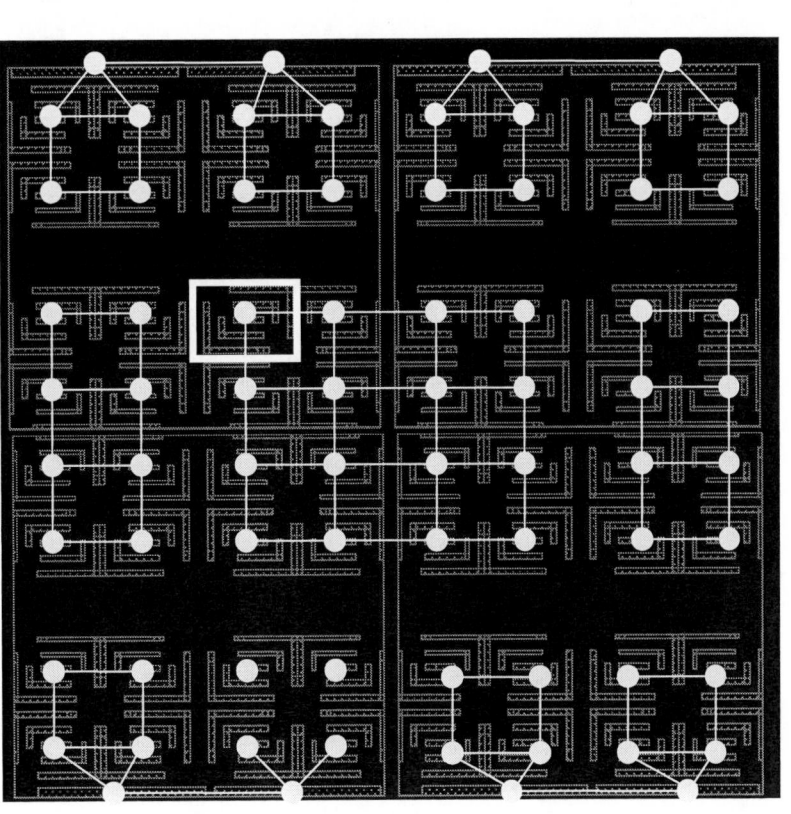

For coloring speed, for predetermined cutting/symmetry, enables incremental fixes
- removes these nodes from the global network

L. Barnes, et. al. Bacus 2009

©Synopsys 2009

How well does distribution work?

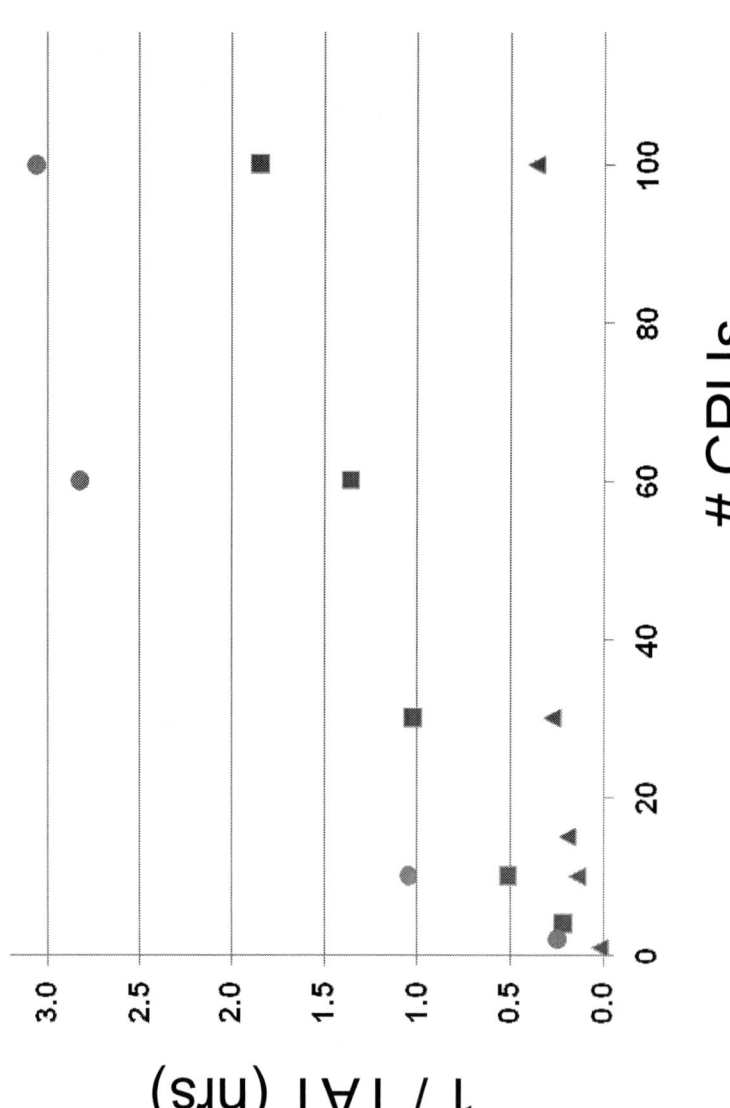

Good scalability seen up to 60-100 cpus

L. Barnes, et. al. Bacus 2009

Triple Patterning tests: Penrose Tiling

Penrose tiling overlaid on 800 year old Islamic decoration

- Penrose Tiling was proposed by the British Mathematician Roger Penrose in 1970.

- A pattern is created using a limited number of tiles, yet completely covers a surface with an infinitely repeatable pattern showing no translation symmetry.

- Rhombus Penrose tiling has 5-fold symmetry and has been shown to be 3-colorable.[1]

- It is also believed to underlie some examples of 800 year old Islamic art, such as this example from the Seljuk Mama Hatun Mosque in Tercan, Turkey.

[1] *Rhombic Penrose Tilings can be 3-Colored*, Sibley and Wagon The American Mathematical Monthly Vol 107 No. 3 (Mar 2000) pp251-253

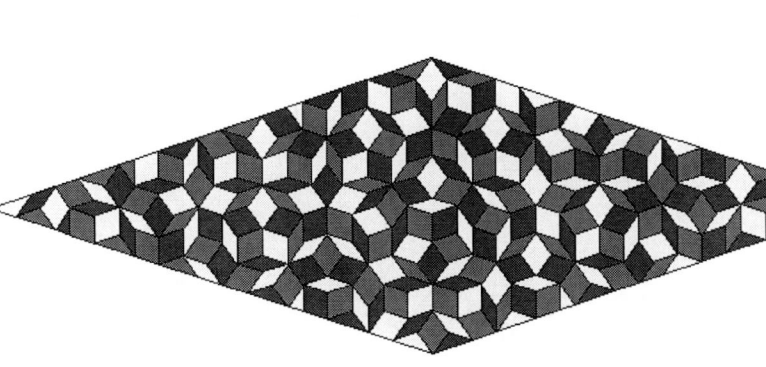

3-colored Rhombus Tiling

C. Cork, et. al. PMJ 2008

©Synopsys 2009

Triple patterning for 16nm contacts

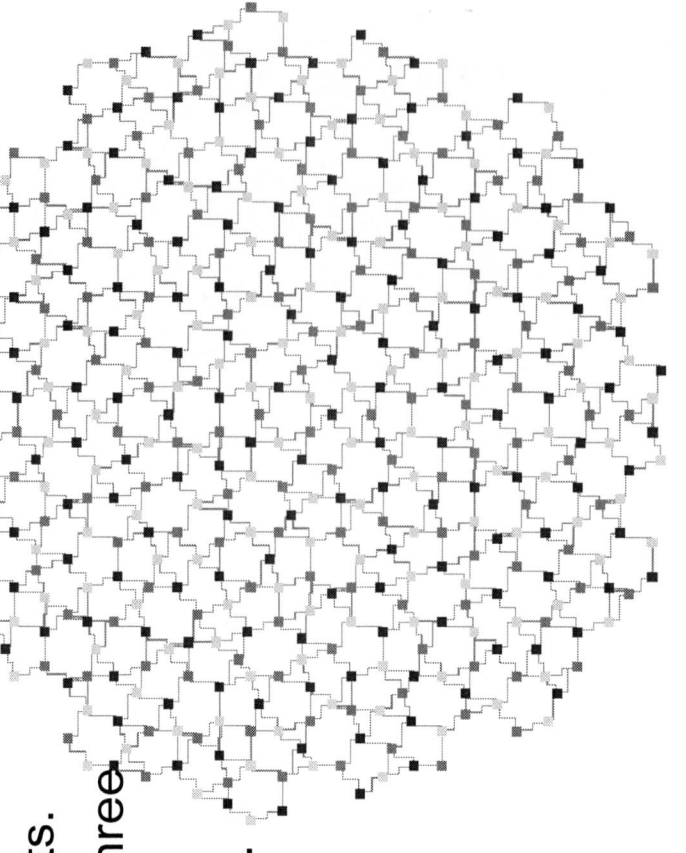

- Allowing odd cycles can give significant pattern density advantages for contacts.
- Certain layout restrictions allow only three mask layers need to be used.
- 3-coloring is an NP-complete problem.
- Limits applicability by network size.

Penrose Rhombic Tiling showing Multiple odd cycle loops. Requires only 3 colors for compliance.

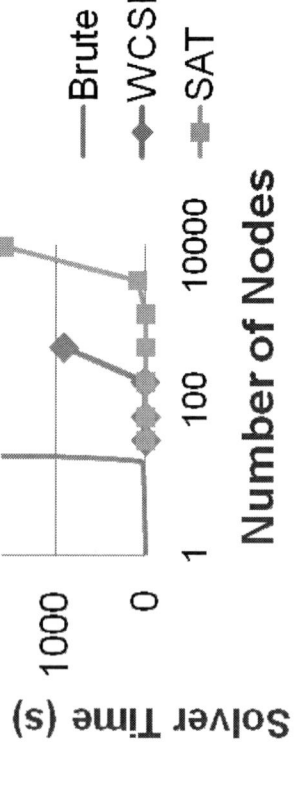

Computation Time vs. Number of Nodes by Algorithm

— Brute Force
◆ WCSP
■ SAT

C. Cork, et. al. PMJ 2008

Predictable Success

©Synopsys 2009

Outline

- Introduction & DPT goals
- Decomposition and coloring
- **Design**
- OPC & Verification
- Wafer results
- Conclusions

DPT for Memory Tests – Metal 1

Memory Array

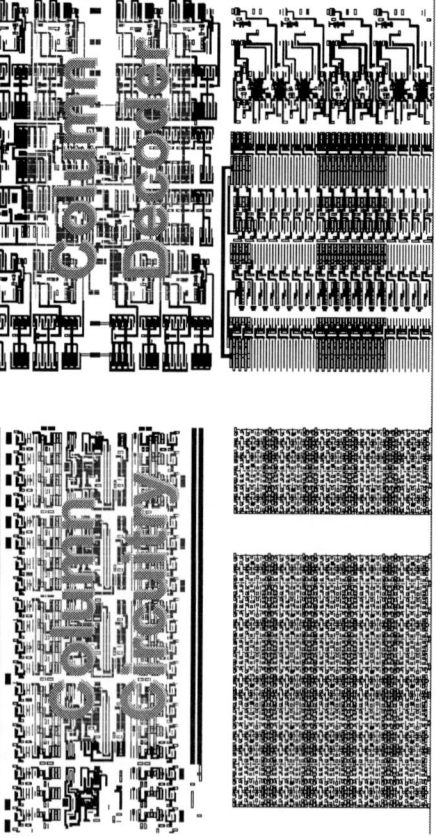

- This layout is currently manufacturable with single patterning.
- Array shows multiple odd-cycles all at about minimum design space.
- Can double patterning be used to shrink circuit area without significant manual redesign effort?

B-S. Seo, et. al. PMJ 2009

©Synopsys 2009

Design Rule Guidance – memory blocks

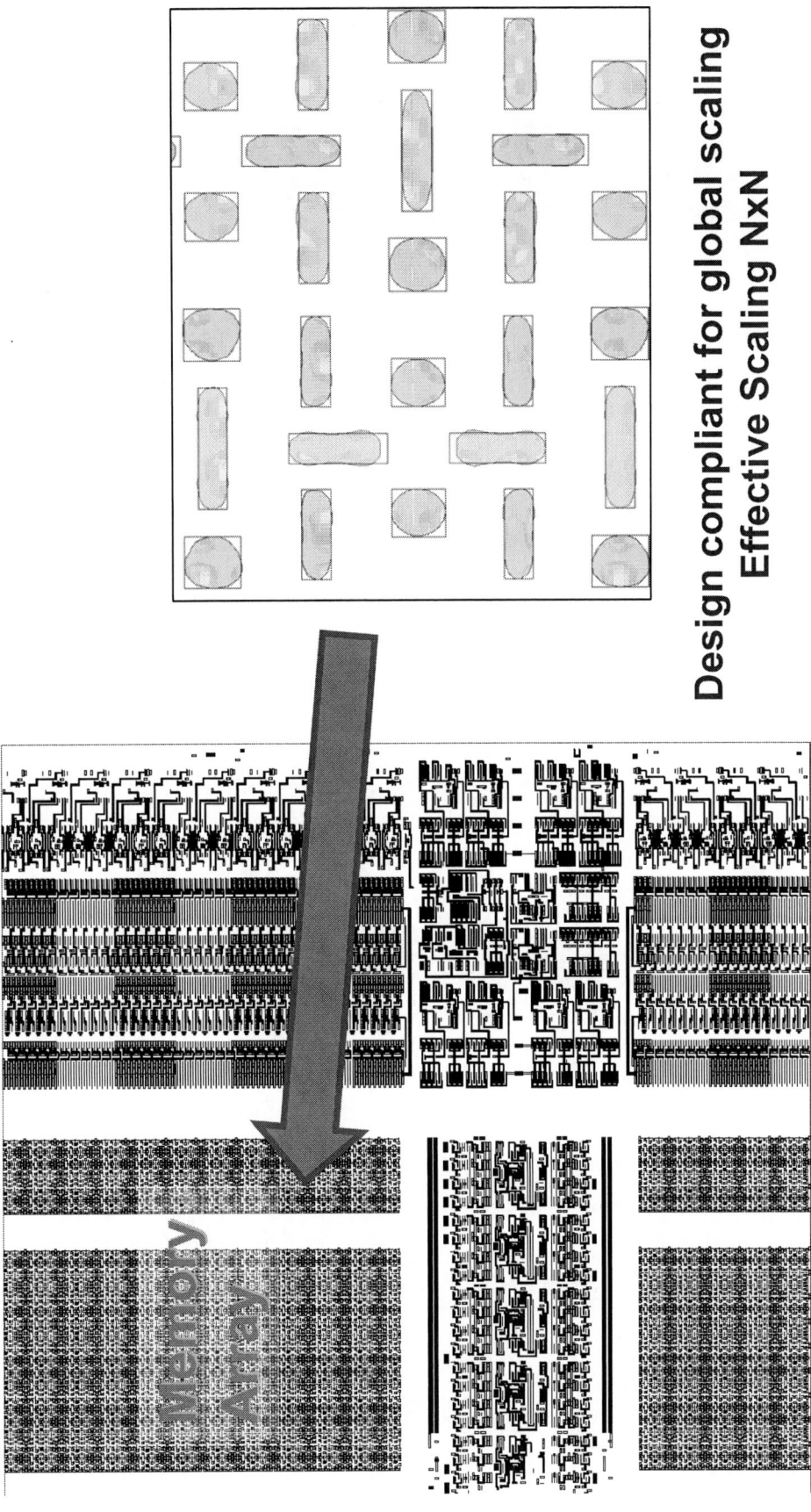

Design compliant for global scaling
Effective Scaling NxN

B-S. Seo, et. al. PMJ 2009

Design Rule Guidance – periphery

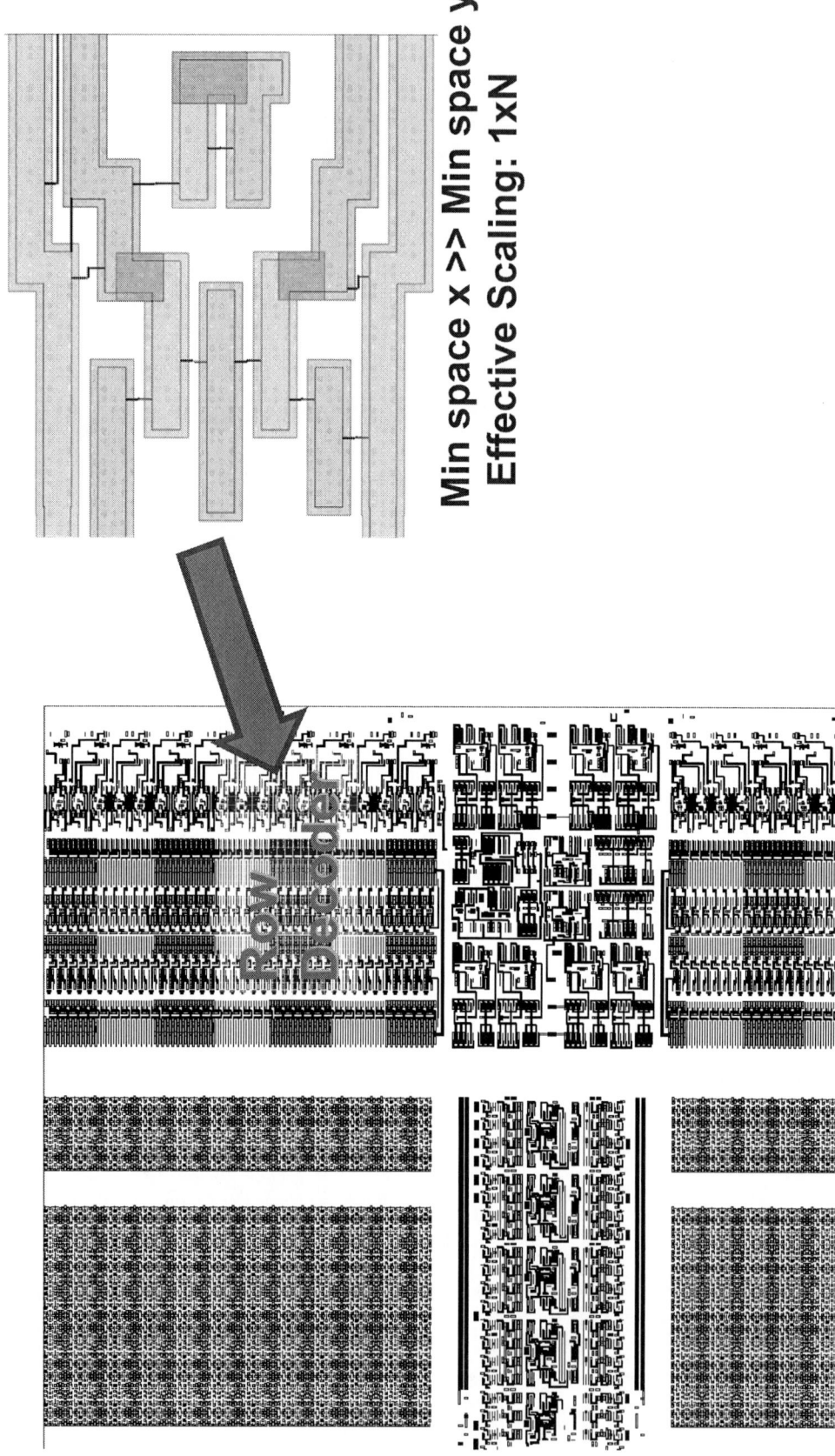

Min space x >> Min space y
Effective Scaling: 1xN

B-S. Seo, et. al. PMJ 2009

Design Rule Guidance – periphery

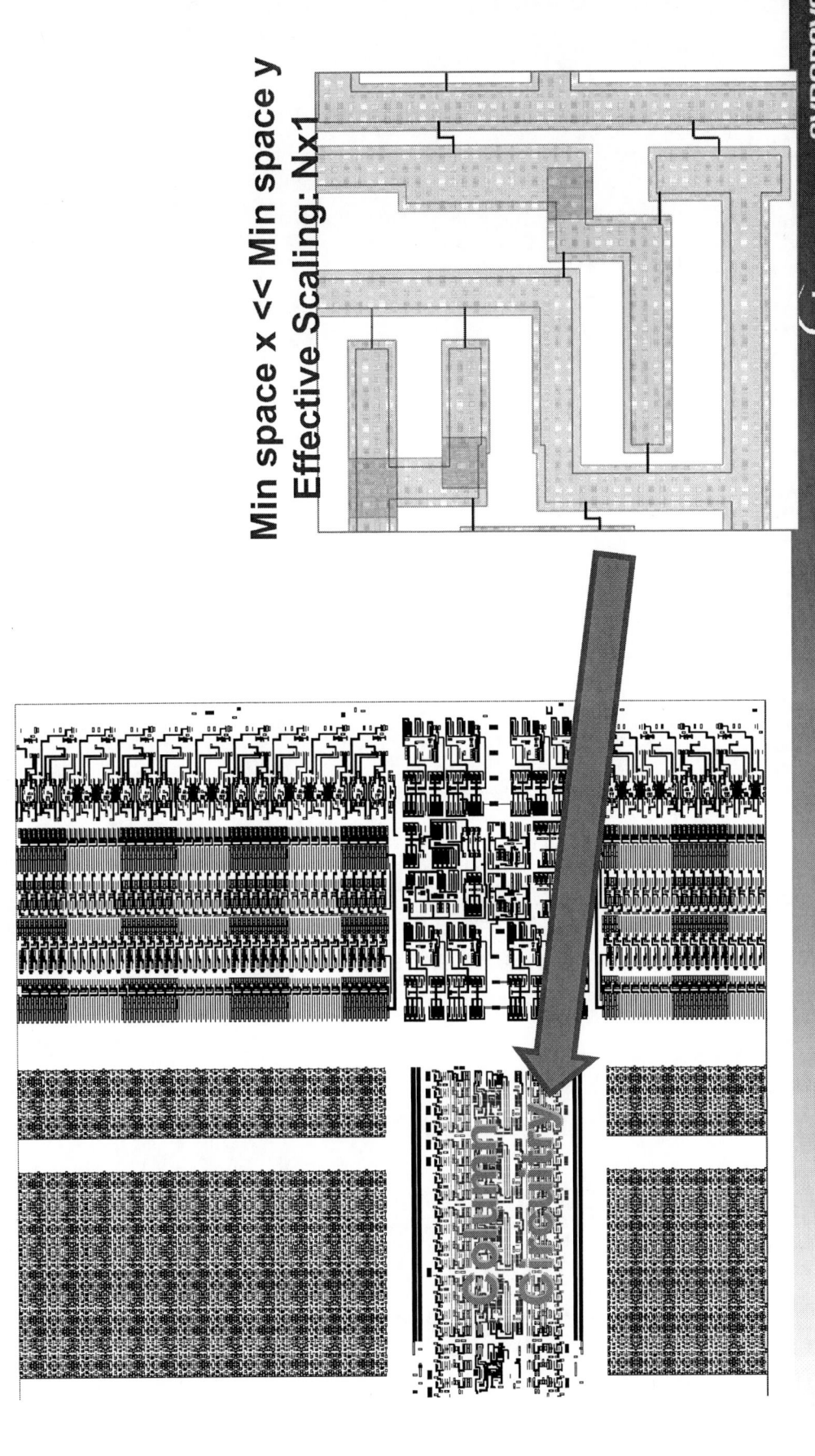

Min space x << Min space y
Effective Scaling: Nx1

column circuitry

B-S. Seo, et. al. PMJ 2009

Design Rule Guidance – periphery

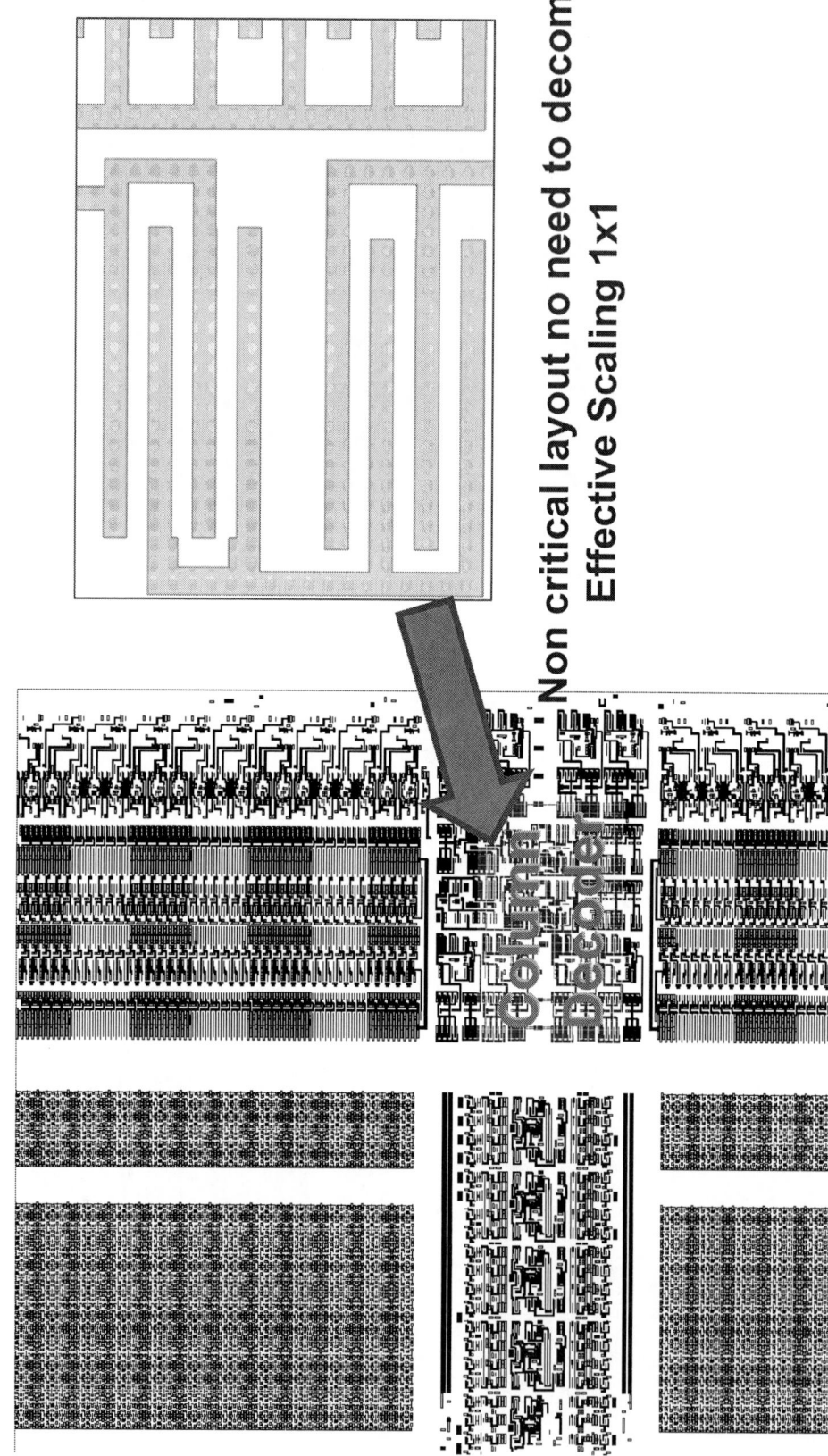

Non critical layout no need to decompose
Effective Scaling 1x1

B-S. Seo, et. al. PMJ 2009

©Synopsys 2009

Short range DPT Compliance errors

Type A: T-Shaped Space

- Can be identified by a vertex having neighbors with small space in both X & Y directions
- Analogous coloring conflicts existed in layouts for alternating phase shift masks

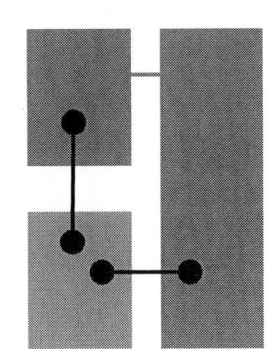

Type B: U-Bend – Short Range

- Feature bends back upon itself
- Here the feature causes color conflict with itself as there is no room to split at the bottom of the U-shape.

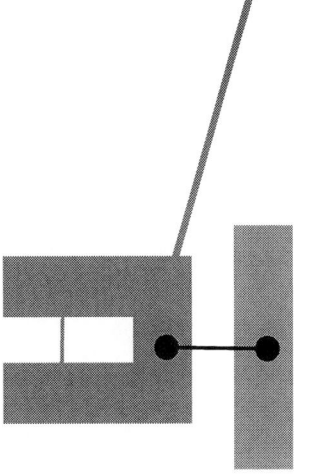

C. Cork, VLSI-TSA 2009

©Synopsys 2009

Longer range DPT compliance issues

Type B: U-Bend – Long Range

- Feature bends back upon itself
- Here the feature causes color conflict because it allows for an odd # of small spaces between features inside the U.

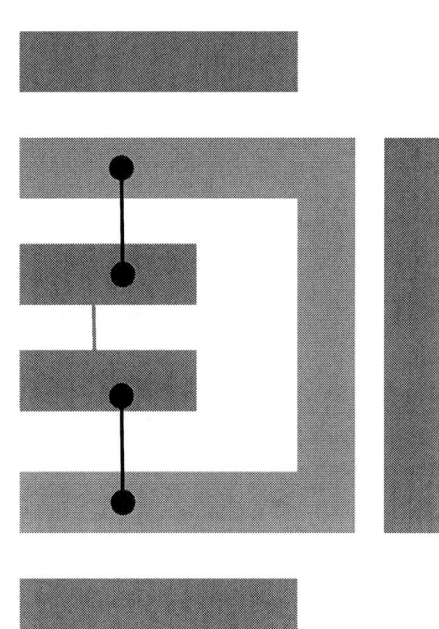

Type C: Jogged Feature

- Jog creates an odd cycle with neighbor.
- Close packing prevents splitting at the jog to resolve the odd cycle.

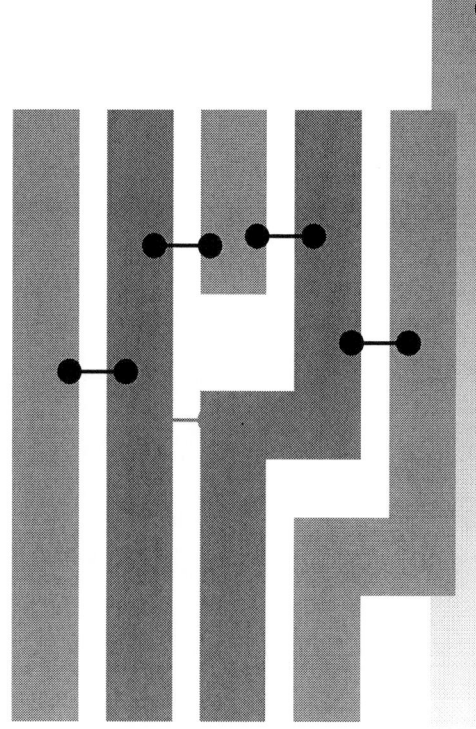

C. Cork, VLSI-TSA 2009

©Synopsys 2009

DRC Deck for DPT errors

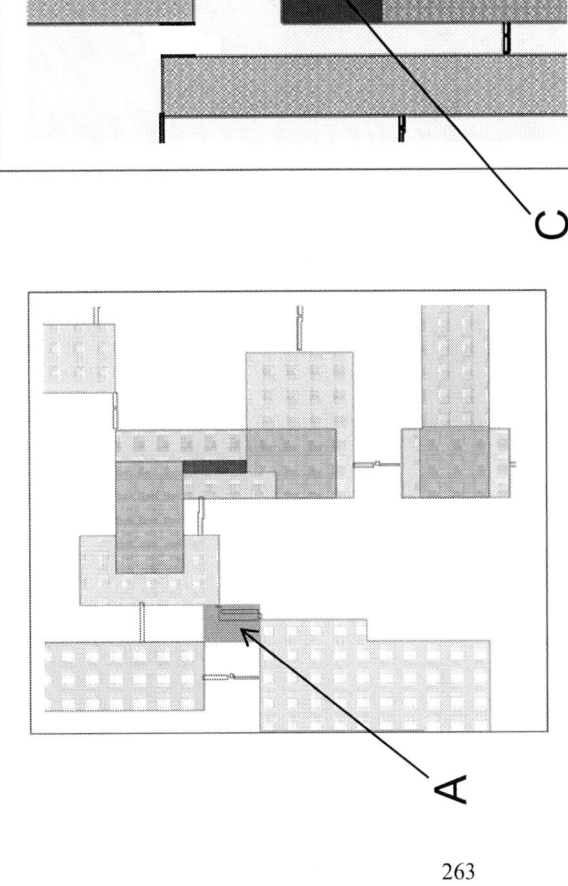

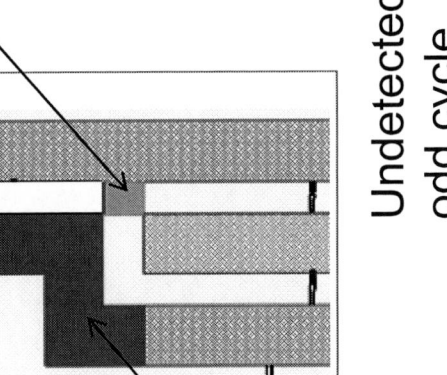

Undetected odd cycle

DRC Deck

- Finds 80-90% of M1 DPT errors.
- It runs 10 – 100 times faster than DPT.
- It can run on design database formats
- Classifies DPT errors by type.

©Synopsys 2009

Outline

- Introduction & DPT goals
- Decomposition and coloring
- Design
- **OPC & Verification**
- Wafer results
- Conclusions

DPT Verification: Overlap Pinches

- Typically from corner rounding and line-end pullback at DPT cut location.

- Worst overlay direction hard to estimate from design intent

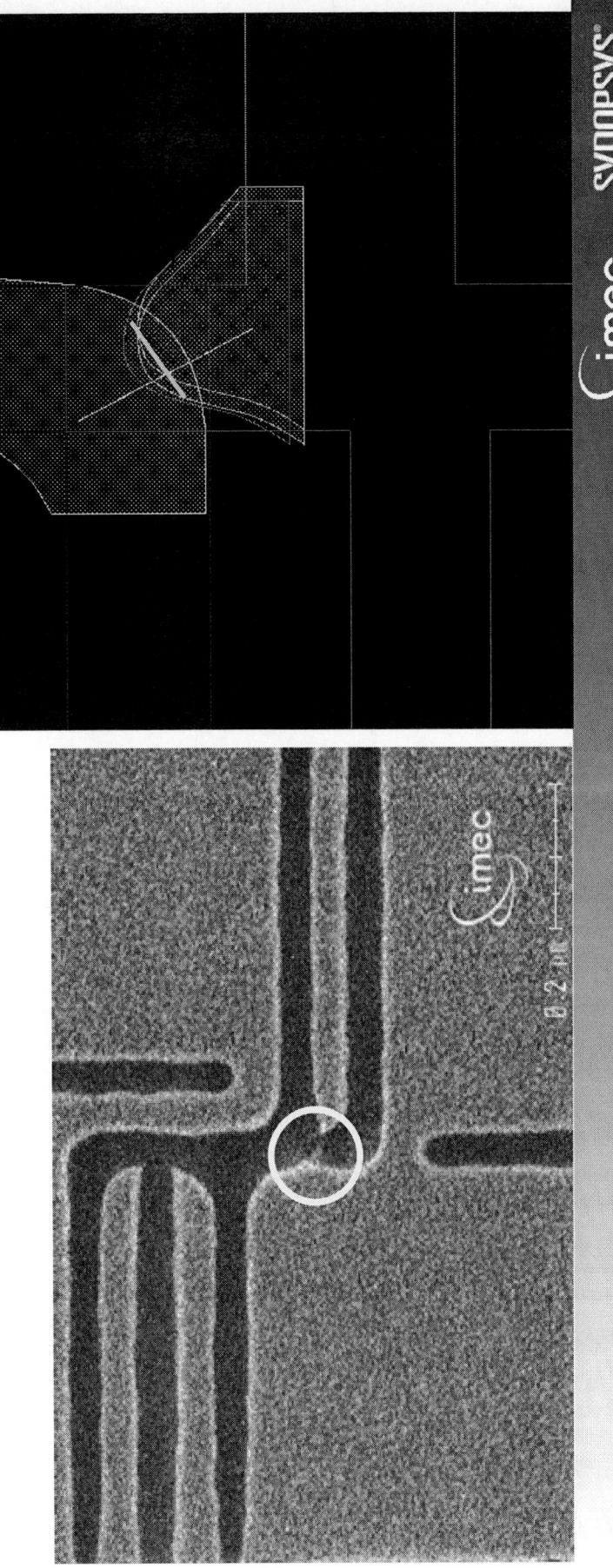

G. Luk-Pat, et. al. Bacus 2008

DPT Pinching: Electrical Impact

- For 40nm nominal line, current density increases 2.8x with CD=30nm and 4.5x with CD=20nm

  <u>Pinching risk at overlap, max overlay error</u>

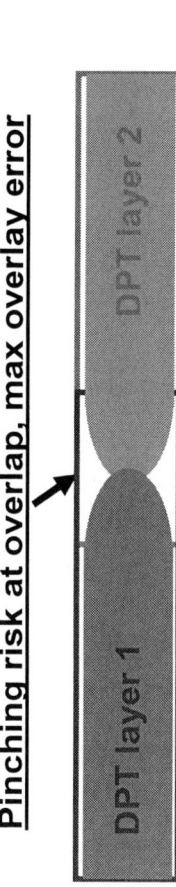

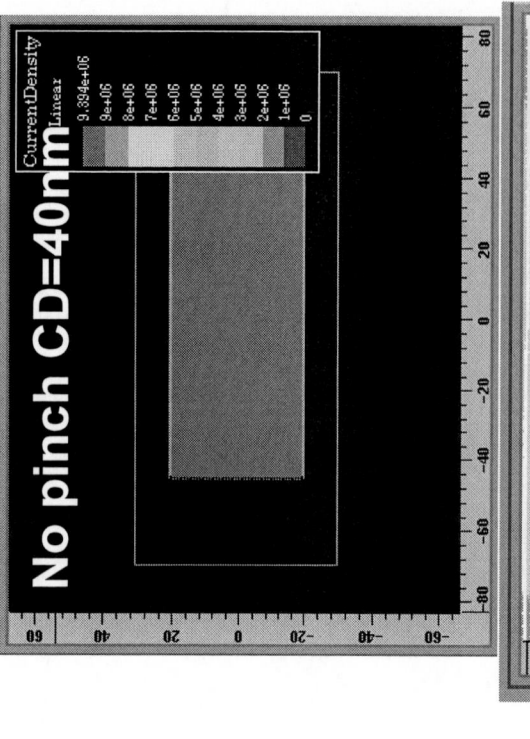

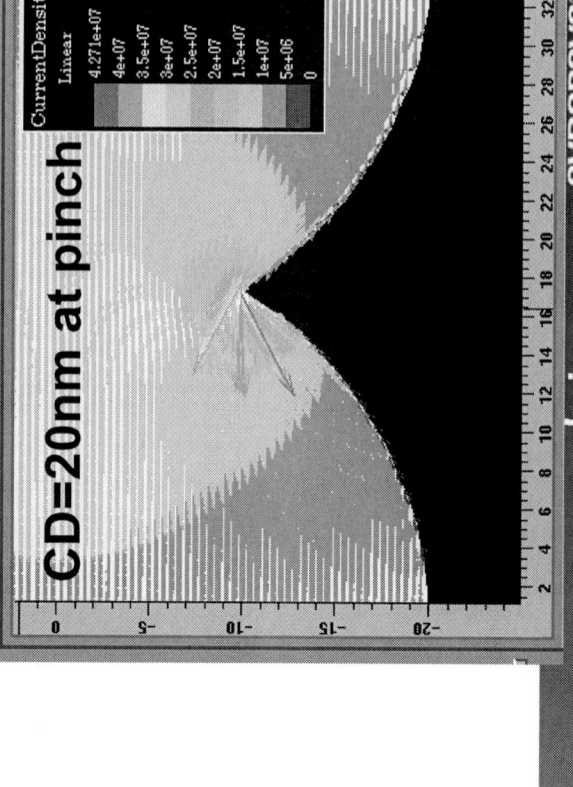

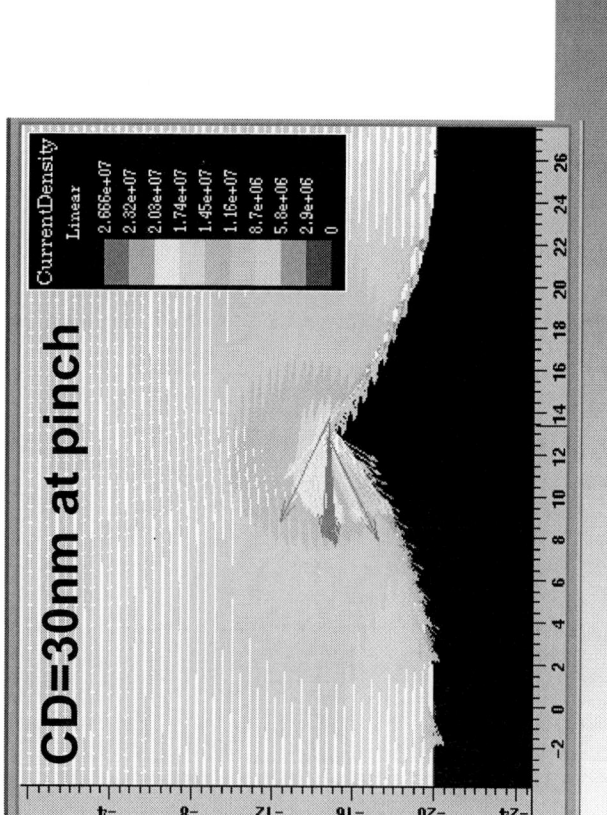

C. Cork, VLSI-TSA 2009

©Synopsys 2009

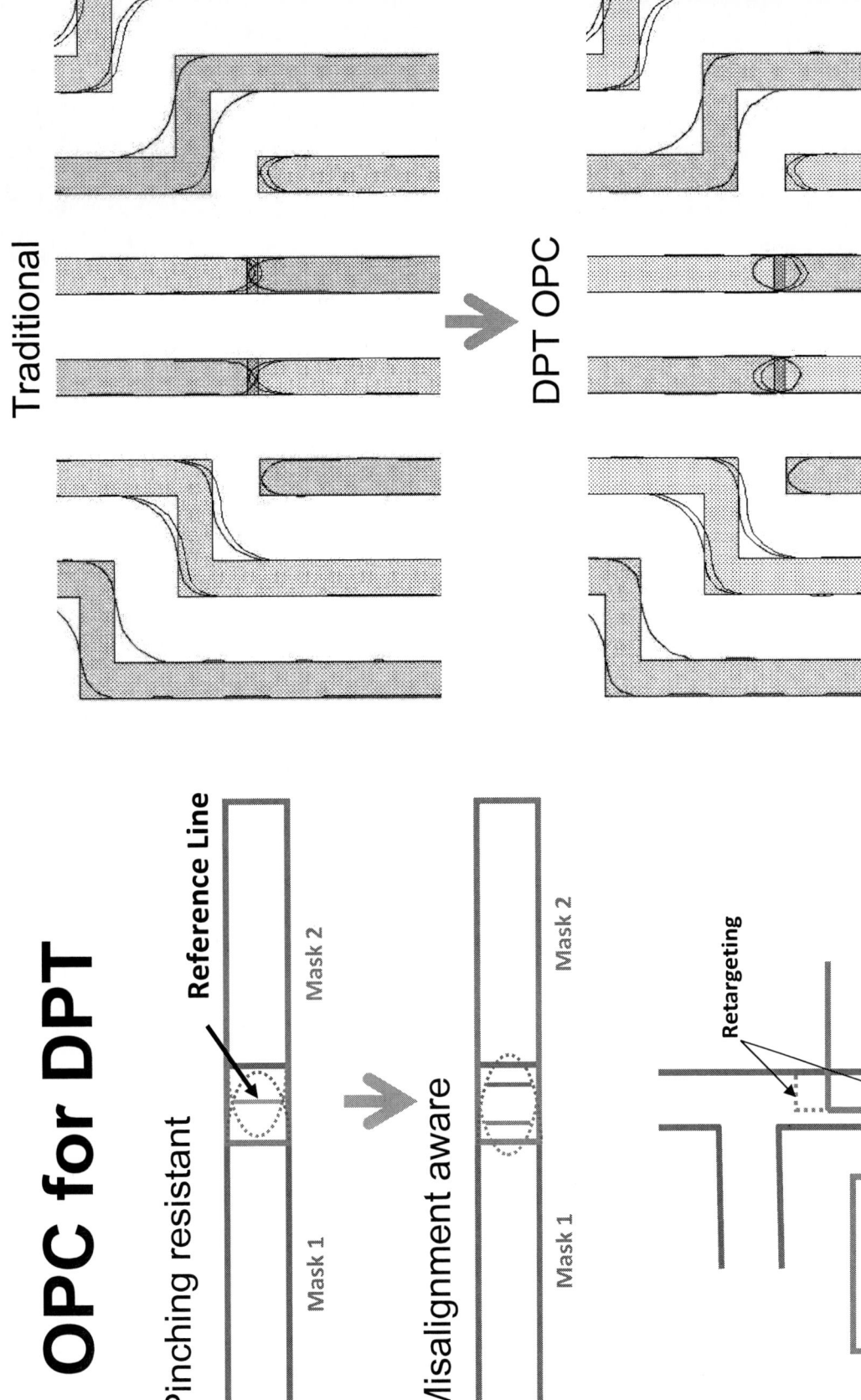

Outline

- Introduction & DPT goals
- Decomposition and coloring
- Design
- OPC & Verification
- **Wafer results**
- Conclusions

Wafer results – 22nm DPT process

Metal1 wafer patterns look good

Wafer Image

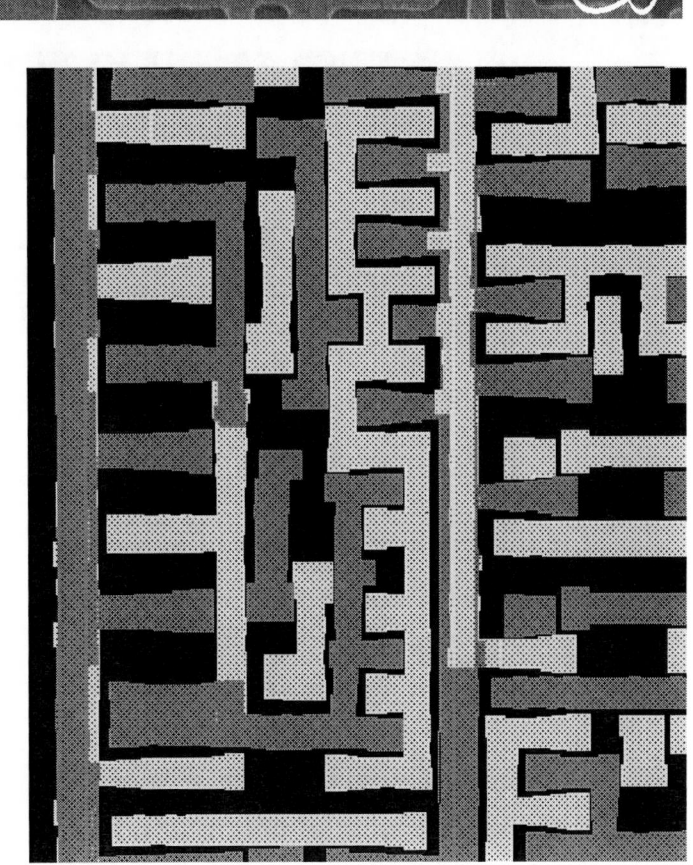

Split Mask Image – Post OPC

Wafer results – DPT splits & overlaps

Small overlaps and tight spaces resolve well

Wafer Image

Split Mask Image – Post OPC

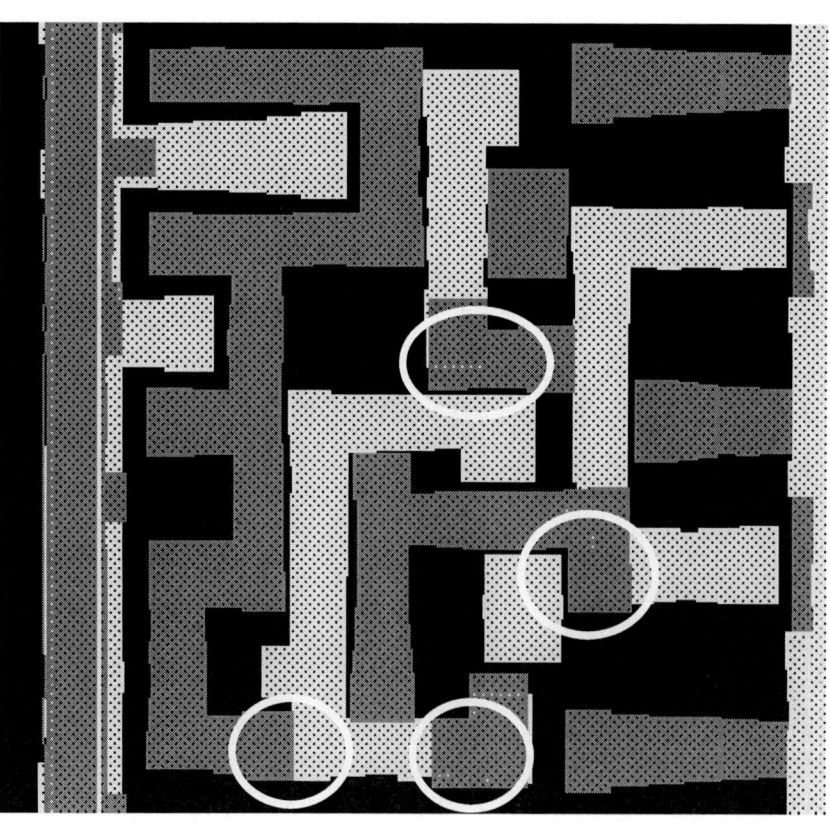

©Synopsys 2009

Wafer results – DPT conflict impact

Wafer results look good except at coloring conflicts

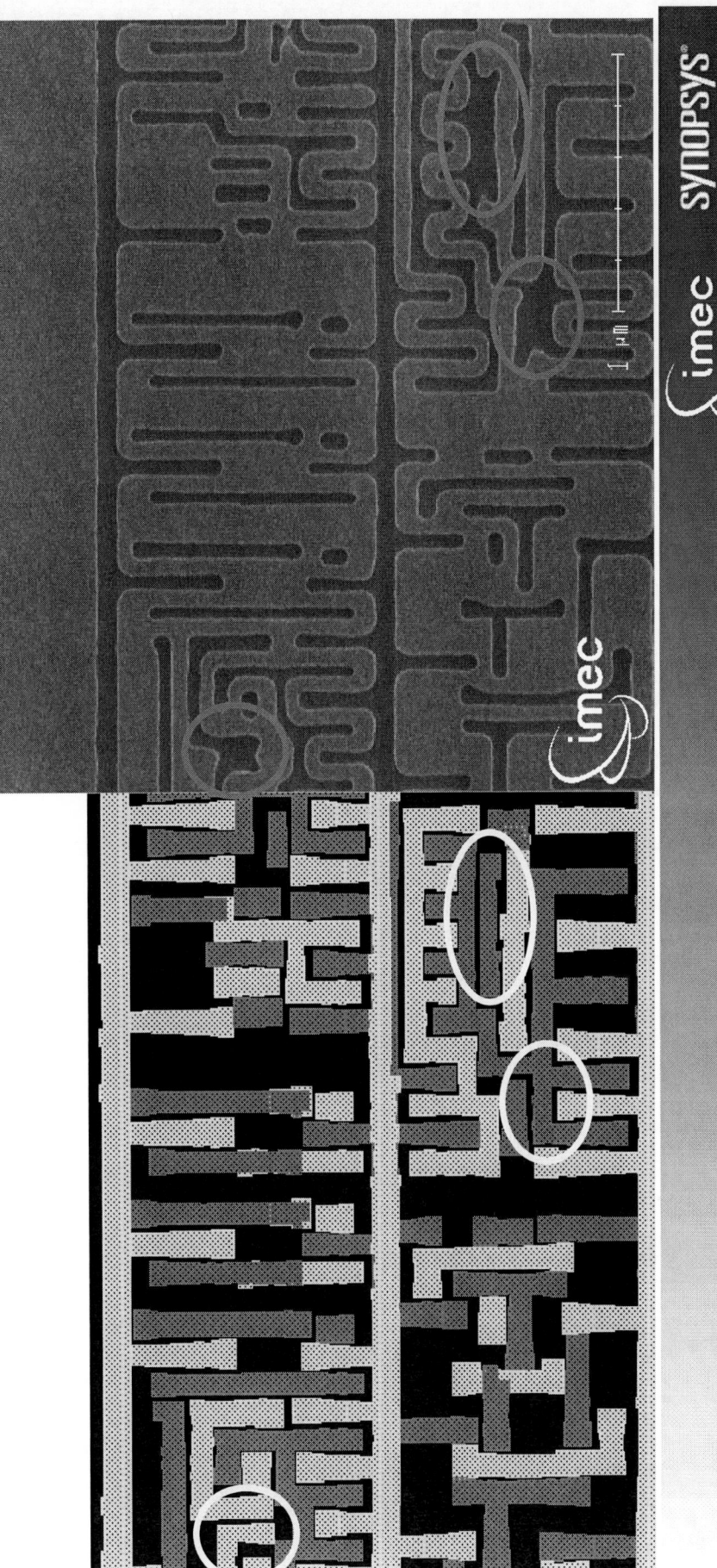

Wafer Image

Split Mask Image – Post OPC

Wafer results – DPT conflict impact

Process margin can be small near min allowed space

Double patterning effects can create sharp failure transitions

17 mj

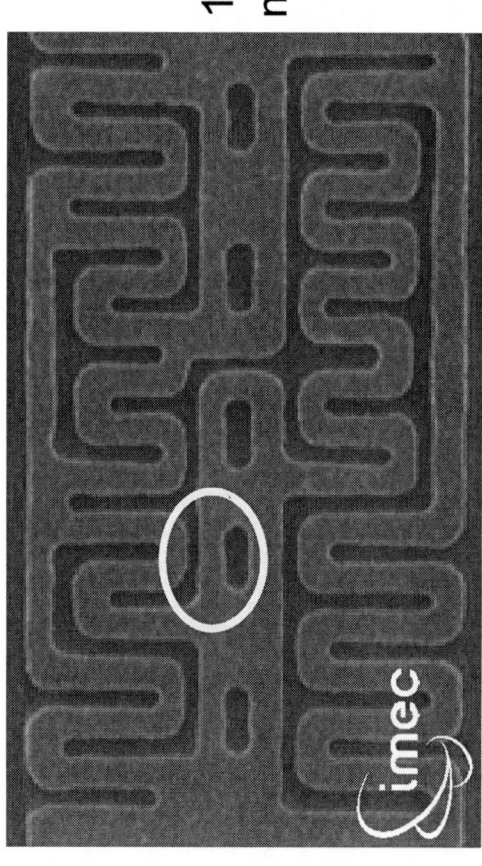

18 mj

©Synopsys 2009

Conclusions

- LELE DPT is a likely RET at 22nm & below
- Intelligent decomposition & coloring benefits product goals
- Accurate, fast full chip coloring is difficult but possible
- Design for DPT compliance essential for optimal density
- DRC for DPT is useful but not a complete solution
- DPT-aware OPC makes aggressive shrinks more manufacturable.
- DPT-aware verification is a requirement.
- Wafer process results show good promise overall

Wafer and Simulation Study Comparing 5 LELE Decomposition Algorithms for Both Compliant and Non-Compliant Layouts.

Germain Fenger,
Pat LaCour,
Alex Tritchkov,
Sergiy Komirenko (MENTOR)
Vincent Wiaux (IMEC)

imec

Mentor Graphics®

Outline

- Fundamentals
  - Separator Definitions
  - Cut Definitions
  - No-cut-zone Definitions
  - Electro-migration Suppression
- Decomposition Comparisons
  - Wafer Investigation
  - Cut Prioritization
  - DP Compliance
- Conclusions

Outline

- Fundamentals
  - Separator Definitions
  - Cut Definitions
  - No-cut-zone Definitions
  - Electro-migration Suppression
- Decomposition Comparisons
  - Wafer Investigation
  - Cut Prioritization
  - DP Compliance
- Conclusions

Fundamentals- Separation Definitions

- **Pitch** is classical definition of resolution limit, but **space** is what fails to image most often

- Rule based space separators are adequate and avoid the increased runtime of model based simulation

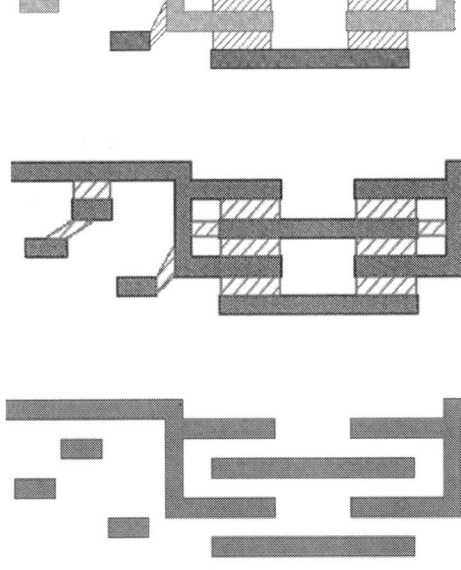
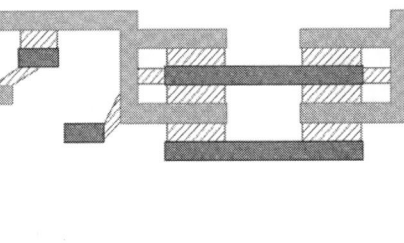

Design Separation Definitions Coloring

Main task for decompositions:
recognize where single patterning cannot resolve and arrange the polygons to two mask

Cut Definition

T-cuts

L-cuts

S-cuts

- Challenge for decomposition occurs when simple sorting of polygons is insufficient. Cuts must be introduced and chosen to enable the decomposition.

- T and L cuts are based on original polygon vertices while S cuts can be introduced anywhere.

No-cut-zone definition

- Process limits are approximated by a noCutZone layer (shown as black lines) where no cuts are allowed.

- Approximation can be based on rules or model simulations

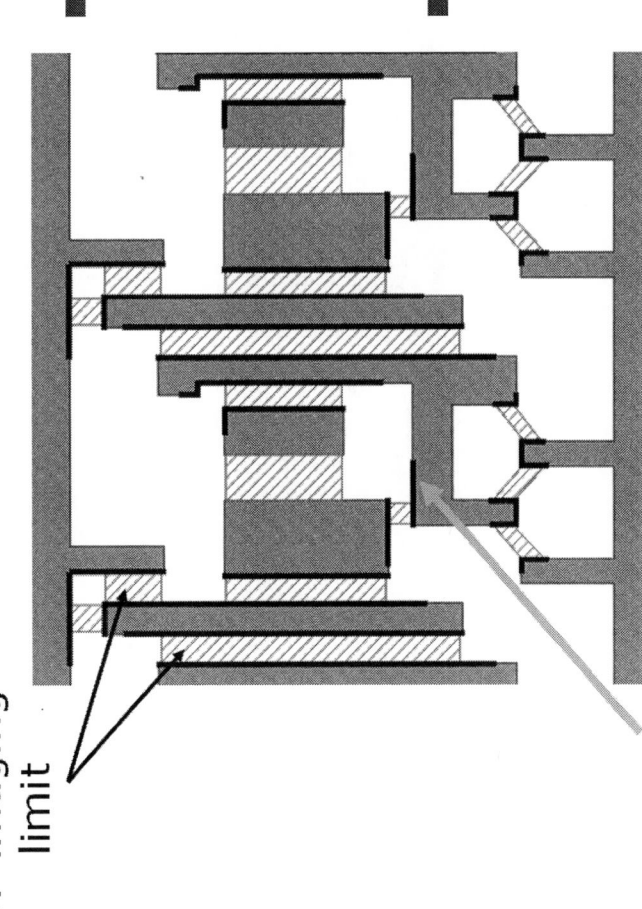

SP imaging limit

noCutZone

Mitigating Electro-migration through Stitching

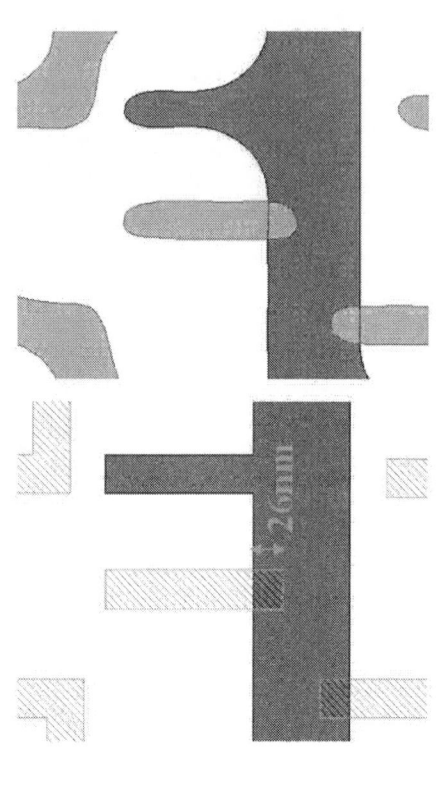

Typical Decomposition Stitching

- DP stitching could potentially introduce electro-migration failures by having sharp corners caused by stitching.

Mitigating Electro-migration through Stitching

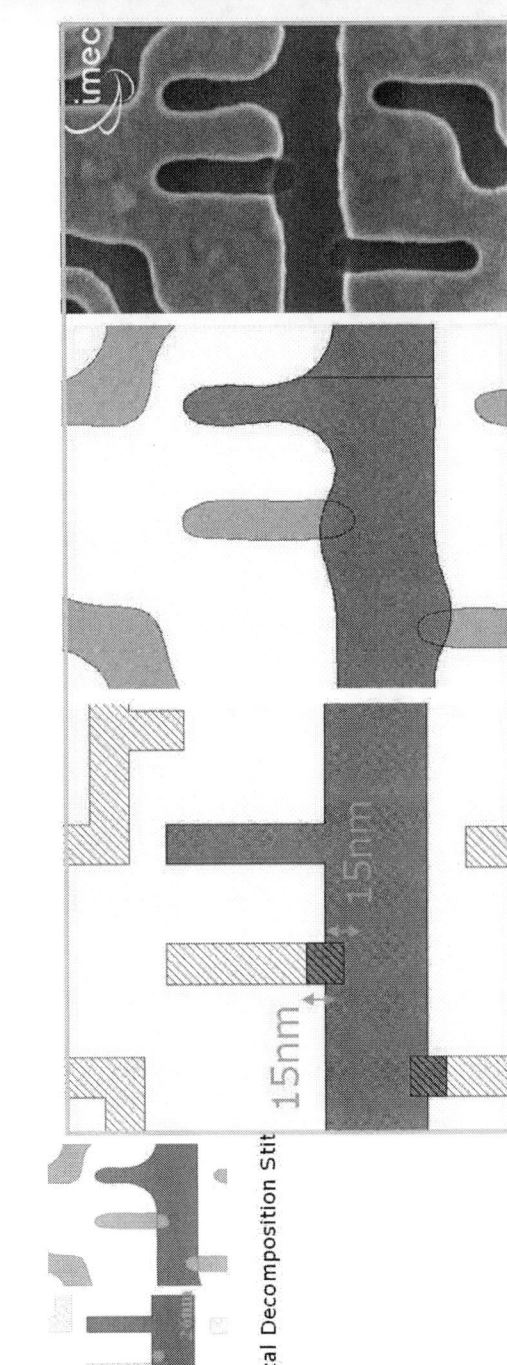

Typical Decomposition Stit...

15nm Stitching tolerated violation area

- DP stitching could potentially introduce electro-migration failures by having sharp corners caused by stitching.
- We have implemented an extended stitching, which rounds the sharp edges and is believed to mitigate electro-migration.

Mitigating Electro-migration through Stitching

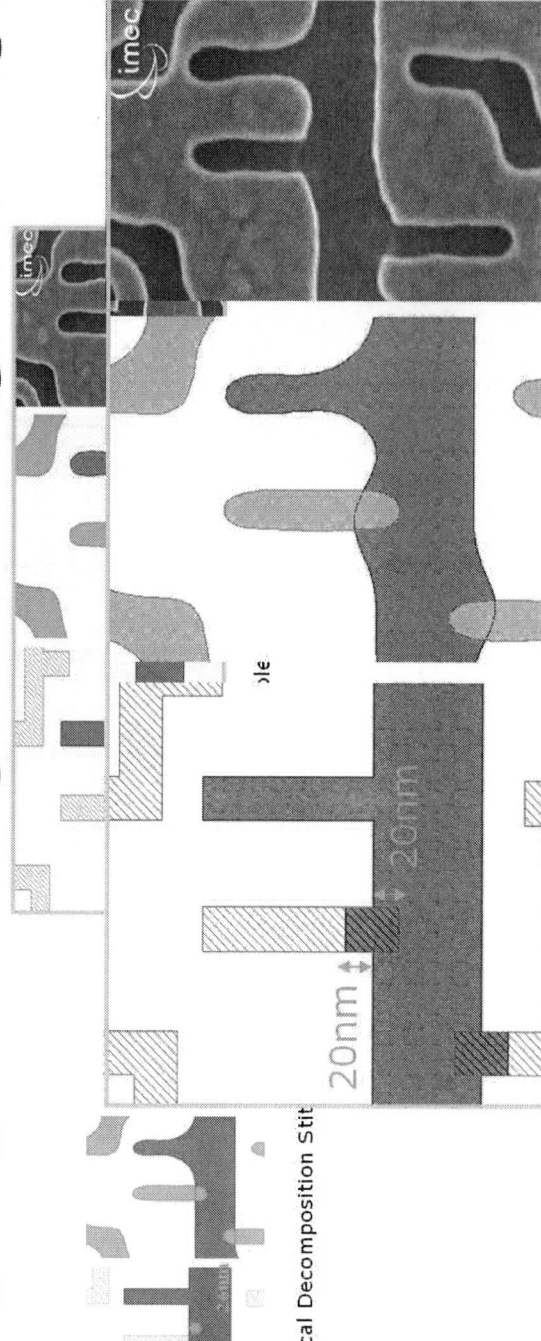

Typical Decomposition Stit...

20nm Stitching tolerated violation area

- DP stitching could potentially introduce electro-migration failures by having sharp corners caused by stitching.
- We have implemented an extended stitching, which rounds the sharp edges and is believed to mitigate electro-migration.

Mitigating Electro-migration through Stitching

Typical Decomposition Stitching

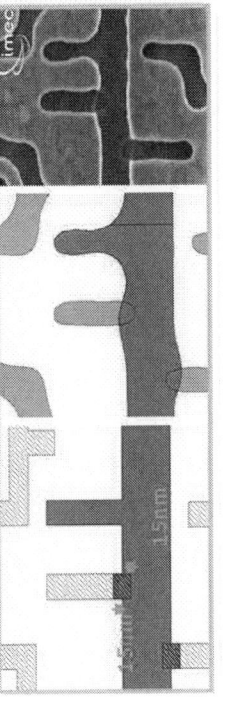

15nm Stitching tolerated violation area

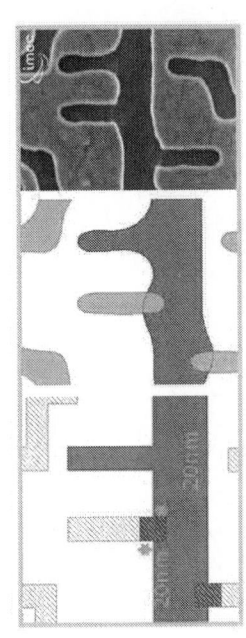

20nm Stitching tolerated violation area

- DP stitching could potentially introduce electro-migration failures by having sharp corners caused by stitching.
- We have implemented an extended stitching, which rounds the sharp edges and is believed to mitigate electro-migration.

Outline

- Fundamentals
  - Separator Definitions
  - Cut Definitions
  - No-cut-zone Definitions
  - Electro-migration Suppression
- Decomposition Comparisons
  - Wafer Investigation
  - Cut Prioritization
  - DP Compliance
- Conclusions

Mentor Graphics imec

Silicon Exploration

- **Motivation:** Evaluate cut selections affect on decomposability when full process knowledge is not simulated as part of decomposition.

- Five different cut prioritizations were investigated to demine the best method for DP decomposition.

- Evaluated compliant and non-compliant designs

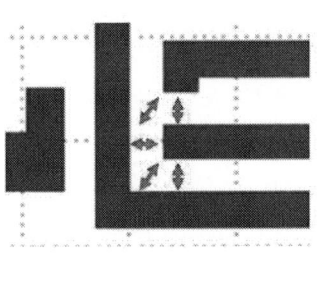

Example of non-compliant layout

Cut Prioritization

- The decomposition application allows user preferences to be set for cut selection.
- The impact of this is to influence cut selection when multiple cuts can solve the required decomposition.
  - E.g. TLS implies that T cut is top preference, followed by L cuts, and then S cuts.

- **The images confirm that forced cuts with non-adaptive stitching causes the L cut to fail before the S cut.**

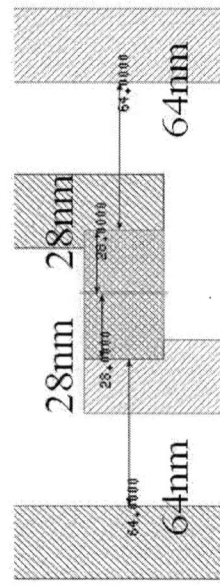

Forced S cut provides best overlap

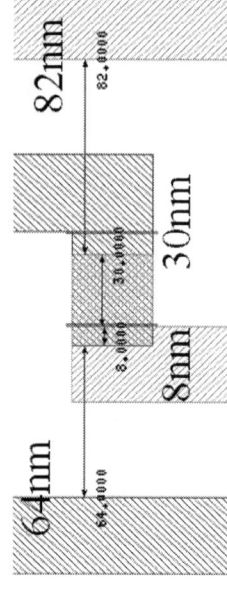

Forced L cut is limited in overlap

Cut Prioritization

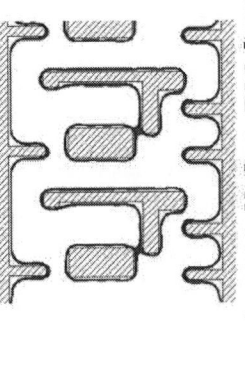

Forced L-cut

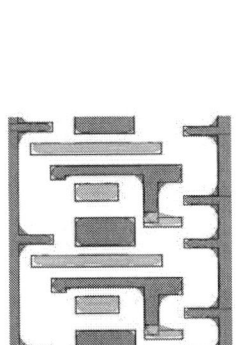

L-Cut silicon

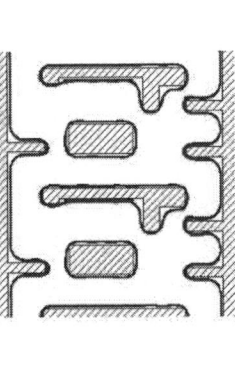

L-cut litho PV bands showing bridge in PW

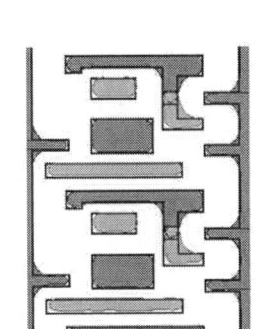

S-cut silicon

Forced S-cut

S-cut litho PV bands

- The results of the S-cut show a clear imaging benefit here, since the stitching reintroduced the original bridge risk when this L-cut was chosen.

Cut Prioritization 2

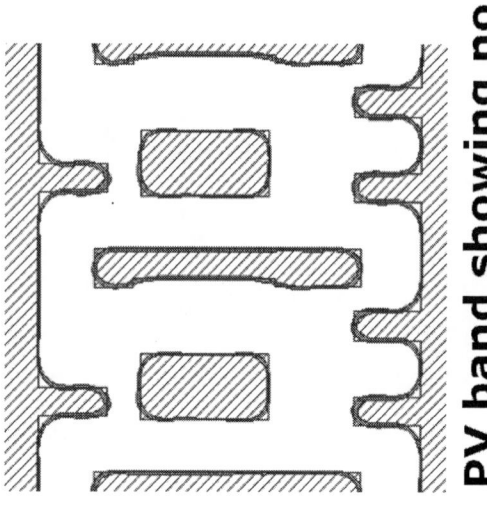

Globally optimized cuts

PV band showing no bridging

- By not forcing a cut, but allowing the cuts to be globally optimized, this L cut provides the most robust solution for this layout

Cut Priorities

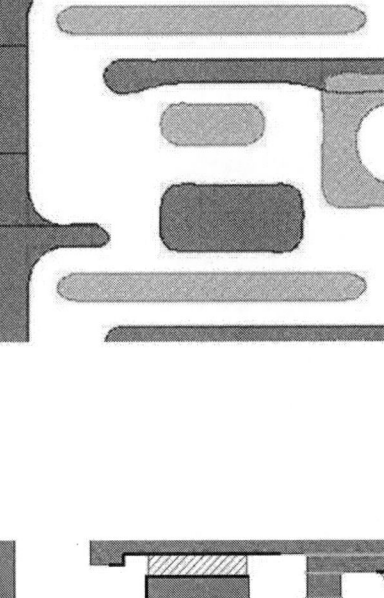

Target, separators and cut selections   **Etch contours**

- Calibre nmDP-L, using a Euclidean metric can decide on the most effective cuts without any user intervention.
  - I.E. Allowing for global optimization rather than forcing a cut prioritization is the best strategy and our algorithms are capable of delivering this automatically

DP Compliance

- The image shows a case where a forced decomposition on a non-compliant layout (includes cycle conflicts) leads to problematic image error. The shapes near the yellow arrow are too close but not decomposed. In this case, the failure is revealed in LE pullback rather than a bridge.

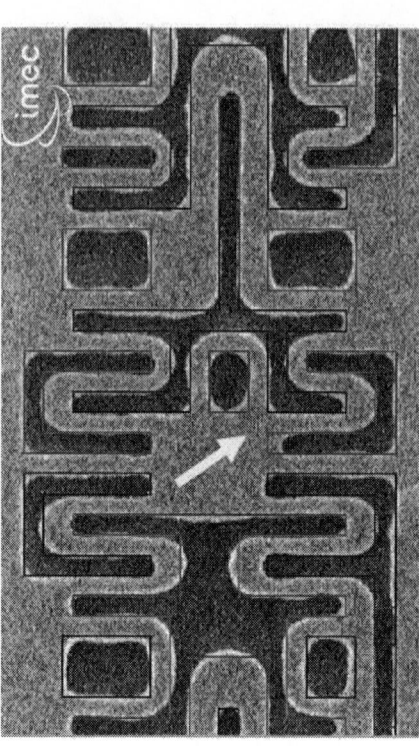

The solution to this problem is to begin the discussion with design about enabling double patterning compliance.

Outline

- Fundamentals
  - Separator Definitions
  - Cut Definitions
  - No-cut-zone Definitions
  - Electro-migration Suppression
- Decomposition Comparisons
  - Wafer Investigation
  - Cut Prioritization
  - DP Compliance
- Conclusions

Conclusions

- We've shown that cut selection does affect decomposition success.

- No single cut type is successful on all layouts and contexts.

- The decomposition solution needs to have automated cut selection which optimizes for the post-decomposition stitching and OPC, preferably without incurring the runtime penalty of additional simulation.

Thank you for your attention

A New Approach Towards Pitch Division for 193 nm Immersion Lithography

Xinyu Gu, Christopher Bates, Younjin Cho, Elizabeth Costner

Tomoki Nagai, Toshiyuki Ogata, C. Grant Willson

Department of Chemical Engineering, The University of Texas at Austin, Austin, TX

Arun K. Sundaresan, Nicholas J. Turro

Department of Chemistry, Columbia University, New York, NY

Robert Bristol

Intel Corporation, Portland, OR

Paul Zimmerman

Intel Assignee to SEMATECH, Austin, TX

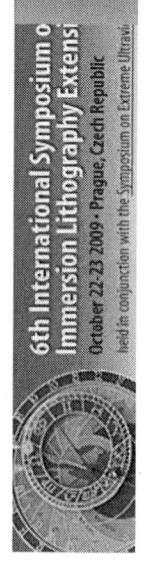

6th International Symposium o
Immersion Lithography Extensi
October 22-23 2009 • Prague, Czech Republic
held in conjunction with the Symposium on Extreme Ultravi

Outline

- Quick Review

- Concept

- Simulation of feasible system

- Experimental verification

- Summary

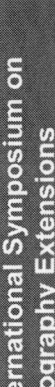

Oct 22nd 2009, 6th International Symposium on Immersion Lithography Extensions

Quick review

- **Double patterning technique**

- **Double exposure technique**

- **Pitch division technique with single exposure**

  Self-Aligned Double Patterning (SADP)

  Dual Tone Development (DTD)

Oct 22nd 2009, 6th International Symposium on
Immersion Lithography Extensions

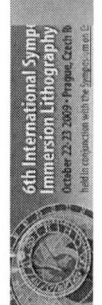

6th International Sympo
Immersion Lithography
October 22-23 2009 • Prague, Czech R
held in conjunction with the Symposium on E

Concept

Mask

Aerial Image

Resist Profile

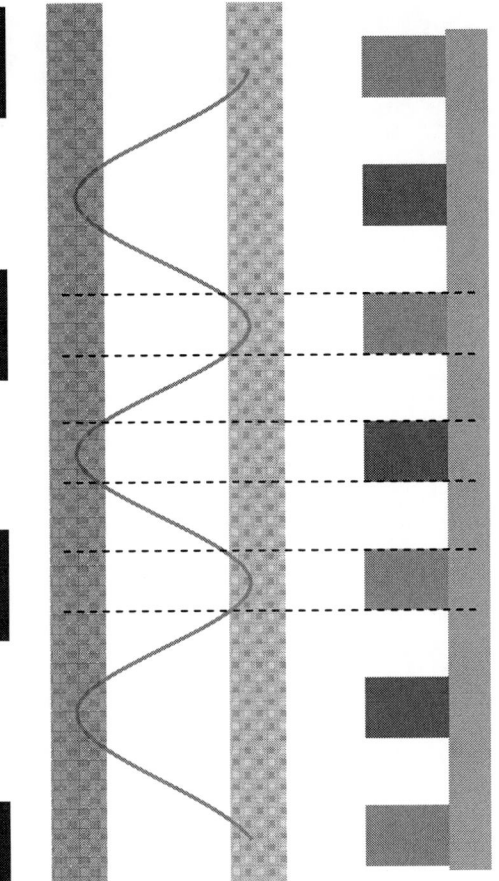

[Acid]

Dose

bake   develop

Film thickness

Dose

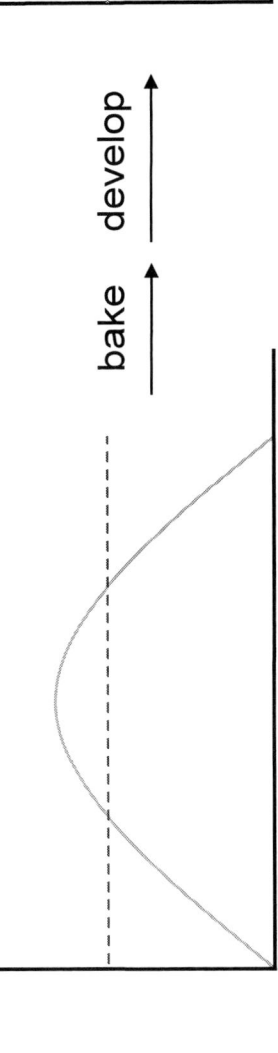

Oct 22nd 2009, 6th International Symposium on Immersion Lithography Extensions

Photoactive system and mechanism

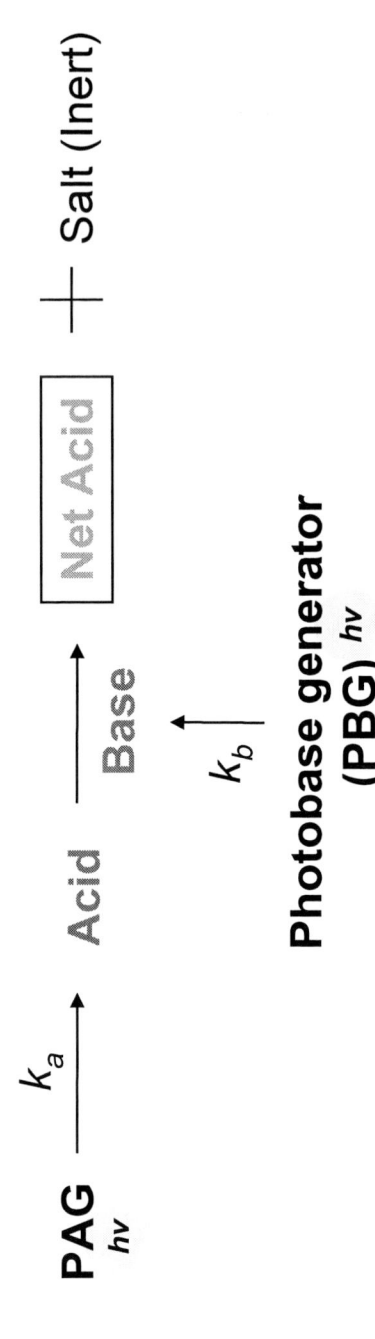

$$PAG \xrightarrow[hv]{k_a} Acid \xrightarrow{} \boxed{Net\ Acid}\ +\ Salt\ (Inert)$$

$$Base \xleftarrow{k_b}$$

Photobase generator (PBG) hv

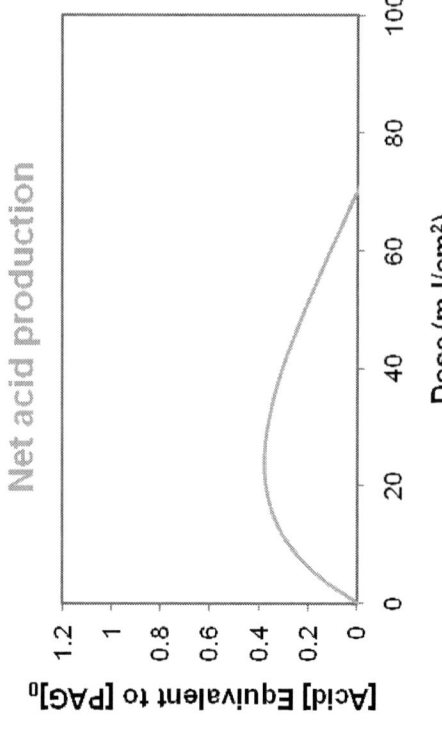

Net acid production

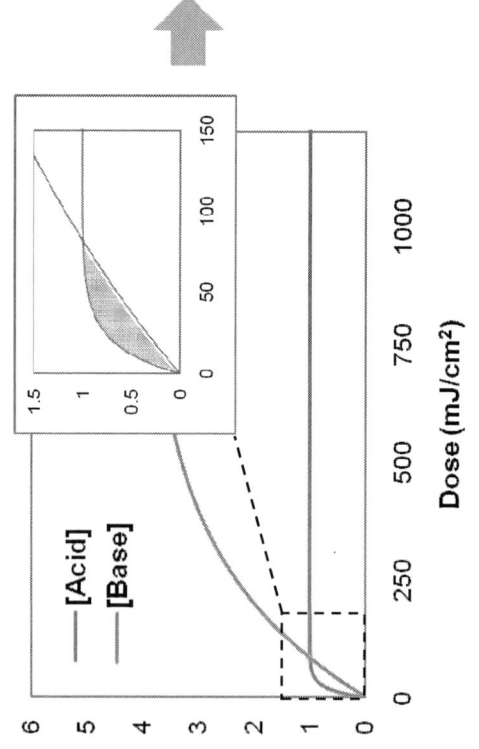

Oct 22nd 2009, 6th International Symposium on Immersion Lithography Extensions

Dependence on formulation and kinetic ratio

$$[Acid] = \boxed{} \cdot \left(1 - e^{-\boxed{} \cdot Dose}\right)$$

$$[Base] = \boxed{} \cdot \left(1 - e^{-\boxed{} \cdot Dose}\right)$$

$[PBG]_0 > [PAG]_0$

$k_b < k_a$

Net acid production

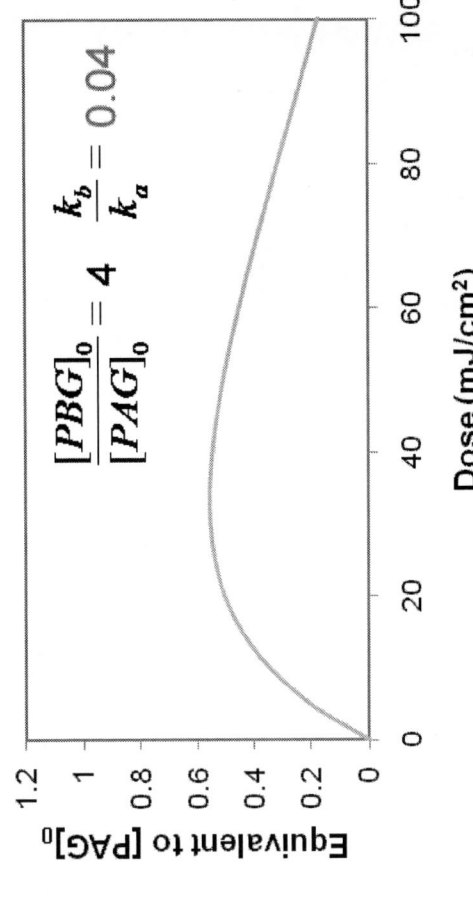

$$\frac{[PBG]_0}{[PAG]_0} = 4 \qquad \frac{k_b}{k_a} = 0.06$$

$$\frac{[PBG]_0}{[PAG]_0} = 4 \qquad \frac{k_b}{k_a} = 0.04$$

Net acid production

Oct 22nd 2009, 6th International Symposium on
Immersion Lithography Extensions

Simulation of pitch division performance

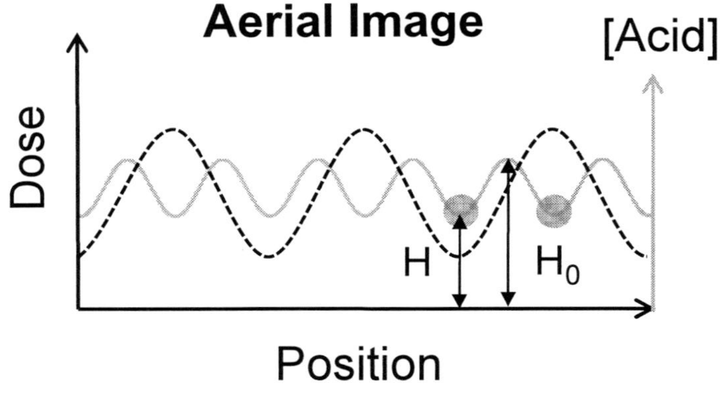

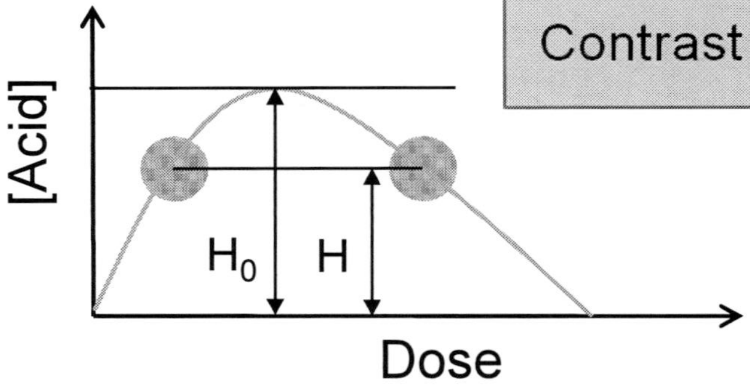

Aerial Image [Acid]

Dose

H_0

H

Position

$$\text{Contrast} = \frac{H_0 - H}{H_0 + H}$$

[Acid]

H_0

H

Dose

Contrast

Net [Acid] = 0

k_b/k_a

$\dfrac{[PBG]_0}{[PAG]_0}$

0.2 0.15 0.1 0.05 0

0.02 0.04 0.06 0.08 0.1 0.12 0.14 0.16 0.18

Oct 22nd 2009, 6th International Symposium on
Immersion Lithography Extensions

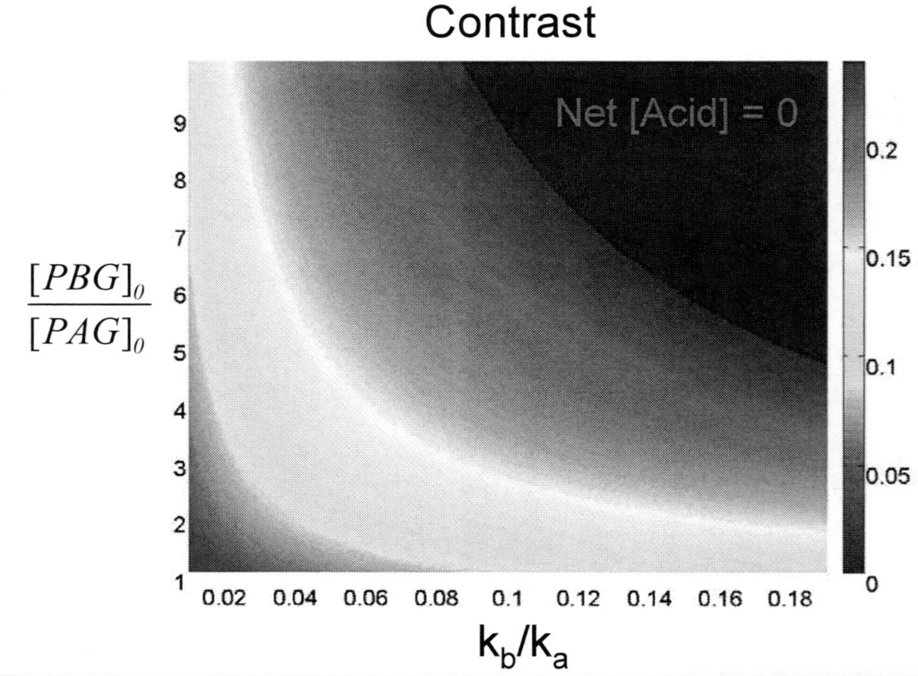

Photolysis of PBG

Houlihan et al, *Macromolecules*, Vol. 21, No. 7, 1998
Cameron et al, *J. Am. Chem. Soc.*, Vol. 113, No. 11, 1991

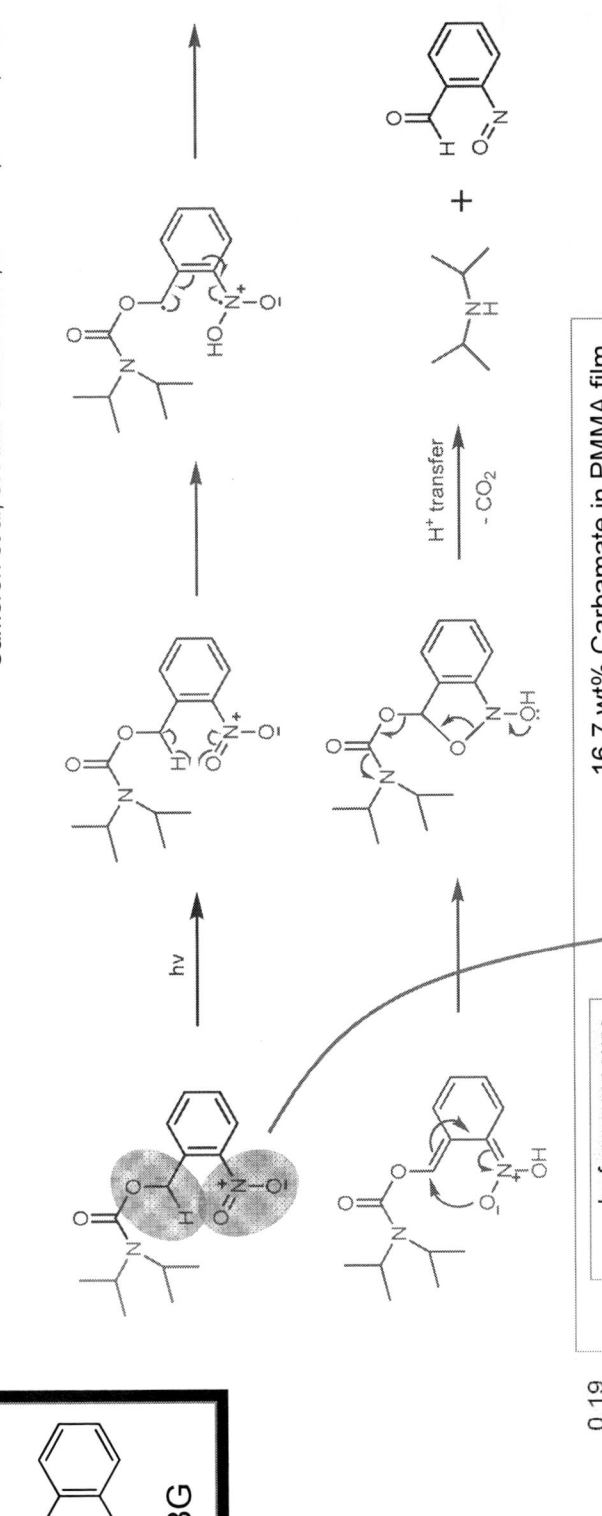

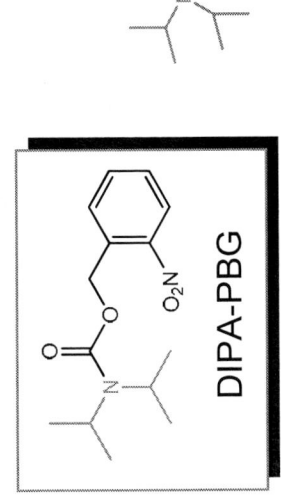

DIPA-PBG

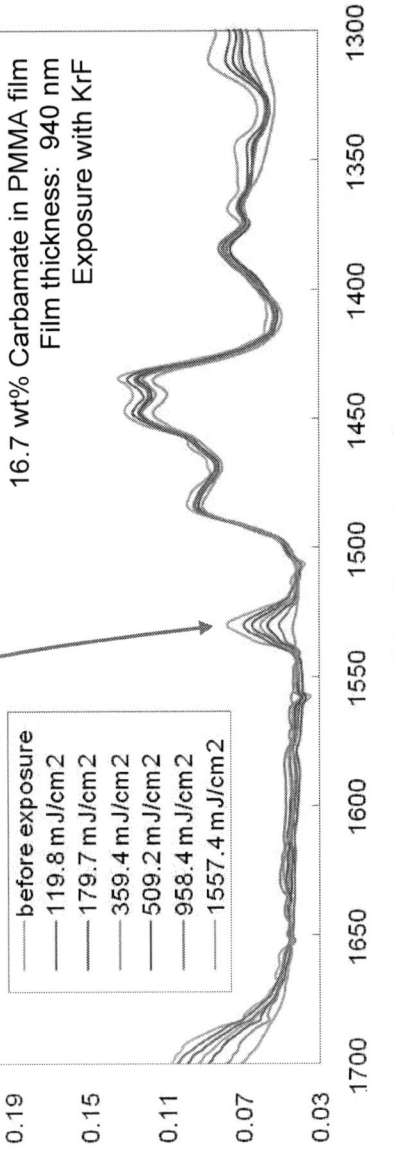

Oct 22nd 2009, 6th International Symposium on
Immersion Lithography Extensions

Rate constant for base generation

NaCl Disk

$$1 - e^{-k_b \cdot Dose} \approx k_b \cdot Dose$$
$$(Dose \to 0)$$

Quartz Disk

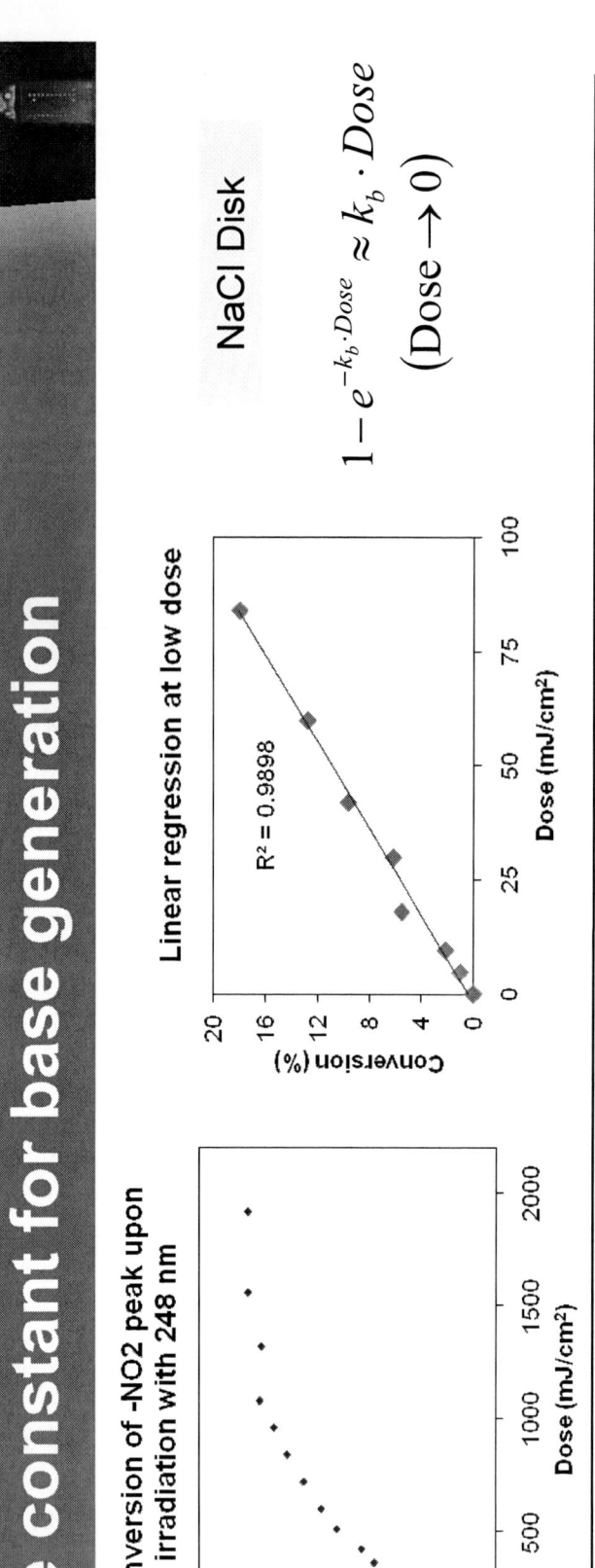

Conversion of -NO2 peak upon irradiation with 248 nm

Linear regression at low dose

$R^2 = 0.9898$

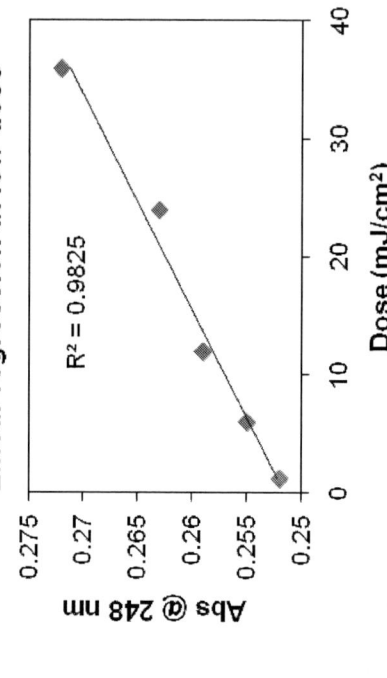

Linear regression at low dose

$R^2 = 0.9825$

Photodarkening @ 248 nm

Considering the photodarkening: $k_b = 0.00243$ (cm²/mJ)

Oct 22nd 2009, 6th International Symposium on Immersion Lithography Extensions

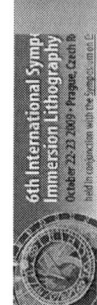

Rate constant for acid generation

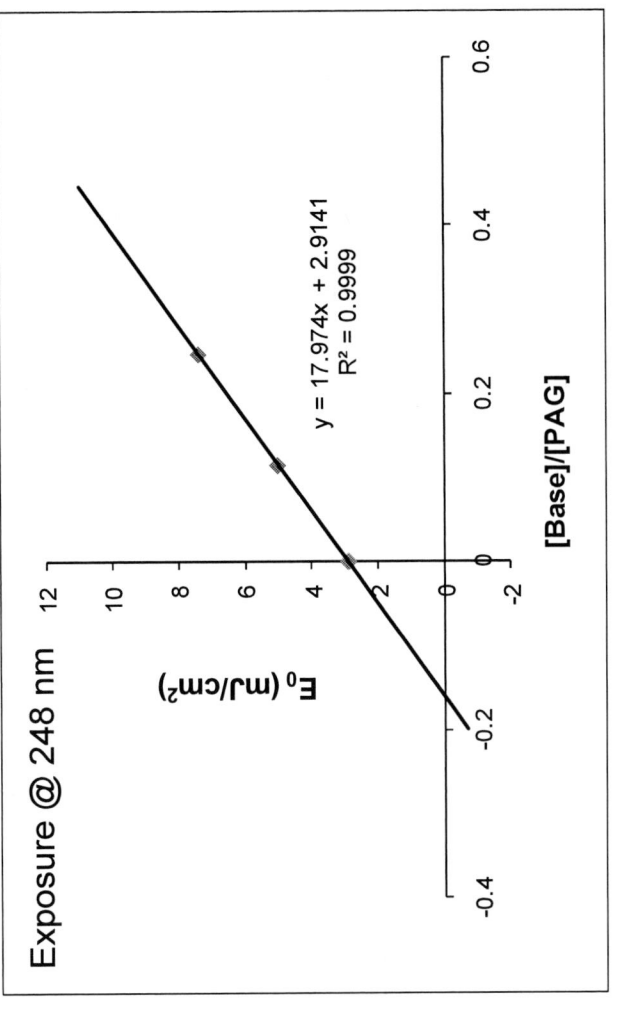

Exposure @ 248 nm

E_0 (mJ/cm²) vs [Base]/[PAG]

$y = 17.974x + 2.9141$
$R^2 = 0.9999$

- 10 wt % **TPS-Tf** in DuPont Generic 193 nm Polymer

- Vary base quencher loading

- Film thickness = 118.9 nm

- Abs @ 248 nm: a = 0.066

$$k_a = \frac{1}{slope} \cdot \frac{a}{(1 - e^{-a})} = 0.0575 \ (cm^2 \cdot mJ^{-1})$$

Oct 22nd 2009, 6th International Symposium on
Immersion Lithography Extensions

Find an appropriate formulation

1. PBG loading < 9.9 wt% to polymer

2. Net [Acid]$_{max}$ equivalent to (> 1 wt%) PAG to polymer

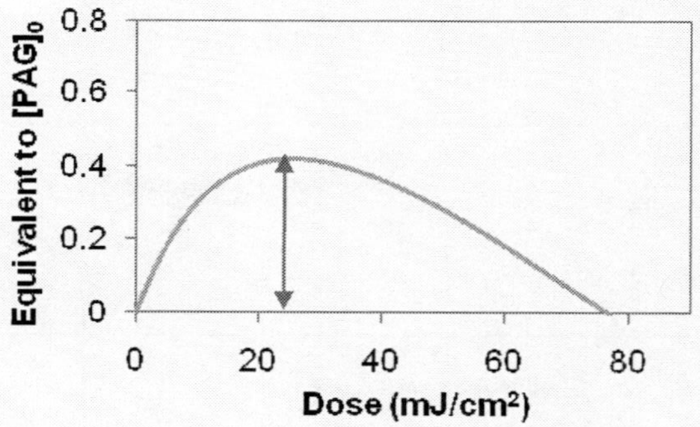

$$\frac{k_b}{k_a} = \frac{0.00243}{0.0575} = 0.0423$$

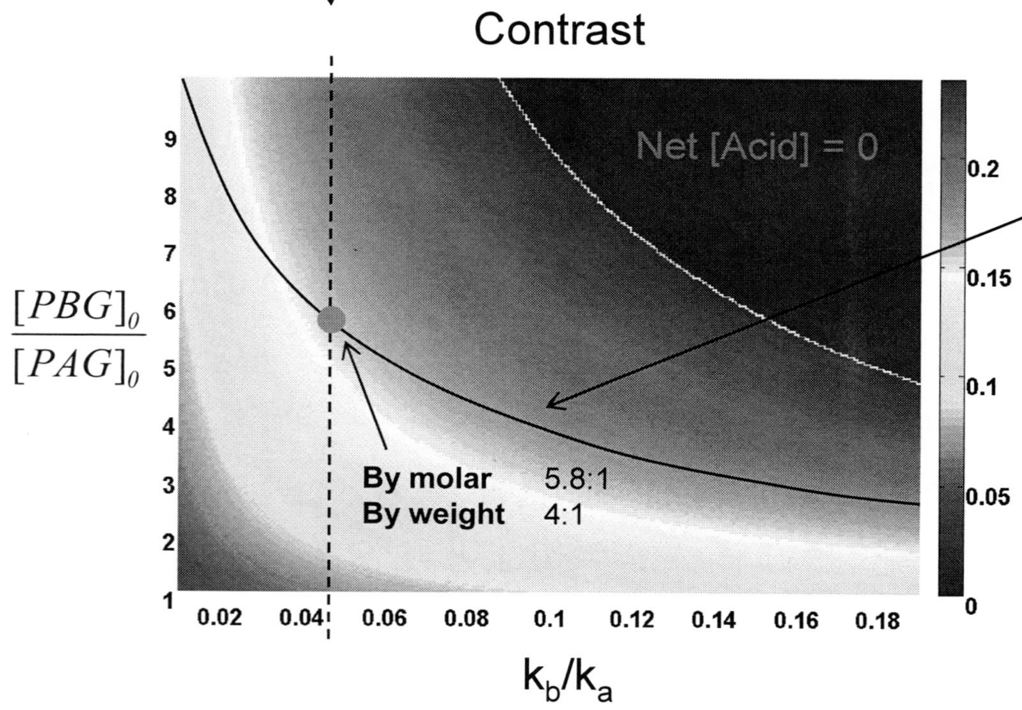

PAG = 2.5 wt%
PBG = 9.9 wt%

Contrast

Net [Acid] = 0

By molar   5.8:1
By weight   4:1

$\dfrac{[PBG]_0}{[PAG]_0}$

k_b/k_a

Equivalent to [PAG]$_0$

Dose (mJ/cm²)

Oct 22nd 2009, 6th International Symposium on
Immersion Lithography Extensions

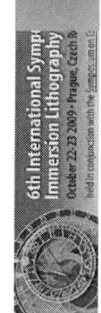

Contrast curves

Formulation

DuPont Polymer	5 wt%	in 2-Heptanone
TPS-Tf	2.5 wt%	to polymer
DIPA-PBG	9.9 wt%	to polymer

Process conditions:

Film Thickness = 111.5 nm

Exposure @ 248 nm

PEB @ various temperatures

Develop in 2.38% TMAH for 1 min

Shrinkage on the relief image

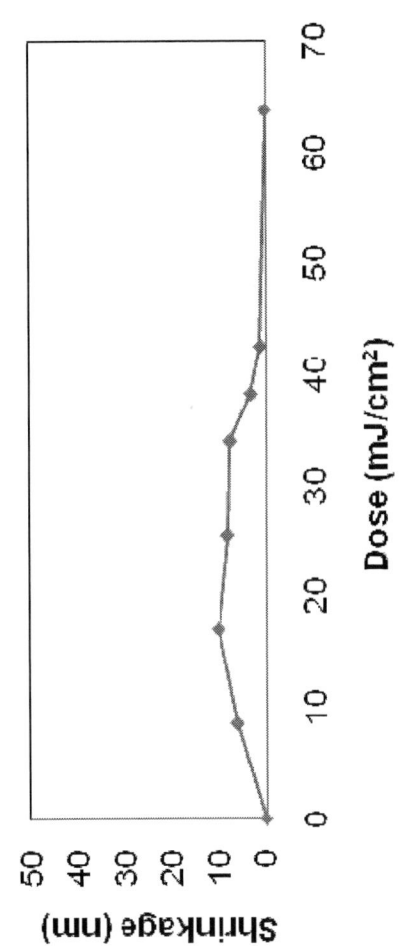

- The film was not developable upon increasing the PEB temperatures.
- The intuitive solution would be to change the formulation since the experimental condition might be deviated from the simulation results.

Oct 22nd 2009, 6th International Symposium on Immersion Lithography Extensions

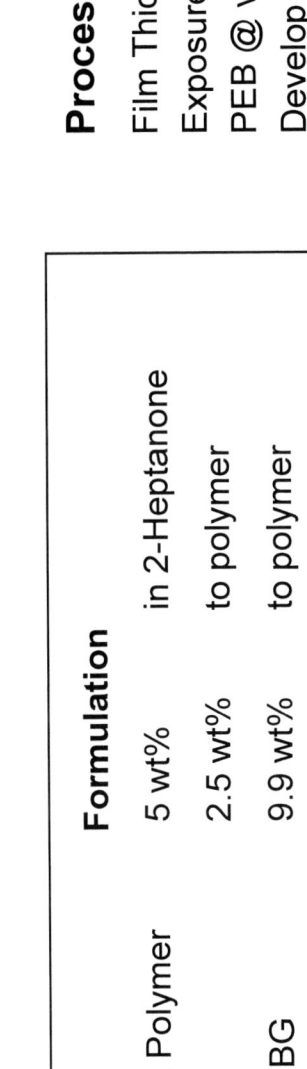

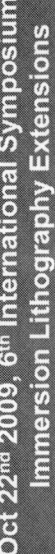

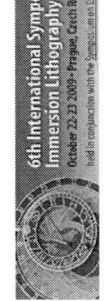

Contrast curves

Formulation

DuPont Polymer	5 wt%	in 2-Heptanone
TPS-Tf	4.5 wt%	to polymer
DIPA-PBG	9.9 wt%	to polymer

Process conditions:

Film Thickness = 111.5 nm
Exposure @ 248 nm
PEB @ various temperatures
Develop in 2.38% TMAH for 1 min

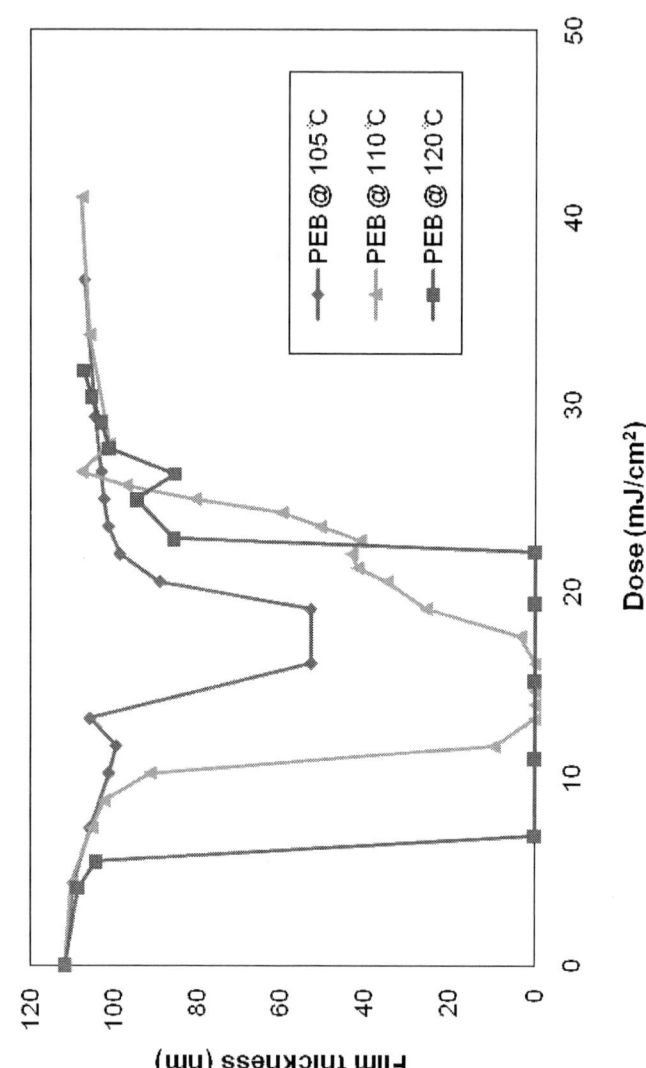

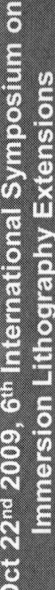

Oct 22nd 2009, 6th International Symposium on
Immersion Lithography Extensions

Contrast curve with 193 nm laser

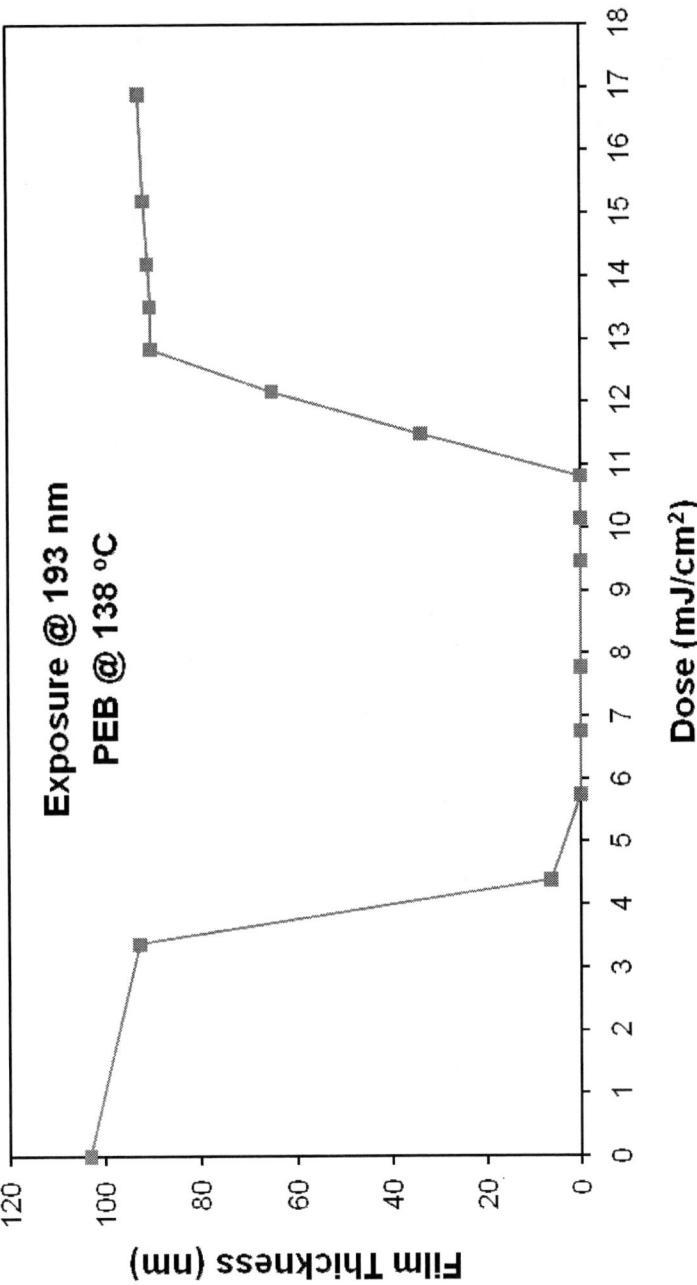

Exposure @ 193 nm
PEB @ 138 °C

Oct 22nd 2009, 6th International Symposium on
Immersion Lithography Extensions

Summary & Future work

- A photoactive system with both PAG and PBG was proposed to achieve pitch division.

- The system is compatible with the current manufacturing tools.

- A simulation was completed to study the influence of rate constants and formulation on the pitch division performance.

- The proof-of-concept was demonstrated.

- The optimization of the formulation and the processing condition is under way.

- The patterning performance of this system on an 193 nm imaging tool is being evaluated.

Oct 22nd 2009, 6th International Symposium on Immersion Lithography Extensions

Acknowledgement

Dr. C. Grant Willson
Dr. Paul Zimmerman

Thanks to all the Willson group members !!!

Undergraduate students

Chuan Shi

Fernando Marzuka

Sponsor

Accelerating the next technology revolution.

Copyright ©2008 SEMATECH, Inc. SEMATECH, and the SEMATECH logo are registered servicemarks of SEMATECH, Inc. International SEMATECH Manufacturing Initiative, ISMI, Advanced Materials Research Center and AMRC are servicemarks of SEMATECH, Inc. All other servicemarks and trademarks are the property of their respective owners.

Collaborations

Oct 22nd 2009, 6th International Symposium on Immersion Lithography Extensions

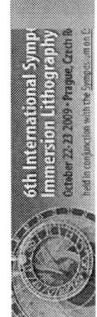

Double Patterning

Coater/Developer Evolution or Revolution?

Anita Viswanathan,

M. Nakano, R. Crowell, S. Scheer

6th International Symposium on Immersion Lithography Extensions, Oct 2009

Outline

- Evolution of CLEAN TRACK Technology

- DP Implications on Coater/Developer Design

- CLEAN TRACK DP Activity Update

- Summary

Evolution of CLEAN TRACK Technology

- Substrate Reflectivity / Surface Reflectivity
- Wafer Edge Defectivity
- Resist Sensitivity Defects

i-line: ADH – Chill – Coat – PAB – WEE – EXP – PEB – Chill – Dev

KrF / ArF: ADH – Chill – BCT – Bake – Chill – Coat – PAB – WEE – Chill – EXP – PEB – Chill – Dev

Immersion: ADH – Chill – BCT – Bake – Chill – Coat – PAB – Chill – TCT – Bake – Chill – WEE – Rinse – Chill – EXP – Rinse – PEB – Chill – Dev

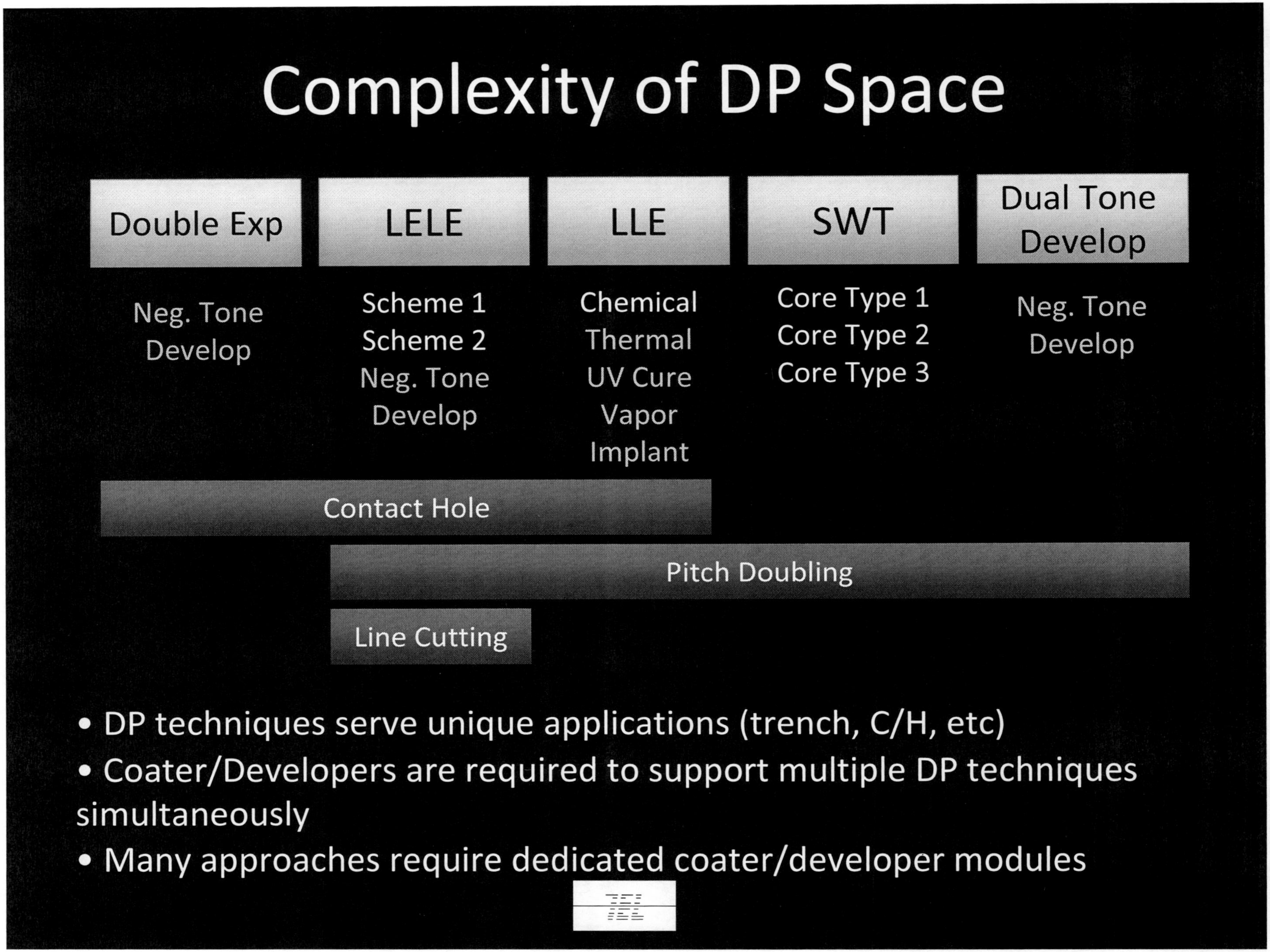

Complexity of DP Space

Double Exp	LELE	LLE	SWT	Dual Tone Develop

Neg. Tone Develop

Scheme 1
Scheme 2
Neg. Tone Develop

Chemical
Thermal
UV Cure
Vapor
Implant

Core Type 1
Core Type 2
Core Type 3

Neg. Tone Develop

Contact Hole

Pitch Doubling

Line Cutting

- DP techniques serve unique applications (trench, C/H, etc)
- Coater/Developers are required to support multiple DP techniques simultaneously
- Many approaches require dedicated coater/developer modules

Throughput Demand from DP

Single Exposure LITHO-ETCH   19 processing steps

LITHO-ETCH-LITHO-ETCH   44 processing steps
LITHO-LITHO-ETCH   39 processing steps

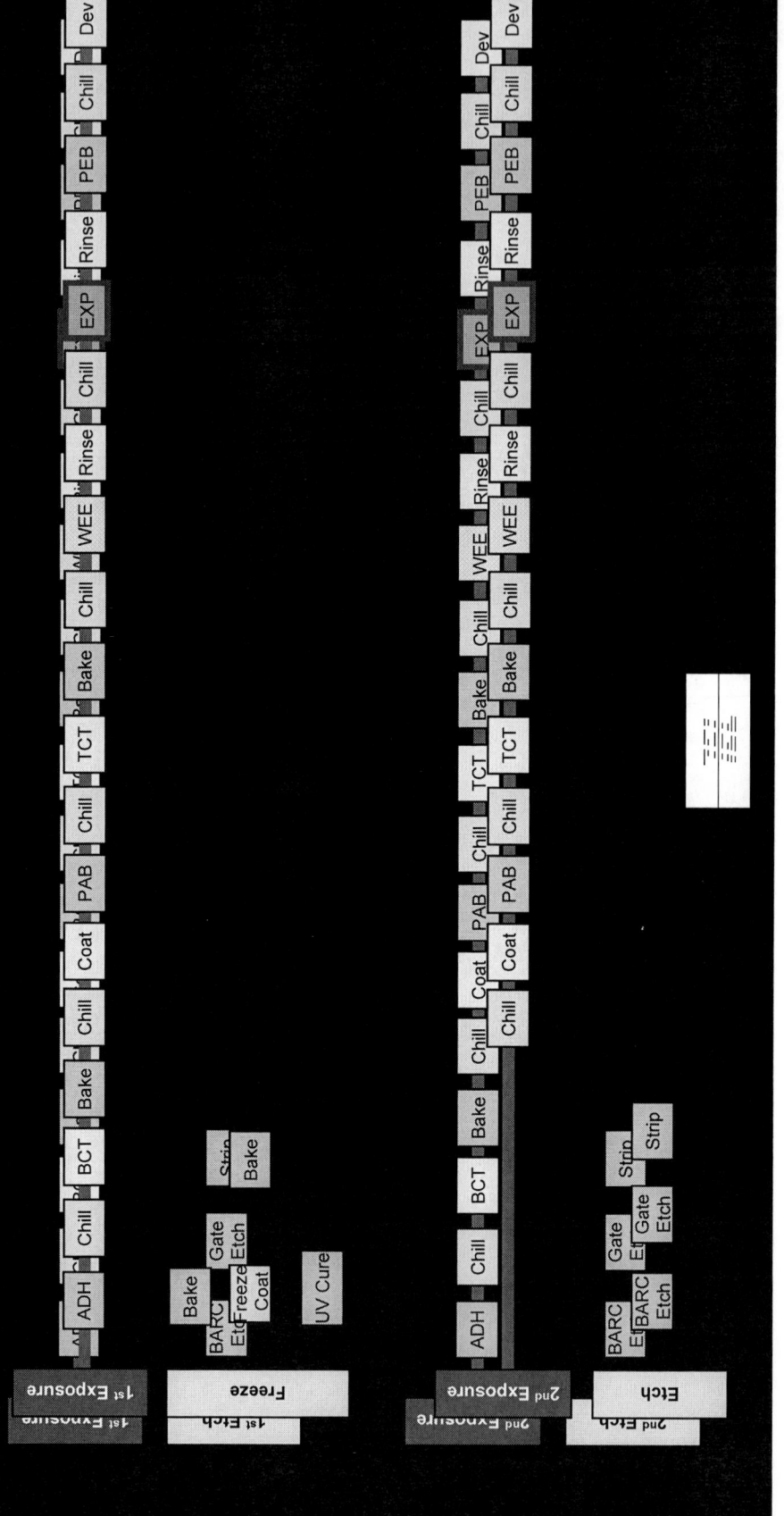

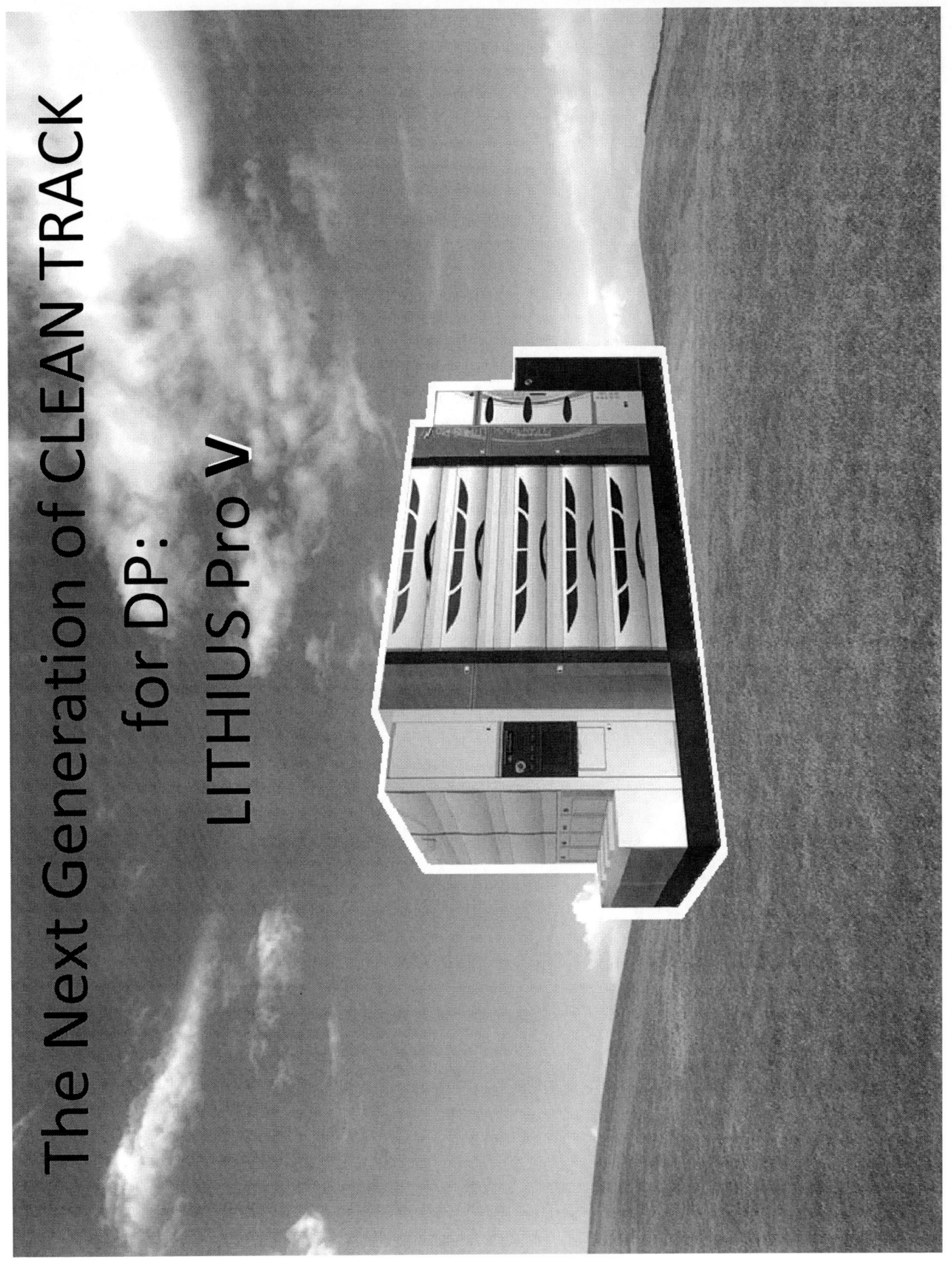

The Next Generation of CLEAN TRACK for DP:

LITHIUS Pro V

Key Coater/Developer DP Concerns

Cost of Ownership

- Process Costs
  - SWT
  - Dual Tone DP
  - Tool Productivity
  - OEE

Process Performance

- Side Wall Technology
- PR Slimming
- LWR Reduction
- Cross Line C/H
- Pattern Collapse
- Negative Tone Develop

TEL Resist Core SWT Overview

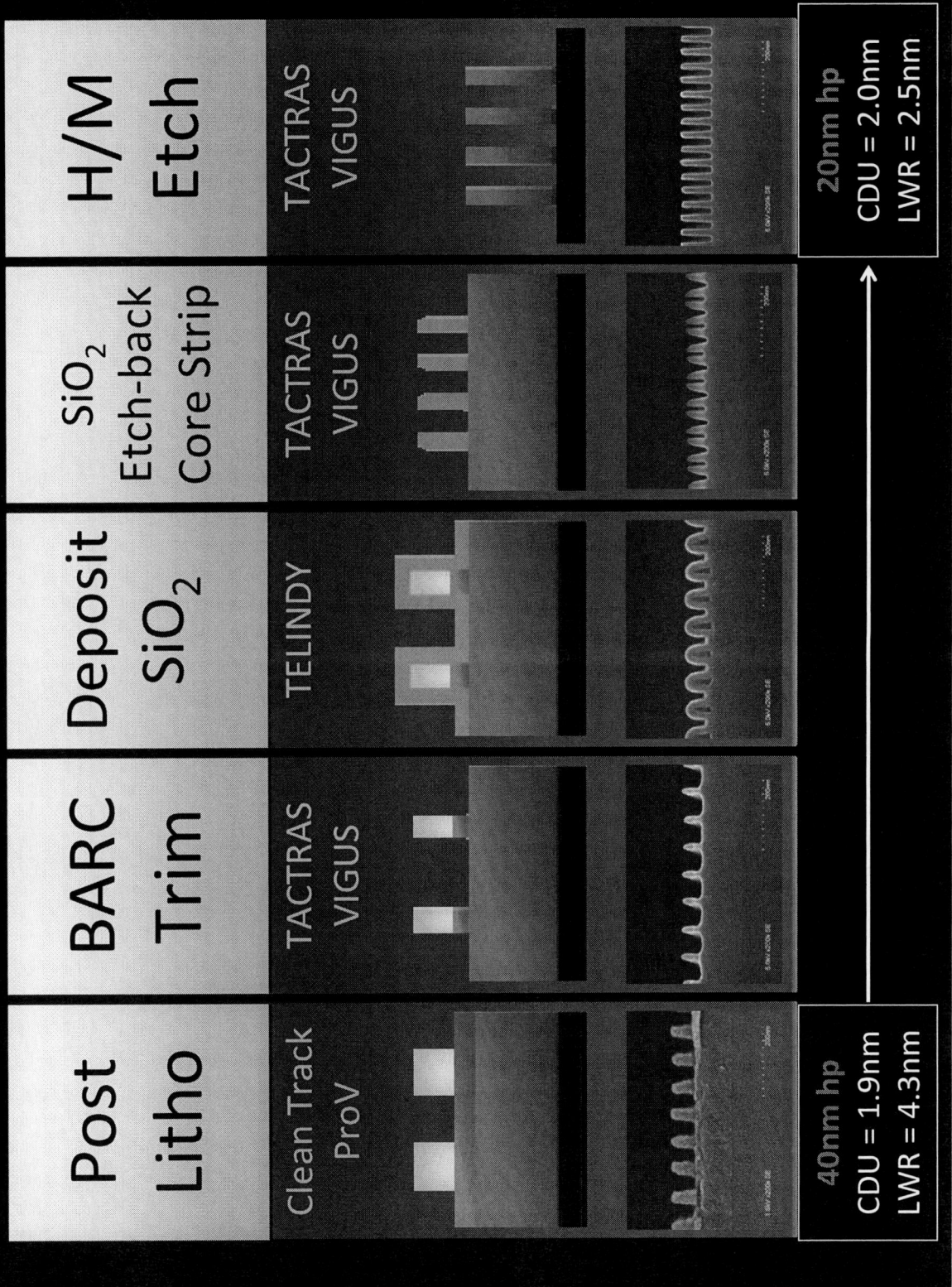

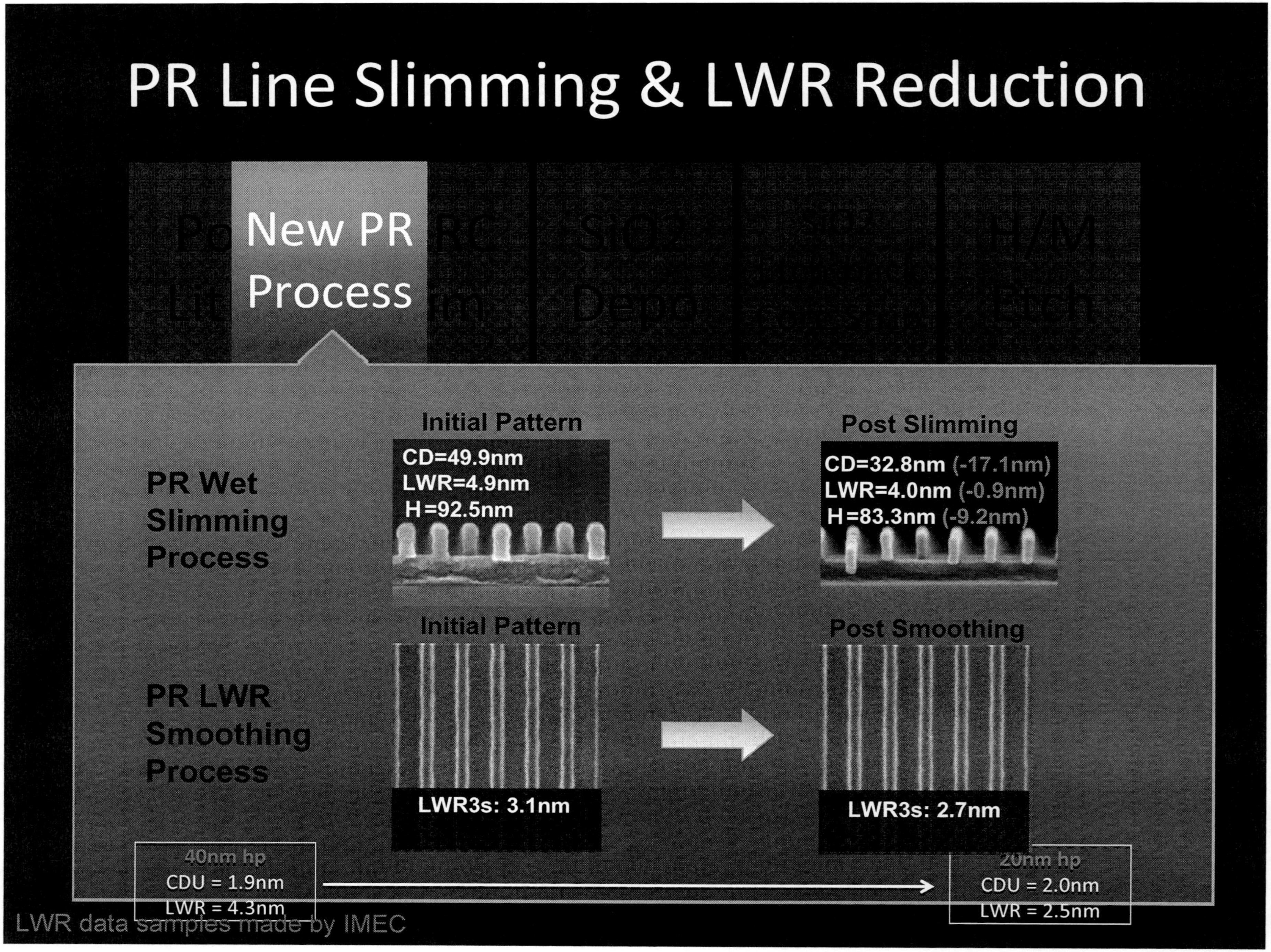

PR Line Slimming & LWR Reduction

New PR Process

PR Wet Slimming Process

Initial Pattern
CD=49.9nm
LWR=4.9nm
H=92.5nm

Post Slimming
CD=32.8nm (-17.1nm)
LWR=4.0nm (-0.9nm)
H=83.3nm (-9.2nm)

PR LWR Smoothing Process

Initial Pattern
LWR3s: 3.1nm

Post Smoothing
LWR3s: 2.7nm

40nm hp
CDU = 1.9nm
LWR = 4.3nm

20nm hp
CDU = 2.0nm
LWR = 2.5nm

LWR data samples made by IMEC

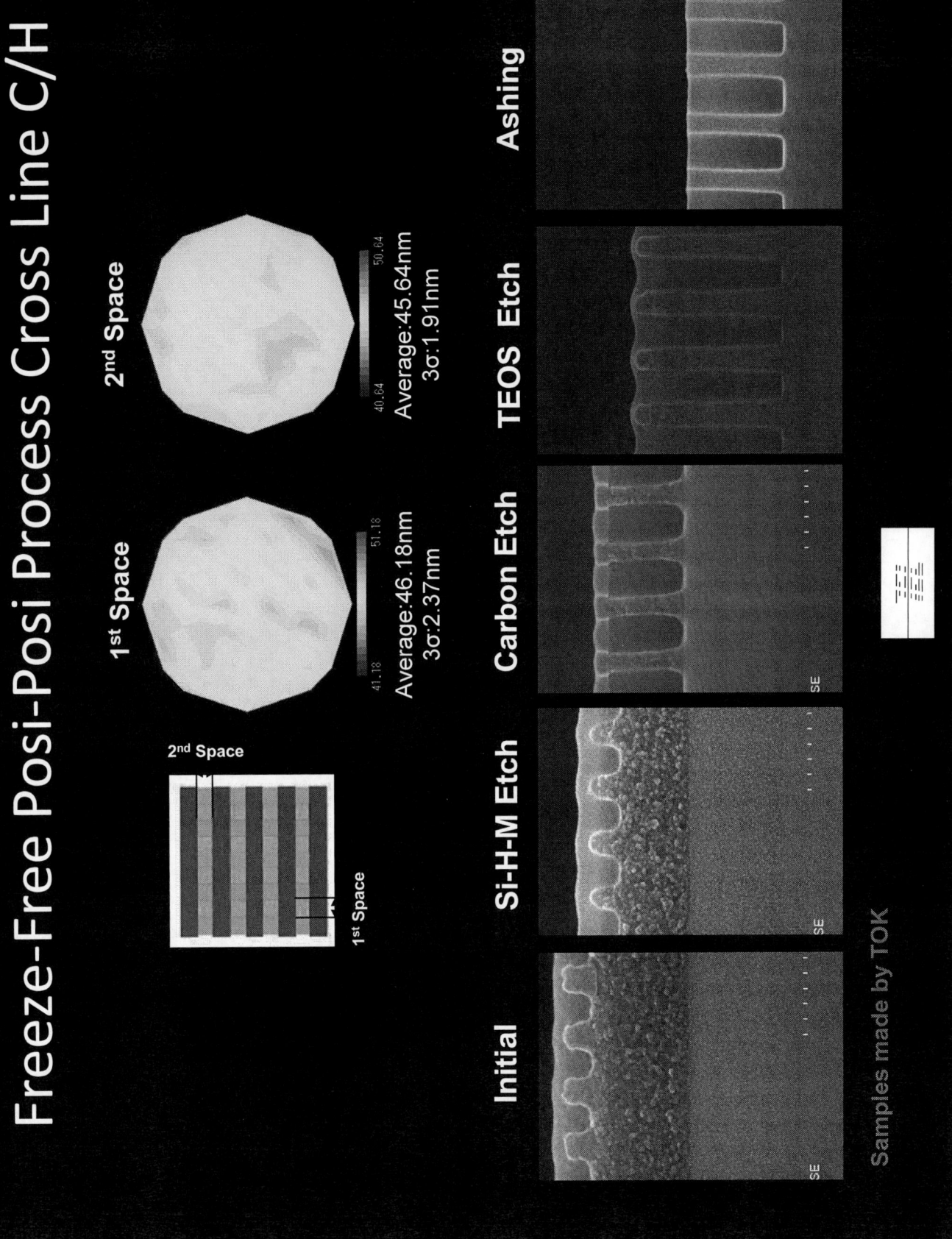

TEL FIRM Impact on Pattern Collapse
LLE L/S Patterning

Coat freeze

DIW Rinse

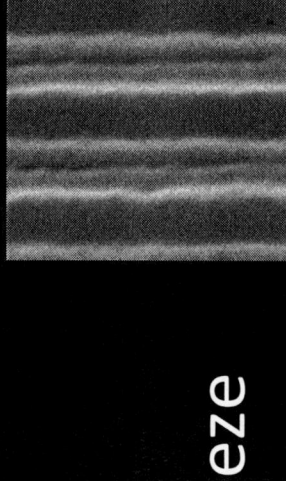

~170 defects
due to pattern collapse

Surfactant Rinse

~0 defects
due to pattern collapse

TEL FIRM has demonstrated effectiveness across multiple LLE processes.

Data courtesy of IMEC

Reference: Towards 26nm hp: Advances in Litho Process Litho; P. Wong

Negative Tone Development

Process A

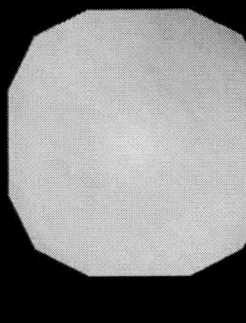

CDU 3σ = 1.39%
DEV = 1 unit/waf

Process B

CDU 3σ = 2.75%
DEV = 1 unit/waf

Process C

CDU 3σ = 3.2%
DEV = 0.45 unit/waf

Process A —1st Trial Defect Data

Measurement Area	LS		RESIST		BARC	
Classification	LS	Cts	RESIST	Cts	BARC	Cts
	Bridge	0	Ball	0	Ball	0
	Deformation	0	Residue (Circle)	0	Residue (Circle)	2
	Residue on LS	0	Residue (Non-Circle)	4	Residue (Non-Circle)	0
	Other	0	Other	0	Other	0
	TOTAL	0	TOTAL	4	TOTAL	2
Defect Counts [cts/waf]	0		4		2	
Defect Density [cts/cm²]	0.000		0.026		0.030	

Key Coater/Developer DP Concerns

Process Performance

- Spacer
- PR Slimming
- LWR Reduction
- Cross Line [illegible]
- Pattern Collapse
- [illegible]

Cost of Ownership

- Process Costs
  - Resist Core SWT
  - Dual Tone DP
- Tool Productivity
  - Overall Equipment Efficiency (OEE)
  - Wafers Per Day (WPD)

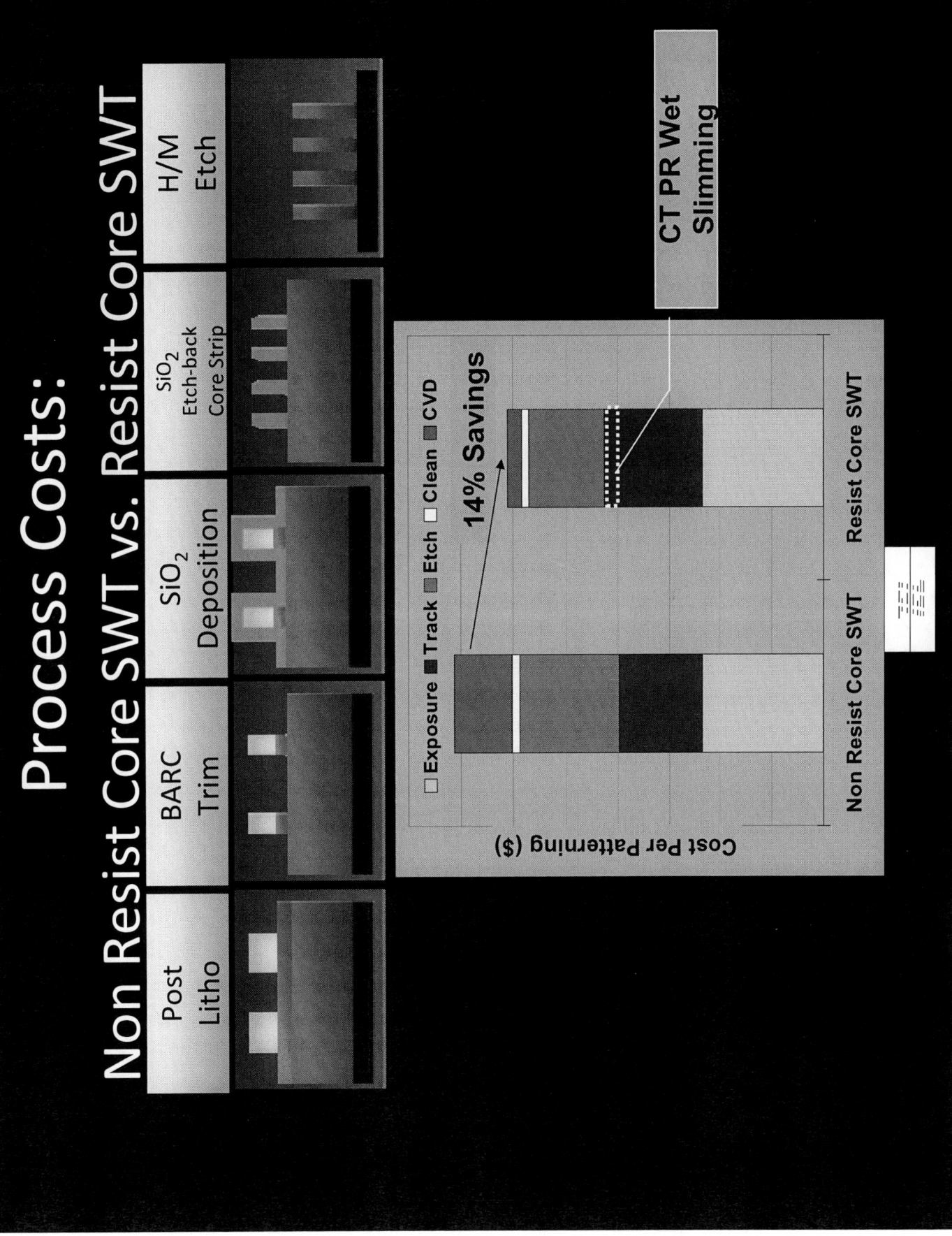

Process Costs: Dual-Tone Development

Overview

Current Status

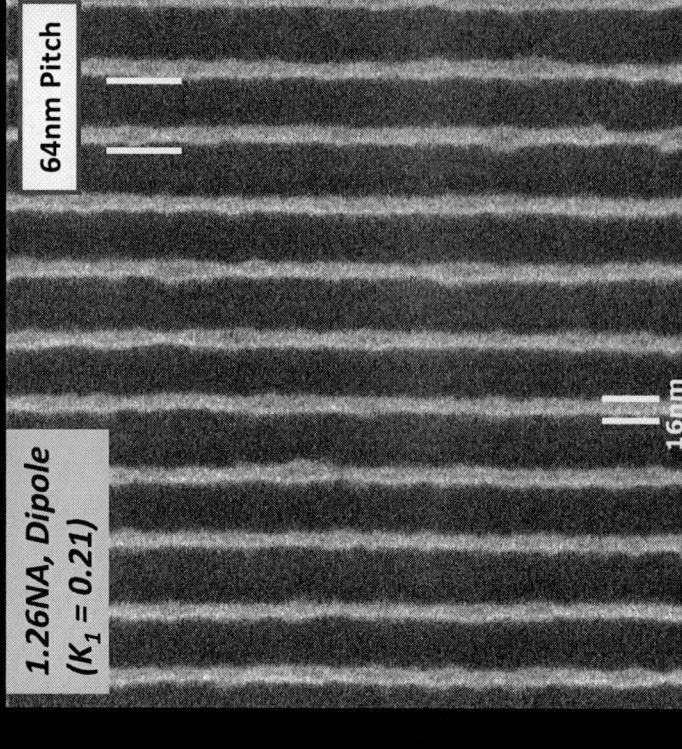

64nm Pitch

1.26NA, Dipole
($K_1 = 0.21$)

16nm

Mask Pitch: 128nm

- Demonstrated image modulation

- Resolution below single exposure & develop 1.35NA scanner capability

Reference: Exploration of ...Dual tone
Development; C. Fonseca

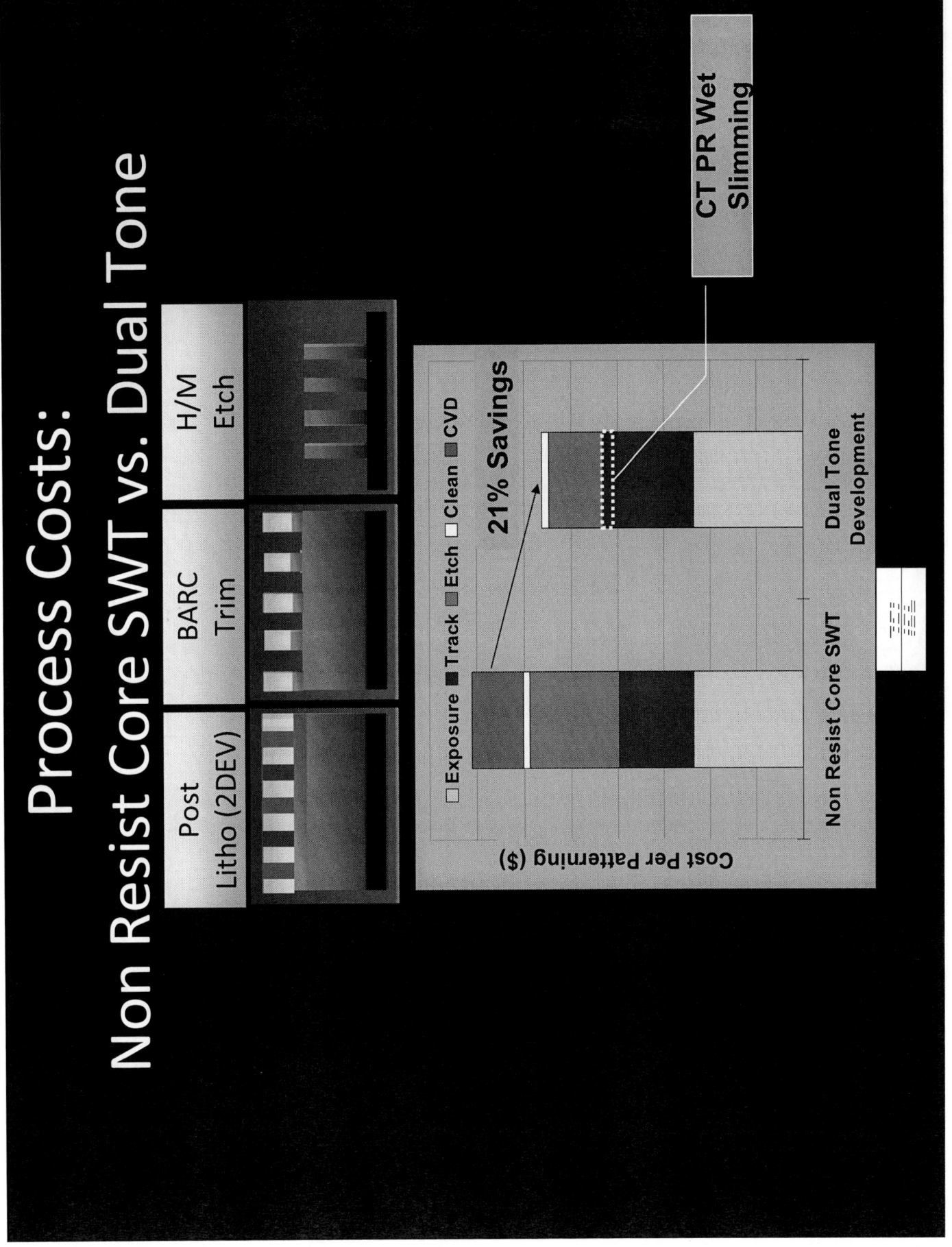

Tool Productivity: LITHIUS Pro V DP Process Capability

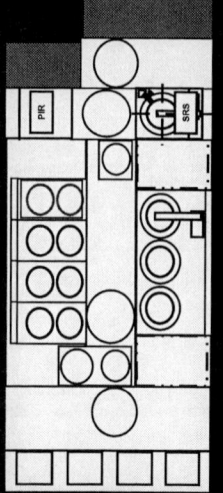

New Process

- Wet Slimming Capability (PR Pattern Trimming)
- Freezing Process Capability for LLE
- Negative Tone Development Process Capability
- FIRM Process for Defect reduction
- Smoothing Process for LWR Improvement (Future Upgrade)

OEE

- FOUP Exchanger System
- Multi cycle wafer flow control Function
- PRIME Cascade Transfer Control
- Additional OEE improvement items

Throughput / Footprint

- High T.P 250wph capability and Flexible Configuration
- LITHIUS Pro vs LITHIUS Pro V 25% Greater Throughput with only 10% Footprint Increase

Tool Productivity: Overall Equipment Efficiency (OEE) Improvement

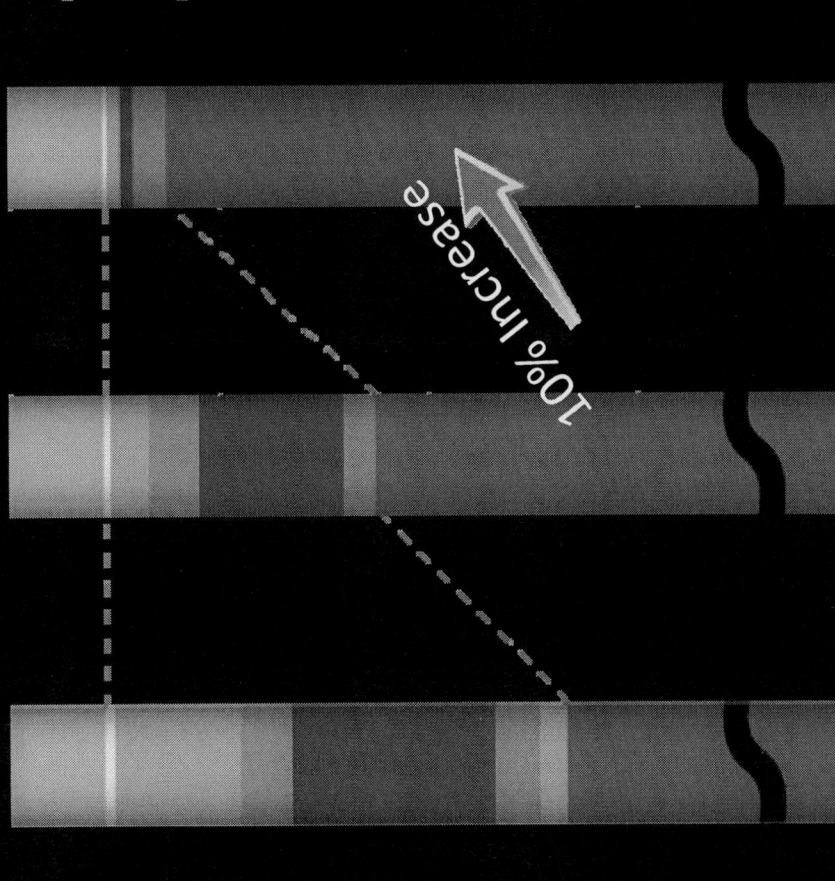

Others (Scanner Issue)

Availability Efficiency Loss (USD & SD)

Rate Efficiency Loss Interval Cleaning & Lot Change

Operational Efficiency Loss (ET & Resist)

OEE (Litho Cluster)

10% Increase

Conventional   LITHIUS Pro (current)   LITHIUS Pro V

Tool Productivity: Wafers Per Day (WPD) Improvement

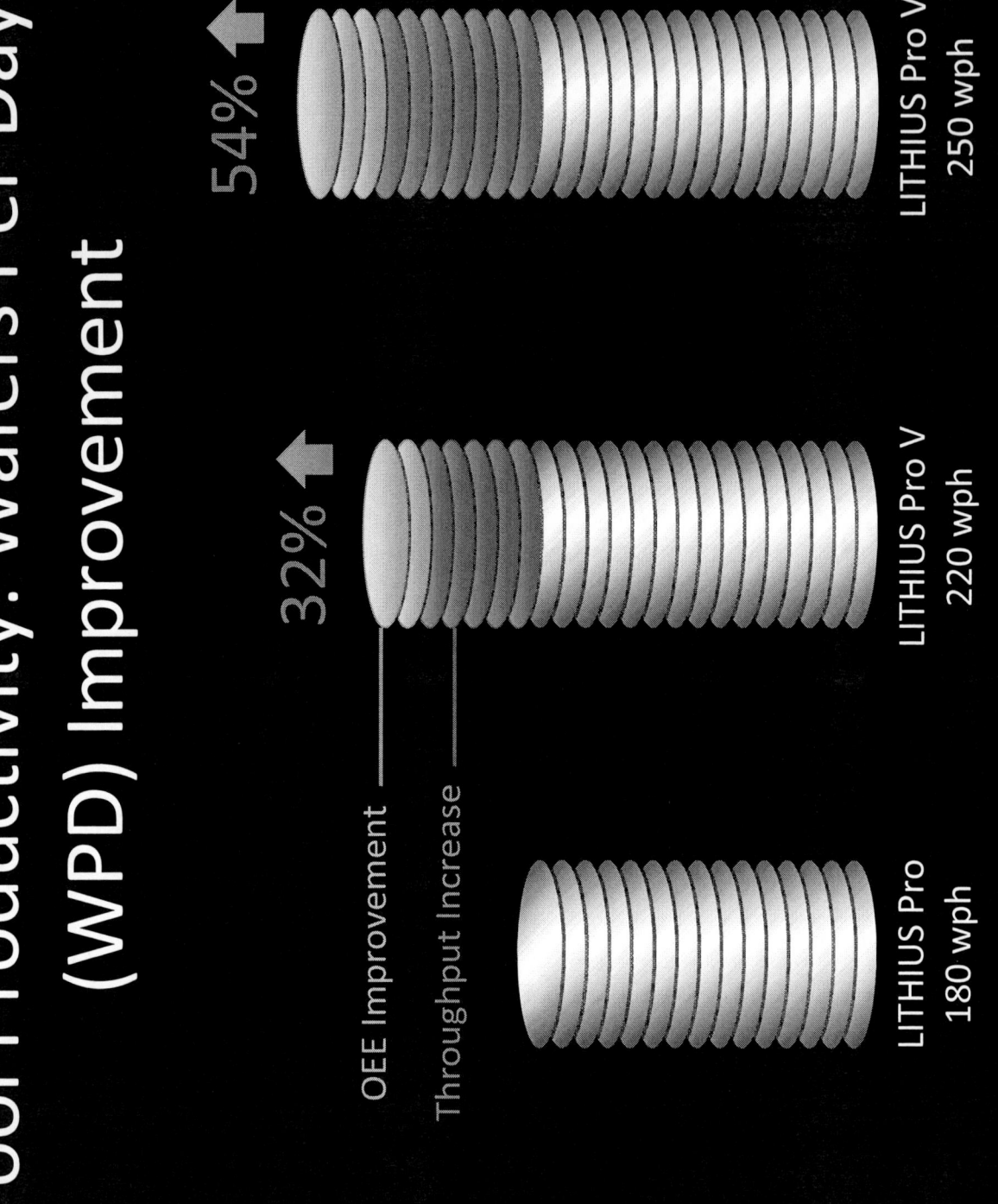

Summary

- DP processes are chosen by technology performance, customer device / layer type, and integration strategy, creating a complex space

- TEL's objective is to be a partner, regardless of the DP technology used, in helping our customers to:

  – Establish world class process performance

  – Achieve the highest productivity with lowest process cost

Acknowledgements

- TEL:
  - S. Shimura, T. Kawasaki, S. Natori, A. Hara, K. Hontake, T. Niwa, C. Fonseca, K. Nafus, H. Yaegashi
- IMEC
  - P. Wong, R. Gronheid

imec

Challenges Building a 22nm Node 6T-SRAM Cell Using Immersion Lithography

M. Ercken, E. Altamirano-Sanchez, C. Baerts, S. Brus, J. De Backer, M. Demand, C. Delvaux, N. Horiguchi, S. Locorotondo, T. Vandeweyer, A. Veloso, S. Verhaegen

Outline

- Introduction
- Process optimization
  - Fin patterning
  - Gate patterning
  - Implant litho
- Conclusions
- Acknowledgements

Monique Ercken
© imec 2009

6th Int. Symp. On Immersion Lithography Extentions
Oct. 22-23'09, Prague – Czech Republic

imec

Outline

- Introduction
- Process optimization
  - Fin patterning
  - Gate patterning
  - Implant litho
- Conclusions
- Acknowledgements

6th Int. Symp. On Immersion Lithography Extentions
Oct. 22-23'09, Prague – Czech Republic

Monique Ercken
© imec 2009

imec

Goal of this work

- Why SRAM ? ⇨ technology lead vehicle for scaling embedded memory
  - Key role in the introduction of new technology nodes

- Why FinFET ? ⇨ most promising candidate for scaling beyond 32nm node

- Industry asks for continuous shrinking device sizes, where we are currently facing a lithographic scaling limit
  - Immersion litho stops @ 1.35NA
  - EUV litho still faces technical challenges and commercial timing issues
  - ⋏ Significant impact on patterning (litho & etch) development

- Work presented here shows patterning development stages for setting front-end part of 22nm node 6T-SRAM cell.
  - 193nm immersion lithography
  - Starting point = patterning processes set-up for a 32nm node SRAM device

Monique Ercken
© imec 2009

6th Int. Symp. On Immersion Lithography Extentions
Oct. 22-23'09, Prague – Czech Republic

imec

FYI: design and targets for 22nm node

- Cell size = 0.099um^2
- Fin level (▨)
  - Min. Pitch = 90nm
  - CD target = 17nm
- Gate level (▬)
  - Min. Pitch = 110nm
  - CD target = 35nm
- Fin and gate are designed unidirectional and perpendicular to each other ⇨ 2D structures still present
- Both levels are exposed on ASML XT:1900Gi
  - 1.35NA
  - Optimized exposure settings in combination with polarized light

Litho-friendly SRAM

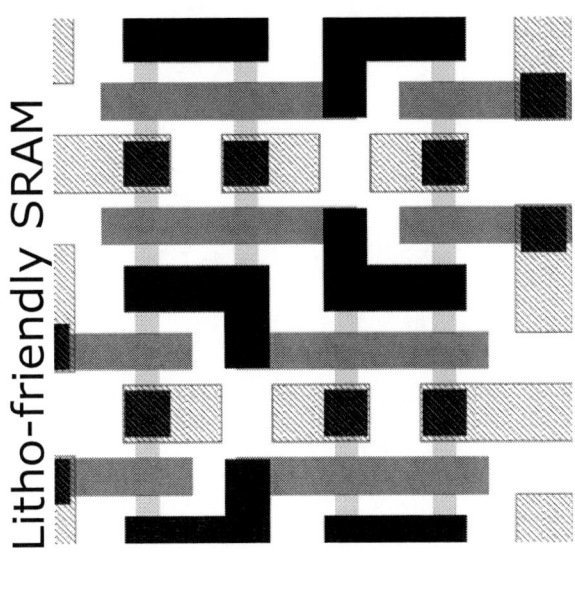

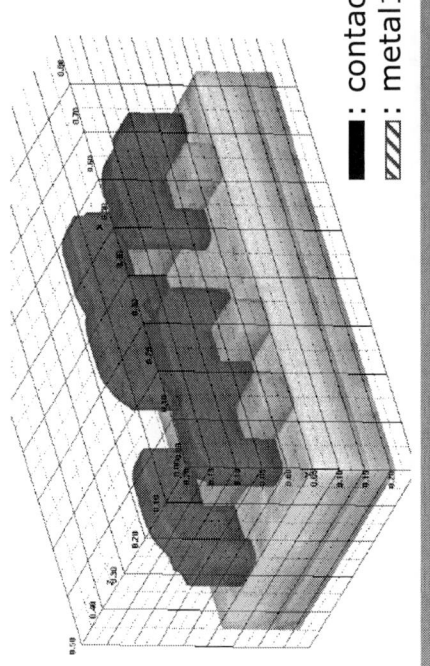

■ : contact level
▨ : metal1 level

6th Int. Symp. On Immersion Lithography Extentions
Oct. 22-23'09, Prague – Czech Republic

Monique Ercken
© imec 2009

imec

Outline

- Introduction
- **Process optimization**
  - Fin patterning
  - Gate patterning
  - Implant litho
- Conclusions
- Acknowledgements

Monique Ercken
© imec 2009

6th Int. Symp. On Immersion Lithography Extentions
Oct. 22-23'09, Prague – Czech Republic

imec

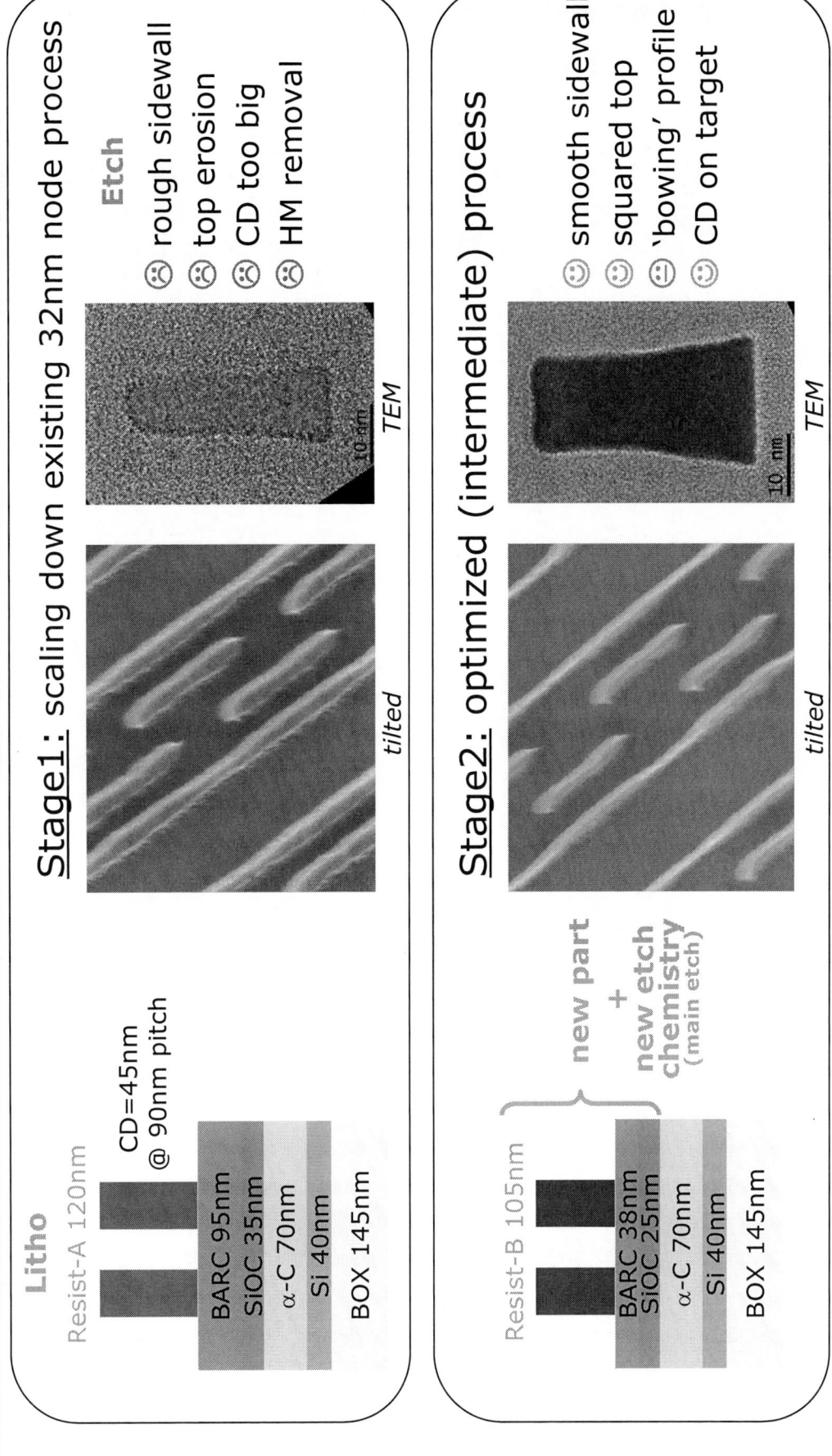

Fin patterning: model-based OPC (litho+etch)

Stage1/2: OPC model 1   **Stage3:** OPC model 2

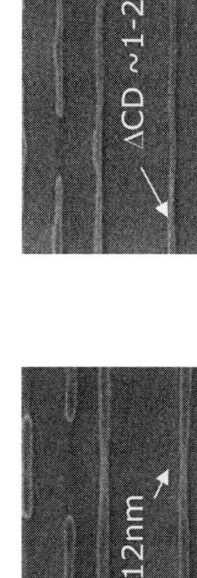

etch

litho

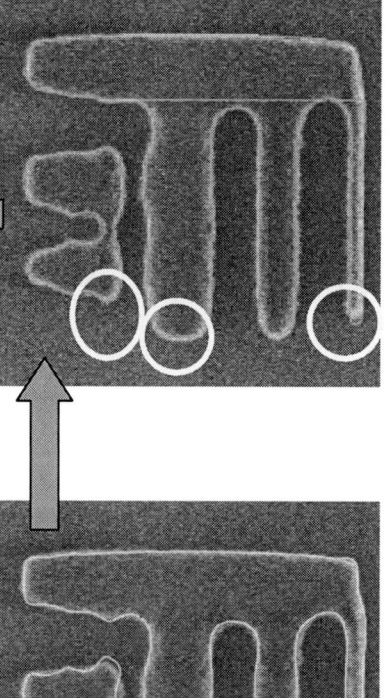

e.g. specific OPC calibration structure

Issue of model 1: necking prohibiting creation of small enough CDs

Reason: difference in resist coverage for mask used to set-up OPC model with device mask

Differences between model 1 and 2:

- More line-end measurements taken into account
- Through dose calibration (Mentor Graphics)
- Usage of Design Gauge (Hitachi)
- OPC calibration structures on appropriate device mask

Result:

- Improved line-end and through pitch behavior

Monique Ercken
© imec 2009

6ᵗʰ Int. Symp. On Immersion Lithography Extentions
Oct. 22-23'09, Prague – Czech Republic

imec

Fin patterning: process optimization

Litho

Resist-B 105nm

BARC 38nm
SiOC 25nm
APF 70nm
Si 40nm

BOX 145nm

Stage4: optimized process

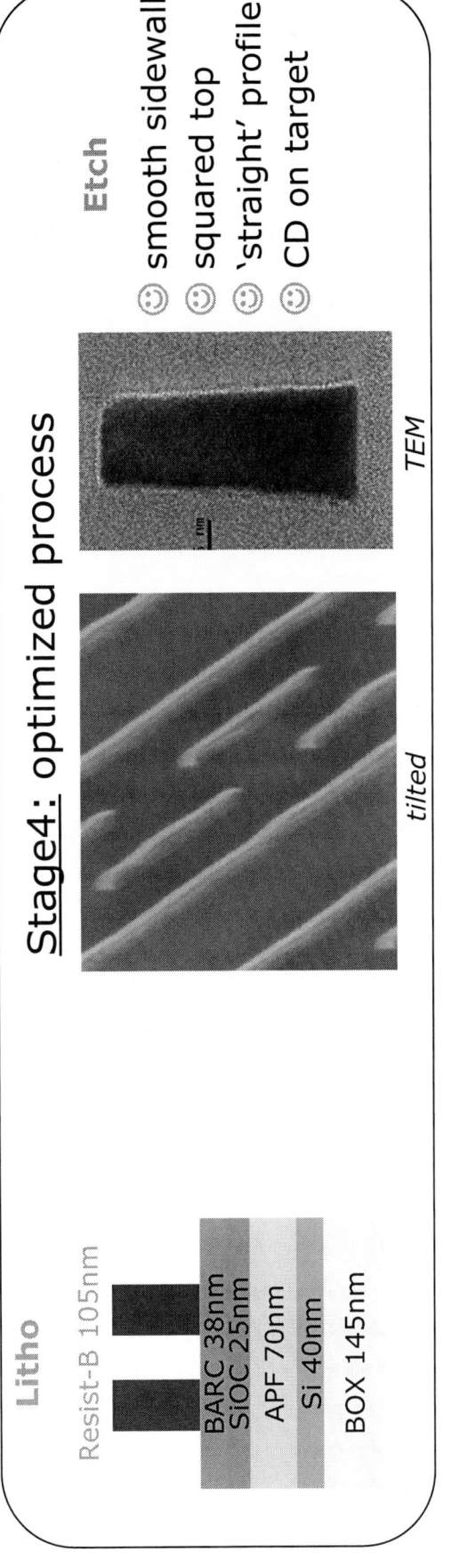

tilted

TEM

Etch

- 😊 smooth sidewall
- 😊 squared top
- 😊 'straight' profile
- 😊 CD on target

Achieved by further improved Si over-etch (optimized plasma chemistry and condition)

Monique Ercken
© imec 2009

6th Int. Symp. On Immersion Lithography Extentions
Oct. 22-23'09, Prague – Czech Republic

imec

Fin patterning: CD performance

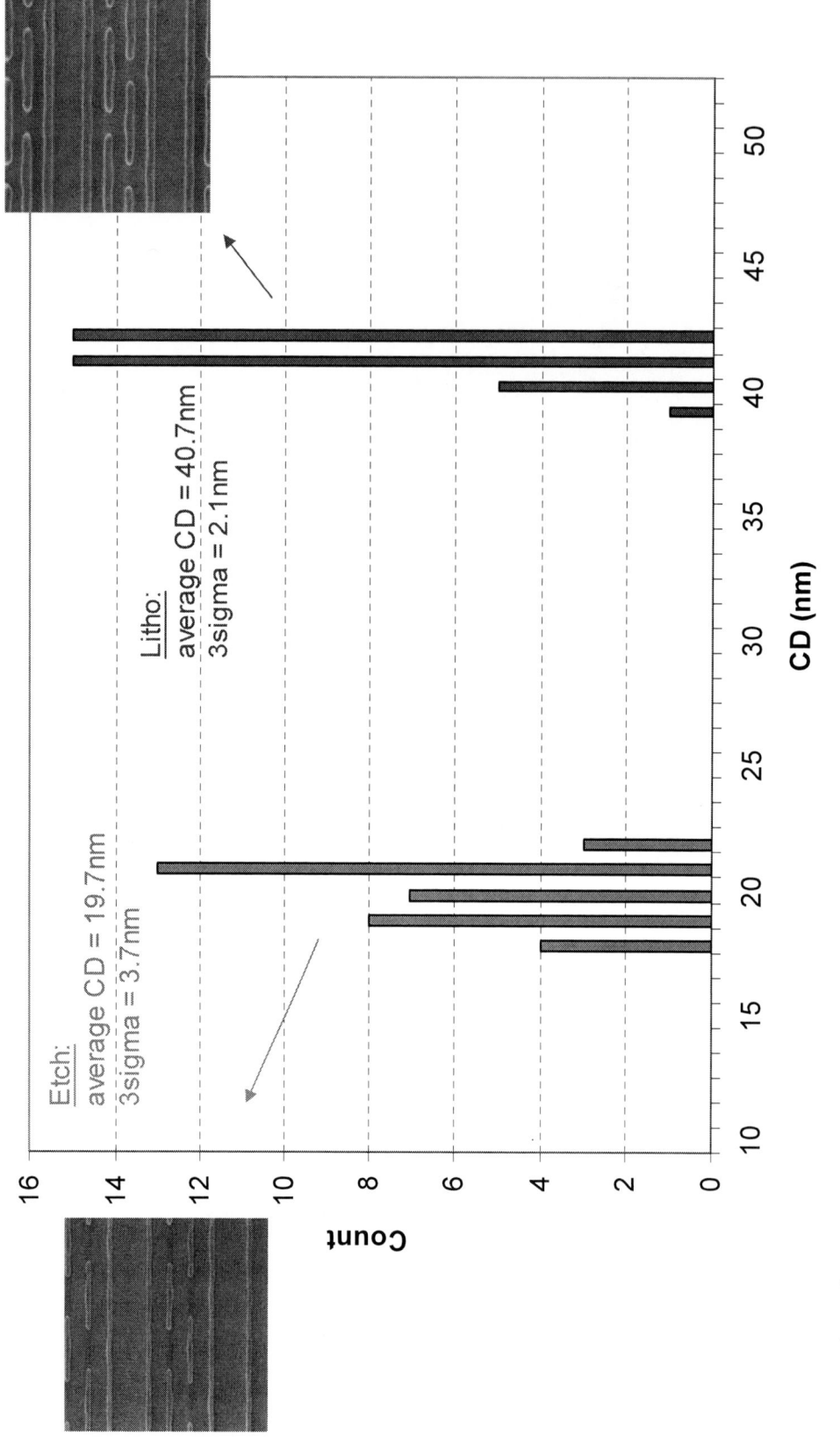

Etch:
average CD = 19.7nm
3sigma = 3.7nm

Litho:
average CD = 40.7nm
3sigma = 2.1nm

CD (nm)

Count

~1 lot / 4 wafers / 9 points per wafer

Monique Ercken
© imec 2009

6th Int. Symp. On Immersion Lithography Extentions
Oct. 22-23'09, Prague – Czech Republic

Gate patterning: process optimization

Stage1: scaling down existing SC32 (150nm pitch) process

Single mask
Double expose

Litho

Etch

☹ Gap CD too big

Overlap needed to protect underlying fin during implants !

Stage2: process optimization

Double mask
Double expose

Line ends touching

Thinner BARC + more selective BARC etch

☺ Gap CD control ↑, but random

☹ Rounded line ends
⇨ varying L_g

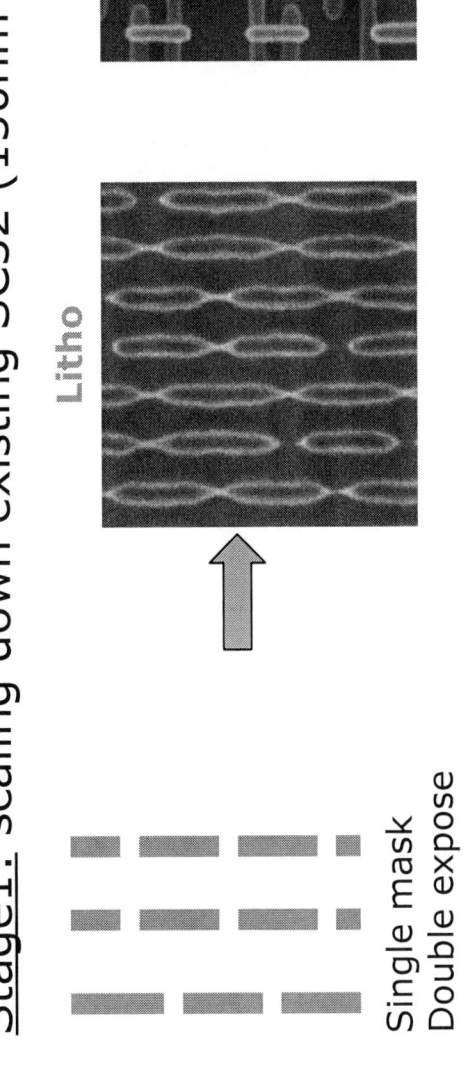
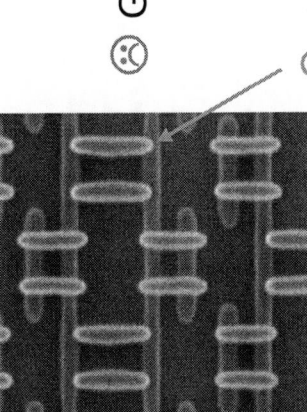

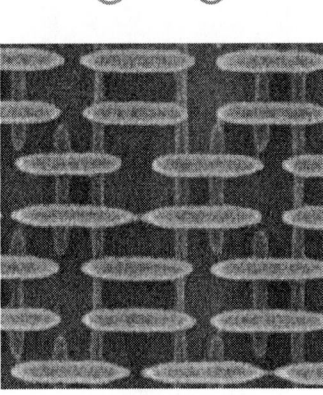

6th Int. Symp. On Immersion Lithography Extentions
Oct. 22-23'09, Prague – Czech Republic

Monique Ercken
© imec 2009

imec

Gate patterning: process optimization

Stage3: process optimization ⇨ introduction of line cut approach

Double mask
Double **patterning (~LELE)**

38nm BARC / 105nm Resist-B

Hard Mask

- 30nm Si3N4
- 35nm SiOC
- 70nm α-C
- 30nm oxide

FIN
80nm
40nm
90nm
150nm BOX

Gate stack: high-k/metal/80nm poly

Sequence of different steps
1. LF photo1 : lines
2. Lines etched in nitride
3. DF photo2: short trenches

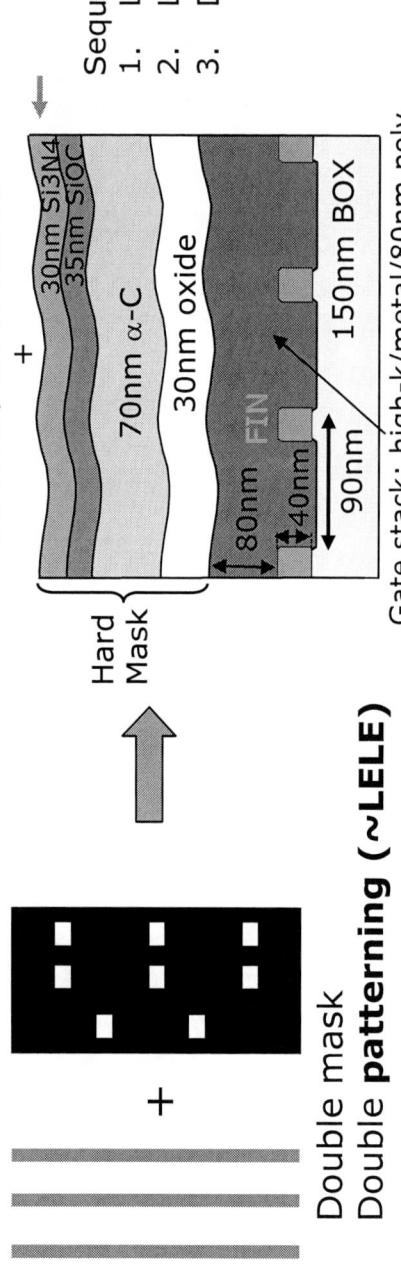

1. LF photo1: lines
CD = 40nm
top-down

2. Lines etched in nitride
tilted

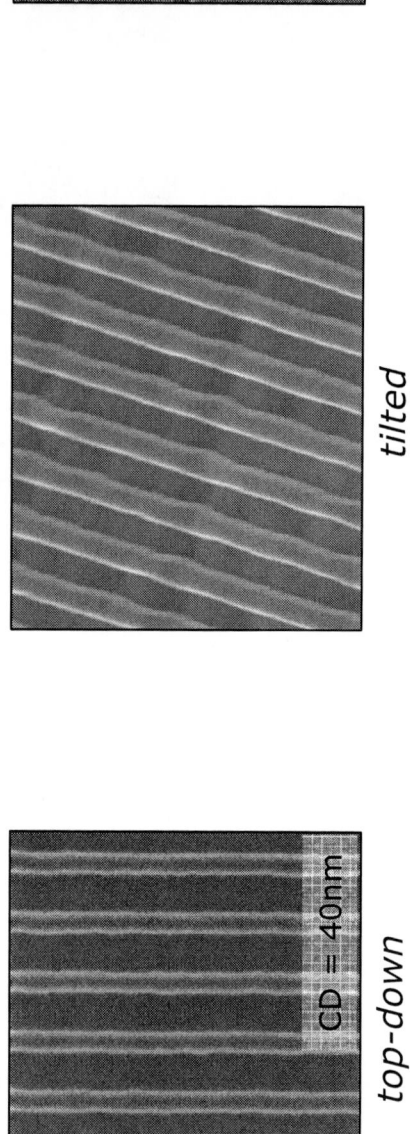

3. DF photo2: trench
CD = 35nm
top-down

Monique Ercken
© imec 2009

6th Int. Symp. On Immersion Lithography Extentions
Oct. 22-23'09, Prague – Czech Republic

imec

Gate patterning: process optimization

Stage3: process optimization ⇨ introduction of line cut approach

38nm BARC / 105nm Resist-B

+

Double mask
Double **patterning (~LELE)**

Hard Mask
- 30nm Si3N4
- 35nm SiOC
- 70nm α-C
- 30nm oxide

80nm FIN
40nm
90nm
150nm BOX

Gate stack: high-k/metal/80nm poly

Sequence of different steps
1. LF photo1 : lines
2. Lines etched in nitride
3. DF photo2: short trenches
4. Cut of nitride lines
5. Transfer into underlying stack

4. Cut of nitride lines

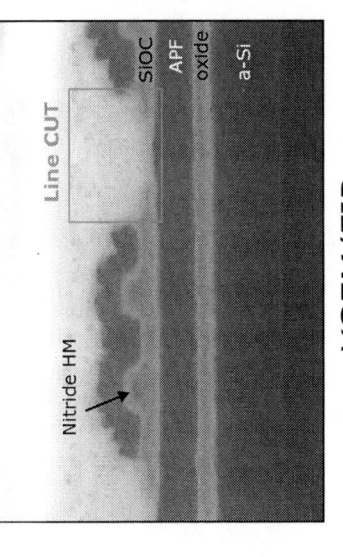

Line CUT
SiOC
APF
oxide
a-Si
Nitride HM

XSEM/FIB

5. Transfer into underlying stack

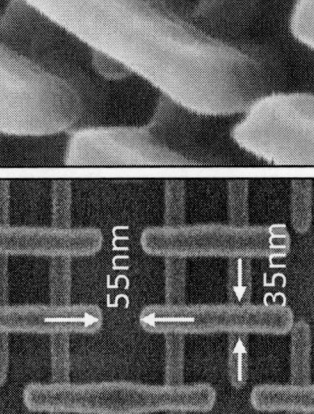

55nm
35nm

top-down

tilted

☺ gap CD control

☹ squared line ends

Monique Ercken
© imec 2009

6th Int. Symp. On Immersion Lithography Extentions
Oct. 22-23'09, Prague – Czech Republic

imec

Implant litho

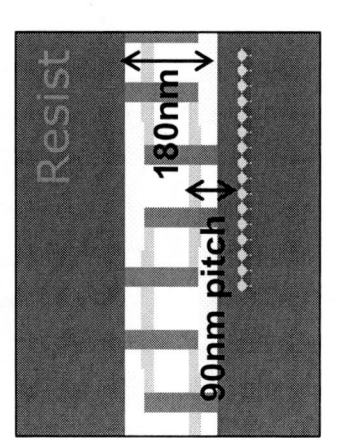

Resist

180nm

90nm pitch

- Fin pitch = 90nm → CD ~ 17nm → Gap ~70nm
- Very strict requirements for overlay control
  - Spec: |m|+3sigma < 15nm
- IMEC's current std. tool for less critical layers (~ implants): |m|+3sigma ~15-20nm (AT:750S; 248nm)

➢ More advanced XT:1250 (193nm dry)

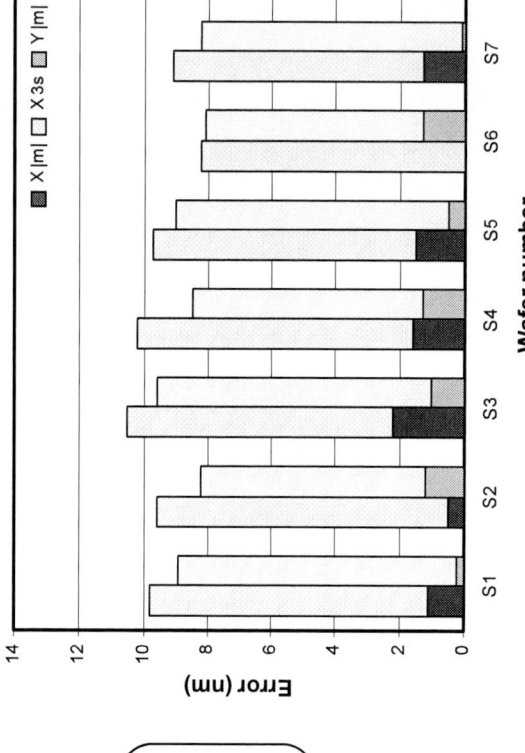

Move to XT:1250 needed, not for resolution purpose, but improved overlay performance !!!

Monique Ercken
© imec 2009

6th Int. Symp. On Immersion Lithography Extentions
Oct. 22-23'09, Prague – Czech Republic

imec

Implant litho

Resist

180nm

90nm pitch

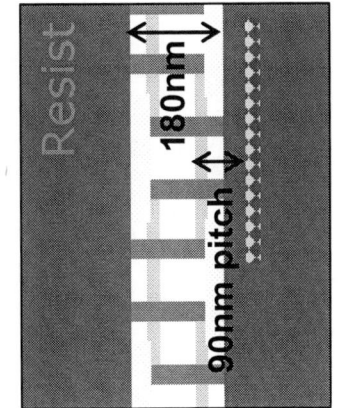

230nm thick resist needed to ensure enough stopping power !

1. Std. 193nm resist

TopCD=212nm

BottomCD=Scum

☹ Resist scum
☹ Underlying fin not enough protected

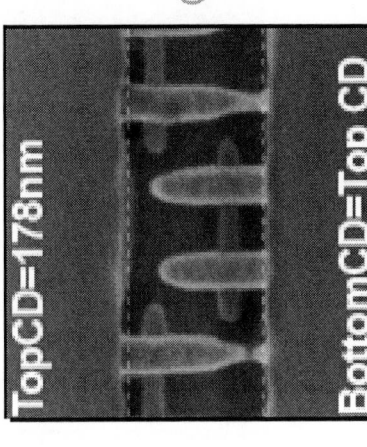

2. Optimized 193nm resist

TopCD=178nm

BottomCD=Top CD

☺ Cleared bottom

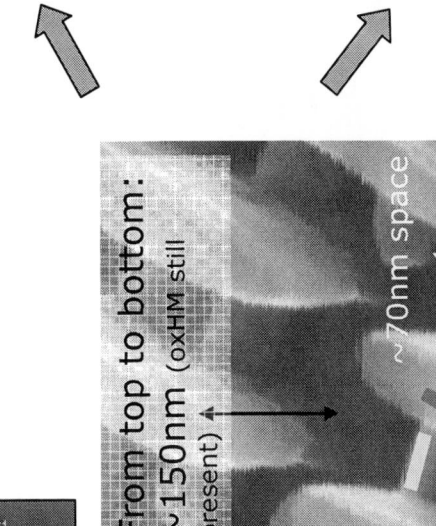

From top to bottom:
~150nm (oxHM still present)

~70nm space

Overlay is becoming very critical *tilted*

= topo to put implant litho on

Monique Ercken
© imec 2009

6ᵗʰ Int. Symp. On Immersion Lithography Extentions
Oct. 22-23'09, Prague – Czech Republic

Implant litho: but ...

248nm

193nm

resist

gate

fin

XSEM

Undercut in resist ⇨
fin enough protected ?

Implant under 7-25° tilt

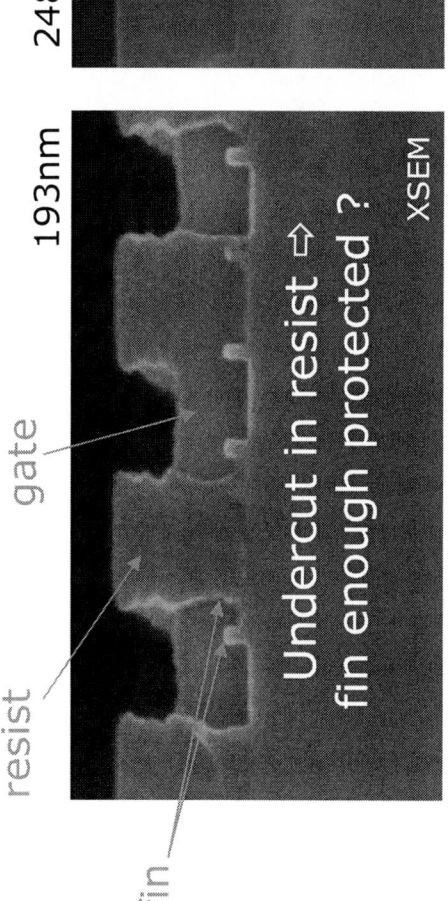

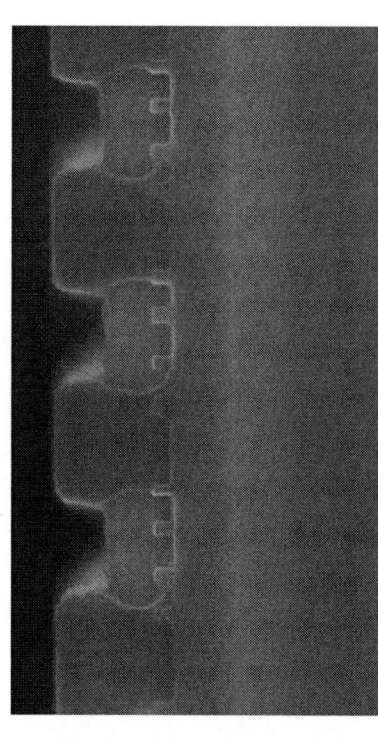

Simulation by Sentaurus Lithography
(using non-optimized resist model)

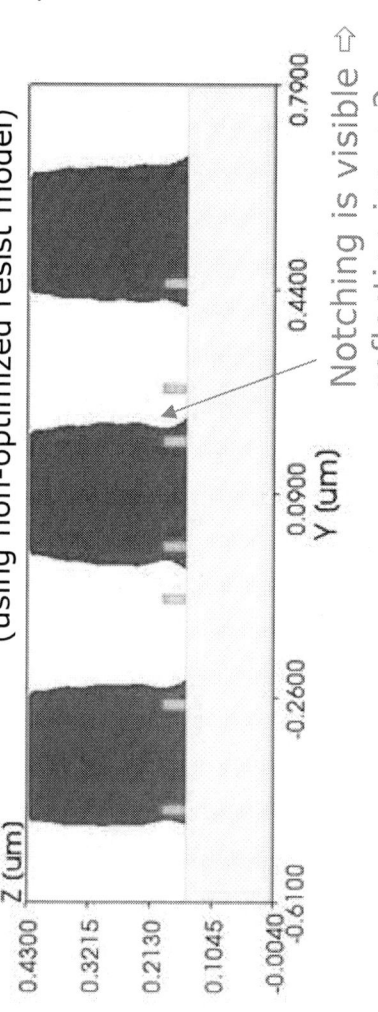

Notching is visible ⇨
reflection issue ?

Solutions:

- Resist(-TARC) only option ?
- Including BARC
  - Wet-developable ?
  - Dry (etch step needed) ?

Monique Ercken
© imec 2009

6th Int. Symp. On Immersion Lithography Extentions
Oct. 22-23'09, Prague – Czech Republic

imec

Outline

- Introduction
- Process optimization
  - Fin patterning
  - Gate patterning
  - Implant litho
- Conclusions
- Acknowledgements

Monique Ercken
© imec 2009

6th Int. Symp. On Immersion Lithography Extentions
Oct. 22-23'09, Prague – Czech Republic

imec

Conclusions

- This work resulted in the first electrically functional 22nm node SRAM cell, with the contact and metal1 level exposed on the ASML EUV alpha-demo tool.
  - Done with initial patterning processes (stage 1 for fin/stage 2 for gate) + implant litho @ 248nm
  - Back-end patterning optimization presented @ 2009 International Symposium on EUV (M. Goethals et al.)
  - Device results to be presented @ IEDM'09 (A. Veloso et al.)

- 'Simple' scaling of existing patterning processes not straightforward anymore
- Optimized fin and gate patterning available
  - Straighter and smaller fins with smoother sidewalls
  - Optimized fin OPC
  - Better control over gate-to-fin overlap

- Implant litho is a concern
  - Stricter overlay control needed for more reliable results
  - Dedicated 193nm resist and BARC development is required !!!
  - Further scaling ?

Monique Ercken
© imec 2009

6th Int. Symp. On Immersion Lithography Extentions
Oct. 22-23'09, Prague – Czech Republic

Acknowledgements

- IMEC pilot line
- Serge Biesemans, Hans Lebon, Mireille Maenhoudt, Werner Boullart, Geert Vandenberghe, Philippe Absil and Kurt Ronse
- ASML
- LAM

Monique Ercken
© imec 2009

6th Int. Symp. On Immersion Lithography Extentions
Oct. 22-23'09, Prague – Czech Republic

imec

Freeform illumination sources:
Source mask optimization for 22 nm node SRAM

Joost Bekaert

IMEC: B. Laenens, S. Verhaegen, L. Van Look, D. Trivkovic, F. Lazzarino, G. Vandenberghe
ASML/Brion: P. van Adrichem, R. Socha, M. Mulder, S. Baron, M.-C. Tsai, K. Ning, S. Hsu, A. Bouma, E. van der Heijden, K. Schreel, R. Carpaij, M. Dusa
ZEISS: J. Zimmermann, P. Gräupner, C. Hennerkes

Outline

- **Motivation and goal**

- Source-Mask Optimization flow

- SMO for the CH & ME1 layer of 22 nm node SRAM

- FlexRay wafer data

- Conclusion

6th International Symposium on
Immersion Lithography Extensions
Prague, 2009

J. Bekaert
© imec 2009

Source Mask Optimization (SMO)

What is Source Mask Optimization?

Clip

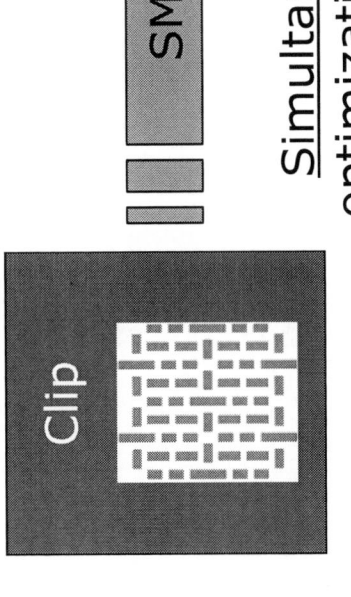

SMO

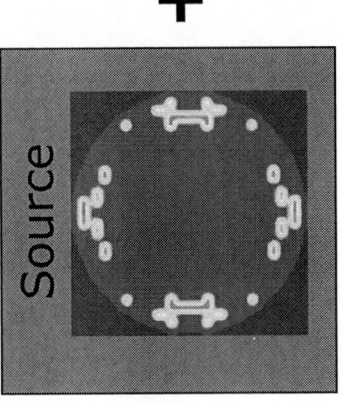

Source

+

OPCed clip

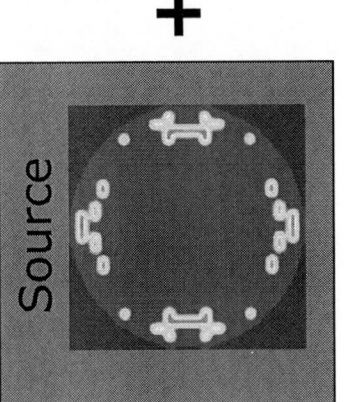

<u>Simultaneous</u> co-optimization of mask and source for given input design.

Goal of Source Mask Optimization:

→ extend the limits of 193 immersion lithography (for SP & DP) by using the optimal source for your specific design

- Increase yield (through PW, MEEF)
- Create margin allowing for some further downscaling

6th International Symposium on
Immersion Lithography Extensions
Prague, 2009

J. Bekaert
© imec 2009

Motivation and goal of this work

This exercise:

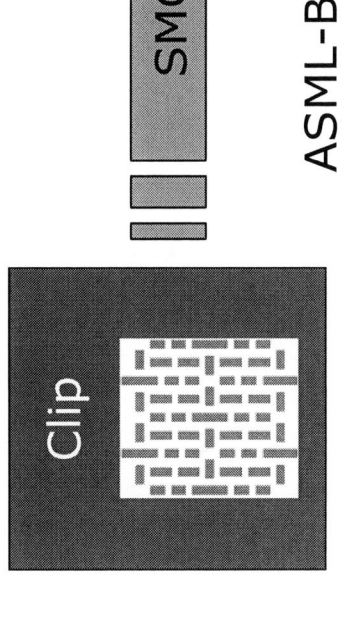

Clip

SRAM clip

SMO

ASML-BRION
Tachyon SMO

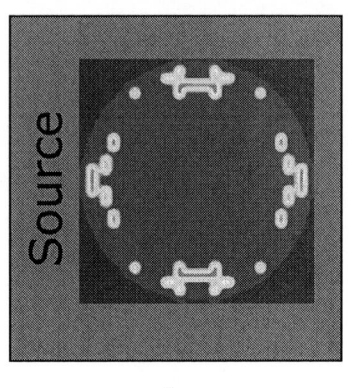

Source

Freeform source
(Pixelated DOE's / FlexRay)

+

OPCed clip

Clips on SMO
mask

Motivation of this work:

- Go through the **full SMO flow**, from design input to printed wafer data.

- Evaluate **benefit of freeform source** shapes over standard source shapes for an advanced use-case.

- Preparation towards full **FlexRay functionalities**

J. Bekaert
© imec 2009

6th International Symposium on
Immersion Lithography Extensions
Prague, 2009

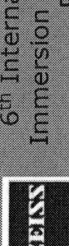  

- Motivation and goal
- **Source-Mask Optimization flow**
- SMO for the CH & ME1 layer of 22 nm node SRAM
- FlexRay wafer data
- Conclusion

6th International Symposium on
Immersion Lithography Extensions
Prague, 2009

J. Bekaert
© imec 2009

imec   ASML   BRION   ZEISS

Tachyon SMO cost-function

The EPE (edge placement error, contour-based) is minimized within a process variation band determined by defocus, dose and mask CD variations.

EPE evaluation at regular placed evaluation points

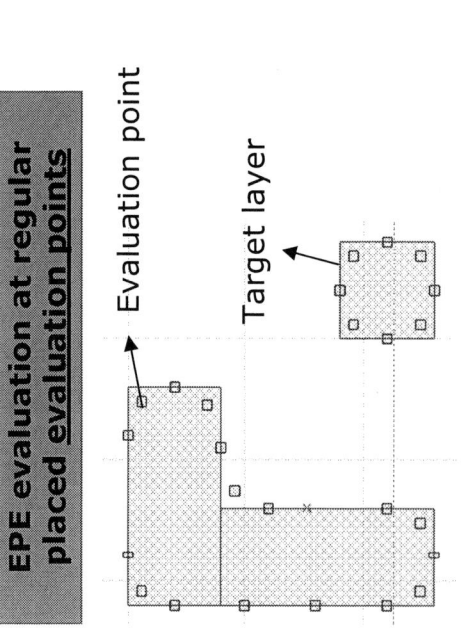

Evaluation point

Target layer

Evaluation through user defined process window and mask error conditions

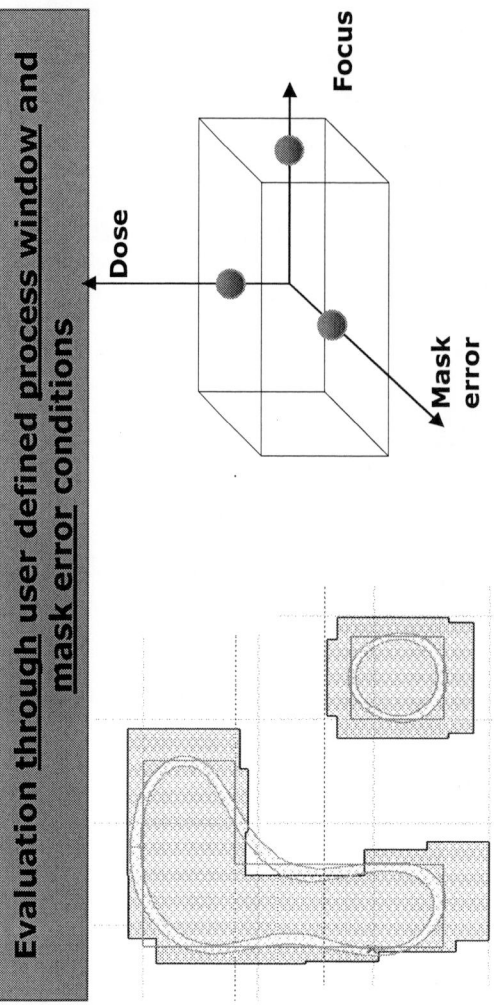

Focus

Dose

Mask error

→ Optimal source has lowest contour EPE

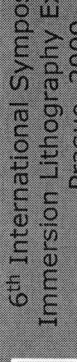

J. Bekaert
© imec 2009

6th International Symposium on
Immersion Lithography Extensions
Prague, 2009

Types of illumination source shapes

In the SMO-setup, the user chooses the type of source to be optimized:

Standard source	Parametric source	Freeform source
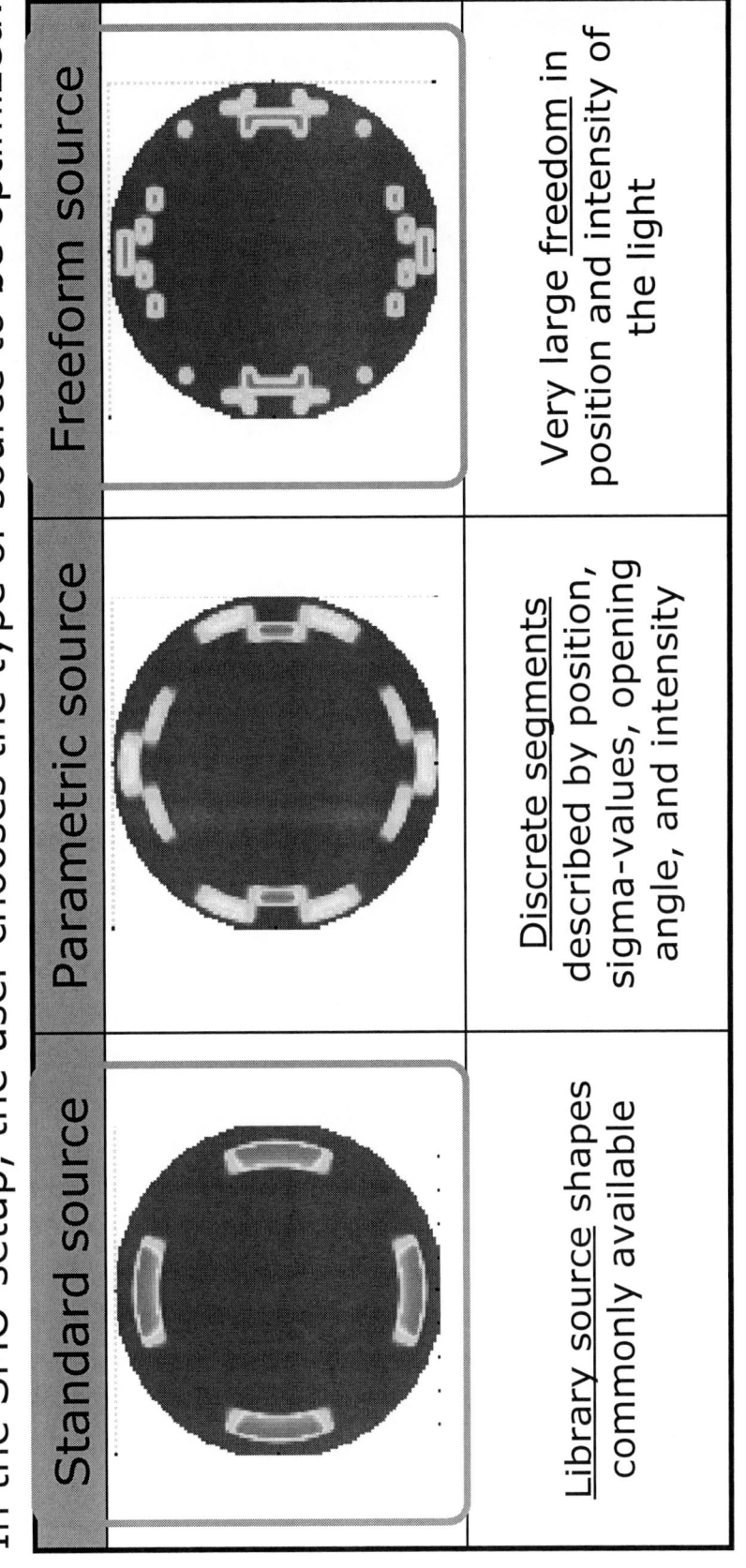		
Library source shapes commonly available	Discrete segments described by position, sigma-values, opening angle, and intensity	Very large freedom in position and intensity of the light

In this work we focus on *freeform* and *standard* source optimization

6th International Symposium on
Immersion Lithography Extensions
Prague, 2009

J. Bekaert
© imec 2009

- Motivation and goal

- Source-Mask Optimization flow

- SMO for the CH & ME1 layer of 22 nm node SRAM
  - SMO input: SRAM clip & settings
  - Calculation of optimal sources
  - Verification of the freeform source
  - Verification on wafer: freeform vs standard source

  Pixelated DOEs made and installed

  SMO-mask made

- FlexRay wafer data

- Conclusion

6th International Symposium on
Immersion Lithography Extensions
Prague, 2009

J. Bekaert
© imec 2009

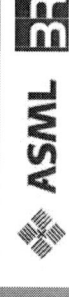

Input design:
SRAM Contact layer

Optimization 1:

Contact layer of 22 nm node SRAM – 0.099 µm² bit cell size

Unit cell = 4 times bit cell

Pitch 90 nm (V), 110 nm (H)
$=> k_1 = 0.314$

bit cell

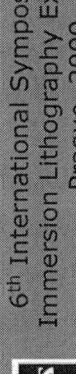

6th International Symposium on
Immersion Lithography Extensions
Prague, 2009

J. Bekaert
© imec 2009

Input design:
SRAM Contact layer

Optimization 1:

Contact layer of 22 nm node SRAM – 0.099 μm² bit cell size

Unit cell = 4 times bit cell

Pitch 90 nm (V), 110 nm (H)
=> k_1 = 0.314

Initial SMO predicted this
clip too aggressive
=> Split design for DPT

Note: Split 1 = Split 2
=> SMO valid for both DP steps

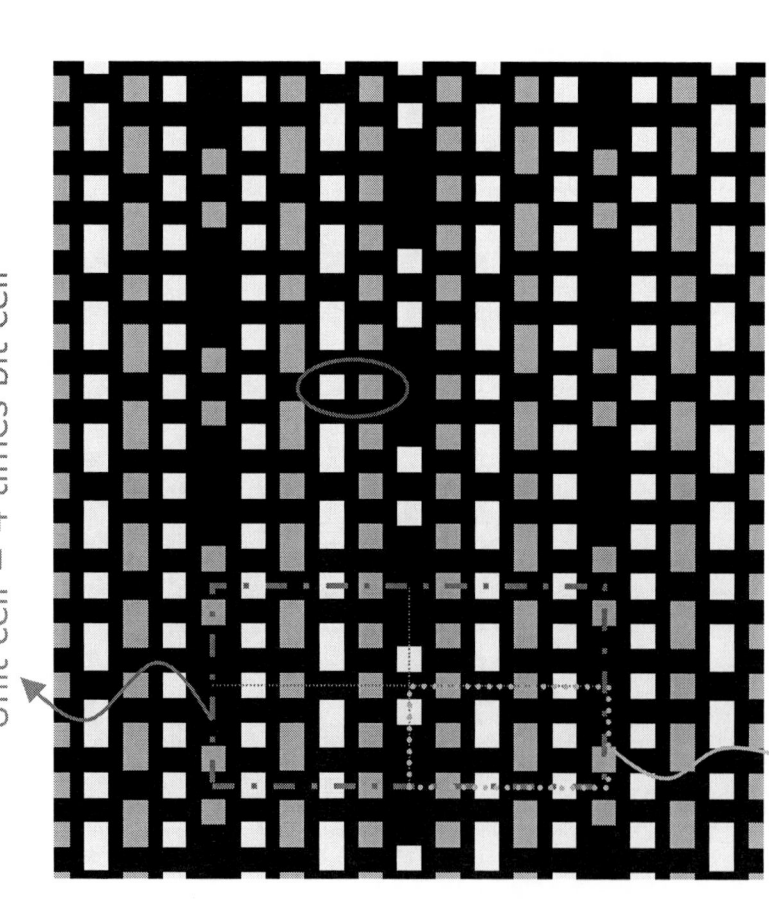

bit cell

6th International Symposium on
Immersion Lithography Extensions
Prague, 2009

J. Bekaert
© imec 2009

Input design:
SRAM Contact layer

Optimization 1:

Contact layer of 22 nm node SRAM – 0.099 µm² bit cell size

Unit cell = 4 times bit cell

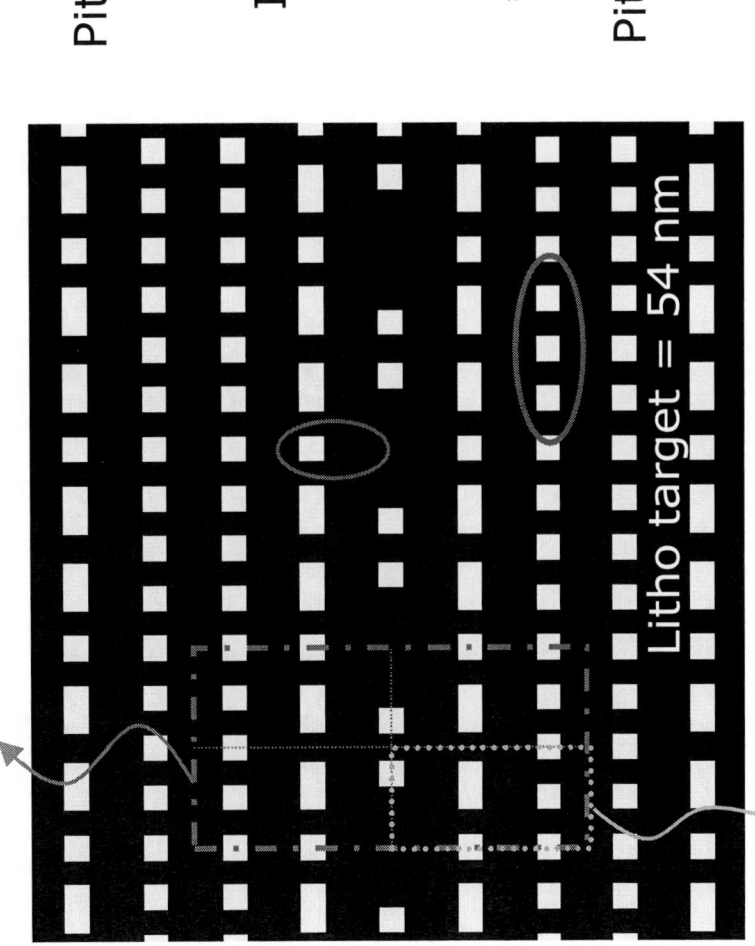

Litho target = 54 nm

bit cell

~~Pitch 90 nm (V), 110 nm (H)~~
~~=> k_1 = 0.314~~

Initial SMO predicted this
clip too aggressive
=> Split design for DPT

Note: Split 1 = Split 2
=> SMO valid for both DP steps

Pitch 180 nm (V), 110 nm (H)
=> k_1 = 0.384

6th International Symposium on
Immersion Lithography Extensions
Prague, 2009

J. Bekaert
© imec 2009

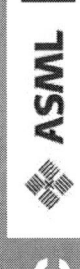

Input design: SRAM Metal layer

Optimization 2:

Metal1 layer of 22 nm node SRAM – 0.099 µm² bit cell size

Pitch 110 nm (H) => k_1 = 0.384
But: small gaps down to 42 nm

Unit cell = 4 times bit cell

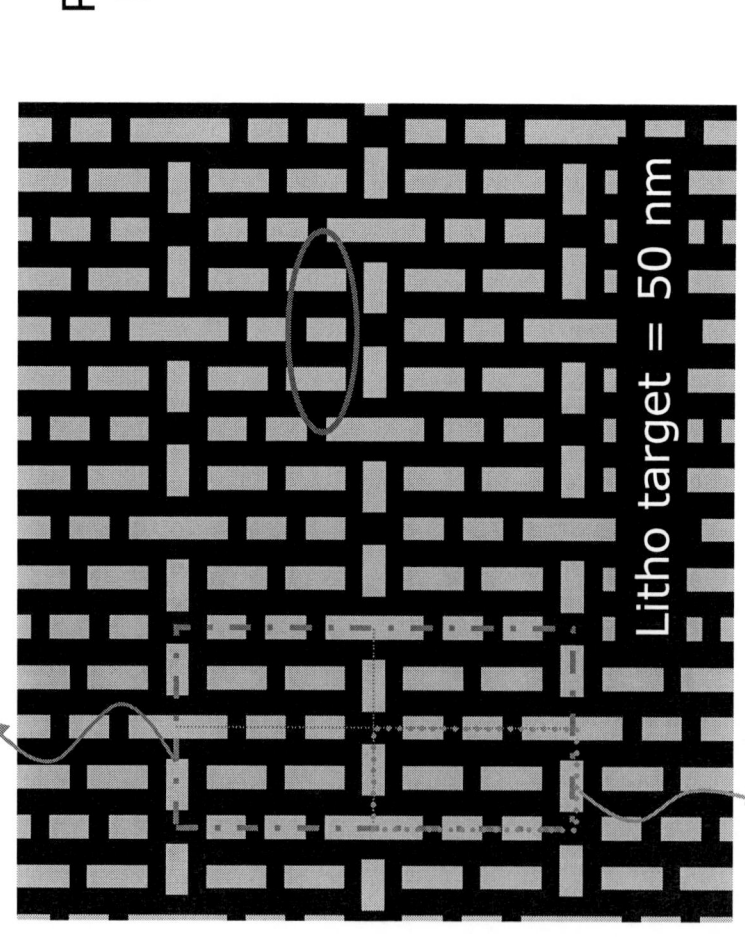

Litho target = 50 nm

bit cell

6th International Symposium on
Immersion Lithography Extensions
Prague, 2009

J. Bekaert
© imec 2009

SMO settings

The simulations (as well as the experiments) were performed using the following stack:

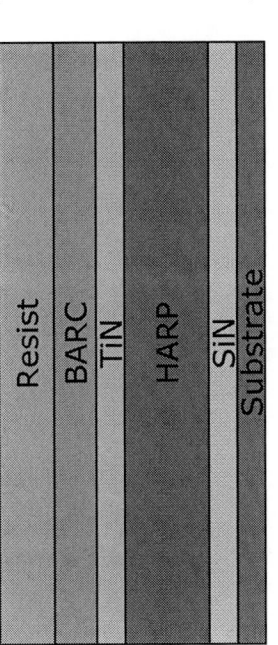

Representative back-end stack

Since no calibrated resist model was available, the calculation of the optimal sources and OPC was done using a simplified model using 12 nm blur.

After investigations of a variety of options, it was found that the better results were predicted using the following settings:

- 6% attPSM mask
- Light field mask in combination with negative tone development
- XY polarized light

6th International Symposium on
Immersion Lithography Extensions
Prague, 2009

J. Bekaert
© imec 2009

- Motivation and goal

- Source-Mask Optimization flow

- SMO for the CH & ME1 layer of 22 nm node SRAM
  - SMO input: SRAM clip & settings
  - Calculation of optimal sources
    - Verification of the freeform source (pixelated DOE)
    - Wafer-based verification: freeform vs standard source

- FlexRay wafer data

- Conclusion

6th International Symposium on
Immersion Lithography Extensions
Prague, 2009

J. Bekaert
© imec 2009

Calculation of optimal sources
Standard

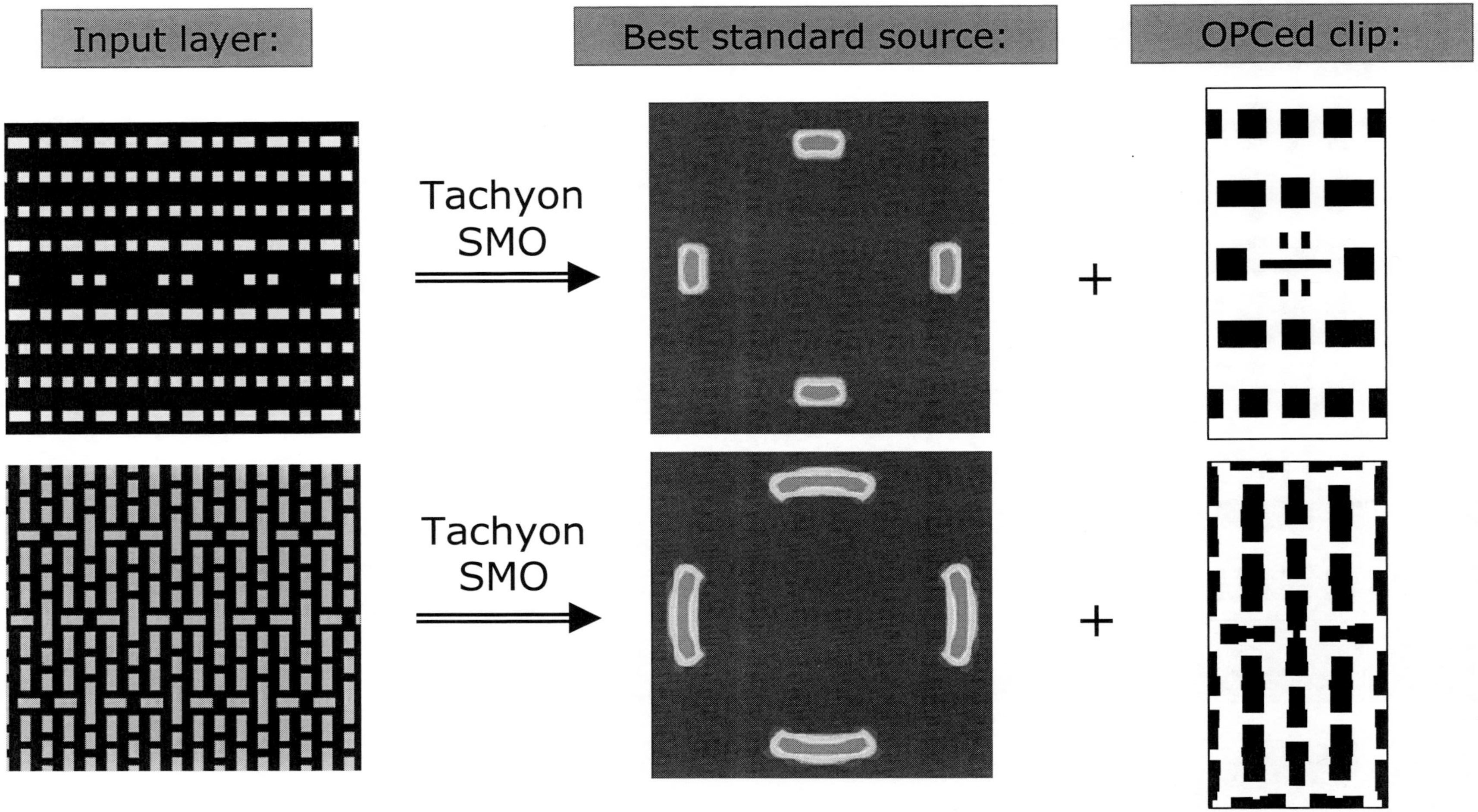

Input layer:

Best standard source:

OPCed clip:

Tachyon SMO

+

Tachyon SMO

+

J. Bekaert
© imec 2009

6th International Symposium on
Immersion Lithography Extensions
Prague, 2009

Calculation of optimal sources
Freeform

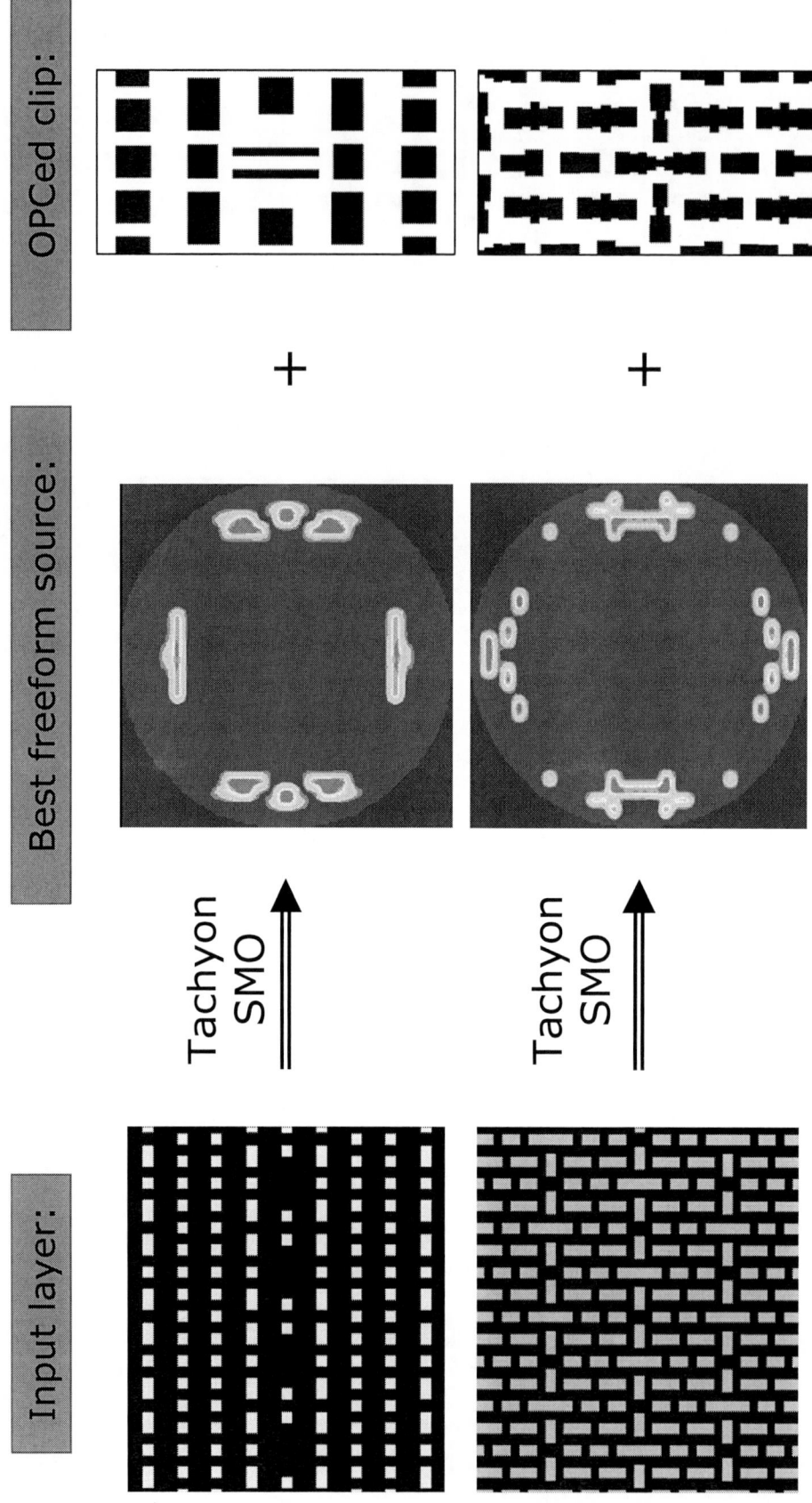

Input layer:

Tachyon SMO

Best freeform source:

+

OPCed clip:

J. Bekaert
© imec 2009

6th International Symposium on
Immersion Lithography Extensions
Prague, 2009

- Motivation and goal

- Source-Mask Optimization flow

- SMO for the CH & ME1 layer of 22 nm node SRAM
  - SMO input: SRAM clip & settings
  - Calculation of optimal sources
  - Verification of the freeform source (pixelated DOE)
  - Wafer-based verification: freeform vs standard source

- FlexRay wafer data

- Conclusion

6th International Symposium on
Immersion Lithography Extensions
Prague, 2009

J. Bekaert
© imec 2009

Pixelated sources: set versus get

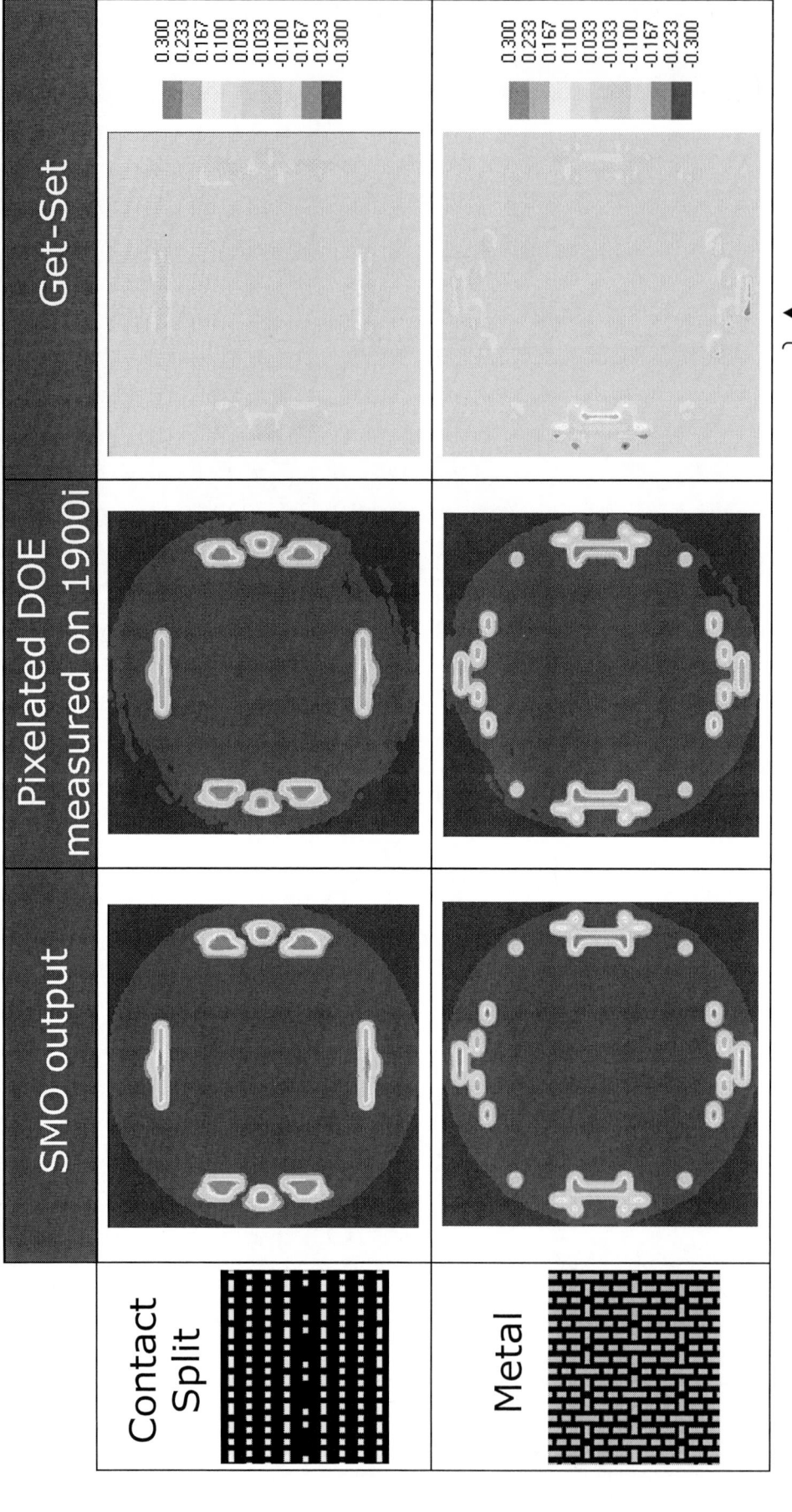

- This result is with limited DOE calibration.
- The differences are easily absorbed in a model calibration that will be carried out in common cases

J. Bekaert
© imec 2009

6th International Symposium on
Immersion Lithography Extensions
Prague, 2009

Verification of the freeform source

Freeform source shapes from fabricated pixelated DOE's are very close to designed source shapes

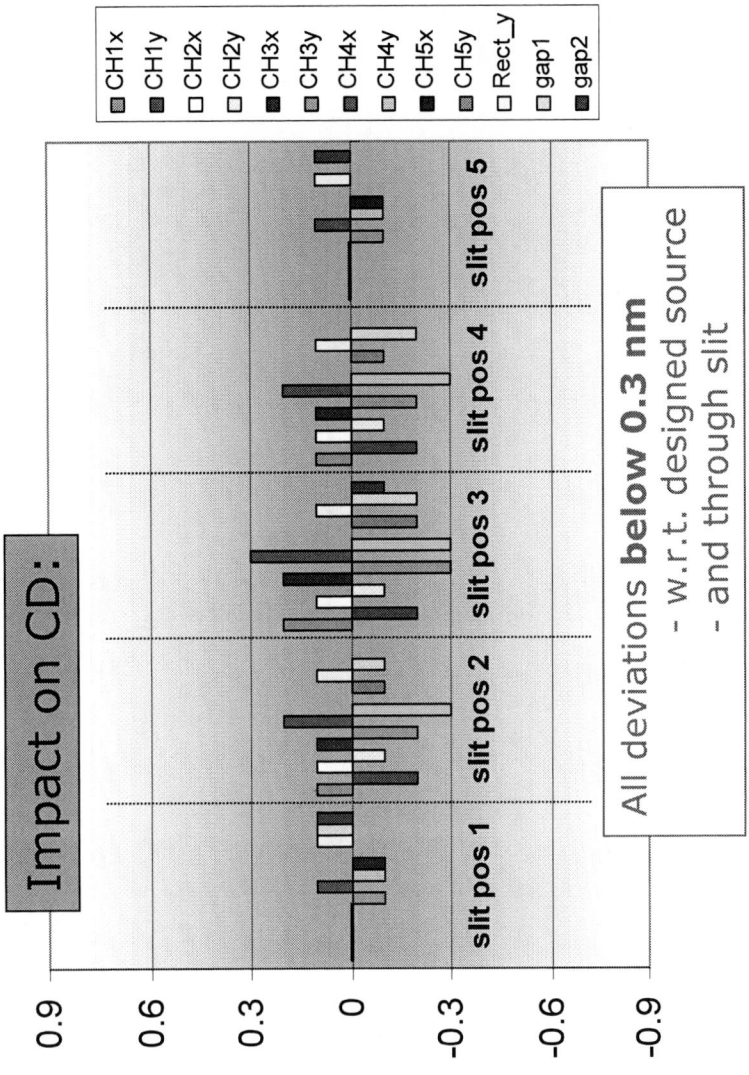

Impact on CD:

All deviations **below 0.3 nm**
- w.r.t. designed source
- and through slit

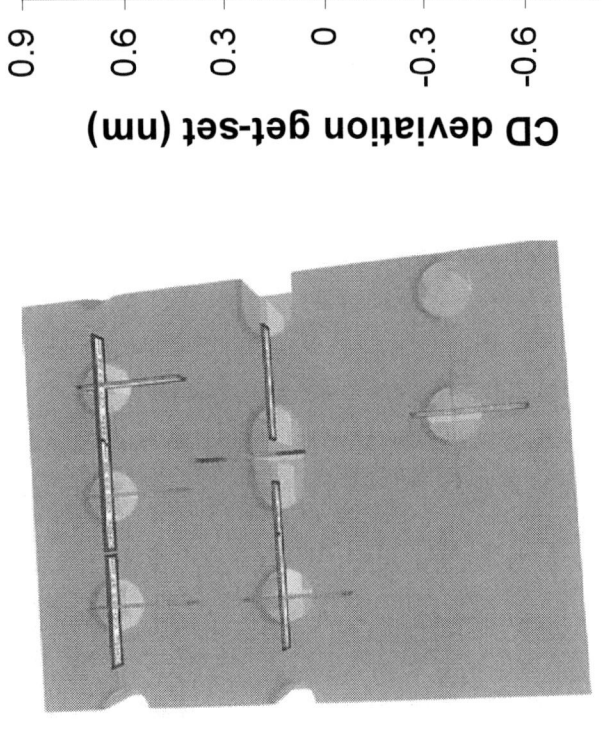

Evaluated 13 cutlines (Prolith)

6th International Symposium on
Immersion Lithography Extensions
Prague, 2009

J. Bekaert
© imec 2009

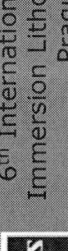

- Motivation and goal

- Source-Mask Optimization flow

- SMO for the CH & ME1 layer of 22 nm node SRAM
  - SMO input: SRAM clip & settings
  - Calculation of optimal sources
  - Verification of the freeform source (pixelated DOE)
  - Wafer-based verification: freeform vs standard source

- FlexRay wafer data

- Conclusion

6th International Symposium on
Immersion Lithography Extensions
Prague, 2009

J. Bekaert
© imec 2009

Contact and metal images at best focus and best dose

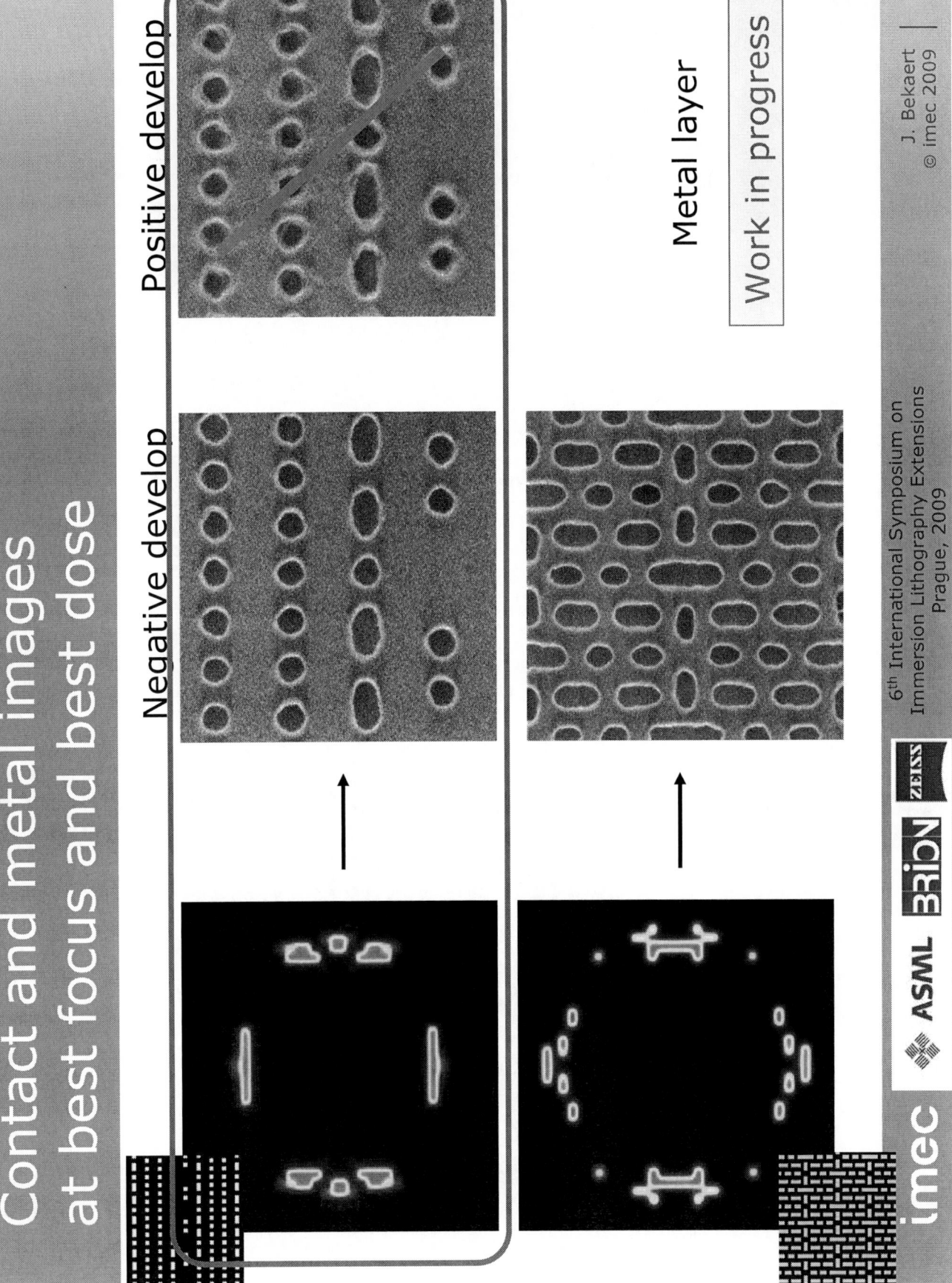

Negative develop Positive develop

Metal layer

Work in progress

6th International Symposium on Immersion Lithography Extensions Prague, 2009

J. Bekaert
© imec 2009

Through focus performance @ best dose

Contact layer – design split – k_1=0.384

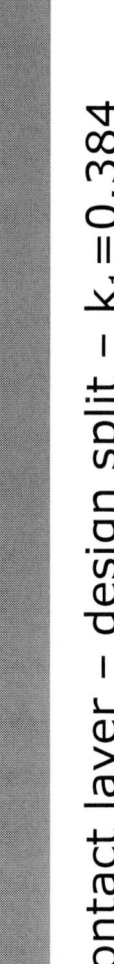

Best focus

−56 nm defocus −48 nm defocus best dose +40 nm defocus +48 nm defocus

Freeform

Standard

6th International Symposium on
Immersion Lithography Extensions
Prague, 2009

J. Bekaert
© imec 2009

SEM metrology

Six metrology sets are defined based on symmetry reasons

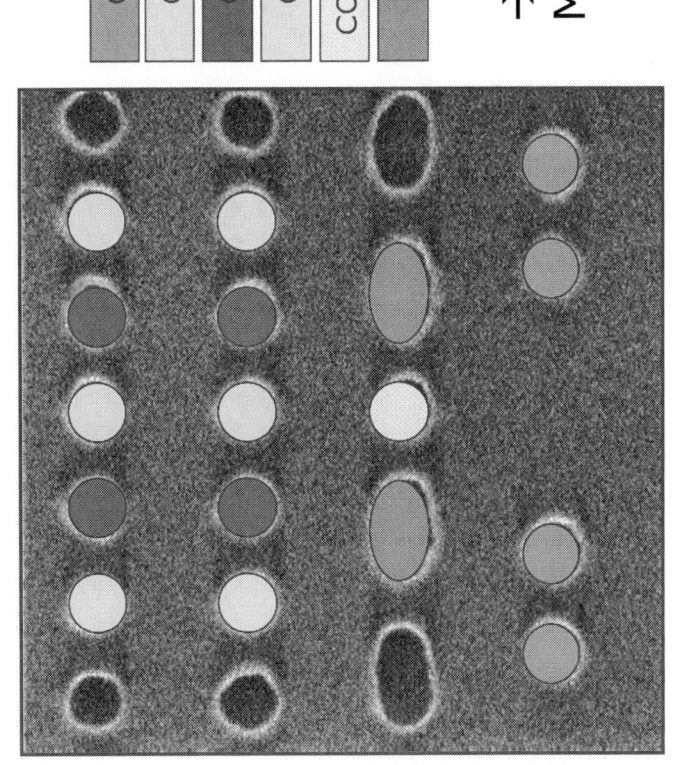

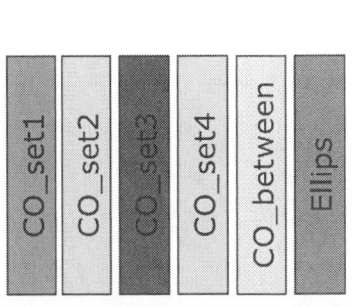

CO_set1
CO_set2
CO_set3
CO_set4
CO_between
Ellips

→ For each set EL, DoF and MEEF are measured on wafer.

All targets are 54 nm with spec of +-4 nm.
(EllipsX = 104 +- 4 nm)

6th International Symposium on
Immersion Lithography Extensions
Prague, 2009

J. Bekaert
© imec 2009

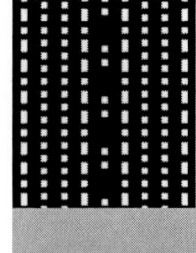

PW window results for freeform and standard source

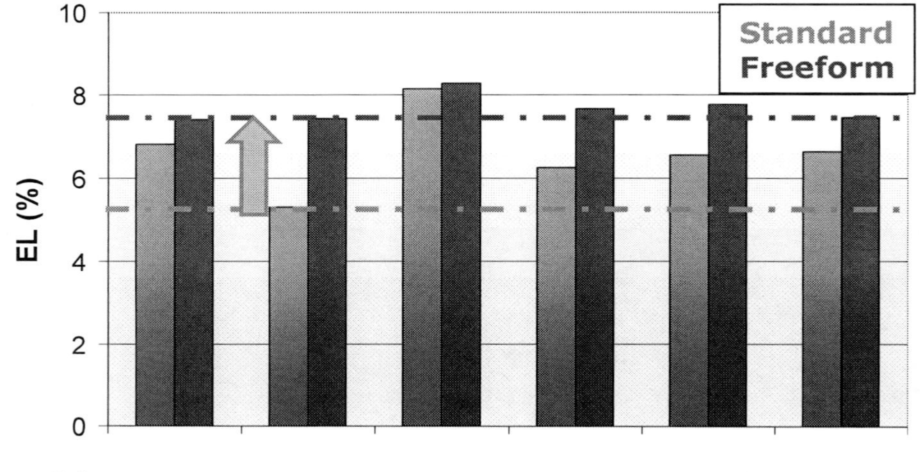

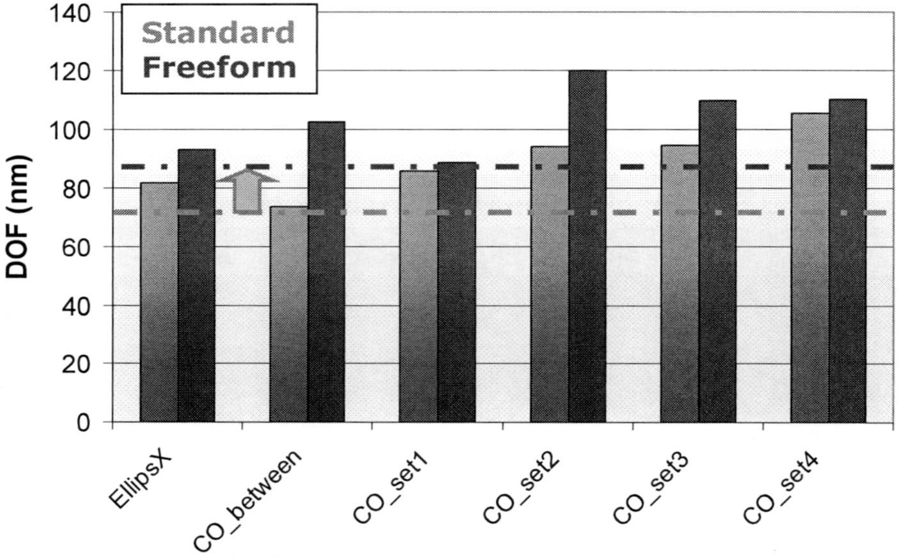

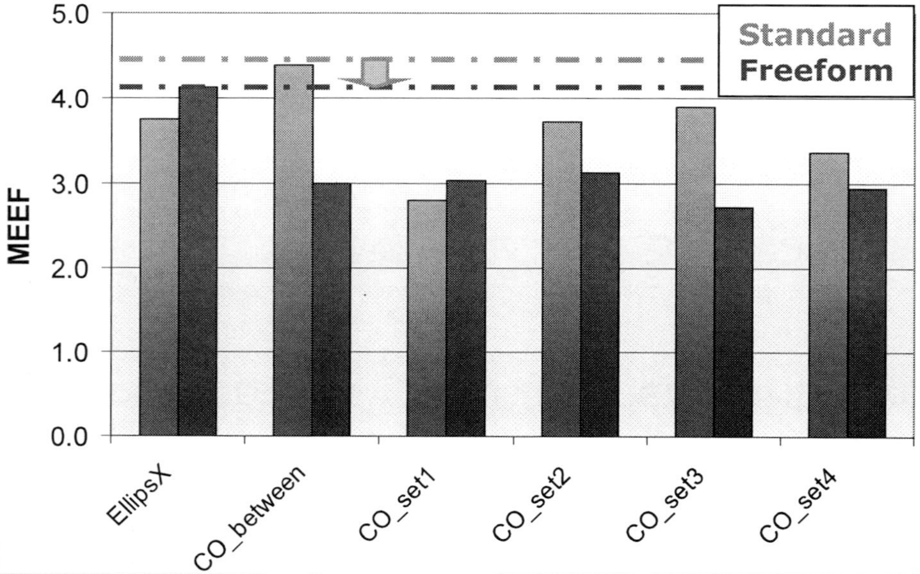

	Standard	Freeform
Lowest EL (%)	5	7
Lowest DOF (nm)	74	89
Max. MEEF	4.4	4.1

→ Overall, freeform source performs better than standard source

J. Bekaert
© imec 2009

Immersion Lithography Extensions
onal Symposium on
Prague, 2009

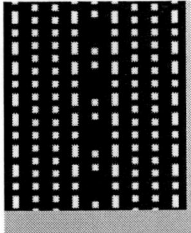

PW window results for freeform and standard source (2)

Feature with the lowest contrast:

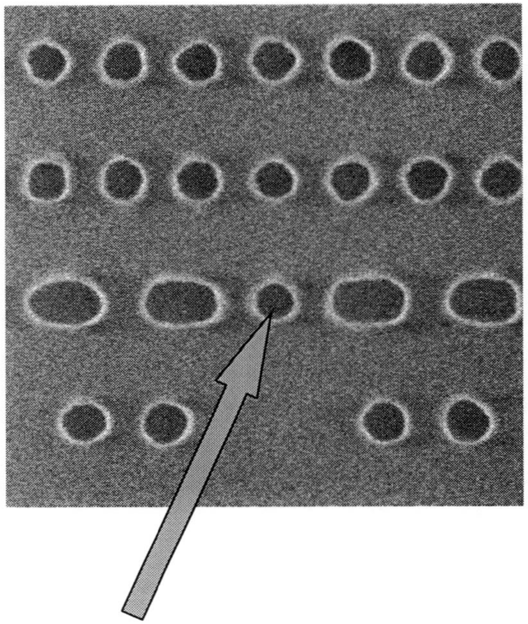

Hot spot

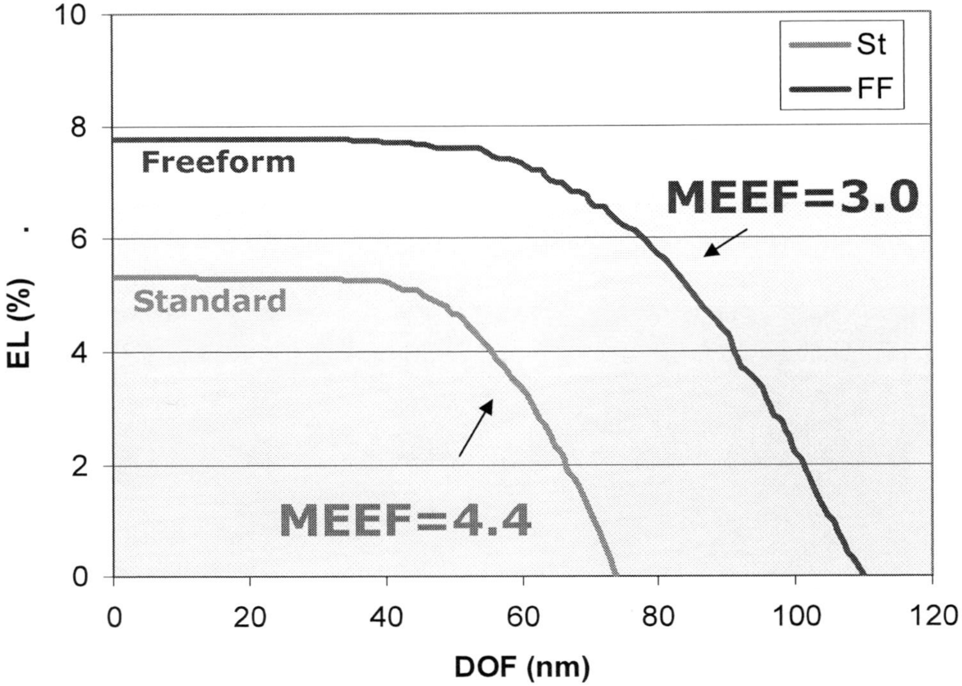

→ Clear benefit from freeform to standard source for the most difficult feature.

6th International Symposium on
Immersion Lithography Extensions
Prague, 2009

J. Bekaert
© imec 2009

Result after double patterning:
Litho – etch – litho – etch (LELE)

Image after etch of the contact pattern into TiN hardmask.

P180

Full contact layer of 22 nm SRAM after LELE.

<u>Note:</u>
initial result
• Etch process not optimized

P110

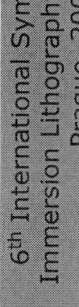

6th International Symposium on Immersion Lithography Extensions Prague, 2009

J. Bekaert
© imec 2009

imec ASML BRION ZEISS

Result after double patterning:
Litho – etch – litho – etch (LELE)

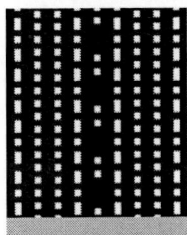

Image after etch of the contact pattern into TiN hardmask.

P90

P90

Full contact layer of 22 nm SRAM after LELE.

P110

<u>Note:</u>
initial result
- Etch process not optimized
- Alignment/overlay not optimized

6th International Symposium on
Immersion Lithography Extensions
Prague, 2009

J. Bekaert
© imec 2009

Co-patterning of standard logic cell with freeform source optimized for SRAM

Pitch for logic/periphery matched to SRAM pitch.
OPC for freeform source.

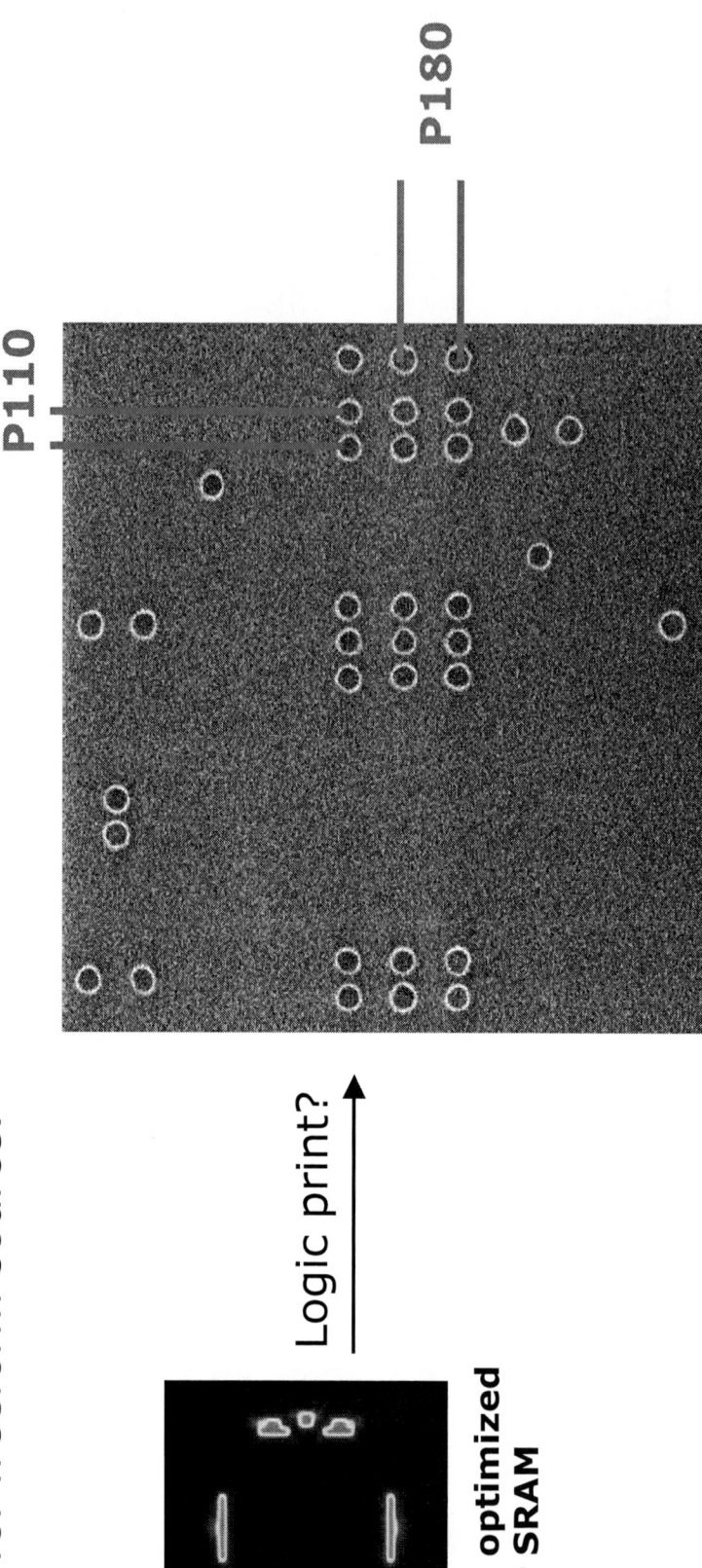

P110

P180

Logic print?

Source optimized for SRAM

→ Standard logic cell can be printed with SRAM-freeform source.

6th International Symposium on
Immersion Lithography Extensions
Prague, 2009

J. Bekaert
© imec 2009

Outline

- Motivation and goal

- Source-Mask Optimization flow

- SMO for the CH & ME1 layer of 22 nm node SRAM

- **FlexRay wafer data**
  - The FlexRay illuminator
  - FlexRay versus DOE: comparison of imaging performance — FlexRay experiments

- Conclusion

6th International Symposium on
Immersion Lithography Extensions
Prague, 2009

J. Bekaert
© imec 2009

FlexRay Optical Design 'Replacing DOE with MMA'
→ *For Instant Freeform Illumination Sources*

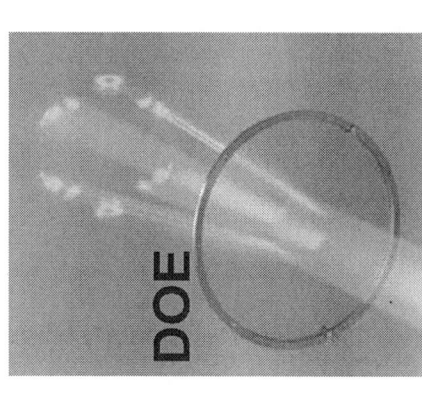

Existing Illumination

Changing Pupils by Exchanging
Diffractive Optical Element (DOE)

→ *DOE order lead time*

FlexRay Illumination

Changing Pupils with
Micro-Mirror-Array (MMA)

→ *Any Pupil – Any Time*

Optional available on
XT:19x0i and NXT:1950i

Immersion Symposium, Prague 2009

FlexRay versus DOE
Source and wafer images

Images at BF/BD, from 1950i scanner before and after FlexRay install.

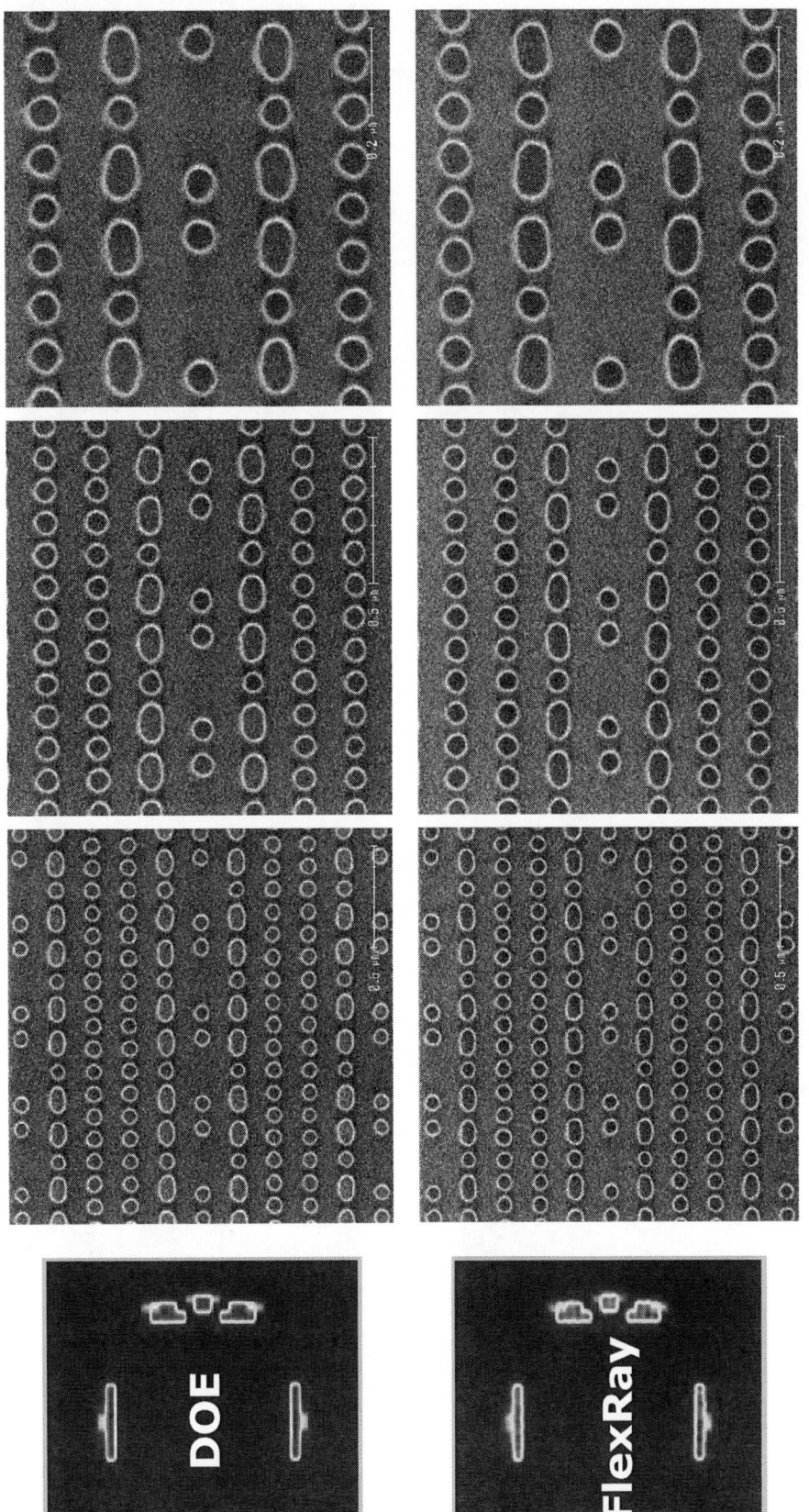

Results obtained on first FlexRay scanner, within first 24 hours of operation.

6th International Symposium on
Immersion Lithography Extensions
Prague, 2009

J. Bekaert
© imec 2009

FlexRay versus DOE
Process window comparison

Measured process windows, before and after FlexRay install.

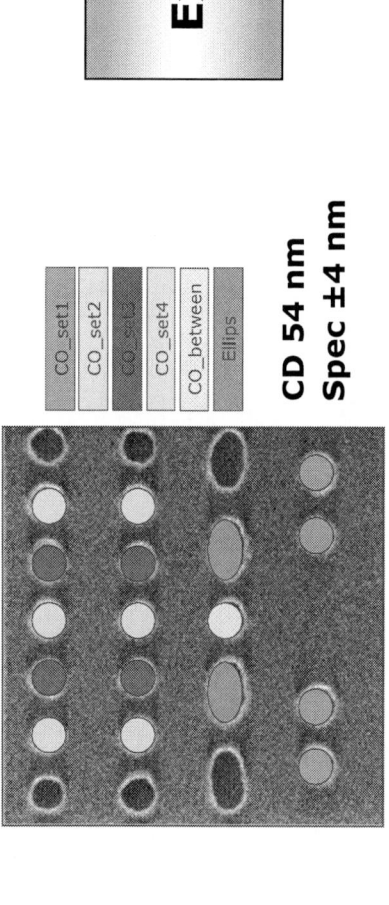

CD 54 nm
Spec ±4 nm

FlexRay versus DOE:
Experimental process windows
are identical

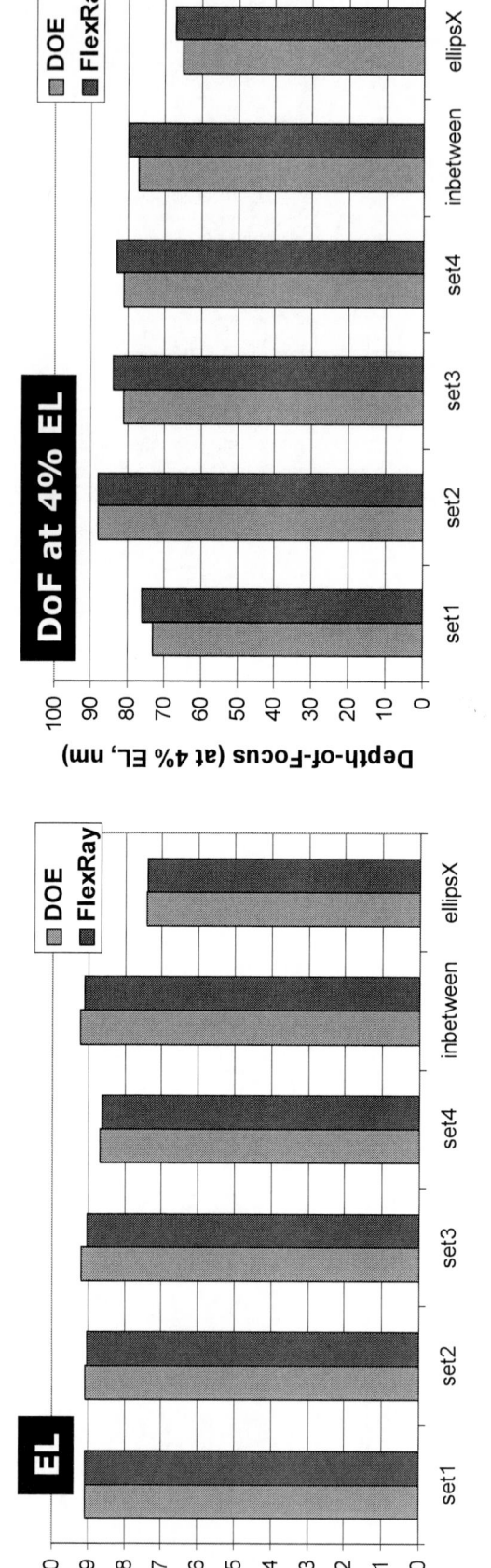

6th International Symposium on
Immersion Lithography Extensions
Prague, 2009

J. Bekaert
© imec 2009

FlexRay
Exploratory work on further device downscaling

Same contact design (split), but downscaled by 20% from 0.099 μm² to 0.078 μm² bit cell size.

(min. pitch 110 => 100 nm)

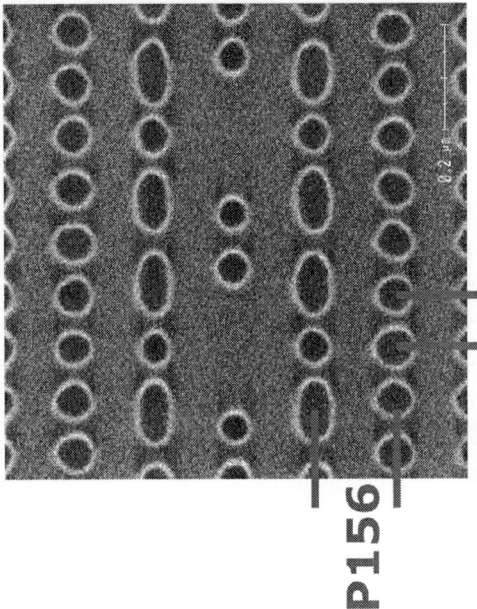

P156

P100

Wafer print

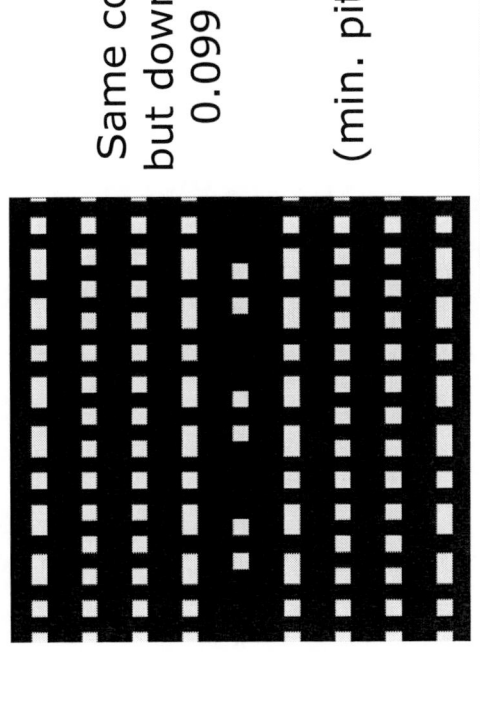

Tachyon SMO

Designed

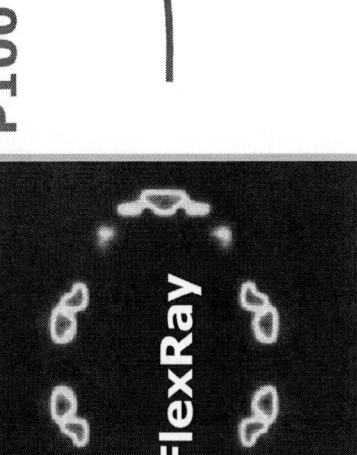

FlexRay

→ FlexRay facilitates exploratory work on downscaling of patterns, with huge source variety in very limited time.

6th International Symposium on Immersion Lithography Extensions
Prague, 2009

J. Bekaert
© imec 2009

- Motivation and goal

- Source-Mask Optimization flow

- SMO for the CH & ME1 layer of 22 nm node SRAM

- FlexRay wafer data

- Conclusion

J. Bekaert
© imec 2009

6th International Symposium on
Immersion Lithography Extensions
Prague, 2009

Summary and conclusions

- We went through a full source-mask optimization flow, and experimentally investigated the benefit of freeform sources to standard sources for a particular 22 nm node SRAM cell (Contact + Metal layer)

- Optimized freeform and standard sources were simulated using Tachyon SMO. The optimized freeform sources and a dedicated SMO-mask were fabricated.

- The measured source shape of the freeform DOE's is in very good agreement with the ordered source shape. This demonstrates the fabrication capabilities and setup of freeform DOE's.

- The gain of freeform to standard source is confirmed for this particular SRAM case. In particular, clear improvement was found for a hotspot in the contact layer.

- FlexRay exposures on a fully operational ASML 1950i scanner have been performed. Identical imaging performance was found for illumination by FlexRay and DOE. FlexRay facilitates exploratory work with freeform sources.

6th International Symposium on
Immersion Lithography Extensions
Prague, 2009

J. Bekaert
© imec 2009

Acknowledgements

Peter De Bisschop, Rudi De Ruyter, Jeroen Van de Kerkhove

Jens-Timo Neumann

Alexander Serebryakov, Steve Hansen, Hua-Yu Liu, Orion Mouraille, Jo Finders, Samy Saleh, Andre Engelen

Jan Hendrik Peters, Karsten Bubke, Romy Schlesier, Tom Busche

Daisuke Hibino, Toru Ishimoto, Kohei Sekiguchi, Naoki Yasui

Mario Reybrouck, Shinji Tarutani

6th International Symposium on Immersion Lithography Extensions
Prague, 2009

J. Bekaert
© imec 2009

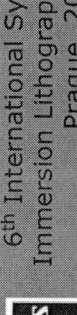

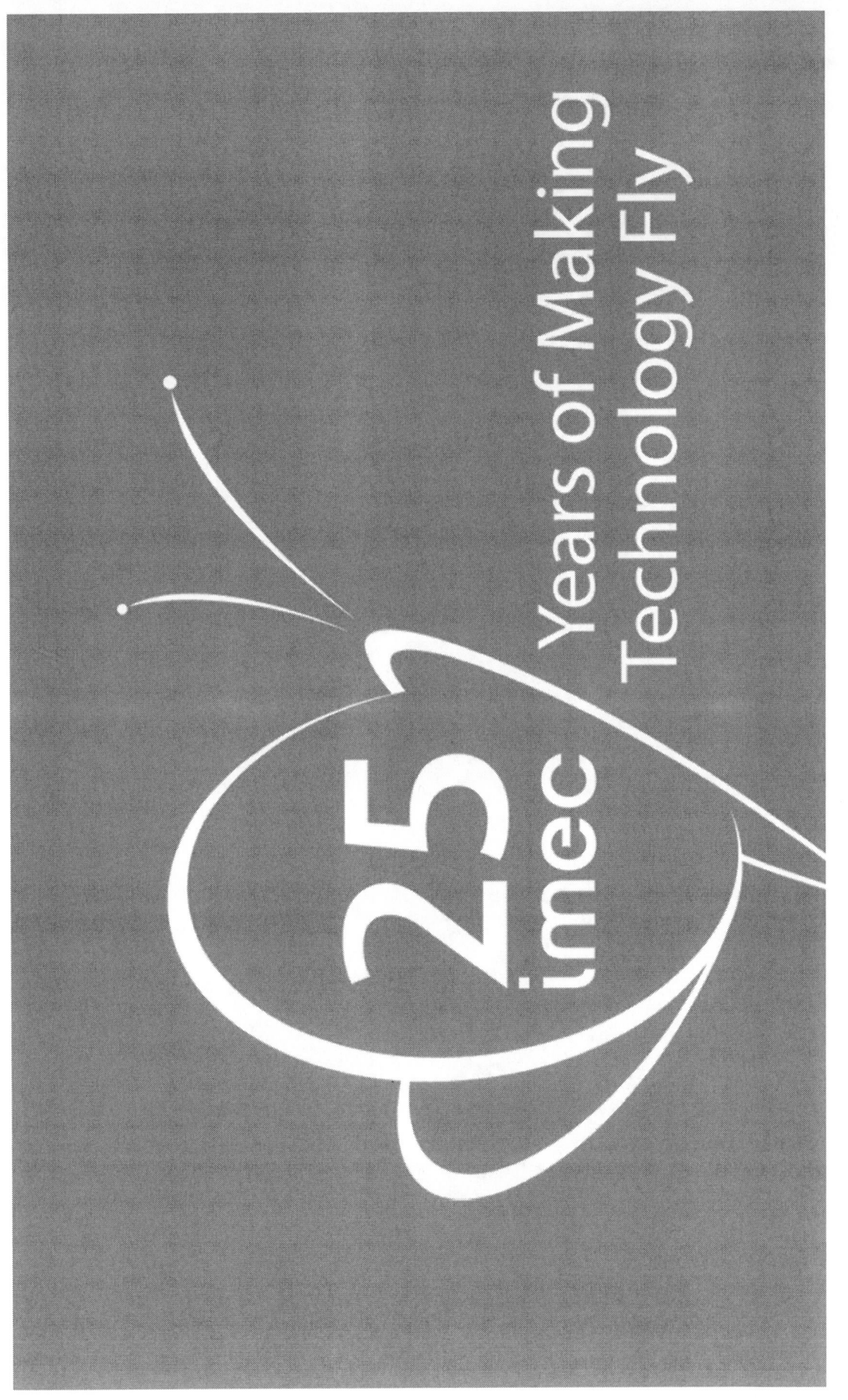

Advances in Modeling and Optical Performance of DOEs Used for OAI in Immersion Lithography

James Carriere, Jared Stack,
Alan Kathman, Marc Himel

10/22/2009

TESSERA

Utilization of DOEs in Scanner Systems

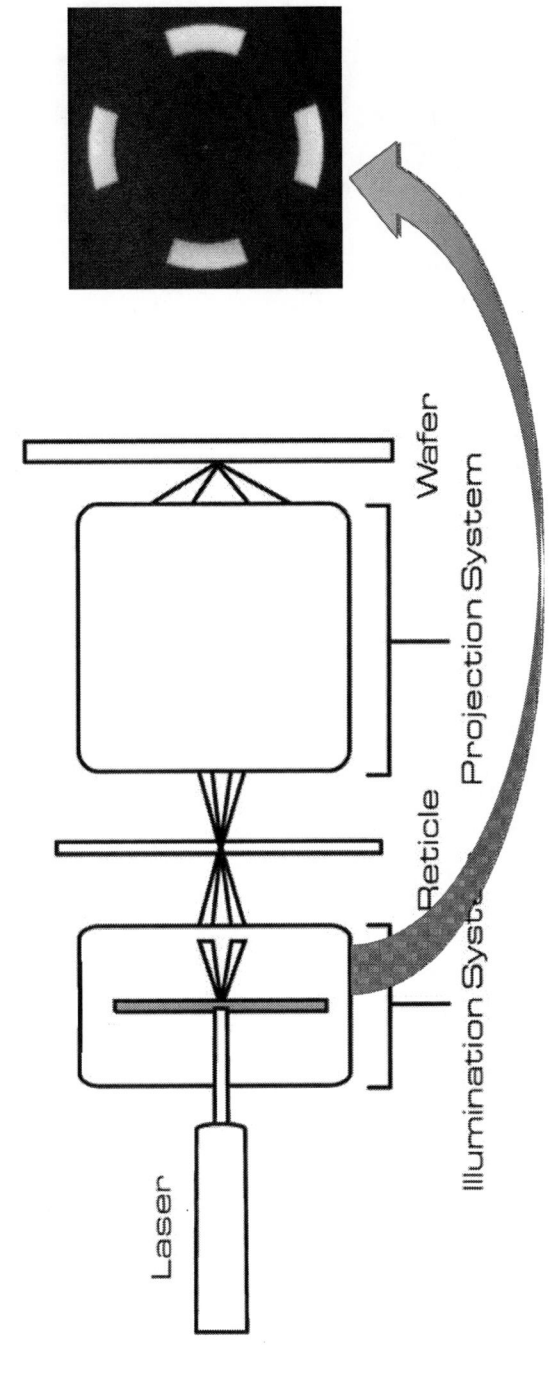

Laser

Wafer

Reticle

Projection System

Illumination System

Tessera Diffractive Optical Elements (DOEs)

- Can create any desired pattern in the pupil plane of the illumination system

- Have been used by the lithography industry since 1996

- Are able to withstand billions of pulses without degrading in performance

- Are stable, proven and require no moving parts

10/22/2009   6th International Symposium on Immersion Lithography Extensions

TESSERA.

Overview

- Manufacture and Usage of DOEs in Scanner Illuminators

- DOE Specifications and Performance

- Pixelated DOEs for Source Mask Optimization (SMO)

- Design and Manufacturing Intrinsic Artifacts

- Predictive Model of DOE Performance

- Recent Advances in Design and Manufacturing

- Availability of Custom DOEs

10/22/2009   6th International Symposium on Immersion Lithography Extensions

TESSERA.

Importance of DOEs to Immersion Lithography

- DOEs are used to reach lower effective k1 values

- Will continue to enable immersion lithography processes
  - Freeform DOEs can extend single patterning for designs whose process windows would otherwise be too small
  - Illumination can be customized for each double patterning layer individually depending on the pattern complexity

- To take full advantage of DOE capabilities, it is important to understand how they are used and fabricated

10/22/2009   6th International Symposium on Immersion Lithography Extensions

Design and Fabrication of DOEs

- DOEs are typically defined by target files with ~200 x 200 pixel resolution

Design & Simulations

10/22/2009   6th International Symposium on Immersion Lithography Extensions

TESSERA

Design and Fabrication of DOEs

- DOEs are typically defined by target files with ~200 x 200 pixel resolution

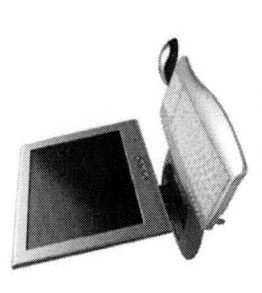

Design & Simulations

10/22/2009   6th International Symposium on Immersion Lithography Extensions

TESSERA.

Design and Fabrication of DOEs

- DOEs are typically defined by target files with ~200 x 200 pixel resolution

Design & Simulations

10/22/2009   6th International Symposium on Immersion Lithography Extensions

Design and Fabrication of DOEs

- DOEs are typically defined by target files with ~200 x 200 pixel resolution

Design & Simulations

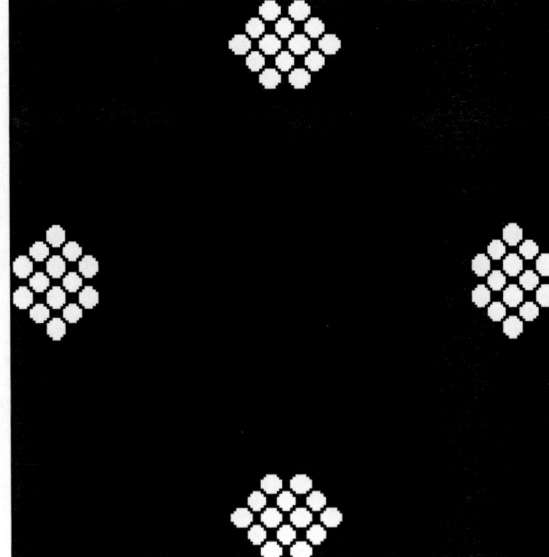

10/22/2009   6th International Symposium on Immersion Lithography Extensions

TESSERA.

Design and Fabrication of DOEs

- DOEs are typically defined by target files with ~200 x 200 pixel resolution

- The intensity of each pixel can be individually controlled to 256 gray levels for a total of 10^7 degrees of freedom

Design & Simulations

10/22/2009   6th International Symposium on Immersion Lithography Extensions

TESSERA.

Design and Fabrication of DOEs

Design & Simulations

- DOEs are typically defined by target files with ~200 x 200 pixel resolution

- The intensity of each pixel can be individually controlled to 256 gray levels for a total of 10^7 degrees of freedom

- The design is transferred to an E-beam written photomask

10/22/2009   6th International Symposium on Immersion Lithography Extensions

TESSERA.

Design and Fabrication of DOEs

- DOEs are typically defined by target files with ~200 x 200 pixel resolution

- The intensity of each pixel can be individually controlled to 256 gray levels for a total of 10^7 degrees of freedom

- The design is transferred to an E-beam written photomask

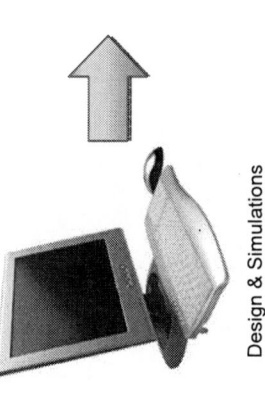

Design & Simulations

E-Beam Photomask

Design and Fabrication of DOEs

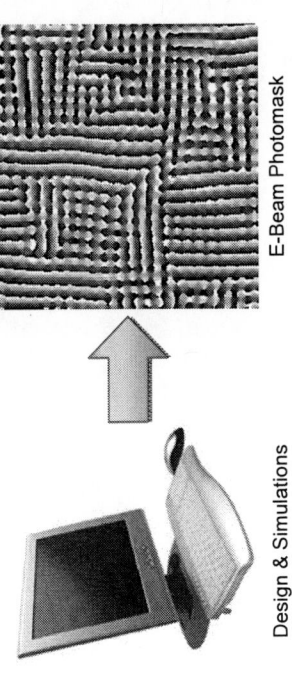

Design & Simulations

E-Beam Photomask

- DOEs are typically defined by target files with ~200 x 200 pixel resolution

- The intensity of each pixel can be individually controlled to 256 gray levels for a total of 10^7 degrees of freedom

- The design is transferred to an E-beam written photomask

10/22/2009   6th International Symposium on Immersion Lithography Extensions

TESSERA.

Design and Fabrication of DOEs

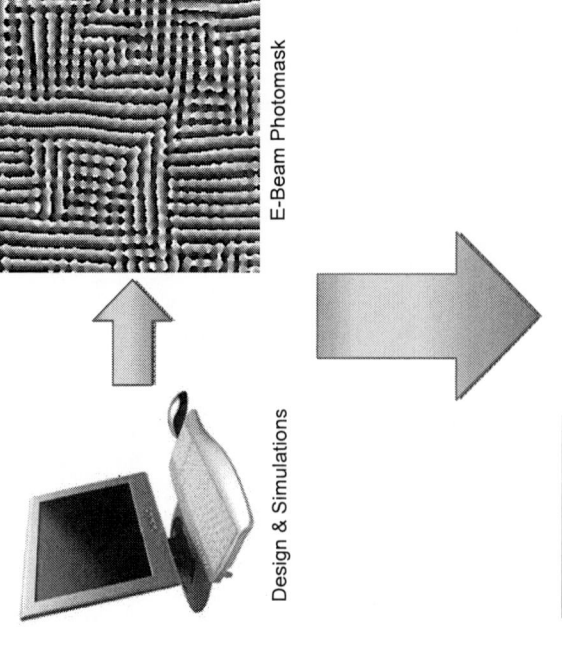

Design & Simulations

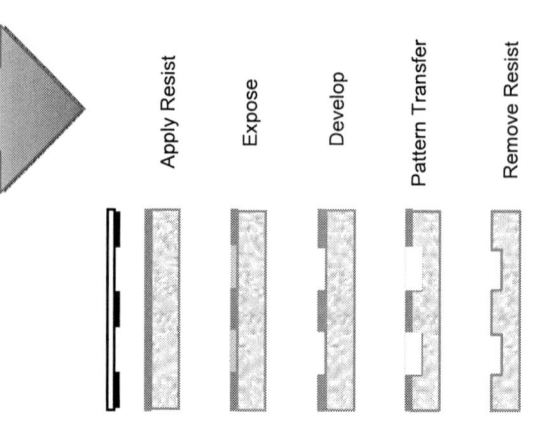

E-Beam Photomask

Apply Resist

Expose

Develop

Pattern Transfer

Remove Resist

- DOEs are typically defined by target files with ~200 x 200 pixel resolution

- The intensity of each pixel can be individually controlled to 256 gray levels for a total of 10^7 degrees of freedom

- The design is transferred to an E-beam written photomask

- Standard lithographic procedures are used to print the pattern in photoresist and transfer etch into a glass substrate

10/22/2009   6th International Symposium on Immersion Lithography Extensions

TESSERA.

Design and Fabrication of DOEs

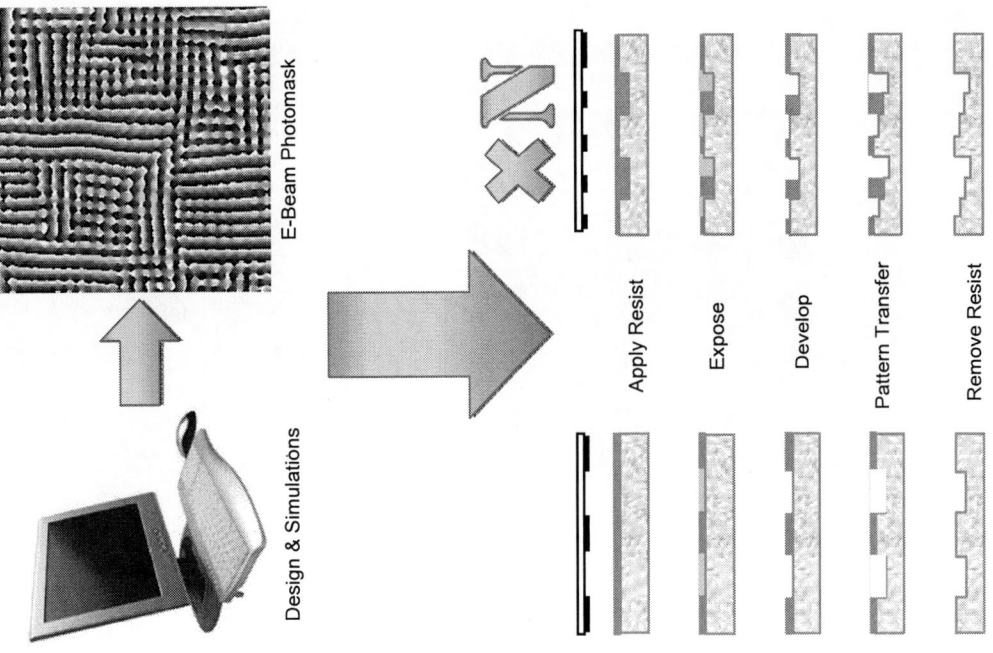

Design & Simulations

E-Beam Photomask

Apply Resist

Expose

Develop

Pattern Transfer

Remove Resist

- DOEs are typically defined by target files with ~200 x 200 pixel resolution

- The intensity of each pixel can be individually controlled to 256 gray levels for a total of 10^7 degrees of freedom

- The design is transferred to an E-beam written photomask

- Standard lithographic procedures are used to print the pattern in photoresist and transfer etch into a glass substrate

10/22/2009   6th International Symposium on Immersion Lithography Extensions

TESSERA.

What Resolution is Needed?

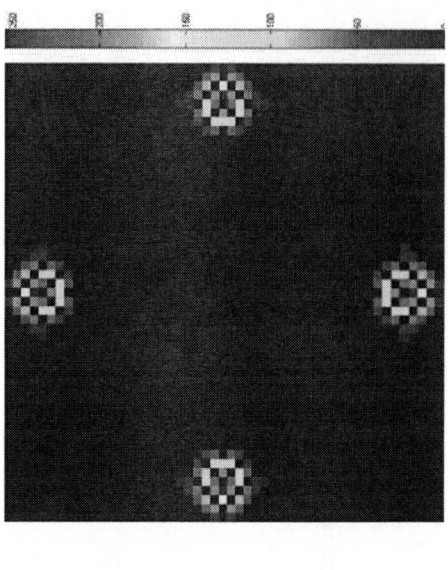

51 x 51 Pixels
Considerable Loss

101 x 101 Pixels
Some Loss

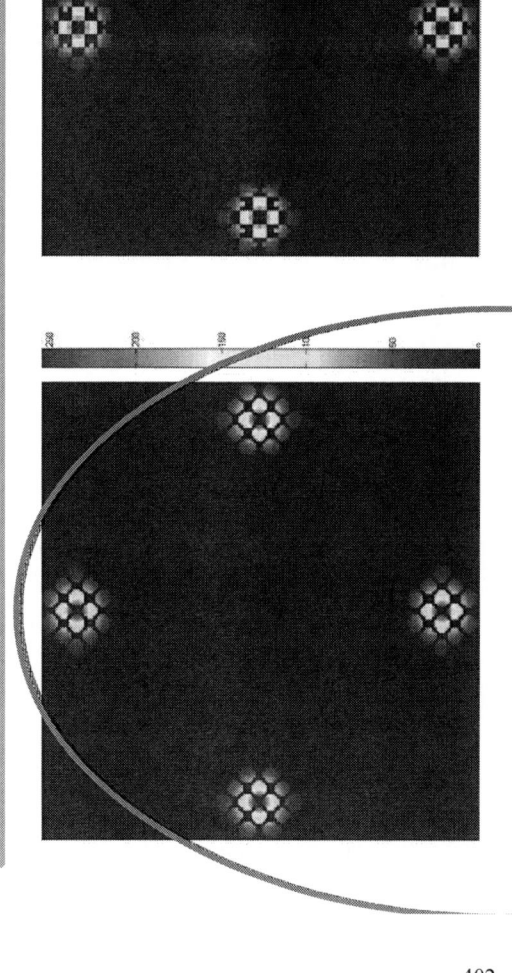

201 x 201 Pixels
Excellent Match

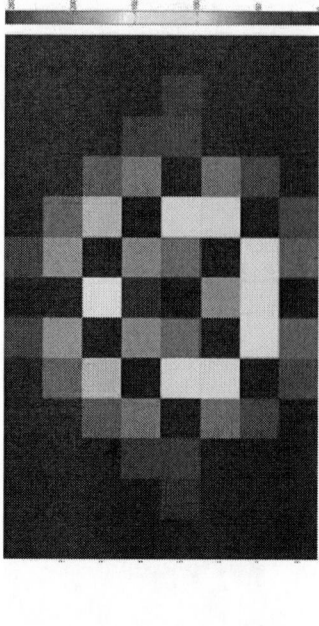

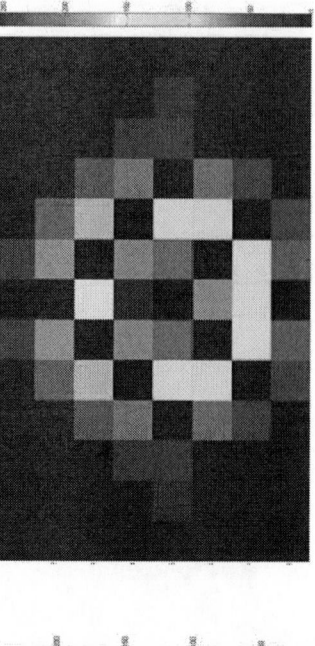

Tessera chooses ~200x200 pixels to maximize the benefits of the DOE illumination

10/22/2009   6th International Symposium on Immersion Lithography Extensions

TESSERA.

Benefits of ~200x200 Pixel Resolution

- Takes full advantage of DOE degrees of freedom

- Resolution exceeds the limits of the illumination system
  - Not the limiting constraint in illumination pattern flexibility

- Higher pixel resolutions have limited impact due to coherence properties of the laser source

- Design complexity is manageable to minimize production lead time

10/22/2009   6th International Symposium on Immersion Lithography Extensions

TESSERA.

Custom DOEs for SMO

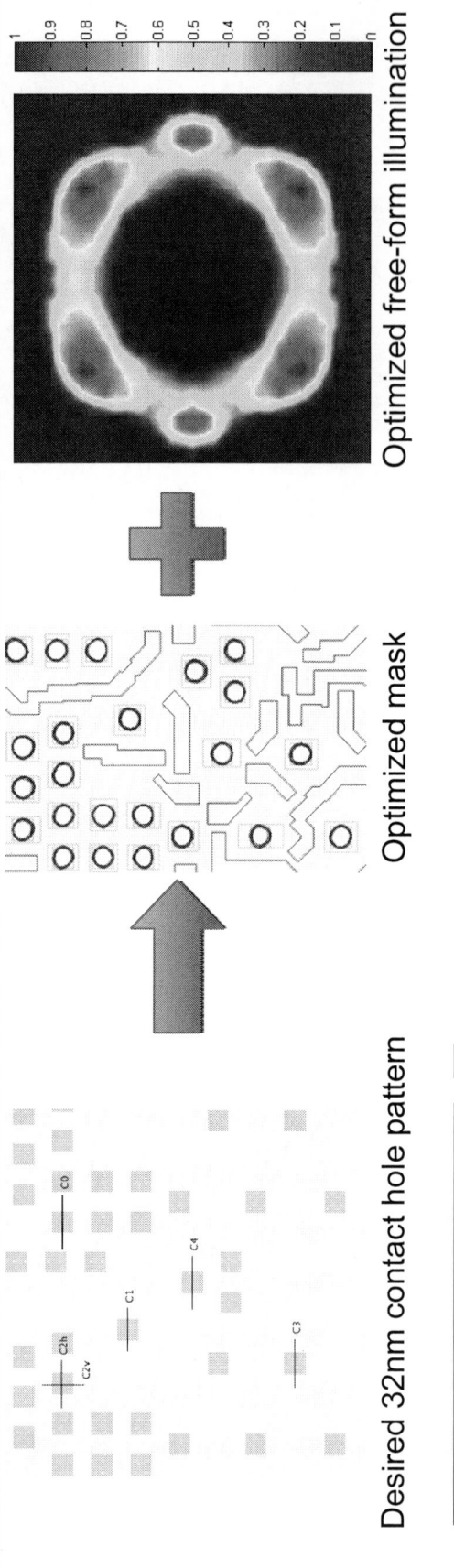

Optimized free-form illumination

Optimized mask

Desired 32nm contact hole pattern

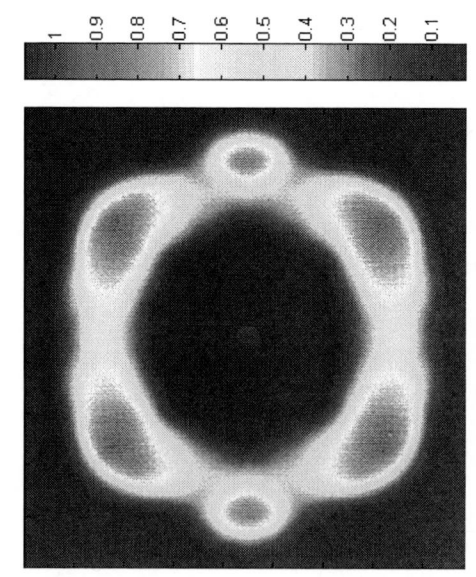

Test data from manufactured free-form illumination part

*Results obtained through collaboration with Cadence Design Systems

10/22/2009   6th International Symposium on Immersion Lithography Extensions

TESSERA.

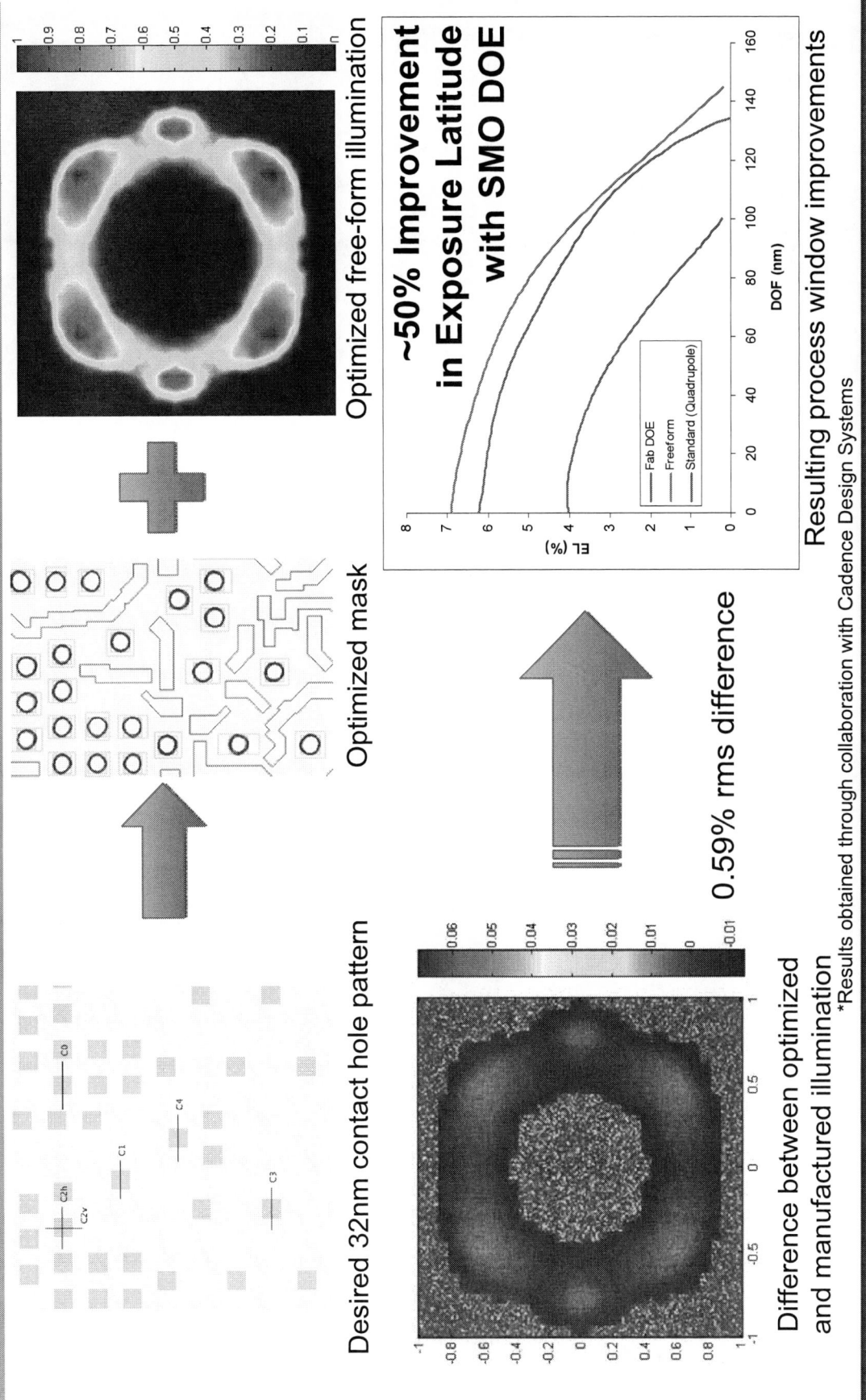

Repeatability of Manufacture

- Nine parts were fabricated to show process repeatability
- RMS variation between parts ranged from 0.12% to 1.0% <u>on a pixel-pixel basis</u>

Part 1

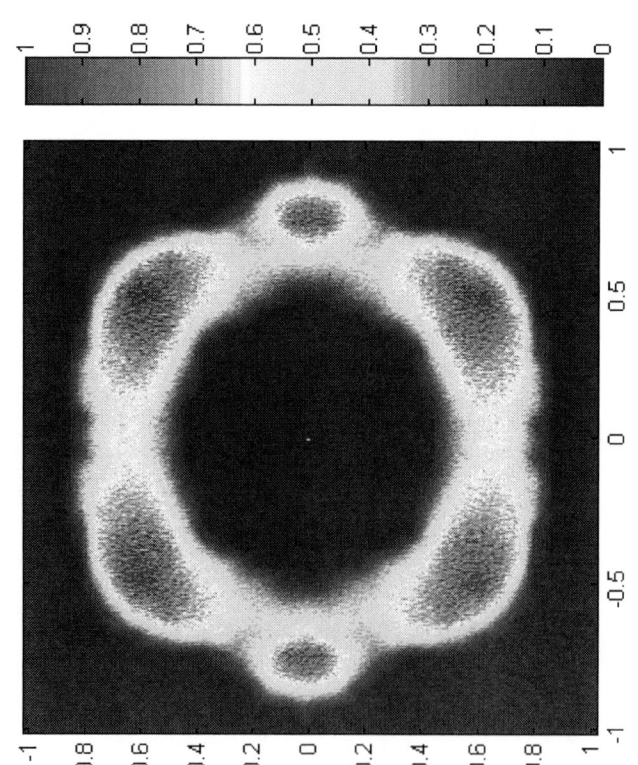

Test Image Data for 9 Parts

Pixel-Pixel Variation Between Parts

10/22/2009   6th International Symposium on Immersion Lithography Extensions   ❖ TESSERA.

Repeatability of Manufacture

- Nine parts were fabricated to show process repeatability
- RMS variation between parts ranged from 0.12% to 1.0% <u>on a pixel-pixel basis</u>

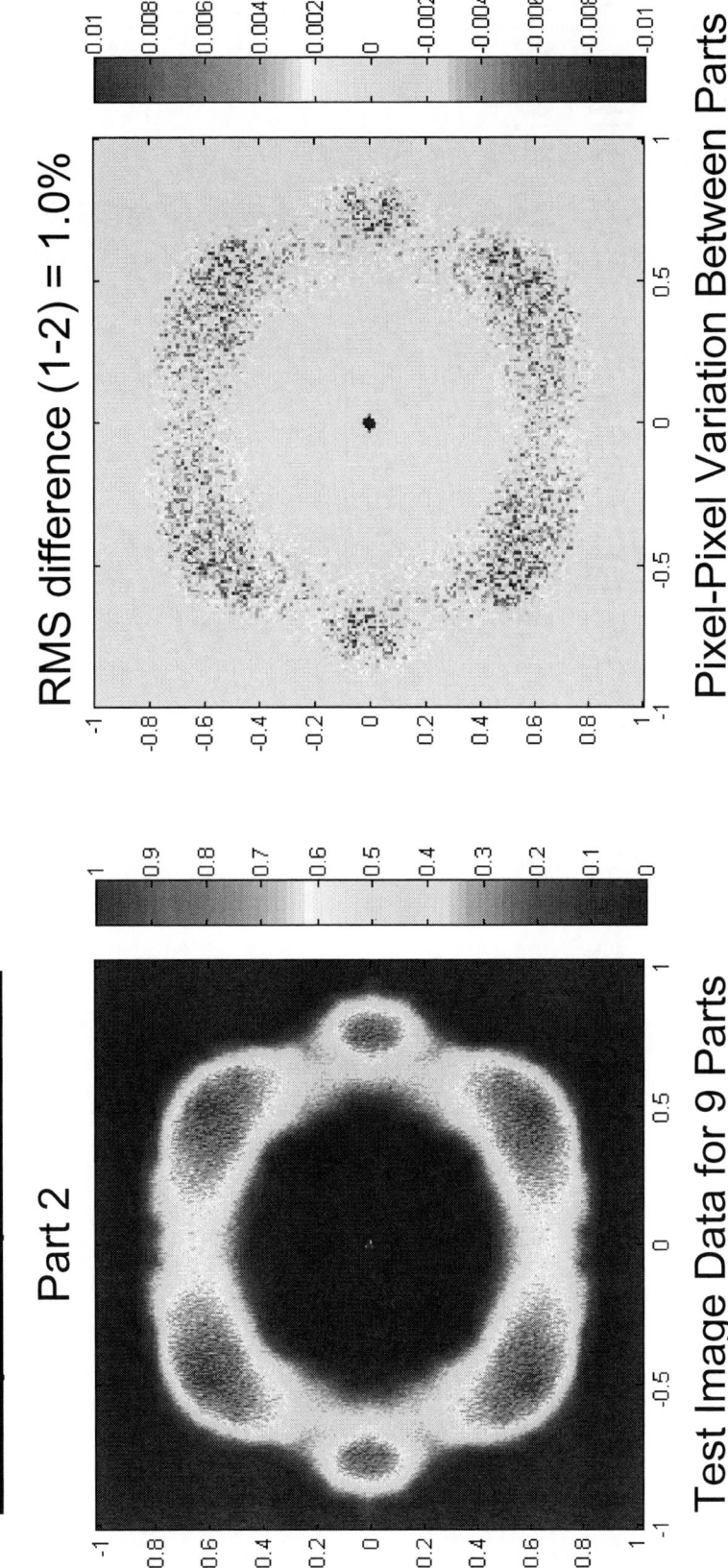

Repeatability of Manufacture

- Nine parts were fabricated to show process repeatability
- RMS variation between parts ranged from 0.12% to 1.0% on a pixel-pixel basis

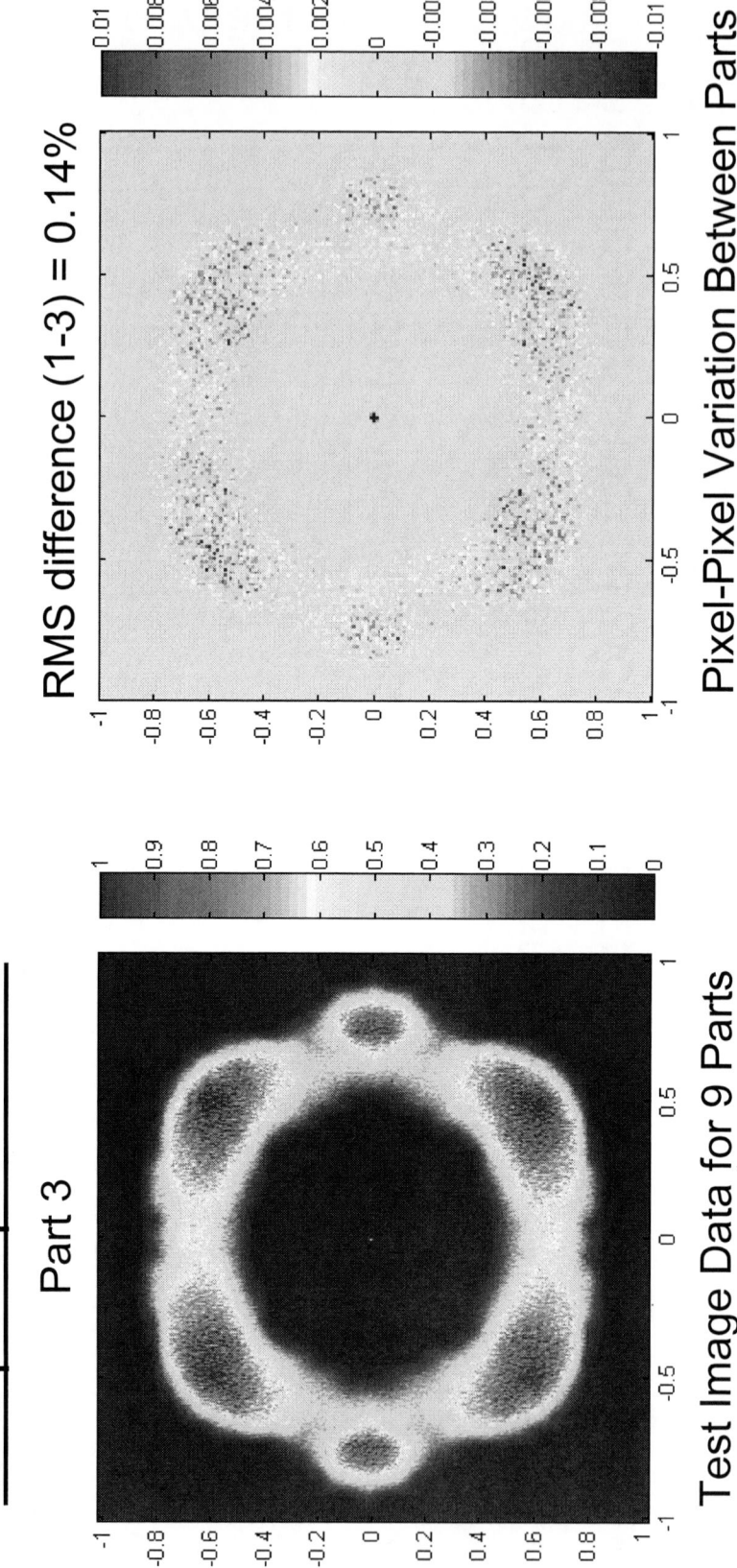

Part 3

Test Image Data for 9 Parts

RMS difference (1-3) = 0.14%

Pixel-Pixel Variation Between Parts

10/22/2009   6th International Symposium on Immersion Lithography Extensions   TESSERA.

Repeatability of Manufacture

- Nine parts were fabricated to show process repeatability
- RMS variation between parts ranged from 0.12% to 1.0% on a pixel-pixel basis

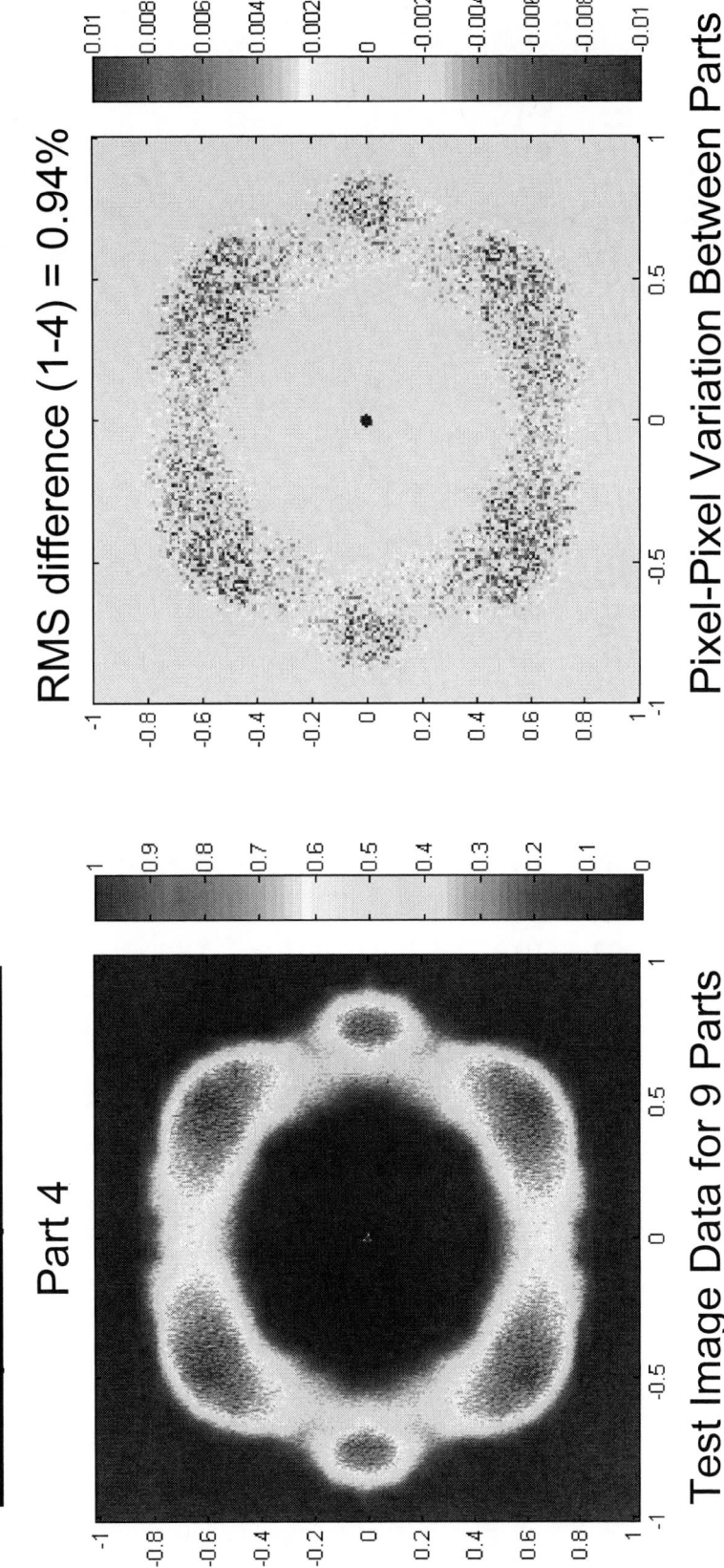

Part 4

Test Image Data for 9 Parts

RMS difference (1-4) = 0.94%

Pixel-Pixel Variation Between Parts

10/22/2009   6th International Symposium on Immersion Lithography Extensions   TESSERA.

Repeatability of Manufacture

- Nine parts were fabricated to show process repeatability
- RMS variation between parts ranged from 0.12% to 1.0% on a pixel-pixel basis

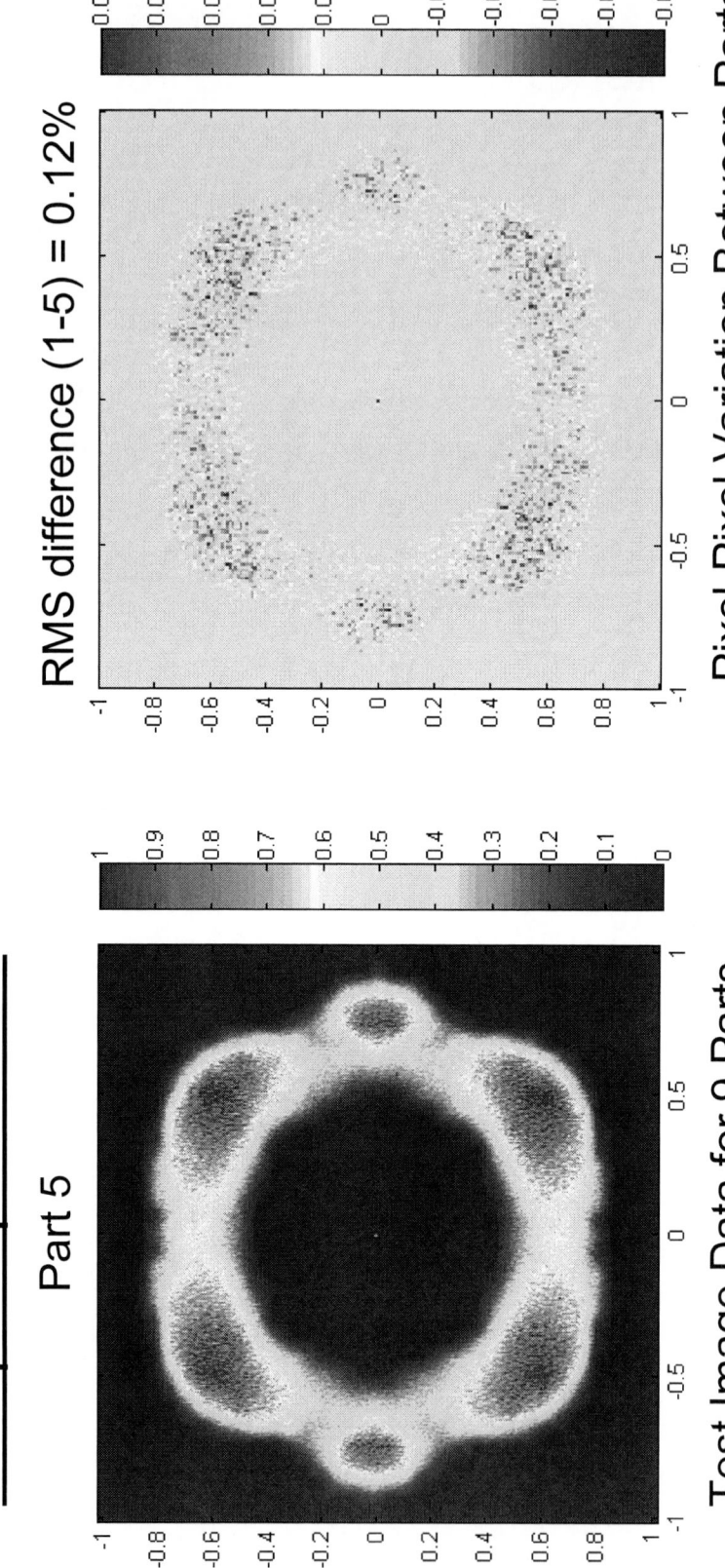

Part 5

Test Image Data for 9 Parts

RMS difference (1-5) = 0.12%

Pixel-Pixel Variation Between Parts

10/22/2009   6th International Symposium on Immersion Lithography Extensions   TESSERA.

Repeatability of Manufacture

- Nine parts were fabricated to show process repeatability
- RMS variation between parts ranged from 0.12% to 1.0% <u>on a pixel-pixel basis</u>

Part 6

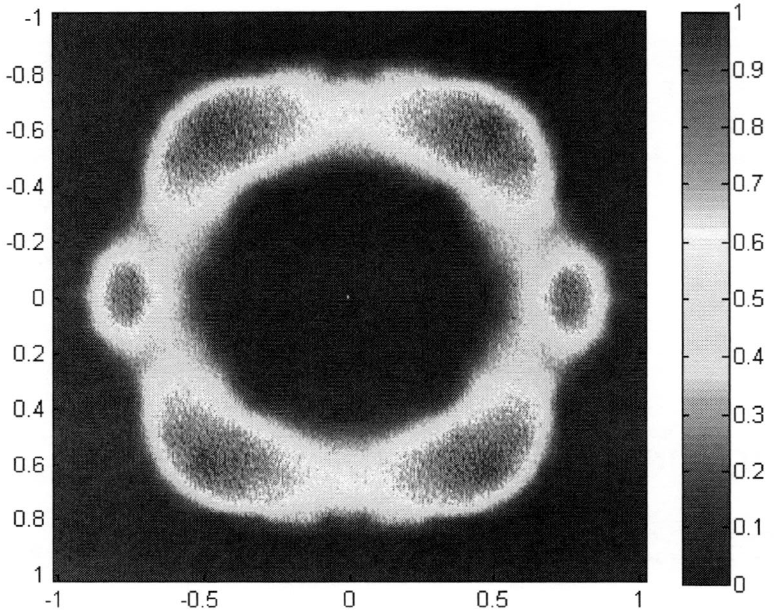

Test Image Data for 9 Parts

RMS difference (1-6) = 0.15%

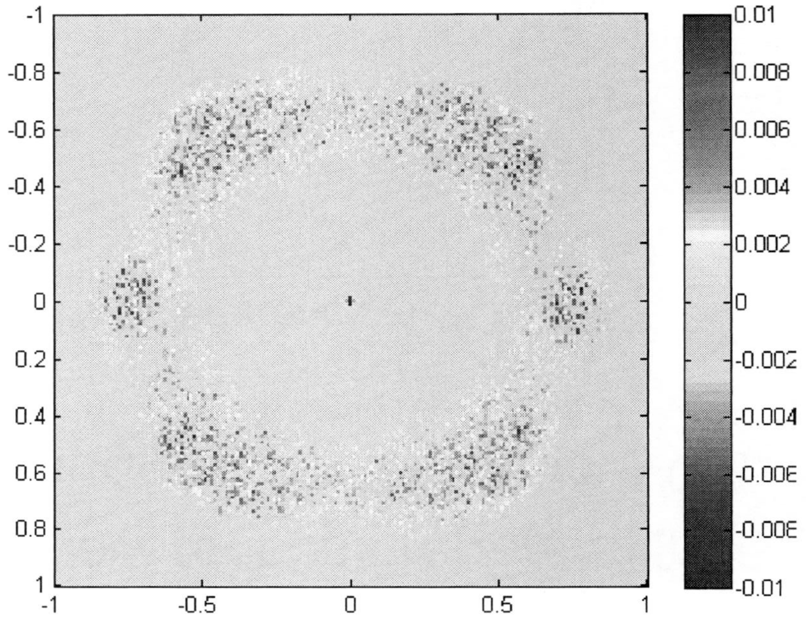

Pixel-Pixel Variation Between Parts

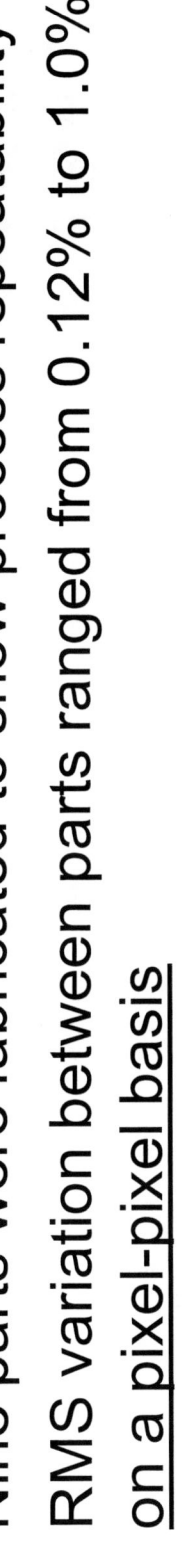

Repeatability of Manufacture

- Nine parts were fabricated to show process repeatability
- RMS variation between parts ranged from 0.12% to 1.0% on a pixel-pixel basis

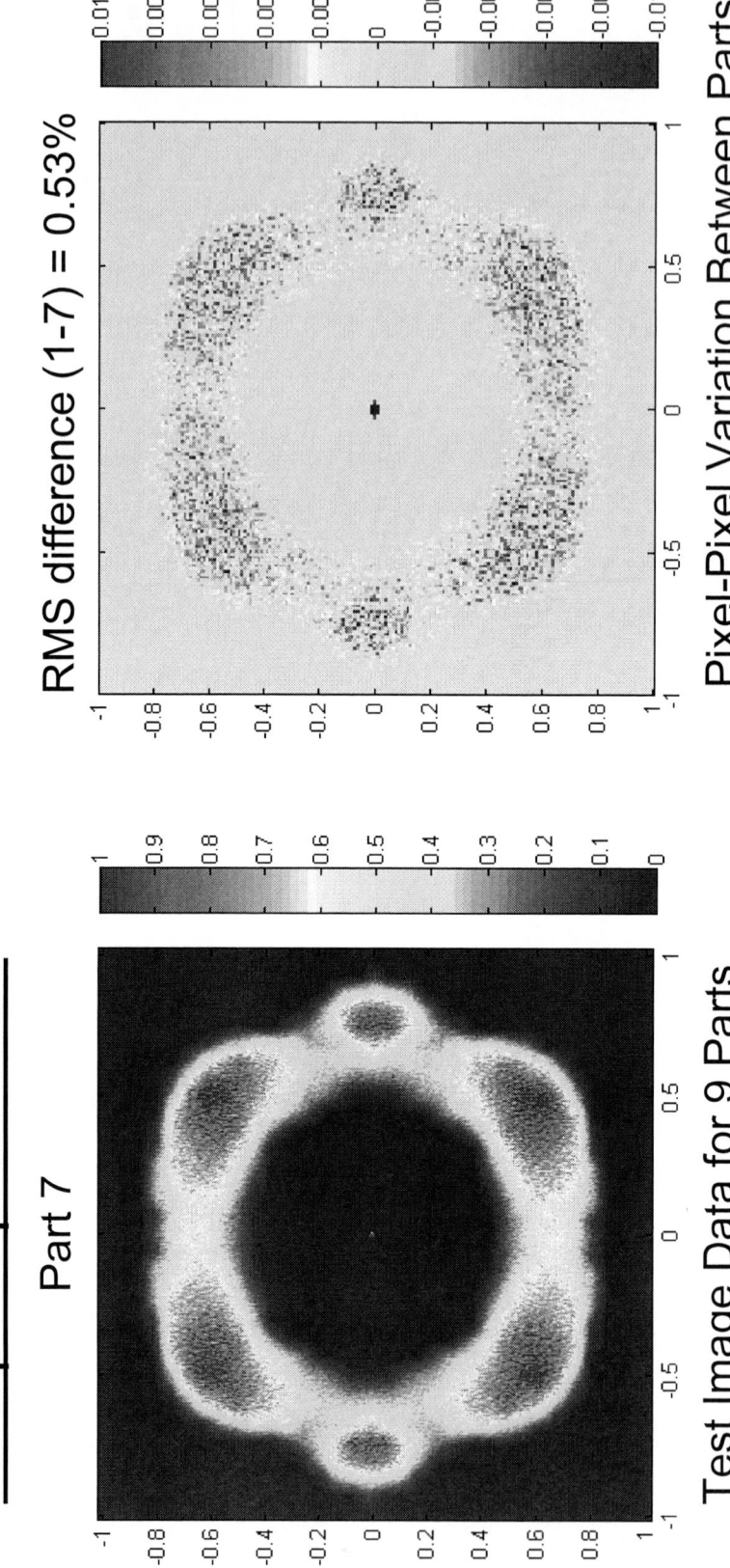

Part 7

Test Image Data for 9 Parts

RMS difference (1-7) = 0.53%

Pixel-Pixel Variation Between Parts

10/22/2009   6th International Symposium on Immersion Lithography Extensions   TESSERA.

Repeatability of Manufacture

- Nine parts were fabricated to show process repeatability
- RMS variation between parts ranged from 0.12% to 1.0% on a pixel-pixel basis

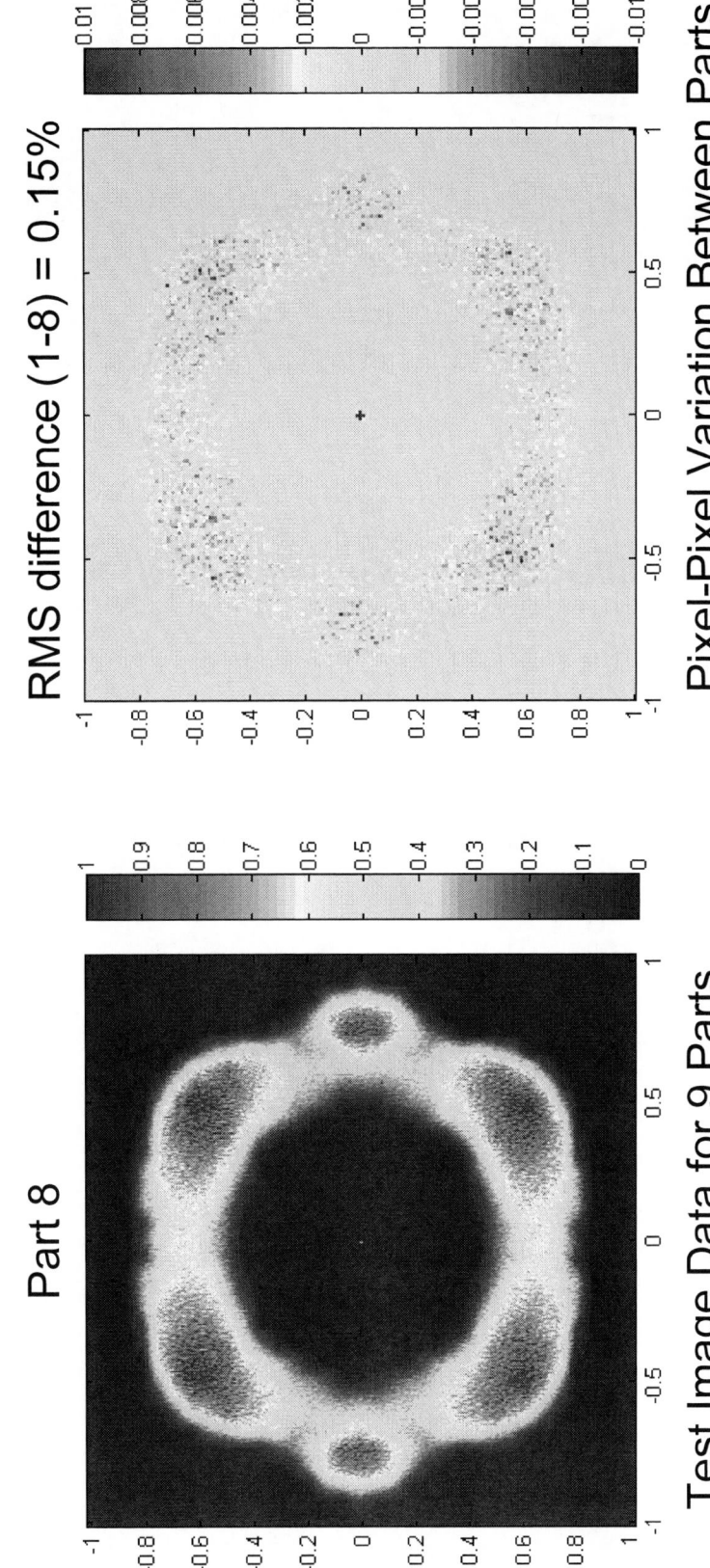

Part 8

Test Image Data for 9 Parts

RMS difference (1-8) = 0.15%

Pixel-Pixel Variation Between Parts

10/22/2009   6th International Symposium on Immersion Lithography Extensions   TESSERA.

Repeatability of Manufacture

- Nine parts were fabricated to show process repeatability
- RMS variation between parts ranged from 0.12% to 1.0% on a pixel-pixel basis

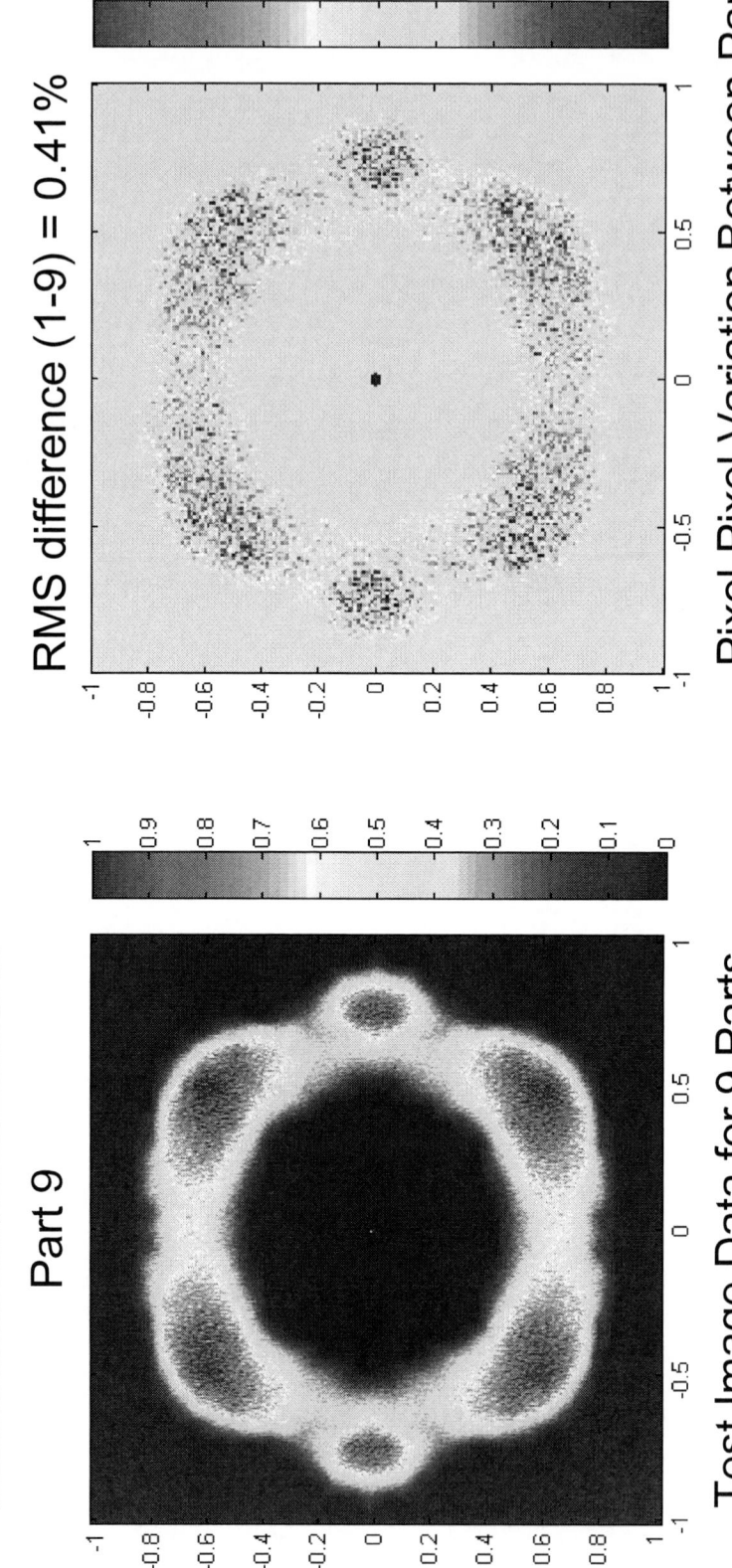

Part 9

Test Image Data for 9 Parts

RMS difference (1-9) = 0.41%

Pixel-Pixel Variation Between Parts

10/22/2009   6th International Symposium on Immersion Lithography Extensions   TESSERA.

Design and Manufacturing Effects of DOEs

Manufactured DOEs do exhibit some small variations from the ideal target

1. Pole Imbalance and H/V Bias

- Energy in poles is not evenly distributed
- Pair-wise horizontal vs. vertical imbalance produces H/V Bias
- Other imbalances produce telecentricity errors
- *Caused by asymmetry in the diffractive microstructure profile*

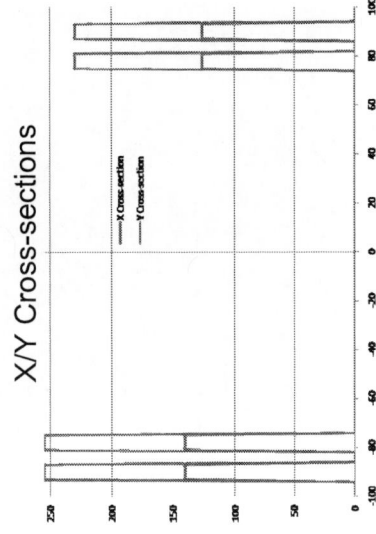

X/Y Cross-sections

Pole imbalance and stray light exaggerated for clarity

10/22/2009   6th International Symposium on Immersion Lithography Extensions   TESSERA.

Design and Manufacturing Effects of DOEs

Manufactured DOEs do exhibit some small variations from the ideal target

1. Pole Imbalance and H/V Bias
   - Energy in poles is not evenly distributed
   - Pair-wise horizontal vs. vertical imbalance produces H/V Bias
   - Other imbalances produce telecentricity errors
   - *Caused by asymmetry in the diffractive microstructure profile*

2. Stray light around poles
   - Increases the effective size of the pole unless the target is biased appropriately
   - *Not the result of manufacturing variations*
   - *Inherent in the physics of DOEs but can be minimized*

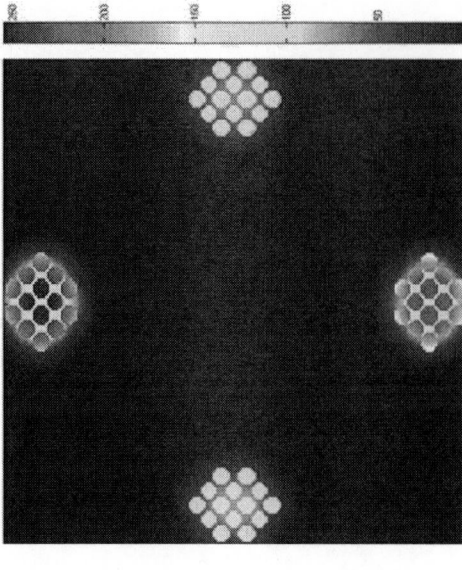

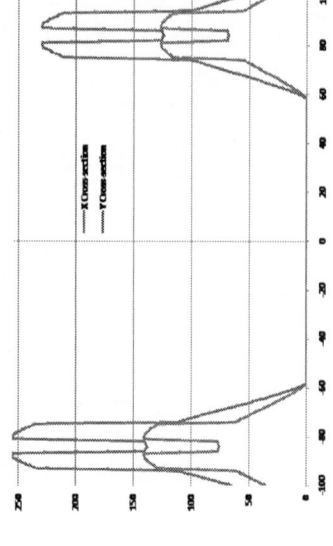

X/Y Cross-sections

Pole imbalance and stray light exaggerated for clarity

TESSERA.

10/22/2009   6th International Symposium on Immersion Lithography Extensions

Design and Manufacturing Effects of DOEs

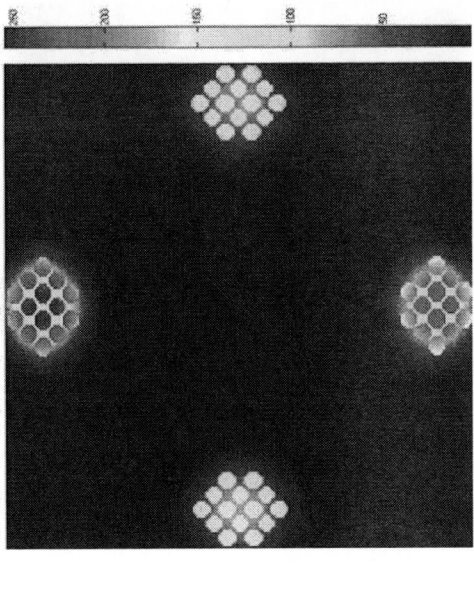  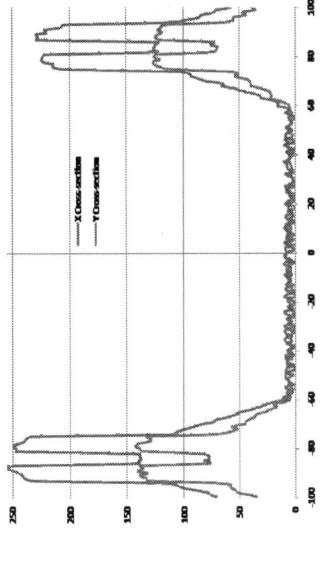

X/Y Cross-sections

Pole imbalance and stray light exaggerated for clarity

Manufactured DOEs do exhibit some small variations from the ideal target

1. Pole Imbalance and H/V Bias

- Energy in poles is not evenly distributed
- Pair-wise horizontal vs. vertical imbalance produces H/V Bias
- Other imbalances produce telecentricity errors
- *Caused by asymmetry in the diffractive microstructure profile*

2. Stray light around poles

- Increases the effective size of the pole unless the target is biased appropriately
- *Not the result of manufacturing variations*
- *Inherent in the physics of DOEs but can be minimized*

3. Constant random background stray light

- Results in a reduction of contrast in the aerial image
- *Caused by random surface defects and micro-roughness*

10/22/2009   6th International Symposium on Immersion Lithography Extensions

Accurate Models Improve the Benefits of SMO

- SMO models currently assume perfect matching to the target illumination

  - Can result in re-spins on both the illumination and OPC reticles

  - Increased cost and time to market

- Combining accurate DOE prediction with existing SMO models can reduce or eliminate the need for re-spins

10/22/2009   6th International Symposium on Immersion Lithography Extensions

TESSERA.

Tessera Prediction Model vs. Test Data

- Tessera's proprietary DOE prediction model can accurately predict both the target pixel and stray light performance

Prediction Model Results

Test Data Results

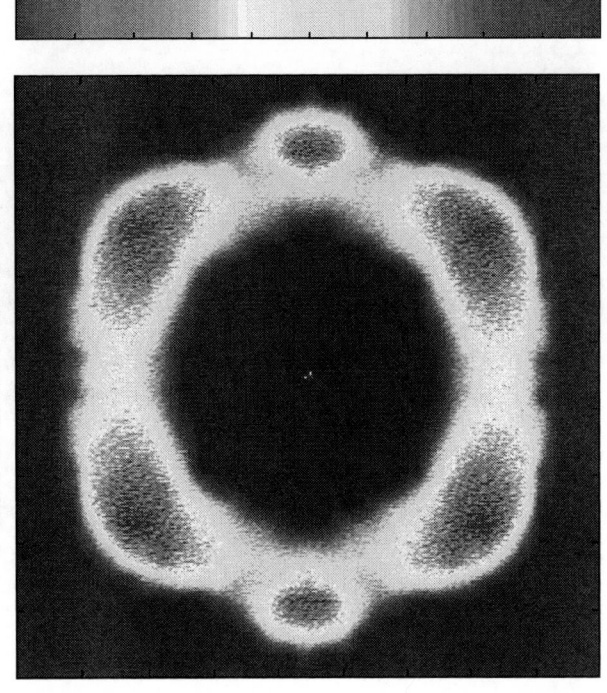

10/22/2009   6th International Symposium on Immersion Lithography Extensions

TESSERA.

Tessera Prediction Model vs. Test Data

- Tessera's proprietary DOE prediction model can accurately predict both the target pixel and stray light performance

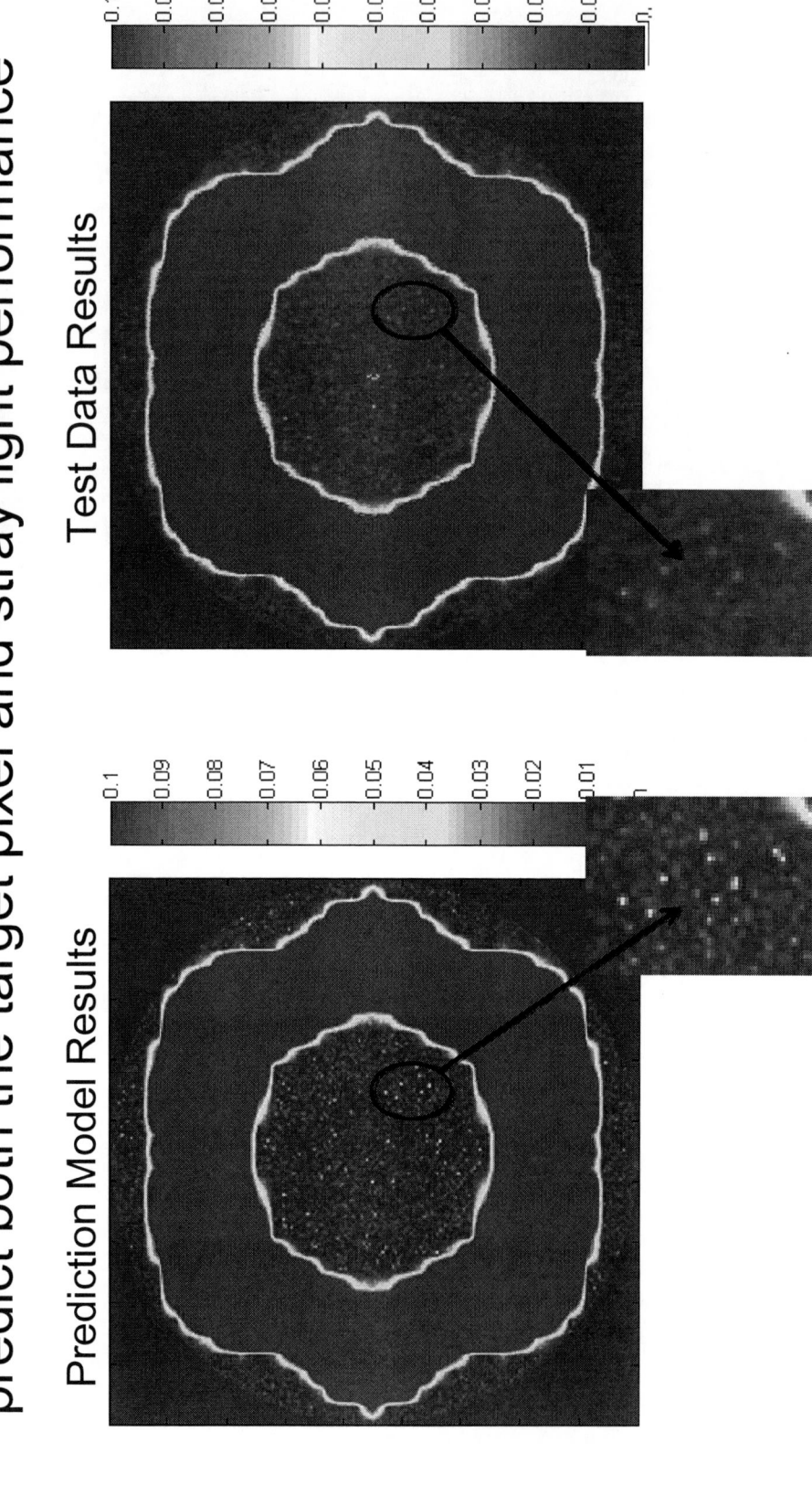

Test Data Results

Prediction Model Results

10/22/2009   6th International Symposium on Immersion Lithography Extensions

TESSERA.

Tessera Prediction Model Cross-Sections

- Cross-sections through the center of the pattern demonstrate model accuracy at all scales

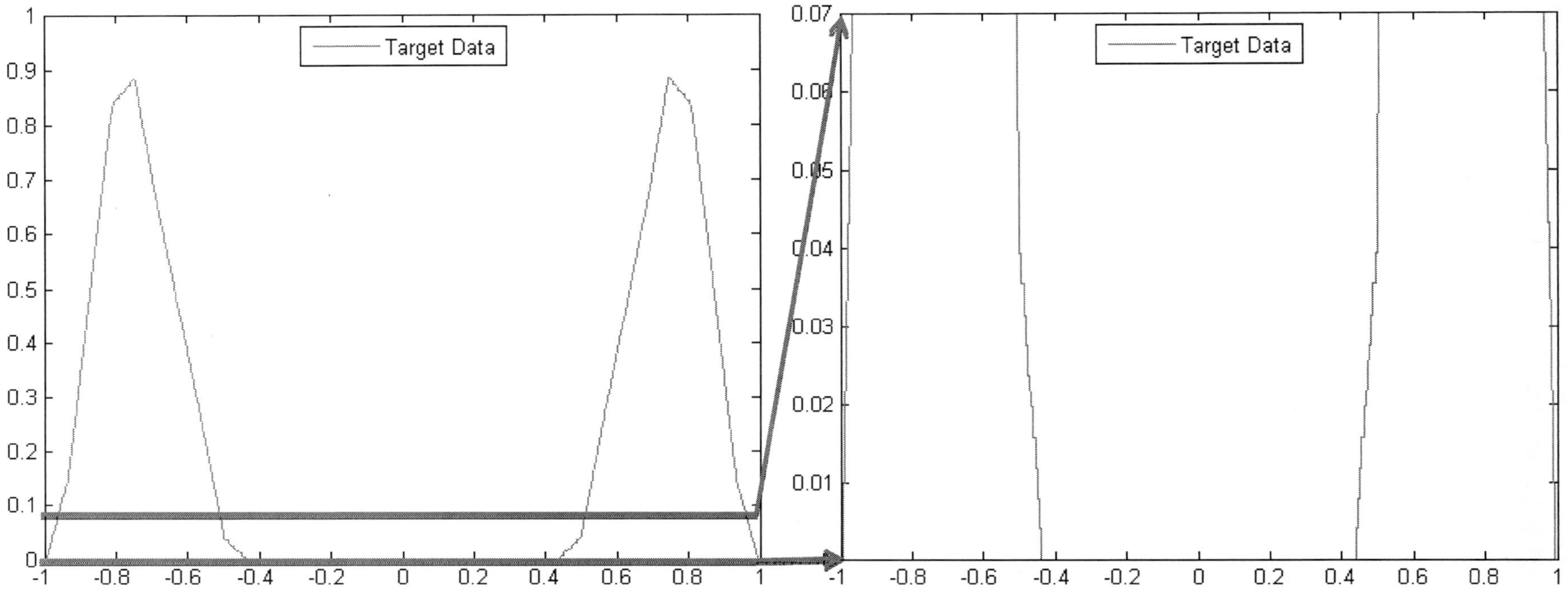

10/22/2009   6th International Symposium on Immersion Lithography Extensions

 TESSERA.

Tessera Prediction Model Cross-Sections

- Cross-sections through the center of the pattern demonstrate model accuracy at all scales

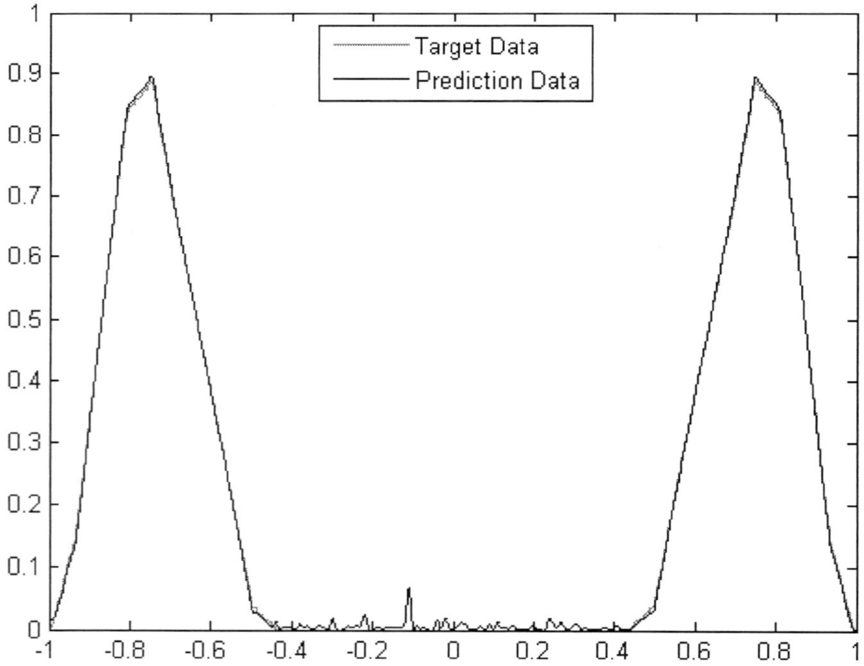

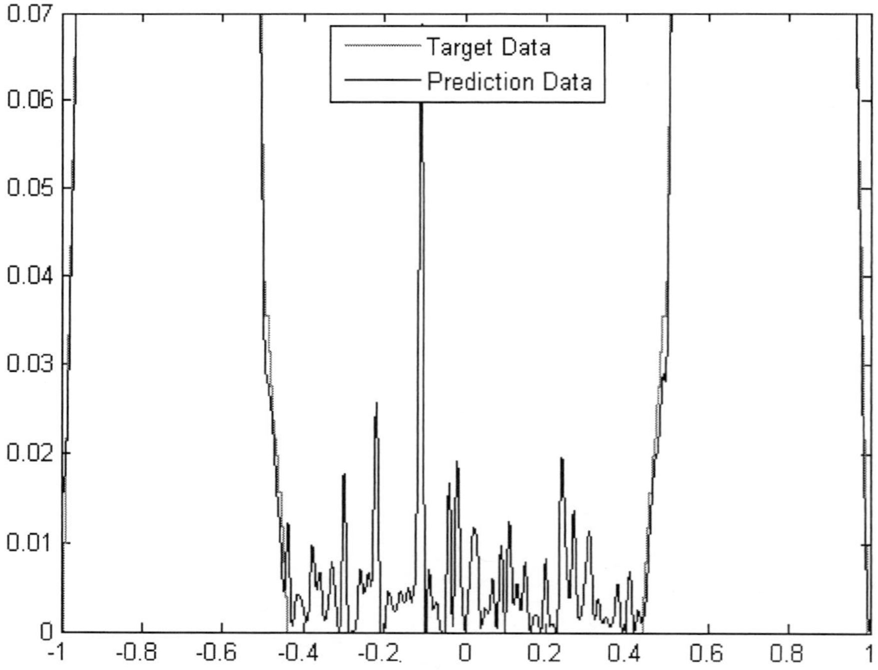

10/22/2009   6th International Symposium on Immersion Lithography Extensions

Tessera Prediction Model Cross-Sections

- Cross-sections through the center of the pattern demonstrate model accuracy at all scales

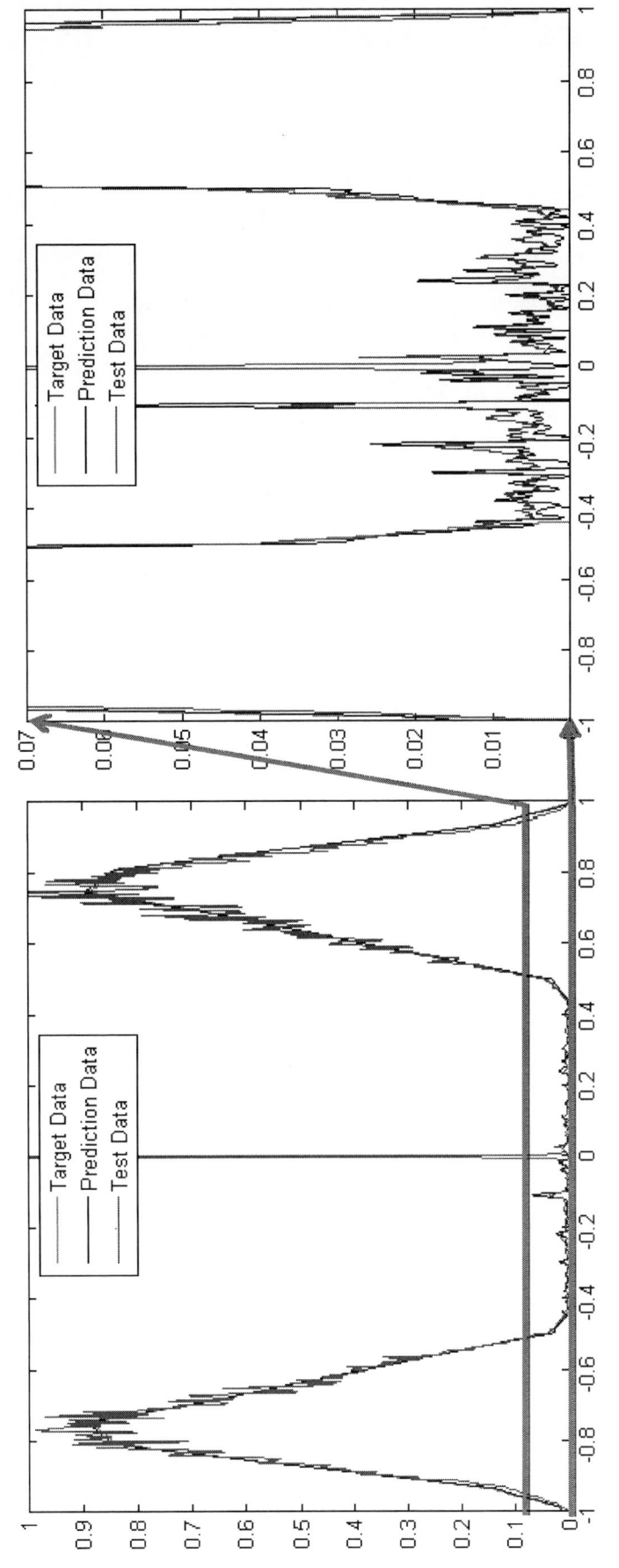

Test data exhibits speckle due to the higher partial coherence of the Tessera test setup compared to a standard scanner source

10/22/2009   6th International Symposium on Immersion Lithography Extensions

TESSERA.

Stray Light Reduction

- Tessera has also developed the ability to reduce the design intrinsic stray light of DOEs

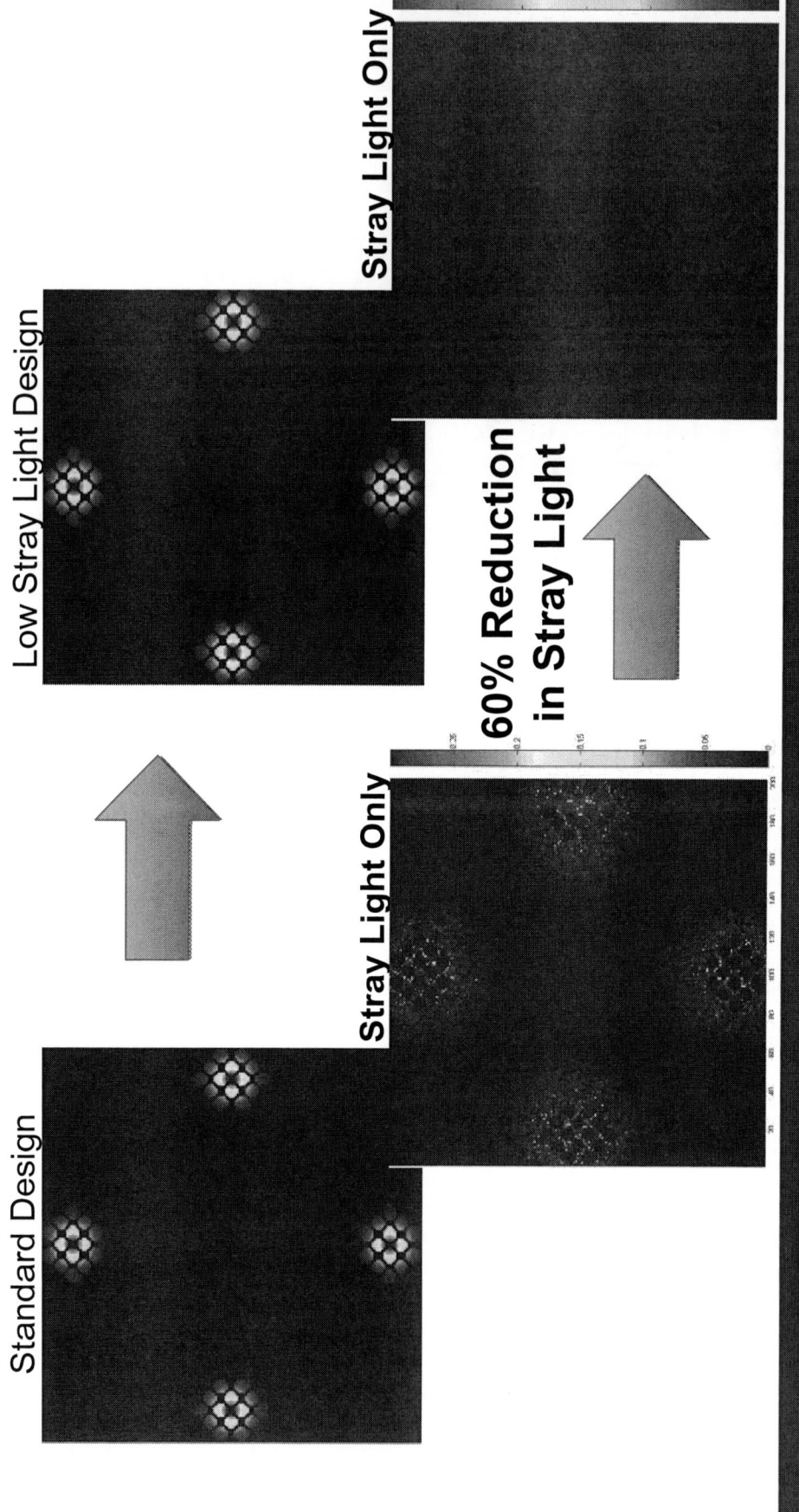

Standard Design

Stray Light Only

Low Stray Light Design

Stray Light Only

60% Reduction in Stray Light

10/22/2009   6th International Symposium on Immersion Lithography Extensions

TESSERA.

DOEs are Ready for Today and the Future

- A proven, stable solution to the challenges facing immersion lithography:
  - SMO, stray light prediction & control, pole imbalance
- Modern DOEs can provide the illumination source you need the first time … every time
- Are a critical enabler of optical lithography for both single and double patterning at 32nm and 22nm
- Can be produced with lead times similar to reticles
- Source-optimized DOEs are available through scanner manufacturers now

10/22/2009   6th International Symposium on Immersion Lithography Extensions

TESSERA.

Technical and Manufacturing Challenges and the Prospect for HVM using ArF Pitch Division

Sam Sivakumar

Intel Corporation

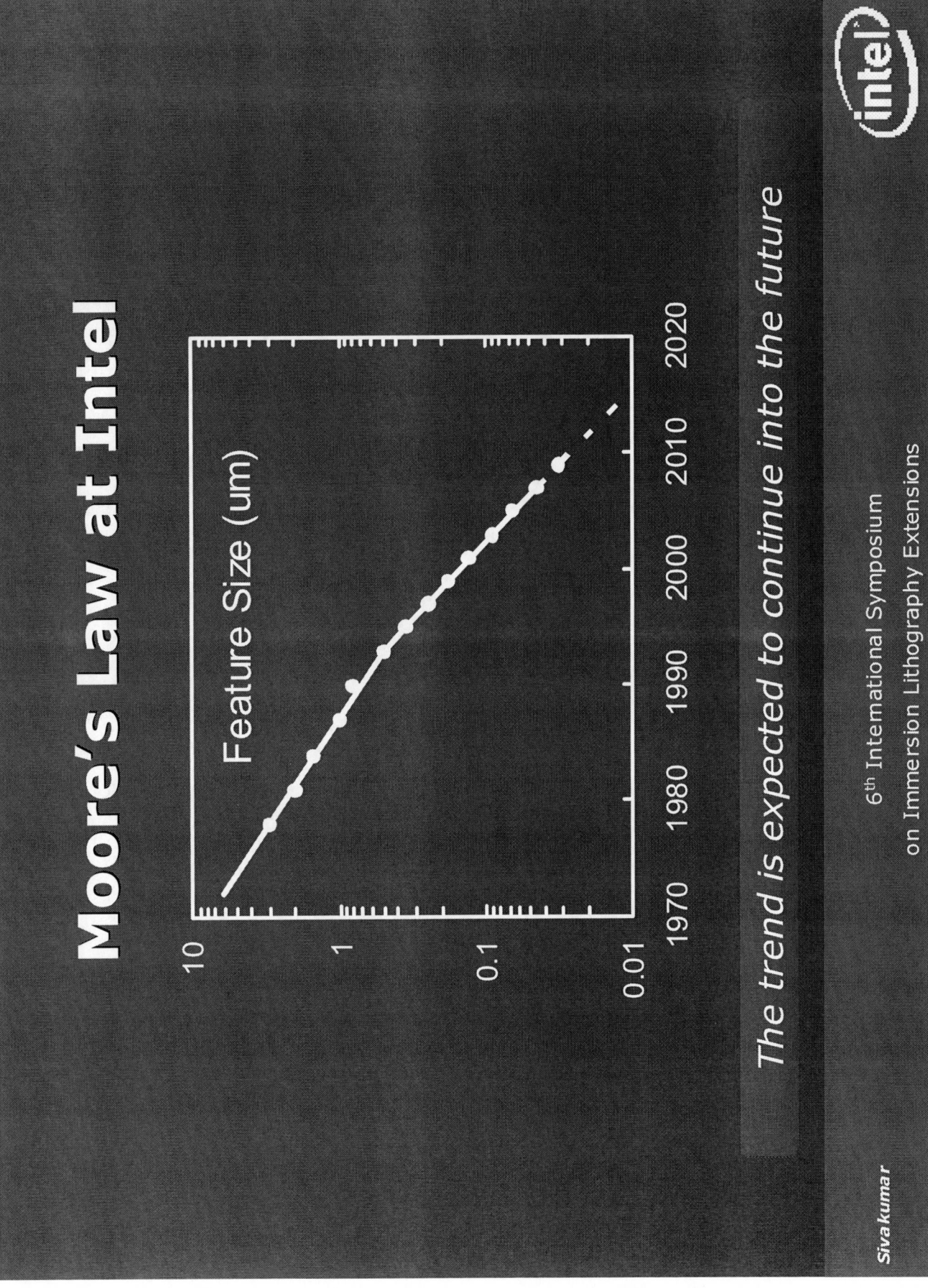

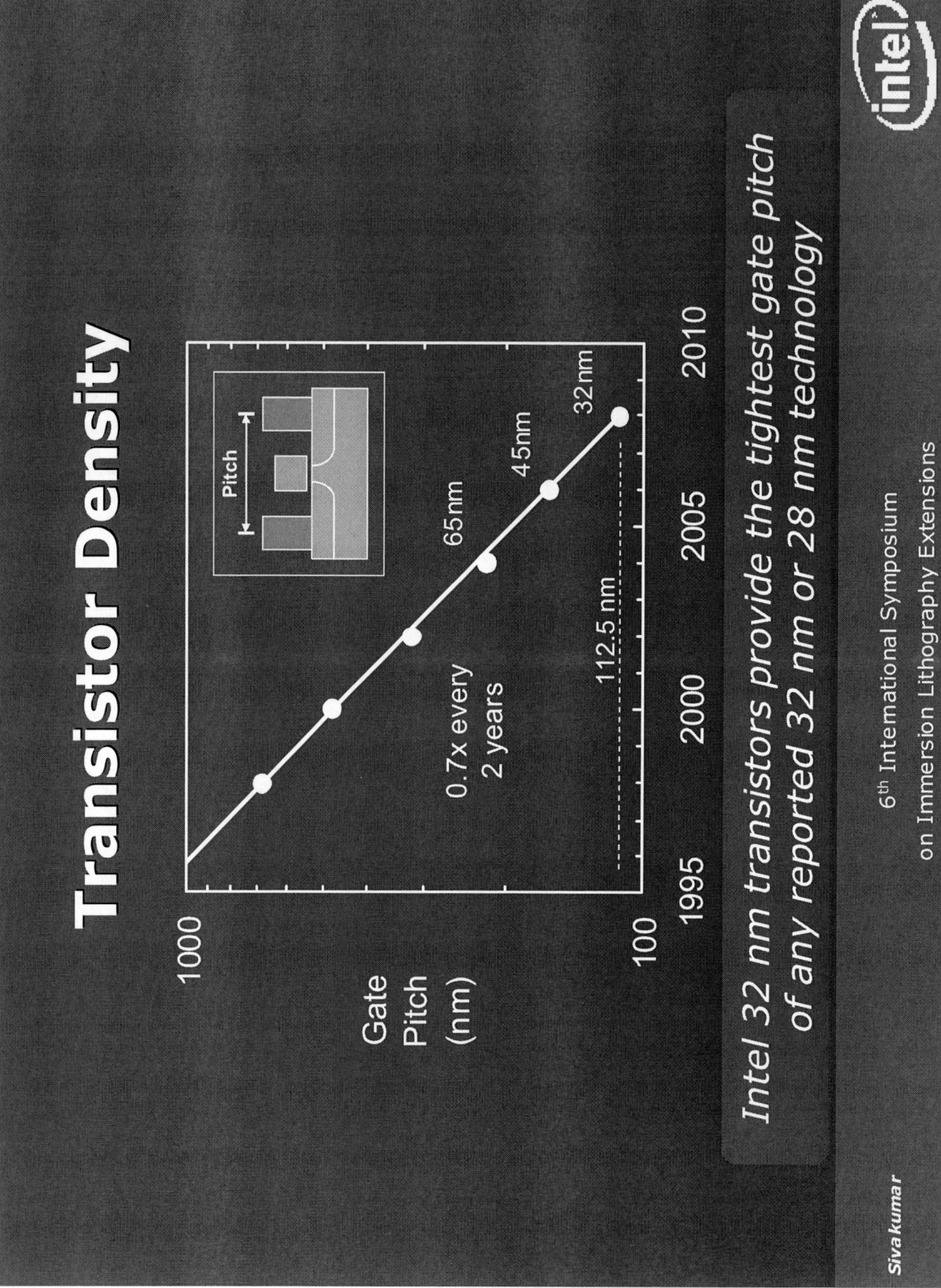

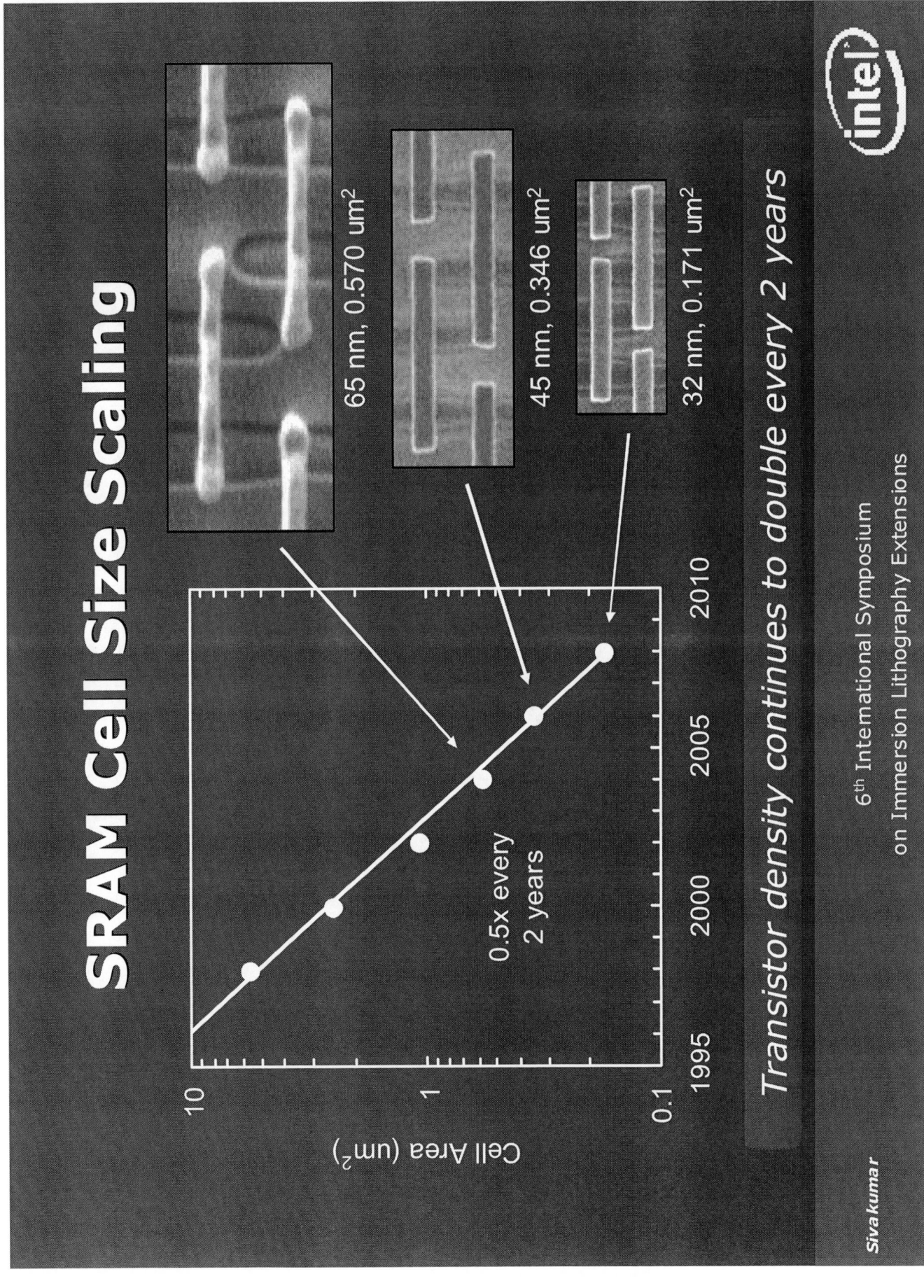

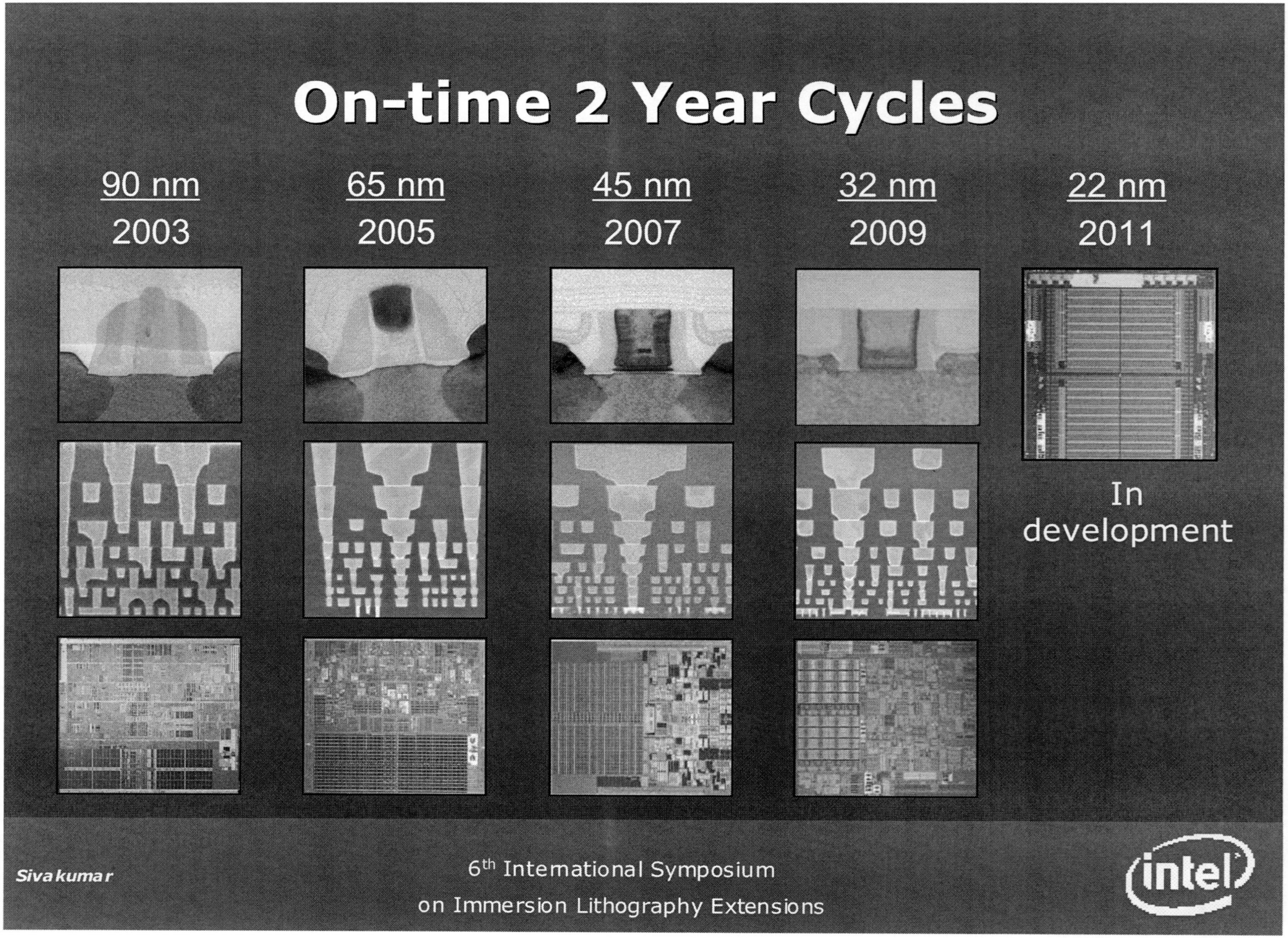

On-time 2 Year Cycles

90 nm 2003	65 nm 2005	45 nm 2007	32 nm 2009	22 nm 2011
				In development

6th International Symposium
on Immersion Lithography Extensions

Sivakumar

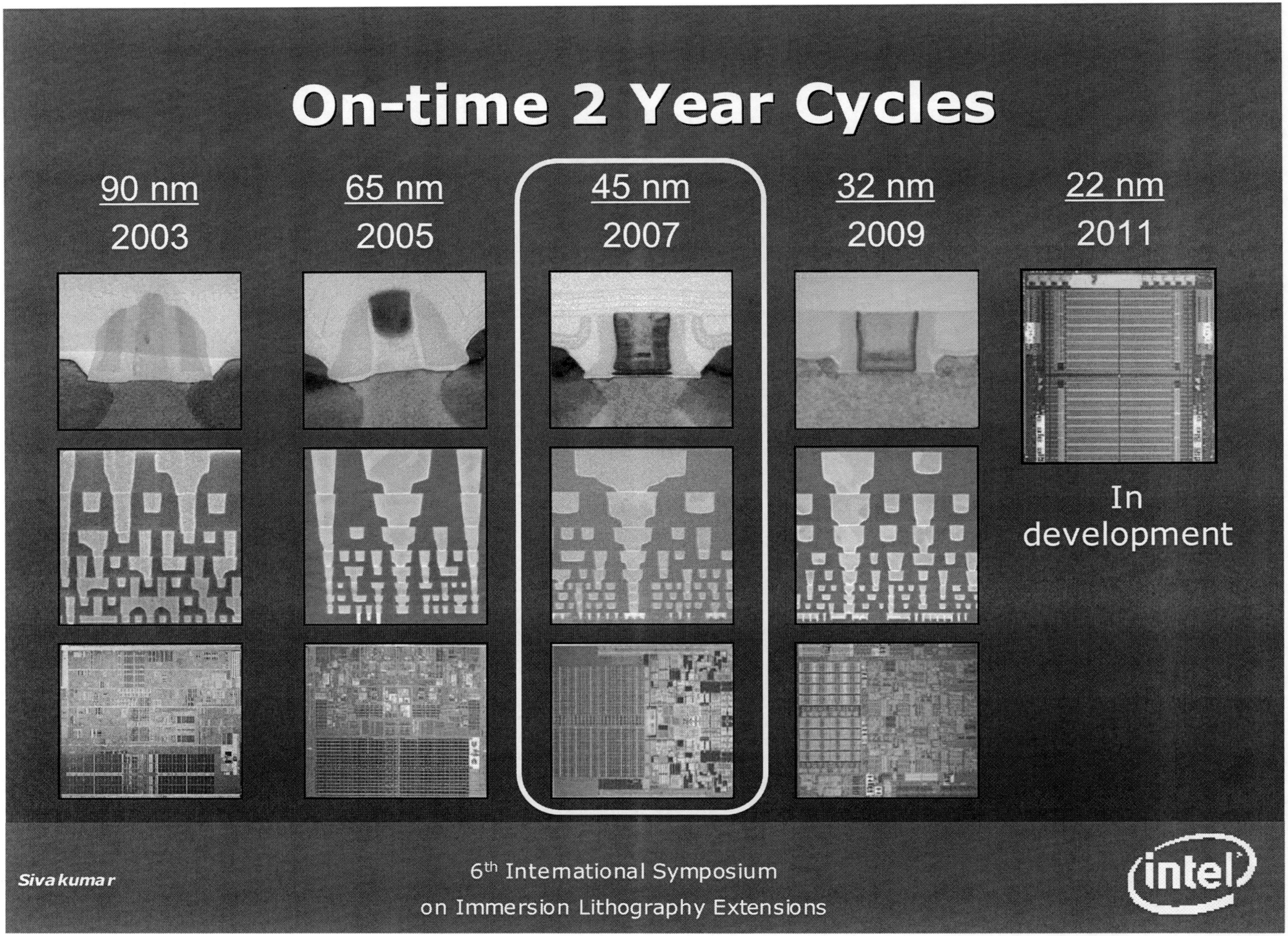

On-time 2 Year Cycles

| 90 nm 2003 | 65 nm 2005 | 45 nm 2007 | 32 nm 2009 | 22 nm 2011 |

In development

intel

6th International Symposium
on Immersion Lithography Extensions

Sivakumar

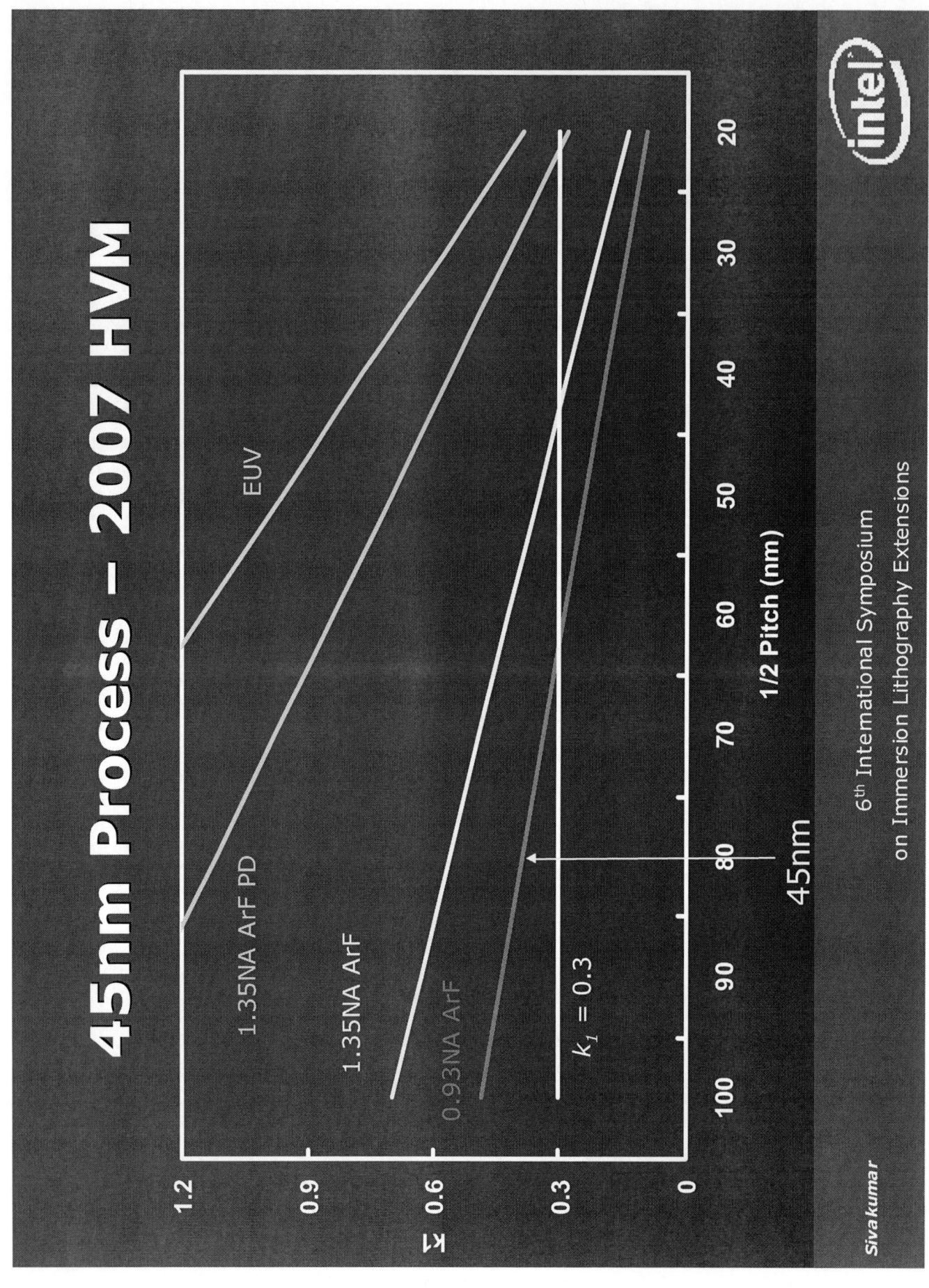

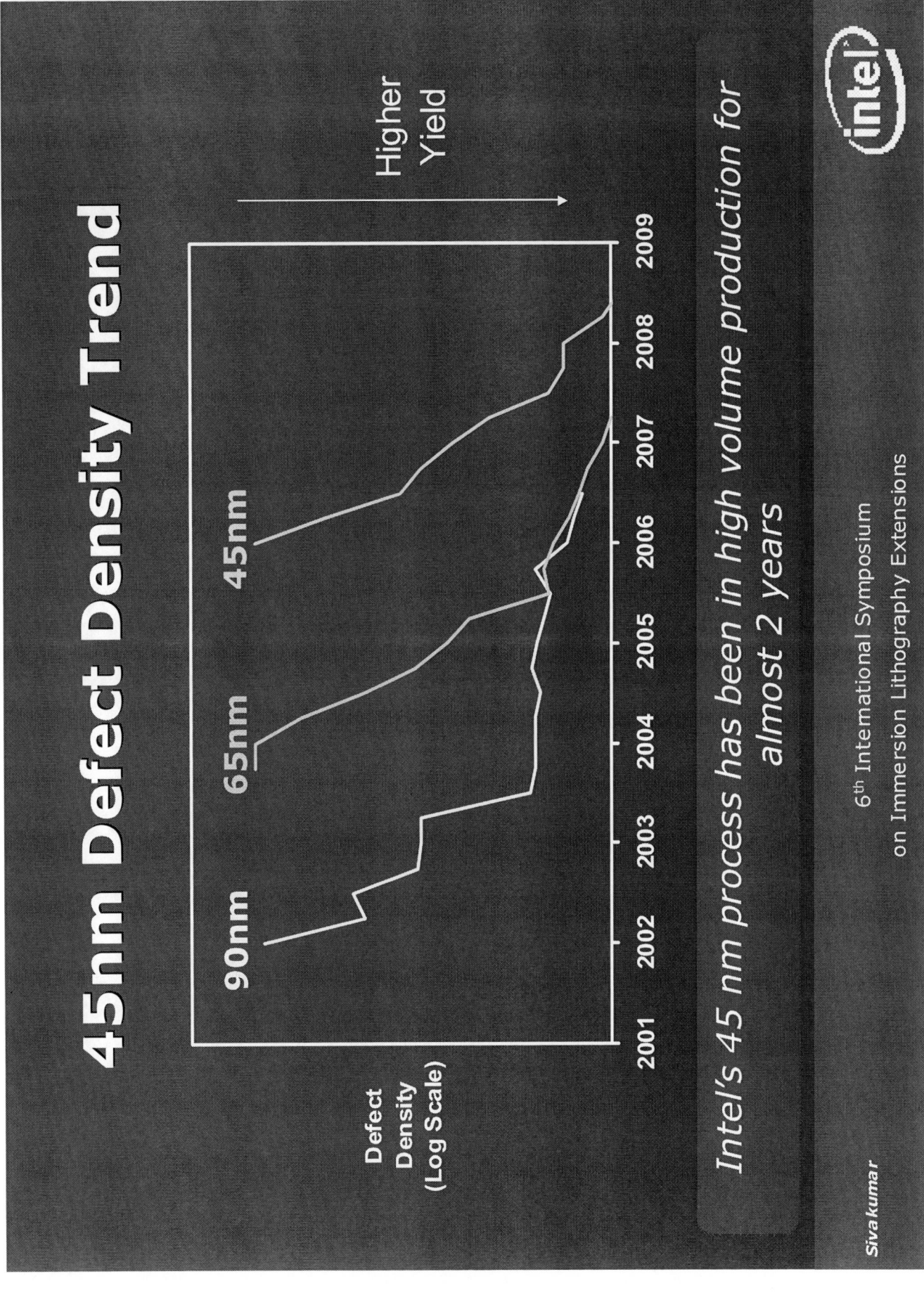

45nm Microprocessor Products

Single Core

Dual Core

Quad Core

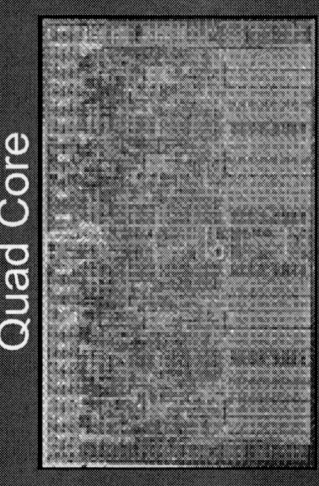

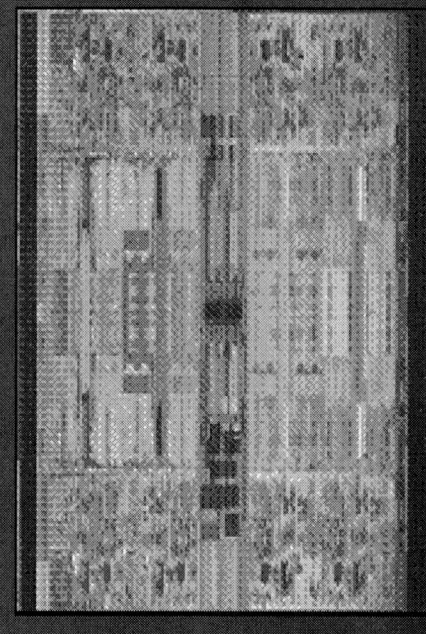

8 Core

6 Core

>200 million 45 nm CPUs shipped to date

6th International Symposium
on Immersion Lithography Extensions

Sivakumar

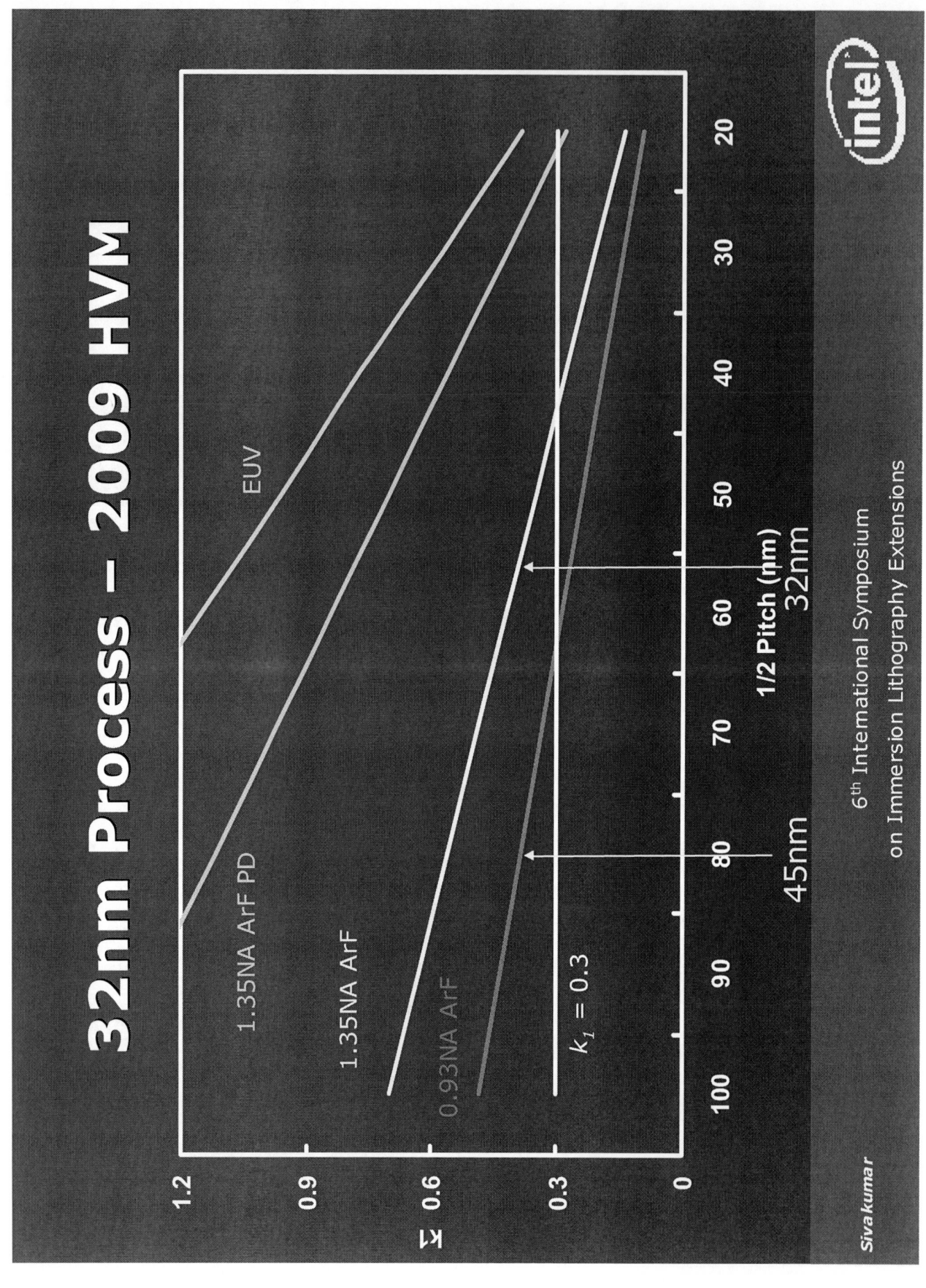

32nm Defect Density Trend

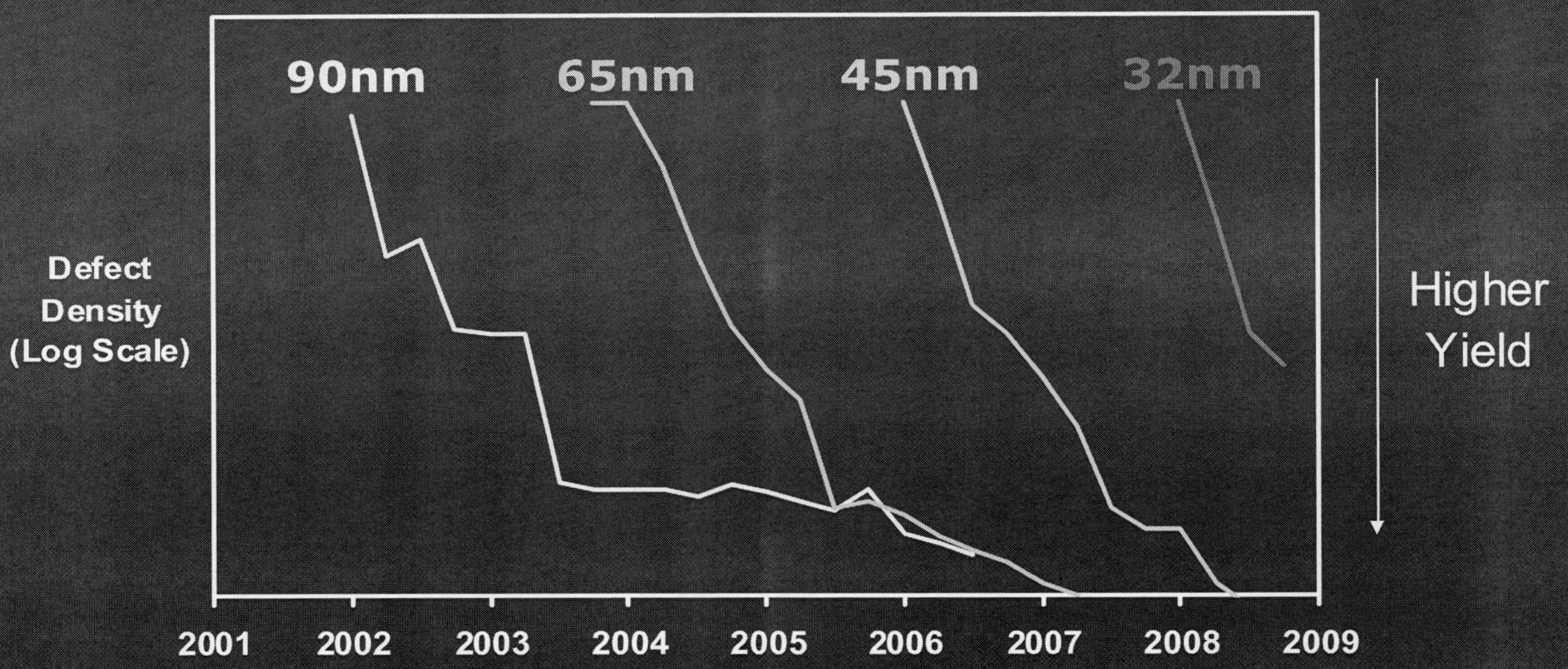

Defect Density (Log Scale)

Higher Yield →

90nm 65nm 45nm 32nm

2001 2002 2003 2004 2005 2006 2007 2008 2009

Intel's 32 nm process is certified and CPU wafers are moving through the factory in support of planned Q4 revenue production

6th International Symposium
on Immersion Lithography Extensions

Sivakumar

32nm Westmere Microprocessor

Dual core Westmere

First in a family of 32 nm microprocessors based upon the Intel® microarchitecture codenamed Nehalem

6th International Symposium
on Immersion Lithography Extensions

Sivakumar

On-time 2 Year Cycles

90 nm 2003	65 nm 2005	45 nm 2007	32 nm 2009	22 nm 2011

In development

intel

6th International Symposium
on Immersion Lithography Extensions

Siva kumar

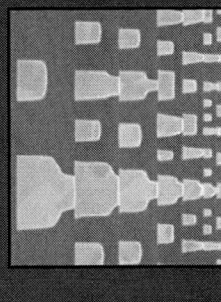

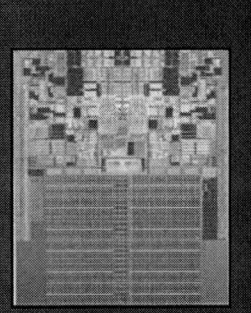

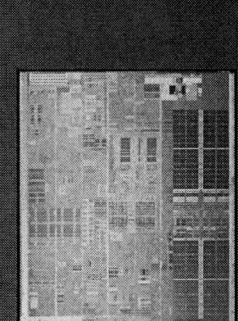

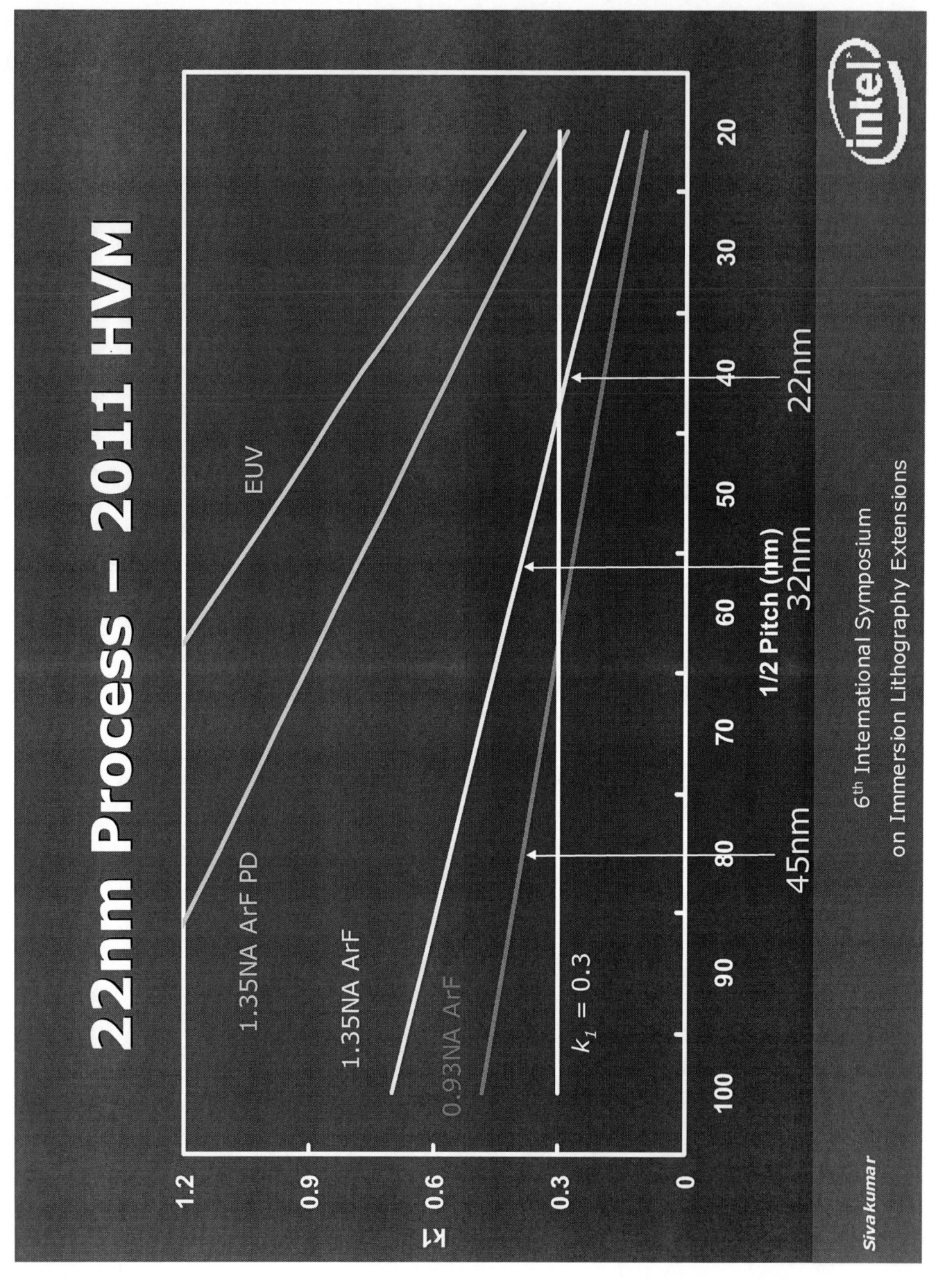

22nm SRAM Test Chip

SRAM, Logic, Mixed-Signal
Test Circuits

SRAM, Logic, Mixed-Signal
Test Circuits

Discrete
Test Structures

*Intel is first in the industry to
demonstrate working 22 nm circuits*

6th International Symposium
on Immersion Lithography Extensions

Sivakumar

22nm SRAM Test Chip

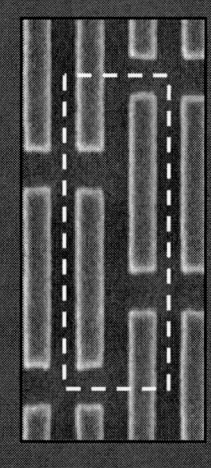

0.092 um^2 SRAM cell
for high density applications

0.108 um^2 SRAM cell
for low voltage applications

*0.092 um2 is the smallest SRAM cell
in working circuits reported to date*

6th International Symposium
on Immersion Lithography Extensions

Sivakumar

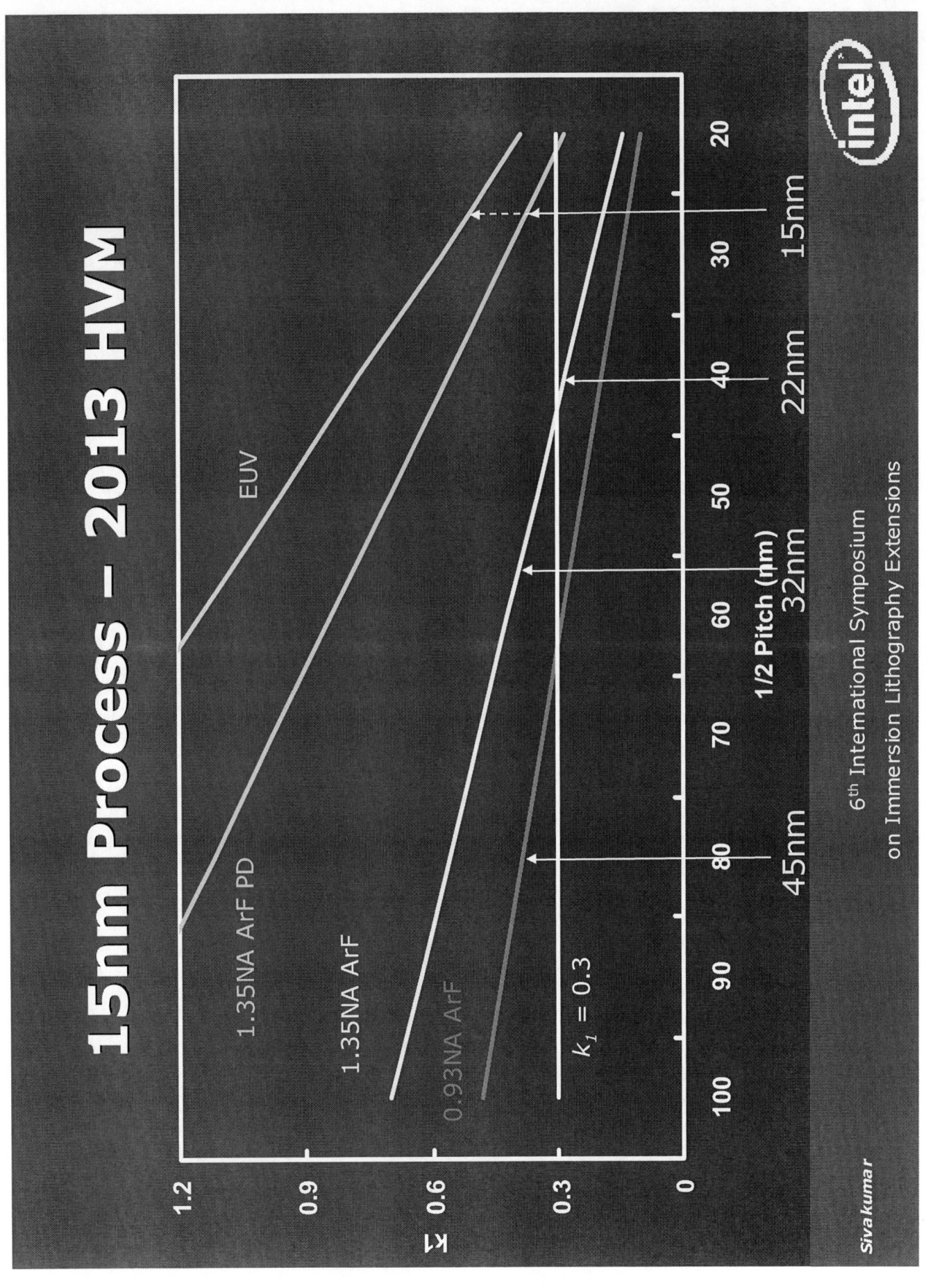

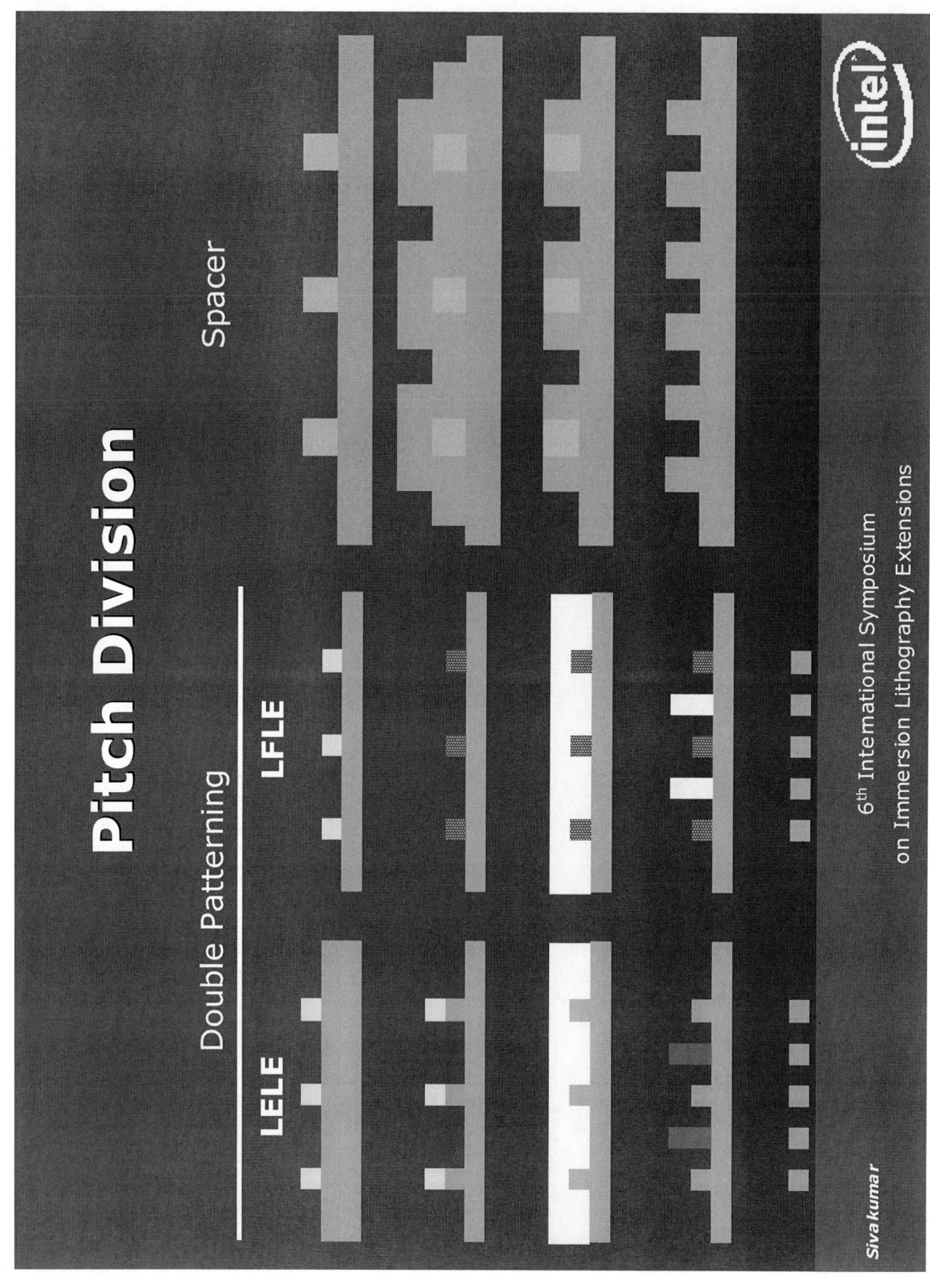

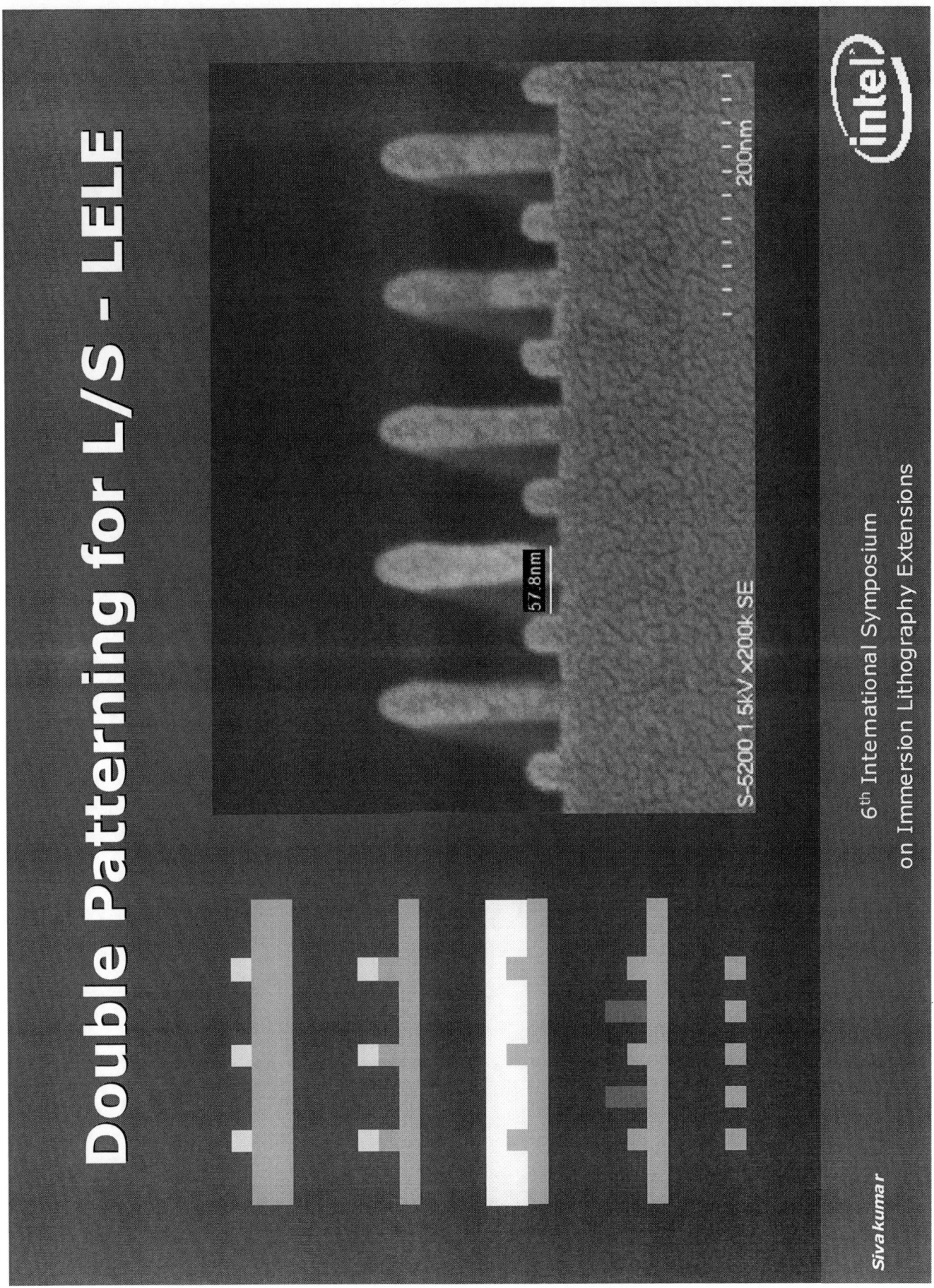

Double Patterning for Vias – LELE

Several different PD options will be picked and used depending on layer-by-layer applicability

6th International Symposium
on Immersion Lithography Extensions

Sivakumar

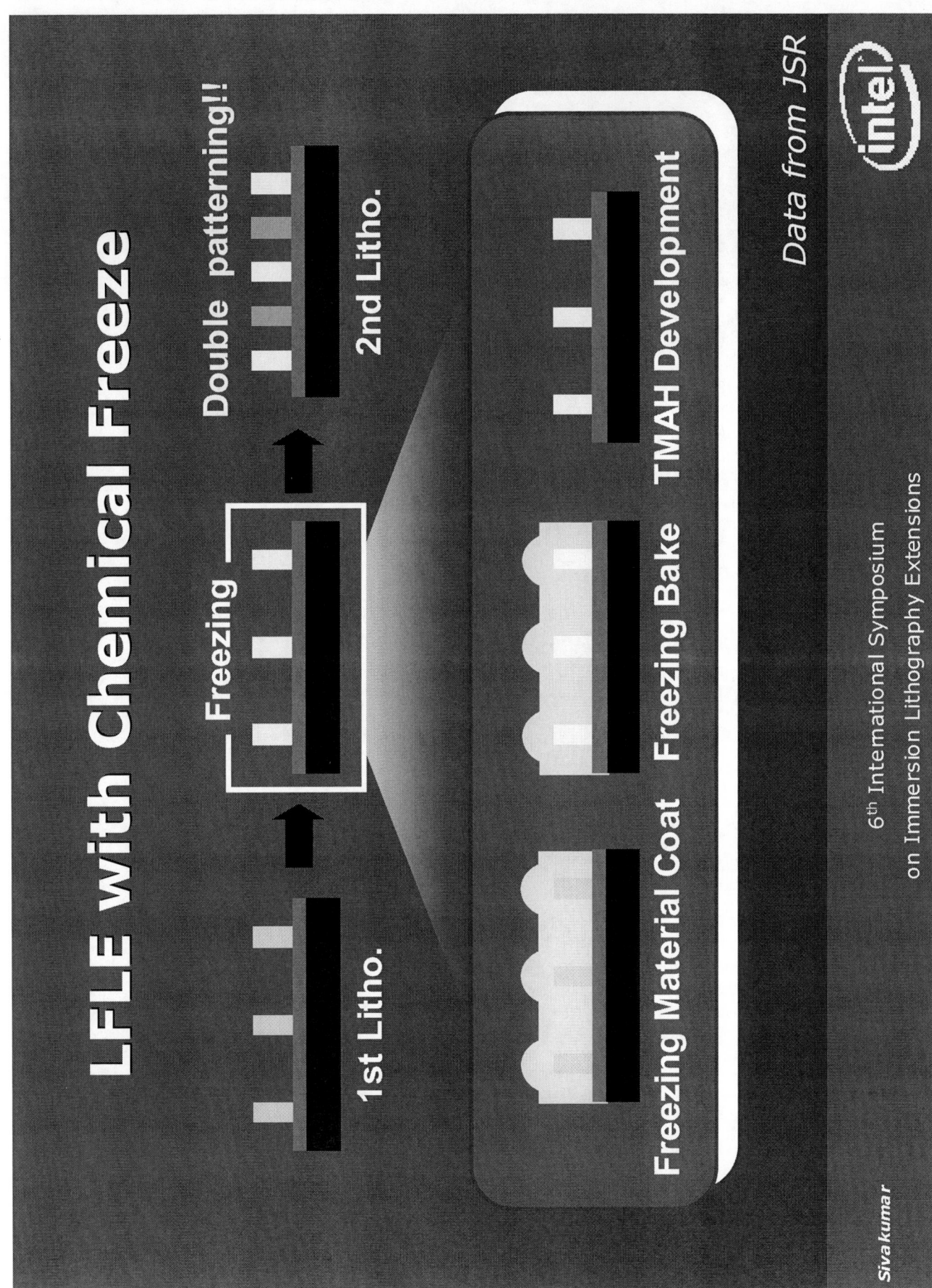

LFLE with Chemical Freeze

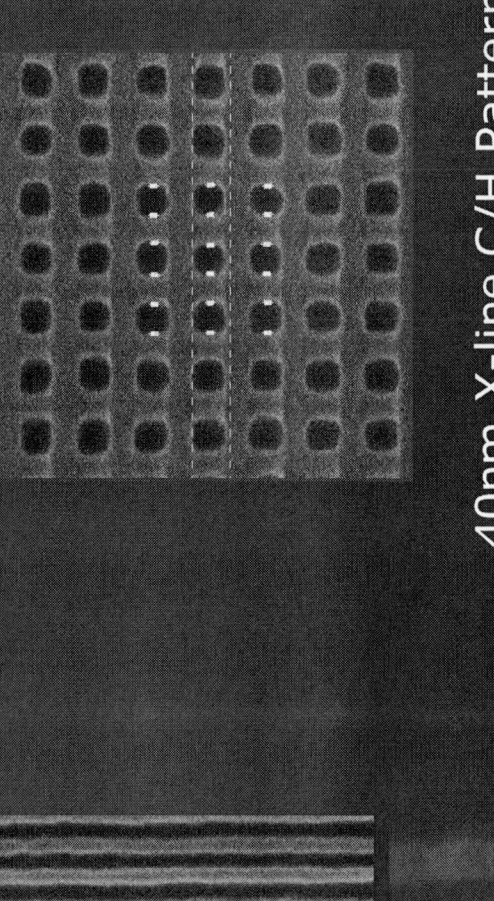

40nm X-line C/H Pattern

Data from JSR

22nm HP Dual Line Pattern

6th International Symposium
on Immersion Lithography Extensions

Sivakumar

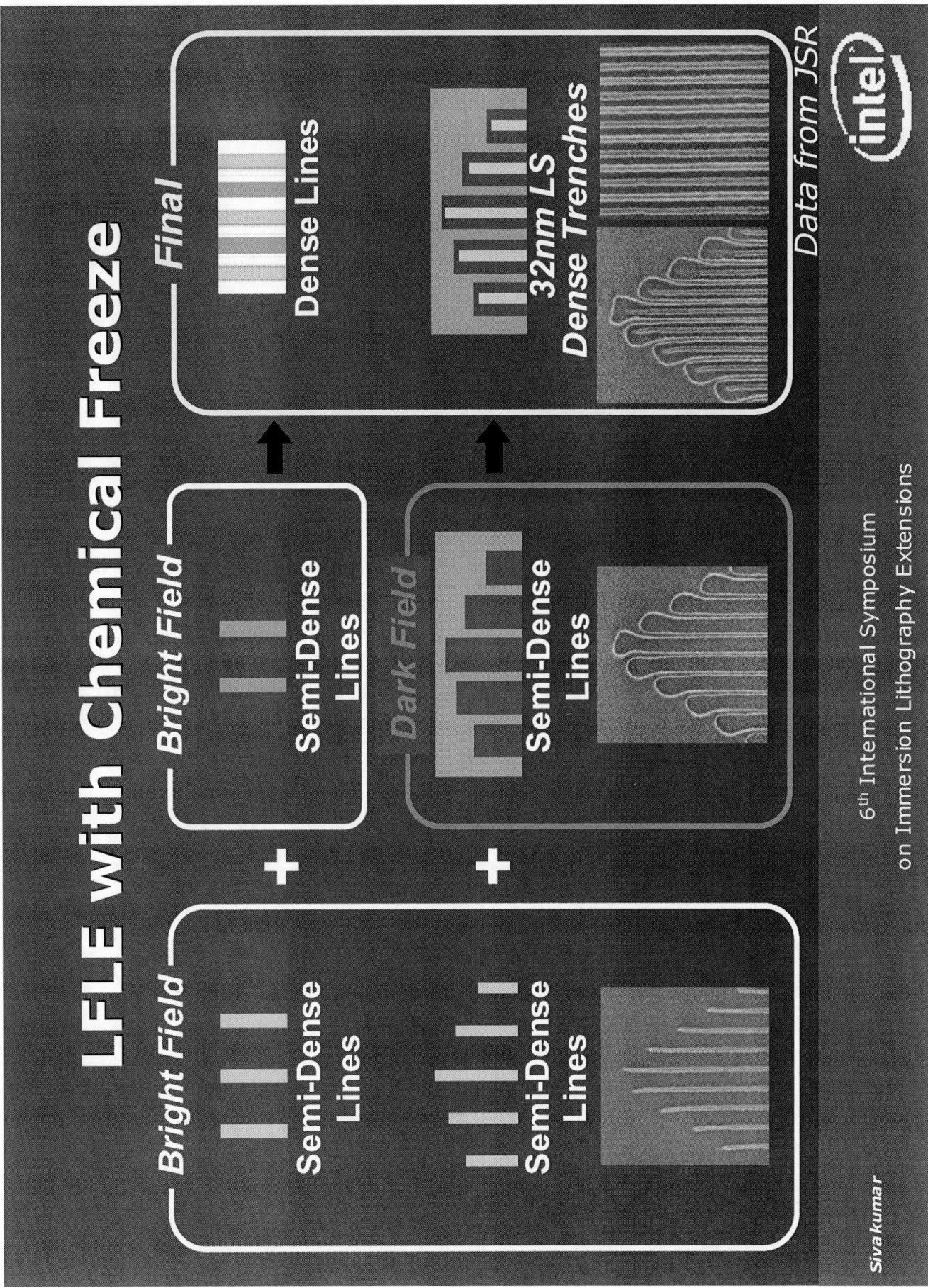

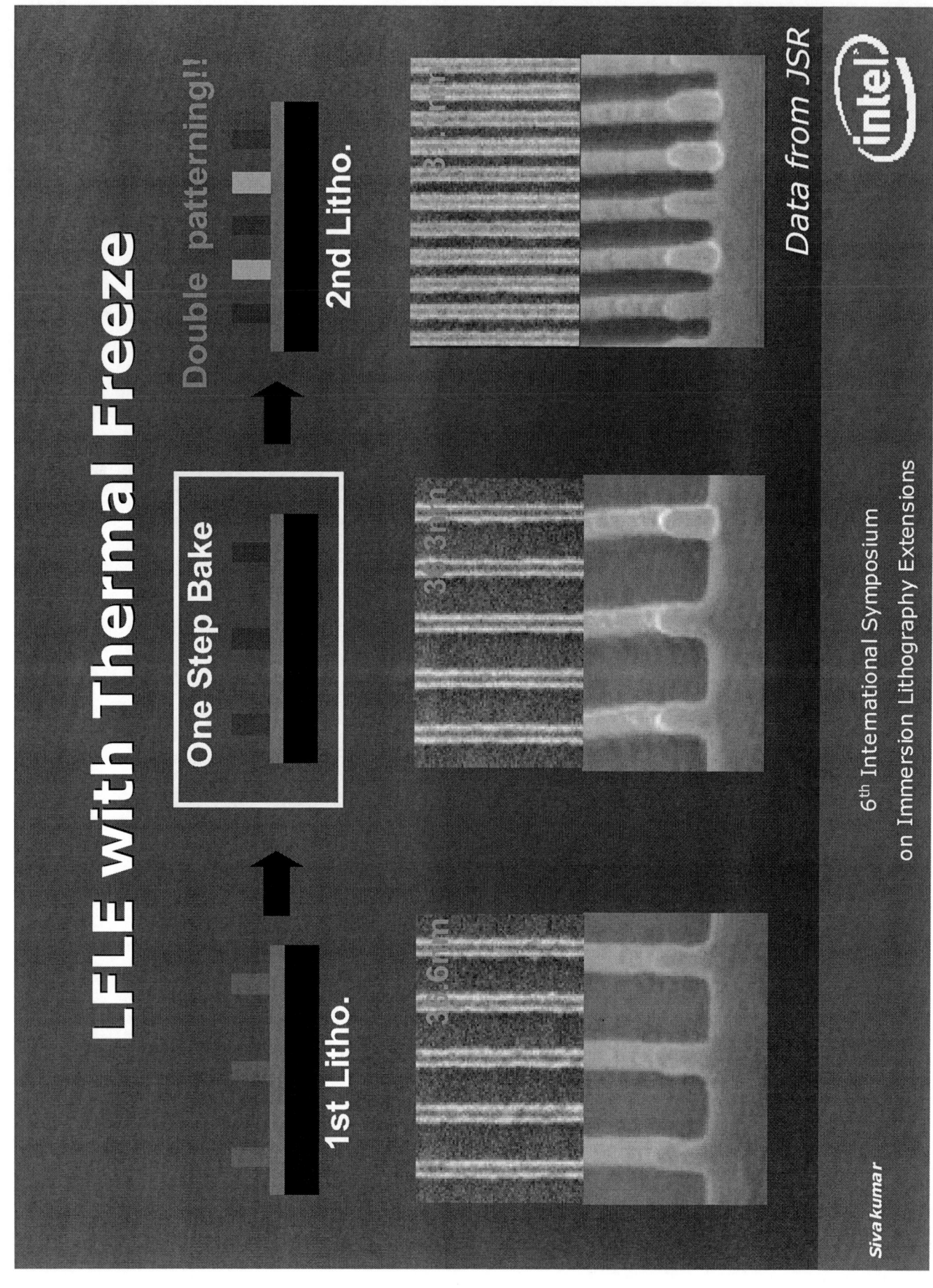

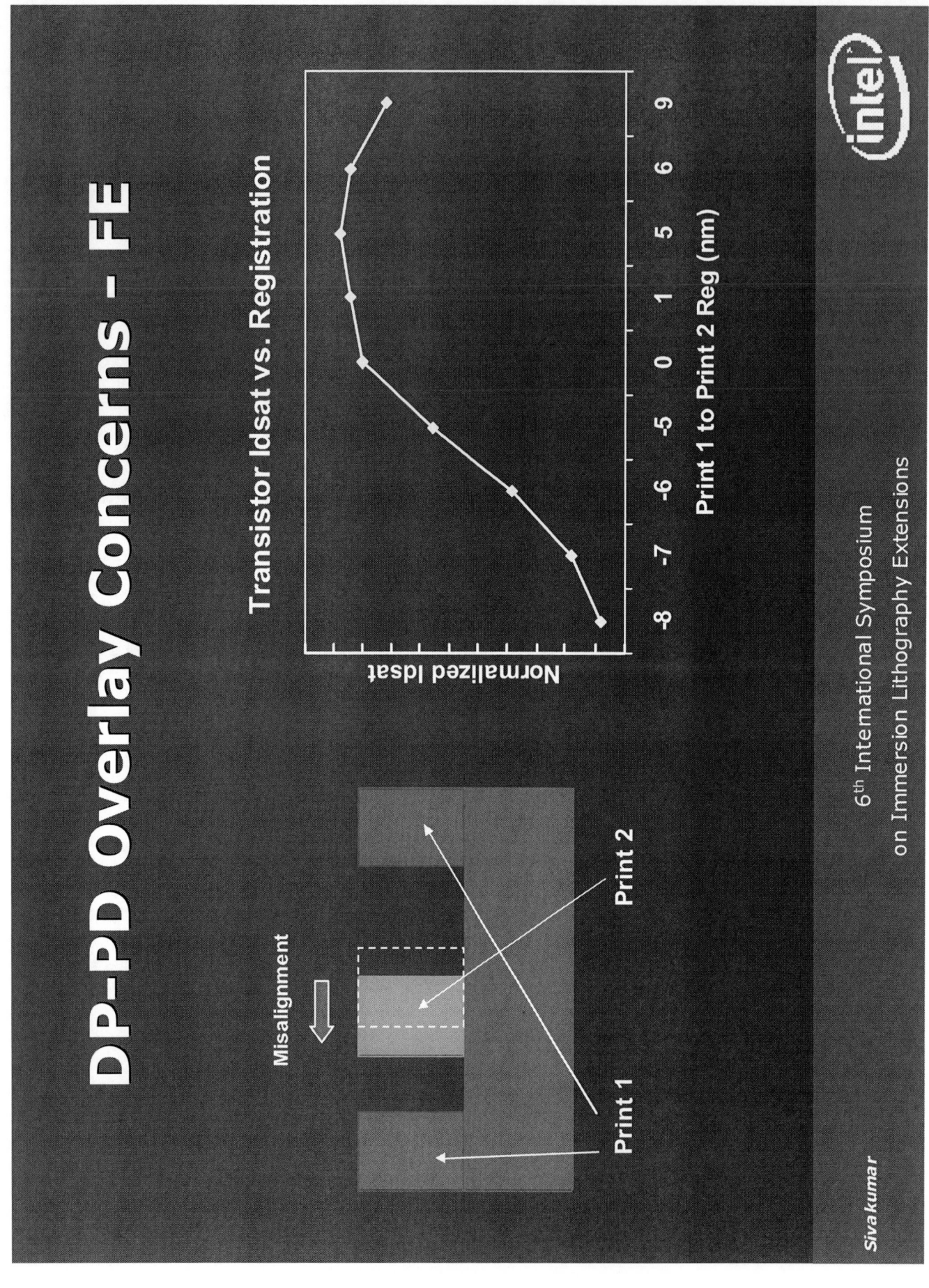

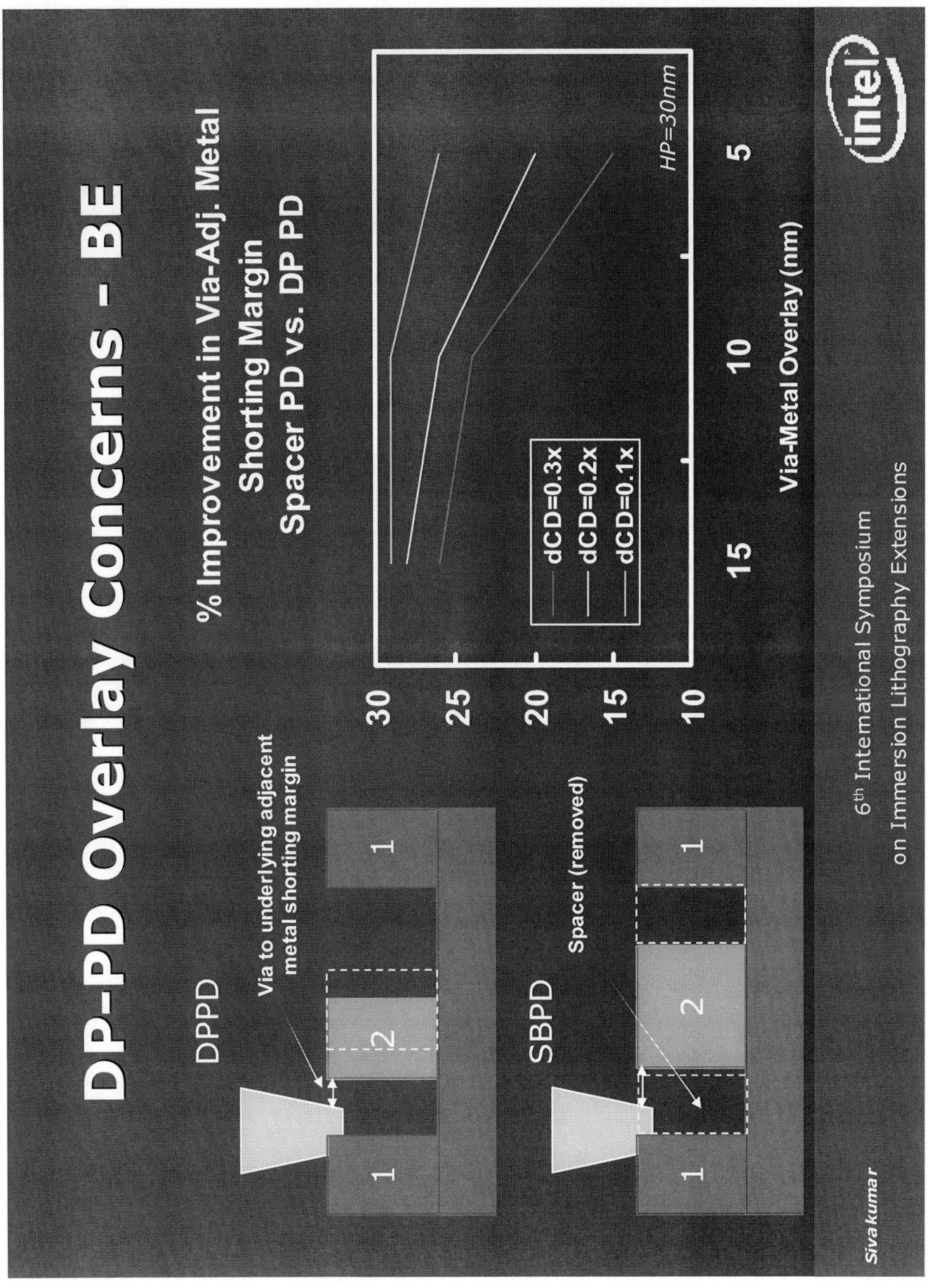

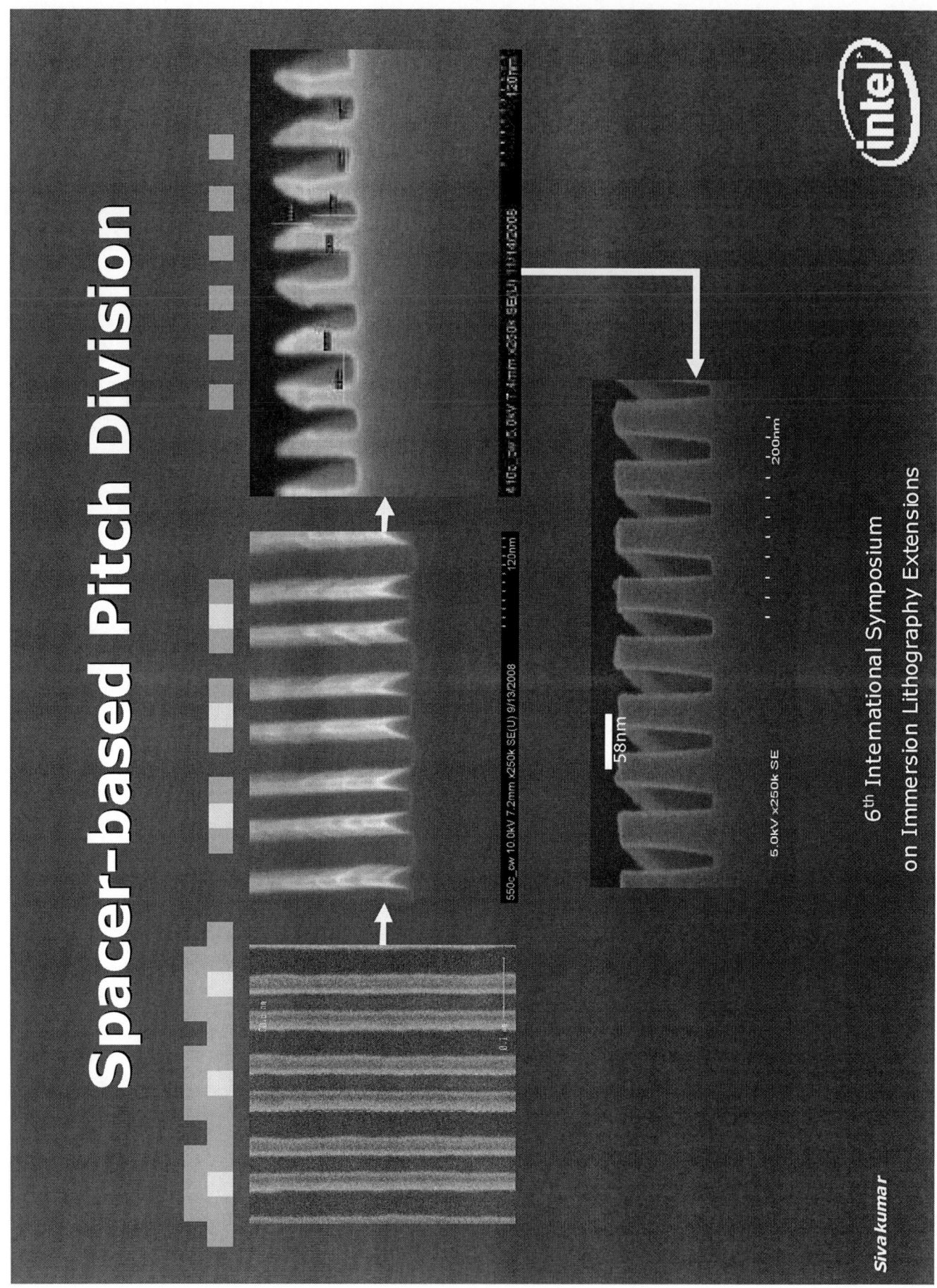

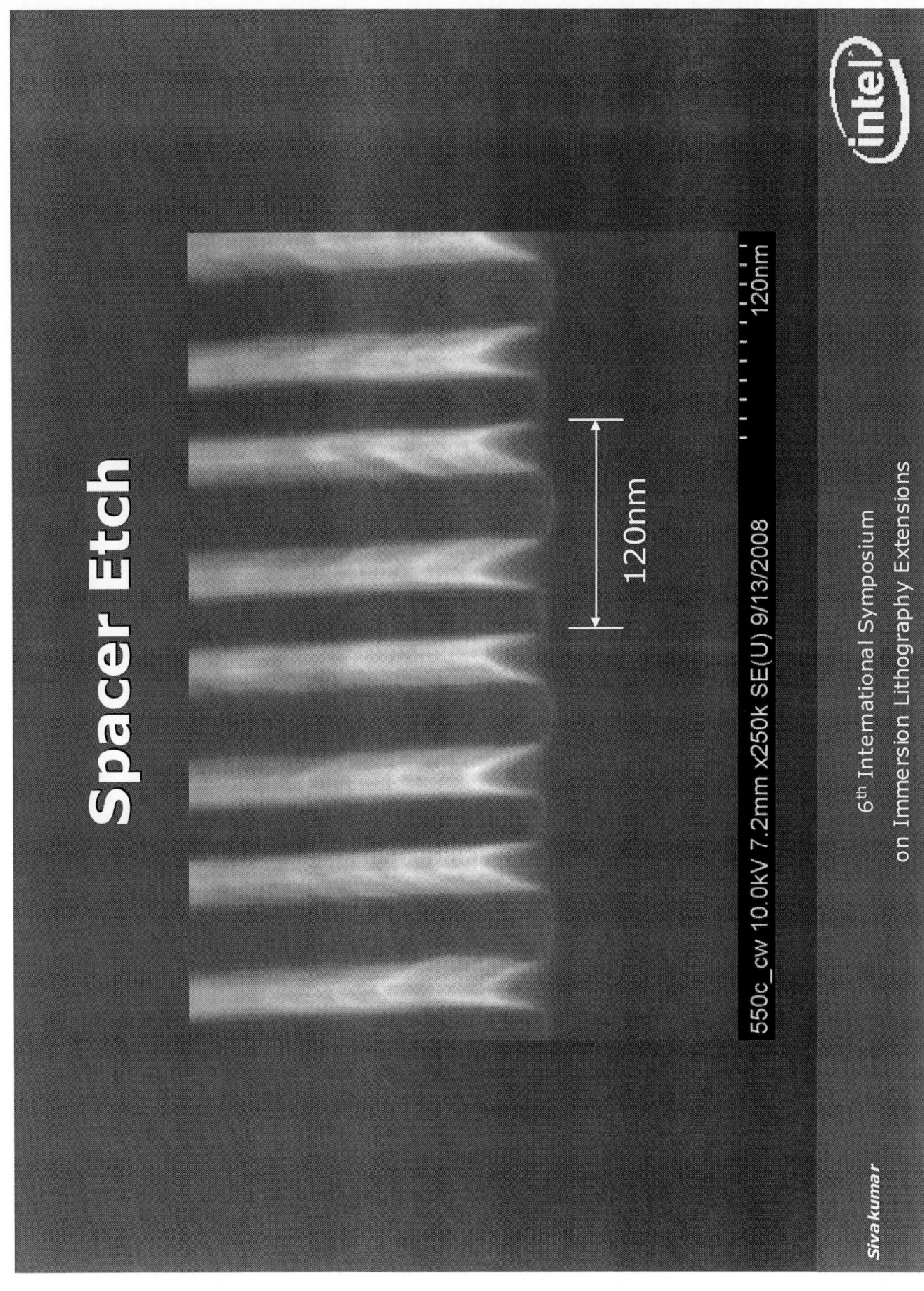

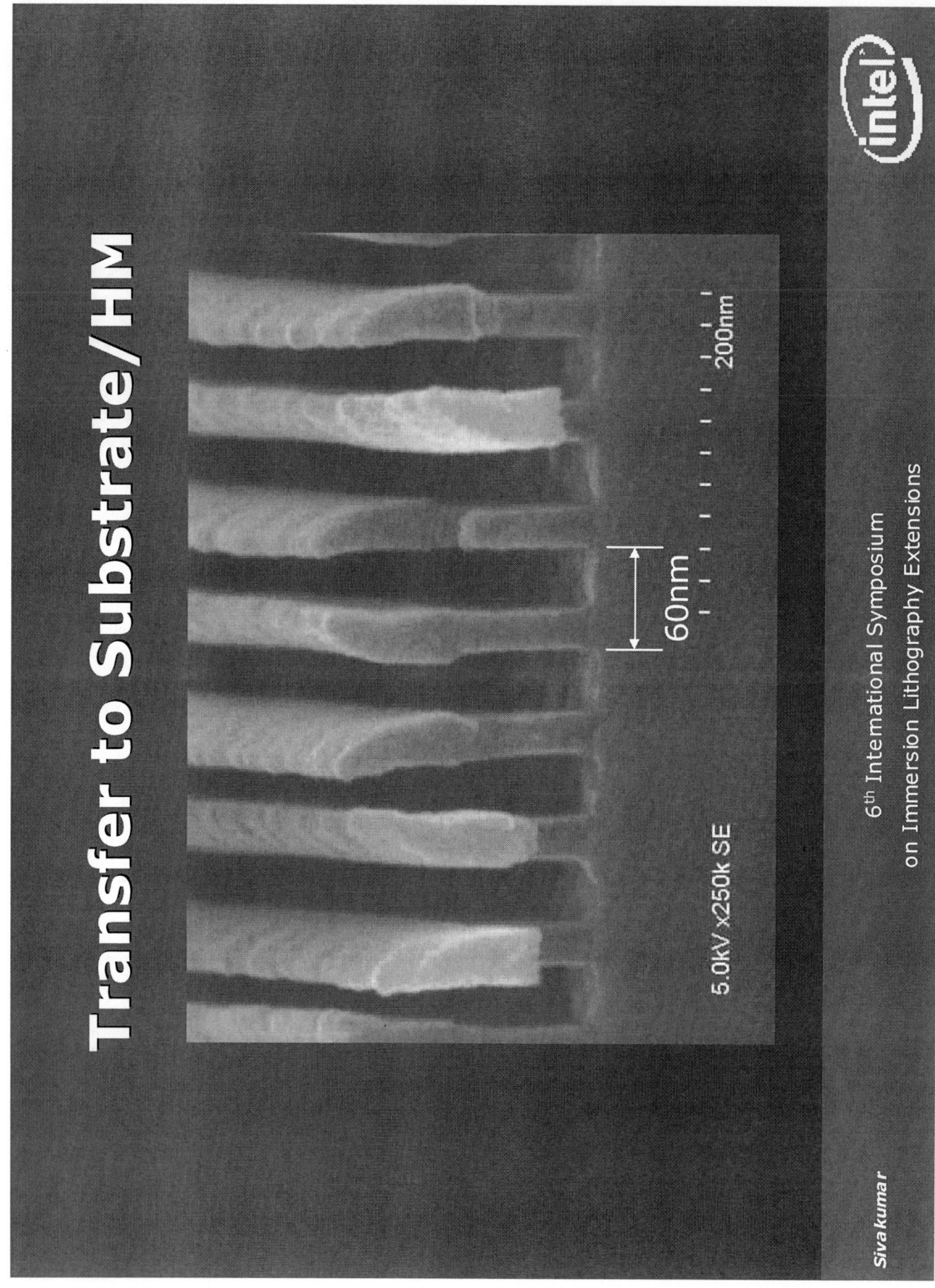

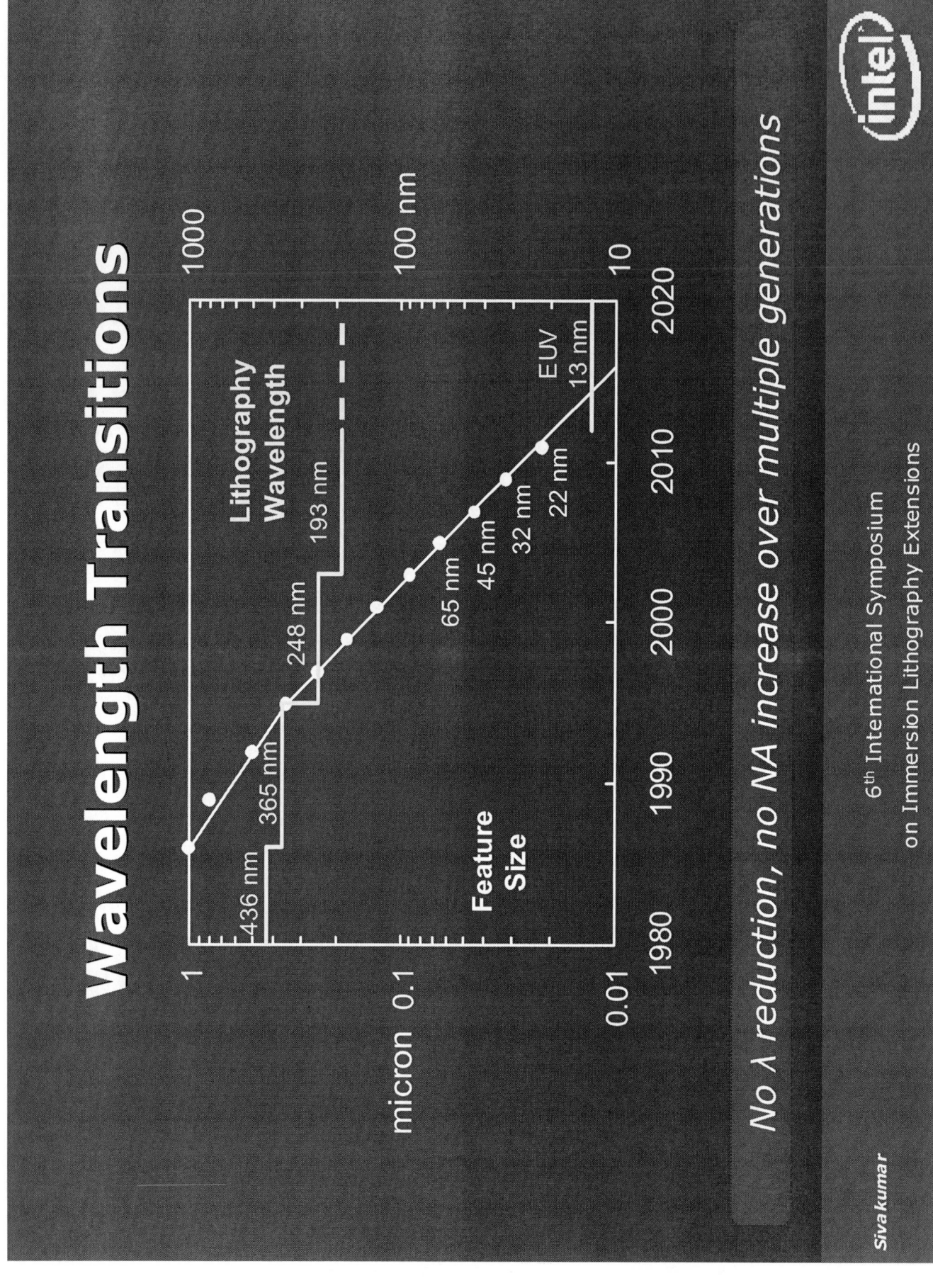

Computational Lithography

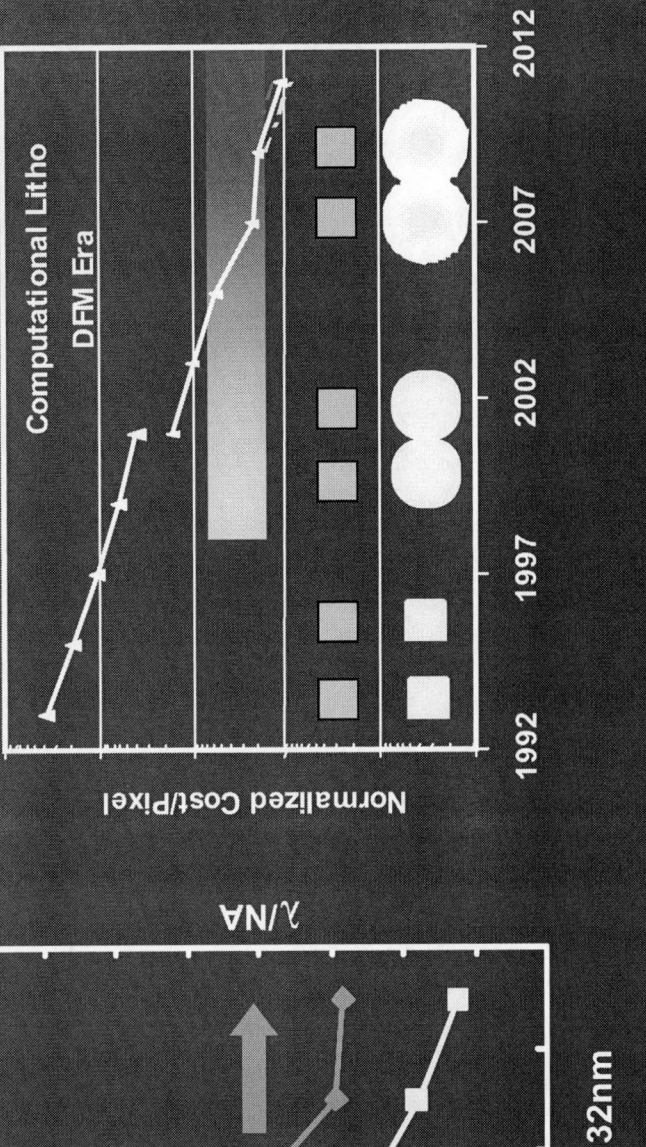

DRs are significantly more complicated with PD
Computational Lithography is key to bridging capability gap

6th International Symposium
on Immersion Lithography Extensions

Siva kumar

Pixelated Masks

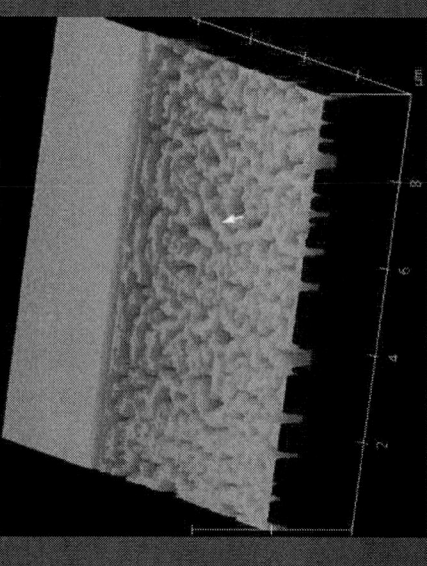

AFM of Mask

Design Layout

Mask

Wafer

6th International Symposium
on Immersion Lithography Extensions

Sivakumar

Optimizing Source and Mask

Source Mask Optimization

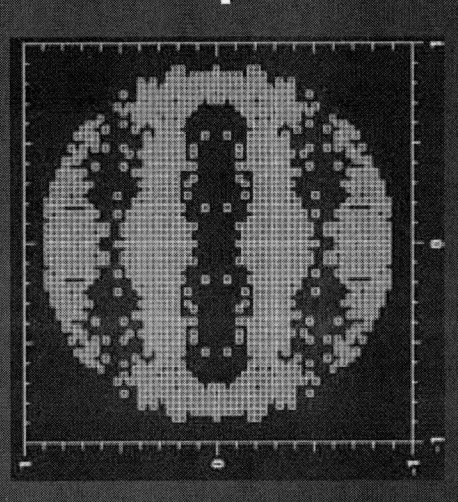

Source Optimization

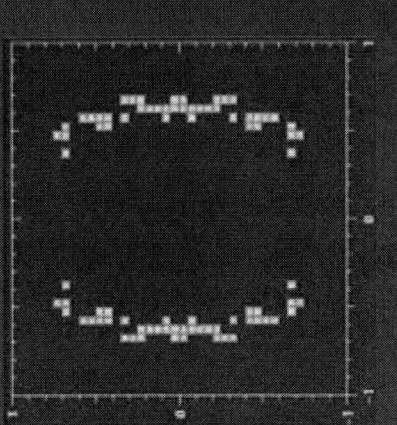

Computational Lithography is a key enabler for extracting maximum resolution from existing technology

6th International Symposium
on Immersion Lithography Extensions

Sivakumar

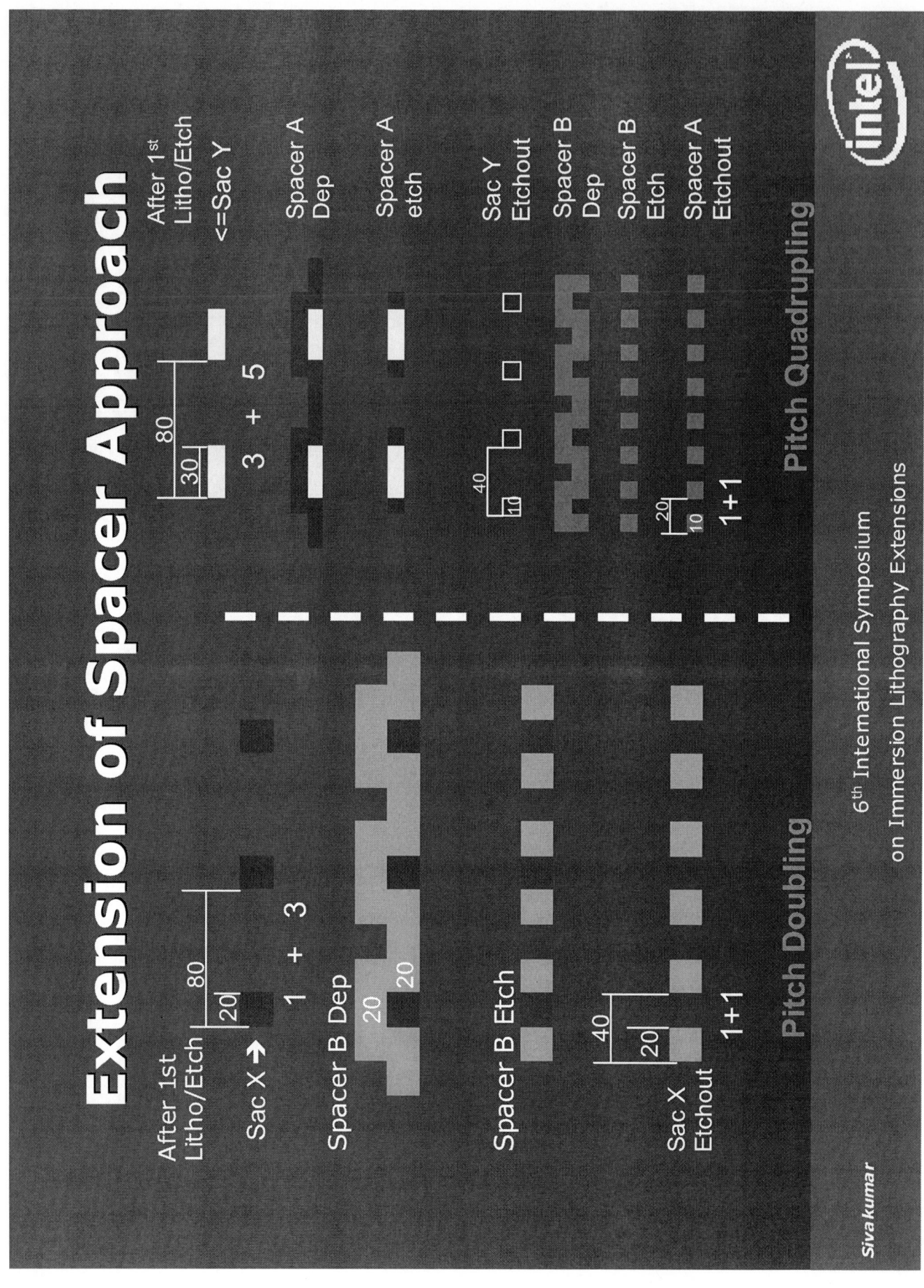

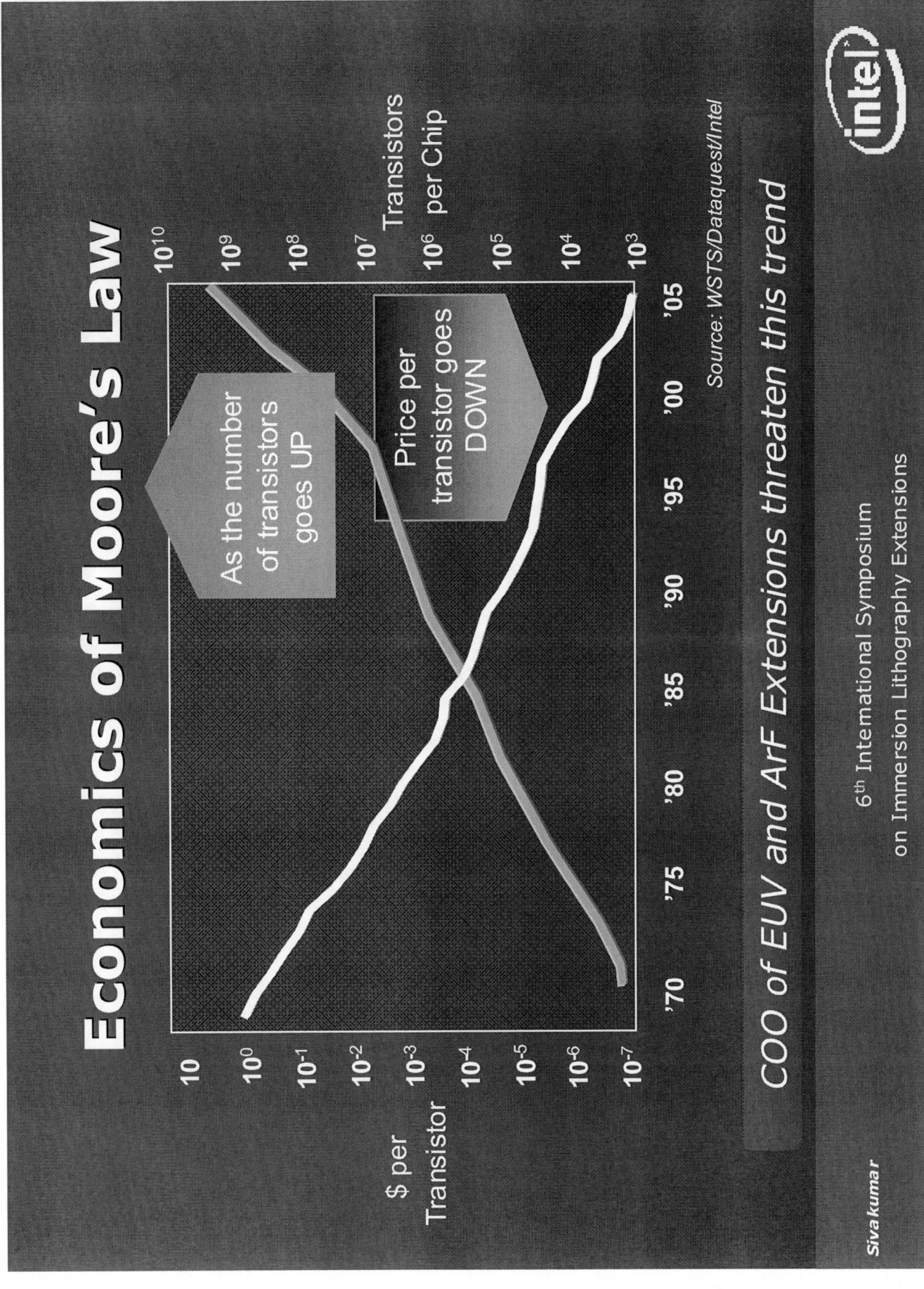

Pitch Division COO

Results – 22 nm (Rigorous, 20,000 wpm)

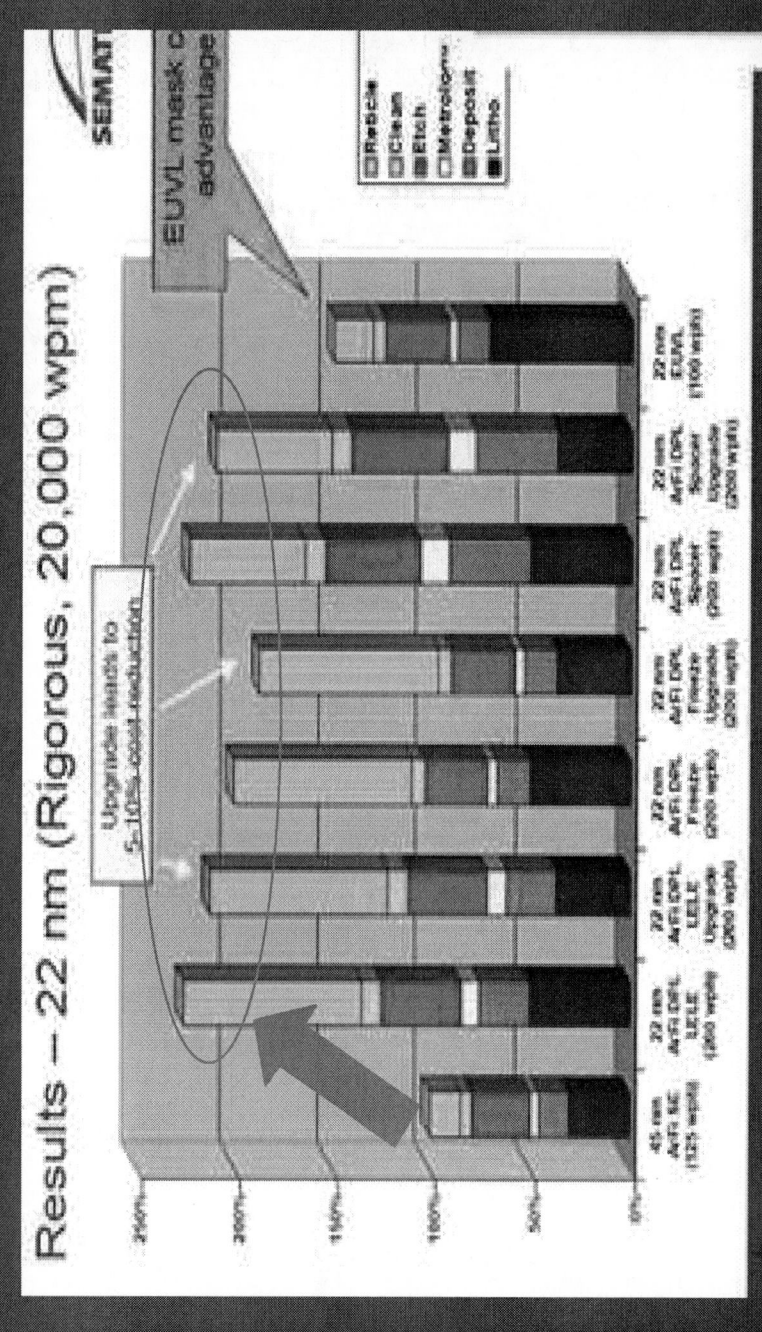

Managing the cost of patterning will be the single biggest challenge facing the 15nm and 11nm nodes

6th International Symposium
on Immersion Lithography Extensions

Sivakumar

Summary

- Intel's process technologies continue to deliver the promise of Moore's Law: higher performing, lower power, and lower cost transistors

- The 15nm logic node will see wide use of ArF pitch division
  - It is expected that all forms of pitch division will be used on a layer-by-layer basis

- Computational lithography will be an increasingly crucial component of the total patterning solution in the pitch-division regime

- Managing the COO of new lithography options is the biggest challenge facing us

Sivakumar

6th International Symposium
on Immersion Lithography Extensions

Acknowledgments

- Lithography staff at the Portland Technology Development Group, Intel Corp., Hillsboro, OR USA

- Mark Bohr, Intel Senior Fellow, Director of Process Architecture and Integration

- Vivek Singh, Intel Fellow, Director of Computational Lithography

- JSR Corp. for sharing several of the resist images shown in this presentation

Sivakumar

6th International Symposium
on Immersion Lithography Extensions

IBM Research Division & Systems and Technology Group

Lithography on the Edge

David Medeiros
IBM

6th International Symposium on Immersion Lithography Extensions

Prague, Czech Republic

23 October 2009

© 2009 IBM Corporation

6th International Symposium on Immersion Lithography Extensions

IBM Research Division & Systems and Technology Group

An Edge

- **A line where an something begins or ends:**

  – A border, a discontinuity, a threshold…

6th International Symposium on Immersion Lithography Extensions

© 2009 IBM Corporation

IBM Research Division & Systems and Technology Group

Scaling Trend – End of an Era?

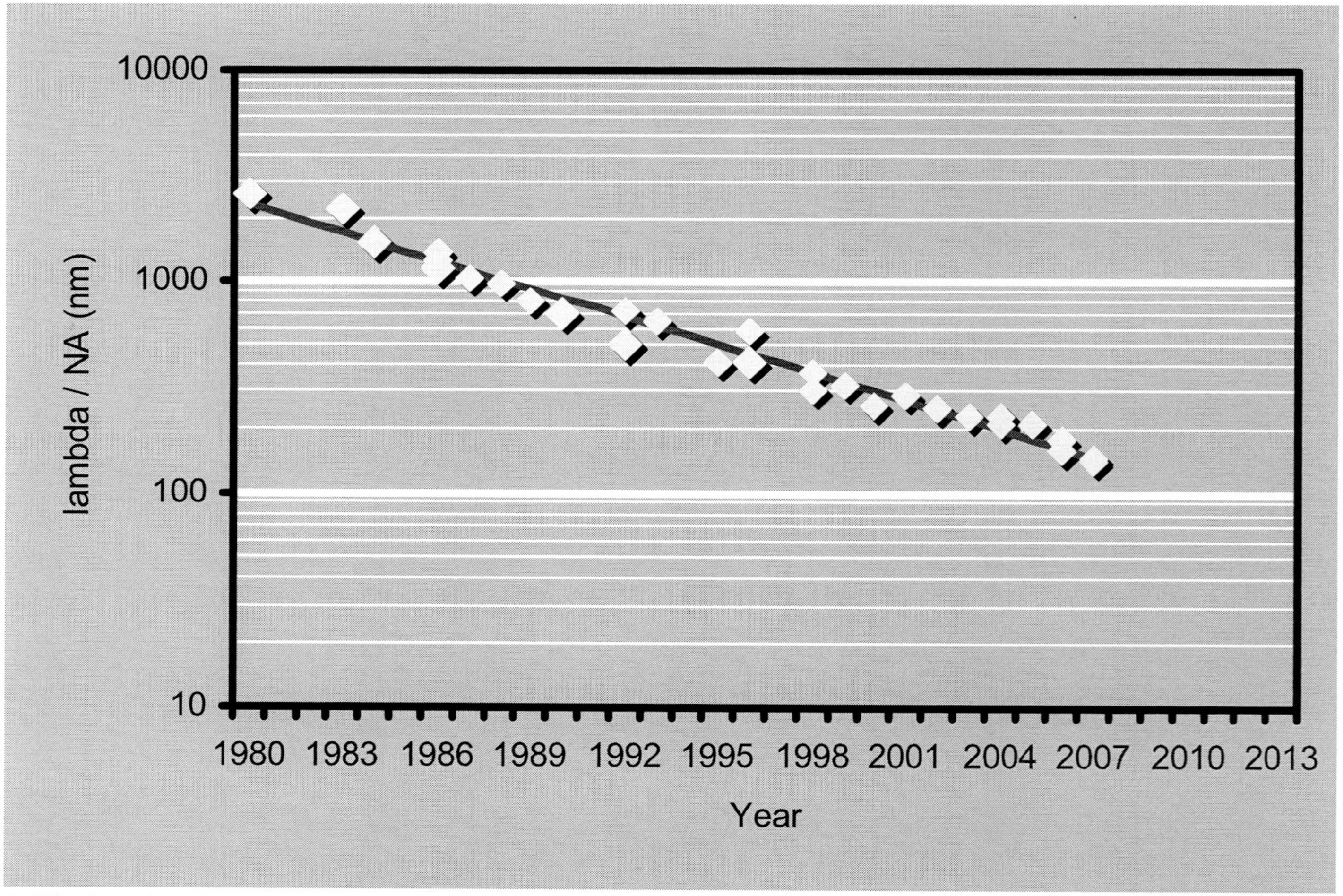

6th International Symposium on Immersion Lithography Extensions

© 2009 IBM Corporation

IBM Research Division & Systems and Technology Group

Edge of a new era?

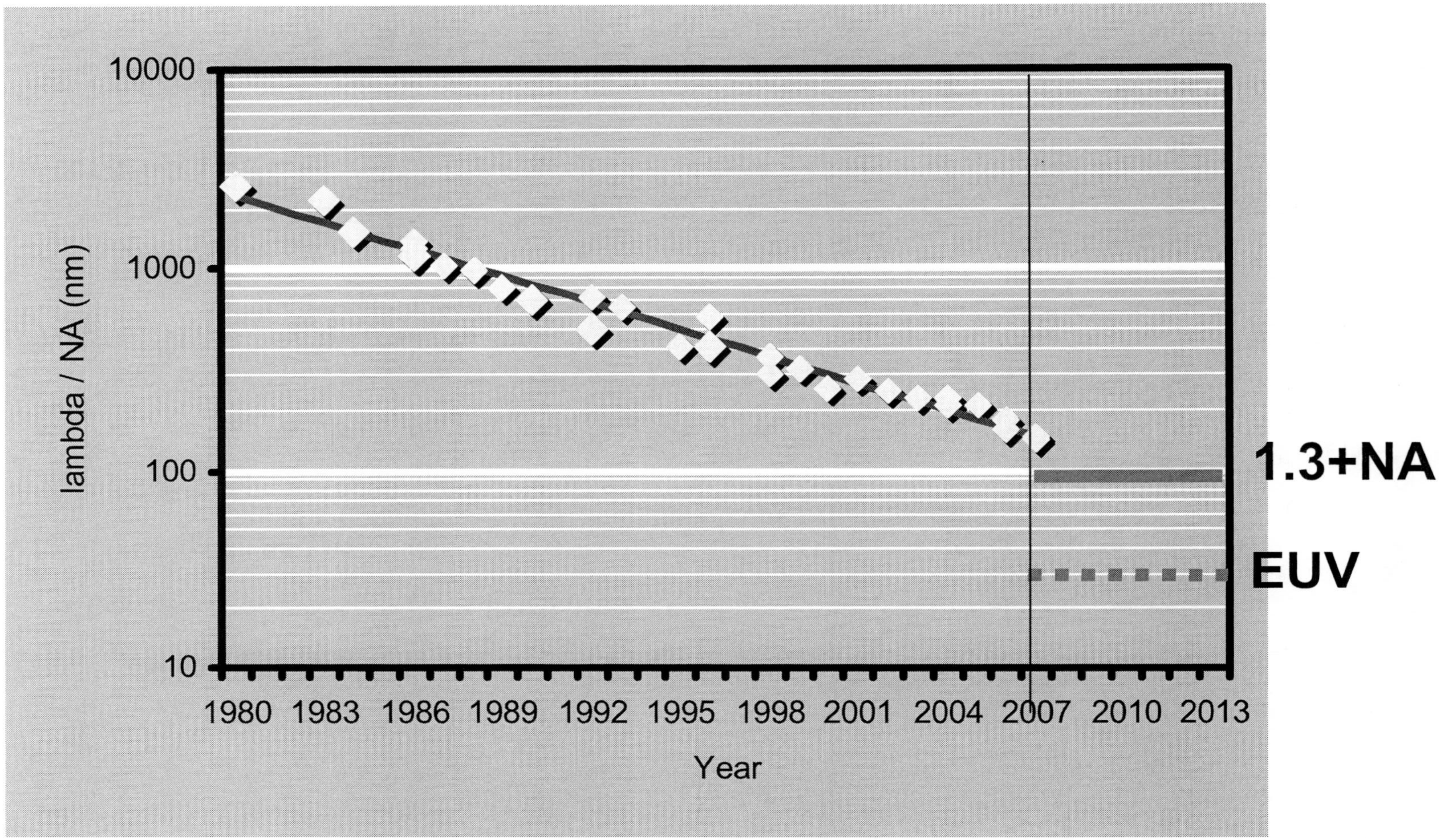

6th International Symposium on Immersion Lithography Extensions

© 2009 IBM Corporation

IBM Research Division & Systems and Technology Group

An Edge

- **A line where an something begins or ends:**
  - A border, a discontinuity, a threshold…

- **A favorable margin:**
  - An advantage, a benefit…

- **What are the challenges of patterning at the edge of the optical lithography capabilities?**

- **What are the implications of trying to move the edge, as with double patterning (DPL)?**

- **What innovations are at our disposal to give us an edge in meeting these challenges?**

6th International Symposium on Immersion Lithography Extensions

© 2009 IBM Corporation

IBM Research Division & Systems and Technology Group

Double Patterning

- **We are on the threshold of extending the utility of ArF litho for logic applications with DPL, which poses challenges in:**

  – Achieving process tolerances for complex designs

    • Overlay

    • CDU

  – Maintaining reasonable mask count

    • More importantly, overall CoO

- **Opportunities – Litho with an Edge**

  – Litho-Litho-Etch (LLE) – Drive down CoO

  – SMO (Source Mask Optimization) – Make the most of k1

6th International Symposium on Immersion Lithography Extensions

© 2009 IBM Corporation

IBM Research Division & Systems and Technology Group

Edge Placement

- **Not an Issue of Demonstration**
  - We know we can print very small
  - Can we meet tolerances?
    - …yield?
    - …profit?

- **Edge Placement Error**
  - How accurately can we print a given set of features? (CD)
  - How precisely can we control the variability of these features (CDU: DoF/EL)
  - How accurately & precisely can we place features together to make useful structures? (OL)

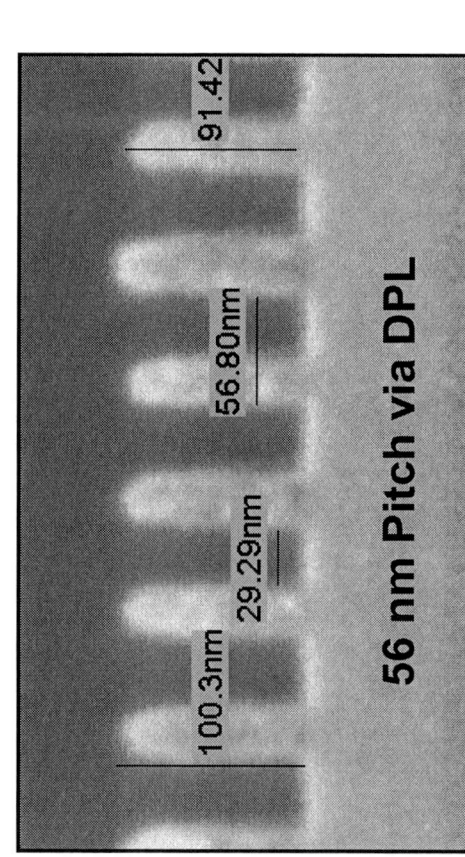

56 nm Pitch via DPL

6th International Symposium on Immersion Lithography Extensions

© 2009 IBM Corporation

IBM Research Division & Systems and Technology Group

Double Patterning Overlay Requirements

- **Sidewall Image Transfer (SIT) aka Self-Aligned Double Patterning (SADP)**

  - "Effective Overlay" ~4-9 nm required

  'Effective OL' or IP dictated by Litho CDU & Dep/Etch Uniformity

  Spacer film

  E1   E1

  CDU_l, CDU_s

  Sidewall Image Transfer

- **Pitch Split Double Patterning (Resist-on-Resist)**

  - Overlay ~ 3-6 nm required

  CDU has significant OL contribution

  CDU_l CDU_s

  L1   L2   L1   L2

  Pitch Split Double Patterning

For example: A.J. Hazelton, et al. "Double-patterning requirements for Optical lithography and prospects for optical extension without Double patterning," JMMM 8(1) 2009

6th International Symposium on Immersion Lithography Extensions

© 2009 IBM Corporation

IBM Research Division & Systems and Technology Group

Overlay

- **4 nm overlay (3σ) in a 25 mm field = 160 ppb**

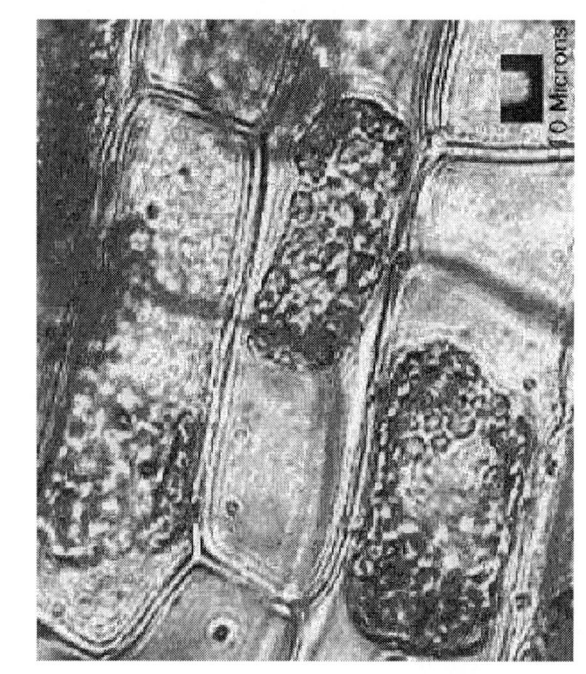

Petrin Tower, Prague ~ 60 m

Plant Cell Width ~ 10 μm

6th International Symposium on Immersion Lithography Extensions

© 2009 IBM Corporation

Overlay

- **Scanner manufacturing have made significant strides in overlay control**
  - Dedication may impact throughput but affords manufacturability

- **Tooling is only part of the equation**
  - Mask Contributions: Image Placement – Systematic, often correctable
  - Nonsystematic Reticle Chucking Offsets
  - Nonsystematic Process Induced Deformations...

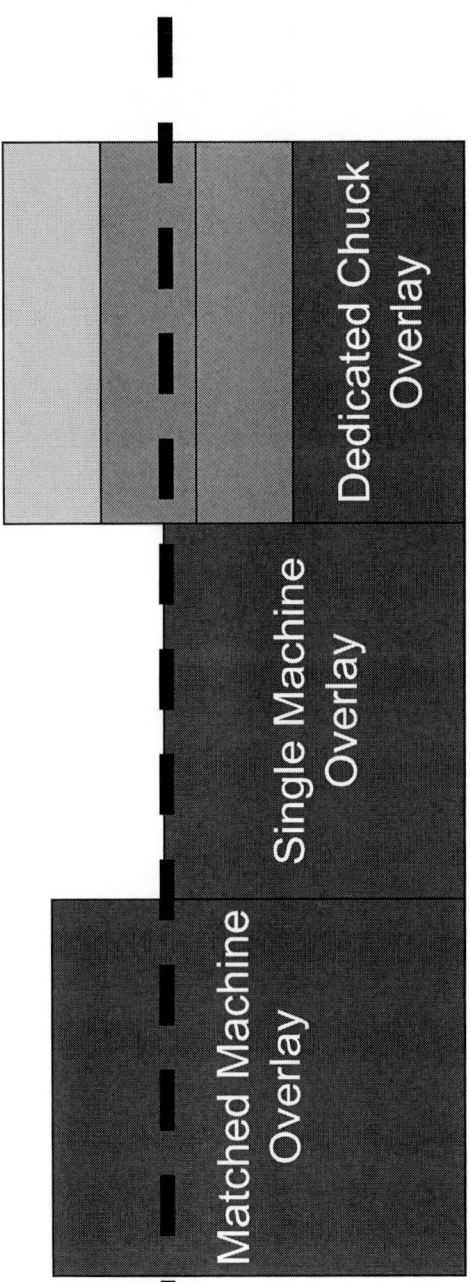

IBM Research Division & Systems and Technology Group

6th International Symposium on Immersion Lithography Extensions

© 2009 IBM Corporation

IBM Research Division & Systems and Technology Group

Overlay Control in DPL

Driving down to manufacturable levels by removal of correctable systematics

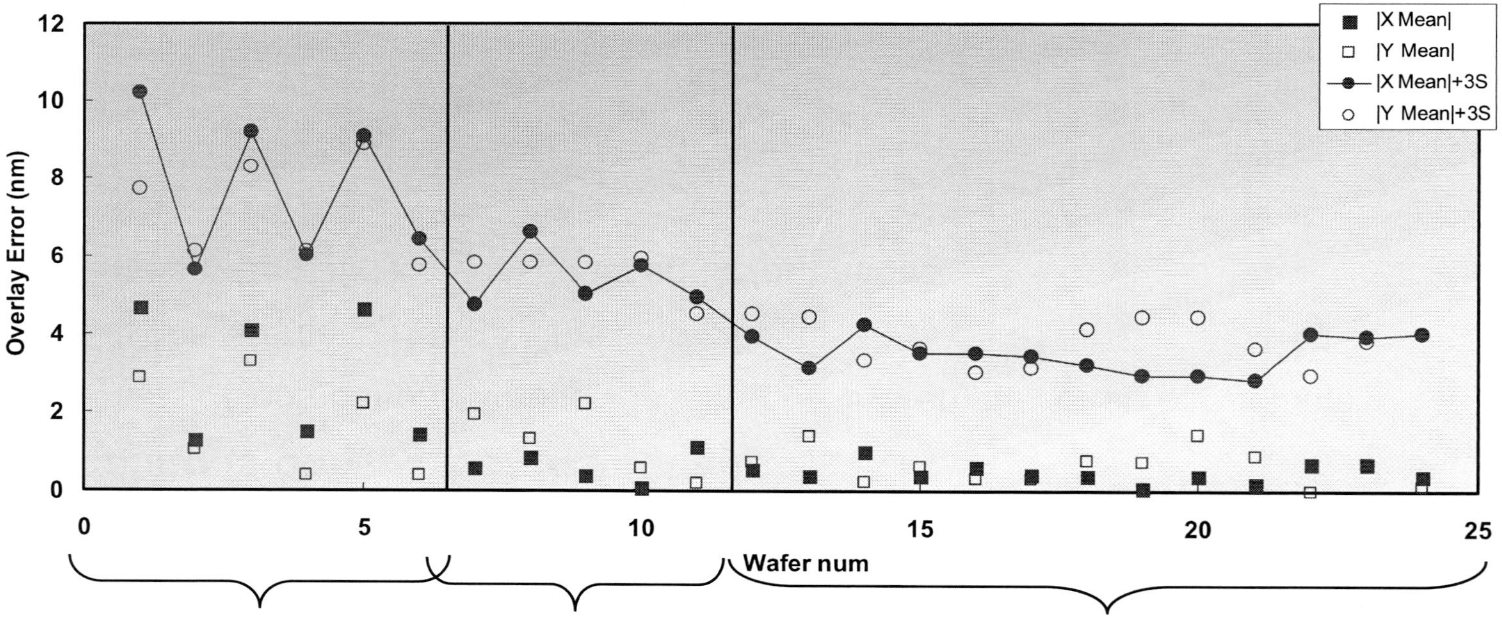

Legend:
- ■ |X Mean|
- □ |Y Mean|
- ● |X Mean|+3S
- ○ |Y Mean|+3S

Y-axis: Overlay Error (nm), 0 to 12

X-axis: Wafer num, 0 to 25

Experiment 1

Experiment 2

Best-Known-Method

Ref: S. Holmes, et al, SPIE (2009) 7273-179
See also: V. Nagaswami, et al, this conference

6th International Symposium on Immersion Lithography Extensions

© 2009 IBM Corporation

IBM Research Division & Systems and Technology Group

Two Primary Line/Space Patterning Methods

- **Sidewall Image Transfer (SIT)**
- **Litho-'Freeze'-Litho-Etch**

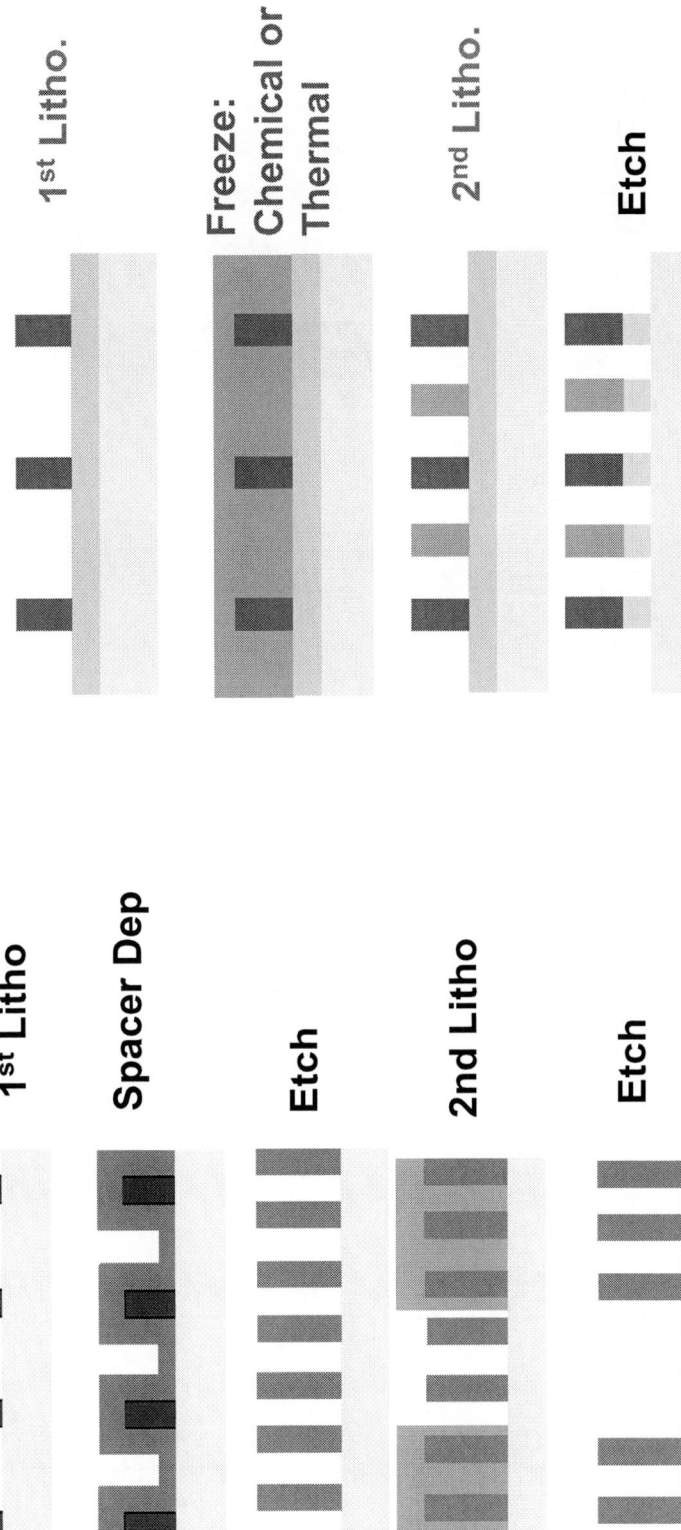

Benefits: CDU, Overlay Insensitive
Challenges: Design Flexibility*, CoO

Benefits: Design Flexibility*
Challenges: Overlay Sensitivity, CoO

6th International Symposium on Immersion Lithography Extensions

© 2009 IBM Corporation

IBM Research Division & Systems and Technology Group

Sidewall Image Transfer

- Complex decomposition, coloring challenges for 2D cases

- Trim layer becomes a challenge as well

Final

Target
Core
Spacer
Trim

M. Burkhardt et al., EIPBN (2009)

6th International Symposium on Immersion Lithography Extensions

© 2009 IBM Corporation

IBM Research Division & Systems and Technology Group

Pitch Split Double Exposure

- Initial target pitch is beyond resolution of 1.35NA Optical Scanners

- After decomposition into separate exposures, the pattern can be imaged using new Double Patterning resist systems.

Target   Decomposition   Bias   OPC   Print   Bias and combine

M. Burkhardt et al., EIPBN (2009)

6th International Symposium on Immersion Lithography Extensions

© 2009 IBM Corporation

CD Uniformity with Optimized Freeze Develop (Steps 1 & 2)

IBM Research Division & Systems and Technology Group

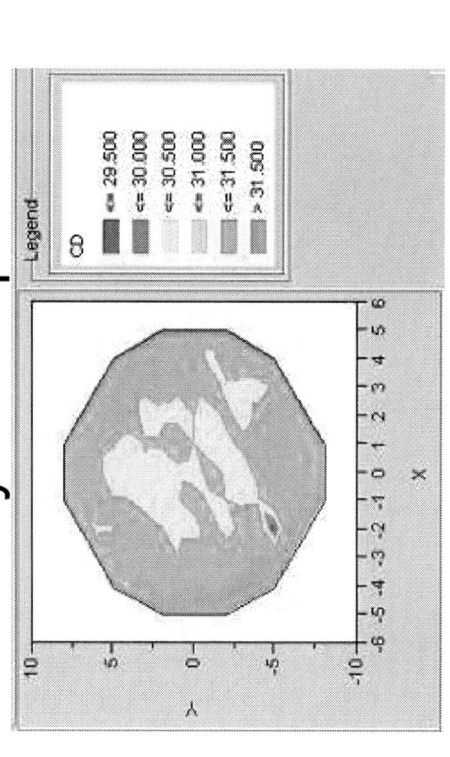

Measured 139 fields

1st Litho + Freeze
ARX2928JN after Freeze
with FZX112

Avg: 28.8nm
3sigma: 1.57nm
Min: 26.9nm
Max: 30.8nm

1st Litho only
ARX2928JN
IBM Develop

Avg: 30.9nm
3sigma: 1.55nm
Min: 29.3nm
Max: 32.3nm

Comparable CDU before and after Freeze   *S. Holmes, et al, SPIE (2009)*

6th International Symposium on Immersion Lithography Extensions

© 2009 IBM Corporation

IBM Research Division & Systems and Technology Group

Intra-field CD before and after DoseMapper Optimization (Steps 1-3)

Level 2 Intrafield CD

Level 1 Intrafield CD

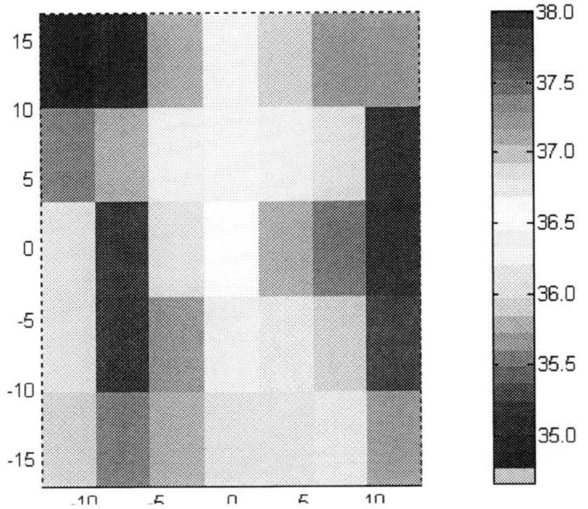

Pre DoseMapper

Level 1
Mean CD: 36.1 nm
3 Sigma: 2.3 nm

Level 2
Mean CD: 34.9 nm
3 Sigma: 2.2 nm

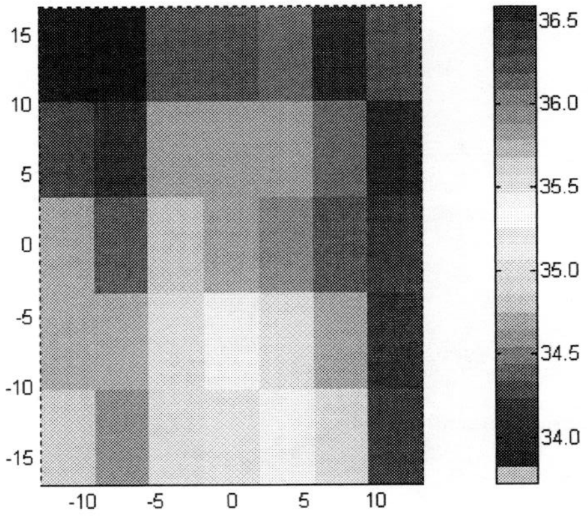

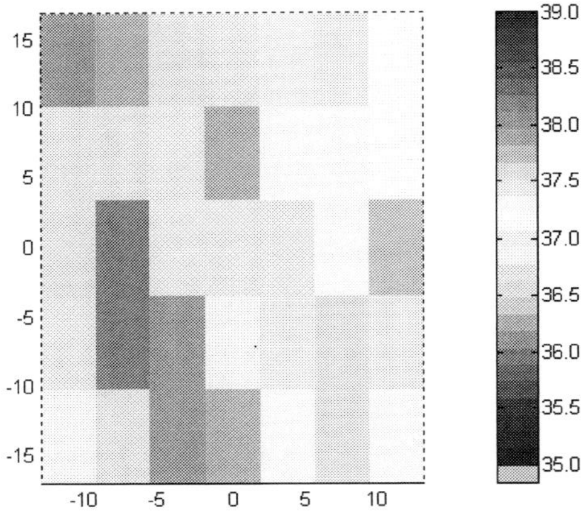

Post DoseMapper

Level 1
Mean CD: 36.8 nm
3 Sigma: 1.5 nm

Level 2
Mean CD: 34.4 nm
3 Sigma: 0.7 nm

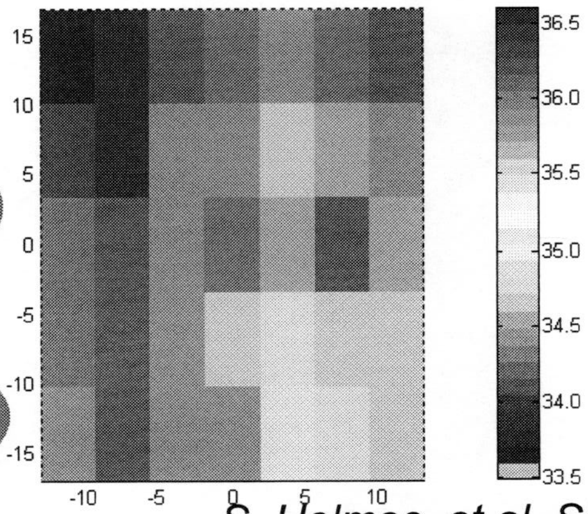

S. Holmes, et al, SPIE (2009)

© 2009 IBM Corporation

6th International Symposium on Immersion Lithography Extensions

Aerial Images for Off-Axis Illumination

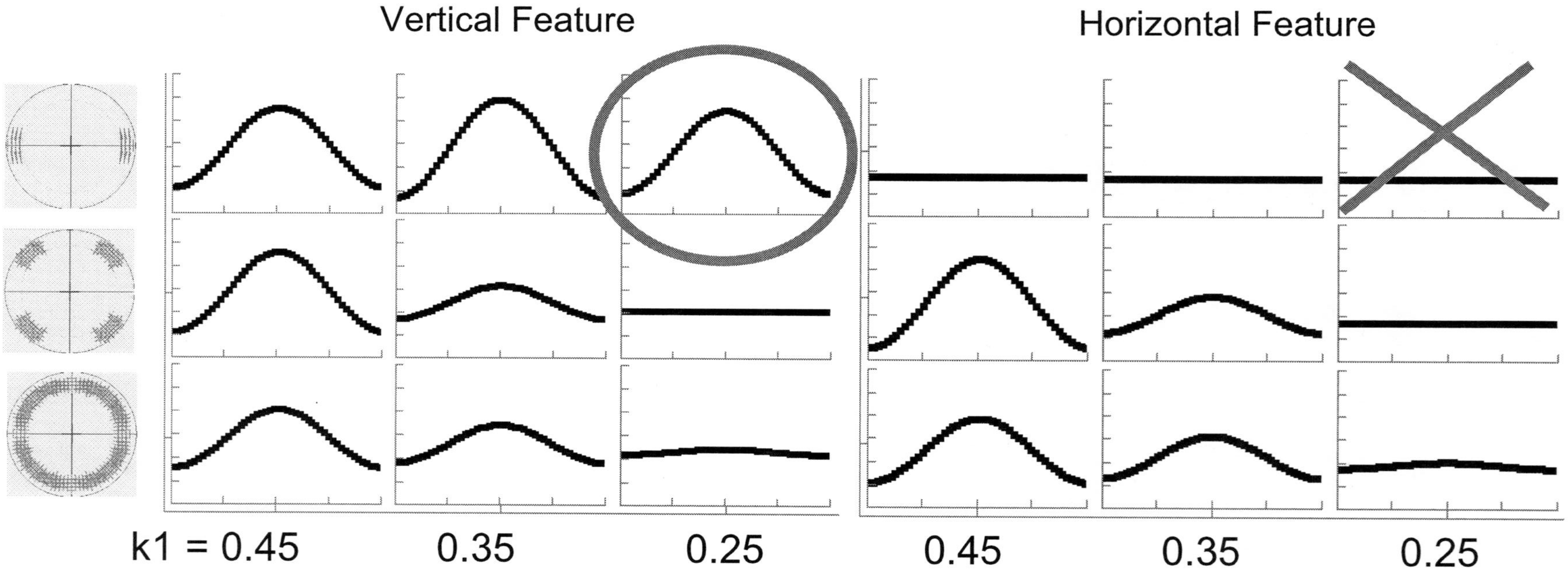

IBM Research Division & Systems and Technology Group

Horizontal Feature

Vertical Feature

	0.25	0.35	0.45
k1 = 0.45			

- Dipole Contrast is highest at low k_1, but can't print wrong orientation at pitch
- Quadrapole prints intermediate structures well but not dense pitch.
- Annular prints multiple orientation but has limited contrast at dense pitch.

6th International Symposium on Immersion Lithography Extensions

© 2009 IBM Corporation

Sensitivity of Illuminator at Active Level: 2D Effects

- Even a pitches greater than minimum, dipole suffers from wrong-way 'ringing', as predicted

- C-Quad allows for improved dual-orientation as well as 2D printability, but limits pitch

M. Colburn, et al. SPIE (2009)

IBM Research Division & Systems and Technology Group

6th International Symposium on Immersion Lithography Extensions

© 2009 IBM Corporation

IBM Research Division & Systems and Technology Group

Printing Wrong Way Gate with Dipole Illuminator: 2D H-Bar

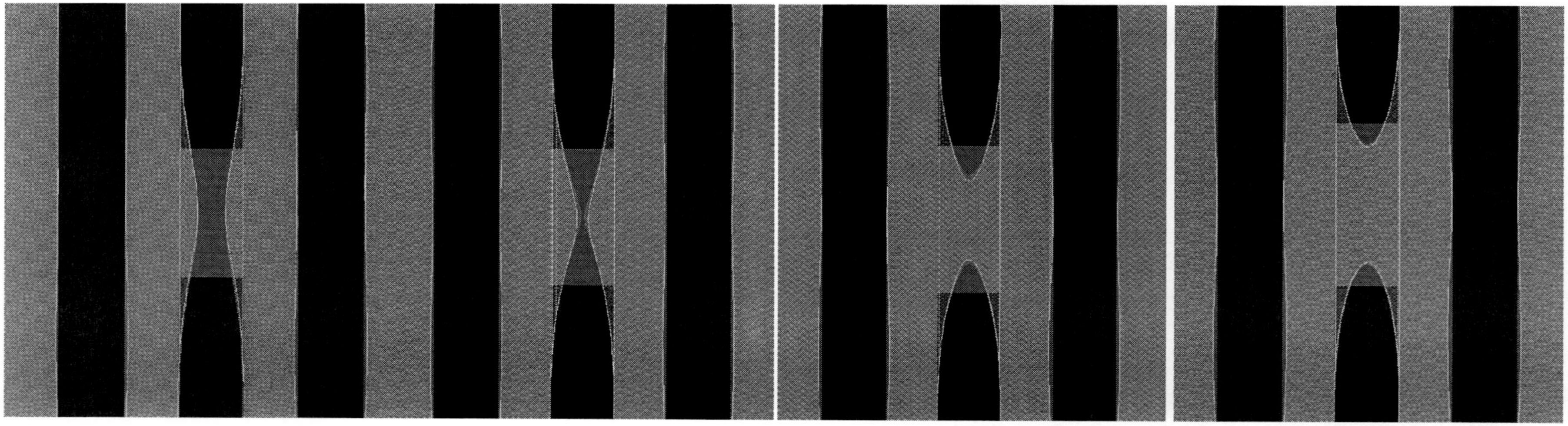

H-Bar:100   110   120   130

- Driving Poly to tight pitch compromises 'wrong-way' printability
- RET must be balanced to accommodate ACLV, WW, Density

M. Colburn, et al. SPIE (2009)

6th International Symposium on Immersion Lithography Extensions

© 2009 IBM Corporation

IBM Research Division & Systems and Technology Group

Double Dipole Litho (DDL) for 2D Printing (*not DPL*)

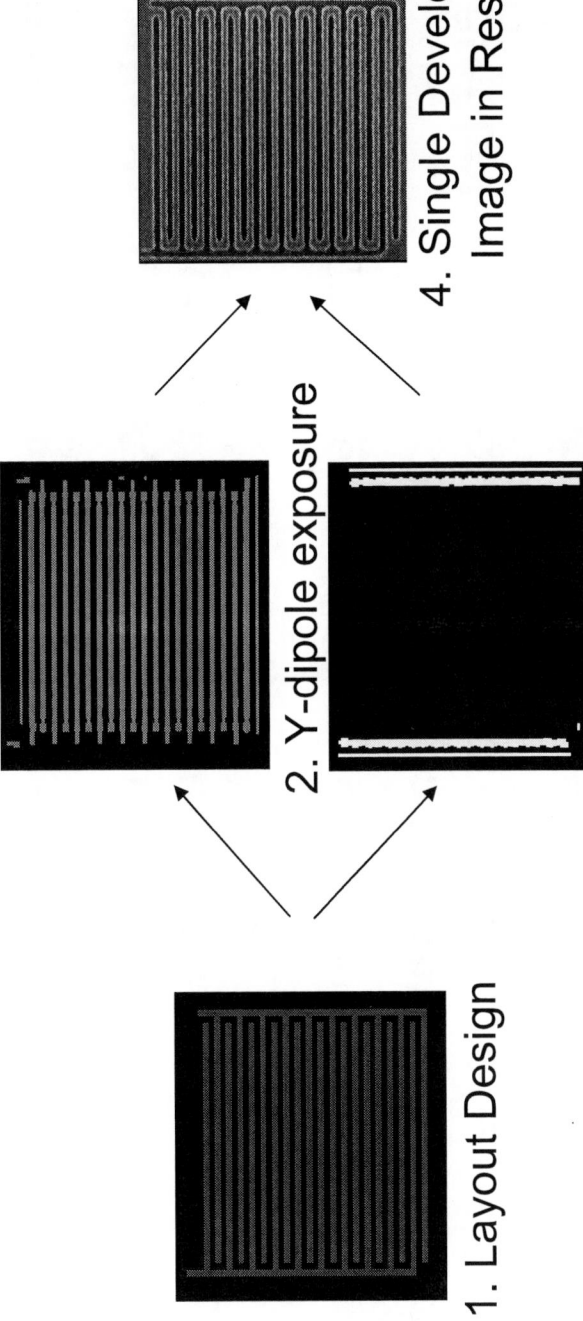

1. Layout Design

2. Y-dipole exposure

3. X-dipole exposure

4. Single Developed Image in Resist

DDL affords increased degree of freedom in the optimization of the print image by separation of design into two masks that generate an image within a single resist.

The sum of the aerial image results in enhanced resolution not afforded by a conventional 1-mask imaging solution.

M. Burkhardt, PMJ 2006

6th International Symposium on Immersion Lithography Extensions

© 2009 IBM Corporation

IBM Research Division & Systems and Technology Group

Scaling Trend: Design Restrictions

- **Shrink of linewidth of pitch imparts a shrink of CD uniformity specification → Image quality needs to improve from one generation to the next**

- **Dipole illumination yields close to perfect imaging for one linewidth/pitch combination.**
  - May be the only option to achieve desired CD uniformity and line edge roughness

- **This implies either unidirectional design / preferred orientation of critical levels or mask proliferation**

- **What are the implications for optimized illumination for each orientation in a design?**

6th International Symposium on Immersion Lithography Extensions

© 2009 IBM Corporation

Mask Count – Hypothetical, for Illustrative Purposes

	Level	45 nm	32 nm	22nm U	22 nm R	15 nm U	15 nm R	11 nm R
	Solution	SE	SE*	SE/DE*	SE/DE*	DPL/SIT	DPL/SIT	DPL/SIT
1	Active	1	1	1	1	3	2	4
2	Gate	1	2	2	2	3	2	4
3	Contact	1	1	2	2	4	3	6
4	Metal 1	1	1	2	2	3	2	4
5	Via 1	1	1	1	1	2	1	2
6	Metal 2	1	1	2	1	3	2	3
7	Via 2	1	1	1	1	2	1	2
8	Metal 3	1	1	2	1	3	2	3
9	Via 3	1	1	1	1	2	1	2
10	Metal 4	1	1	2	1	3	2	3
	Total for 10 Crit. Levels	10	11	16	13	28 !	18	33 !!

IBM Research Division & Systems and Technology Group

Mask Count – Hypothetical, for Illustrative Purposes

Level	45 nm	32 nm	22nm II	22 nm R	15 nm II	15 nm R	11 nm R
							L/SIT
1							4
2							4
3							6
4							4
5							2
6							3
7							2
8							3
9							2
10							3
Total for 10 Crit. Levels	10	11	16	13	28 !	18	33 !!

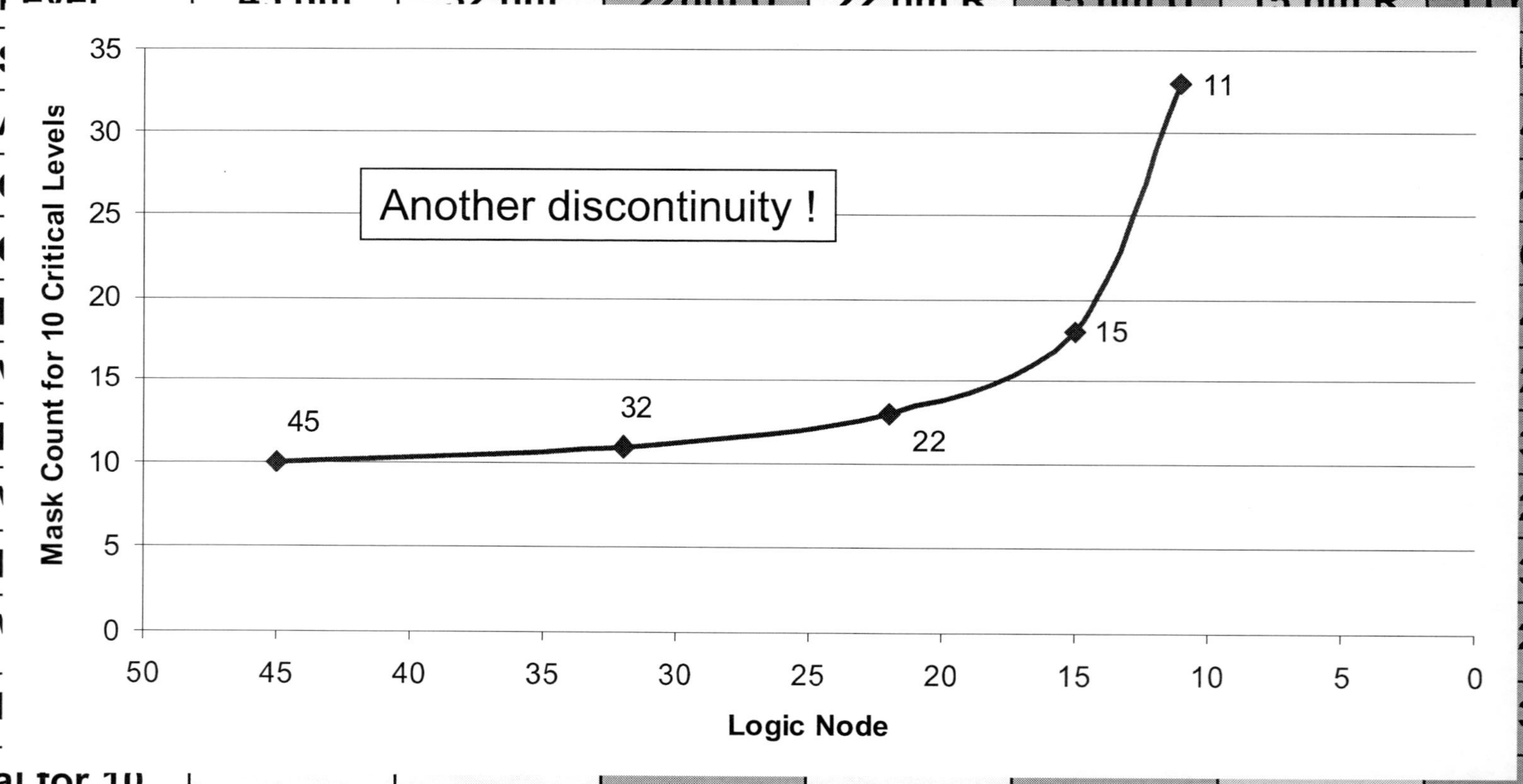

Another discontinuity !

Chart — Mask Count for 10 Critical Levels vs Logic Node

Data points: 45, 32, 22, 15, 11

6th International Symposium on Immersion Lithography Extensions

© 2009 IBM Corporation

IBM Research Division & Systems and Technology Group

Process Steps / CoO

- Mask Count is only one part of the cost of ownership equation
  - Different products require different ground rules: e.g., MPU (HP) vs. Mobile (LP)
  - Depending on wafers/mask, mask count can either be huge component of CoO or relatively small

- Increased complexity and added process steps are always concerns
  - Double patterning adds wafer passes on expensive litho clusters
  - Adds to the etch, deposition, cleans, etc…
  - Adds to the metrology burden (non-value add costs increase)
  - Added capacity adds to capital demands
  - Added steps add to the propensity for yield detractors, defectivity

- Cost effective solutions are required that preclude need for mask & process complexity
  - Litho-litho-etch
  - Source mask optimization

6th International Symposium on Immersion Lithography Extensions

© 2009 IBM Corporation

IBM Research Division & Systems and Technology Group

Source Mask Optimization (SMO)

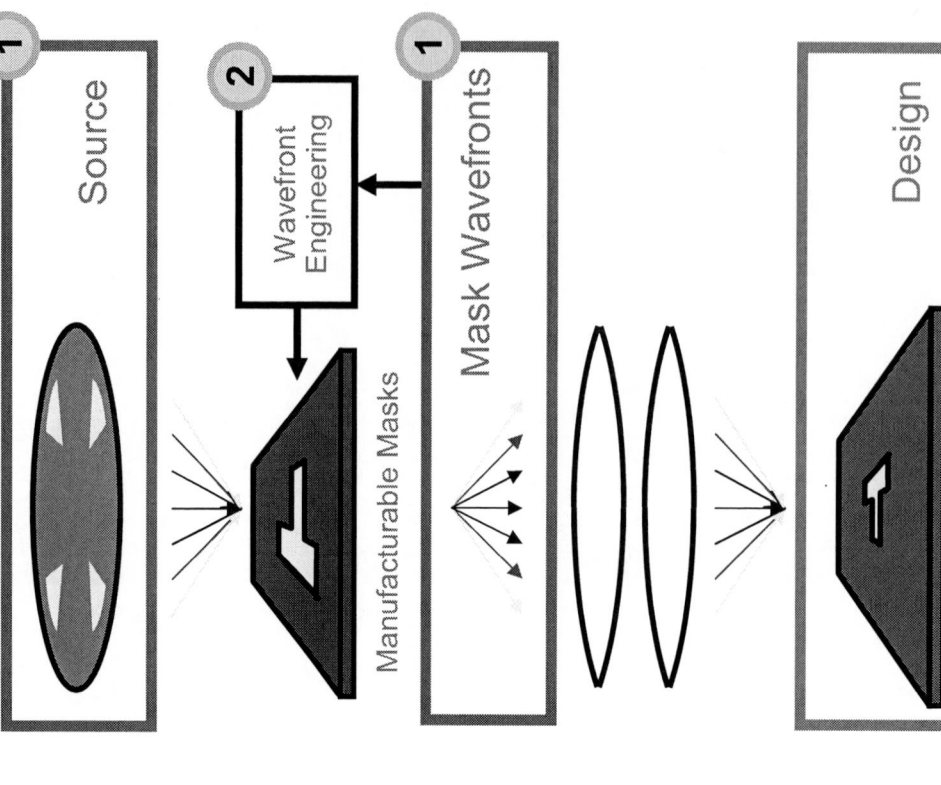

- SMO applies intensive optimization to the fundamental degrees of freedom in the system

- Goal is to find the best set of image-forming waves that can propagate through the exposure tool

- No starting mask or source: only the target & manufacturability tolerances are required.

1 Seek the globally optimum set of image-forming waves (incl. multiple exposures) within constraints (NA, PV, mask...)

2 Create manufacturable masks from the result

T. Inoue, NGL Japan, 2009

6th International Symposium on Immersion Lithography Extensions

© 2009 IBM Corporation

IBM Research Division & Systems and Technology Group

SMO Example: Contact Level

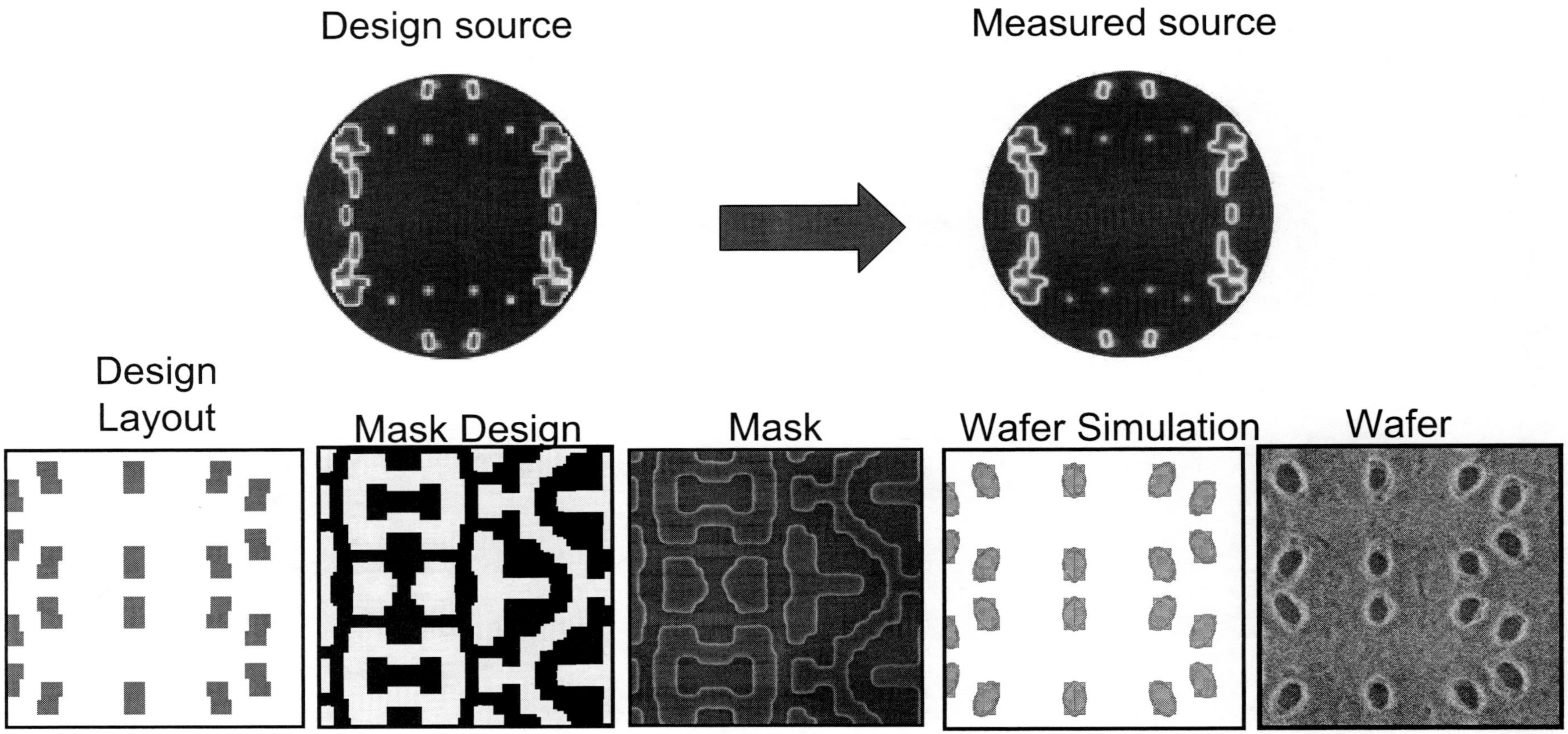

Design source

Measured source

Design Layout

Mask Design

Mask

Wafer Simulation

Wafer

D. Gil, PMJ, 2009 &
T. Inoue, NGL Japan, 2009

6th International Symposium on Immersion Lithography Extensions

© 2009 IBM Corporation

IBM Research Division & Systems and Technology Group

Applications for low k_1 Single Exposure (k_1 ~0.28)

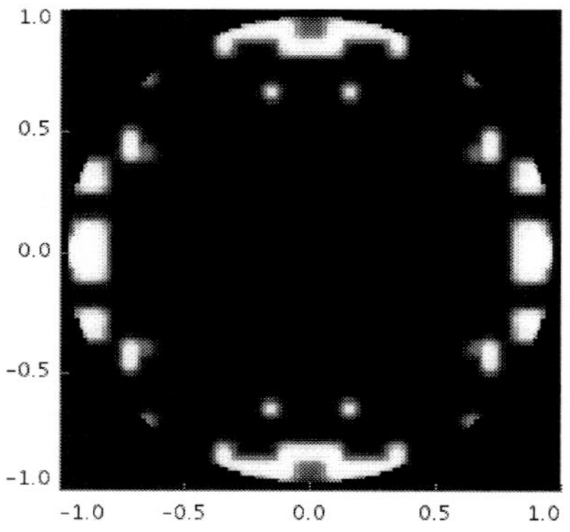

40 nm
Brick wall

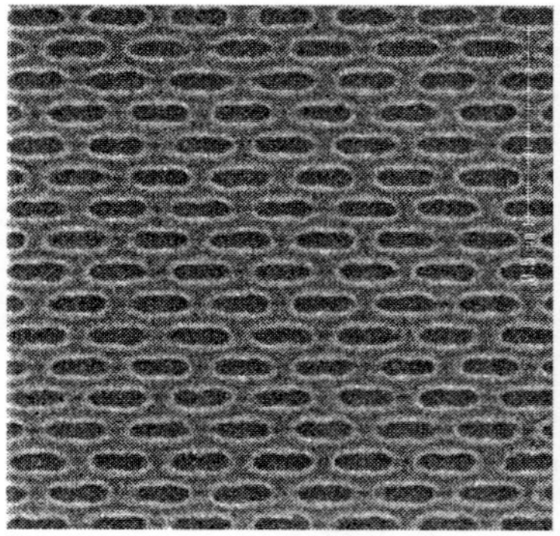

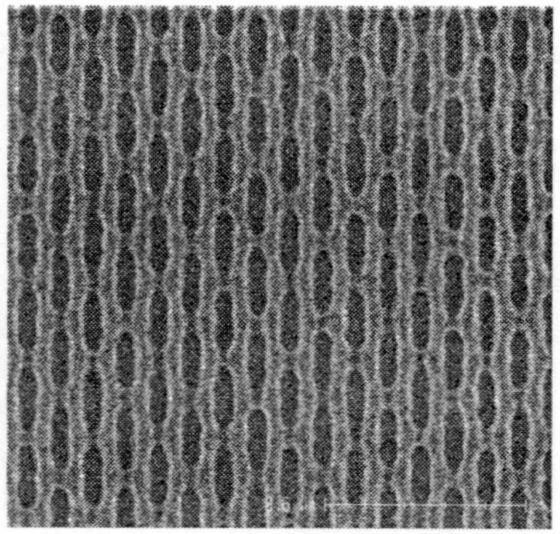

Vertical

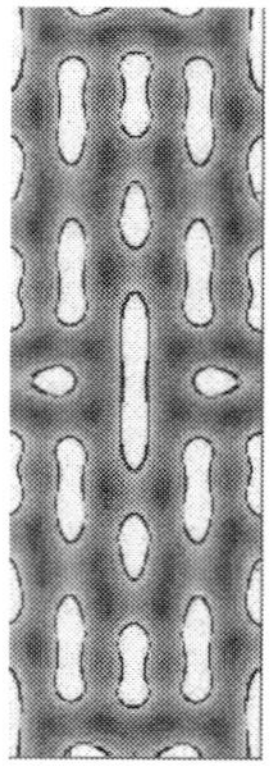

Horizontal

40 nm
1:1 L/S

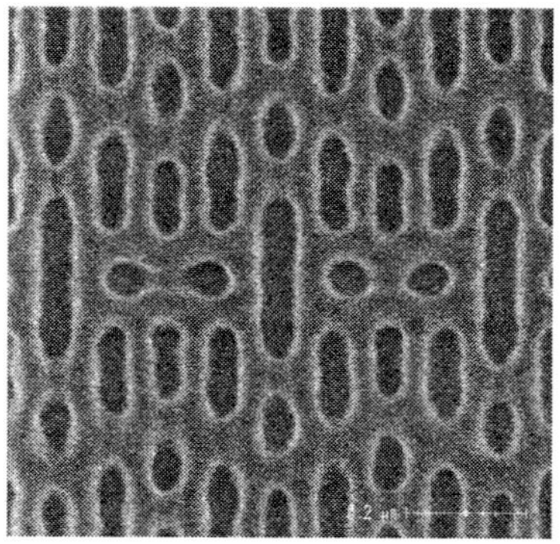

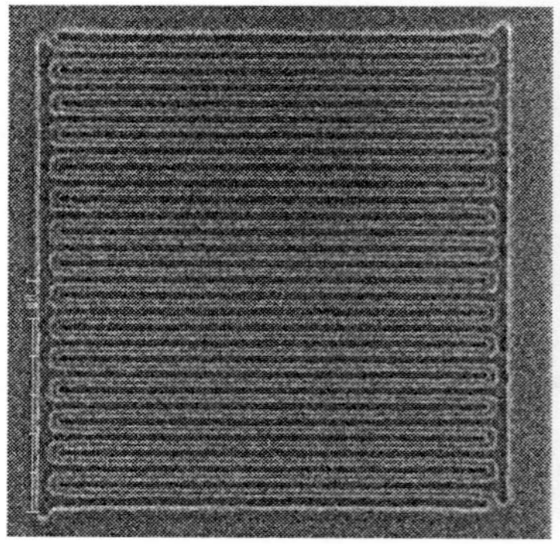

22nm SRAM M1 (0.10 μm²)

On Wafer

Simulation

© 2009 IBM Corporation

6th International Symposium on Immersion Lithography Extensions

IBM Research Division & Systems and Technology Group

Illuminator Trend

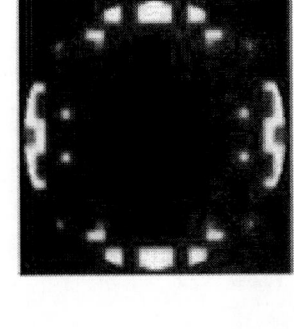

Free Form Source via
Programmable
Illuminator

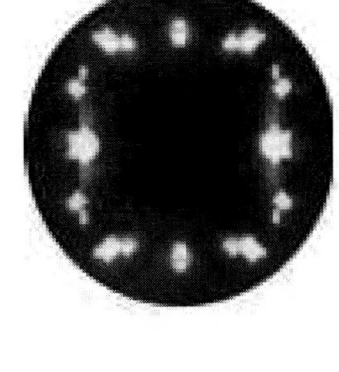

Custom Free Form
DOE

Split Exposures with
Multiple 'Stock' DOEs
or Custom Parametric
'Conventional' DOE

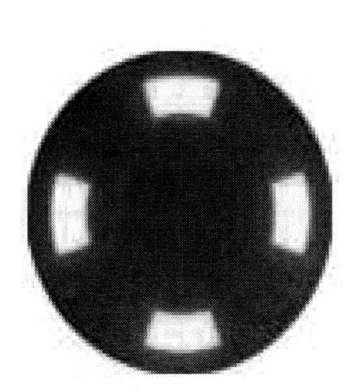

Conventional 'Stock'
Diffractive Optical
Elements (DOE)

6th International Symposium on Immersion Lithography Extensions

© 2009 IBM Corporation

IBM Research Division & Systems and Technology Group

Flexray First Image: SE 22 nm SRAM First Metal

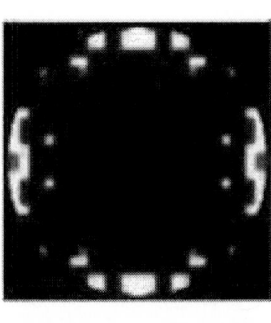

Flexray Source

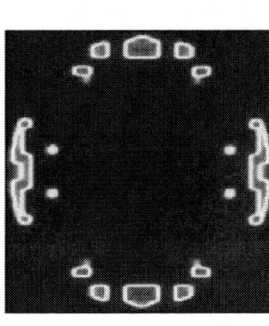

Rendered Source

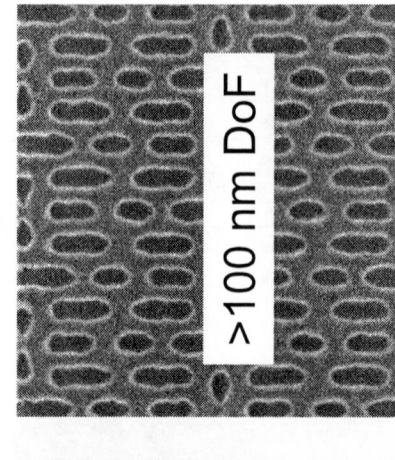

Flexray Wafer

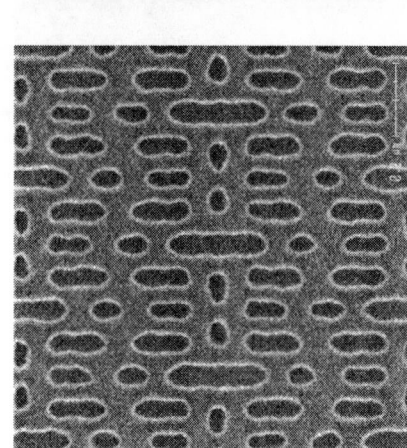

DOE Wafer

>100 nm DoF

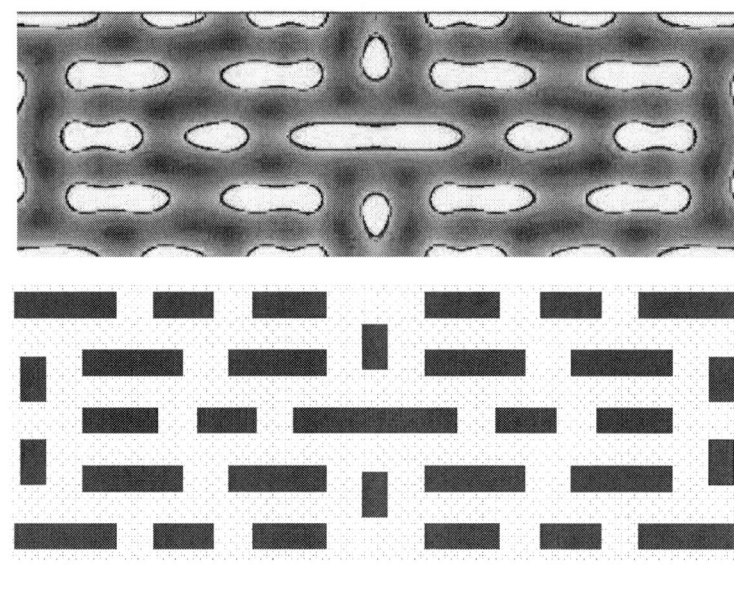

Target

Aerial Image

22nm SRAM M1 (0.100 µm²)
- min 90 nm 'pitch'
- min 40 nm CD

6th International Symposium on Immersion Lithography Extensions

See also: F. de Jong, ASML, this conference

IBM Research Division & Systems and Technology Group

Summary

- **We are entering into a new regime of patterning where unprecedented control in CDU & OL is required to meet process tolerances**

- **Design implications balanced with economics (cost of ownership) will dictate DPL use model and in turn to level of complexity required**

  – Solutions will differ depending on product / market demands

- **Innovations in materials, process controls and computational methods should provide the necessary tools to extend ArF lithography for logic to the 15 nm node.**

6th International Symposium on Immersion Lithography Extensions

© 2009 IBM Corporation

IBM Research Division & Systems and Technology Group

Acknowledgements

- M. Burkhardt, S. Burns, M. Colburn, D. Corliss, A. Gabor, D. Gil, T. Farrell, G. Han, T. Inoue, K. Lai, S. Halle, S. Holmes, S. Mansfield, G. McIntyre, J. Meiring, D. Melville, A. Rosenbluth, K. Tian, G. Gomba…

- ASML, JSR, KLA-Tencor, Mentor Graphics, TEL, Zeiss

- Alliance Partners & CNSE @ SUNY Albany

- Portions of this talk are from work performed by the Research Alliance Teams at various IBM Research and Development Facilities

6th International Symposium on Immersion Lithography Extensions

© 2009 IBM Corporation

IBM Research Division & Systems and Technology Group

International Business Machines
www.ibm.com
davidmed@us.ibm.com

6th International Symposium on Immersion Lithography Extensions

© 2009 IBM Corporation

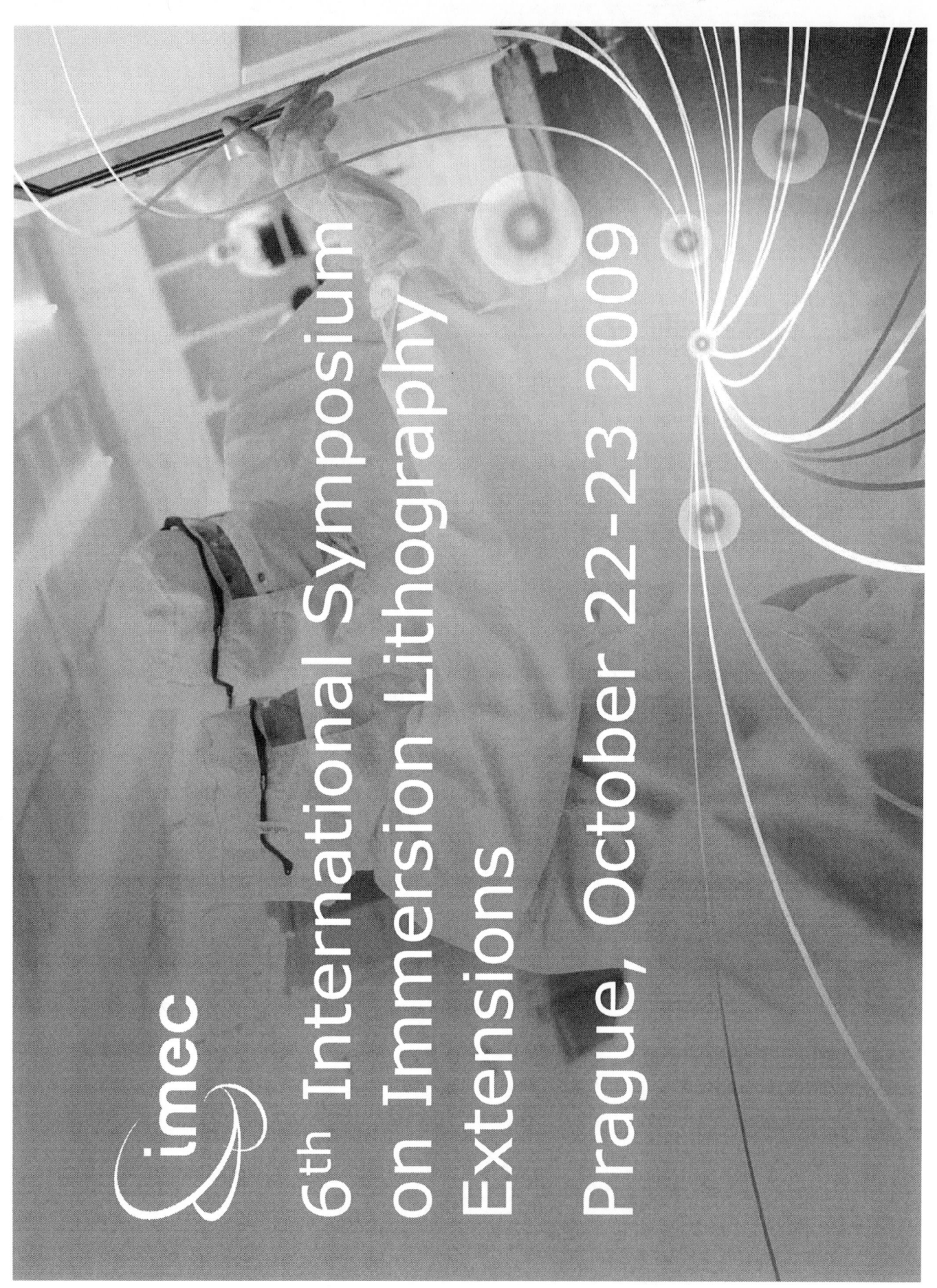

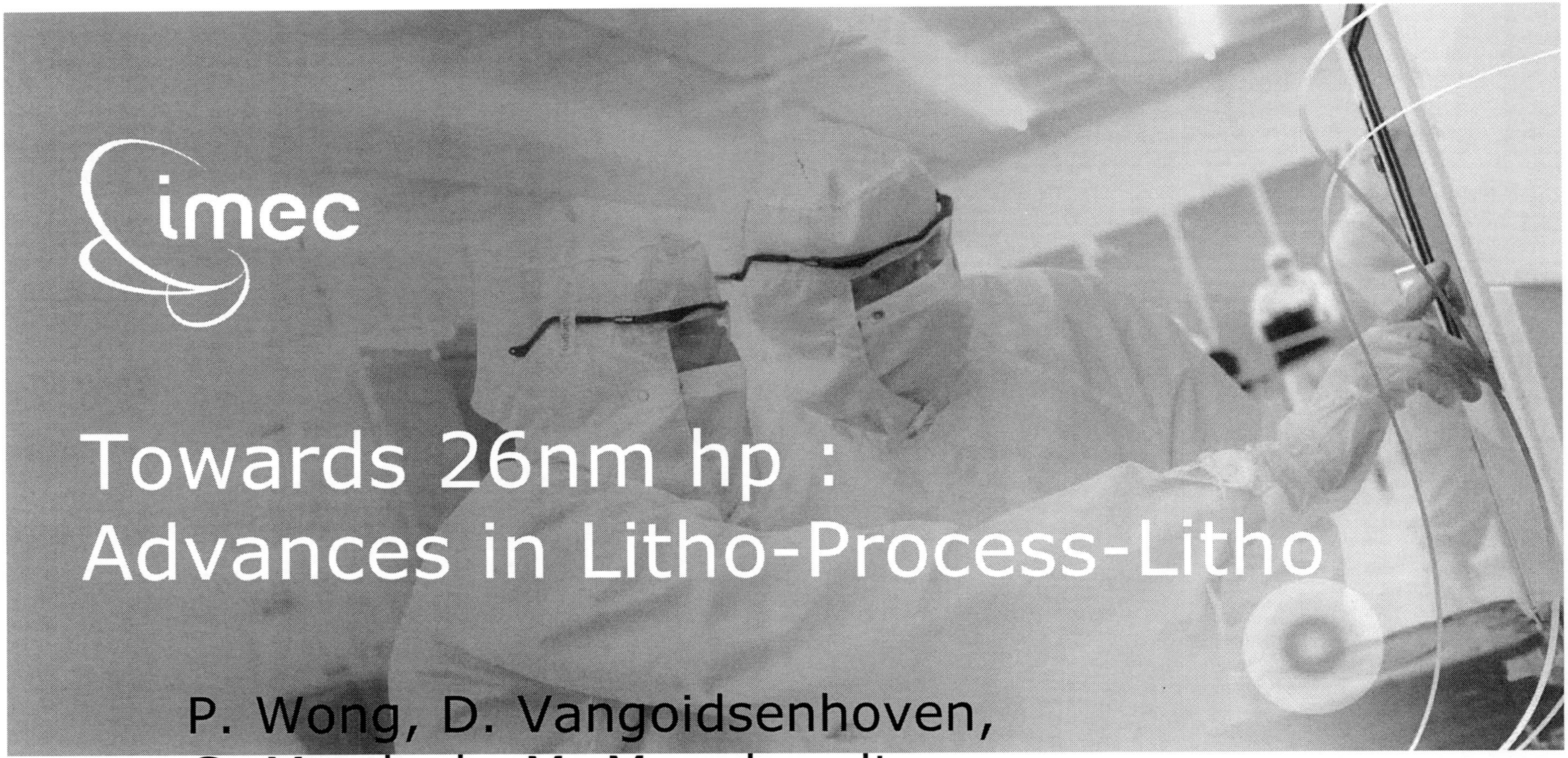

Towards 26nm hp :
Advances in Litho-Process-Litho

P. Wong, D. Vangoidsenhoven,
G. Murdoch, M. Maenhoudt,
S. Verhaegen, V. Wiaux

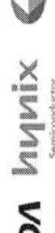
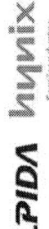
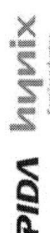

Outline

- Introduction

- Litho-Process-Litho-Etch Process Development Update

  1. Coat freezing layer

  2. Posi/Posi. Resist

  3. Thermal freeze

- Conclusions

Patrick Wong
Imec 2009

6th International Symposium on
Immersion Lithography Extensions
October 22-23 2009 · Prague, Czech Republic

imec

Outline

> Introduction

- Litho-Process-Litho-Etch Process Development Update
  1. Coat freezing layer
  2. Posi/Posi. Resist
  3. Thermal freeze

- Conclusions

Patrick Wong
Imec 2009

6th International Symposium on
Immersion Lithography Extensions
October 22-23 2009 • Prague, Czech Republic

imec

Double patterning process schemes

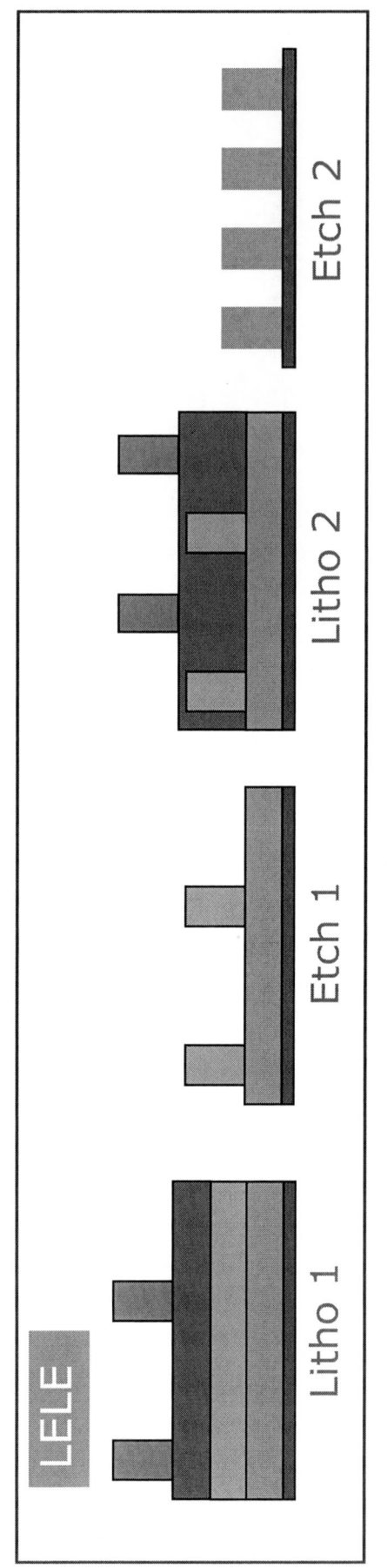

LELE

Litho 1 · Etch 1 · Litho 2 · Etch 2

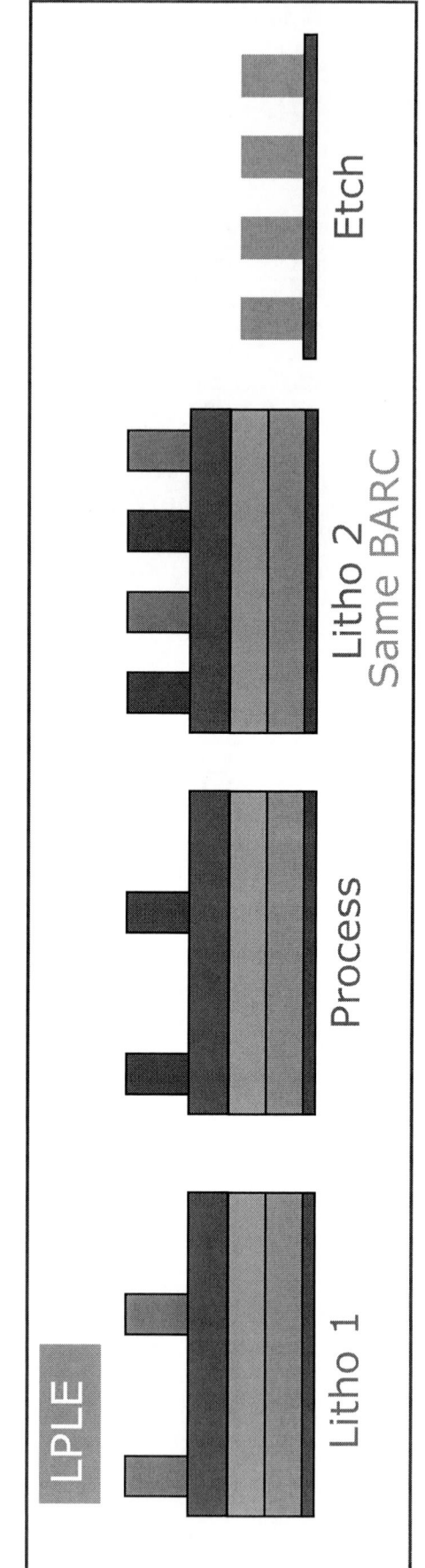

LPLE

Litho 1 · Process · Litho 2 Same BARC · Etch

Patrick Wong
Imec 2009

6th International Symposium on
Immersion Lithography Extensions
October 22-23 2009 · Prague, Czech Republic

Processing cost comparison
double patterning process schemes

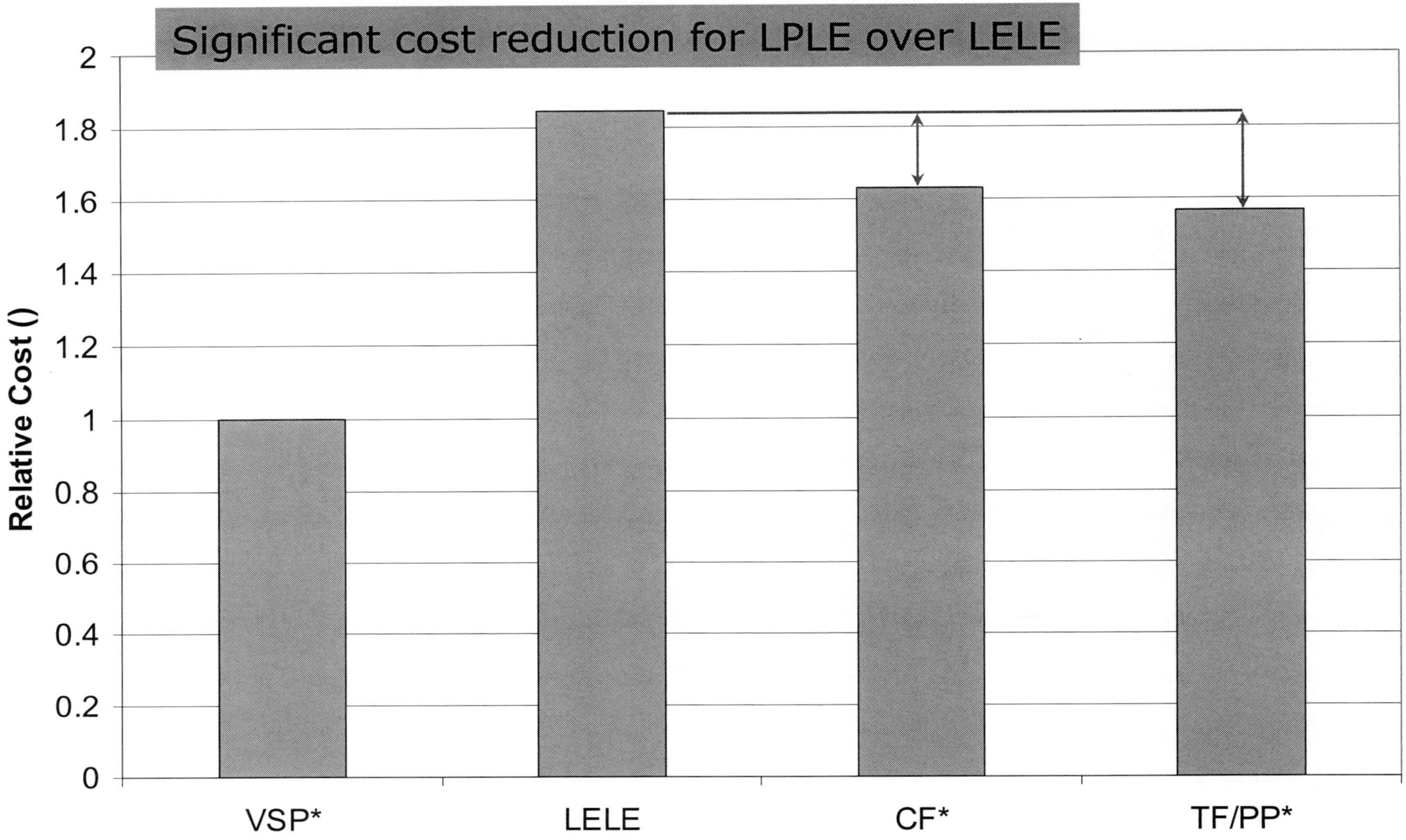

Significant cost reduction for LPLE over LELE

Relative Cost ()

VSP*   LELE   CF*   TF/PP*

* : VSP = virtual single patterning; CF = Coat Freeze; TF = Thermal Freeze; PP = Posi posi

6th International Symposium on
Immersion Lithography Extensions
October 22-23 2009 • Prague, Czech Republic

Patrick Wong
Imec 2009 | 6

imec

Litho-Process-Litho-Etch Alternatives

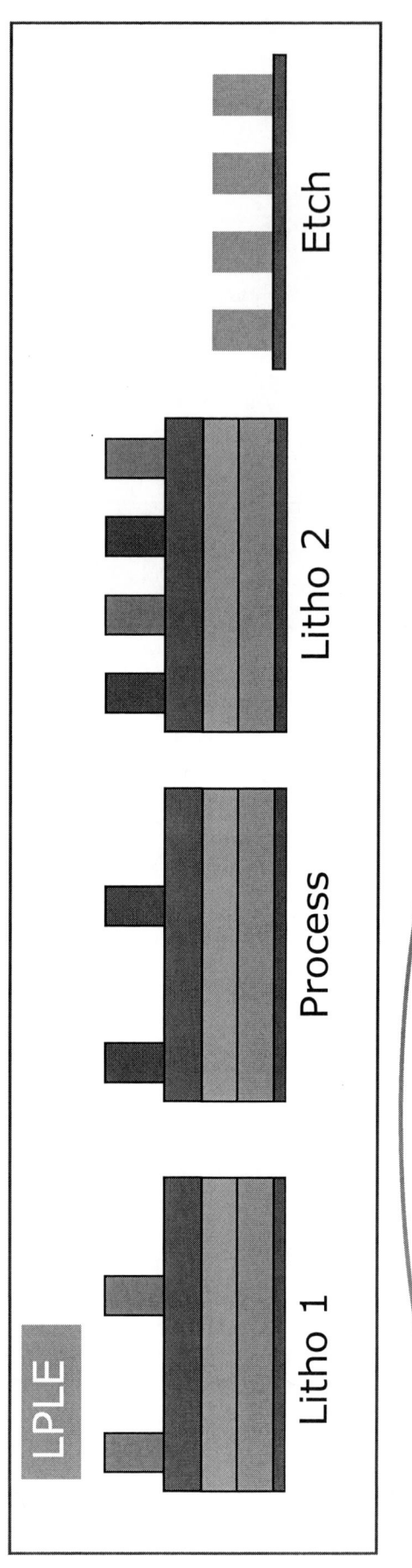

LPLE

Litho 1   Process   Litho 2   Etch

- Coat freezing layer
- Positive/Positive resist   Possible in Litho cluster
- Thermal freeze
- Vapour freeze
- Implant freeze
- ...

Patrick Wong
Imec 2009

**6th International Symposium on
Immersion Lithography Extensions**
October 22-23 2009 · Prague, Czech Republic

imec

Outline

✓ Introduction

▲ Litho-Process-Litho-Etch Process Development

Update

▲ Coat freezing layer

2. Posi/Posi. Resist

3. Thermal freeze

● Conclusions

Patrick Wong
Imec 2009

6th International Symposium on
Immersion Lithography Extensions
October 22-23 2009 · Prague, Czech Republic

Coat freezing alternative

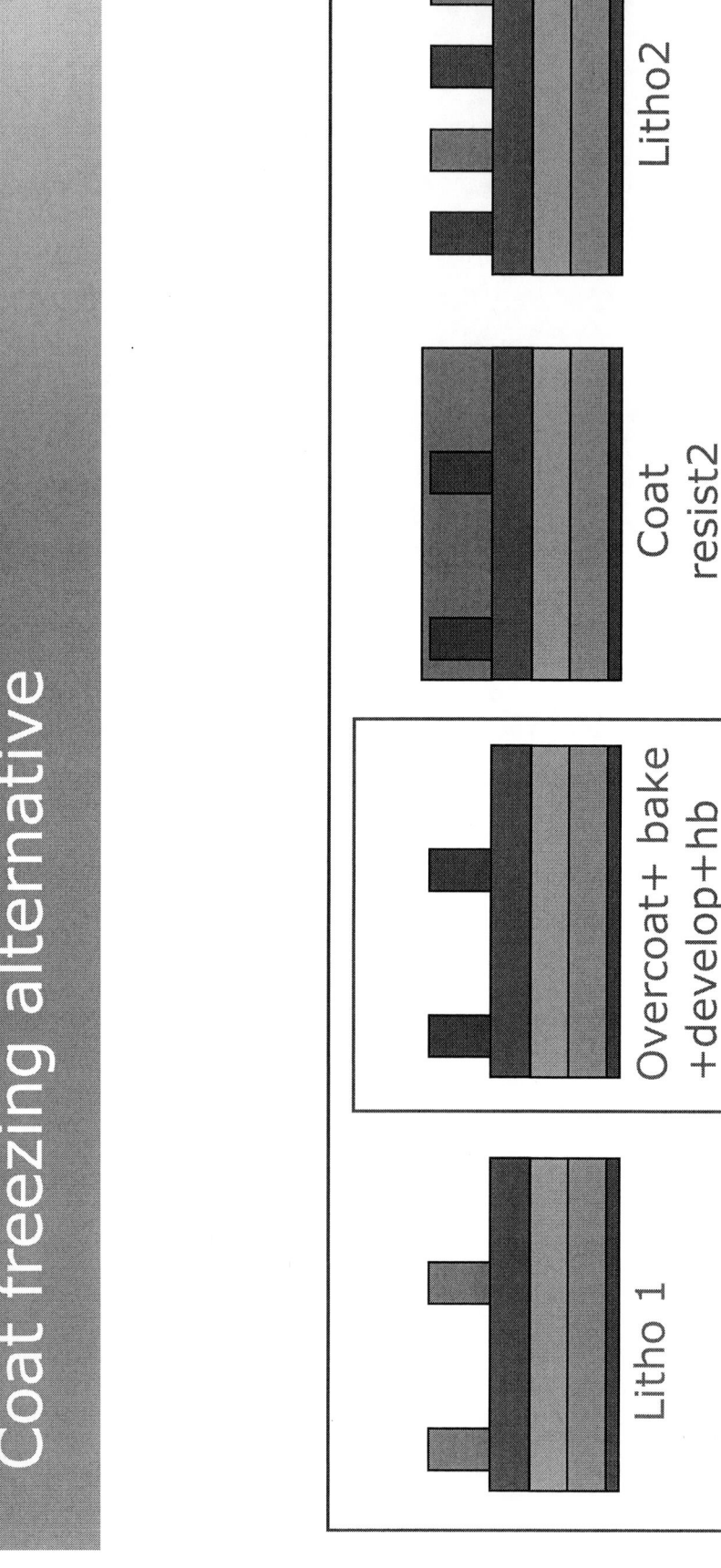

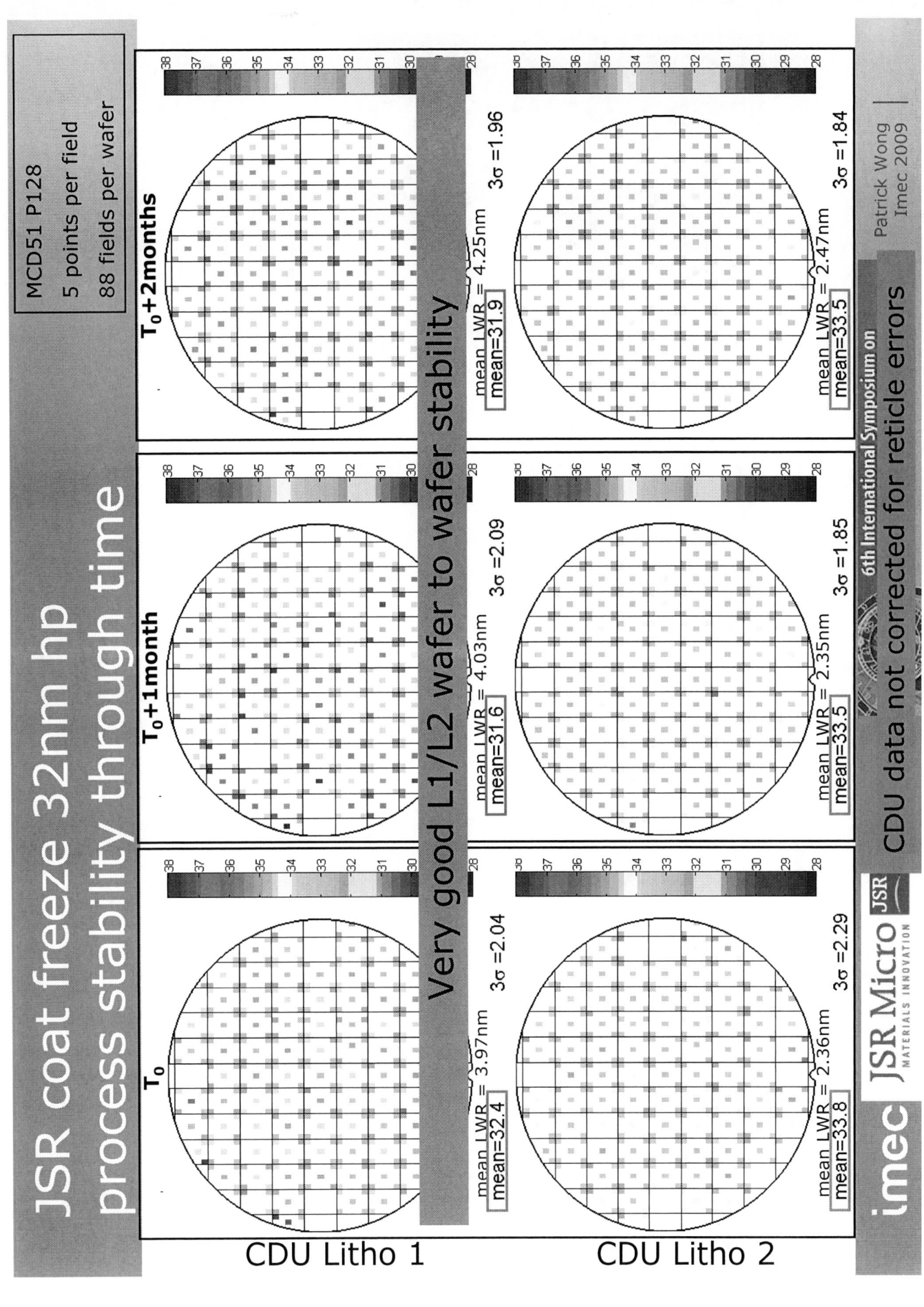

JSR coat freeze
Double Patterning Defectivity

DPT Defect Breakdown

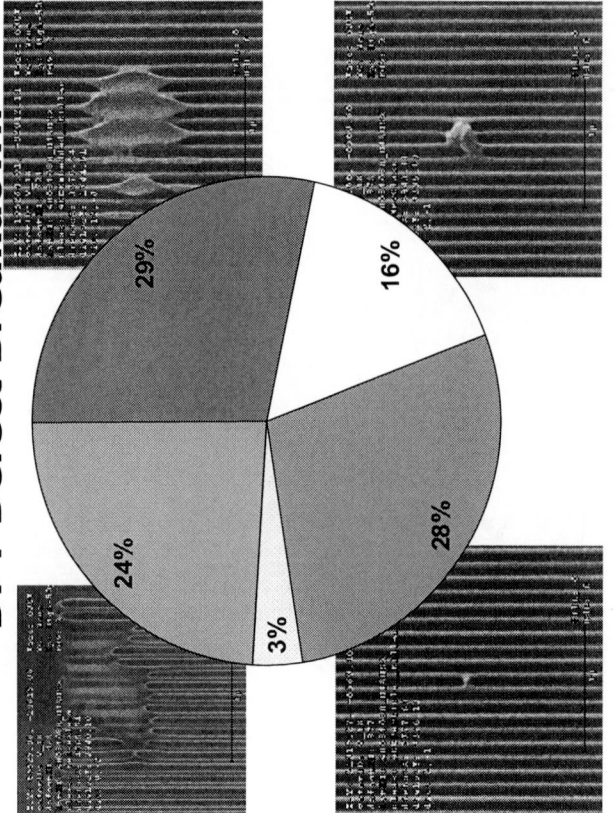

- Immersion
- Bridging
- Residues
- Particles
- Protrusion

29%
24%
3%
28%
16%

- Number of double patterning defects comparable to average number of single patterning defects
  - ❖ more residues and bridging
  - ❖ less immersion defects

Relative Comparison of SPT with DPT Defects

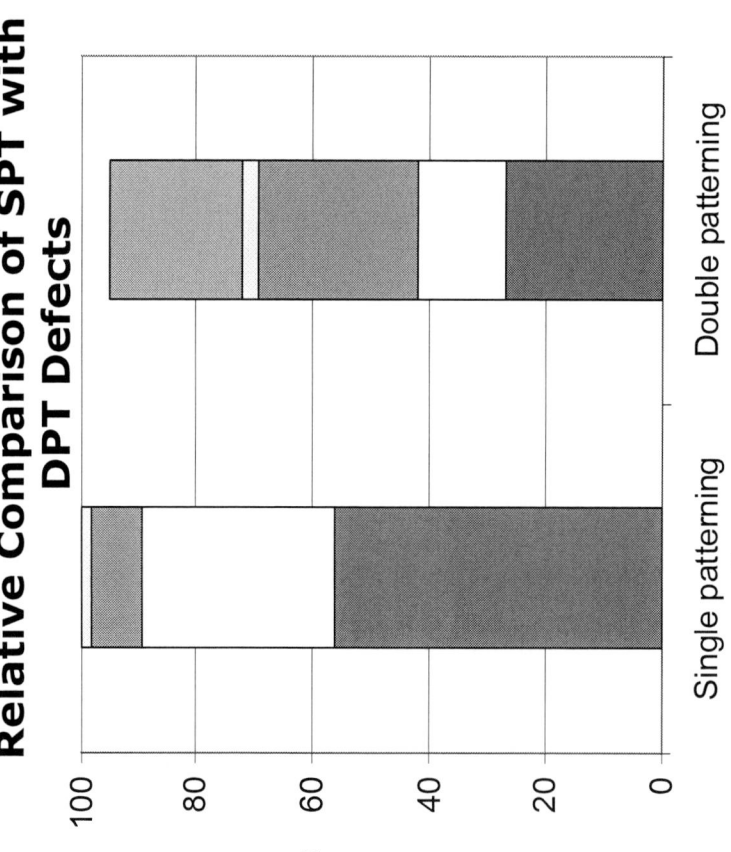

%

Single patterning monitor

Double patterning

Patrick Wong
Imec 2009

6th International Symposium on
Immersion Lithography Extensions
October 22-23 2009 • Prague, Czech Republic

JSR coat coat freeze : 2D Feasibility

1D — Memory

Dipole, 26hp

2D — Logic

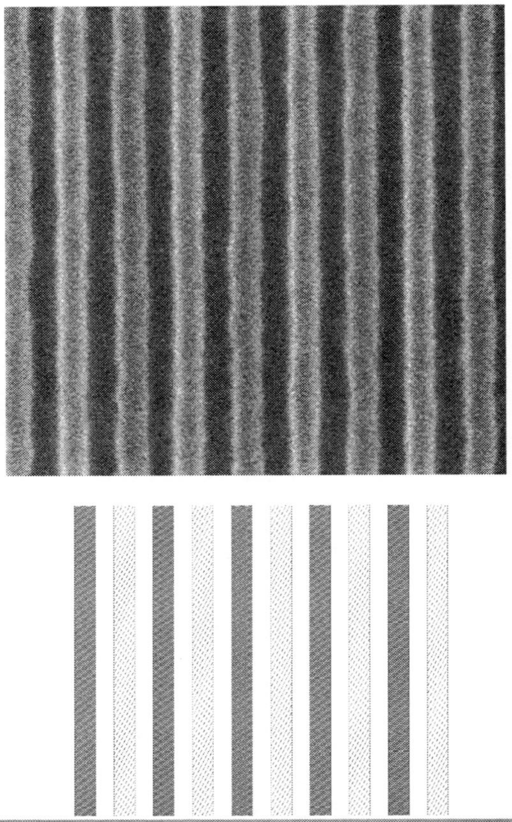

Annular, 32hp

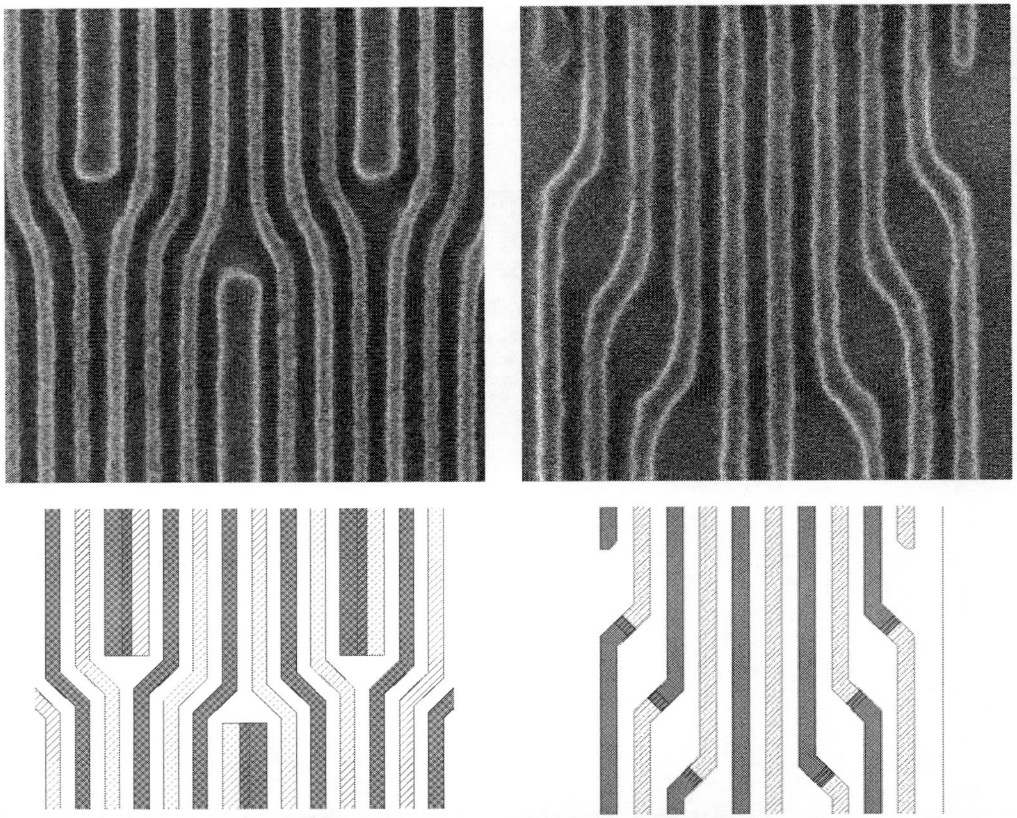

Pattern collapse : a general issue @ 26nm hp

- **LPL specific issues**
  - Line 2 is patterned on developed BARC
  - OVL errors can result in space asymmetry
  - Aggressive pitch in resist

Capillary Forces acting:

$$F_{net} = \frac{\gamma \cos \theta}{S} \times \Delta H_r, L$$

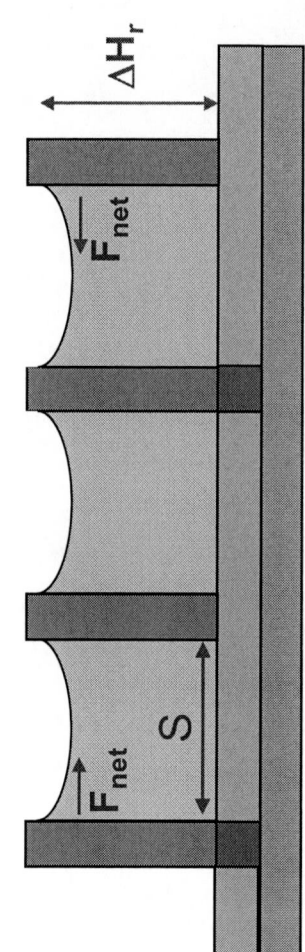

26nm hp with DIW
Rinse example

>50% sites suffer from pattern collapse

- **Solutions:**
  - Lowering the resist film thickness
  - Reduce surface tension (γ) * cos(contact angle (θ))
    product : **surfactant rinse**
  - Increase resist adhesion on substrate : immersion dedicated BARC

Patrick Wong
Imec 2009

6th International Symposium on Immersion Lithography Extensions
October 22-23 2009 · Prague, Czech Republic

imec

JSR coat freeze 26nm hp
CDU performance

MCD39 P104
1 point per field
311 fields per wafer

CDU Litho L2

3σ =1.46

mean LWR = 3.4nm
mean=27.2

CDU Litho L1

3σ =1.90

mean LWR = 4.1nm
mean=28.2

0 flyers due to pattern collapse

6th International Symposium on
Immersion Lithography Extensions
October 22-23 2009 · Prague, Czech Republic

Patrick Wong
Imec 2009

imec   JSR Micro
JSR MATERIALS INNOVATION

Outline

- Introduction

- Litho-Process-Litho-Etch Process Development Update
  - Coat freezing layer
  - Posi/Posi. Resist
  - 3. Thermal freeze

- Conclusions

Patrick Wong
Imec 2009

6th International Symposium on Immersion Lithography Extensions
October 22-23 2009 · Prague, Czech Republic

imec

Posi posi alternative

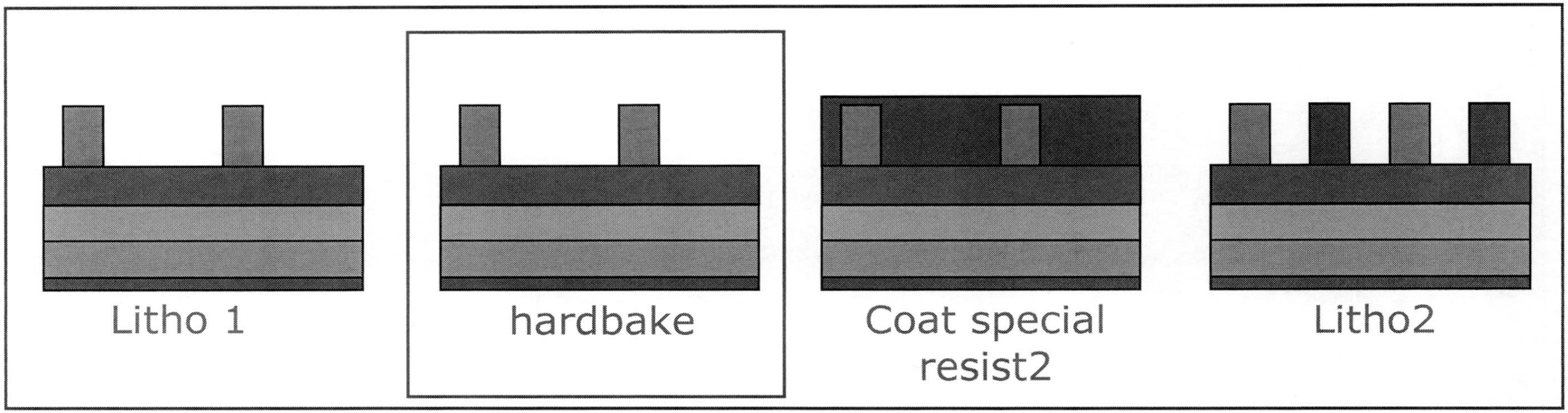

Litho 1

hardbake

Coat special resist2

Litho2

Patrick Wong
Imec 2009

6th International Symposium on Immersion Lithography Extensions
October 22-23 2009 · Prague, Czech Republic

imec

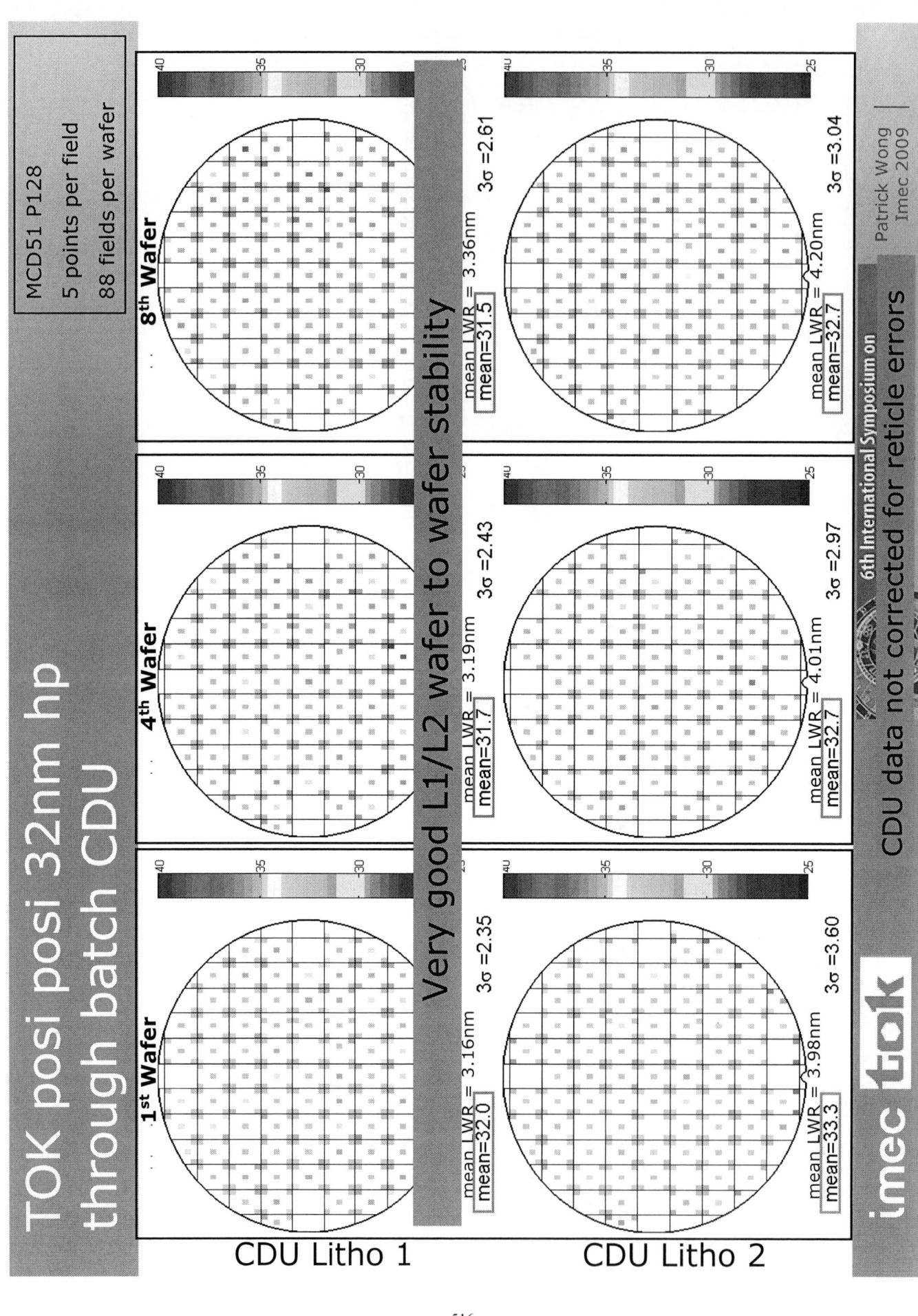

Posi Posi 26nm hp
CDU performance

MCD39 P104
1 point per field
311 fields per wafer

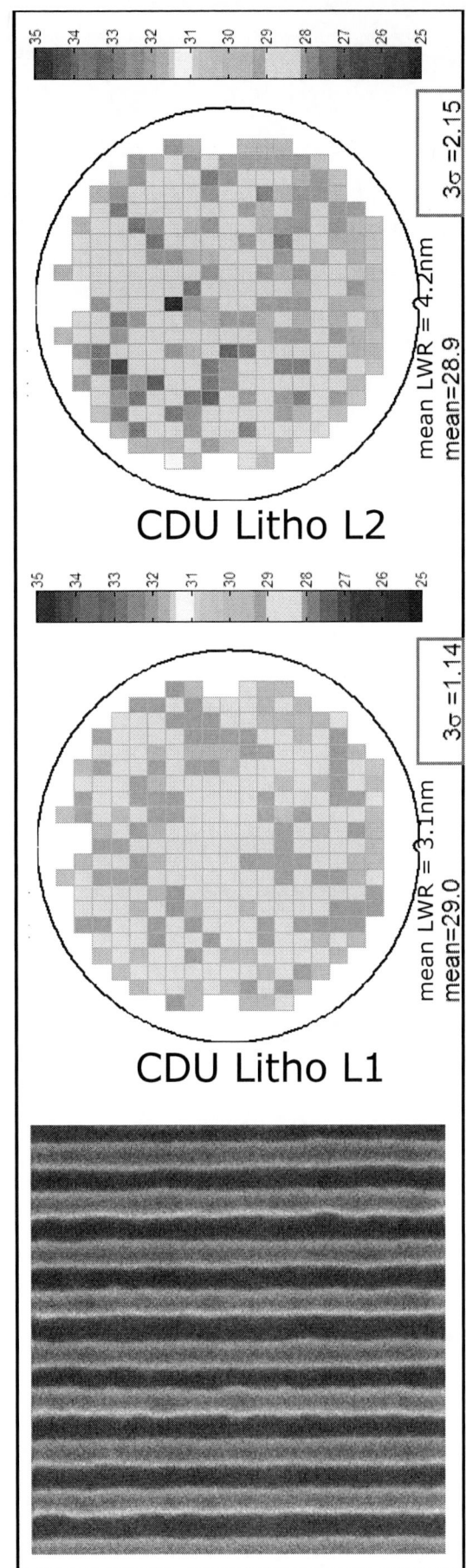

CDU Litho L1

mean LWR = 3.1nm
mean=29.0

$3\sigma = 1.14$

CDU Litho L2

mean LWR = 4.2nm
mean=28.9

$3\sigma = 2.15$

~10 flyers due to pattern collapse

Patrick Wong
Imec 2009

6th International Symposium on
Immersion Lithography Extensions
October 22-23 2009 · Prague, Czech Republic

Outline

- Introduction
- Litho-Process-Litho-Etch Process Development Update
  - Coat freezing layer
  - Posi/Posi. Resist
  - Thermal freeze
- Conclusions

6th International Symposium on
Immersion Lithography Extensions
October 22-23 2009 • Prague, Czech Republic

Patrick Wong
Imec 2009

imec

Thermal freezing alternative

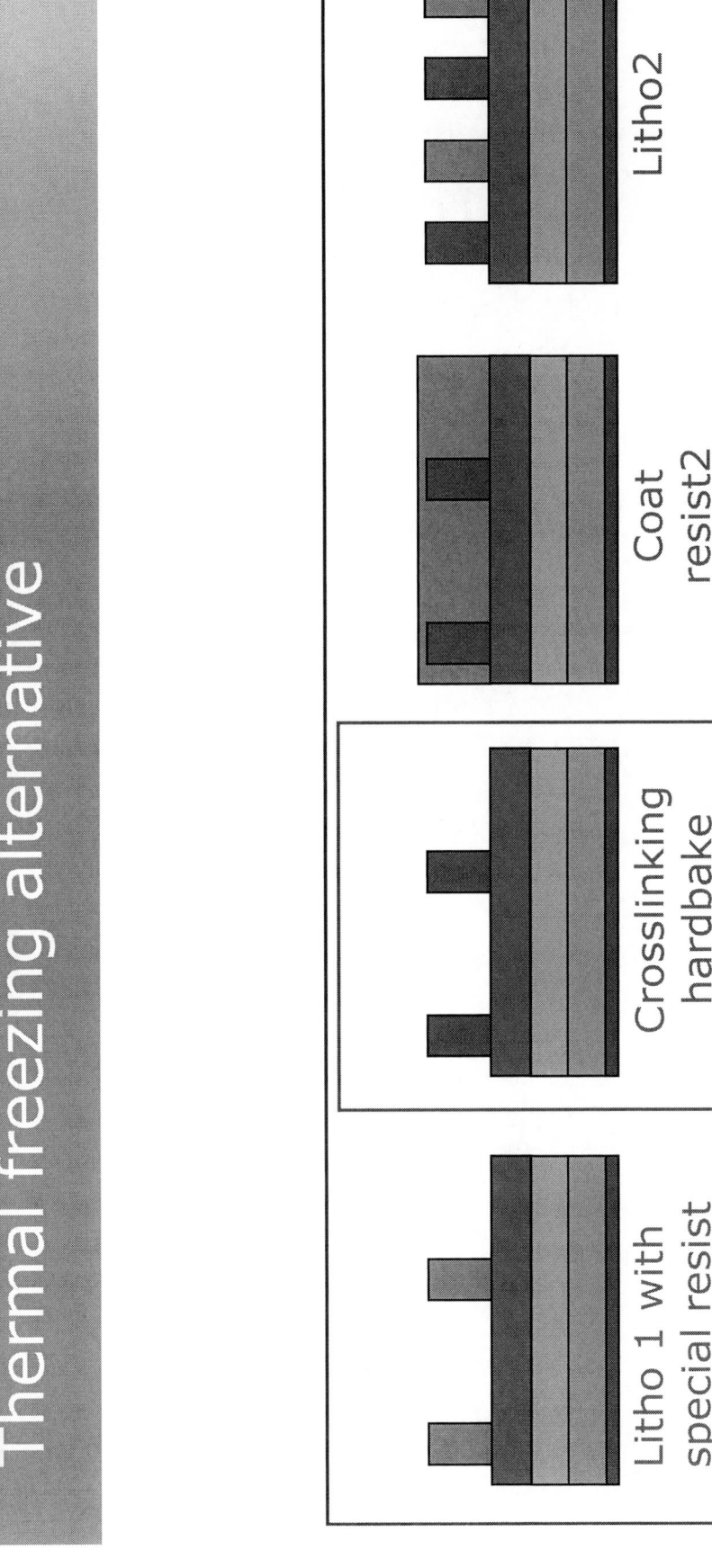

Litho 1 with special resist

Crosslinking hardbake

Coat resist2

Litho2

Patrick Wong
Imec 2009

6th International Symposium on
Immersion Lithography Extensions
October 22-23 2009 · Prague, Czech Republic

imec

JSR thermal freeze 26nm hp
CDU result

MCD39 P104
1 point per field
311 fields per wafer

CDU Litho L2

3σ =1.32

mean LWR = 2.9nm
mean=26.0

CDU Litho L1

3σ =1.67

mean LWR = 4.0nm
mean=27.8

Flyers @ wafer edge due to coat non uniformity (wrong viscosity)

Patrick Wong
Imec 2009

6th International Symposium on
Immersion Lithography Extensions
October 22-23 2009 • Prague, Czech Republic

Outline

✓ Introduction

✓ Litho-Process-Litho-Etch Process Development Update
  ✓ Coat freezing layer
  ✓ Posi/Posi. Resist
  ✓ Thermal freeze

⋀ Conclusions

6th International Symposium on Immersion Lithography Extensions
October 22-23 2009 • Prague, Czech Republic

Patrick Wong
Imec 2009

imec

26nm hp CDU performance comparison

MCD39 P104
1 point per field
311 fields per wafer

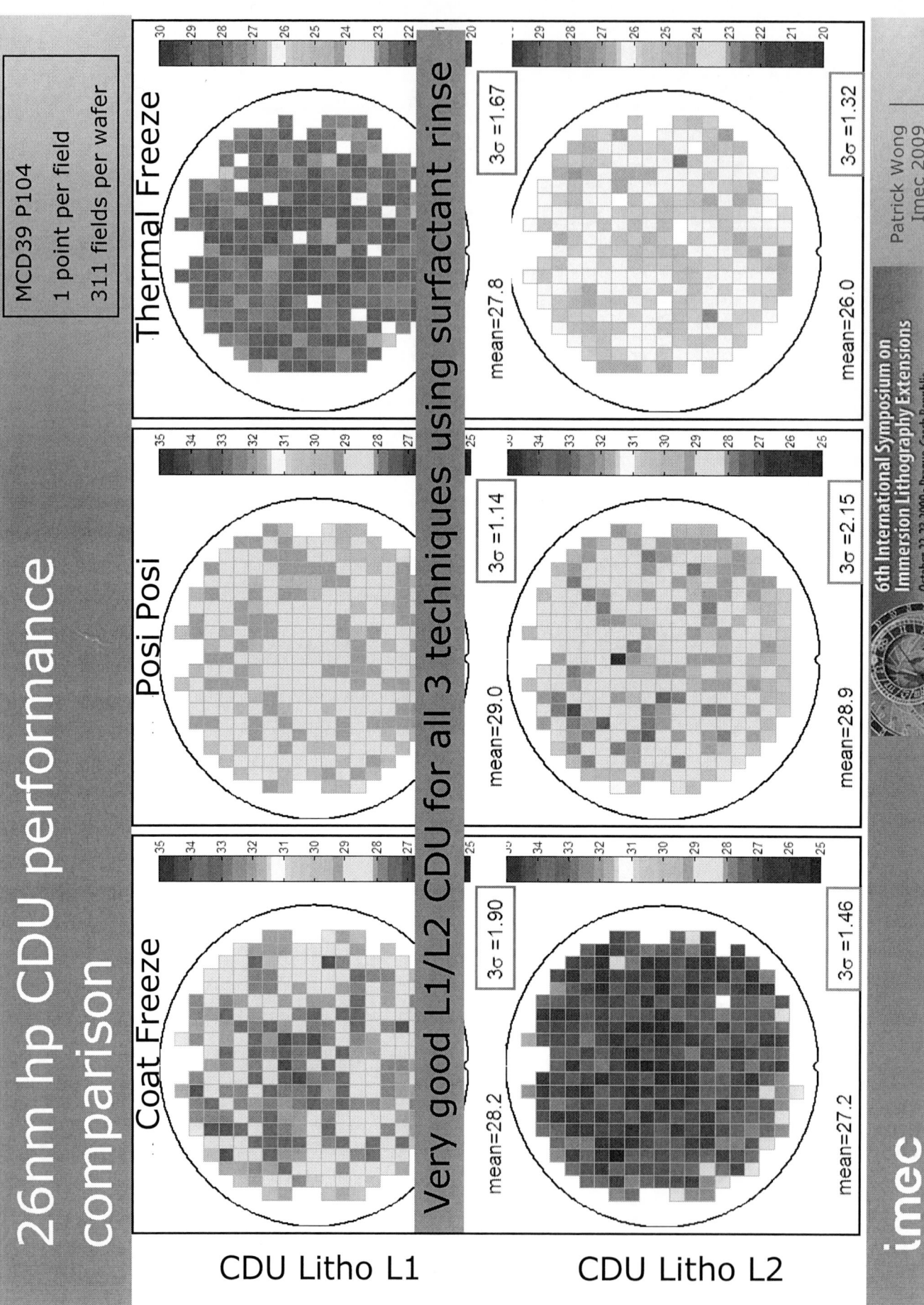

Coat Freeze

CDU Litho L1: mean=28.2, 3σ=1.90

CDU Litho L2: mean=27.2, 3σ=1.46

Posi Posi

CDU Litho L1: mean=29.0, 3σ=1.14

CDU Litho L2: mean=28.9, 3σ=2.15

Thermal Freeze

CDU Litho L1: mean=27.8, 3σ=1.67

CDU Litho L2: mean=26.0, 3σ=1.32

Very good L1/L2 CDU for all 3 techniques using surfactant rinse

Patrick Wong
Imec 2009

6th International Symposium on
Immersion Lithography Extensions
October 22-23 2009 - Prague, Czech Republic

imec

Conclusions

- ## 32nm hp process maturity

  - Cost effective alternative processes show good litho performance, defectivity ok

  - Excellent wafer to wafer process stability demonstrated through batch and through time

- ## 26nm hp imaging capability

  - 26nm hp imaging demonstrated for 3 alternative techniques with very good CDU for L1 & L2

  - Pattern collapse issue effectively solved by surfactant rinse for coat freeze, posi posi and thermal freeze

- Thermal freeze and posi posi most attractive techniques with respect to process complexity and CoO

6th International Symposium on
Immersion Lithography Extensions
October 22-23 2009 • Prague, Czech Republic

Patrick Wong
Imec 2009

imec

Acknowledgements

- Litho: P. Foubert, R Gronheid.
- Etch: S. Locorotondo, M. Demand
- JSR : H. Tanaka, K. Hoshiko
- TOK : R. Takasu
- AZ : R. Collett
- Sokudo: C. Rosslee, Y. Theroude
- TEL: S. Hatakeyama, K. Nafus, N. Bradon, M. McCarthy
- Hitachi: T. Ishimoto, N. Yasui, D. Hibino

Patrick Wong
Imec 2009

6th International Symposium on Immersion Lithography Extensions
October 22-23 2009 · Prague, Czech Republic

imec

Outlines

- **Introduction**
  - State of the art
  - Goal and challenges

- **Hardening treatments evaluation**
  - HBr plasma
  - Surface Curing Agent process
  - Thermal Cure Process

- **Low temperature deposition process**

- **Etch transfer feasability**
  - Impact of pattern height
  - Impact of deposition conformity

- **Conclusion and perspectives**

6th International Symposium on Immersion Lithography Extensions

S. Gaugiran   22/10/09

© CEA 2009. All rights reserved
Any reproduction in whole or in part on any medium or use of the information contained herein
is prohibited without the prior written consent of CEA

Electronic Materials

State of the art

Self Aligned Double Patterning

- SADP used for 64GB NAND flash memory by Samsung
  - Only option for 32nm hp flash memory last year
  - Extension to 22nm hp patterning
  - Higher CoO but overlay spec released

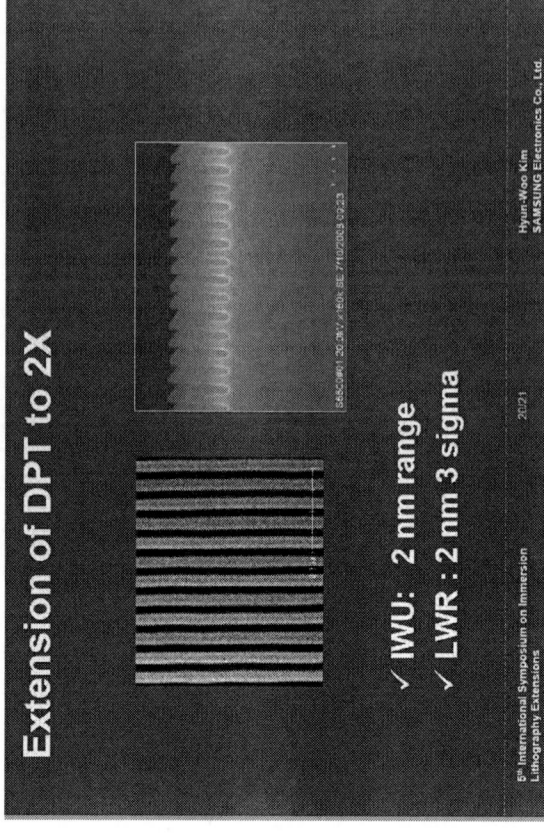

Extension of DPT to 2X

- ✓ IWU: 2 nm range
- ✓ LWR : 2 nm 3 sigma

Hyun-Woo Kim
SAMSUNG Electronics Co., Ltd.

5th International Symposium on Immersion Lithography Extensions

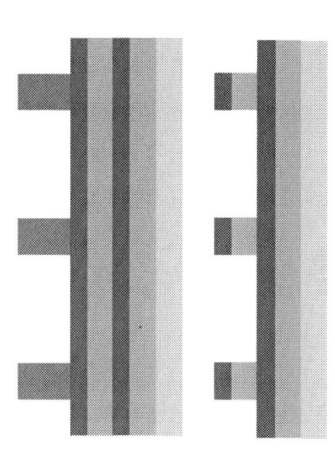  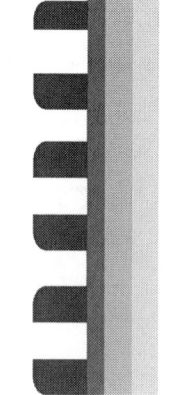 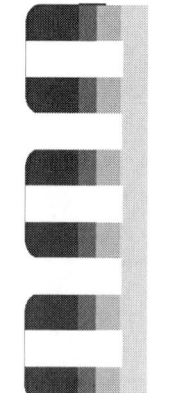

Electronic Materials

6th International Symposium on Immersion Lithography Extensions

S. Gaugiran   22/10/09

Goals and challenges

- Spacer deposition directly on the top of resist patterns (no preliminary transfer into an underlayer)

- Resist hardening process development
  - Thermal stability
  - Pattern shape conservation

- Low temperature oxide deposition (200°C) process development:
  - Conformality
  - Uniformity

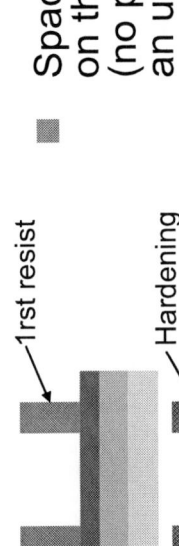

Self Aligned Double Patterning

CVD spacers on resist

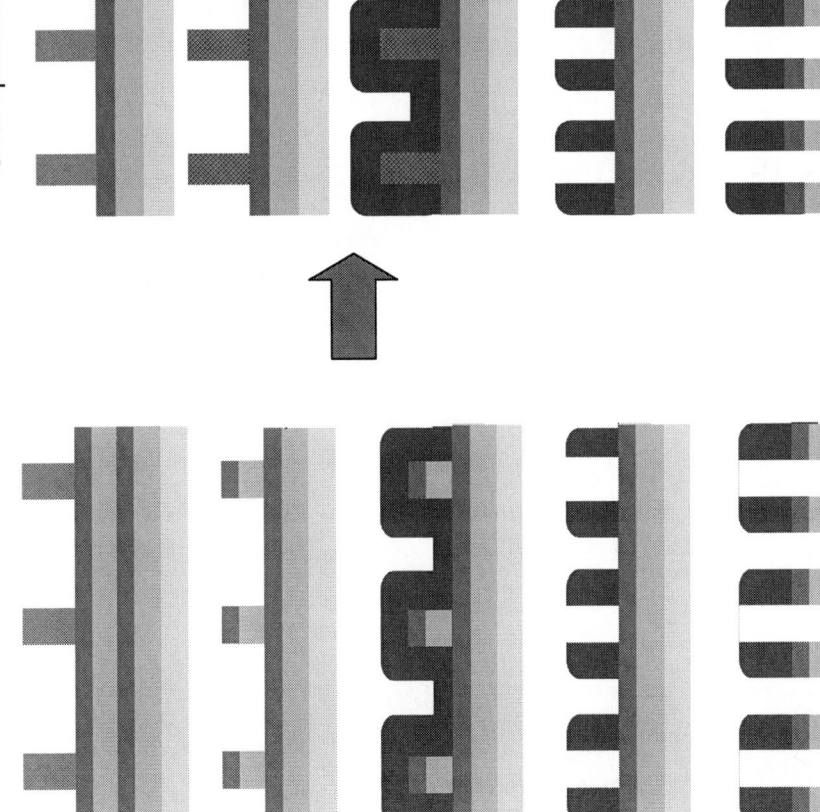

1rst resist

Hardening process

CVD spacer deposition

spacer Etch

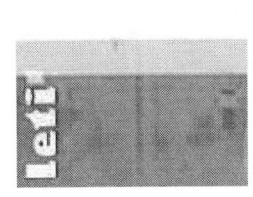

6th International Symposium on Immersion Lithography Extensions

Electronic Materials

S. Gaugiran   22/10/09

© CEA 2009. All rights reserved
Any reproduction in whole or in part on any medium or use of the information contained herein is prohibited without the prior written consent of CEA

Process conditions

- **Track Sokudo RF3:**
  - Stack: Barc AR26 32nm or carbon mask +DARC
  - Developer TMAH

- **Scanner Nikon NSR307-E NA=0.85**
  - Illumination dipolar 20°, polarization
  - Mask= 6% attenuated PSM, brightfield
  - Target: 45L/S P180nm

- **LAM Versys**
  - HBr plasma cure

- **Applied material Producer:**
  - PECVD Oxide deposition
  - Capacitive parallel plate plasma discharge
  - Deposition temperature: 200°C

- **DPS from Applied Materials:**
  - Etch transfer

Si

SiO$_x$

Carbon mask

Si bulk

6th International Symposium on Immersion Lithography Extensions

S. Gaugiran   22/10/09

Any reproduction in whole or in part on any medium or use of the information contained herein
is prohibited without the prior written consent of CEA

© CEA 2009. All rights reserved

Electronic Materials

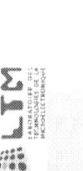

Hardening treatments

Cure by HBr plasma	Surface Curing Agent	Thermal Cure Process
– In the etch reactor	– In track: coating process	– In track: hard bake
– Well know for resist etch resistance increase	– Initially developped for LFLE process	– Initially developped for LFLE process
– Resist surface graphitisation (amorphous carbon like layer)	– Resist pattern surface modification	– Crosslink of the resist patterns
=>*Reference process*		

Electronic Materials

6th International Symposium on Immersion Lithography Extensions

S. Gaugiran   22/10/09

© CEA 2009. All rights reserved
Any reproduction in whole or in part on any medium or use of the information contained herein
is prohibited without the prior written consent of CEA

Hardening by HBr plasma

TGA

After HBr

16% loss at 200°C

- 16% weight loss at 200°C
- Stong top loss and CD loss

Before HBr — CD=57nm
After HBr — CD=45nm

after HBr
before HBr

6th International Symposium on Immersion Lithography Extensions

S. Gaugiran

22/10/09

© CEA 2009. All rights reserved
Any reproduction in whole or in part on any medium or use of the information contained herein is prohibited without the prior written consent of CEA

Hardening by HBr plasma

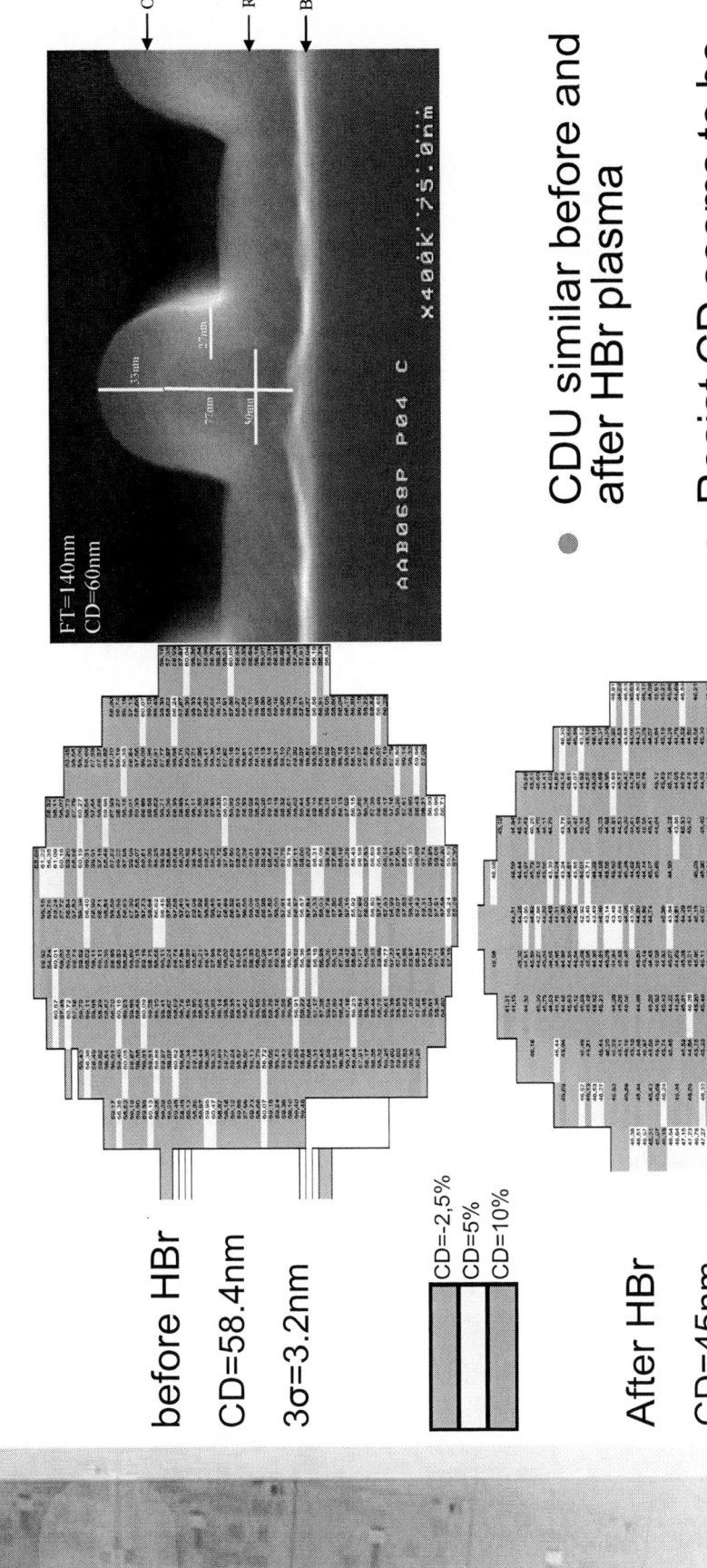

before HBr

CD=58.4nm

3σ=3.2nm

CD=-2,5%
CD=5%
CD=10%

After HBr

CD=45nm

3σ =2.9nm

- CDU similar before and after HBr plasma

- Resist CD seems to be preserved after deposition

FT=140nm
CD=60nm

Oxide
Resist
Barc

X400K 75.0nm

6th International Symposium on Immersion Lithography Extensions

© CEA 2009. All rights reserved
Any reproduction in whole or in part on any medium or use of the information contained herein
is prohibited without the prior written consent of CEA

S. Gaugiran   22/10/09

Hardening by SCA process

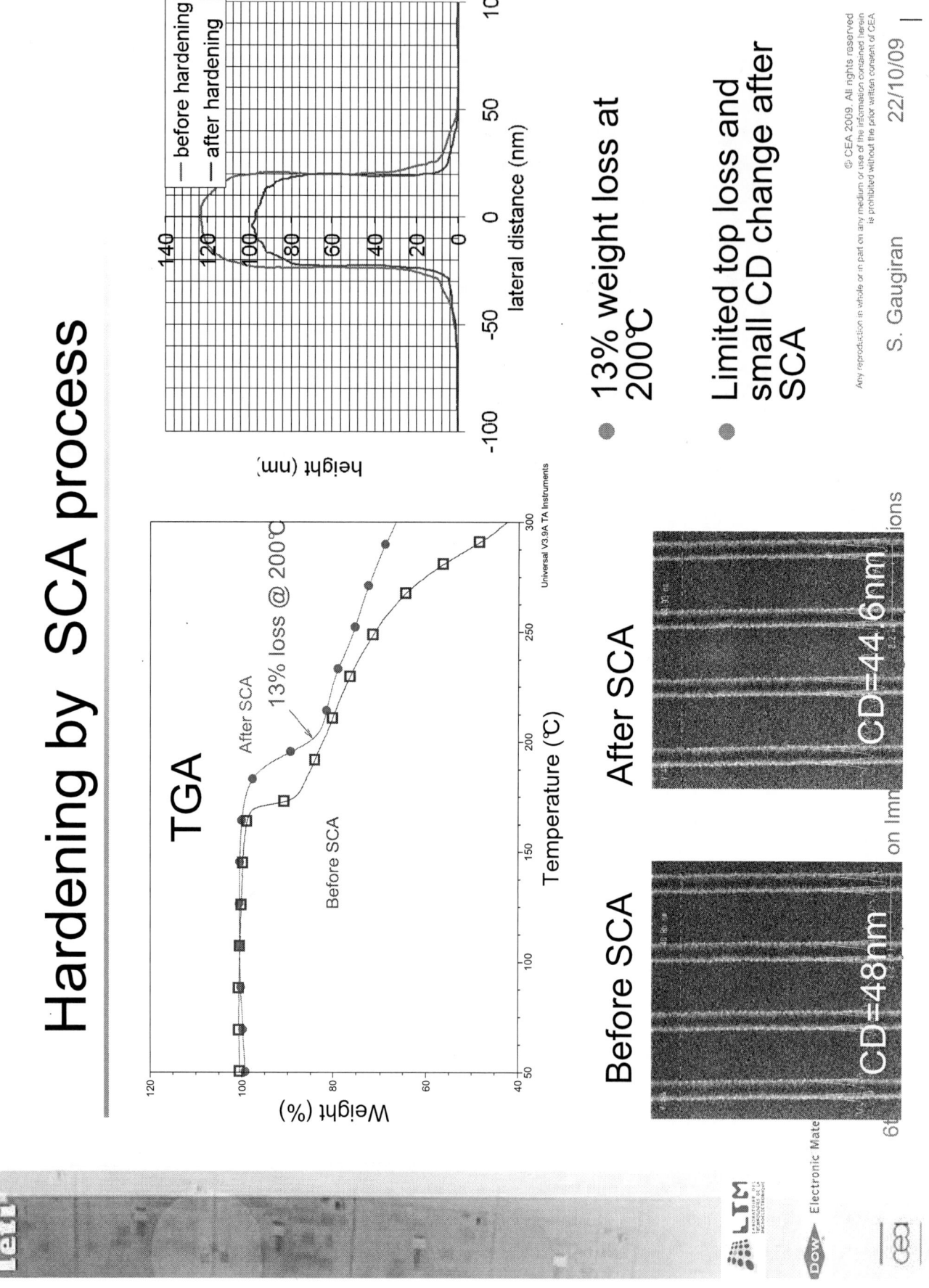

- 13% weight loss at 200°C

- Limited top loss and small CD change after SCA

Hardening by SCA process

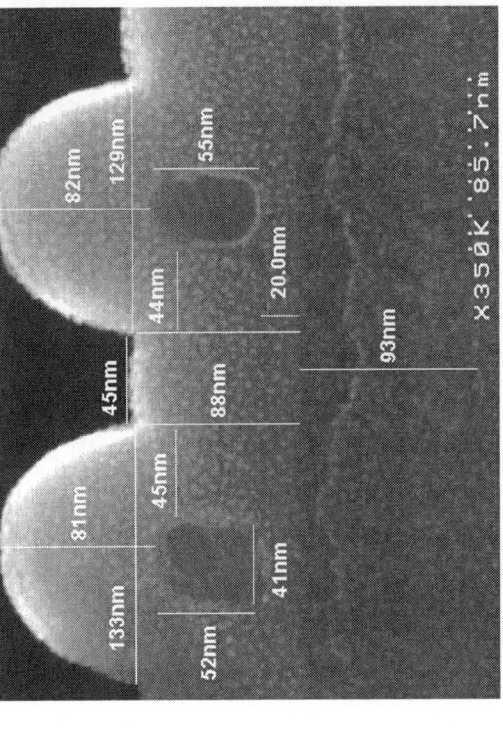

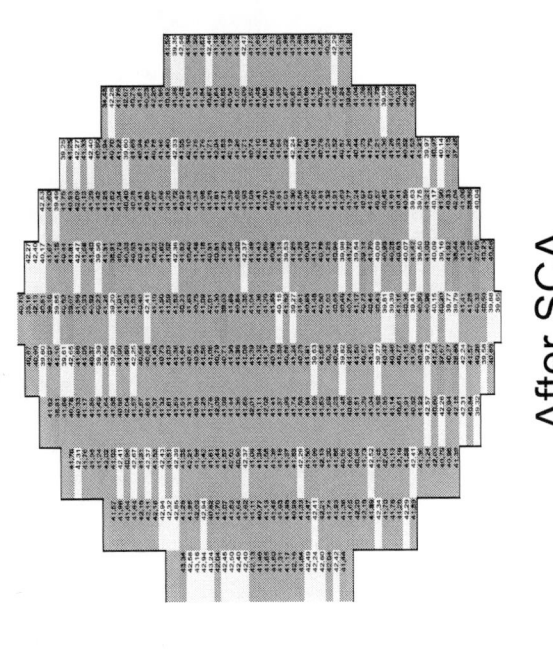

After SCA

CD=41.2nm

3σ=2.7nm

- Low CD dispersion after SCA process

- Resist CD seems to be preserved after deposition

- Strong top loss after deposition

6th International Symposium on Immersion Lithography Extensions   S. Gaugiran   22/10/09

Hardening by TCR process

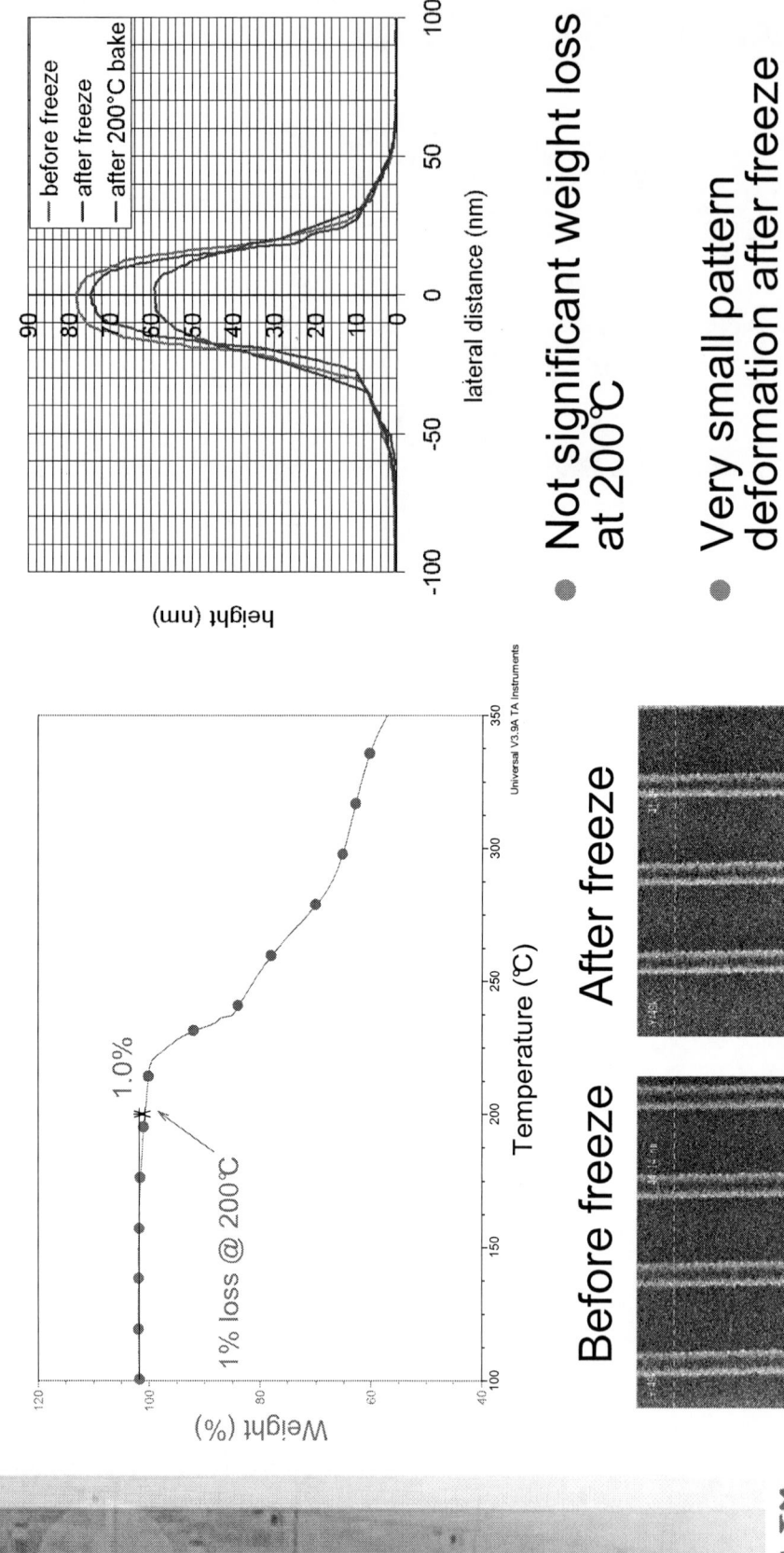

- Not significant weight loss at 200°C

- Very small pattern deformation after freeze

- Pattern reflow during 200°C bake

Before freeze   After freeze

CD=44nm   CD=41.5nm

S. Gaugiran   22/10/09

6th International Symposium on Immersion Lithography Extensions

Hardening by TCR process

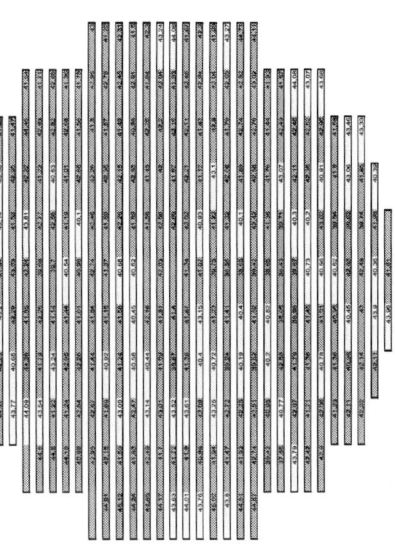

S5500 2.0kV x250k SE

- CD dispersion higher than previously

- Resist patterns altered during deposition

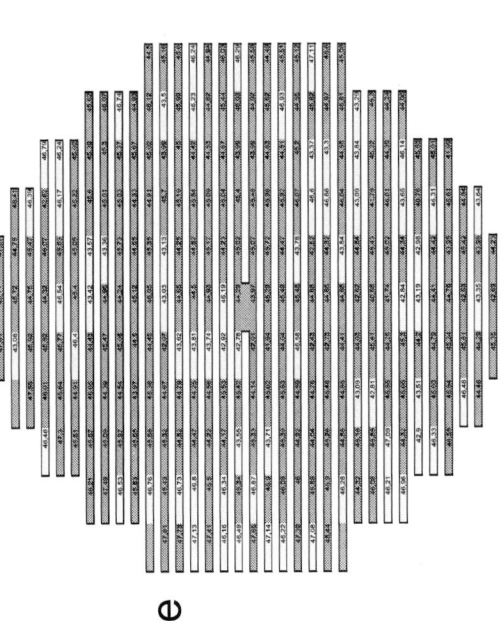

Before freeze

CD=45nm

3σ=3.9nm

After freeze

CD=42nm

3σ=4nm

6th International Symposium on Immersion Lithography Extensions

S. Gaugiran   22/10/09

© CEA 2009. All rights reserved
Any reproduction in whole or in part on any medium or use of the information contained herein is prohibited without the prior written consent of CEA

Conclusion on hardening treatments

	HBr	SCA	TCR
Weight loss	=	=	+
Pattern preservation during hardening	-	=	=
CDU	=	+	-
Pattern preservation during deposition	+	=	-

6th International Symposium on Immersion Lithography Extensions

S. Gaugiran   22/10/09

Outlines

- **Introduction**
  - State of the art
  - Goal and challenges

- **Hardening treatments evaluation**
  - Hardening by HBr plasma
  - Hardening by Surface Curing Agent process
  - Hardening by Thermal Cure Process

- **Low temperature deposition process**

- **Etch transfer feasability**
  - Impact of pattern height
  - Impact of deposition conformity

- **Conclusion and perspectives**

6th International Symposium on Immersion Lithography Extensions

S. Gaugiran   22/10/09

Any reproduction in whole or in part on any medium or use of the information contained herein is prohibited without the prior written consent of CEA.
© CEA 2009. All rights reserved

Low temperature deposition process

Deposition time	6s	7s	10s
CD spacer	18nm	30nm	43nm
Top views	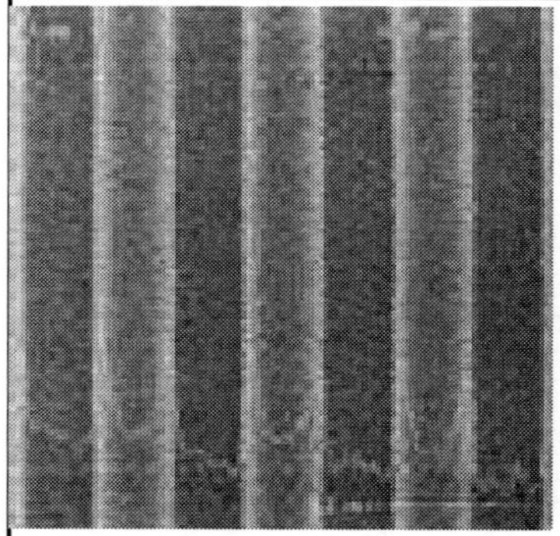		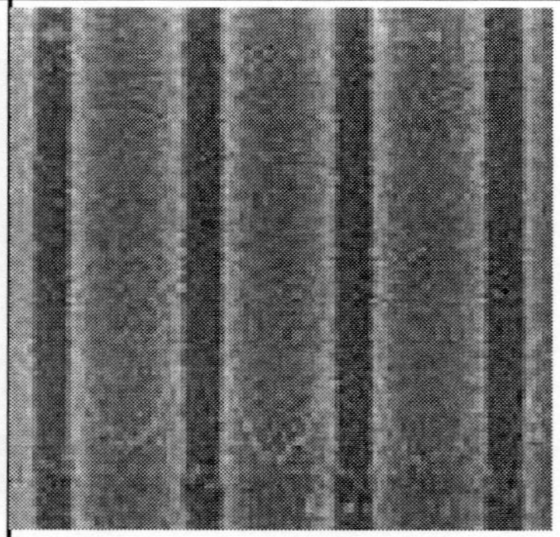
Cross sections	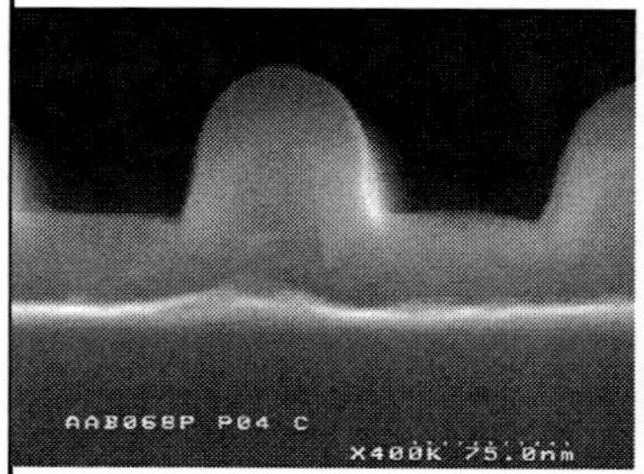	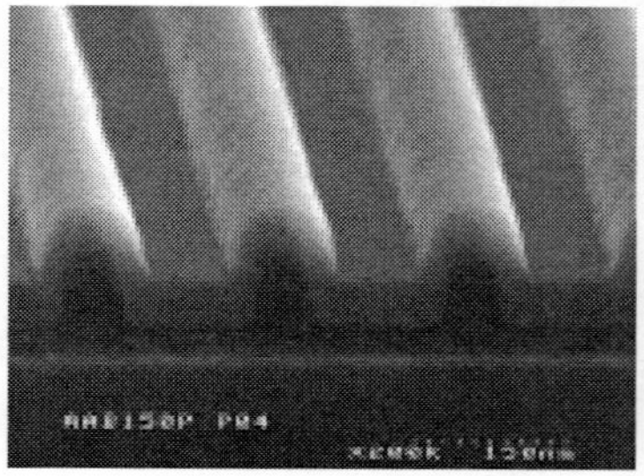	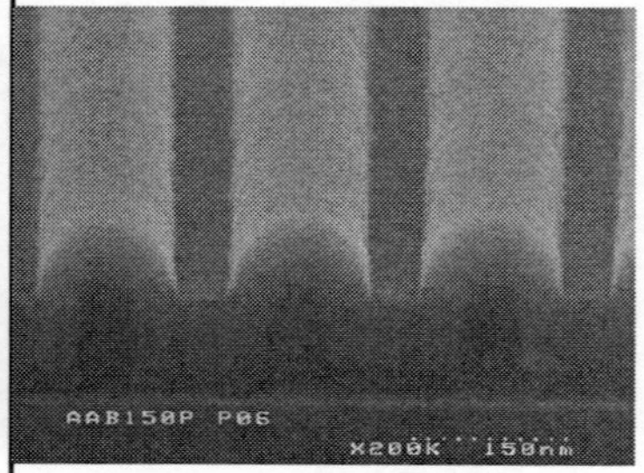
Conformity	80%	61%	48%

6th International Symposium on Immersion Lithography Extensions

S. Gaugiran

22/10/09

Etch Transfer feasability

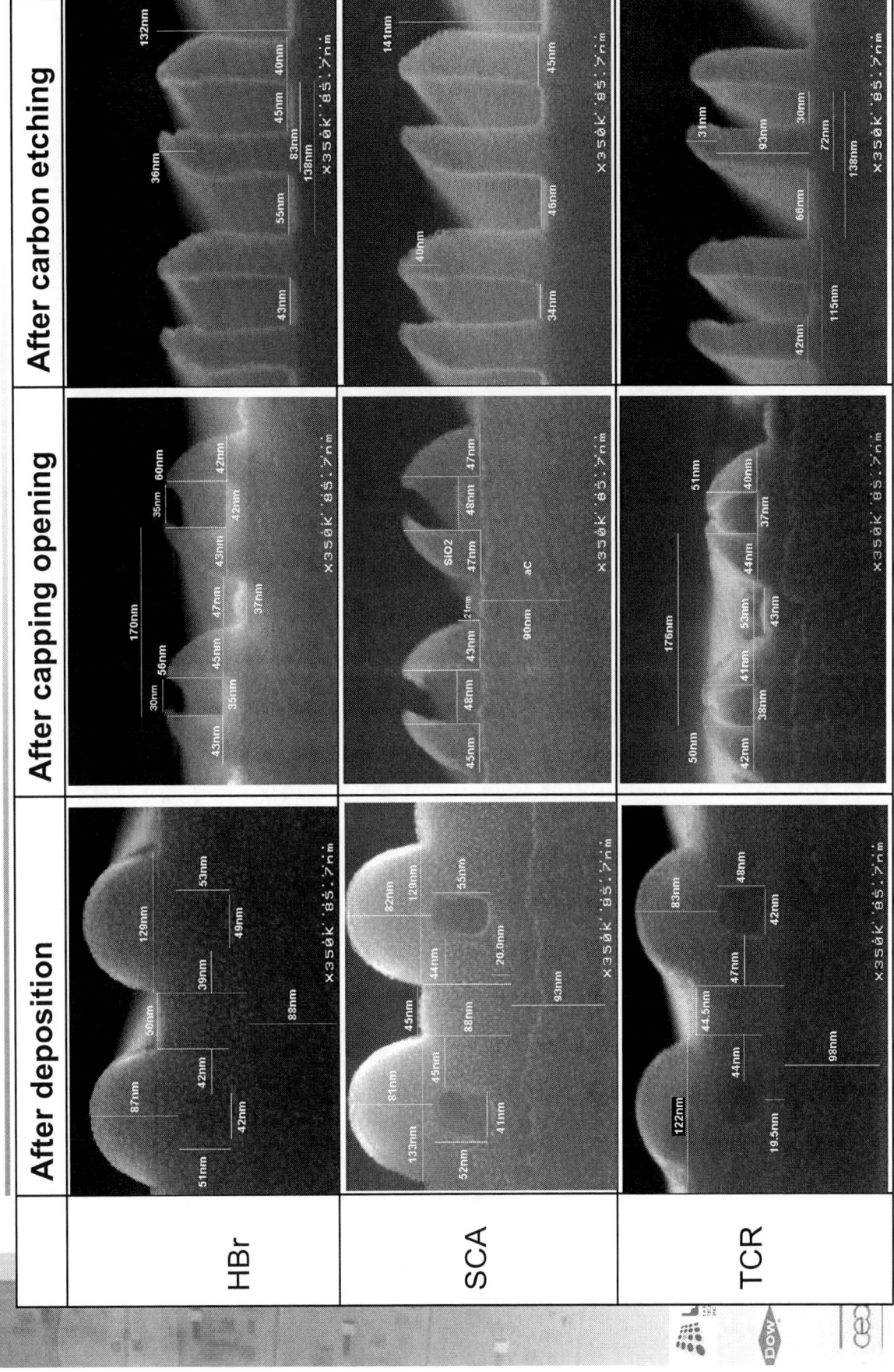

Impact of resist pattern height

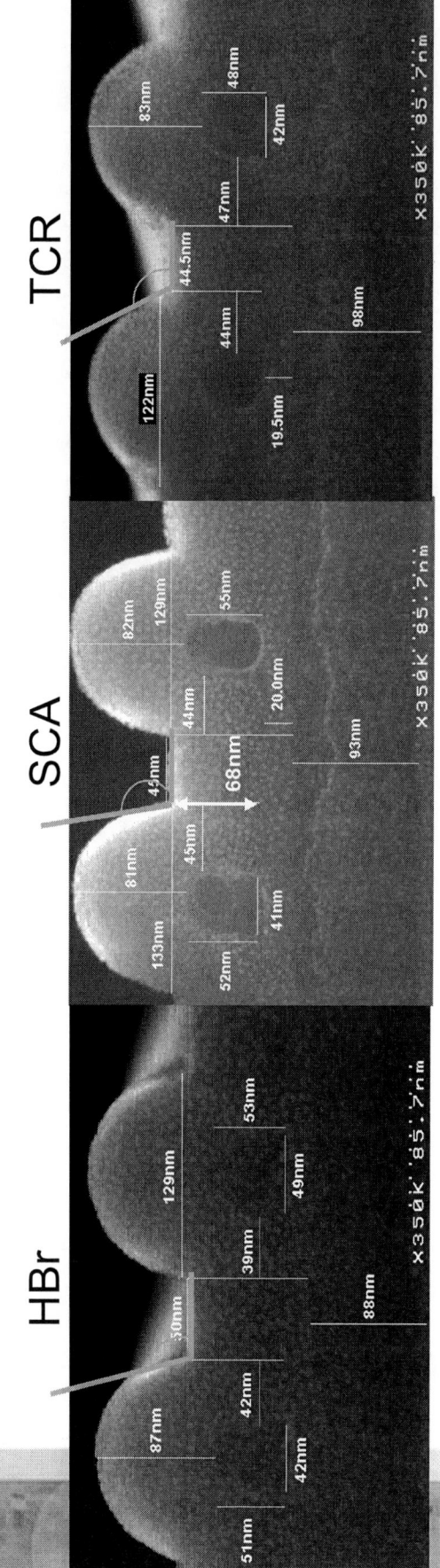

HBr   SCA   TCR

	Resist pattern height after spacer etch	CD conservation through etch
HBr	60nm	+
SCA	62nm	+
TCR	50nm	-

=> Pattern height must be increased

6th International Symposium on Immersion Lithography Extensions

S. Gaugiran   22/10/09

Impact of deposition conformity

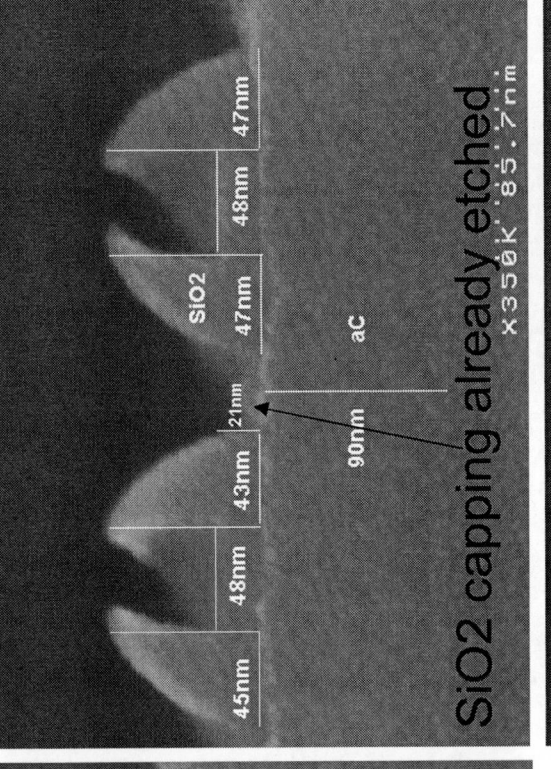

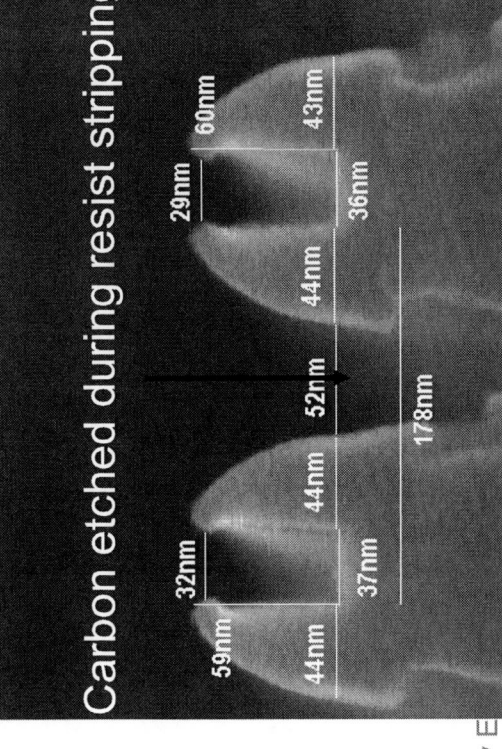

Carbon etched during resist stripping

SiO2 capping already etched

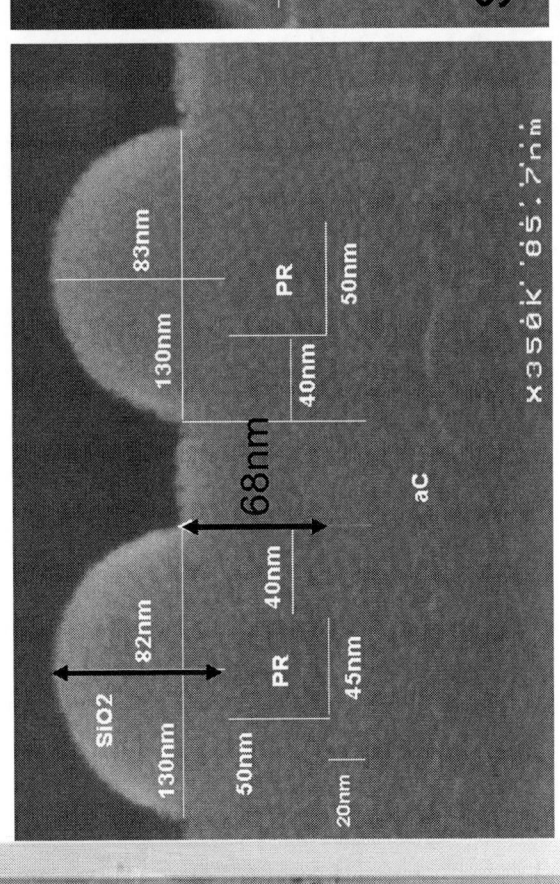

- If SiO2 capping is opened during the spacer etch, the carbon mask is destroyed during the resist strip

$$T_{\text{oxide top}} - T_{\text{oxide bottom}} < T_{\text{SiO2}}$$

6th International Symposium on Immersion Lithography E

Conclusion and perspectives

- Feasibility of pattern doubling process on resist is demonstrated on 45nm dense lines with 3 different hardening approaches (HBr cure, SCA and TCR)

- Surface curing agent process gives the best results in terms of transfer quality (due to the initial resist height and low top loss value)

- HBr cure is a robust hardening treatment and do not show any pattern degradation during deposition

6th International Symposium on Immersion Lithography Extensions

S. Gaugiran

22/10/09

Any reproduction in whole or in part on any medium or use of the information contained herein is prohibited without the prior written consent of CEA

© CEA 2009. All rights reserved

Conclusion and perspectives

- Key points for process optimization have been highlighted:
  - Resist pattern height must be increased
    - => Balance between pattern height/pattern collapse
    - => Top loss limitation during hardening treatment
  - Spacer deposition conformity
    - => Reduce deposition rate

- Next step
  - CDU/ LER evaluation at each process step
  - Impact of the pitch on spacer CD

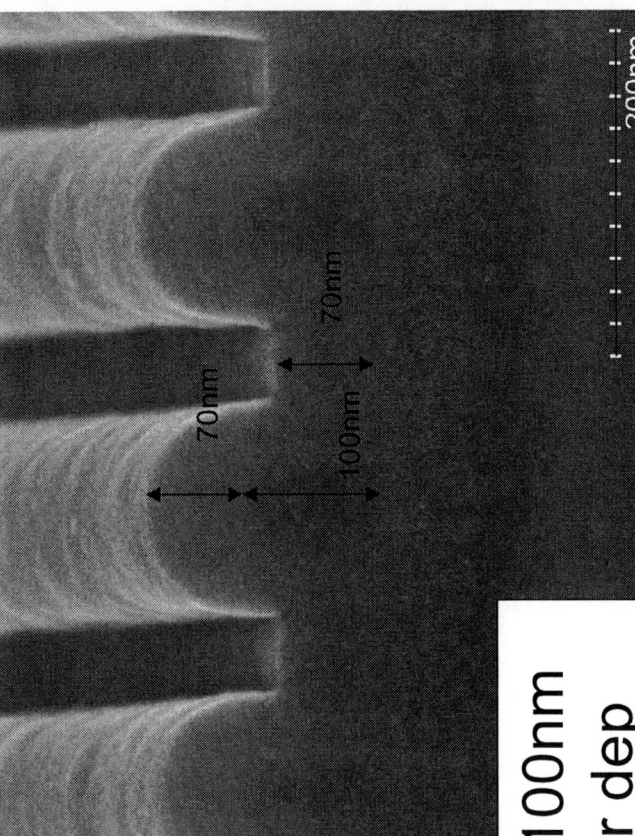

HBr 100nm after dep

6th International Symposium on Immersion

Acknowledgments

- S. Derrough, J. Simon, L. Mage, L. Vandroux, C. Tallaron, M.L. Villani

- European Community for the fundings in the frame of the MD3 project

6th International Symposium on Immersion Lithography Extensions

S. Gaugiran

22/10/09

Any reproduction in whole or in part on any medium or use of the information contained herein is prohibited without the prior written consent of CEA
© CEA 2009. All rights reserved

NIKON CORPORATION
Precision Equipment Company

6th International Symposium on Immersion Lithography Extensions

Immersion Scanner Nikon NSR-S620 for Double Patterning

Yuichi Shibazaki, Masato Hamatani, Kazuhiro Hirano, Jun Ishikawa, Yasuhiro Iriuchijima, and Soichi Owa

October 23, 2009

Nikon

6th International Symposium on Immersion Lithography Extensions   S.Owa

NIKON CORPORATION
Precision Equipment Company

Outline

- ◆ *Streamlign platform*

- ◆ Technology
  - ◆ *Bird's Eye Control*
  - ◆ *Stream Alignment*
  - ◆ *Modular² Structure*

- ◆ Conclusion

October 23, 2009

6th International Symposium on Immersion Lithography Extensions   S.Owa

Outline

Streamlign platform

◆ Technology
- Bird's Eye Control
- Stream Alignment
- $Modular^2$ Structure

◆ Conclusion

NIKON CORPORATION
Precision Equipment Company

October 23, 2009

6th International Symposium on Immersion Lithography Extensions   S.Owa

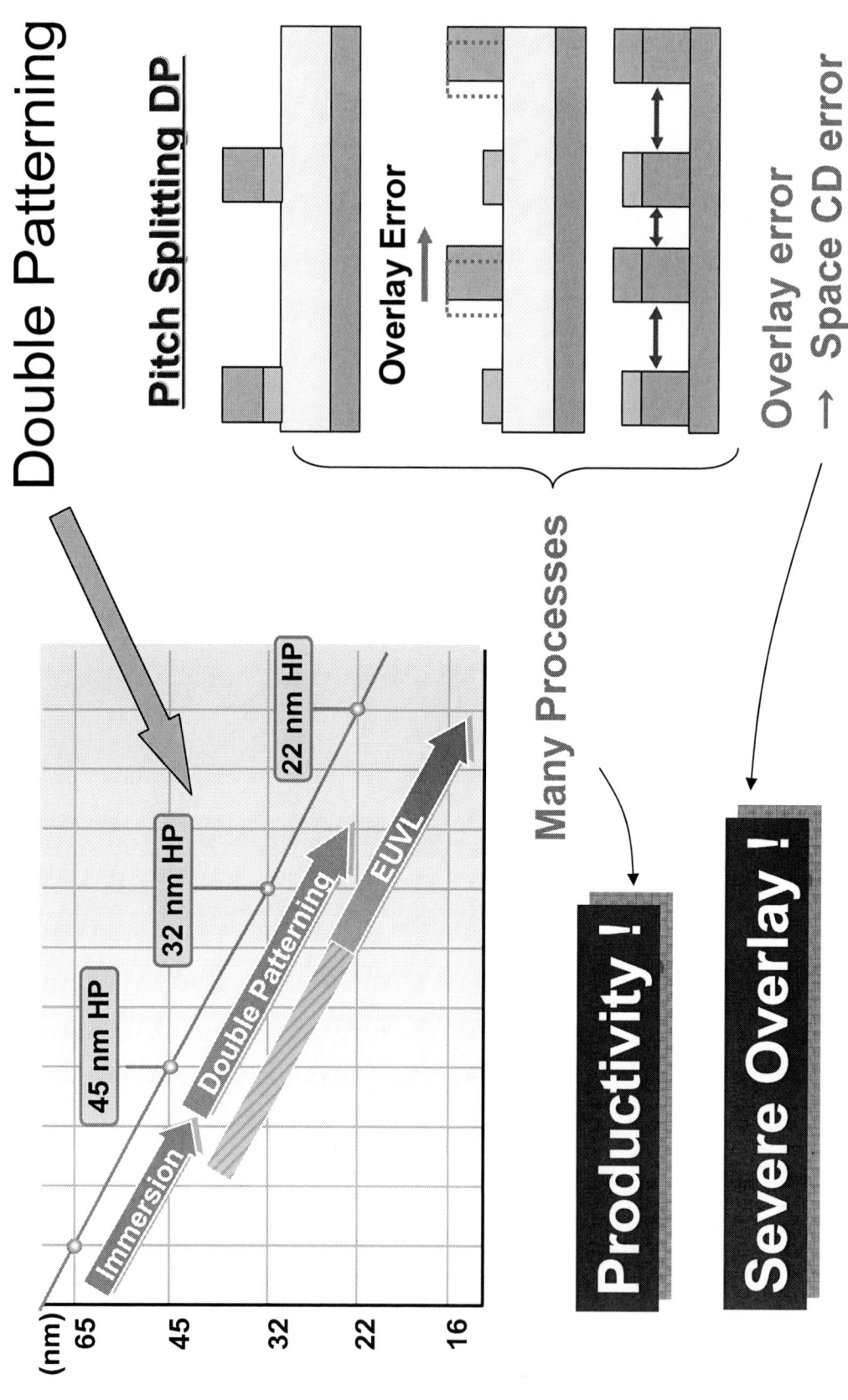

Requirements for Litho Tools

NIKON CORPORATION
Precision Equipment Company

Double Patterning

Pitch Splitting DP

Overlay Error

Overlay error
→ Space CD error

Many Processes

Productivity !

Severe Overlay !

October 23, 2009

6th International Symposium on Immersion Lithography Extensions   S.Owa

(nm)
65
45
32
22
16

45 nm HP
32 nm HP
22 nm HP

Immersion

Double Patterning

EUVL

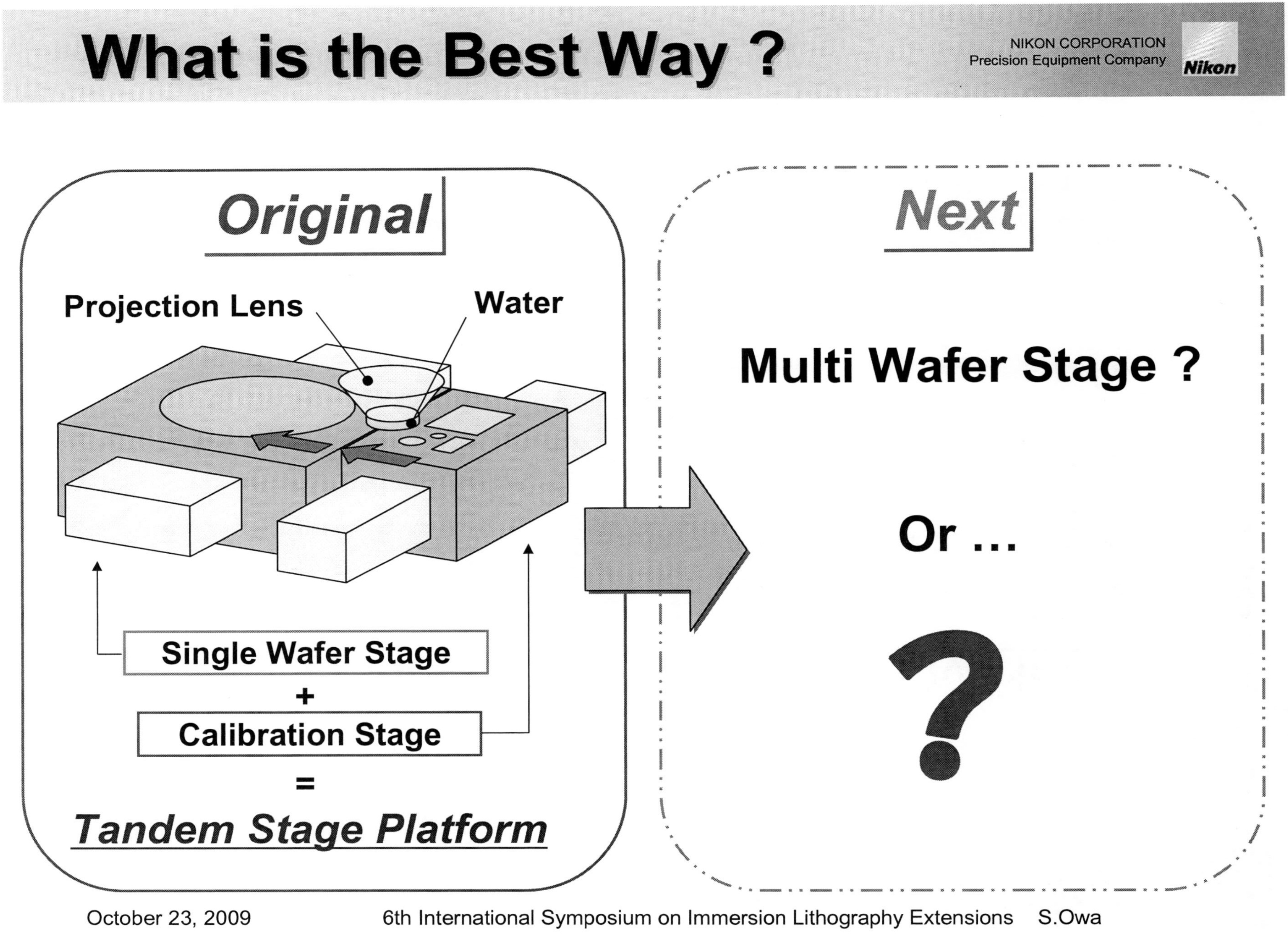

NIKON CORPORATION
Precision Equipment Company

What is the Best Way ?

Original

Projection Lens

Water

Single Wafer Stage
+
Calibration Stage
=
Tandem Stage Platform

Next

Multi Wafer Stage ?

Or ...

October 23, 2009

6th International Symposium on Immersion Lithography Extensions　S.Owa

Overlay Advantage of Single Stage

NIKON CORPORATION
Precision Equipment Company

Nikon

Budget of Single Machine Overlay

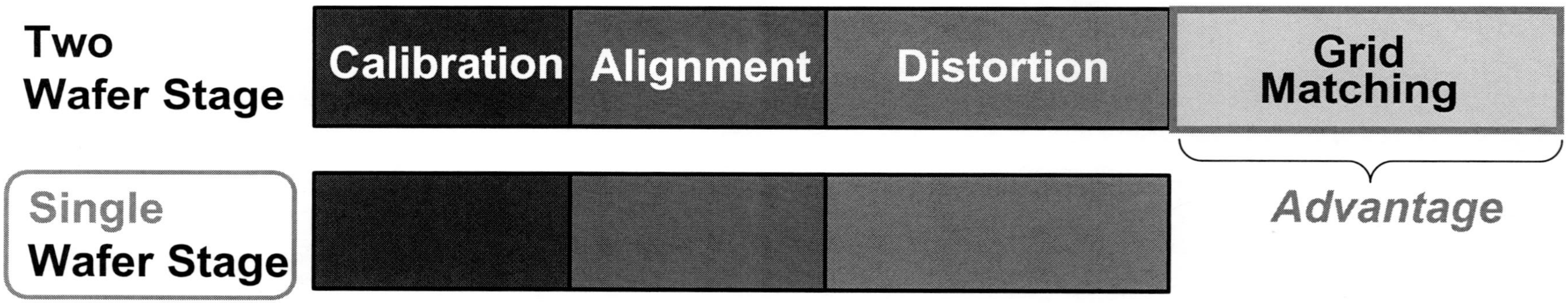

Two
Wafer Stage

Calibration	Alignment	Distortion	Grid Matching

Single
Wafer Stage

Advantage

Single stage has key overlay advantage

October 23, 2009

6th International Symposium on Immersion Lithography Extensions   S.Owa

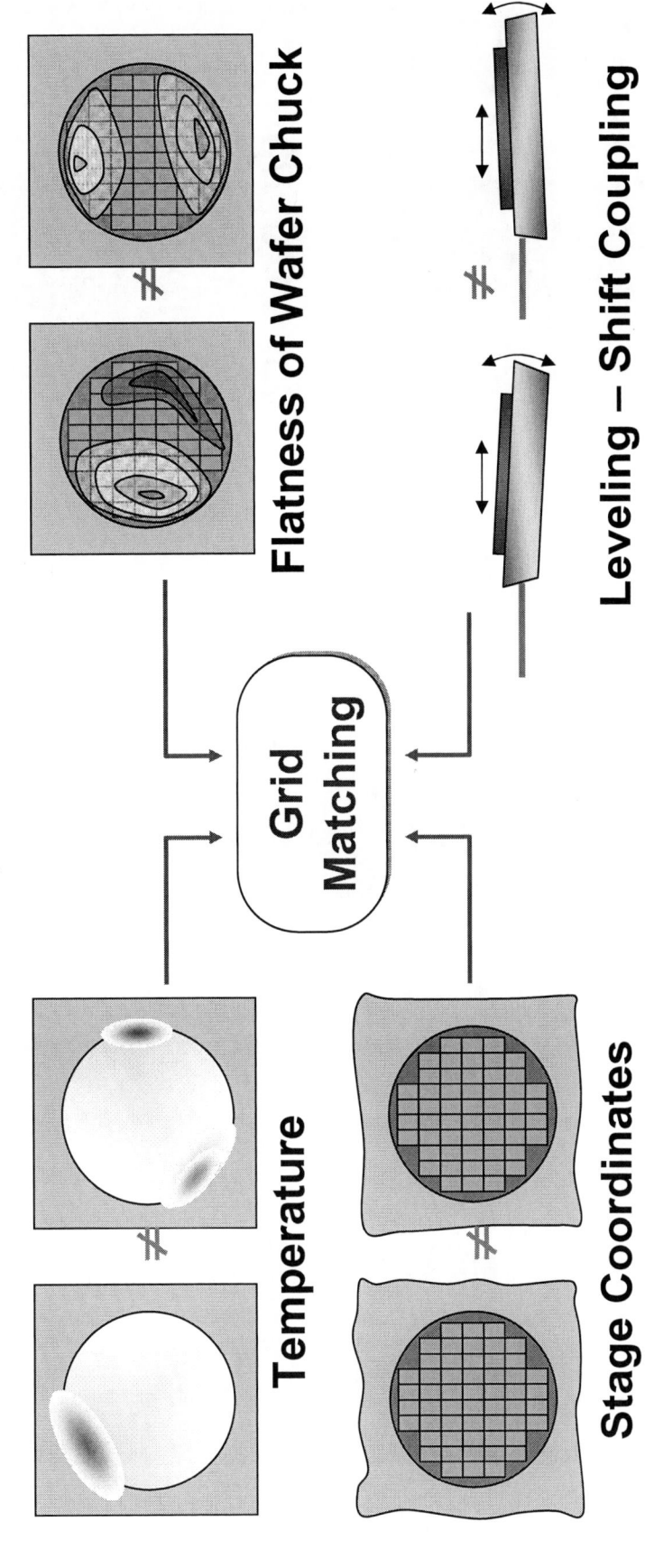

Overlay Advantage of Single Stage

NIKON CORPORATION
Precision Equipment Company

Flatness of Wafer Chuck

Leveling – Shift Coupling

Grid Matching

Temperature

Stage Coordinates

These cannot be perfectly corrected

October 23, 2009   6th International Symposium on Immersion Lithography Extensions   S.Owa

NIKON CORPORATION
Precision Equipment Company

Streamlign Platform

for 2 nm self overlay

Bird's Eye Control

for 200 wph / 125 shots throughput

Stream Alignment

NSR-S620
Challenge for High Accuracy and High Productivity for Double Patterning

October 23, 2009   6th International Symposium on Immersion Lithography Extensions   S.Owa

Outline

- *Streamlign platform*
- **Technology**
  - *Bird's Eye Control*
  - *Stream Alignment*
  - *Modular^2 Structure*
- Conclusion

NIKON CORPORATION
Precision Equipment Company

October 23, 2009   6th International Symposium on Immersion Lithography Extensions   S.Owa

The Dominant Factor of Overlay Error

NIKON CORPORATION
Precision Equipment Company

◆ We are near the limit of interferometer metrology

We need a breakthrough

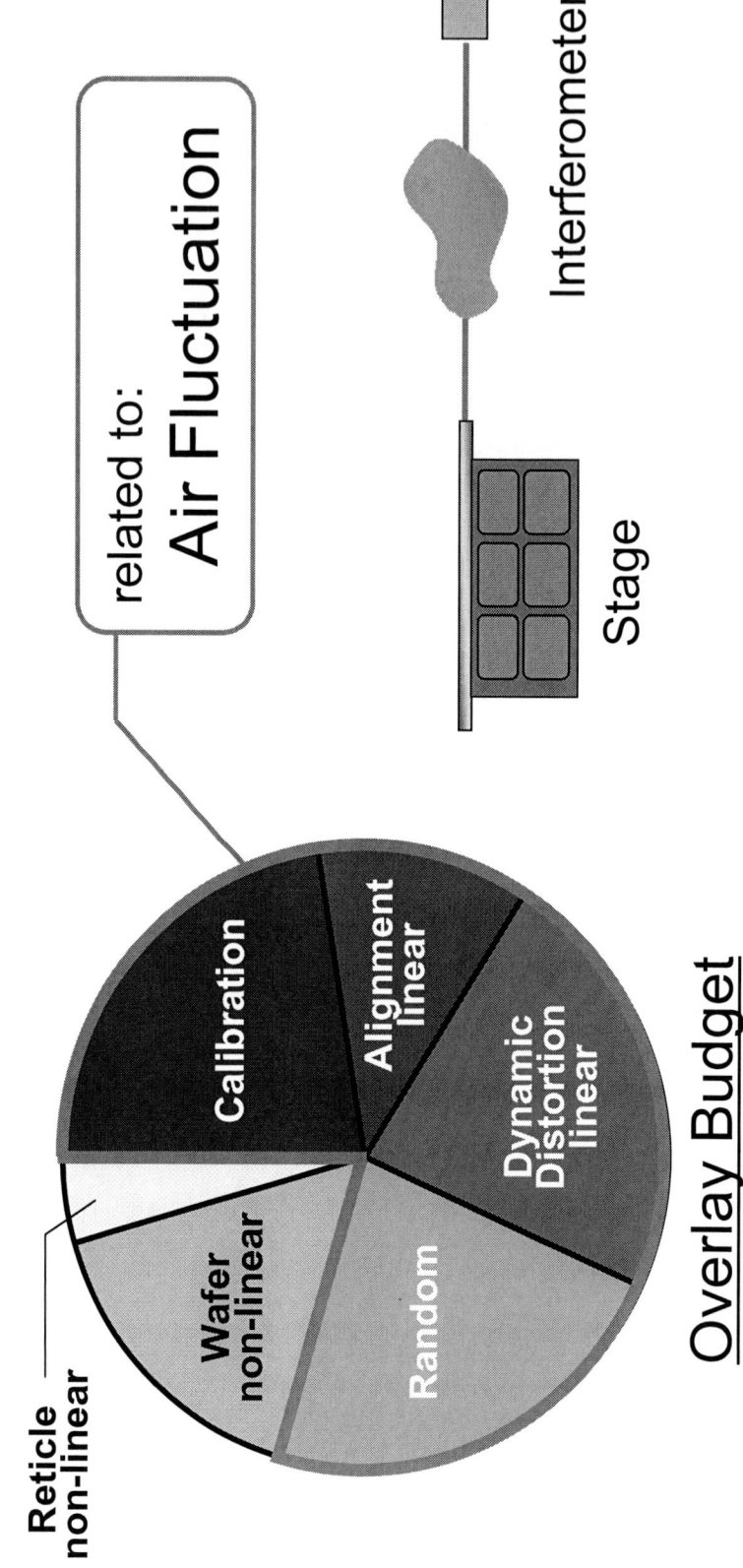

related to:
Air Fluctuation

Interferometer

Stage

Reticle non-linear

Wafer non-linear

Random

Dynamic Distortion linear

Alignment linear

Calibration

Overlay Budget

October 23, 2009

6th International Symposium on Immersion Lithography Extensions   S.Owa

Breakthrough : Bird's Eye Control

NIKON CORPORATION
Precision Equipment Company

Nikon

◆ Use encoders with interferometers

◆ View the top of the Wafer Stage from above

like a Bird

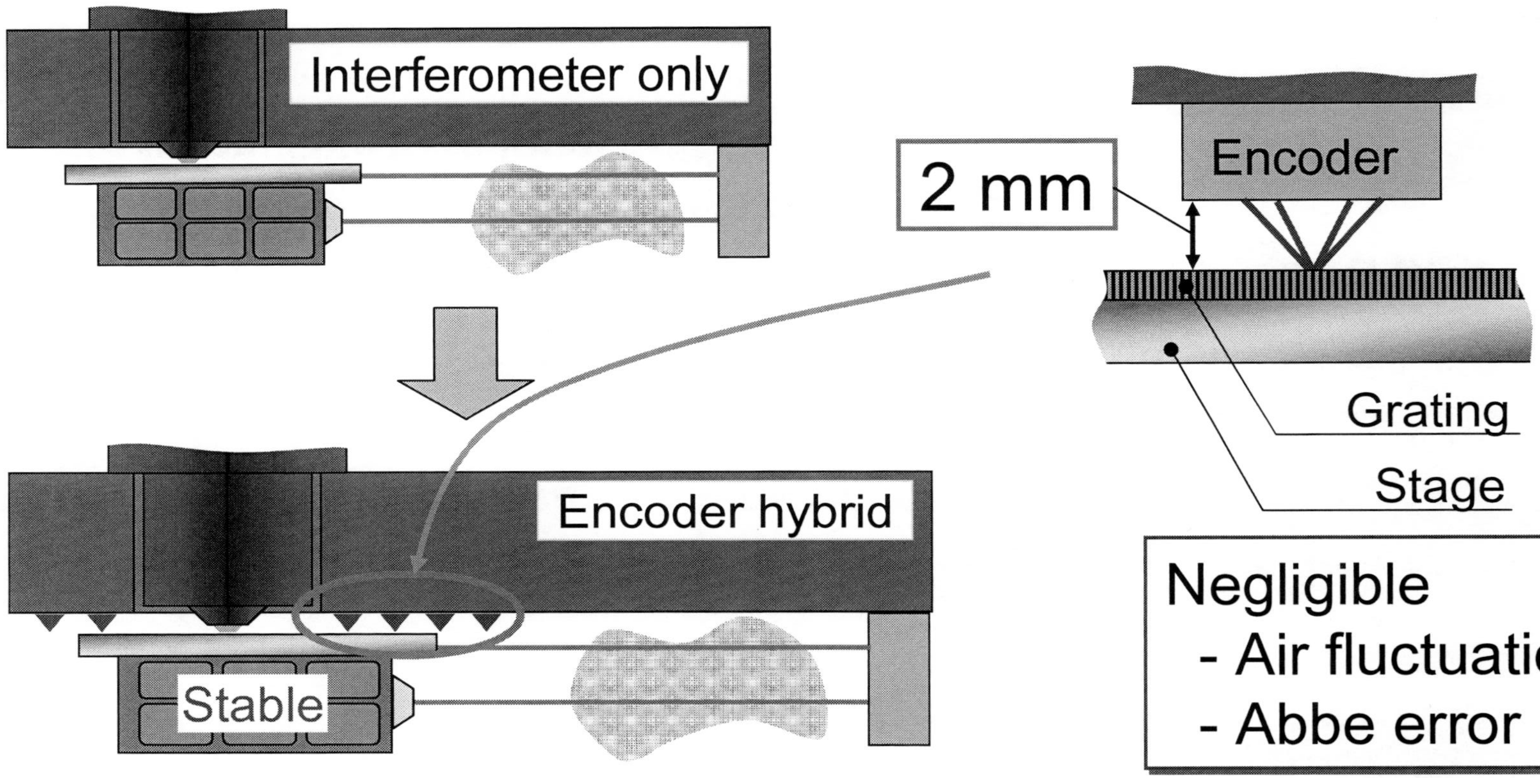

Encoder

Grating

Stage

2 mm

Negligible
- Air fluctuation
- Abbe error

Interferometer only

Encoder hybrid

Stable

October 23, 2009

6th International Symposium on Immersion Lithography Extensions　S.Owa

NIKON CORPORATION
Precision Equipment Company

Hybrid Metrology with Interferometer

- **Easy calibration**
  - Grating is calibrated by interferometer
- **Intelligent correction**
  - Compensation of low order drift and other factors
- **Stability**
  - Insensitive to water droplets or particles
  - Insensitive to thermal drift

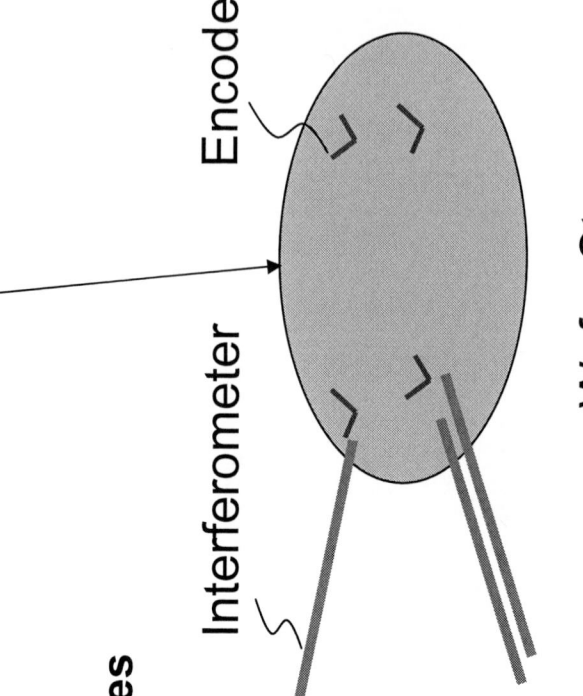

Grating can't be perfect

Encoder

Interferometer

Wafer Stage

	IF	Enc	Hybrid
Linearity	Good	Bad	Good
Repeatability	OK	Good	Good
Long time stability	Good	OK	Good

October 23, 2009

6th International Symposium on Immersion Lithography Extensions   S.Owa

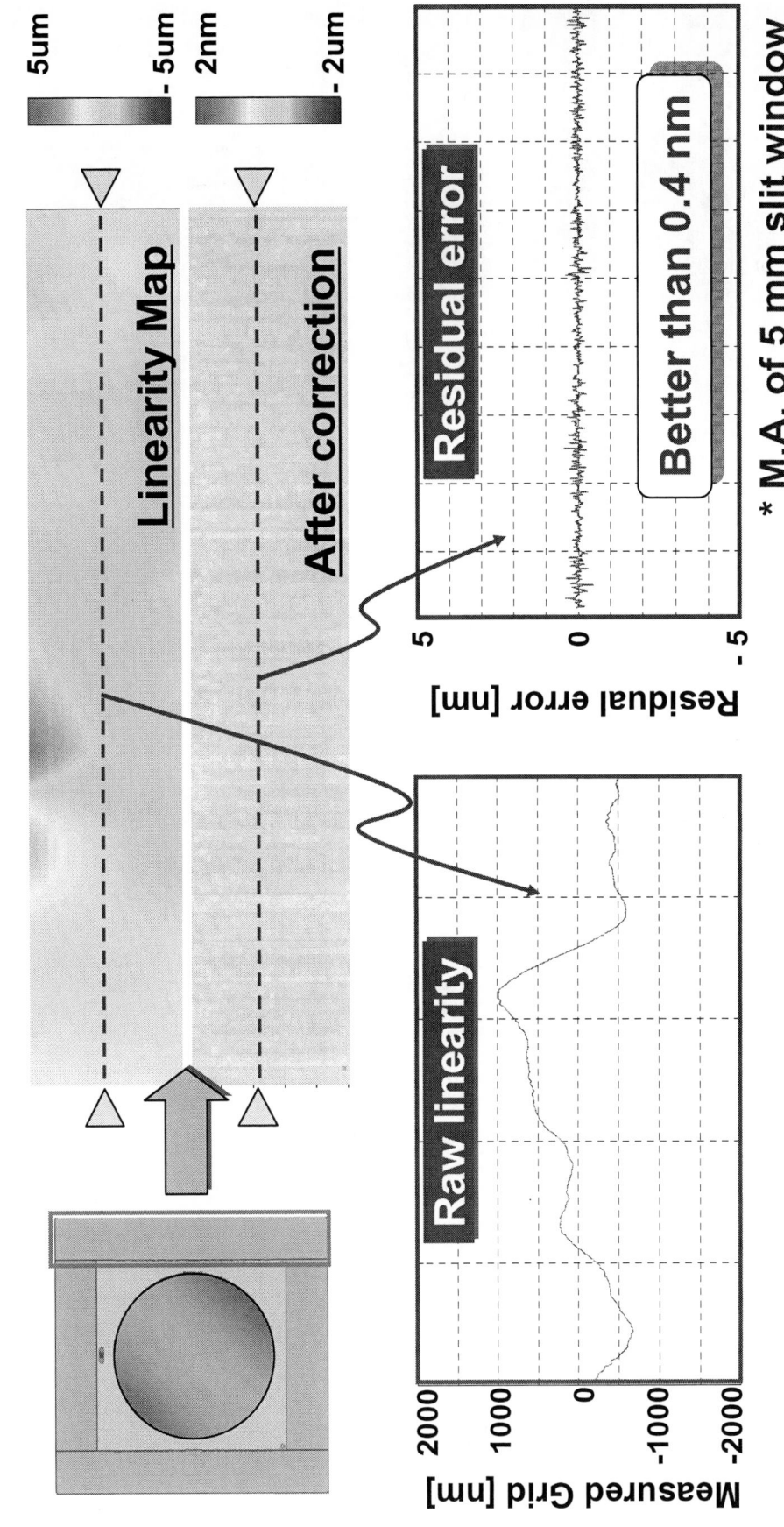

Results of Grating Calibration

NIKON CORPORATION
Precision Equipment Company

Nikon

◆ Grating accuracy is corrected by interferometer

5um

- 5um

2nm

- 2um

Linearity Map

After correction

Residual error

Better than 0.4 nm

* M.A. of 5 mm slit window

Residual error [nm]

5

0

- 5

Raw linearity

Measured Grid [nm]

2000

1000

0

-1000

-2000

October 23, 2009

6th International Symposium on Immersion Lithography Extensions   S.Owa

NIKON CORPORATION
Precision Equipment Company

Repeatability

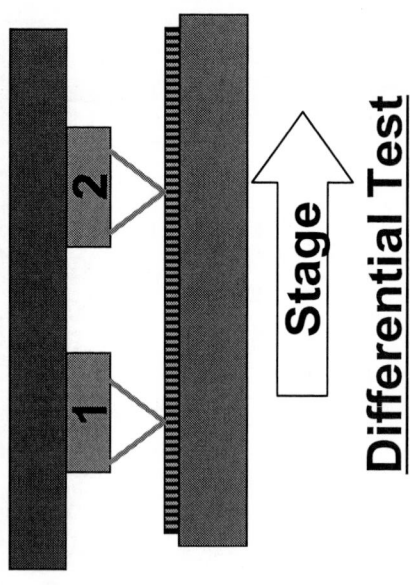

Stage

Differential Test

◆ Not affected by
   ◆ Air fluctuation
   ◆ High scan speed (pressure change)
◆ Fast interpolation calculation

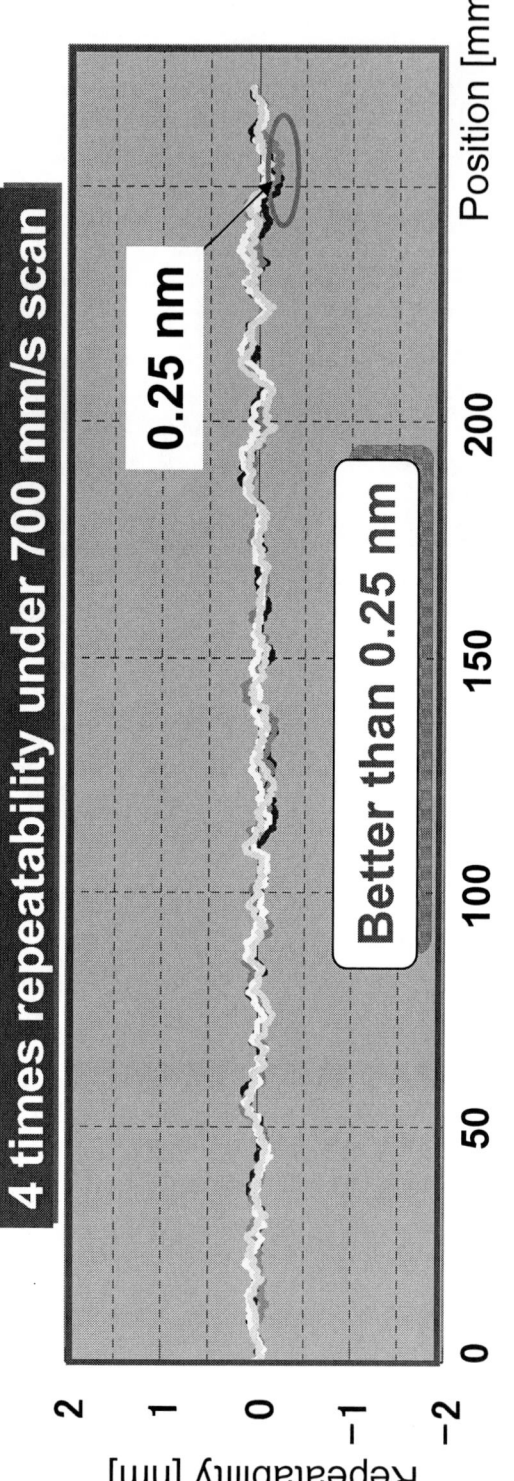

4 times repeatability under 700 mm/s scan

0.25 nm

Better than 0.25 nm

Repeatability [nm]

Position [mm]

* M.A. of 5 mm slit window

October 23, 2009

6th International Symposium on Immersion Lithography Extensions   S.Owa

NIKON CORPORATION
Precision Equipment Company

Outline

- *Streamlign platform*
- **Technology**
  - *Bird's Eye Control*
  - **Stream Alignment**
  - *Modular^2 Structure*
- Conclusion

October 23, 2009   6th International Symposium on Immersion Lithography Extensions   S.Owa

Alignment Points Should Increase for DP

NIKON CORPORATION
Precision Equipment Company

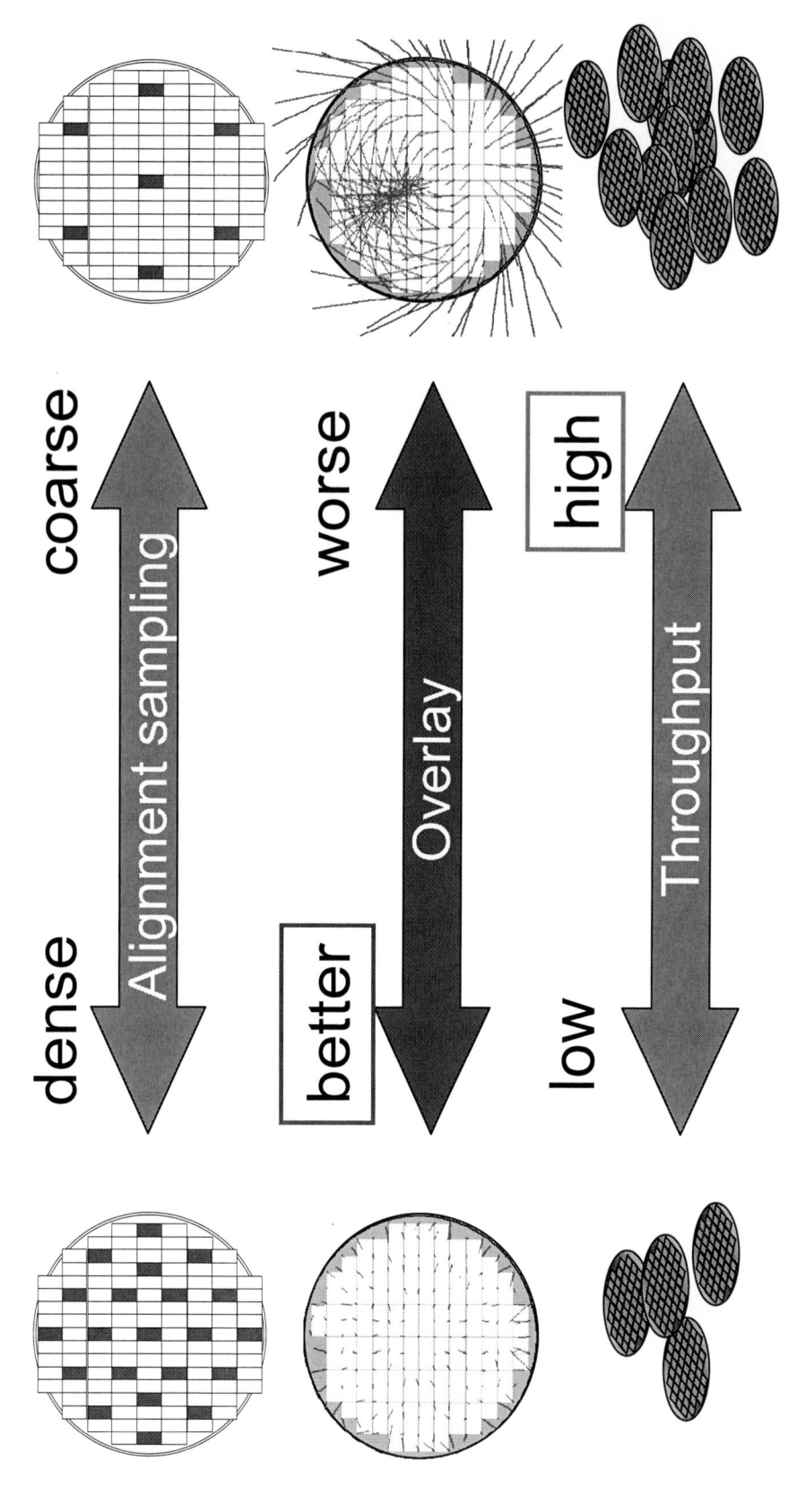

Cycle time

Exposure / Stepping   Alignment / Other

How to reduce ??

6th International Symposium on Immersion Lithography Extensions   S.Owa

October 23, 2009

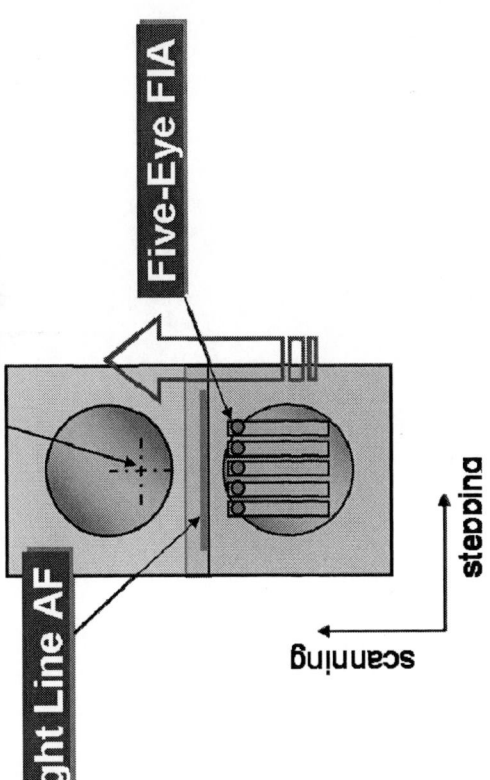

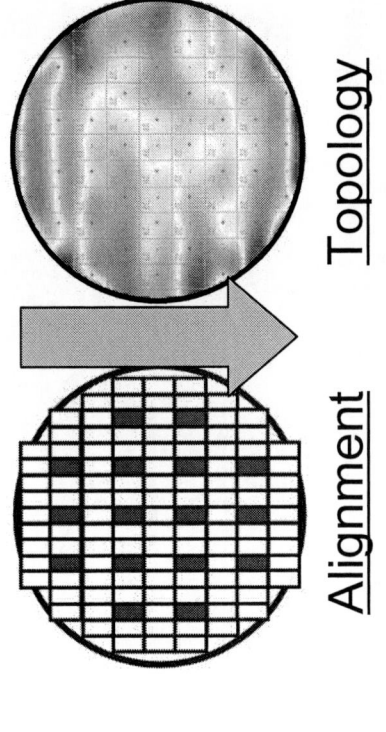

Breakthrough : Stream Alignment

NIKON CORPORATION
Precision Equipment Company

Nikon

Exposure Center

Five-Eye FIA

Straight Line AF

scanning

stepping

Alignment   Topology

- ◆ Newly developed
  - ◆ *Five-Eye FIA*
    (5 alignment microscopes)
  - ◆ *Straight Line Auto Focus*
    which covers the full wafer diameter
- ◆ Straight and short
  trajectory for alignment

Exposure / Stepping

Alignment

67% reduced !

October 23, 2009   6th International Symposium on Immersion Lithography Extensions   S.Owa

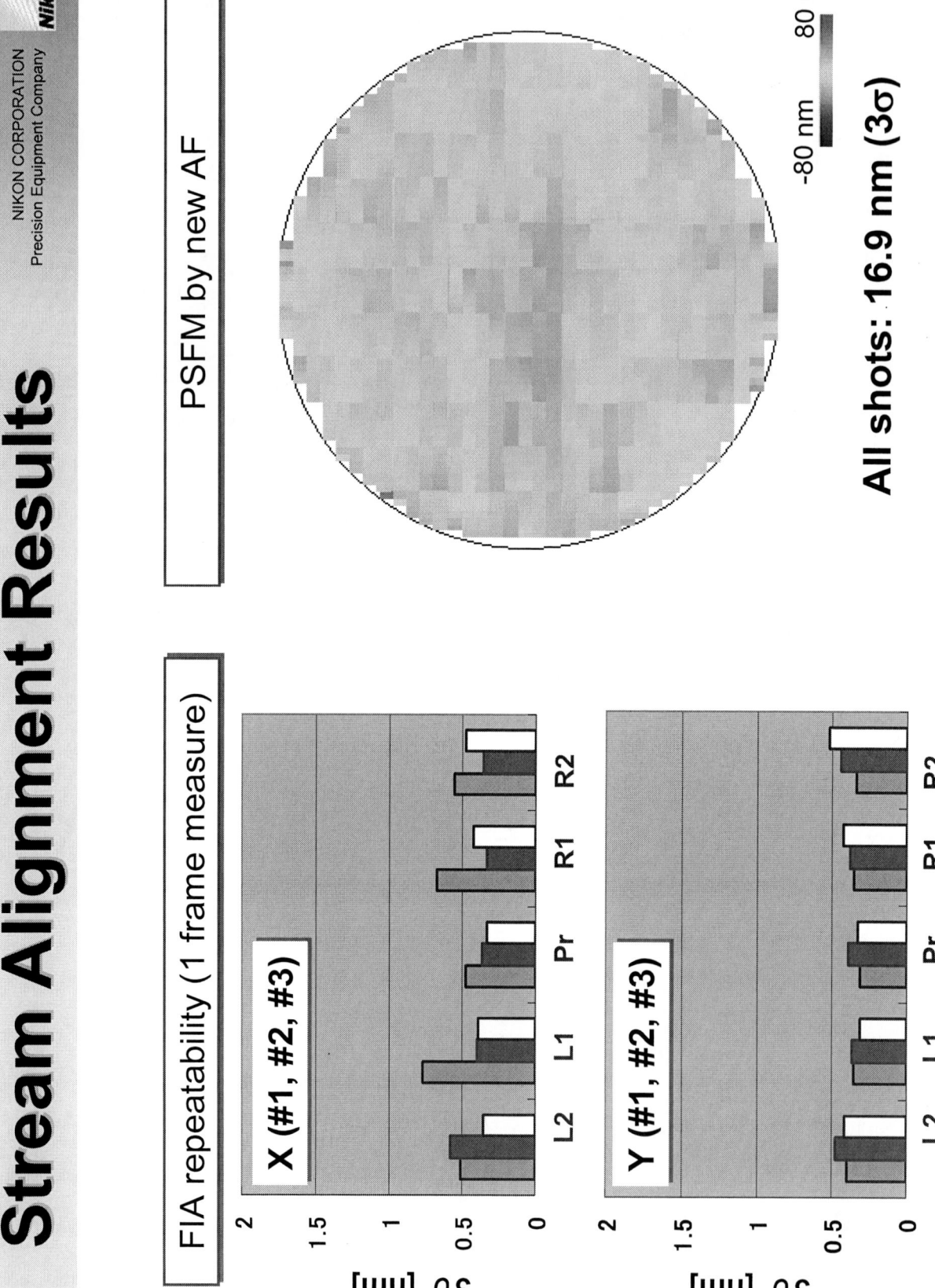

Outline

- *Streamlign platform*

- ◆ **Technology**
  - ◆ *Bird's Eye Control*
  - ◆ *Stream Alignment*

- ◆ **Modular² Structure**

- ◆ Conclusion

October 23, 2009

6th International Symposium on Immersion Lithography Extensions   S.Owa

NIKON CORPORATION
Precision Equipment Company

Nikon

Hierarchical Structure : Modular² Structure

NIKON CORPORATION
Precision Equipment Company

◆ Easy setup and maintenance
◆ Excellent dynamics

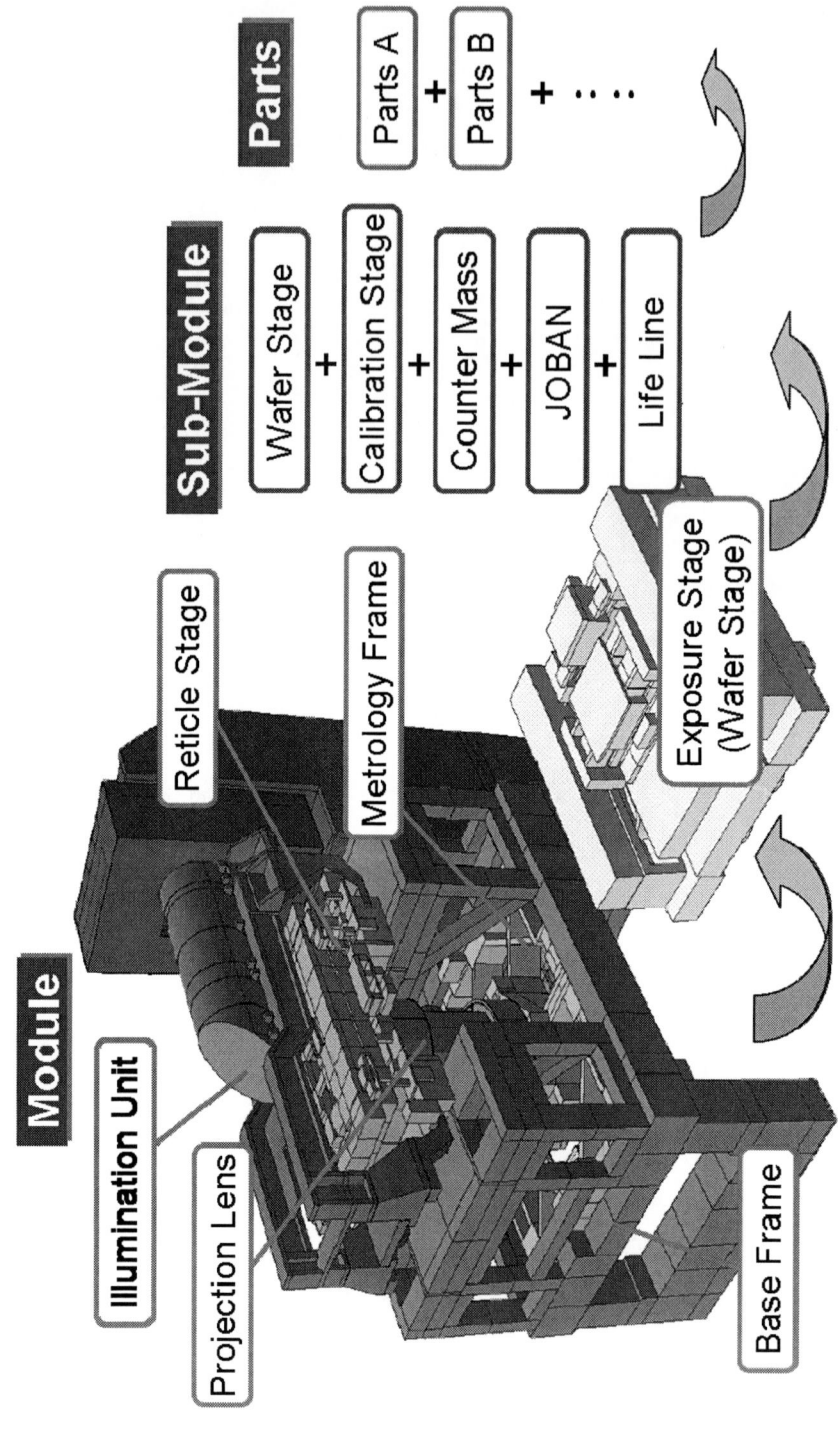

Module
- Reticle Stage
- Illumination Unit
- Metrology Frame
- Projection Lens
- Base Frame

Sub-Module
- Wafer Stage + Calibration Stage + Counter Mass + JOBAN + Life Line

Exposure Stage (Wafer Stage)

Parts
- Parts A + Parts B + ...

October 23, 2009

6th International Symposium on Immersion Lithography Extensions   S.Owa

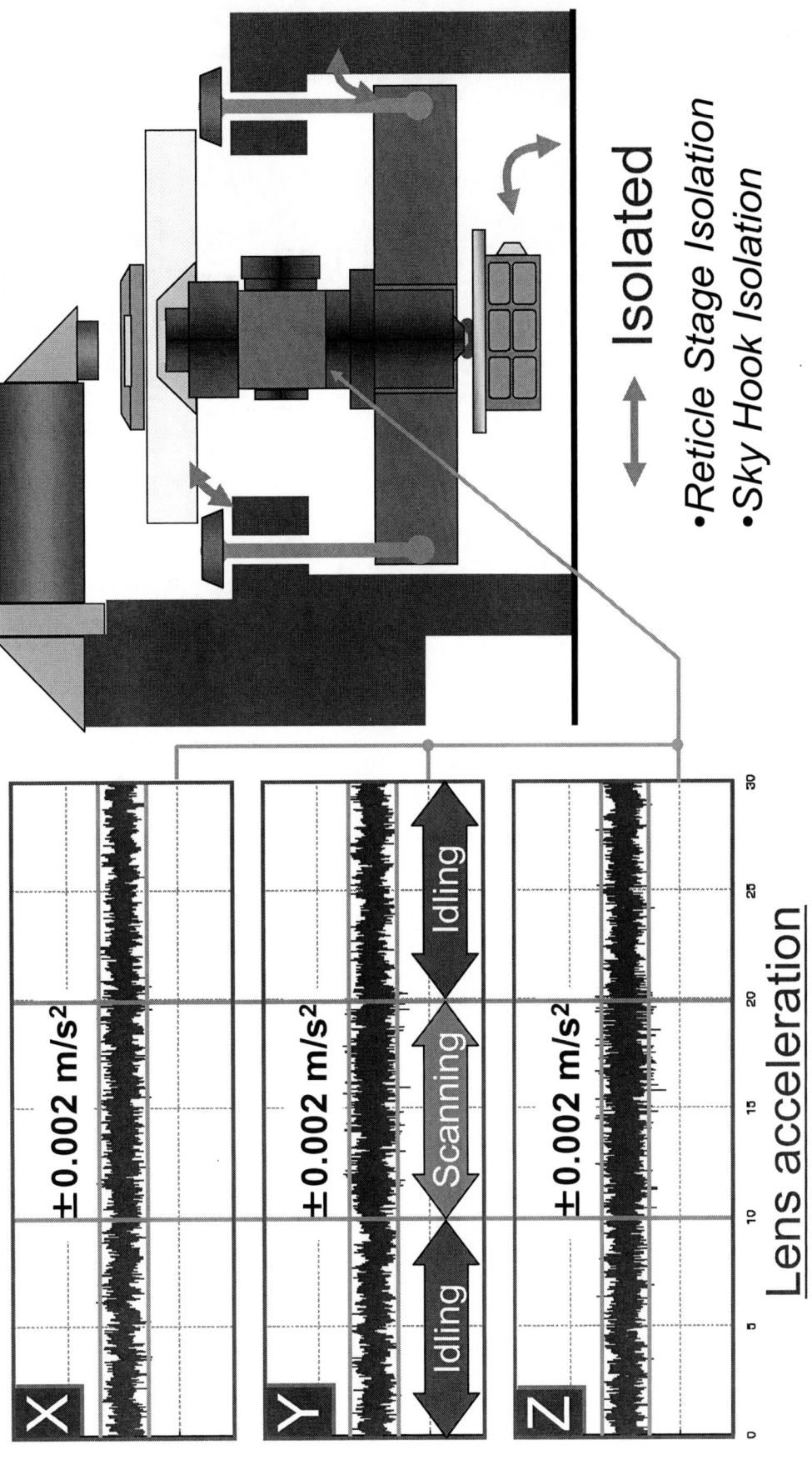

Advances in Stage Dynamics

NIKON CORPORATION
Precision Equipment Company

◆ Higher scan speed

◆ Higher acceleration

◆ Shorter settling time

Reticle Stage

Light and stiff
High efficiency motor
No wires or tubes

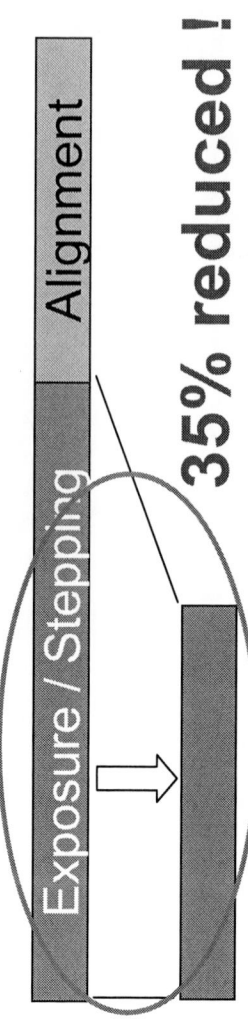

Alignment

Exposure / Stepping

35% reduced !

Wafer Stage

Max: 700 mm/s

October 23, 2009   6th International Symposium on Immersion Lithography Extensions   S.Owa

Synchronization Accuracy: 700 mm/s

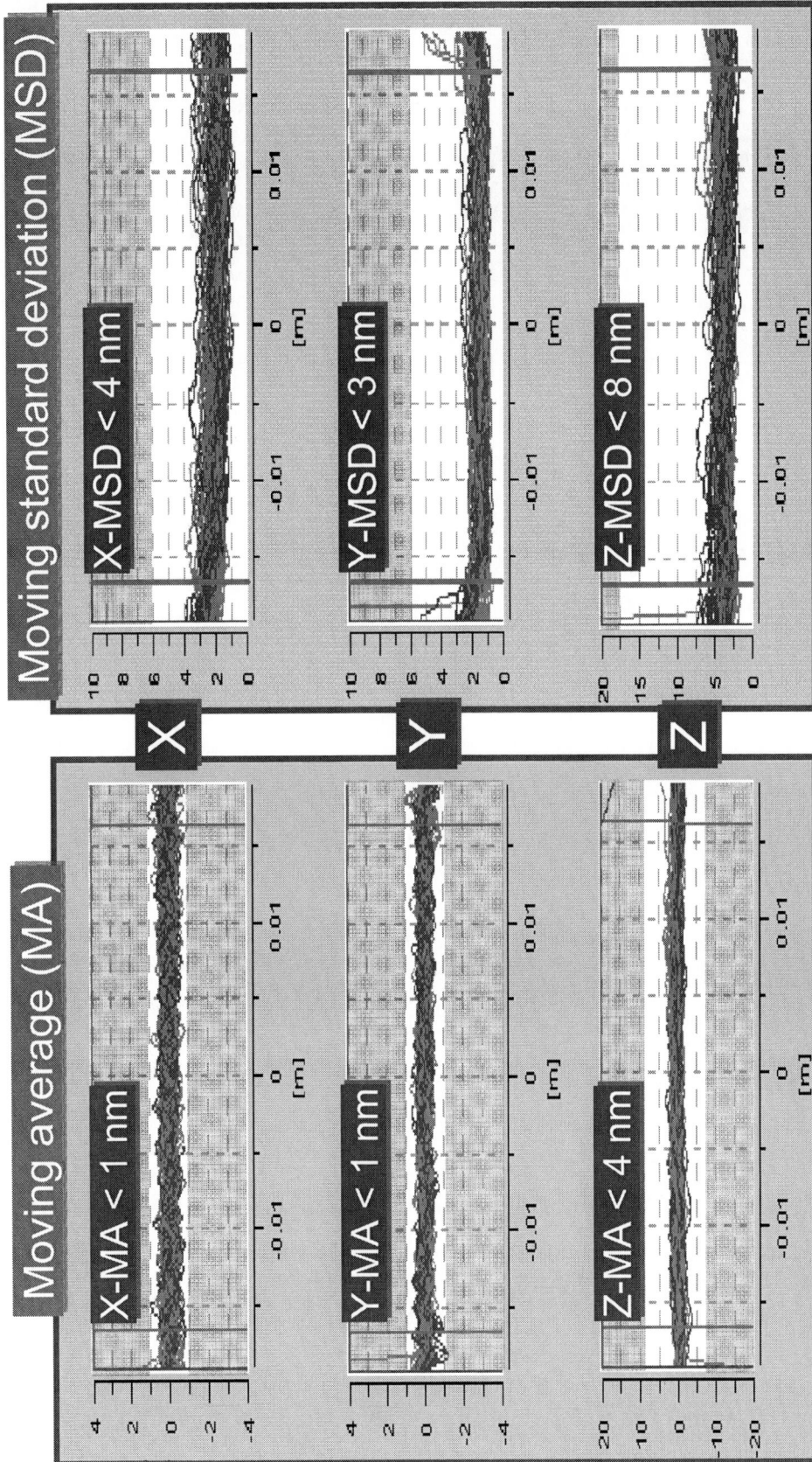

Already achieved the necessary performance

October 23, 2009   6th International Symposium on Immersion Lithography Extensions   S.Owa

NIKON CORPORATION
Precision Equipment Company

Outline

◆ *Streamlign platform*

◆ *Technology*

　◆ *Bird's Eye Control*

　◆ *Stream Alignment*

　◆ *Modular² Structure*

◆ **Conclusion**

October 23, 2009　　　6th International Symposium on Immersion Lithography Extensions　　S.Owa

NIKON CORPORATION
Precision Equipment Company

Conclusion

Nikon's new *Streamlign* platform enables…

- ◆ Enhanced overlay by
  Bird's Eye Control and multi-point alignment.

- ◆ Enhanced throughput by
  Stream Alignment and advanced stage dynamics.

- ◆ Easy setup, maintenance and improved dynamics by
  Modular² *Structure*.

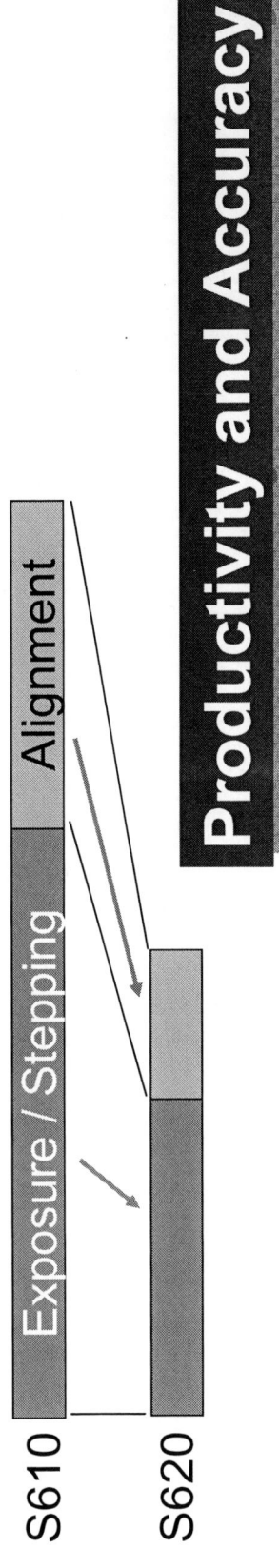

Productivity and Accuracy !

October 23, 2009   6th International Symposium on Immersion Lithography Extensions   S.Owa

NIKON CORPORATION

ASML

Improvements in Exposure Systems and Applications
Enabling Ultimate Extension with Immersion Lithography

F. de Jong, B. Vleeming, M. Mulder, A. Engelen and J. Mulkens

NXT Immersion System for 1.35 NA Optical Extensions

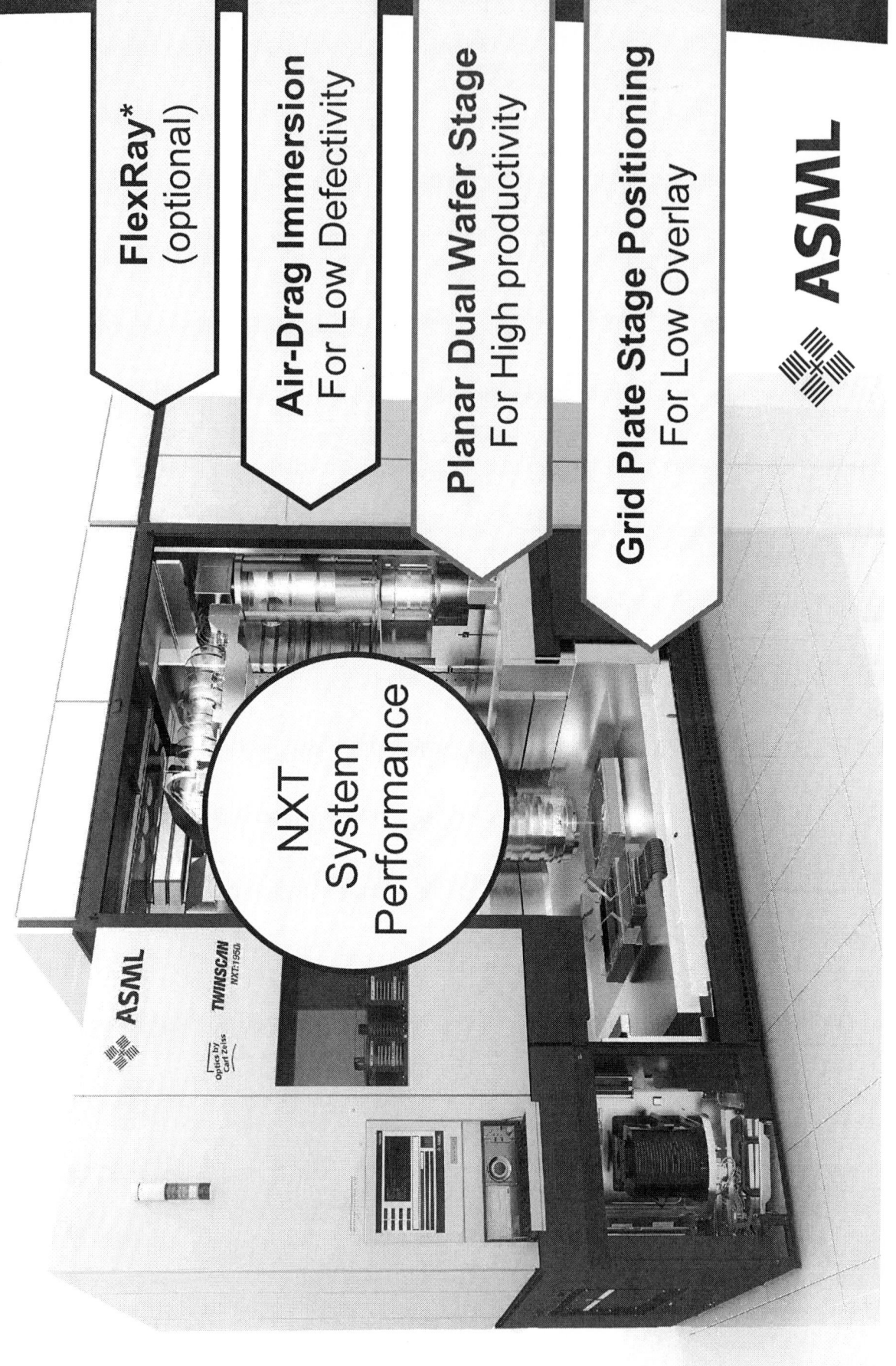

FlexRay* (optional)

Air-Drag Immersion For Low Defectivity

Planar Dual Wafer Stage For High productivity

Grid Plate Stage Positioning For Low Overlay

NXT System Performance

ASML

TWINSCAN NXT:1950i

Optics by Carl Zeiss

NXT:1950i system enables overlay required for double patterning

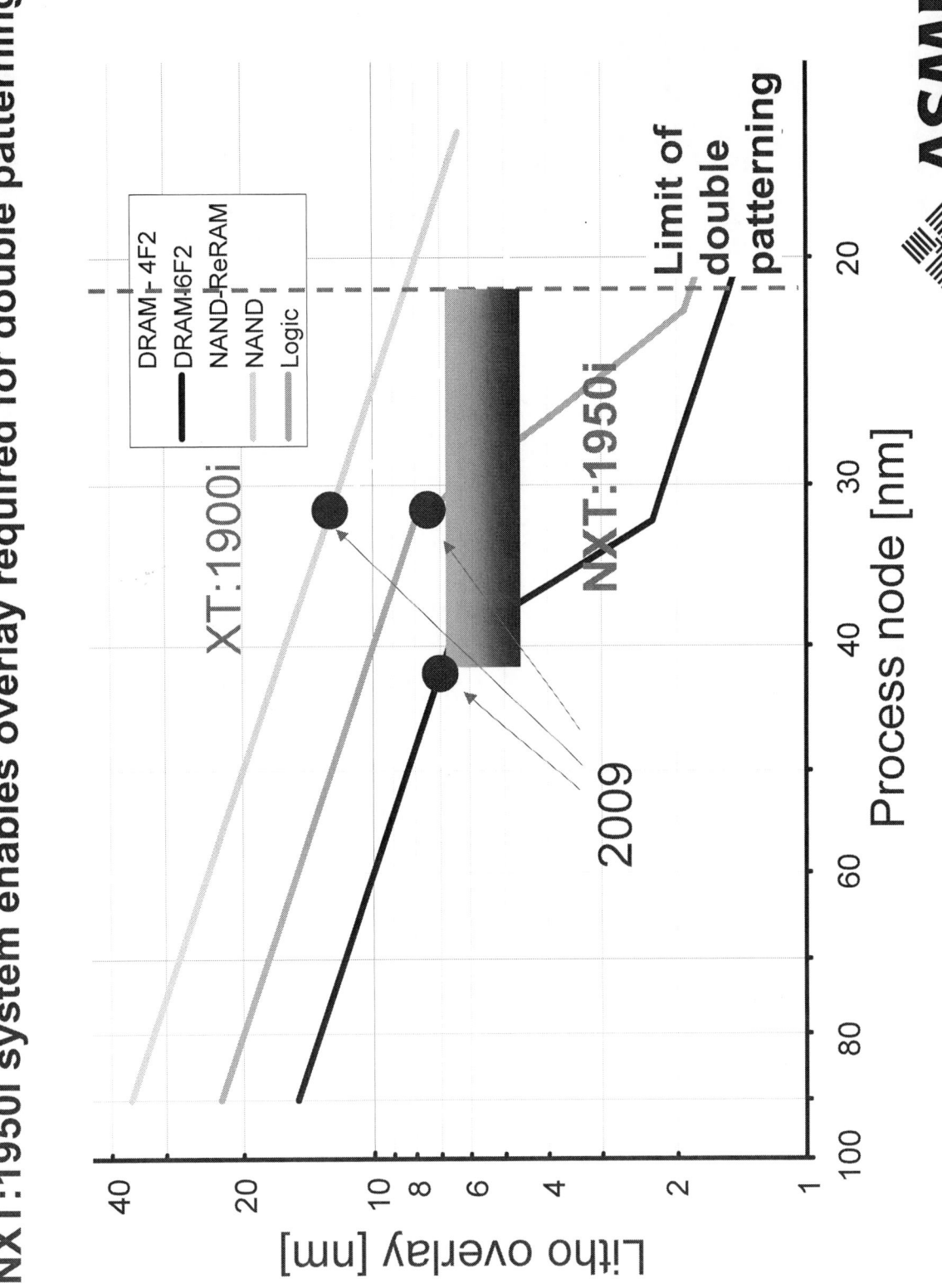

6th International Symposium on Immersion Lithography Extensions

Superior accuracy by grid plate measurement system

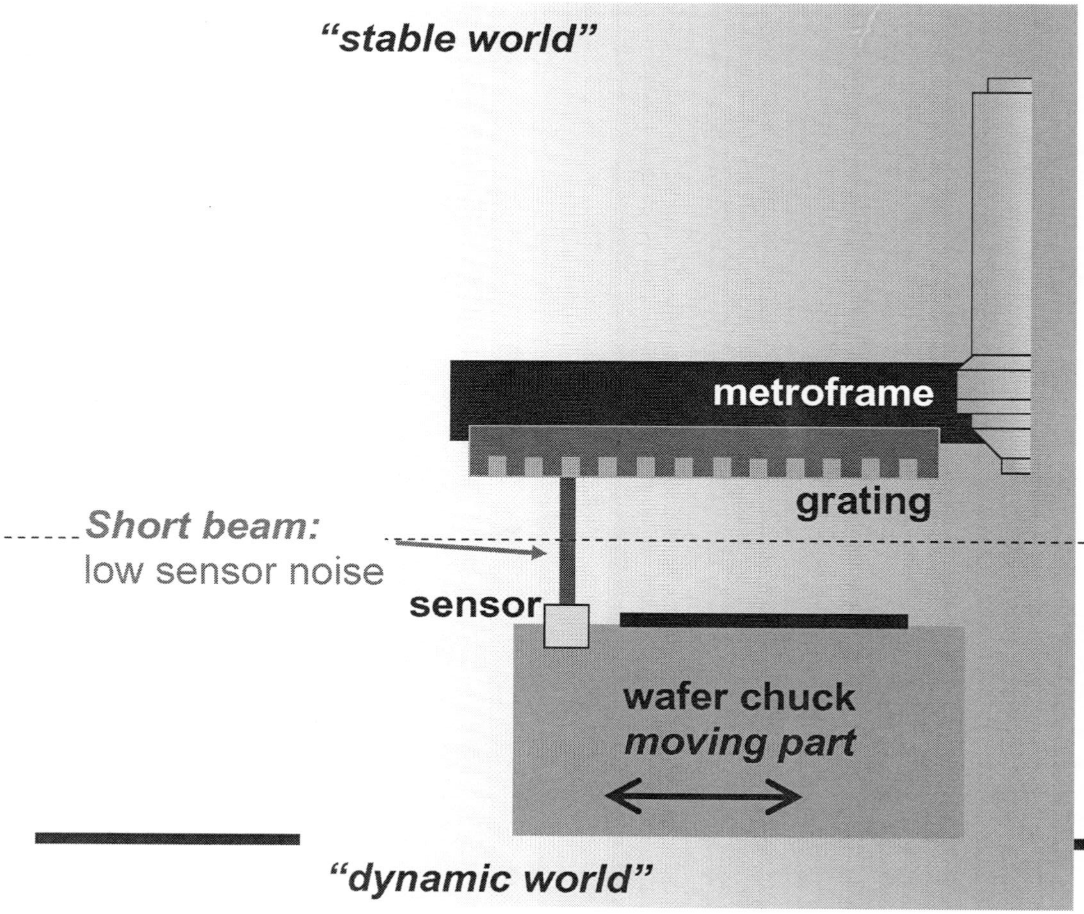

"stable world"

Design for :
maximum stability

Low expansion materials
Temperature conditioning

Short beam:
low sensor noise

metroframe

grating

sensor

wafer chuck
moving part

"dynamic world"

Design for :
fast in line correction

Correct sensor drift every
wafer (using redundancy in
only 4 sensors)

6th International Symposium on Immersion Lithography Extensions

Interferometer versus encoder

- noise.....

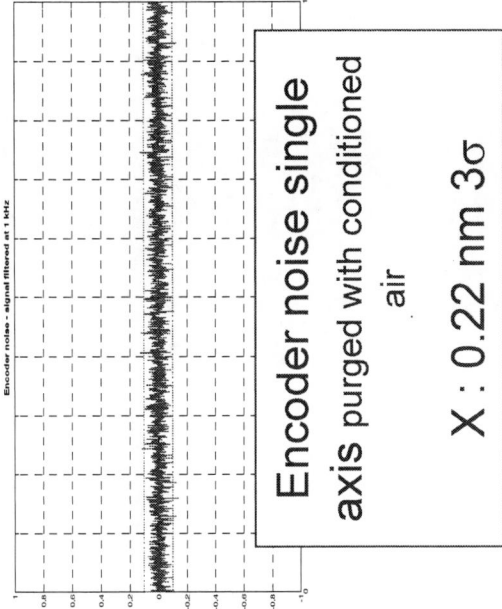

Encoder noise single axis purged with conditioned air

X : 0.22 nm 3σ

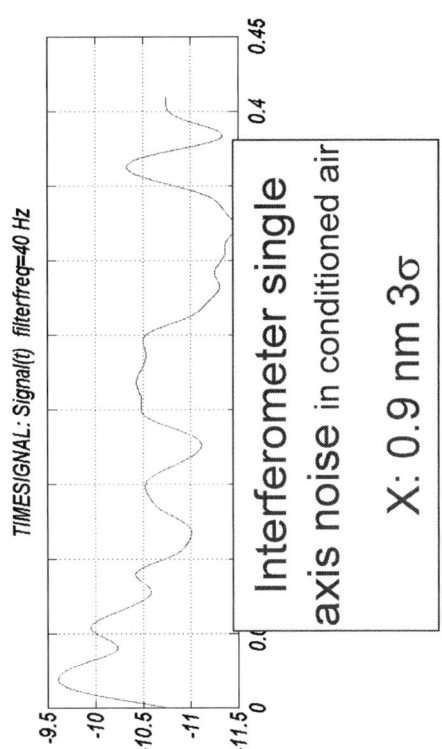

TIMESIGNAL: Signal(t) filterfreq=40 Hz

Interferometer single axis noise in conditioned air

X: 0.9 nm 3σ

- thermal sensitivity in horizontal direction (large range) :

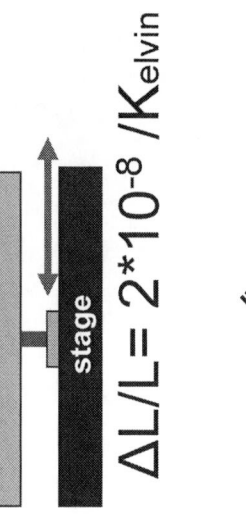

$\Delta L/L = 2*10^{-8}$ /K_{elvin}

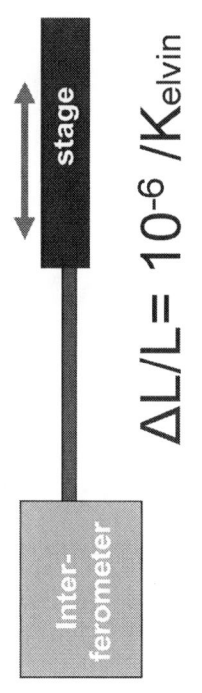

$\Delta L/L = 10^{-6}$ /K_{elvin}

ASML

6th International Symposium on Immersion Lithography Extensions

Lower & uniform alignment repro for NXT
Measurement on NXT is insensitive to stage position

interferometers

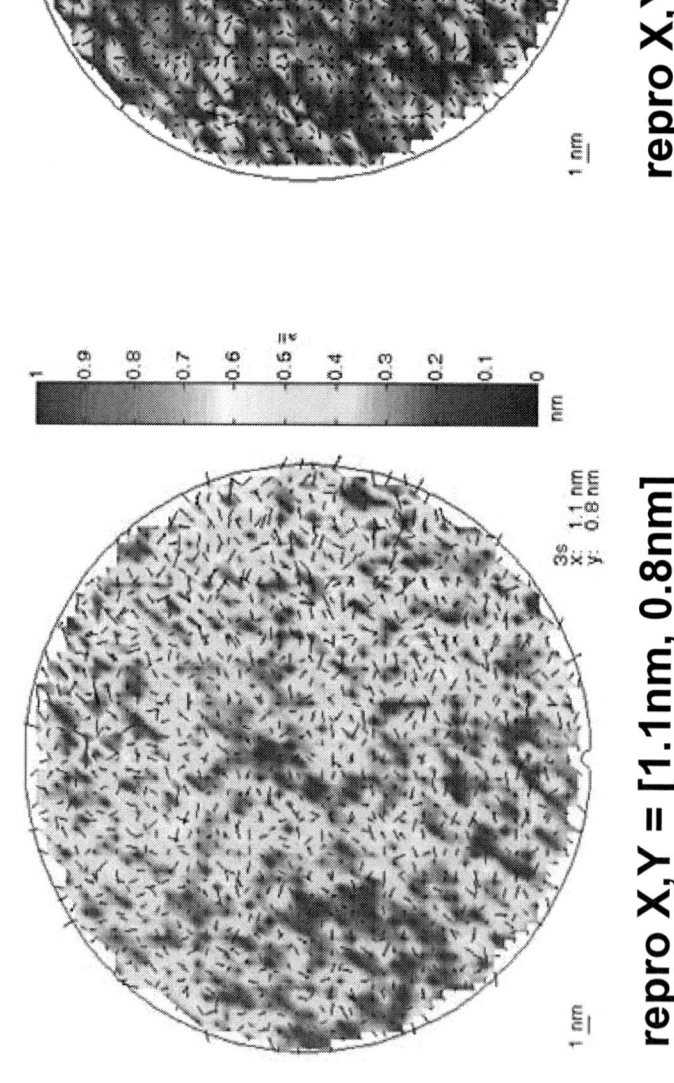

repro X,Y = [1.1nm, 0.8nm]

grid plate measurement

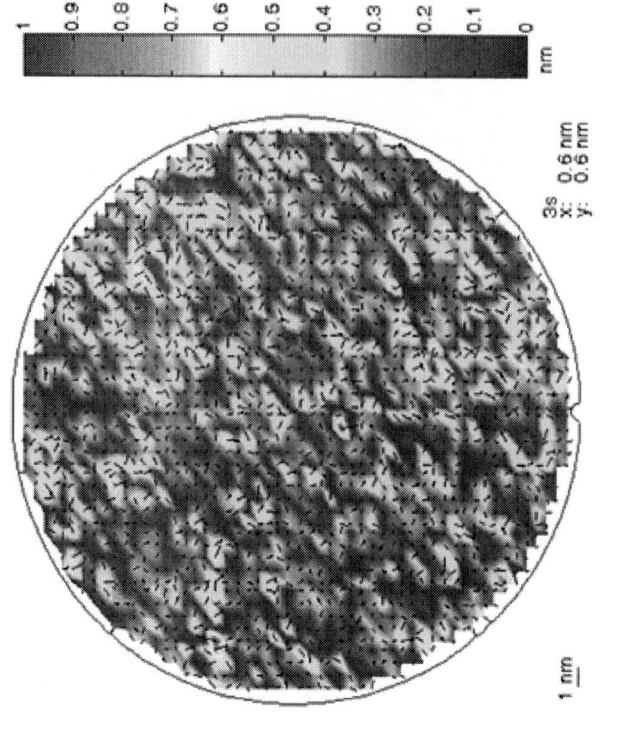

repro X,Y = [0.6nm, 0.6nm]

measured using zero shift test

ASML

6th International Symposium on Immersion Lithography Extensions

Demonstrated Grid Stability < 0.25 nm/week

Measurement period one week (156 hours)

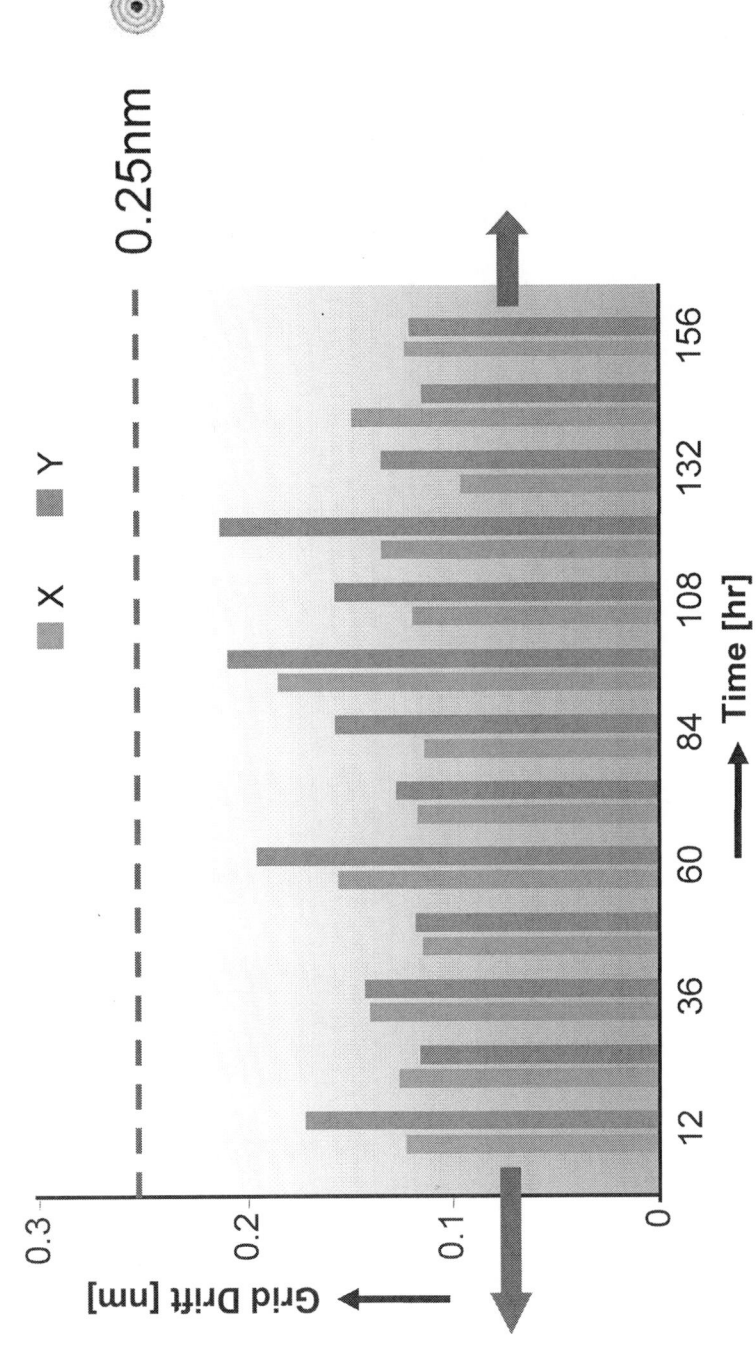

continuous wafer read out with one wafer on chuck

6th International Symposium on Immersion Lithography Extensions

Grid Plate Encoder Design

- Encoders reduce positioning noise with a factor 4

- Positioning system divided in stable and dynamic world

  - Stable world shown to be stable over one week

  - Dynamic world only needs to be stable within life of a wafer

6th International Symposium on Immersion Lithography Extensions

NXT Immersion System for 1.35 NA Optical Extensions

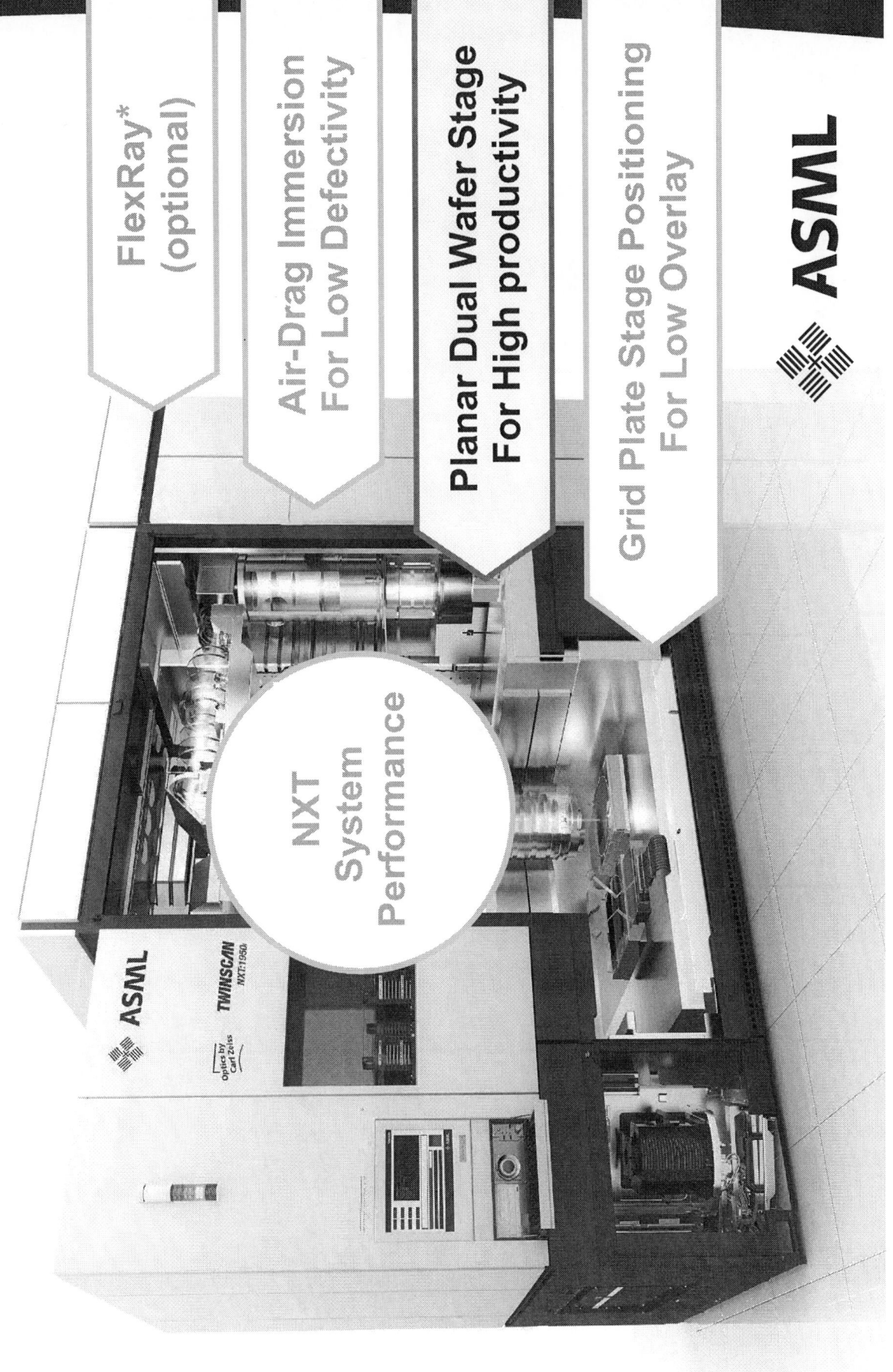

FlexRay* (optional)

Air-Drag Immersion For Low Defectivity

Planar Dual Wafer Stage For High productivity

Grid Plate Stage Positioning For Low Overlay

NXT System Performance

ASML

TWINSCAN NXT:1950

Optics by Carl Zeiss

Dual stages: maximize throughput & measurement time

TWINSCAN delivers both throughput and performance

Squeezed sequential operation

swap | scan time | step time | measure | swap

Swap & Reticle Align times are minimal!

NXT dual stage

metrology operation

wafer metrology focus + align 16 pairs | chuck swap

1. Overlay performance
   - metrology corrects wafer grid

2. High throughput
   - increased acceleration
   - chuck swap without H-bar
   - chuck swap without closing disk

wafer exposure operation

scan time | step time | chuck swap

Time needed to expose a single 300mm wafer ⟶

6th International Symposium on Immersion Lithography Extensions

NXT Immersion System for 1.35 NA Optical Extensions

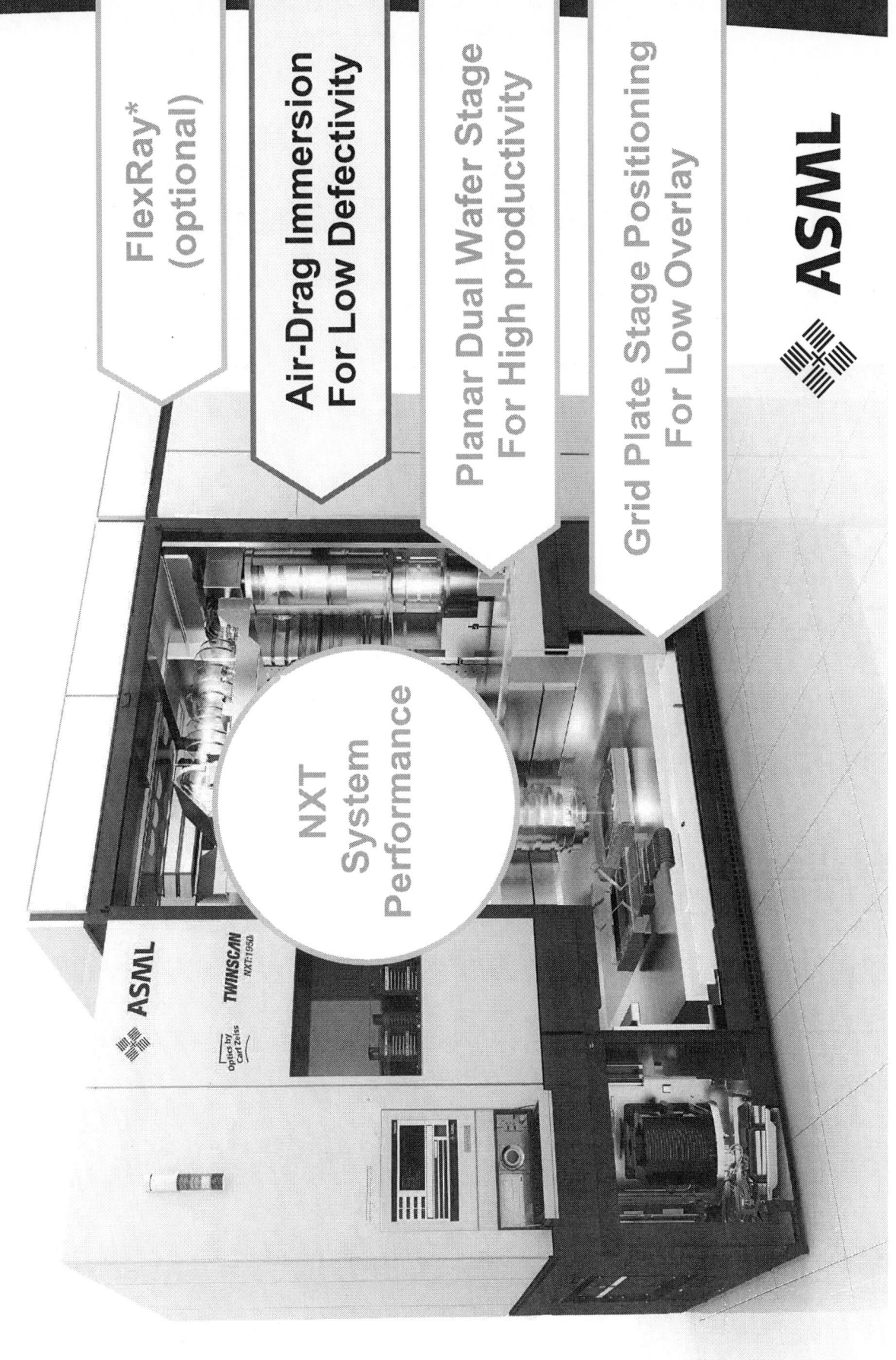

FlexRay* (optional)

Air-Drag Immersion For Low Defectivity

Planar Dual Wafer Stage For High productivity

Grid Plate Stage Positioning For Low Overlay

NXT System Performance

ASML

Patterned Defects Result: 2.2 defects/wafer
→ With new type Immersion Hood

1.2 defect/wafer
Immersion

1 defect/wafer
Printing Particles

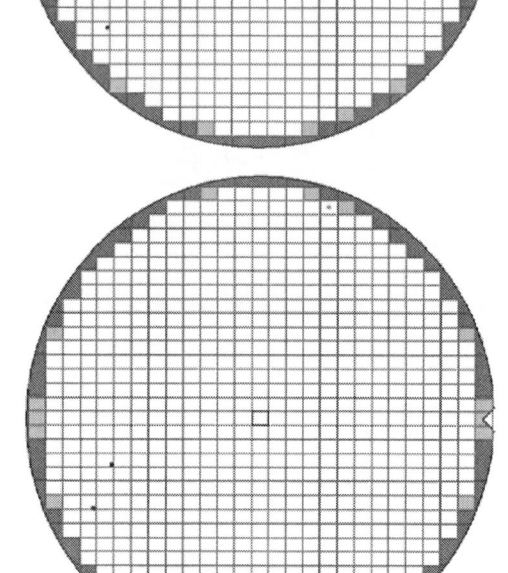

Stacked results of 5 wafers

•Arc29/6001/TCX041

ASML

New type: Air-drag

Water extraction
via 'large holes'

6th International Symposium on Immersion Lithography Extensions

NXT Immersion System for 1.35 NA Optical Extensions

FlexRay* (optional)

Air-Drag Immersion For Low Defectivity

Planar Dual Wafer Stage For High productivity

Grid Plate Stage Positioning For Low Overlay

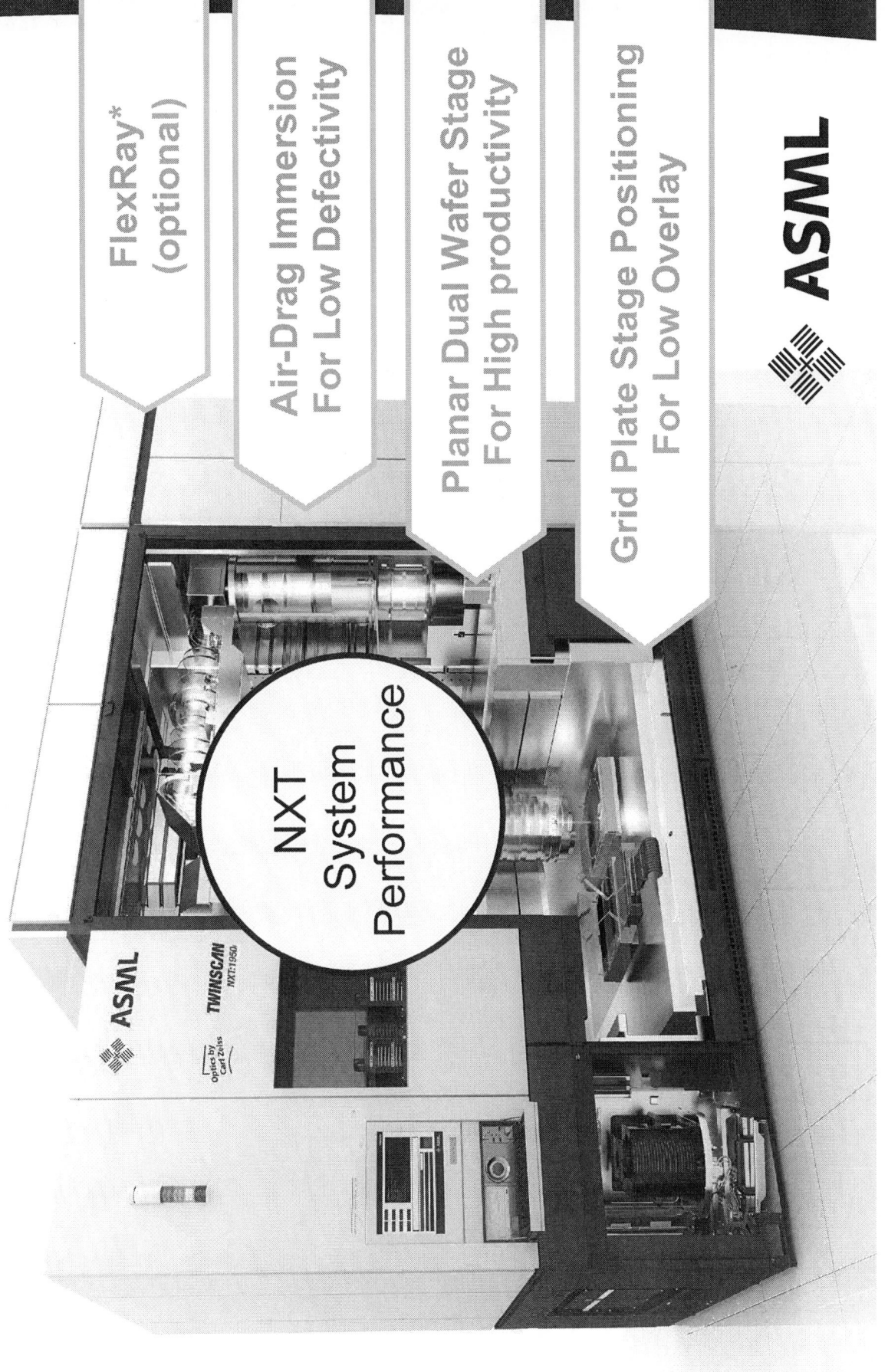

NXT System Performance

ASML

Full field focus uniformity ~ 25nm

Chuck 1

$3\sigma = 25.7$ nm

Chuck 2

$3\sigma = 24.2$ nm

6th International Symposium on Immersion Lithography Extensions

CD uniformity for 40-nm lines
Intra-field 0.7 nm (3σ) and intra-field 1.8 nm (3σ)

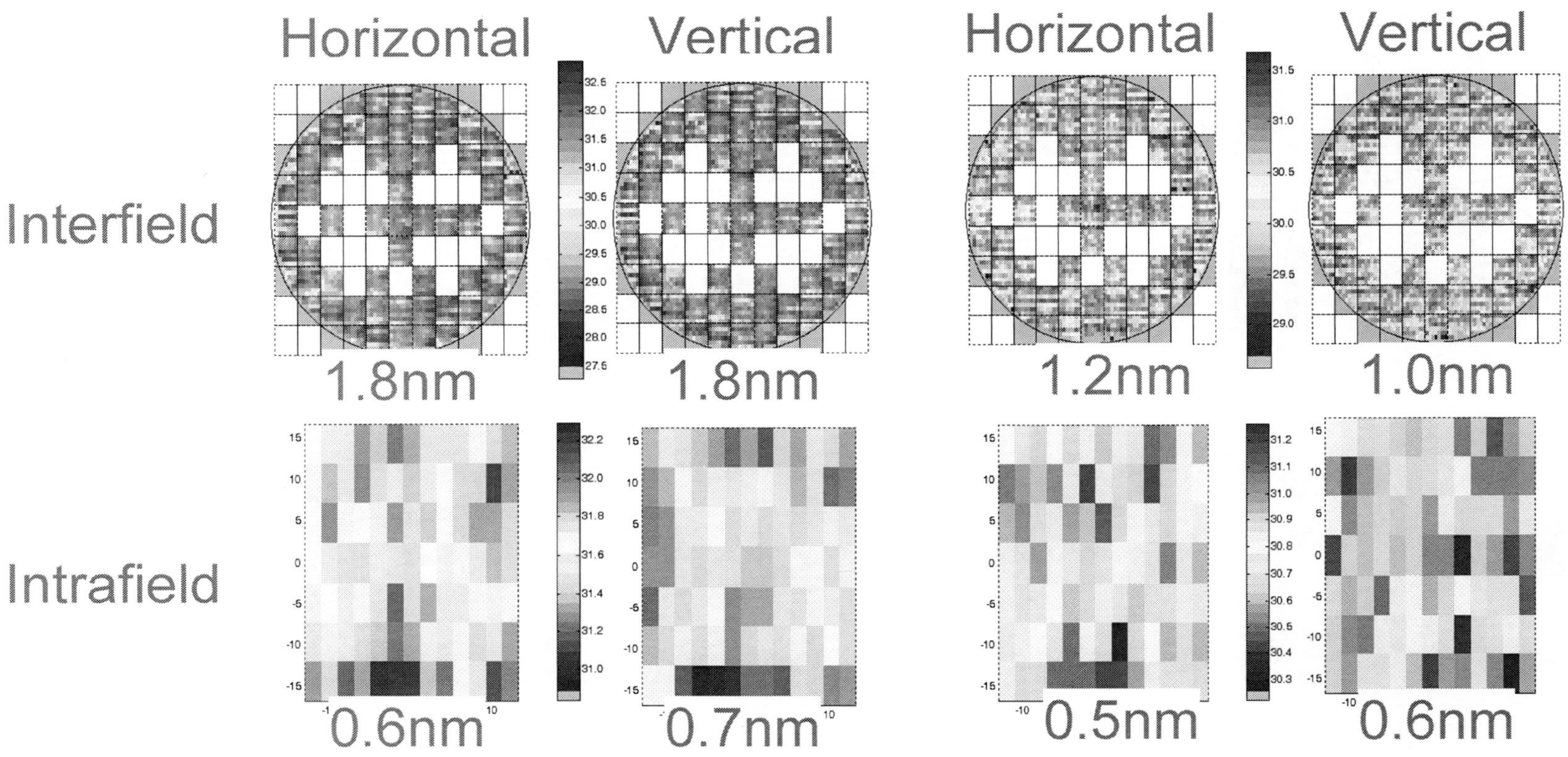

40nm isolated lines

40nm dense lines

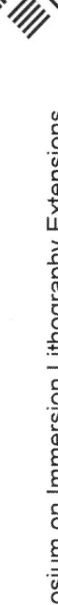

Horizontal   Vertical   Horizontal   Vertical

Interfield

1.8nm   1.8nm   1.2nm   1.0nm

Intrafield

0.6nm   0.7nm   0.5nm   0.6nm

ASML

6th International Symposium on Immersion Lithography Extensions

Lot Overlay 1.7 nm → long term overlay below 2.5 nm

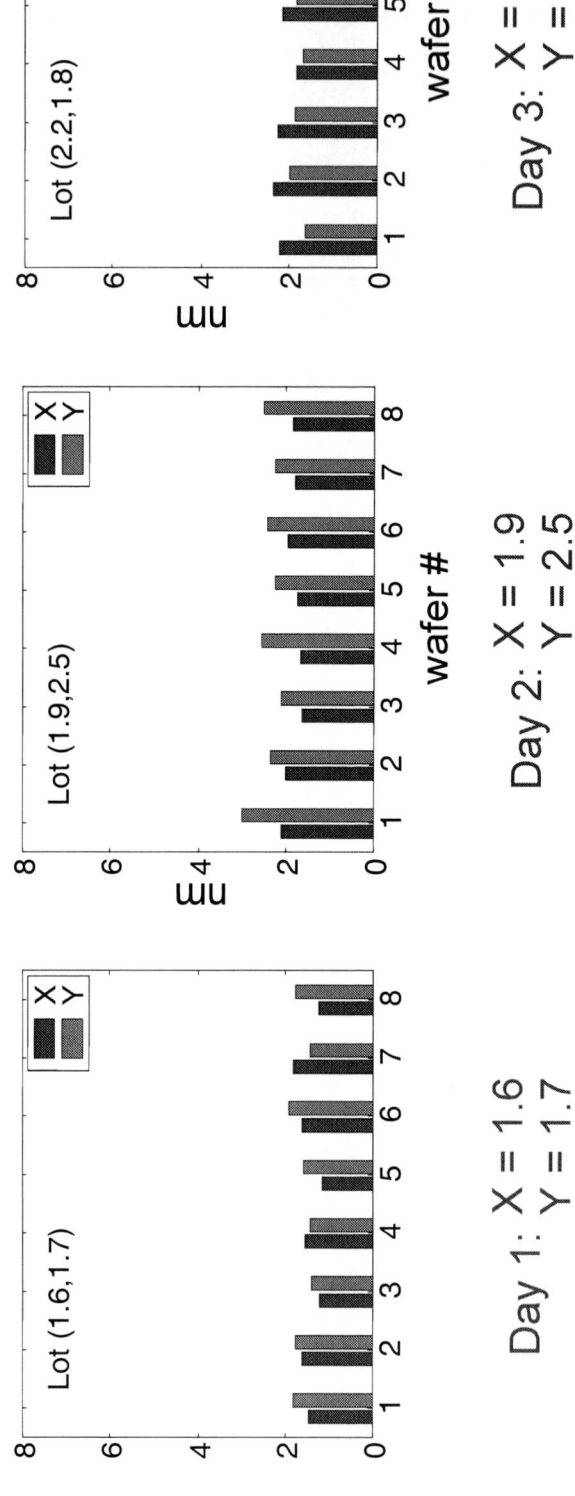

Lot (1.6,1.7)

Day 1: X = 1.6 Y = 1.7

Lot (1.9,2.5)

Day 2: X = 1.9 Y = 2.5

Lot (2.2,1.8)

Day 3: X = 2.2 Y = 1.8

Dedicated Chuck Overlay

6th International Symposium on Immersion Lithography Extensions

Matched Machine Overlay Overlay NXT to XT 5.8nm

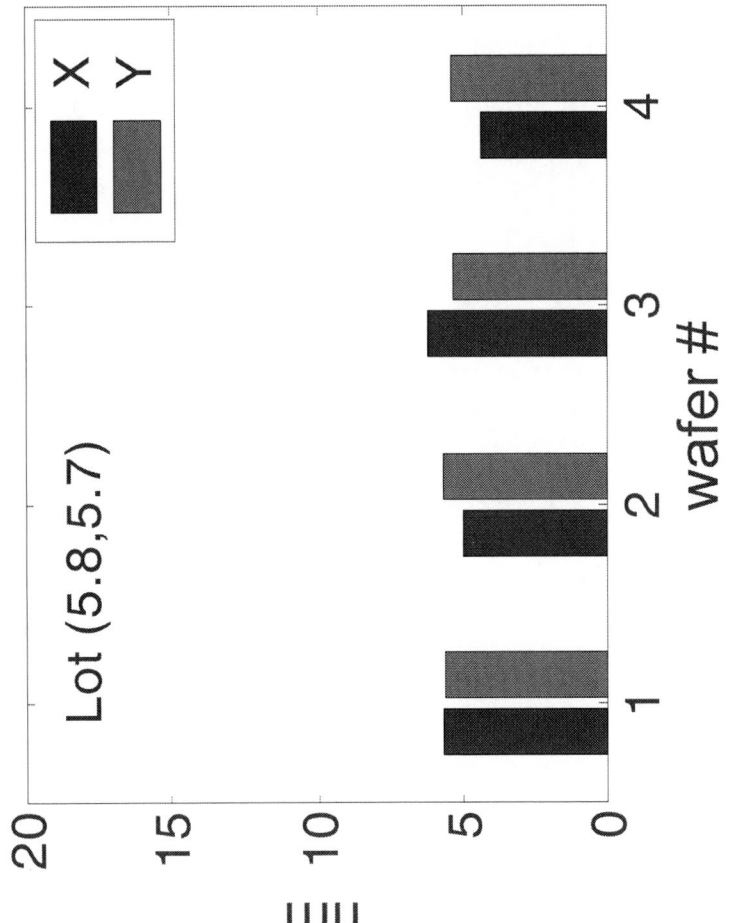

6th International Symposium on Immersion Lithography Extensions

LELE Double Patterning on product stack: 2.7nm

Production-like process control loop:

Intra layer overlay

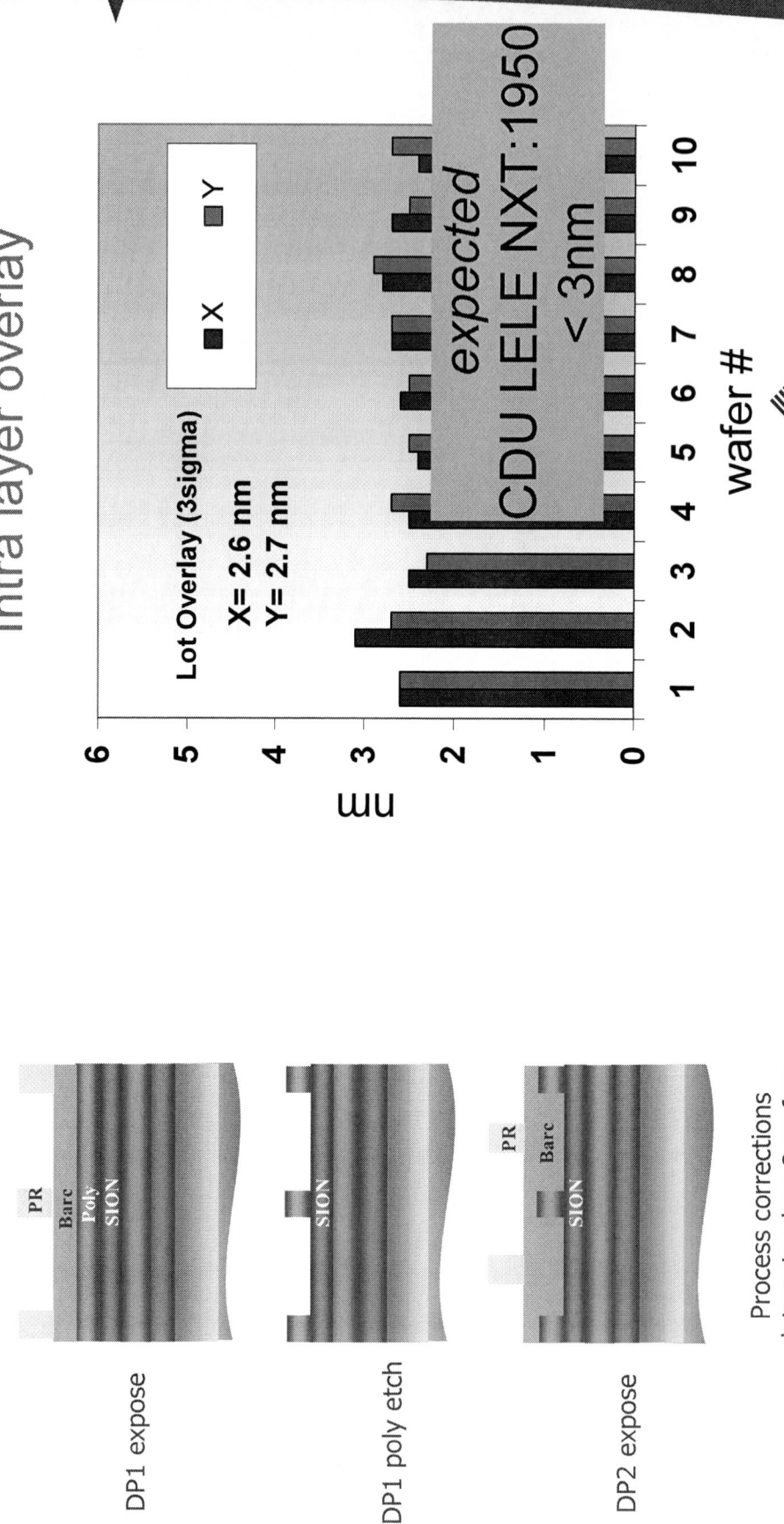

Lot Overlay (3sigma)

X= 2.6 nm
Y= 2.7 nm

■ X   ■ Y

expected
CDU LELE NXT:1950
< 3nm

wafer #

DP1 expose

DP1 poly etch

DP2 expose

Process corrections
determined on 2 wafers

6th International Symposium on Immersion Lithography Extensions

ASML

NXT Immersion System for 1.35 NA Optical Extensions

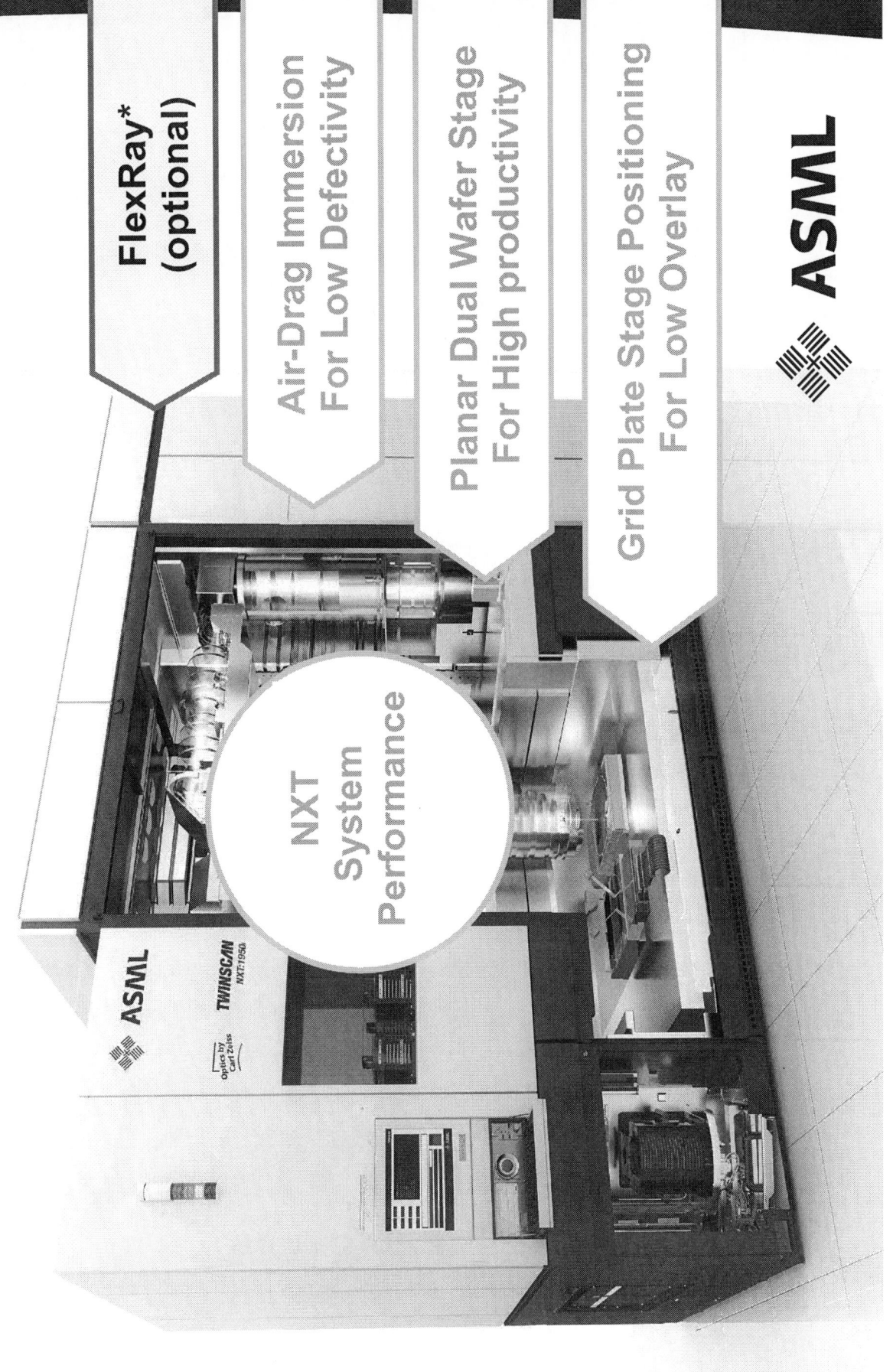

FlexRay* (optional)

Air-Drag Immersion For Low Defectivity

Planar Dual Wafer Stage For High productivity

Grid Plate Stage Positioning For Low Overlay

NXT System Performance

ASML

TWINSCAN NXT:1950i

Optics by Carl Zeiss

ASML

Flexray enables instant freeform illumination

Advantages freeform illumination explained by J. Bekaert yesterday.

A short reminder:

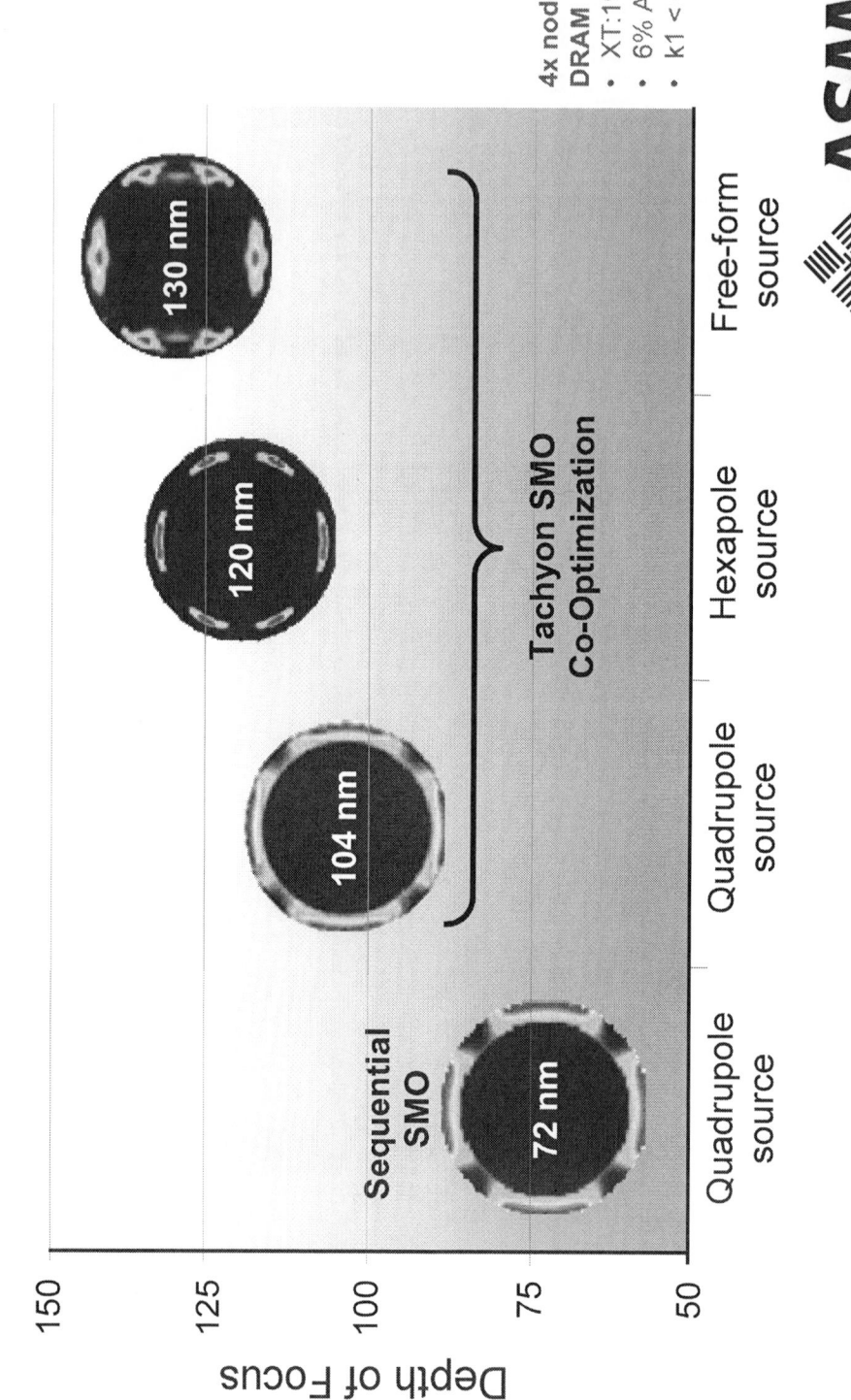

4x node
DRAM Contact
- XT:1900i, 1.35NA
- 6% AttPSM
- k1 < 0.35

6th International Symposium on Immersion Lithography Extensions

Imaging results in XT:1950i Scanner
→ 22nm SRAM Metal with FlexRay Pupil   *(J. Bekaert et al.)*

Target

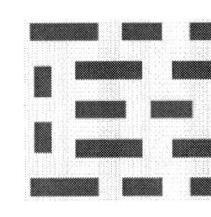

pupil

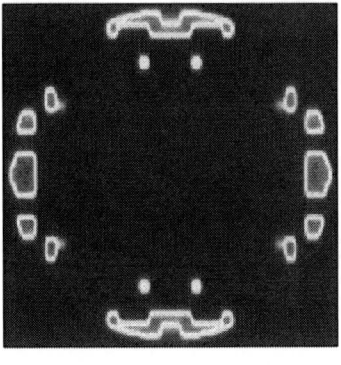

Result on wafer
With DOE

Result on wafer
with flexray

A single exposure with:
- smallest pitch 90 nm
- smallest feature width 40 nm

6th International Symposium on Immersion Lithography Extensions

Conclusions and Outlook

- ## Now: NXT:1950i immersion system is alive and shipped

  - Overlay of 1.7-nm Demonstrated

    NXT:1950i
    Overlay 2,5 nm
    Throughput 175-200 WPH

    - Long term overlay stability below 2.5-nm
    - On product LELE overlay 2.7nm

  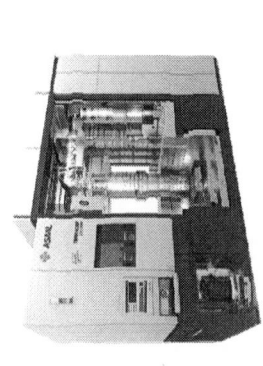

  - Full wafer CDU for 40-nm lines is 1.5-nm

    - Intra-field CDU is 0.7-nm

  - Defects immersion + particles is below 3/wafer

Up next: FlexRay
Any pupil - any time

6th International Symposium on Immersion Lithography Extensions

Thank You

Thanks to FlexRay teams

... and thanks to NXT Project teams

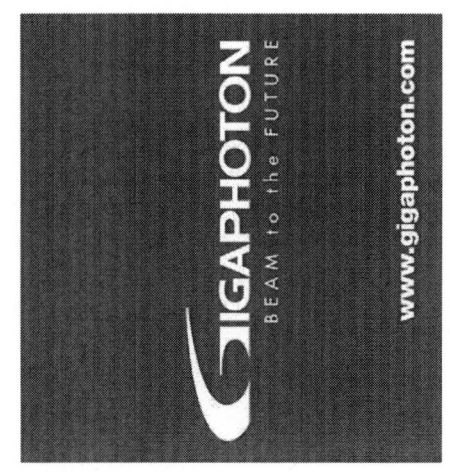

Reliability report of high power injection lock laser light source for double exposure and double patterning ArF immersion lithography

H. Tanaka, H. Tsushima, M. Yoshino, T. Kumazaki, H. Watanabe, S. Matsumoto, H. Umeda, Y. Kawasuji, T. Suzuki, S. Tanaka, A. Kurosu, T. Matsunaga, J. Fujimoto, and H. Mizoguchi

Gigaphoton Inc.

BEAM to the FUTURE

www.gigaphoton.com

Contents

➤ **INTRODUCTION**
- ✓ ArF Roadmap
- ✓ GigaTwin advantage
- ✓ ArF Specifications

➤ **FEATURES OF GT62A-1S xE**
- ✓ Approach to the advanced exposure technology
- ✓ Other Features

➤ **CONCLUSION**

GIGAPHOTON
BEAM to the FUTURE

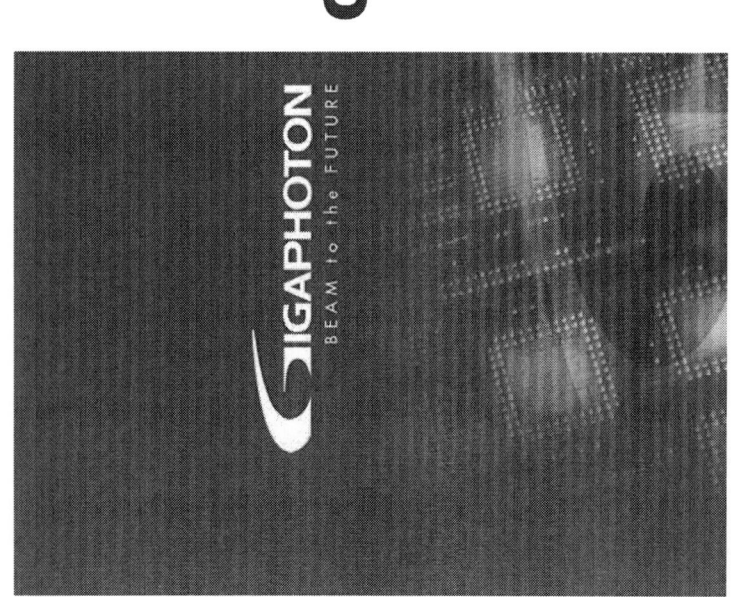

Contents

⋏ **INTRODUCTION**
- ✓ **ArF Roadmap**
- ✓ **GigaTwin advantage**
- ✓ **ArF Specifications**

⋏ **FEATURES OF GT62A-1S xE**
- ✓ **Approach to the advanced exposure technology**
- ✓ **Other Features**

⋏ **CONCLUSION**

INTRODUCTION

➤ 193nm ArF light sources are widely used in semiconductor mass production from the 90 nm node and beyond.

➤ The ArF immersion technology is even spotlighted as the enabling technology for the 45nm node and beyond.

➤ Beyond that, double patterning is considered to be most promising technology to meet the requirement of the next generation 32nm node.

➤ To achieve this, market demands for ArF light source are getting more severe.

For Example :

Higher throughput / Higher reliability / Less running cost

Immersion Symposium 2009

■ Copyright 2009 GIGAPHOTON INC. all rights reserved.

ArF Roadmap

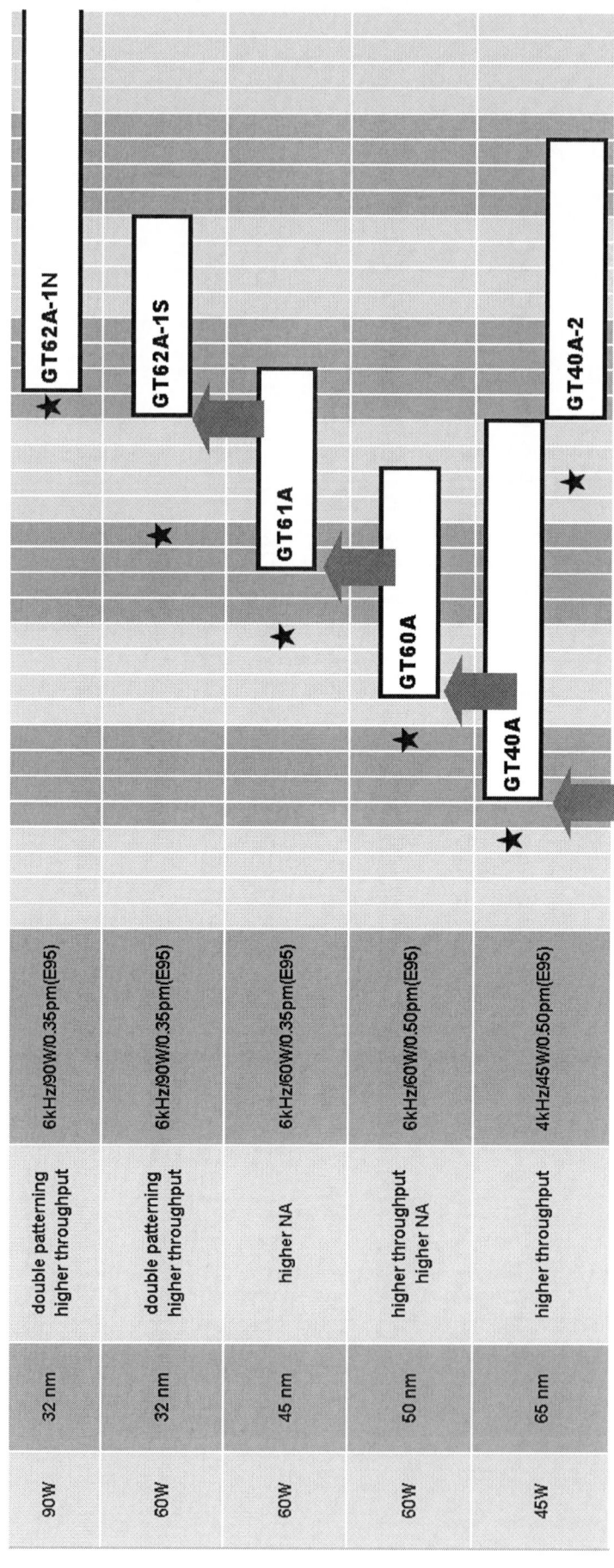

GigaTwin advantage

Flexibility

- Performance improvement
  - ✓ Develop new technology quickly
- Inherited Reliability
  - ✓ Smoothly introduced to a mass-production after installation

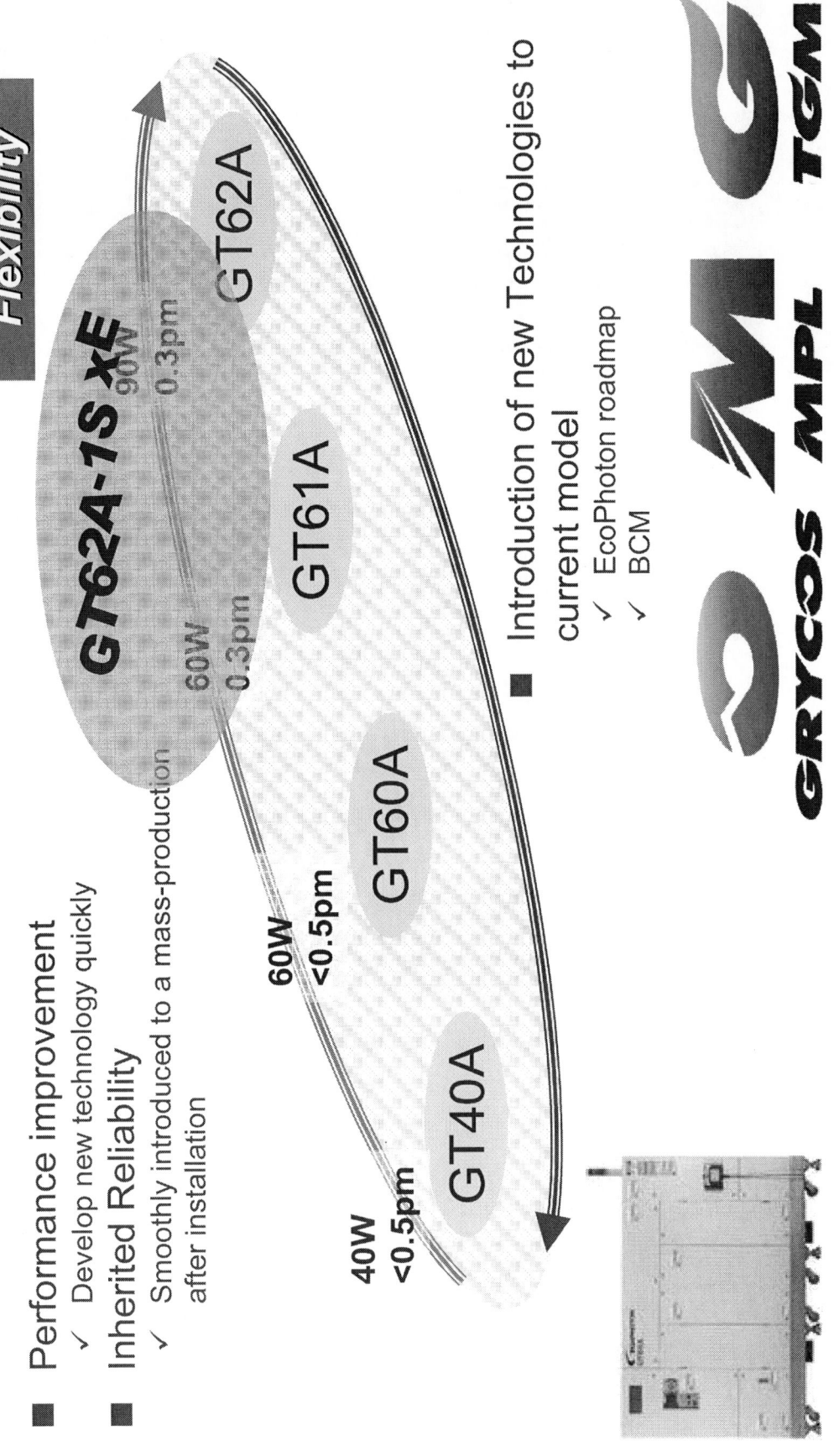

GT62A-1S xE

GT62A
90W
0.3pm

GT61A
60W
0.3pm

GT60A
60W
<0.5pm

GT40A
40W
<0.5pm

- Introduction of new Technologies to current model
  - ✓ EcoPhoton roadmap
  - ✓ BCM

Immersion Symposium 2009

Copyright 2009 GIGAPHOTON INC. all rights reserved.

ArF Specifications

Flexibility

ArF model		GT40A	GT60A	GT61A	GT62A-1S	GT62A-1N
Wavelength	nm	193	193	193	193	193
Power	W	45	60	60	60	90
Pulse energy	mJ	11.25	10	10	10	15
Max. rep rate	Hz	4000	6000	6000	6000	6000
FWHM	pm	0.2	0.2	N.A	N.A	N.A
E95	pm	<0.5	<0.5	0.3	0.3	0.3
Durability (Expected)						
MO Chamber	Bpls	40*	40*	40*	40*	40*
PO Chamber	Bpls	40*	40*	40*	40*	40*
LNM / MO LNM	Bpls	60**	60**	60**	60**	60**
MM	Bpls	30	30	30	30	30
FM / PO FM	Bpls	30	30	30	30	30
PO RM	Bpls	30	30	30	30	30

* GRYCOS technology
** MPL (Multi Positioning LNM)
*** Durability extension @ <90W

GT62A-1S xE is the laser matching the enhancement technology of advanced Exposure Systems. It has the capability of power extension from 60W to 90W.

Immersion Symposium 2009

■ Copyright 2009 GIGAPHOTON INC. all rights reserved.

Contents

GIGAPHOTON BEAM to the FUTURE

➤ **INTRODUCTION**
- ✓ ArF Roadmap
- ✓ GigaTwin advantage
- ✓ ArF Specifications

➤ **FEATURES OF GT62A-1S xE**
- ✓ Approach to the advanced exposure technology
- ✓ Other Features

➤ **CONCLUSION**

FEATURES OF GT62A-1S xE

- ➤ **Approach to the advanced exposure technology**
  - ✓ **Extendable Power**
  - ✓ **Long pulse duration**
- ➤ **Other Features**
  - ✓ **Running cost reduction**
    - ● **Chamber lifetime extension (GRYCOS)**
    - ● **LNM lifetime extension (MPL)**
    - ● **Gas lifetime extension (TGM)**
  - ✓ **Reliability**

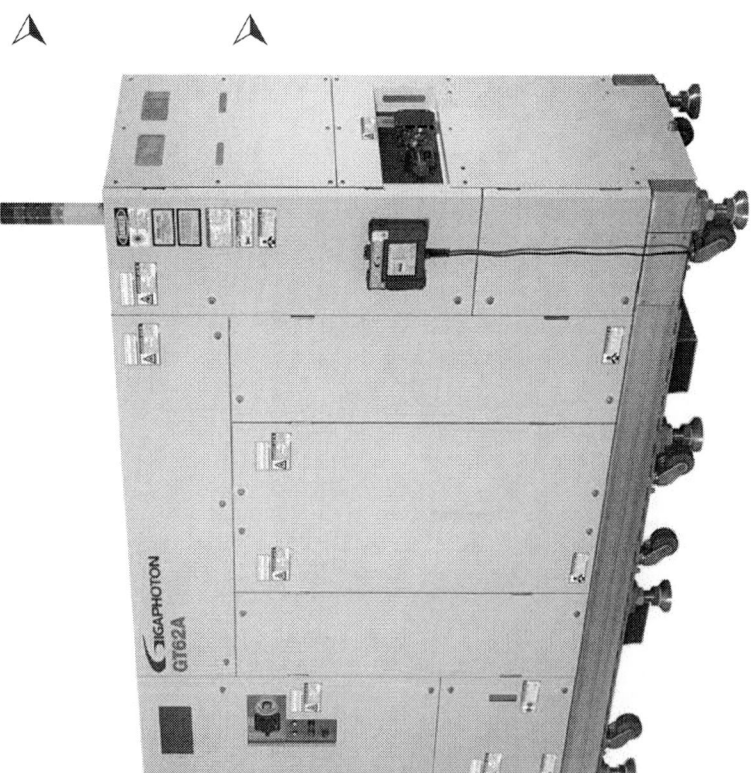

Immersion Symposium 2009

■ Copyright 2009 GIGAPHOTON INC. all rights reserved.

FEATURES OF GT62A-1S xE

⋀ **Approach to the advanced exposure technology**
- ✓ **Extendable Power**
- ✓ **Long pulse duration**

⋀ **Other Features**
- ✓ Running cost reduction
  - ● Chamber lifetime extension (GRYCOS)
  - ● LNM lifetime extension (MPL)
  - ● Gas lifetime extension (TGM)
- ✓ Reliability

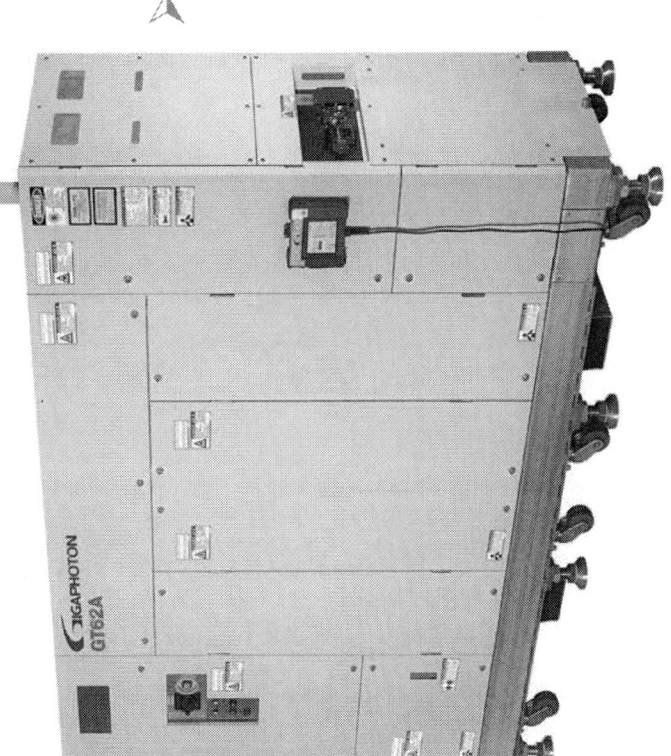

Immersion Symposium 2009

■ Copyright 2009 GIGAPHOTON INC. all rights reserved.

Extendable Power

➤ Illumination Power optimum for Resist Sensitivity is provided.

  ✓ Power extension from 60W to 90W

Feed-back the various technologies of 90W in the 60W laser.

Flexibility of the lithography processes can be increased.

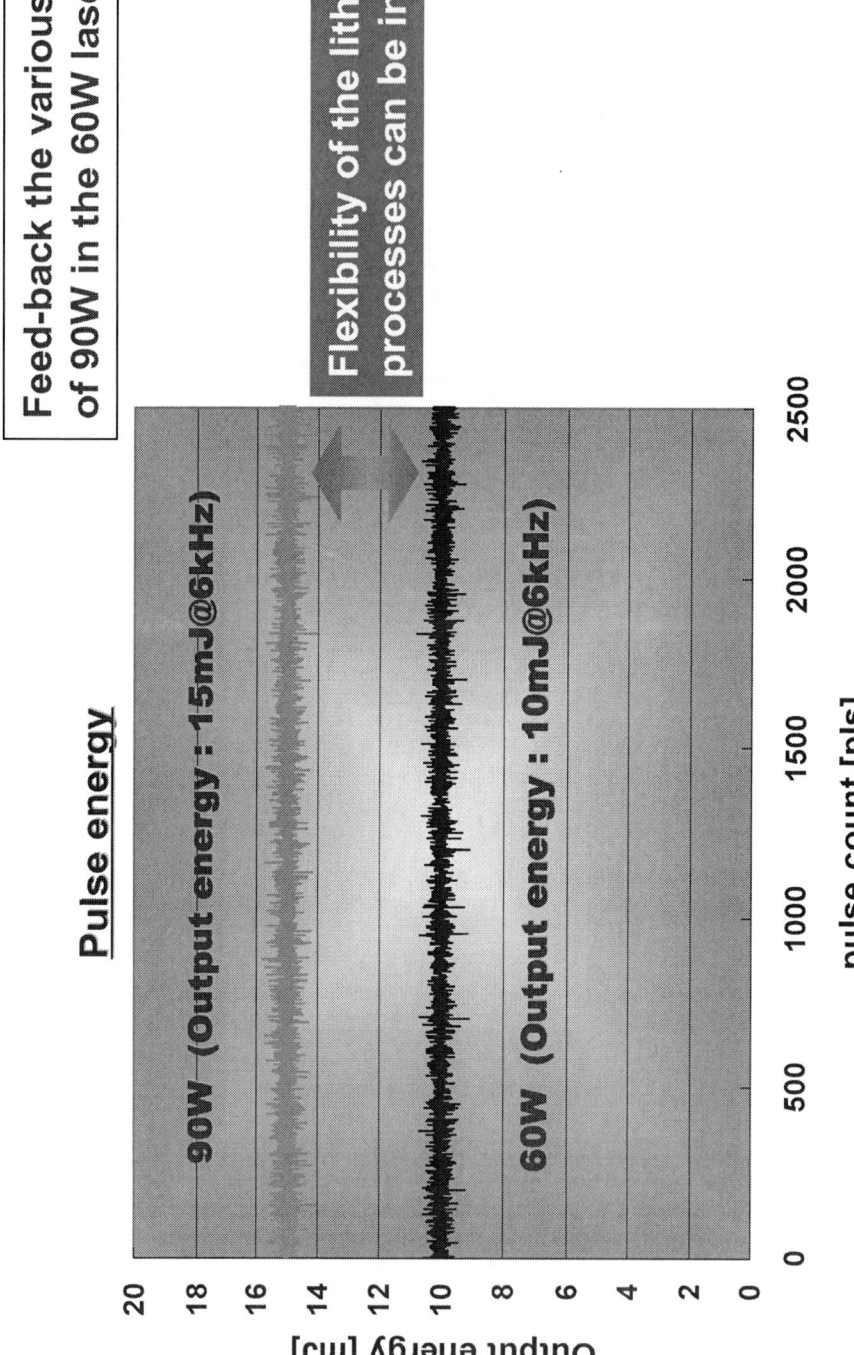

Immersion Symposium 2009

■ Copyright 2009 GIGAPHOTON INC. all rights reserved.

Extendable Power

➤ Beam Quality related to CD variation is kept stable in Power extension.

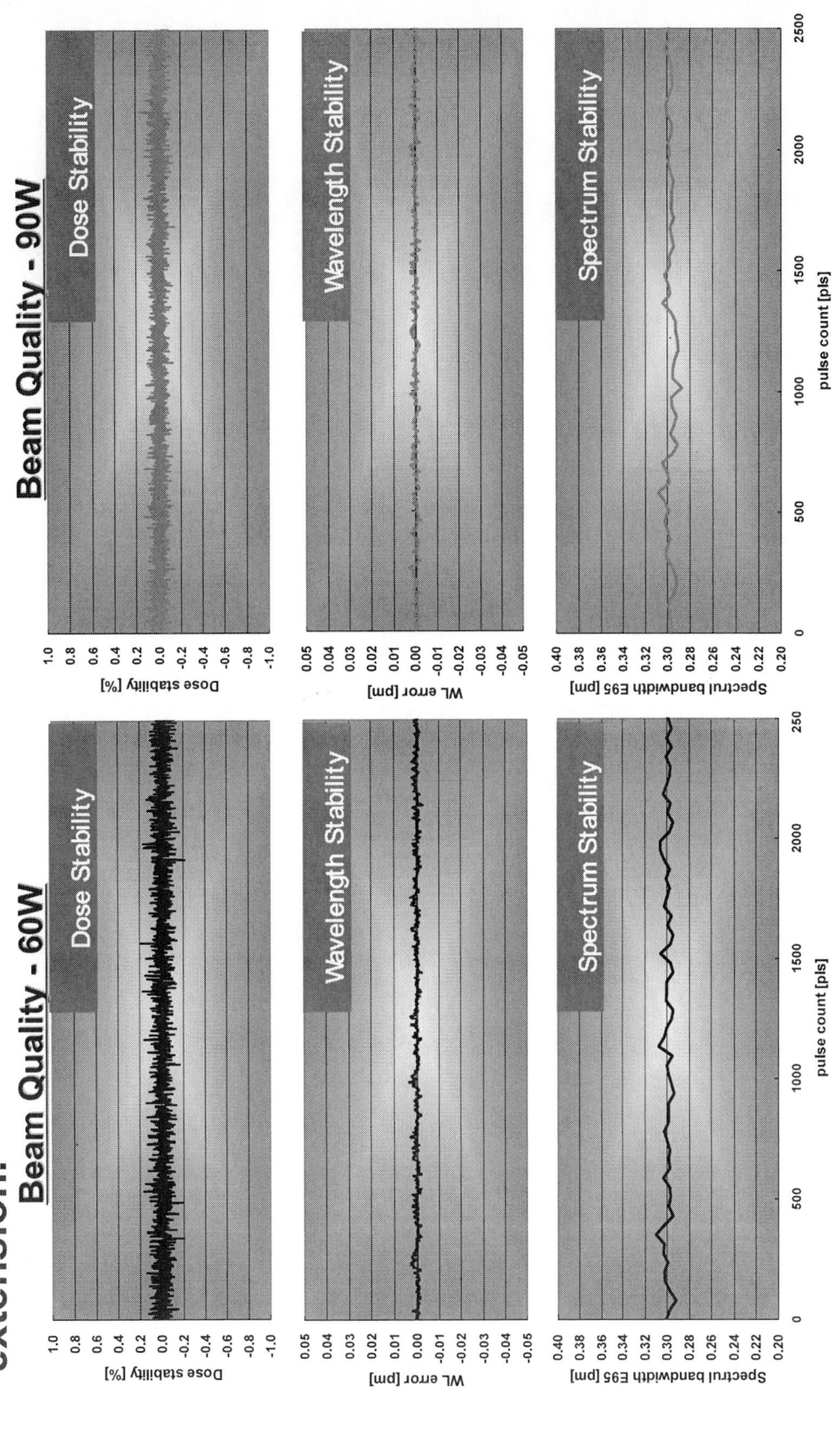

Immersion Symposium 2009

■ Copyright 2009 GIGAPHOTON INC. all rights reserved.

Long pulse duration

➢ GT62A-1S xE contributes optics durability and is able to reduce LER.

- New OPS (Optical Pulse Stretcher) developed for >60W operation
  - 2 stage pulse stretch : Tis = 130ns
- Advantage of New OPS
  - Lower peak power slows down optics deterioration
  - Reduce Spatial / Temporal coherence

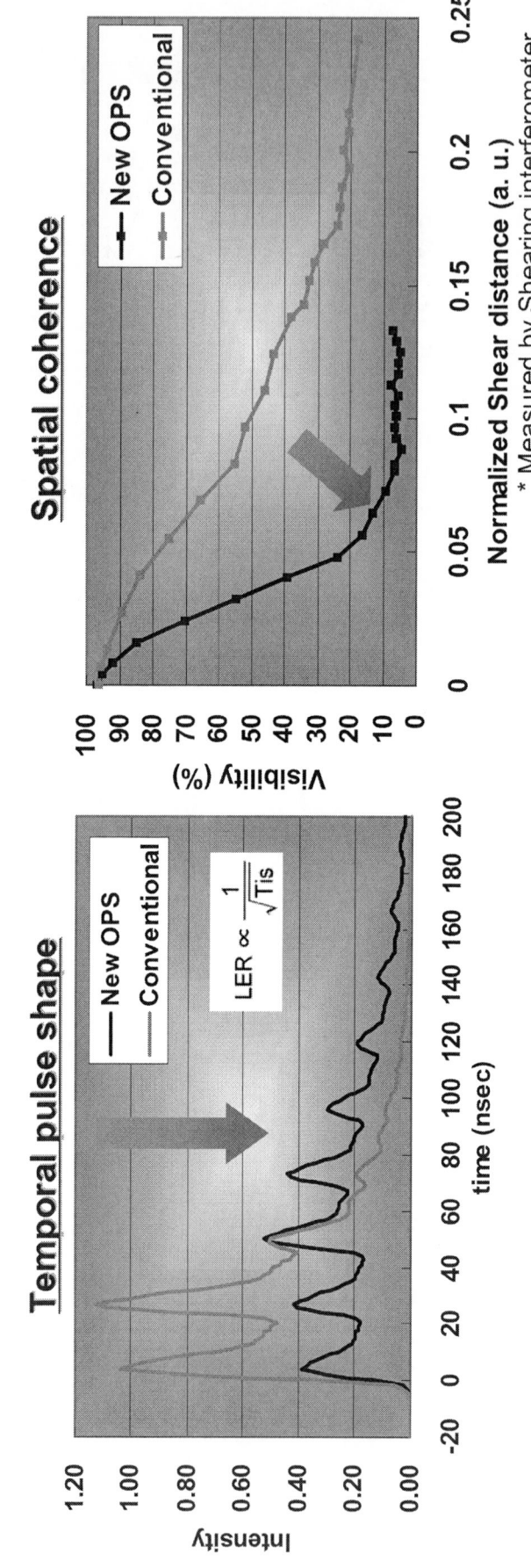

Temporal pulse shape

$$LER \propto \frac{1}{\sqrt{Tis}}$$

Spatial coherence

* Measured by Shearing interferometer
Immersion Symposium 2009

- Copyright 2009 GIGAPHOTON INC. all rights reserved.

FEATURES OF GT62A-1S xE

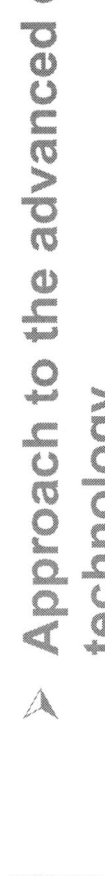

△ Approach to the advanced exposure technology
  - √ Extendable Power
  - √ Long pulse duration

△ **Other Features**
  - √ Running cost reduction
    - Chamber lifetime extension (GRYCOS)
    - LNM lifetime extension (MPL)
    - Gas lifetime extension (TGM)
  - √ Reliability

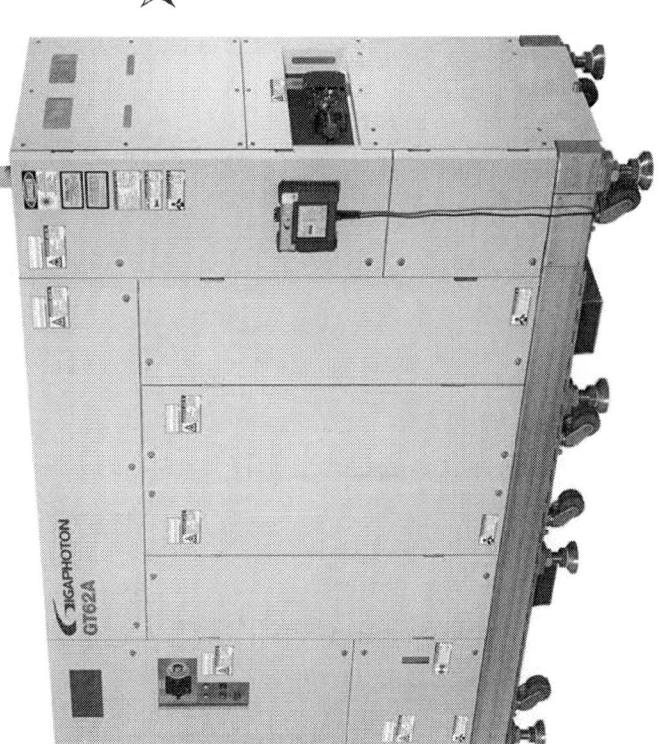

Immersion Symposium 2009

■ Copyright 2009 GIGAPHOTON INC. all rights reserved.

Running cost reduction

➢ Inheriting the GigaTwin platform, GT62A-1S xE features the reduced running costs.

√ Three technologies for running cost reduction are equipped.

Technologies for running cost reduction :

Chamber lifetime extension (Gigaphoton Recycled Chamber Operation System)

LNM lifetime extension (Multi Positioning LNM technology)

Gas lifetime extension (Total Gas Manager)

Immersion Symposium 2009

Copyright 2009 GIGAPHOTON INC. all rights reserved.

Chamber lifetime extension (GRYCOS) :

No impact to Beam Quality

Beam Quality

➢ Each laser chamber can be used up to 40Bpls.

By using a chamber as an oscillator and then an amplifier

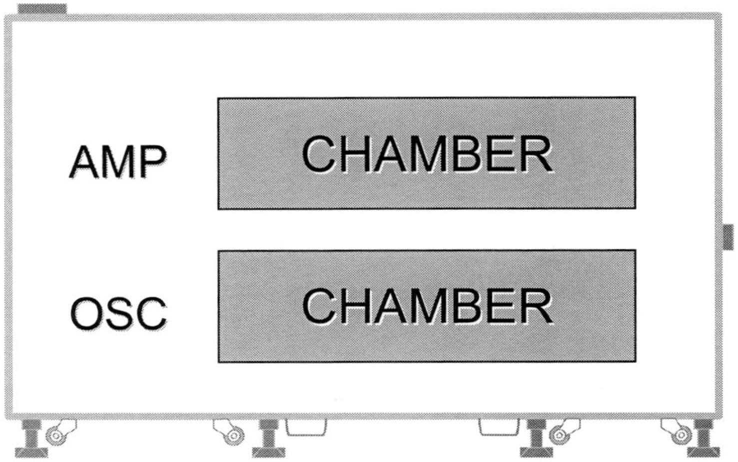

Dose Stability

GRYCOS

Dose stability (%)
1.0 / 0.8 / 0.6 / 0.4 / 0.2 / 0.0 / -0.2 / -0.4 / -0.6 / -0.8 / -1.0

Wavelength Stability

WL error (pm)
0.10 / 0.08 / 0.06 / 0.04 / 0.02 / 0.00 / -0.02 / -0.04 / -0.06 / -0.08 / -0.10

Spectrum Stability

Without E95 control

Spectral Bamdwidth E95 (pm)
1.0 / 0.9 / 0.8 / 0.7 / 0.6 / 0.5 / 0.4 / 0.3 / 0.2 / 0.1 / 0.0

Pulse count [Bpls]
22 23 24 25 26 27 28 29 30

Immersion Symposium 2009

AMP — CHAMBER

OSC — CHAMBER

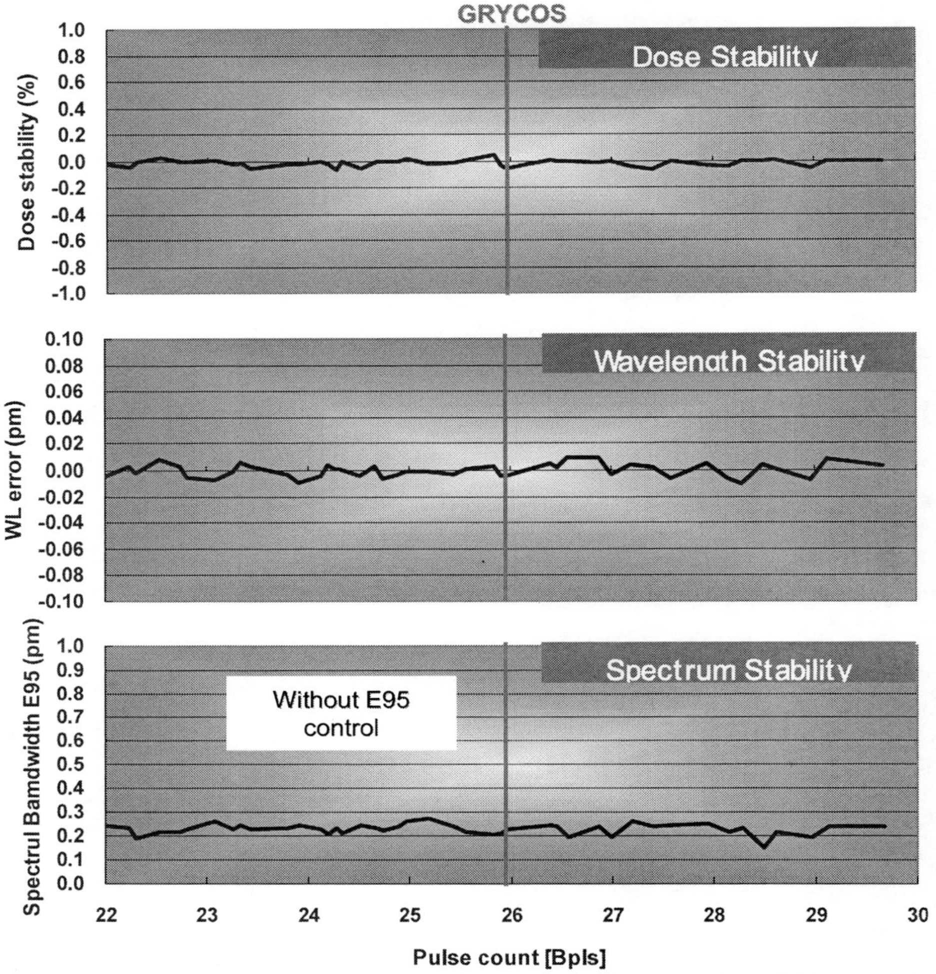

■ Copyright 2009 GIGAPHOTON INC. all rights reserved.

GRYCOS

GIGAPHOTON

LNM lifetime extension (MPL) :

➢ LNM lifetime extends to double (30Bpls ⇒ 60Bpls).

By changing optical path efficiently

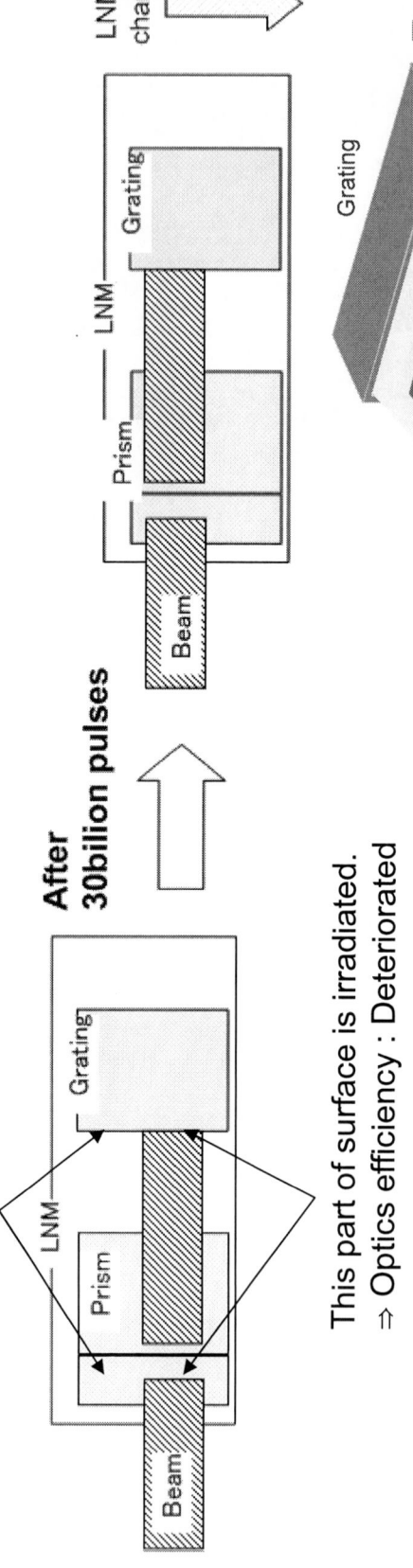

This part of surface is not irradiated.
⇒ Optics efficiency : No deterioration

**After
30billion pulses**

This part of surface is irradiated.
⇒ Optics efficiency : Deteriorated

Immersion Symposium 2009

■ Copyright 2009 GIGAPHOTON INC. all rights reserved.

Gas lifetime extension (TGM) :

➢ Gas refill interval extends remarkably (3days ⇒15days : 24times/year).

By Improving Gas Control
√ Stabilization of fluorine partial pressure
√ Reduction of the amount of impurity

Beam Quality is stable during extended gas lifetime.

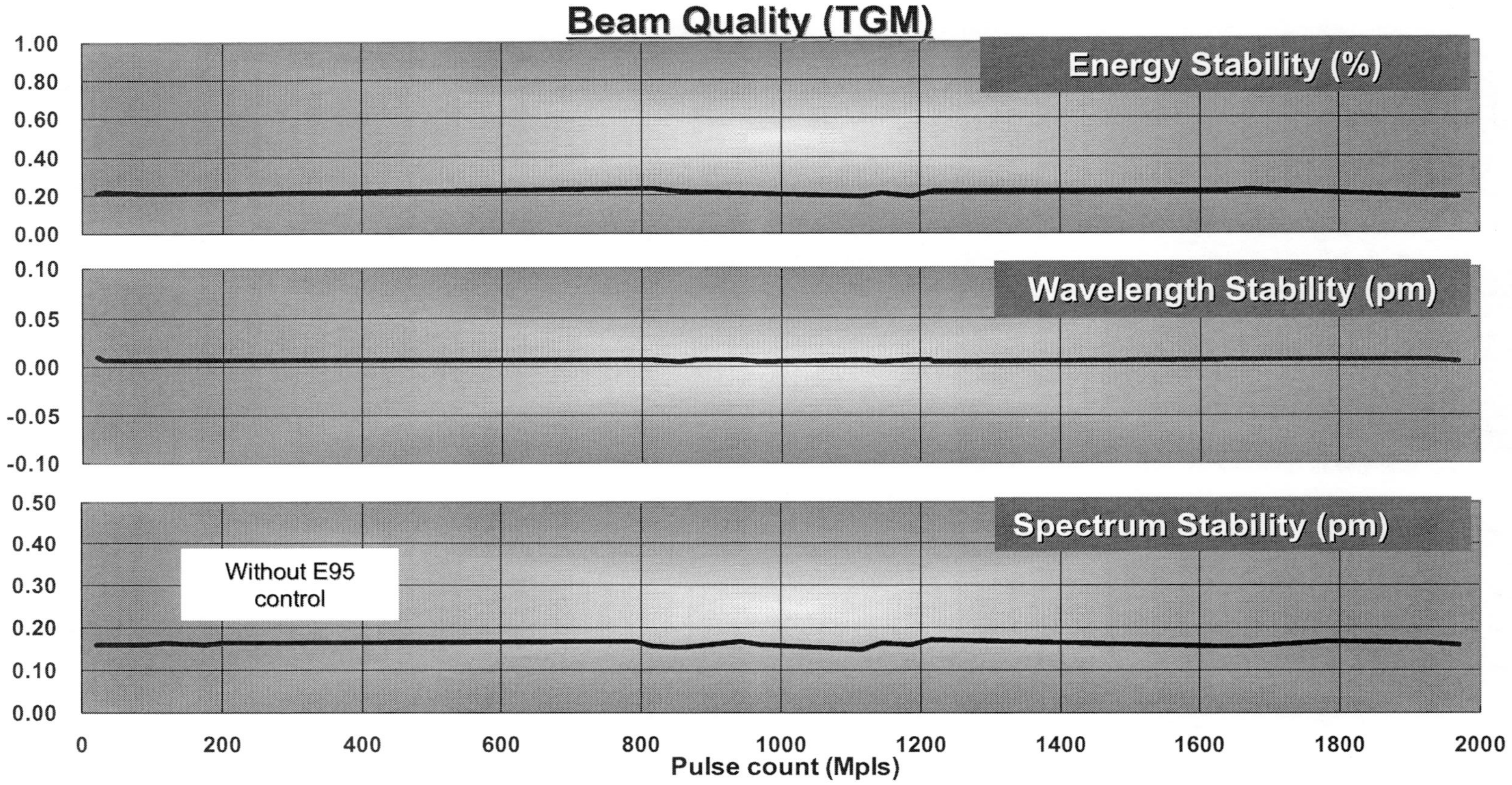

Beam Quality (TGM)

Energy Stability (%)

Wavelength Stability (pm)

Spectrum Stability (pm)

Without E95 control

Pulse count (Mpls)

Immersion Symposium 2009

■ Copyright 2009 GIGAPHOTON INC. all rights reserved.

Reliability

➢ Inheriting the GigaTwin platform, GT62A-1S xE features proven reliability.

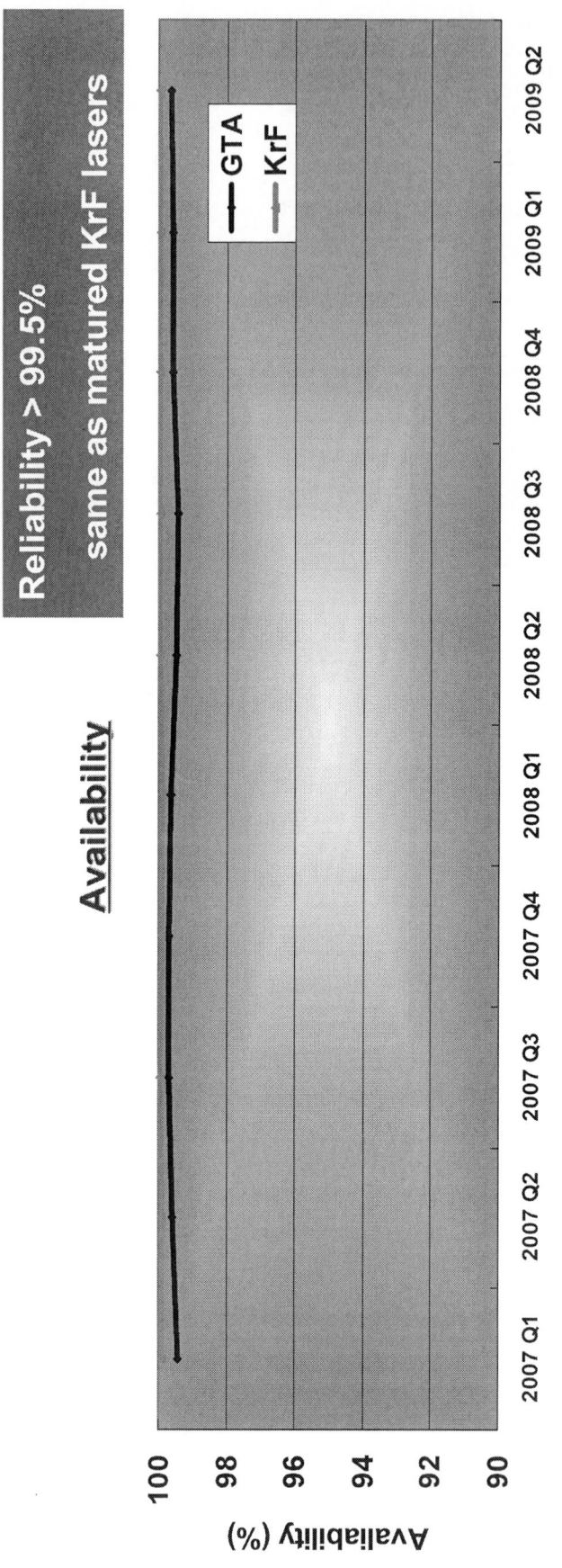

Reliability > 99.5% same as matured KrF lasers

Immersion Symposium 2009

■ Copyright 2009 GIGAPHOTON INC. all rights reserved.

Contents

GIGAPHOTON
BEAM to the FUTURE

➢ **INTRODUCTION**
- ✓ ArF Roadmap
- ✓ GigaTwin advantage
- ✓ ArF Specifications

➢ **FEATURES OF GT62A-1S xE**
- ✓ Approach to the advanced exposure technology
- ✓ Other Features

➢ **CONCLUSION**

CONCLUSION

➤ GT62A-1S xE designed to support the requirement of process parameter flexibility of exposure tool and end customer.

√ Fit to Advanced Exposure Systems like a new illumination system.

√ Provide Illumination Power optimum for Resist Sensitivity

√ Maintain CD variation well in Power extension

√ Contribute to optics durability and is able to reduce LER.

➤ Inheriting the GigaTwin platform, it features the reduced running costs and proven reliability by GRYCOS, MPL, TGM.

Immersion Symposium 2009

■ Copyright 2009 GIGAPHOTON INC. all rights reserved.

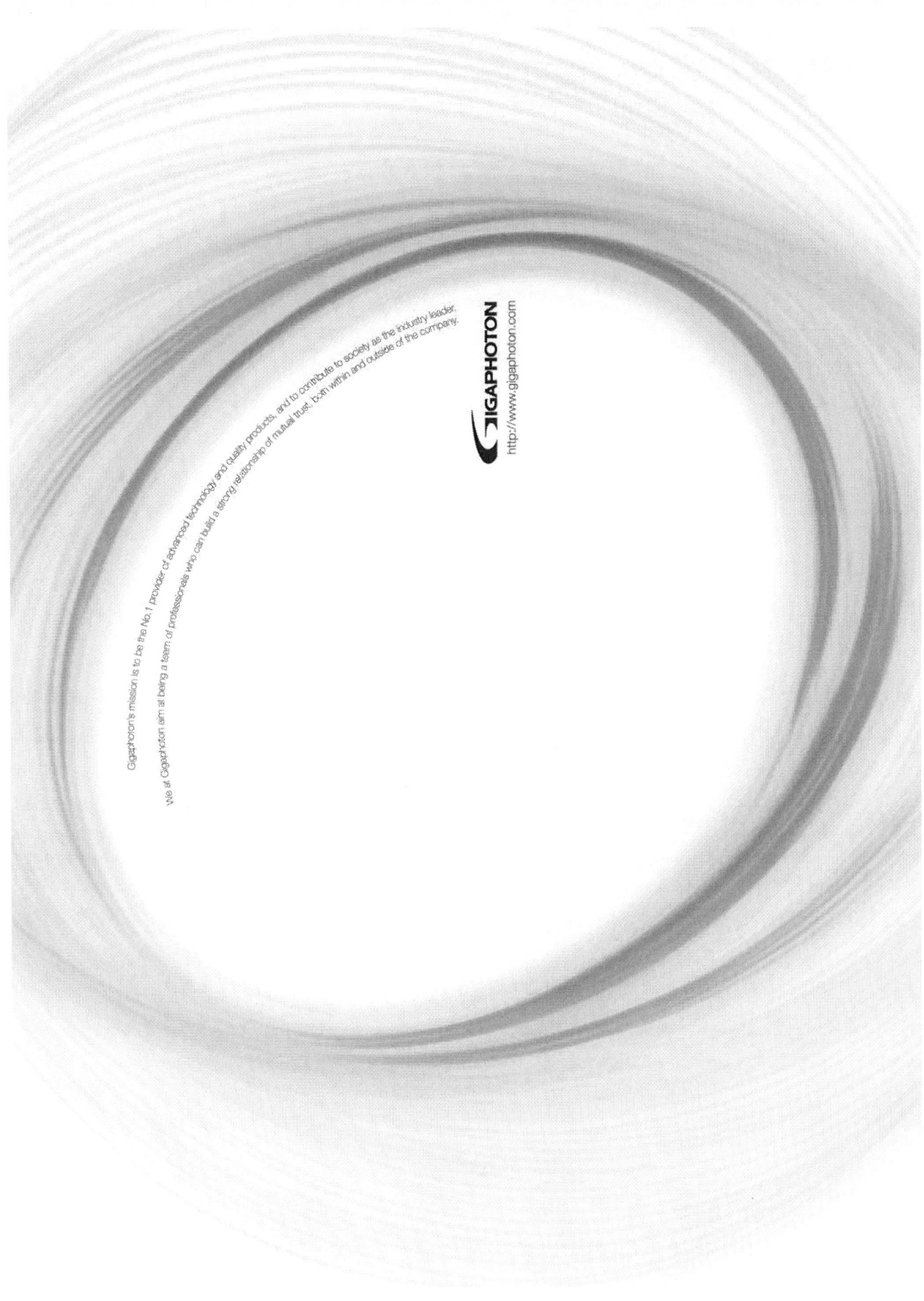

Gigaphoton's mission is to be the No.1 provider of advanced technology and quality products, and to contribute to society as the industry leader.

We at Gigaphoton aim at being a team of professionals who can build a strong relationship of mutual trust, both within and outside of the company.

GIGAPHOTON
http://www.gigaphoton.com

Characterisation of Direct Alignment for LFLE Process

David Laidler, Philippe Leray, Shaunee Cheng (IMEC)

Maya Doytcheva, Manfred Tenner, Richard van Haren (ASML)

Options for DPT Alignment

- For Gate DPT there are two options for aligning DPT2 to DPT1:
  - Option 1: Align DPT2 to DPT1 by indirectly aligning both to STI,
  - Option 2: Align DPT2 directly to DPT1.

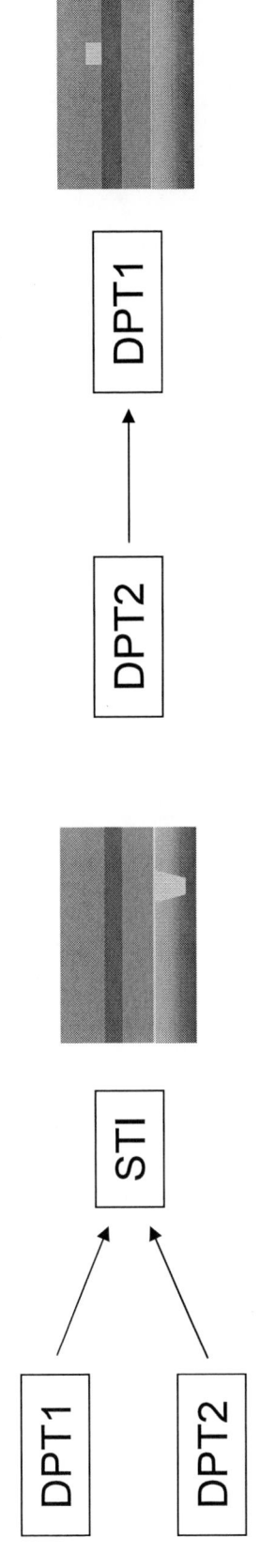

- Indirect alignment of DPT2 to DPT1 has many advantages, however is not possible for a first layer such as STI or if the stack is not transparent.

- To use LFLE for a first layer process we need to be able to align DPT2 directly to DPT1.

- This work looks at the feasibility of direct alignment for LFLE.

David Laidler
imec 2009

Feasibility of Direct DPT2 to DPT1 Alignment for LFLE

Objective:

- Characterize ability to align DPT2 directly to DPT1 frozen and resist coated marks.

- Identify what marks look like after freeze step and 2nd coat.

- Determine if there is sufficient signal strength.

- Recommend optimum mark type and alignment recipe.

- Compare full batch overlay performance for direct alignment of LFLE to baseline "Resist to Resist".

- Demonstrate performance on a real stack where it could be necessary.

David Laidler
imec 2009

imec ASML

JSR LFLE Process Stack and Exposure Conditions

- DPT1:
  - ARC91 38.5nm
  - ARX2928JN 105nm
  - TCX041 90nm
  - Exposure conditions:
    - $NA = 1.0$, $\sigma_o = 0.85$, $\sigma_i = 0.65$, Dipole 40X, Y Polarisation
    - Dose = 51mJ/cm^2, Focus = +0.07µm

- Freeze process:
  - FZX114 150nm

- DPT2:
  - AIM5631JN 90nm
  - Exposure conditions:
    - $NA = 1.0$, $\sigma_o = 0.85$, $\sigma_i = 0.65$, Dipole 40X, Y Polarisation
    - Dose = 35mJ/cm^2, Focus = -0.01µm

- Investigation was performed on blank silicon wafers.

David Laidler
imec 2009

imec ASML

Alignment Adviser
-SMASH Alignment Read Out (SARO) on SMASH Systems

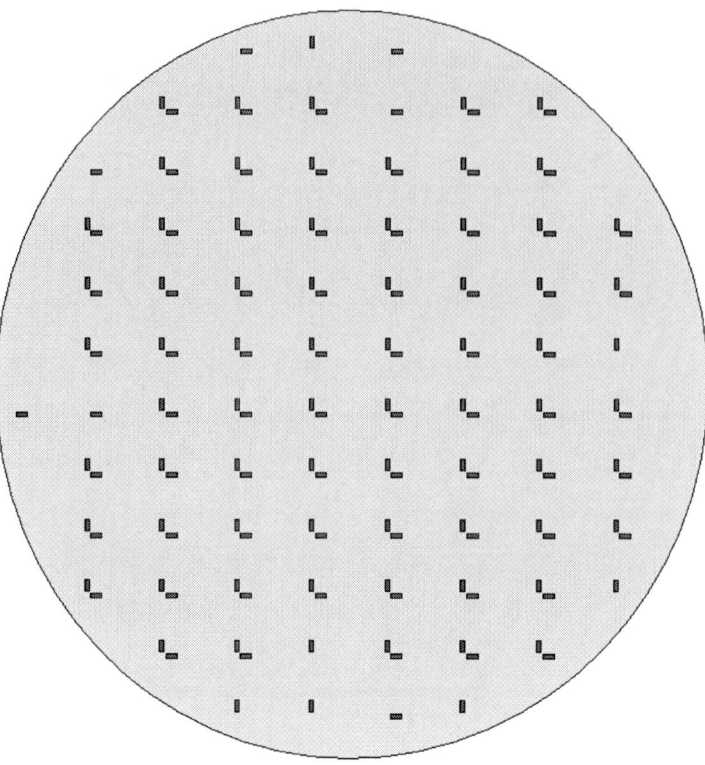

5 nm

- Alignment Adviser readouts were made on a full batch at the following steps in the LFLE process:
  - After DPT1,
  - After freeze,
  - After freeze and 2nd coat at DPT2.

- Allows us to look at any changes in wafer deformation going through the process and also alignment performance.

- Two types of mark were read out:
  - VSPM-AH325374,
  - VSPM-AA157.

- SMASH alignment wavelengths:
  - 633nm (Red),
  - 532nm (Green),
  - 770nm (NIR),
  - 850nm (FIR).

David Laidler
imec 2009

imec ASML

VSPM Scribeline Alignment Mark Designs

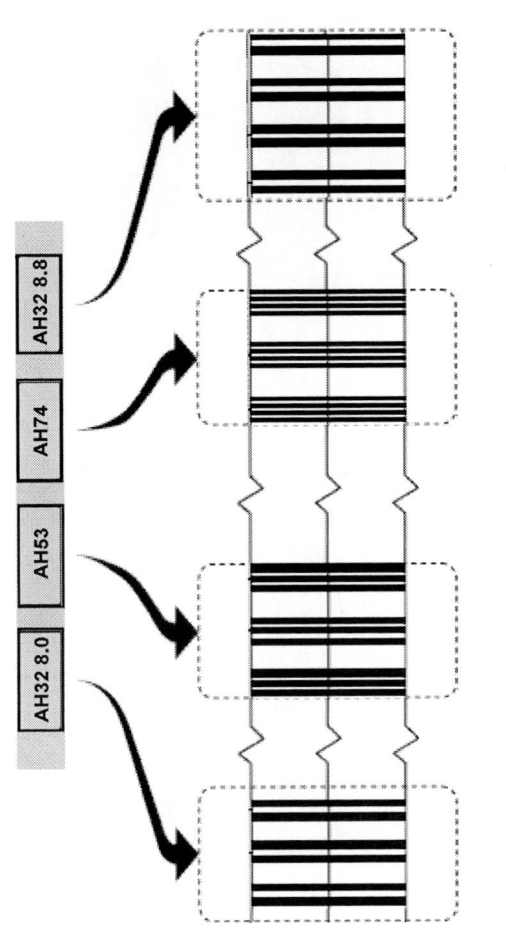

VSPM-AH325374

Standard

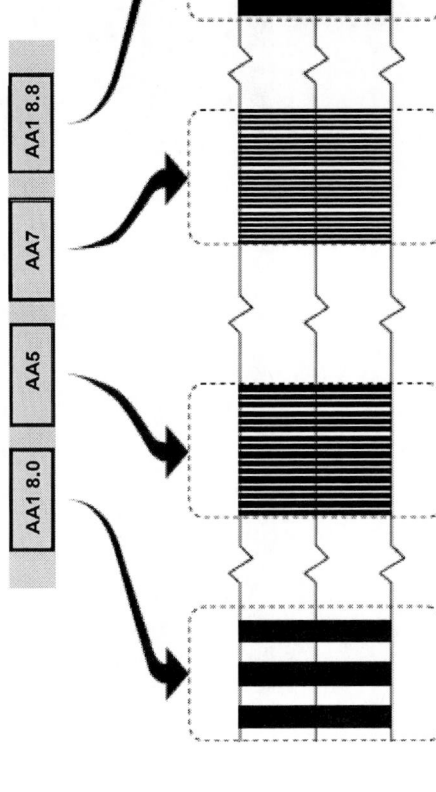

VSPM-AA157

Pure High Order

David Laidler
imec 2009

imec ASML

DPT1 Alignment Marks After Freeze and 2ⁿᵈ Coat at DPT2

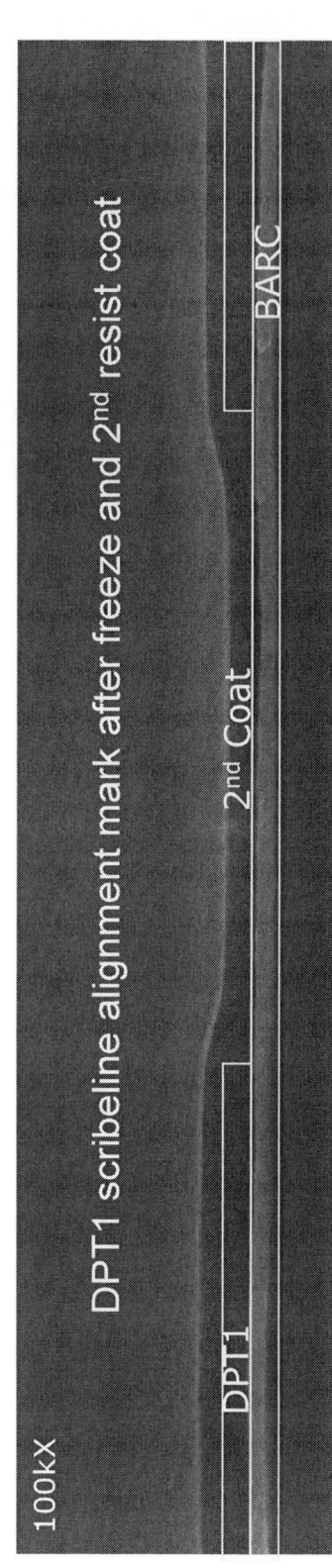

100kX

DPT1 scribeline alignment mark after freeze and 2ⁿᵈ resist coat

2ⁿᵈ Coat

BARC

DPT1

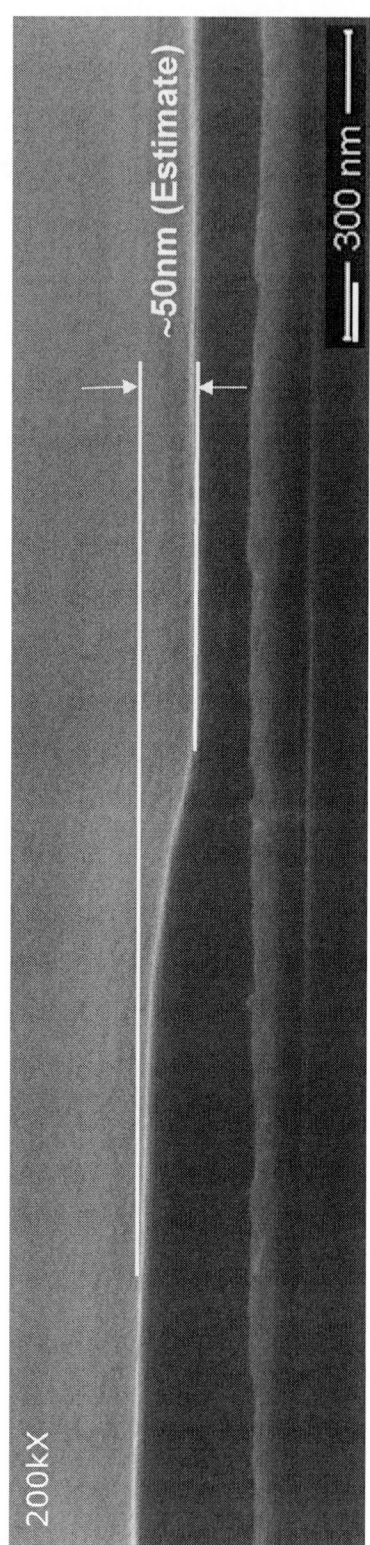

200kX

~50nm (Estimate)

300 nm

David Laidler
imec 2009

DPT1 Alignment Marks After Freeze and 2nd Coat at DPT2

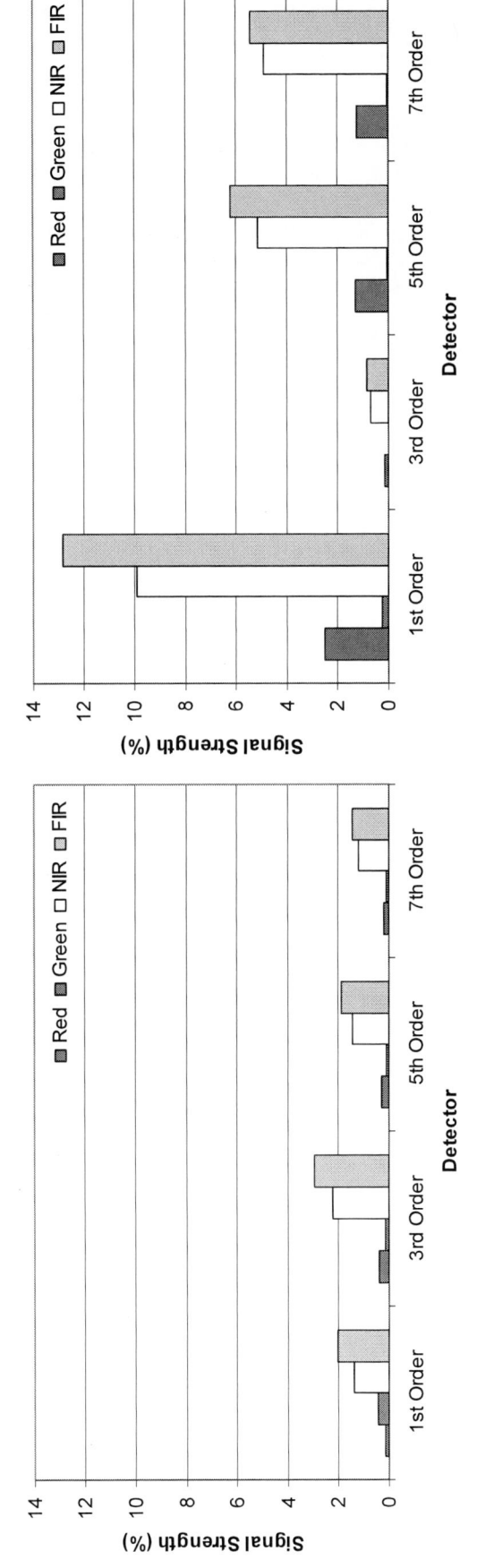

VSPM-AA157 (Pure High Order)

VSPM-AH325374 (Standard)

The VSPM pure high order mark shows higher signal strength than the standard VSPM mark.

Red 5th offers a solution for ATHENA systems.

NIR and FIR 5th offer additional process robustness, due to higher signal strength, for SMASH systems.

David Laidler
imec 2009

imec ❖ ASML

5th Order Signal Strength for Before Freeze, After Freeze and After Freeze and Second Coat at DPT2

- Signal strength for the four alignment wavelengths at different stages of the LFLE process.

- What matters is the performance after freeze and 2nd coat at DPT2.

- Signal strength significantly less after freeze and 2nd coat, but there is still adequate signal strength for Red, NIR and FIR.

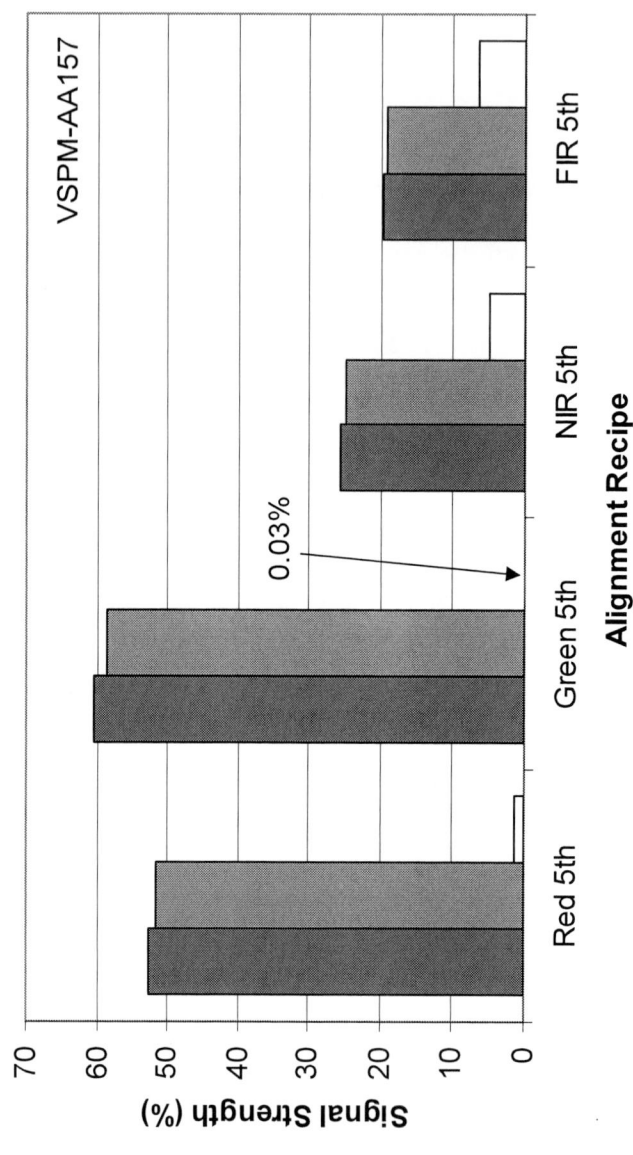

David Laidler
imec 2009

ROPI – Residual Overlay Performance Indicator

- The ROPI number gives a measure of both wafer deformation and alignment performance.

- The performance is consistent for all alignment recipes and through the LFLE process.

- Indicates that there is no deformation of the DPT1 pattern during the LFLE process and that consistent alignment performance is achievable at each step.

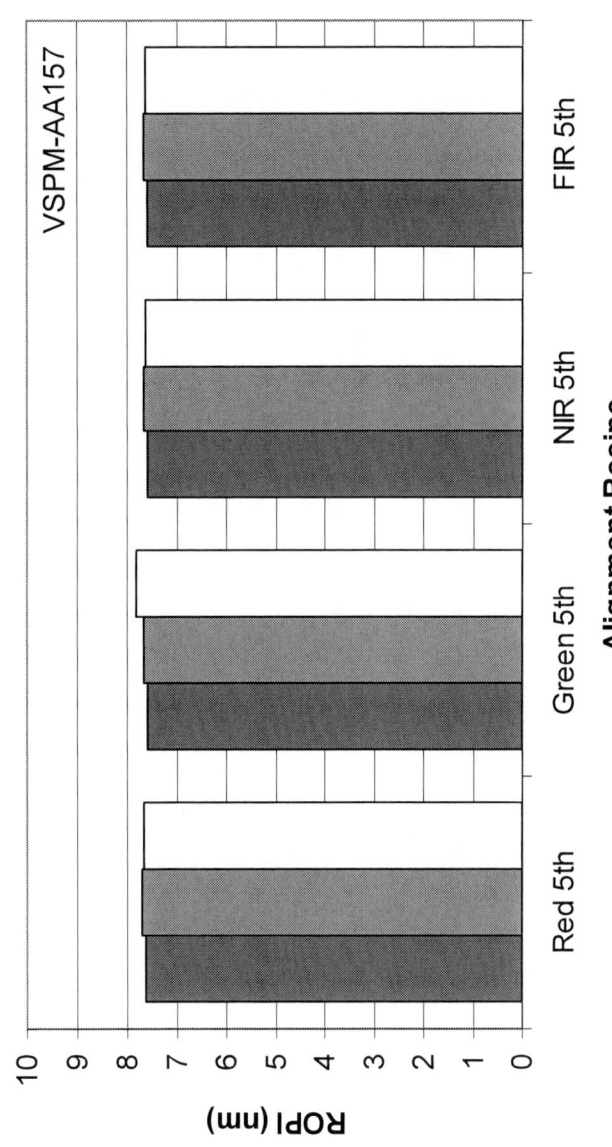

David Laidler
imec 2009

ASML

RPN – Residual Process Noise

- RPN (Residual Process Noise) removes the average wafer signature, in this case immersion related, from the residuals.

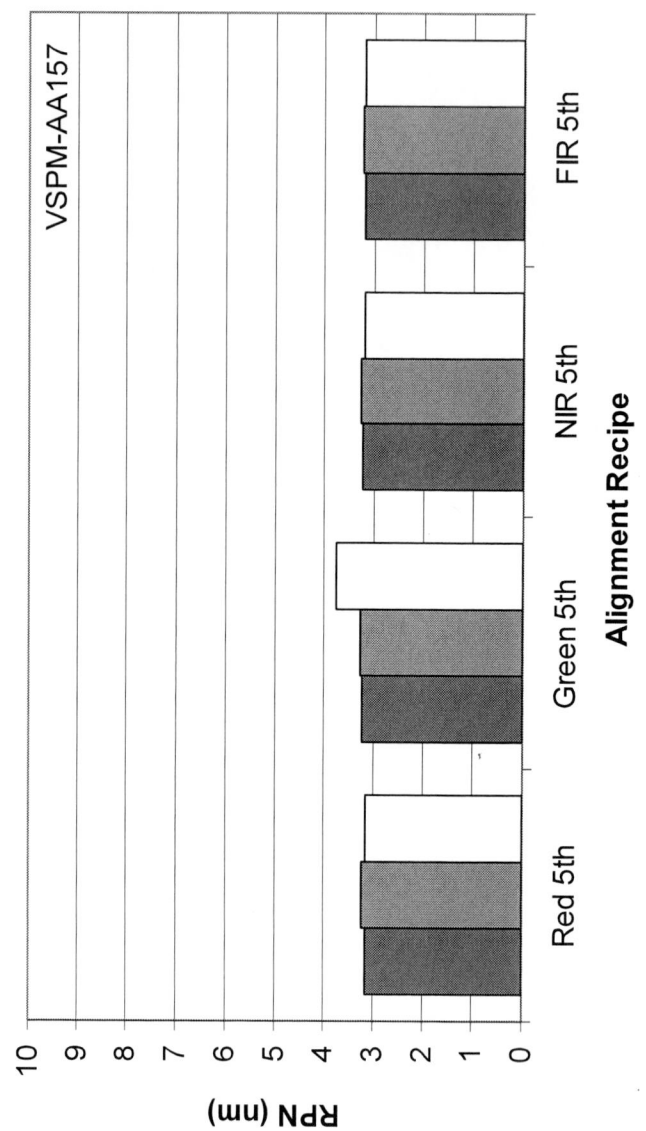

- The RPN shows no alignment performance degradation or wafer deformation as we go through the LFLE process.

David Laidler
imec 2009

Experimental Procedure

- Three 10 wafer batches exposed on the XT:1900Gi, single chuck, using the VSPM-AA157 alignment mark and each using a different alignment recipe:
  - Red 5th,
  - NIR 5th,
  - FIR 5th.

- Additional batch of 10 wafers exposed, again single chuck, but using baseline "Resist to Resist" process as a reference.

- Alignment performed using 16 mark pairs and 6 parameter model.

- All measurements performed on K-T Archer AIM using a Bar-in-Bar structure.

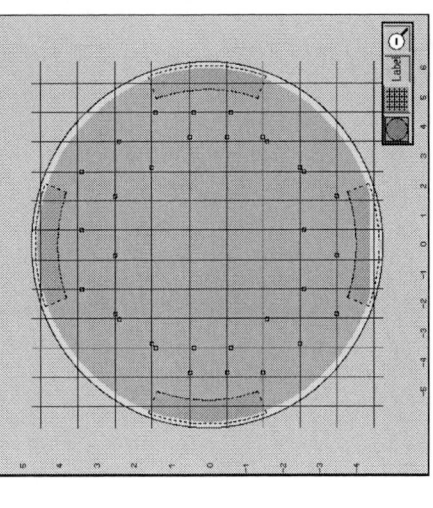

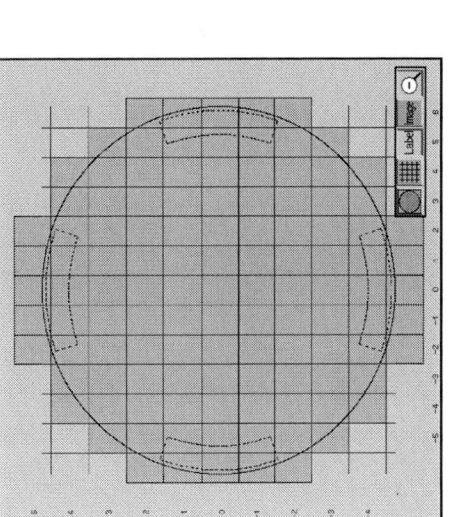

David Laidler
imec 2009

Stability of Correctables and Residuals Through Batch – Red 5th (Pure Grid)

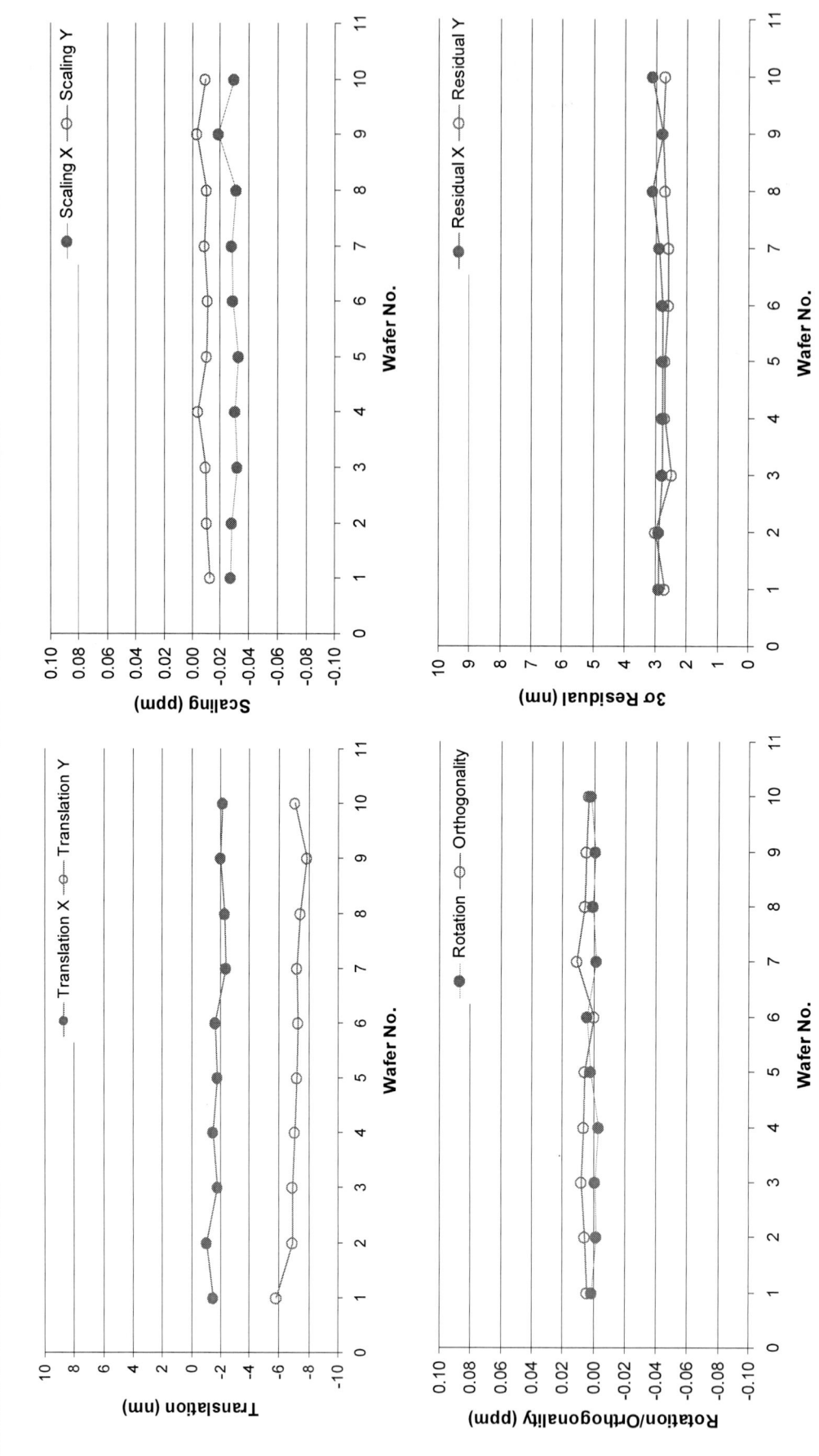

David Laidler
imec 2009

Range of Correctables Across 10 Wafer Batch (Pure Grid)

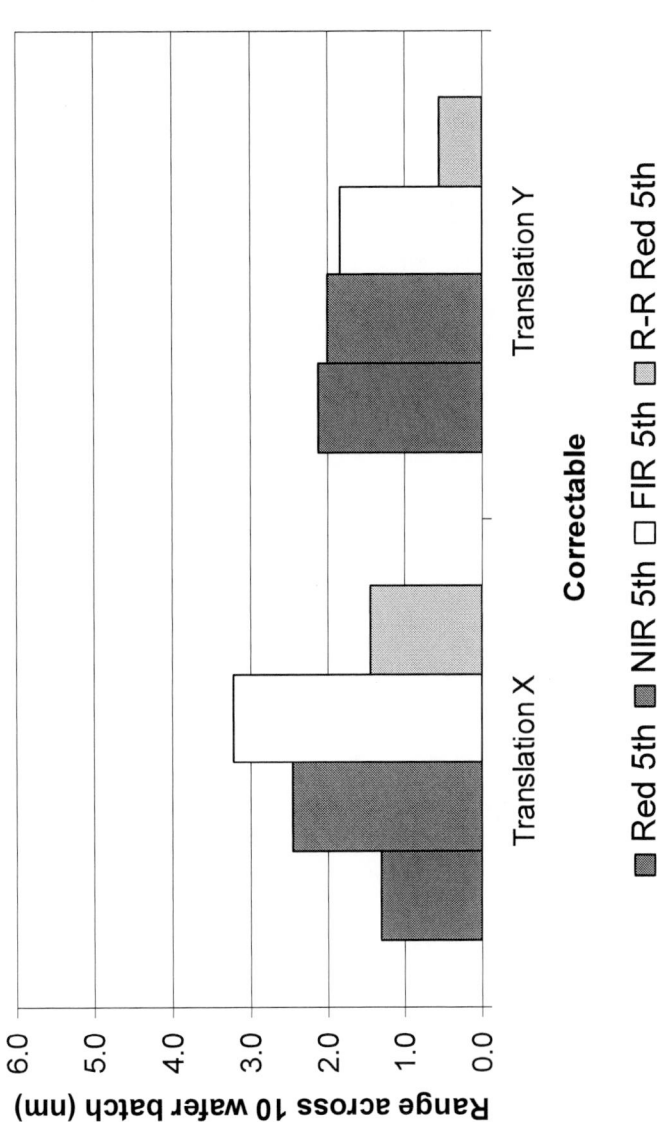

- Typical translation variation through a full batch on an XT:1900Gi is of the order of ±1nm, so all modes are comparable.

David Laidler
imec 2009

Range of Correctables Across 10 Wafer Batch (Pure Grid)

- Variation of correctables through batch comparable to baseline "Resist to Resist" and within noise levels the same for all modes.

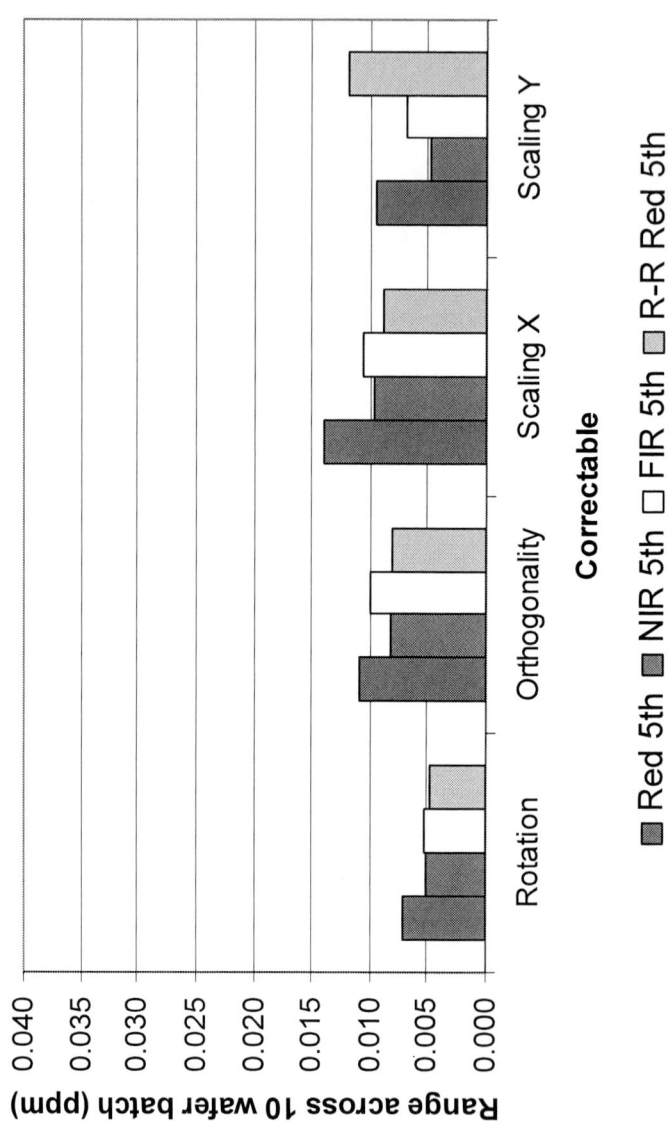

Grid – 0.01ppm (nm/mm) is significant, equivalent to 1.5nm at edge of wafer,
Intrafield – 0.1ppm (nm/mm) is significant, equivalent to 1.6nm at edge of field.

David Laidler
imec 2009

Full Batch Overlay Performance for LFLE Direct Alignment Compared to Baseline "Resist to Resist"

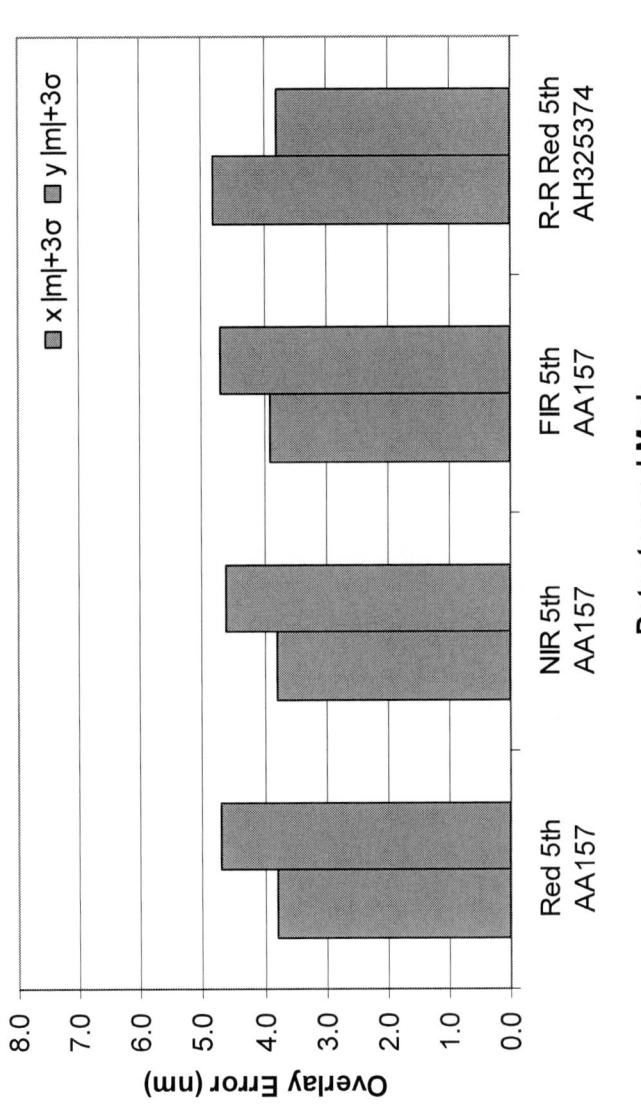

- Full batch overlay performance, 10 wafers per strategy, single chuck with chuck dedication. Nine points per field, all fields and average batch correctables removed.
- Consistent overlay performance has been obtained for Red 5th, NIR 5th and FIR 5th.
- Full batch overlay performance is comparable to that obtained for the baseline "Resist to Resist" process.

David Laidler
imec 2009

imec ASML

LFLE Direct Alignment Performance on <u>Product</u> STI Stack
- Optimum Signal Strength

Using VSPM-AA157 mark type, evaluated optimum signal strength to use on product STI stack:

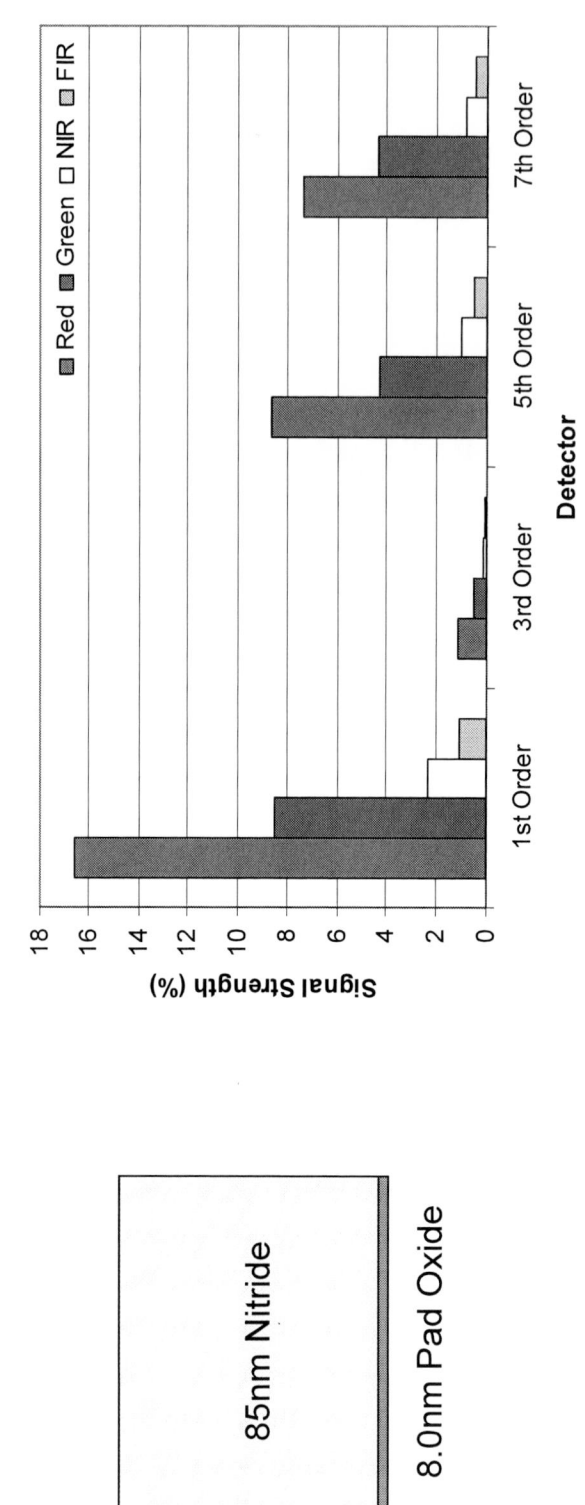

85nm Nitride

8.0nm Pad Oxide

On this particular product STI stack, the optimum wavelength, giving the maximum signal strength was found to be Red (633nm).

David Laidler
imec 2009

imec ❖ ASML

LFLE Direct Alignment Performance on Product STI Stack
- Full Batch Overlay Performance

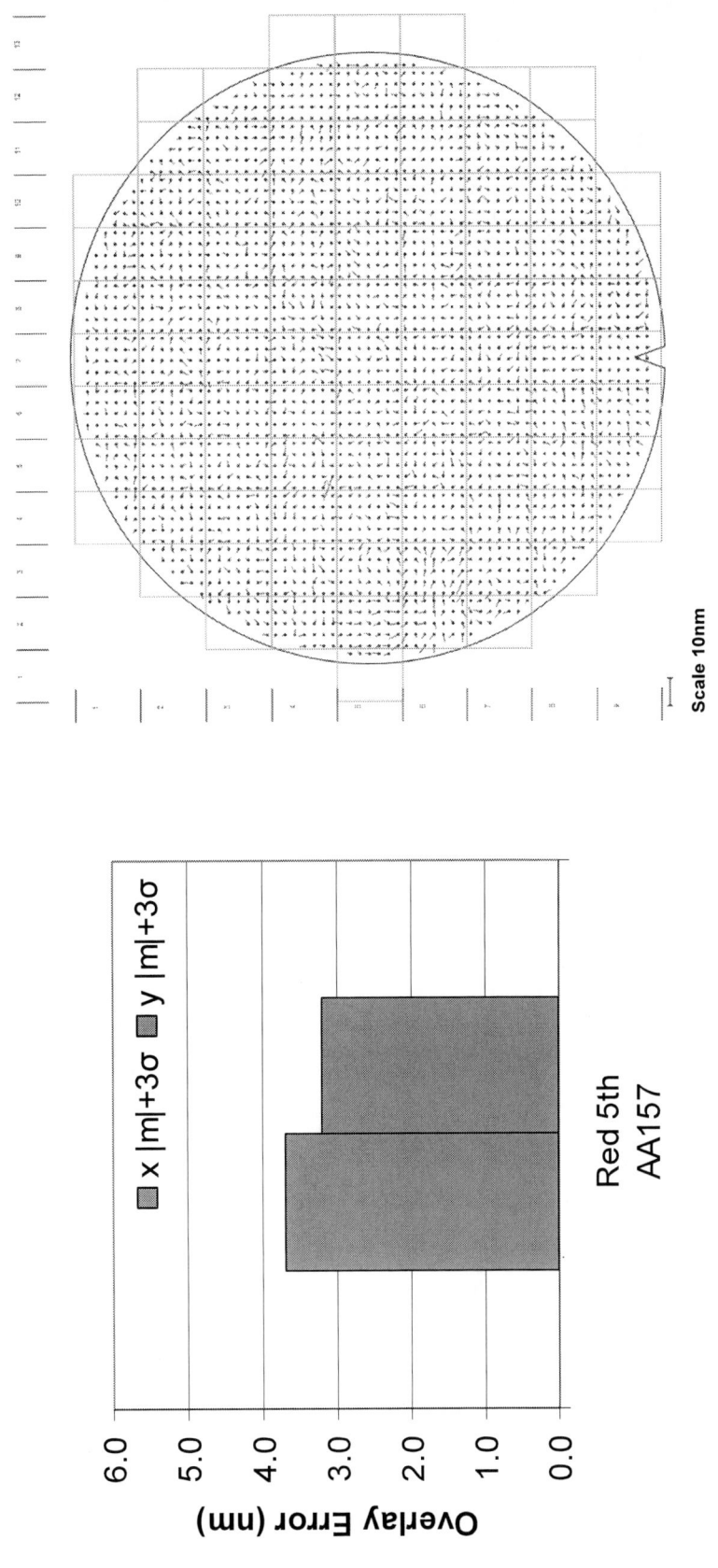

Scale 10nm

Red 5th
AA157

- x |m|+3σ - y |m|+3σ

Overlay Error (nm)

- Full batch overlay performance, 22 wafers with chuck dedication, full wafer coverage, 35 points per field, average correctables removed per chuck.

David Laidler
imec 2009

imec ❖ ASML

Summary

- By selection of the appropriate alignment mark and recipe, sufficient signal strength can be obtained for direct alignment of the JSR LFLE process.

- The VSPM-AA157 mark has been shown to have benefit over the standard VSPM-AH325374 mark.

- Consistent overlay performance has been demonstrated for Red 5th, NIR 5th and FIR 5th.

- Full batch overlay performance is comparable to that obtained for baseline "Resist to Resist".

- For direct alignment of the JSR LFLE process, Red 5th offers a solution for ATHENA and SMASH systems and NIR and FIR 5th offer additional process robustness, for SMASH systems.

- Direct alignment for LFLE has also been demonstrated on a product STI stack, where a direct alignment strategy could be required.

David Laidler
imec 2009

Acknowledgements

- Frieda Van Roey for cross-section work.

David Laidler
imec 2009

imec ASML

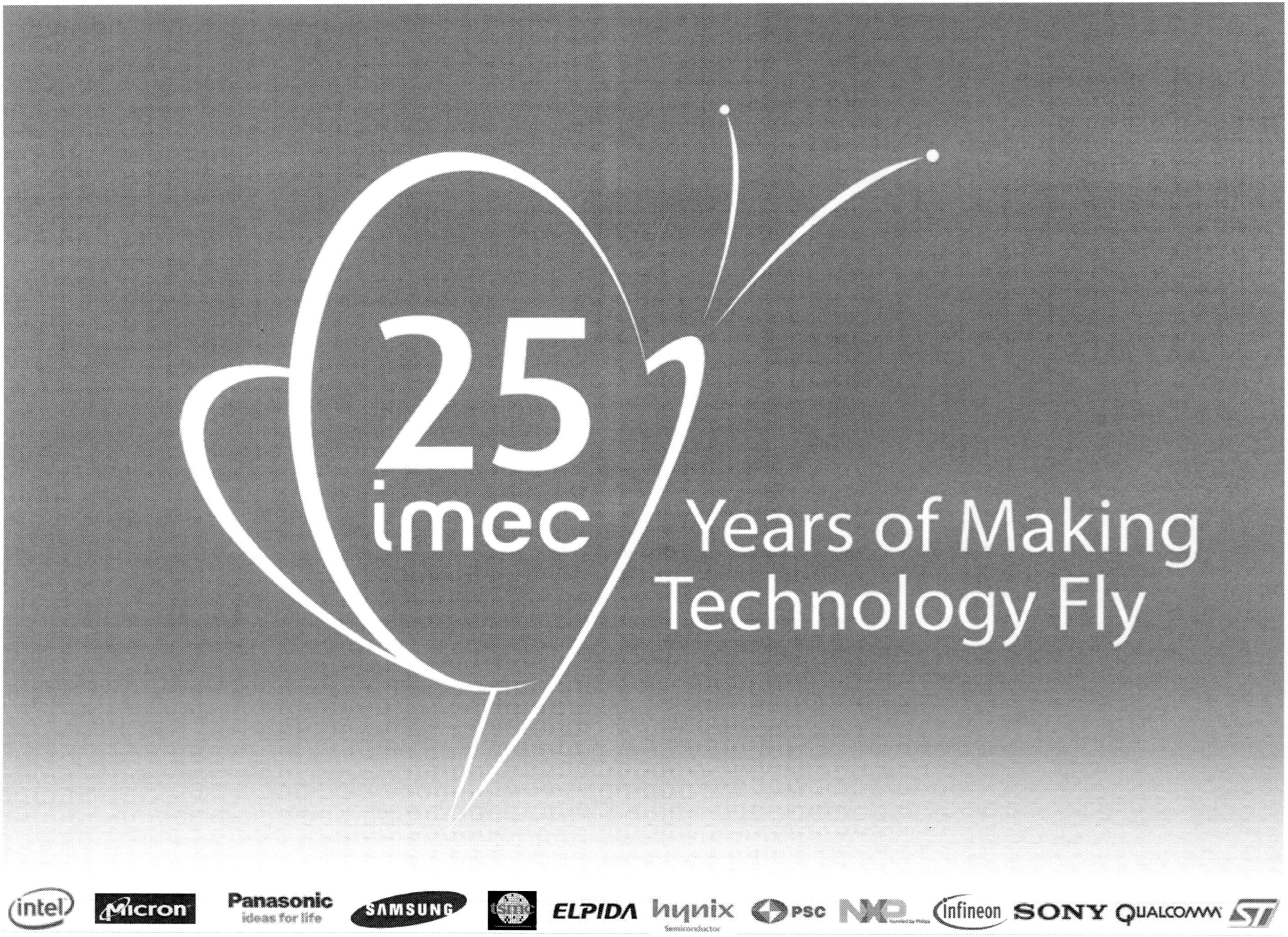
25
imec
Years of Making
Technology Fly
intel
Micron
Panasonic
ideas for life
SAMSUNG
ELPIDA
hynix
Semiconductor
PSC
NXP
Infineon
SONY
QUALCOMM
ST

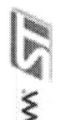

Accelerating the next technology revolution

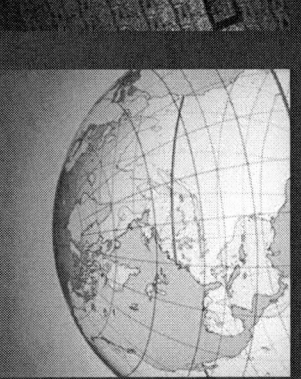

Can Nanoimprint and Maskless Lithography Thrive in a 193i World?

Lloyd C. Litt,
SEMATECH / GLOBAL FOUNDRIES

193nm Immersion Symposium
October 2009, Prague, CZ

Copyright ©2008
SEMATECH, Inc. SEMATECH, and the SEMATECH logo are registered servicemarks of SEMATECH, Inc. International SEMATECH Manufacturing Initiative, ISMI, Advanced Materials Research Center and AMRC are servicemarks of SEMATECH, Inc. All other servicemarks and trademarks are the property of their respective owners.

Overview

- Introduction
  - Incumbent Technology
  - Nanoimprint (NIL)
  - Maskless (ML2)
- Technology Introduction to HVM
  - Market Forces and Supplier Perspective
  - NIL applications
  - ML2 applications
- Conclusions

Lloyd C. Litt, 193nm Immersion Symposium 2009

193i – The Incumbent Technology

- 193i is acknowledged as the incumbent technology for high volume CMOS manufacturing.

- Manufacturing will likely use 193i and double patterning until a more capable and lower cost option (or EUV) becomes available.

- The Litho industry will determine if/when other technologies are adopted :
  - Manufacturers (chip makers)
  - Tool Suppliers
  - Mask Suppliers
  - Resist & Process Suppliers

Lloyd C. Litt, 193nm Immersion Symposium 2009

Nanoimprint Lithography(NIL)

Apply Monomer   **Press & Flash**   **Lift**

TEMPLATE

WAFER

ADVANTAGES
Relatively low cost

3D features possible – MEMS

Defect-tolerant applications –
Disk Storage,
gratings,
micro lens

ISSUES
1X template manufacturing

1X template inspection and repair

Defects

Overlay / magnification

Template lifetime

Lloyd C. Litt, 193nm Immersion Symposium 2009

ML2 – Maskless Lithography

ADVANTAGES

No mask
Flexible patterning
Potential for fast cycle time (RTAT)

E-Beam Source

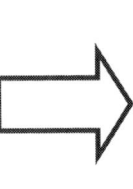

Aperture/Control

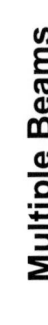

Multiple Beams

ISSUES

Throughput
Multiple Beams (thousands)
Cluster Modules

Stitching
Data transfer

Lloyd C. Litt, 193nm Immersion Symposium 2009

The Big Picture

Lloyd C. Litt, 193nm Immersion Symposium 2009

New Technologies:
Influences on Insertion – "The Big Picture"

- Technical Capability

- Infrastructure

- Technology "Inertia"

- Financial Considerations
  - Depreciation
  - Introduction Timing

Lloyd C. Litt. 193nm Immersion Symposium 2009

Industry Vested Interest in Technology Change

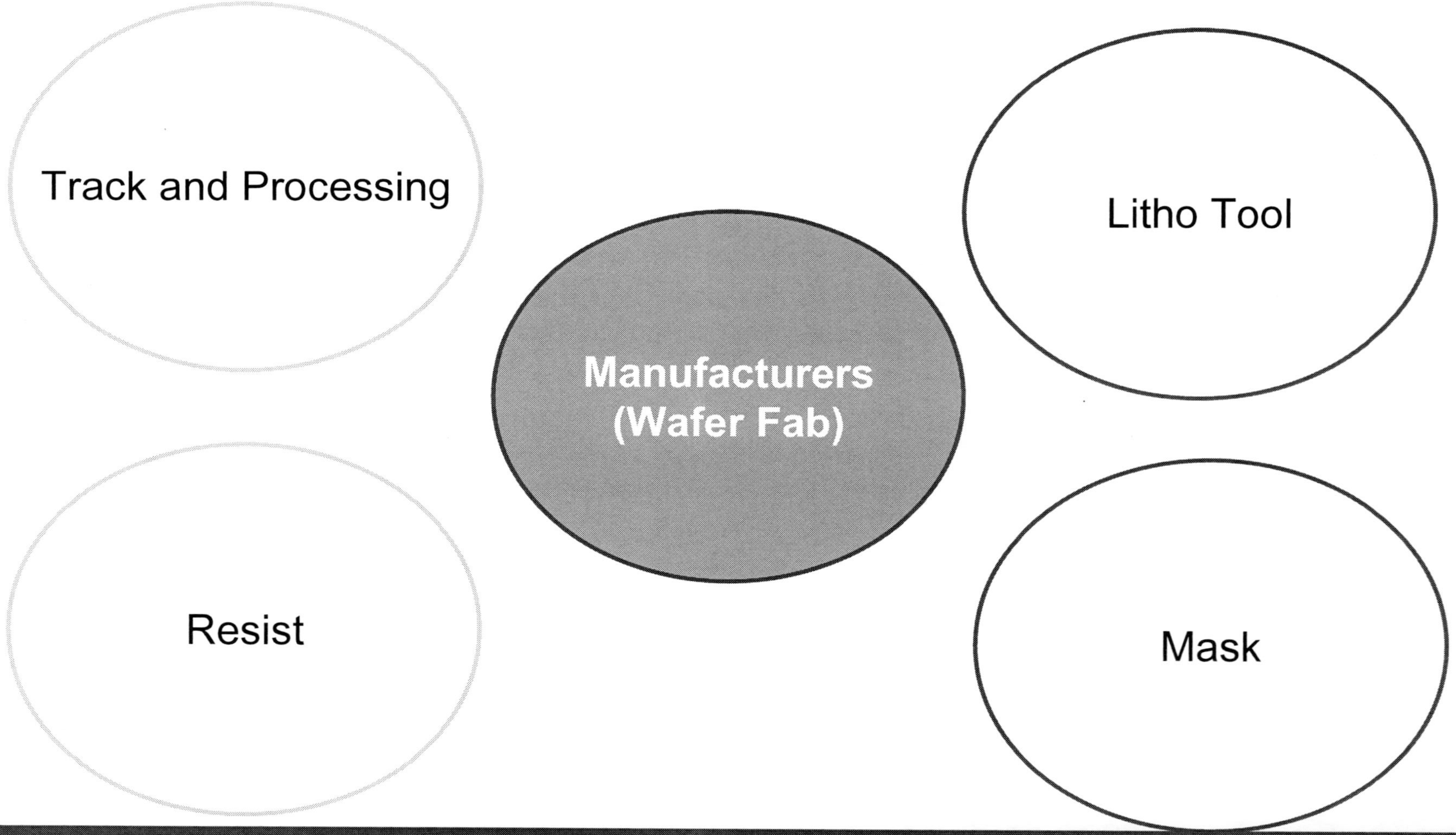

- Litho Tool
- Mask
- Manufacturers (Wafer Fab)
- Track and Processing
- Resist

Lloyd C. Litt, 193nm Immersion Symposium 2009

Tool Suppliers Dilemma
Motivation for Change?

	Double Patterning	EUV	e-Beam	Nano Imprint
Legacy Knowledge				1X Features
Optics			e-Beam optics	1X Features
Hardware		Vacuum	Vacuum	
Capital Investment				

Lloyd C. Litt, 193nm Immersion Symposium 2009

Mask Makers Dilemma
Motivation for Change?

	Double Patterning	EUV	e-Beam	Nano Imprint
Write				1X Features
Inspect				1X Features
Format Substrate		Multilayer plate		Similar
Capital Investment		Inspection		Inspection

Newton's Laws Applied to Litho

> **A technology in manufacturing tends to stay in manufacturing unless acted upon by an external force.**

- External Forces:
  - Technical Capability
  - Cost
  - Infrastructure

- Change in wafer size is a large factor in the timing of technology change acceptance.
  - Retooling for new wafer sizes facilitates technology change.

Sir Isaac Newton, Philosophiæ Naturalis Principia Mathematica, 1687

Lloyd C. Litt, 193nm Immersion Symposium 2009

Lithography Timeline

SEMATECH

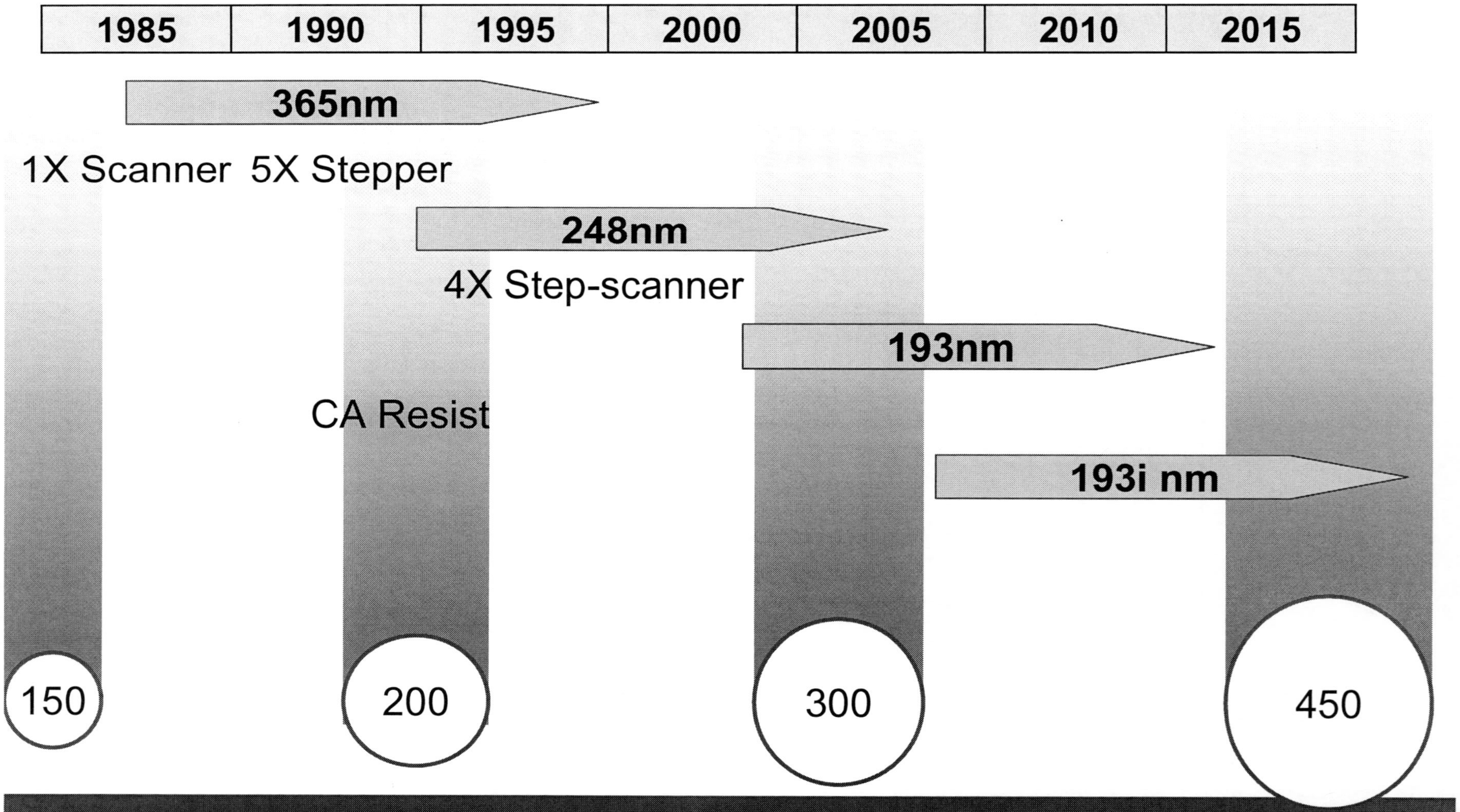

1985	1990	1995	2000	2005	2010	2015

365nm

1X Scanner 5X Stepper

248nm

4X Step-scanner

193nm

CA Resist

193i nm

150   200   300   450

Lloyd C. Litt, 193nm Immersion Symposium 2009

Technology Success Requirements

For any technology to have a chance of success, it requires:

- Strong financial backing.

- A viable, demonstrated technology (including throughput requirements).

- A strong customer "champion" to push the development and be a potential future customer for systems.

- Viable, supportive infrastructure.

SEMATECH CONFIDENTIAL Immersion Symposium 2009

Nanoimprint (NIL)

Lloyd C. Litt, 193nm Immersion Symposium 2009

Chasing the ITRS – NIL Overlay

Imprio is a trademark of Molecular Imprints Inc.

Lloyd C. Litt, 193nm Immersion Symposium 2009

Nanoimprint Capability Mapping
(22nm node)

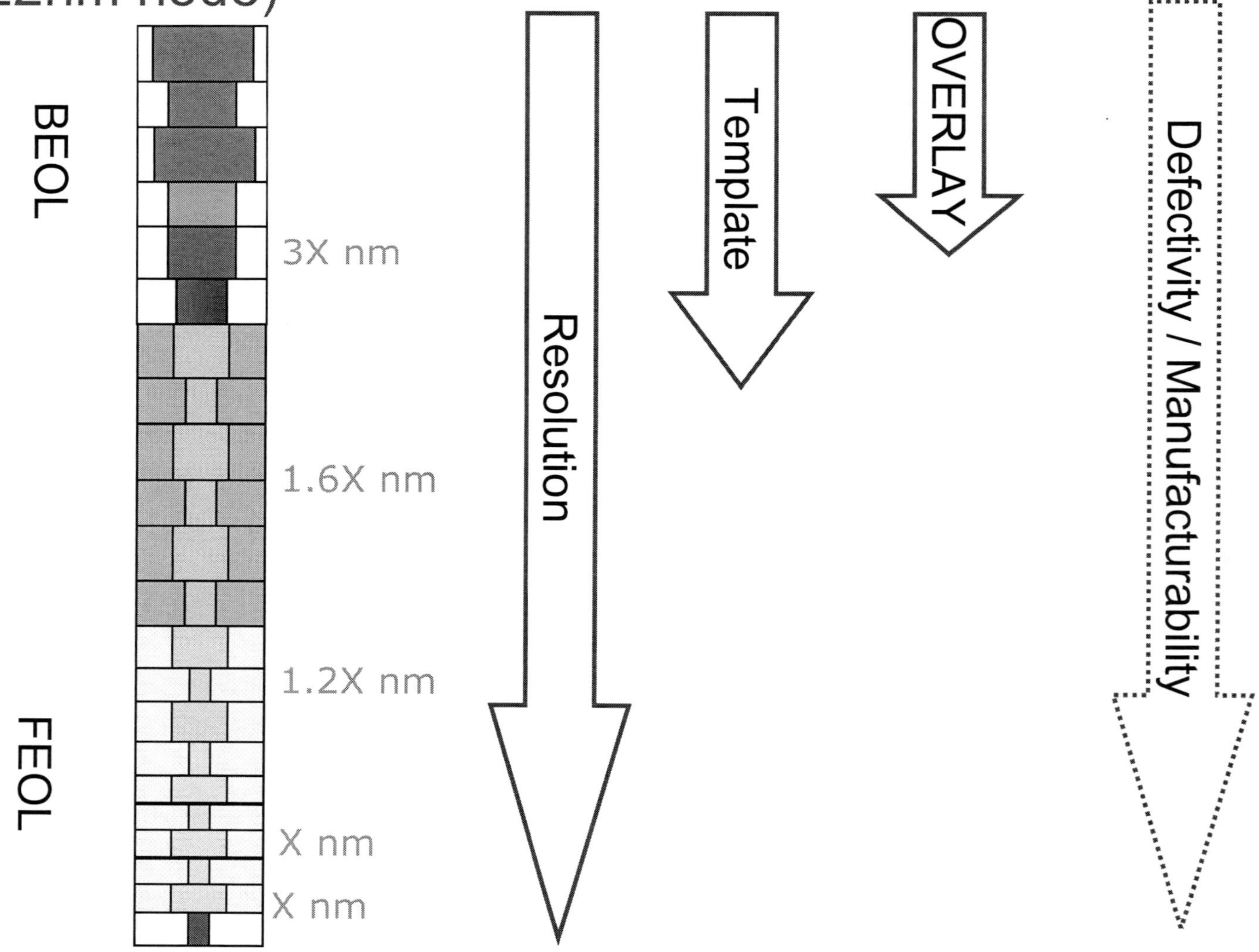

Defectivity / Manufacturability

OVERLAY

Template

Resolution

3X nm

1.6X nm

1.2X nm

X nm

X nm

BEOL

FEOL

Lloyd C. Litt, 193nm Immersion Symposium 2009

Nanoimprint Applications

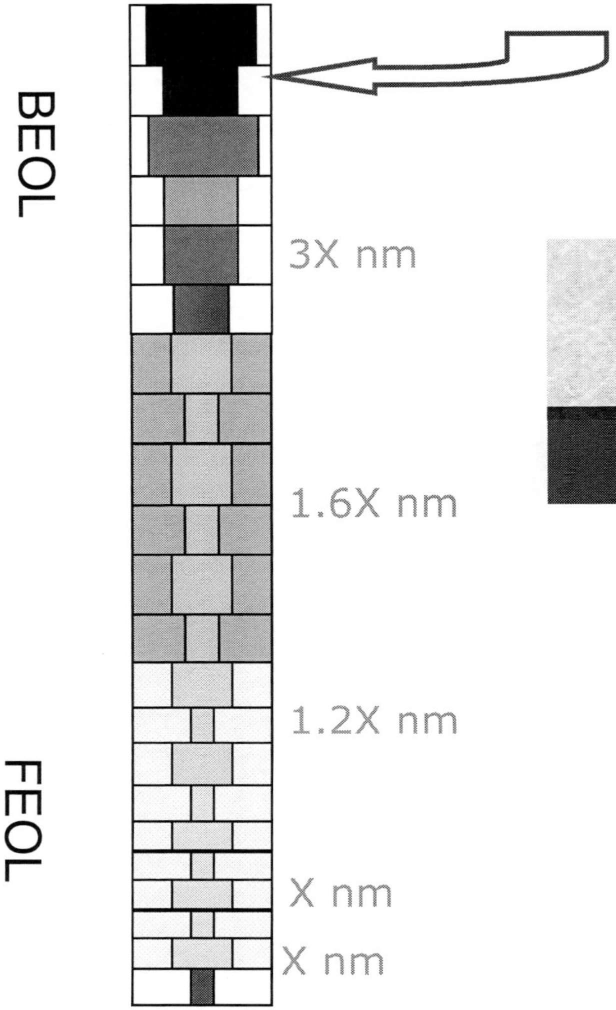

Dual Damascene – Metal & Via

Single Template
Single Litho Step

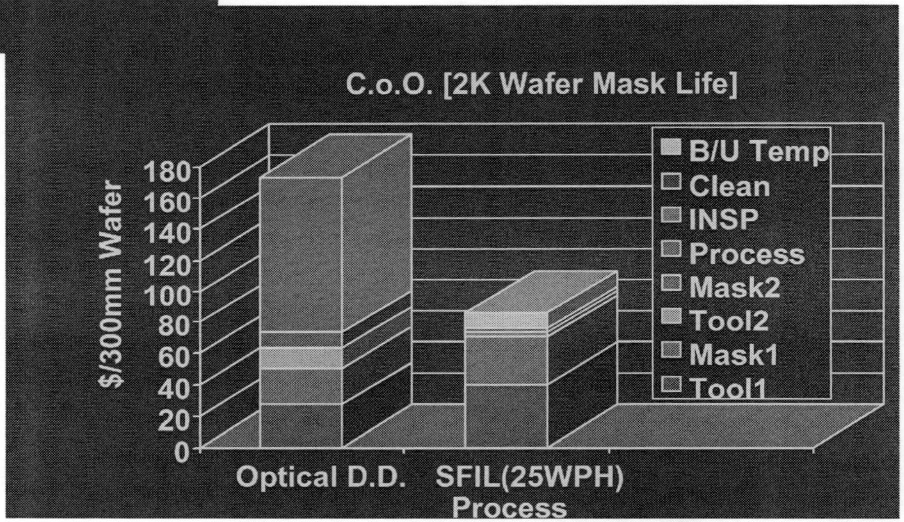

BEOL

FEOL

3X nm

1.6X nm

1.2X nm

X nm

X nm

C.o.O. [2K Wafer Mask Life]

■ B/U Temp
□ Clean
■ INSP
□ Process
■ Mask2
■ Tool2
■ Mask1
□ Tool1

180
160
140
120
100
80
60
40
20
0

$/300mm Wafer

Optical D.D.   SFIL(25WPH) Process

Litt, et al.,Cost Analysis of NIL,NNT06

Lloyd C. Litt, 193nm Immersion Symposium 2009

Maskless Lithography (ML2)

Lloyd C. Litt, 193nm Immersion Symposium 2009

ML2 Technology Applications

Prototype/development
Early resolution capability, process development

Low volume production
Mask cost avoidance, cycle time, NPI

Critical layer mix & match
Mask cost reduction, customization, respins

Risk reduction
Back-up for optical systems

Pattern Generator
Potential to apply technology to reduce mask cost

Lloyd C. Litt, 193nm Immersion Symposium 2009

Design Verification Application

- Reduced cost of entry – more designs
- Reduced cost of early design revisions

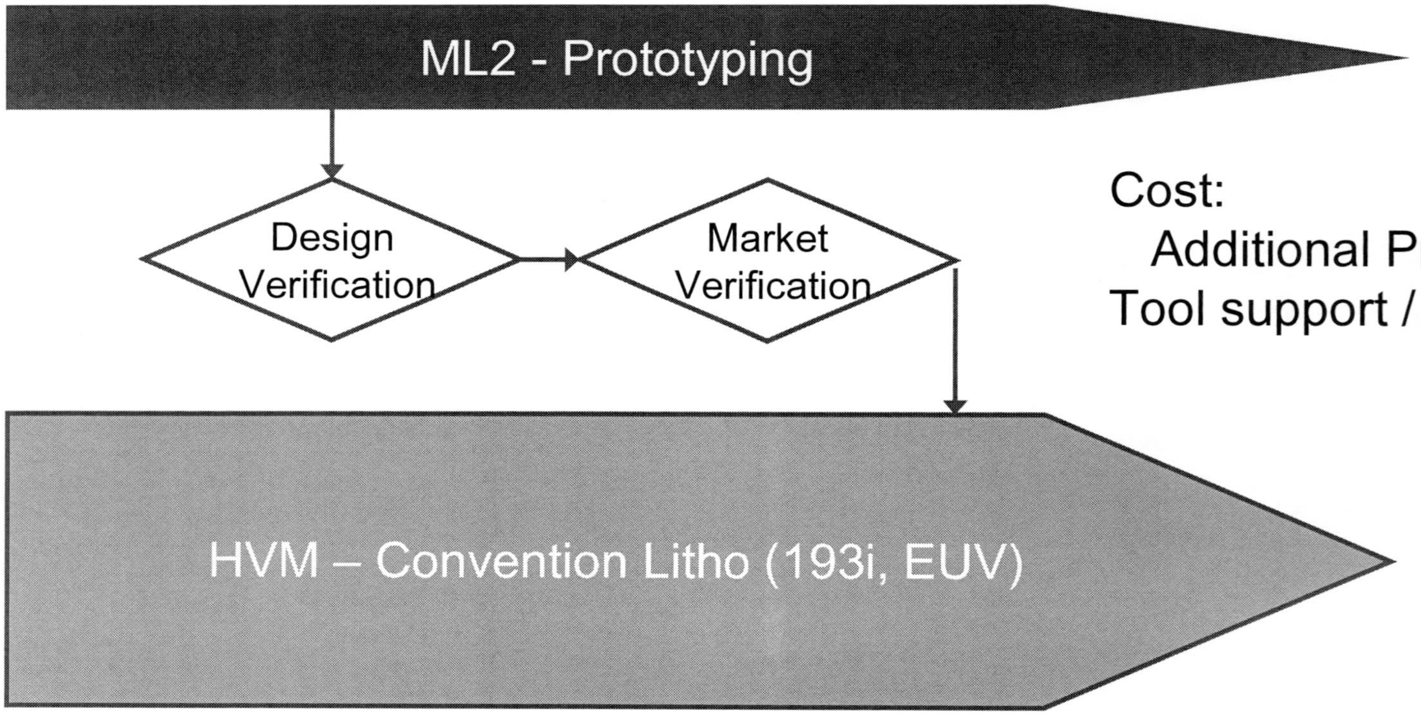

ML2 - Prototyping

Design Verification

Market Verification

Cost:
Additional Process &
Tool support / space, etc.

HVM – Convention Litho (193i, EUV)

Multiple Ebeam(ML2) Capability Mapping

Defectivity / Manufacturability

OVERLAY

Resolution

Via Layers

Best Throughput

3X nm

1.6X nm

1.2X nm

X nm

X nm

BEOL

FEOL

Lloyd C. Litt, 193nm Immersion Symposium 2009

Overview – 1,000 Wafers/Mask

Legend:
- Reticle
- Clean
- Etch
- Metrology
- Deposition
- Litho

Excludes data prep and additional inspection costs.

Normalized values to 45 nm SE

45 nm — **32 nm** — **22 nm**

Technology (wph)

- 45 nm ArFi SE (125)
- 32 nm LELE (180)
- 32 nm Freeze (180)
- 32 nm HI Spacer (180)
- 32 nm HI ArFi SE (120)
- 32 nm EUVL (50)
- 22 nm LELE (200)
- 22 nm Freeze (200)
- 22 nm Spacer (200)
- 22 nm HI LELE (135)
- 22 nm EUVL (100)
- 22 nm ML2 (50)
- 22 nm ML2 (100)

Y-axis: 0%, 50%, 100%, 150%, 200%, 250%, 300%, 350%, 400%, 450%, 500%

SEMATECH

Lloyd C. Litt, 193nm Immersion Symposium 2009

Process COO Implications of ML2

*Relative costs approximated by size scale of boxes

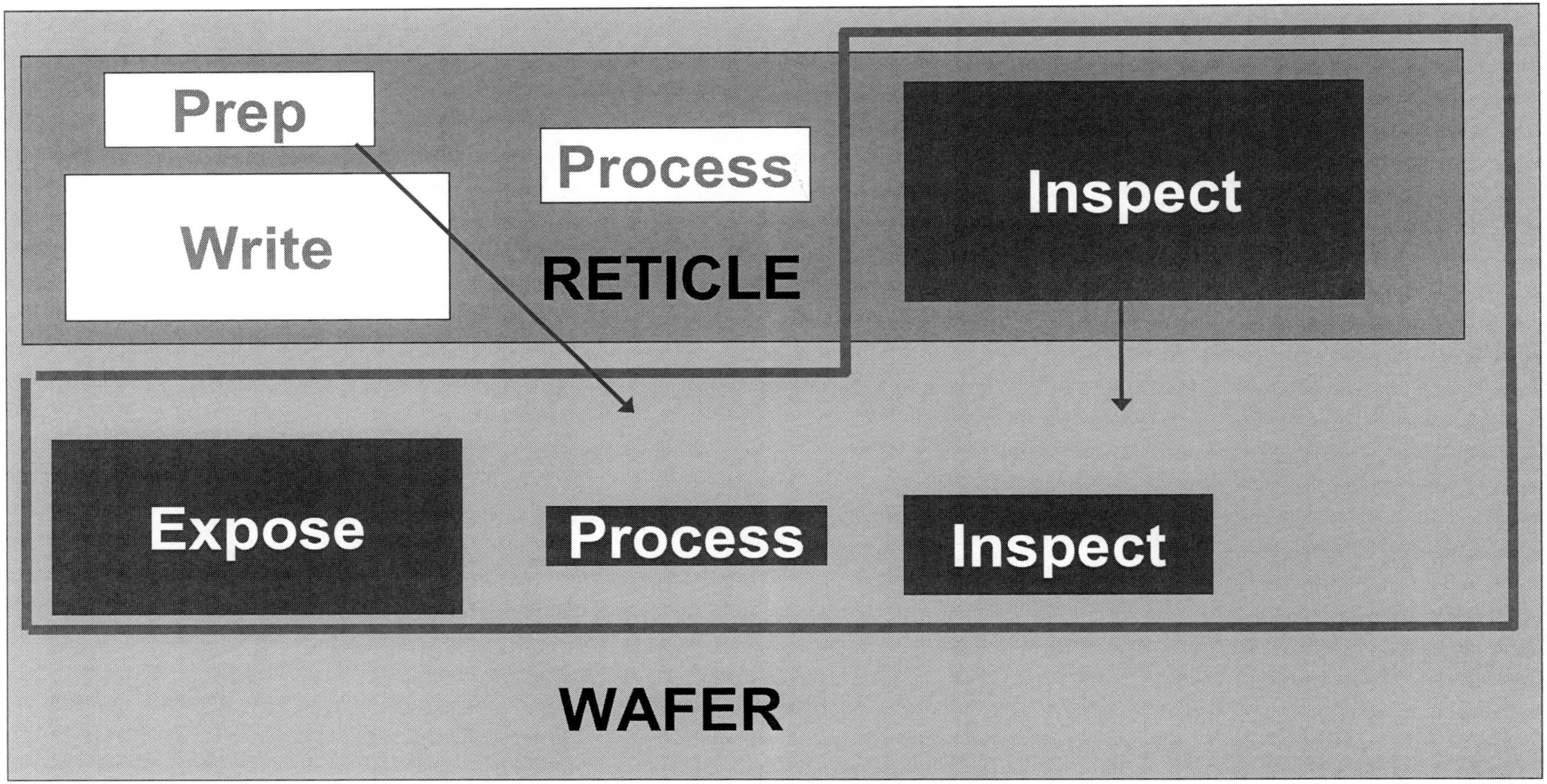

RETICLE

Prep

Write

Process

Inspect

WAFER

Expose

Process

Inspect

Lloyd C. Litt, 193nm Immersion Symposium 2009

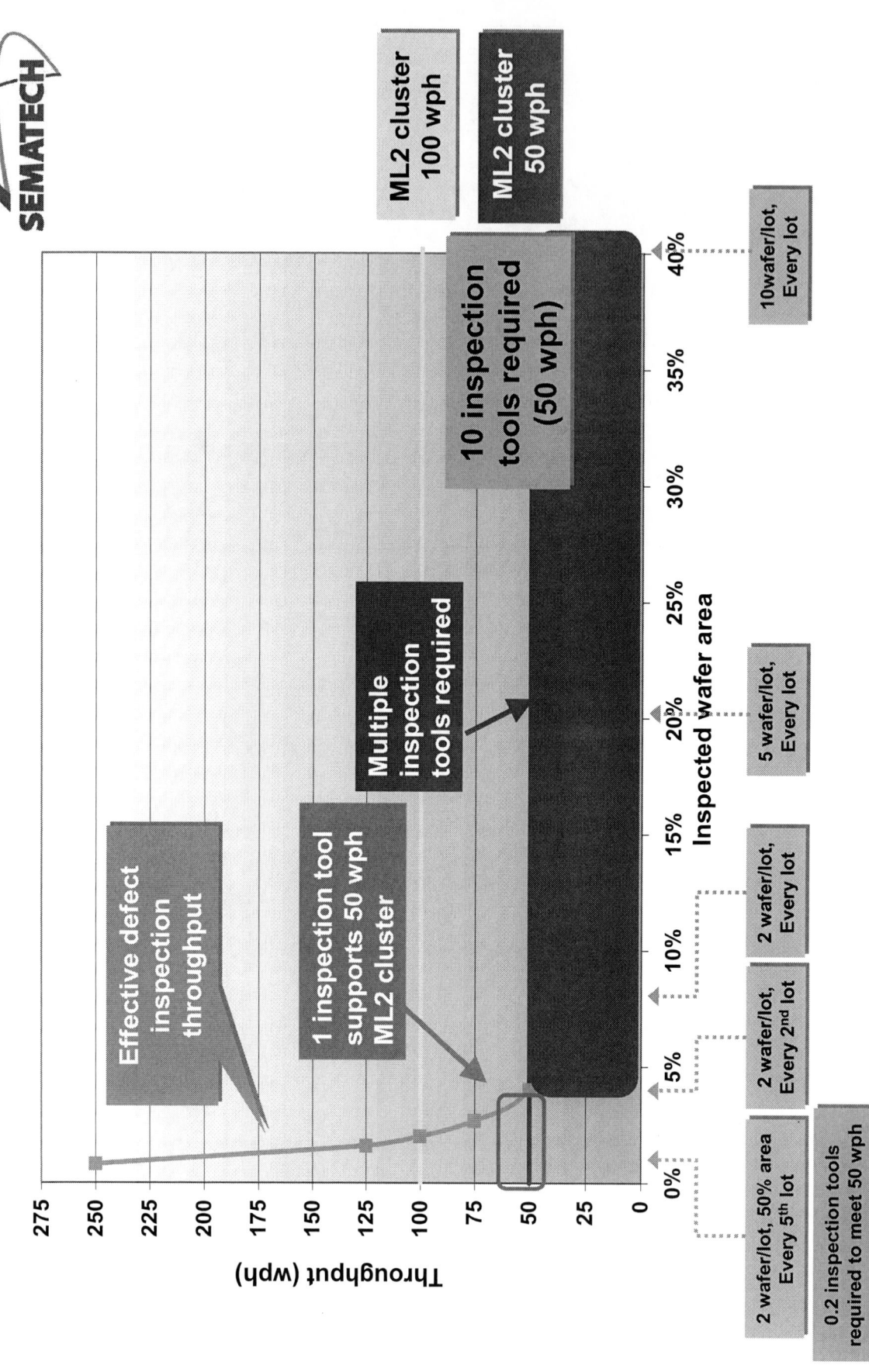

Summary and Conclusions

- NIL and ML2 have unique advantages but additional development effort is required for both alternatives to demonstrate performance.

- There are applications where ML2 and NIL can provide cost effective solutions in manufacturing.
  - Most promising is prototyping and early process development

- The timing of NIL or ML2 introduction could be critical.
  - Wafer size change is a primary opportunity

Lloyd C. Litt, 193nm Immersion Symposium 2009

Acknowledgements

- Matt Malloy
- Jongwook Kye
- Greg Hughes
- Andrea Wuest
- Grant Willson

Lloyd C. Litt, 193nm Immersion Symposium 2009

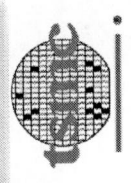

Readiness of Multiple E-Beam Maskless Lithography (MEB ML2)

Speaker: S.C.Wang

Jack J.H. Chen, T.Y. Fang, S.J. Lin, S.M. Chang, Faruk Krecinic, Wen Wang, and Burn J. Lin

Taiwan Semiconductor Manufacturing Co., LTD, Taiwan

Bert-Jan Kampherbeek, Guido de Boer, Bart Schipper, Paul Scheffers, Christiaan van den Berg, and Marco Wieland

MAPPER Lithography B.V., The Netherlands

Contact: jhchen@tsmc.com

The Proven Path to Success ᔆᴹ

23 Oct 2009, 6th International Symposium on Immersion Lithography Extensions

© 2009 TSMC, Ltd

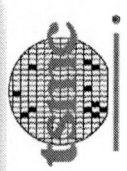

Why Maskless?

- "The limit of lithography will not be in resolution but in economy."

  – Dr. Burn J. Lin, in 1987

- "The devil is in the mask!"

  – Dr. Burn J. Lin, in 2007

Source from Proc. of SPIE Vol. 6520-02, (2007)

23 Oct 2009, 6th International Symposium on Immersion Lithography Extensions

The Proven Path to Success ℠

© 2009 TSMC, Ltd

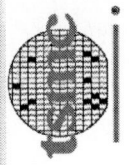

ITRS Roadmap (2007)

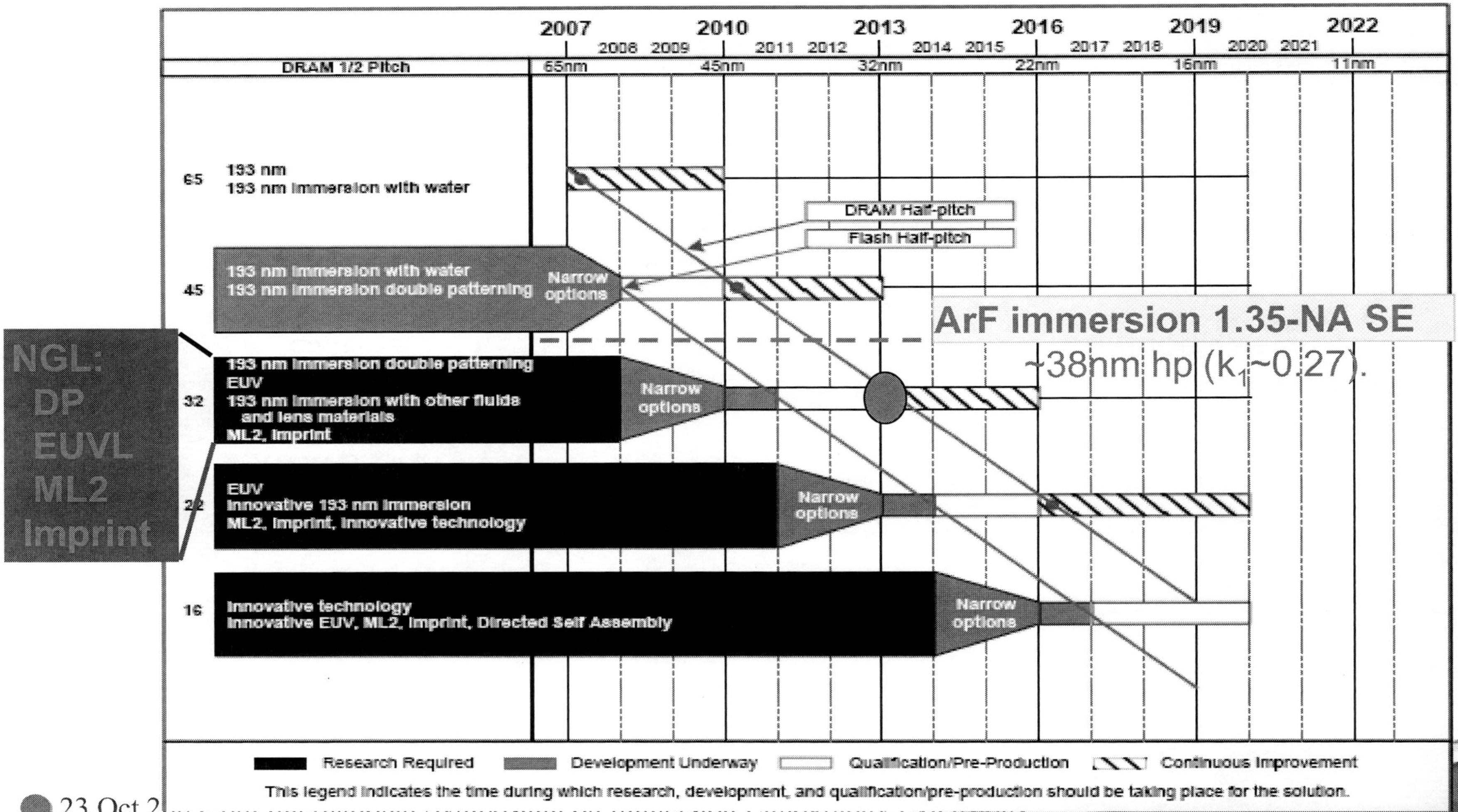

The Proven Path to Success

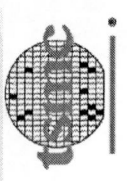

Considerations for NGL

- **Cost**
  - Comparable to existing single exposure
  - >100WPH at similar or less than one scanner footprint

- **Extensibility**
  - Resolution & Throughput
  - Extensible from 22nm node and at least next two nodes

- **Mask**
  - Remove/Relax mask making challenges

- **Patterning performance**
  - CDU & LWR
  - SMO & MMO with existing optical lithography

- **Defectivity**
  - Low defect density
  - Inspection solution!

23 Oct 2009, 6th International Symposium on Immersion Lithography Extensions

The Proven Path to Success ℠

© 2009 TSMC, Ltd

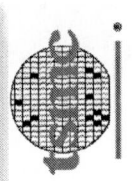

Major Challenges of NGLs

- Fundamentally, masks are too expensive and too difficult for <32nm-HP node and beyond

- **Double Patterning by ArFi**: Double masks/processes cost! Design rule restrictions!

- **EUV**: ML mask defect, inspection and source power.

- **Nanoimprint**: 1X, 3D template is too tough, defect and overlay

- **MEB ML2**: Throughput is a concern! However, it has a lot of advantages
  - No mask cost & mask induced troubles,
  - Remove design rule constraints,
  - Lowest cost if throughput can be > 10wph, or >100WPH by cluster
  - Cost (mainly from electronics) trend down by Moore's law,
  - MEB column is much cheaper than optical lenses

23 Oct 2009, 6th International Symposium on Immersion Lithography Extensions

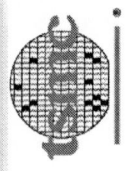

The MAPPER Technology

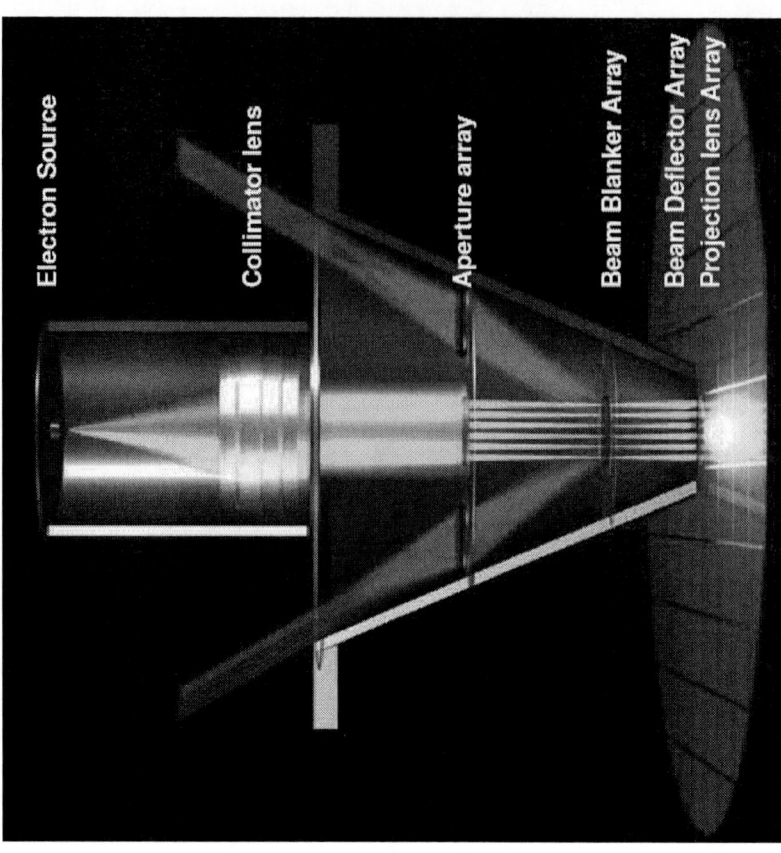

Electron Source

Collimator lens

Aperture array

Beam Blanker Array

Beam Deflector Array
Projection lens Array

- Single electron source split in 13,000 Gaussian beams

- V_{acc} = 5keV

- Apertures are imaged on substrate through 13,000 micro lenses

- MEMS-stacked static electric lenses

- Optical-switched CMOS-MEMS blanker array

* Infomation from MAPPER Lithography.

The Proven Path to Success ℠

- 23 Oct 2009, 6th International Symposium on Immersion Lithography Extensions

© 2009 TSMC, Ltd

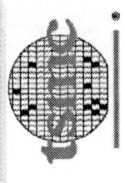

45nm images by Pre-Alpha Tool (Q4, '08)

HSQ thickness = 40nm

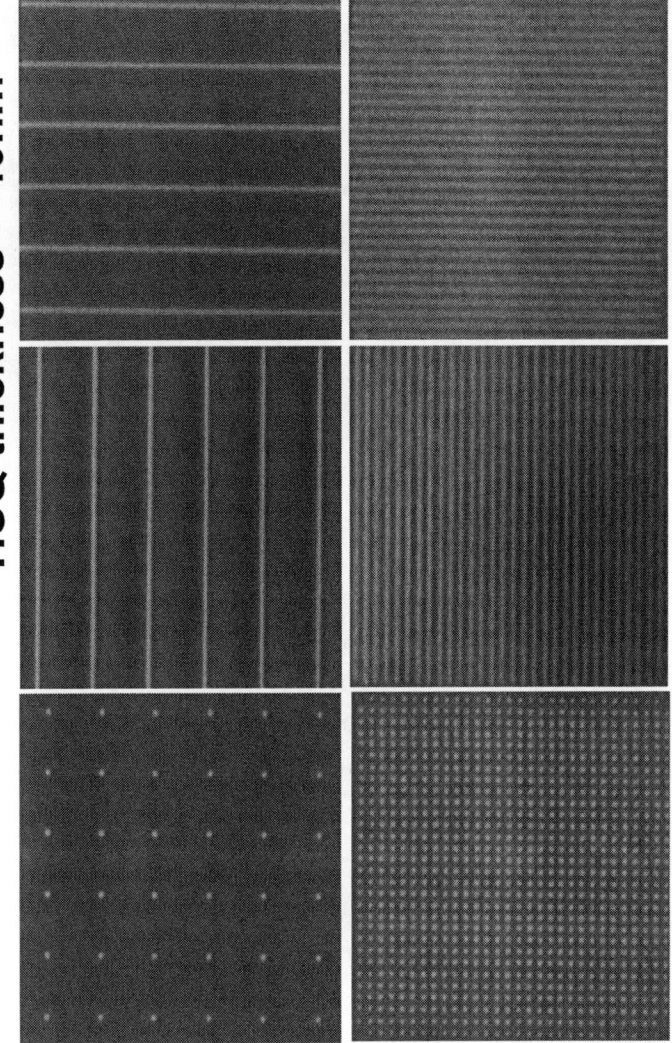

pattern	CD [nm]		CD Mean-to-target [nm]		CDu [nm]	
	Measured	Required	Measured	Required	Measured	Required
Dots dense	43.4	45	1.6	3.2	2.5	4.5
Dots isolated	46.4	45	1.4	3.2	2.8	4.5
Lines_Horizontal dense	42.8	45	2.2	3.2	1.9	4.5
Lines_Horizontal isolated	42.1	45	2.9	3.2	3.0	4.5
Lines_Vertical dense	44.9	45	0.1	3.2	2.8	4.5
Lines_Vertical isolated	46.5	45	1.5	3.2	2.9	4.5

MAPPER Pre-Alpha Tool 110beams @ 5keV,

23 Oct 2009, 6th International Symposium on Immersion Lithography Extensions

The Proven Path to Success

Cluster concept for 100WPH tool

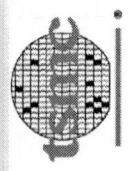

HVM clustered production tool:

- >13,000 beams per chamber (10WPH)
- 10WPH x 5 x 2 = 100WPH
- Footprint ~ArF scanner < 2/3 EUV scanner
- In-line to track

Interface to track

1 m

MAPPER single column tool
Upgrade to 13,000 beams
for 10WPH

Courtesy by MAPPER,

Proc. of SPIE 2009,Vol. 7271, 72710O

The Proven Path to Success℠

23 Oct 2009, 6th International Symposium on Immersion Lithography Extensions

© 2009 TSMC, Ltd

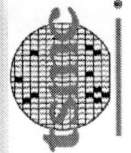

MAPPER Pre-Alpha Tool @ TSMC

- **Tool configuration**
  - 110x Gaussian beams @ 5keV
  - Raster scan by individual beam, with MEMS blanker array controlled by 110x optical data channels
  - 300mm wafer stage & loadlock interface
  - Resolution start from 45nm HP, will upgrade to 32nm HP.

- **Possibly upgrade to 10WPH on the same platform**

23 Oct 2009, 6th International Symposium on Immersion Lithography Extensions

The Proven Path to Success SM

© 2009 TSMC, Ltd

45nm HP resolution & CDU correction

- Individual beam current can be measured by using Faraday cup,

- Correlation of CD vs beam current shows the possibility of correction CDU by apply different dosage offset

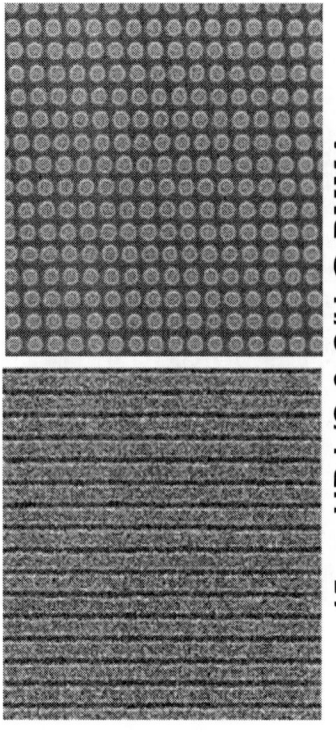

45nm HP L/S & C/H @ PMMA

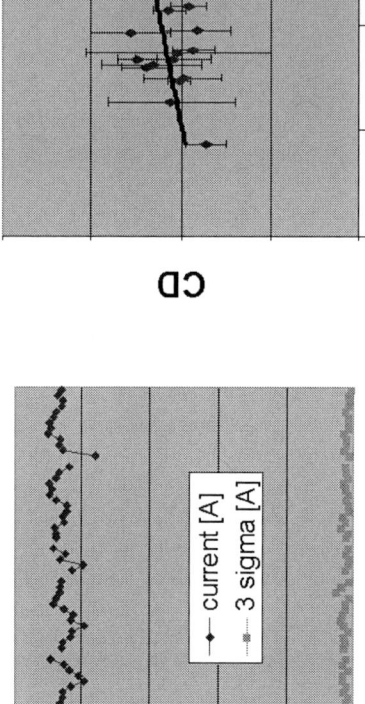

CD vs beam current @45nm HP L/S

Individual Beam Current

23 Oct 2009, 6th International Symposium on Immersion Lithography Extensions

The Proven Path to Success SM

© 2009 TSMC, Ltd

30nm HP resolution @5keV

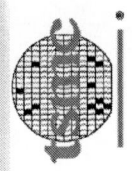

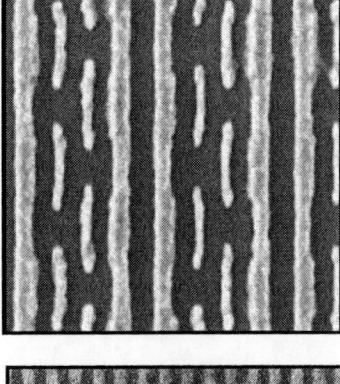
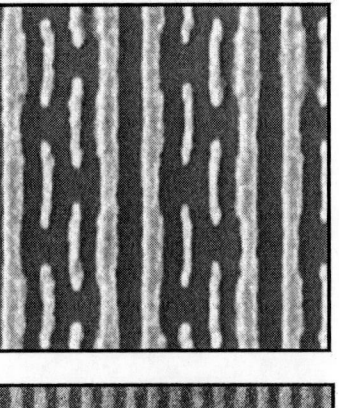

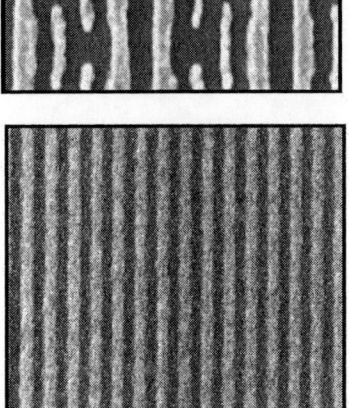

Elionix ESM9000

- Mimic 5keV writing experiments at spot size ~ 25nm were done by a SEM-converted writer in NTU IRND Lab.

- Manual processes, in poor environmental control.

- **HSQ – Negative resist** @40nm thickness

- **ZEP520A – Positive resist** @40nm thickness

23 Oct 2009, 6th International Symposium on Immersion Lithography Extensions

The Proven Path to Success ℠

© 2009 TSMC, Ltd

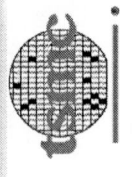

E-Beam Proximity Correction Verification

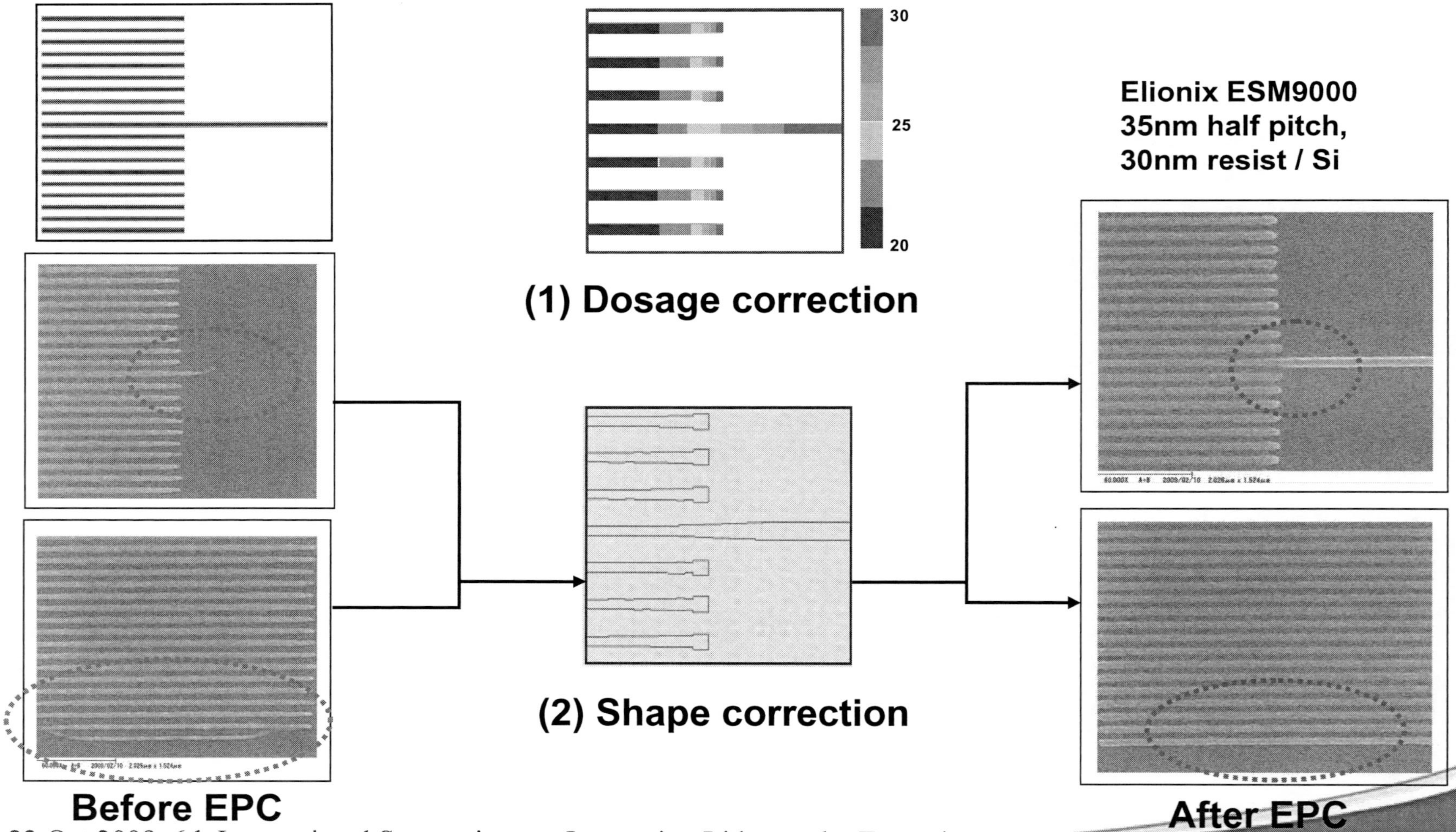

Elionix ESM9000
35nm half pitch,
30nm resist / Si

(1) Dosage correction

(2) Shape correction

Before EPC

After EPC

23 Oct 2009, 6th International Symposium on Immersion Lithography Extensions

The Proven Path to Success℠

© 2009 TSMC, Ltd

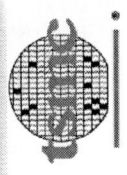

Proximity Effect Correction

Test Clip:

32nm Logic clips

Conditions:

HSQ thickness 40nm
Beam size = 35 nm
Scanning pixel = 2.25 nm

Reference: SPIE 7271_54 (2009)

W/O EPC

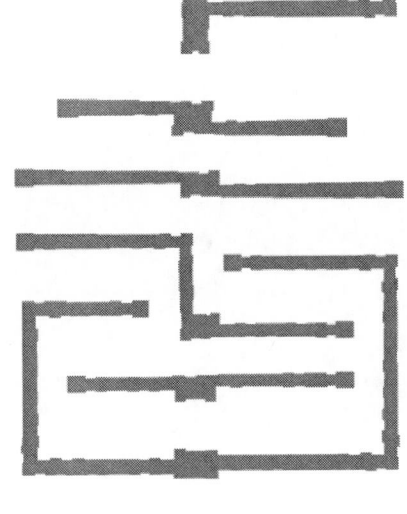

W/ EPC

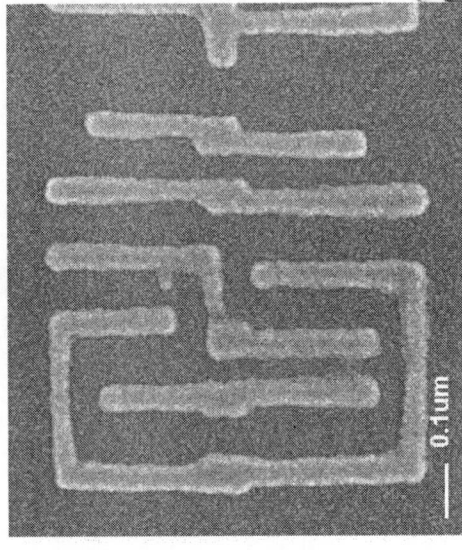

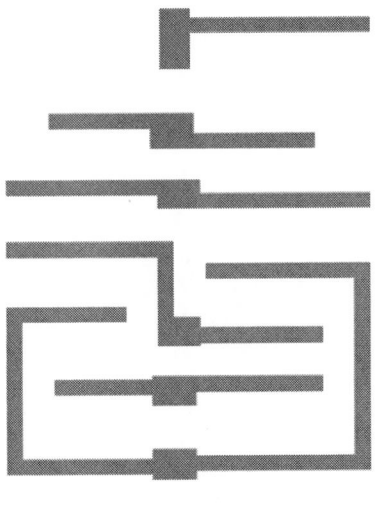

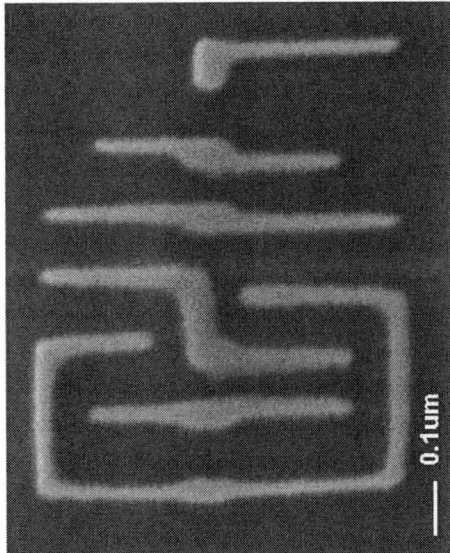

by MOSES

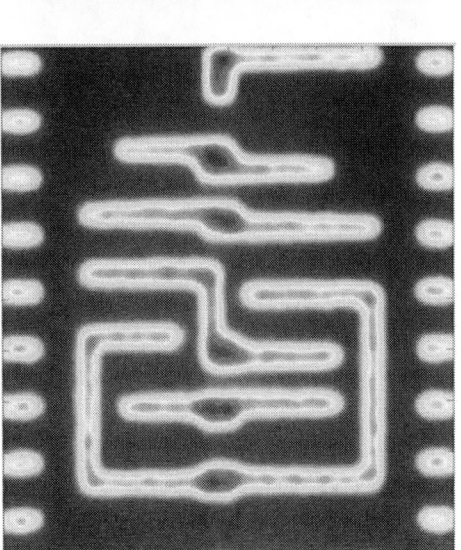

23 Oct 2009, 6th International Symposium on Immersion Lithography Extensions

The Proven Path to Success℠

© 2009 TSMC, Ltd

Chemical Amplified Resist @ 5keV

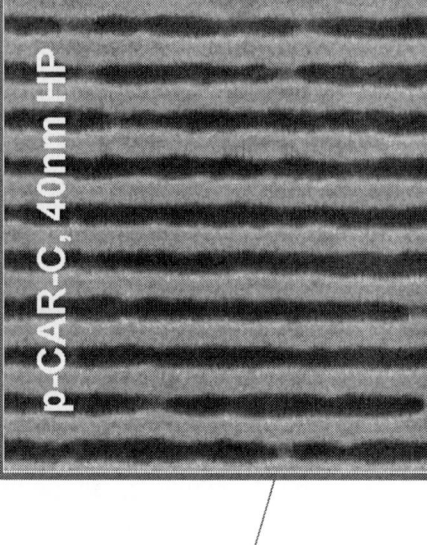

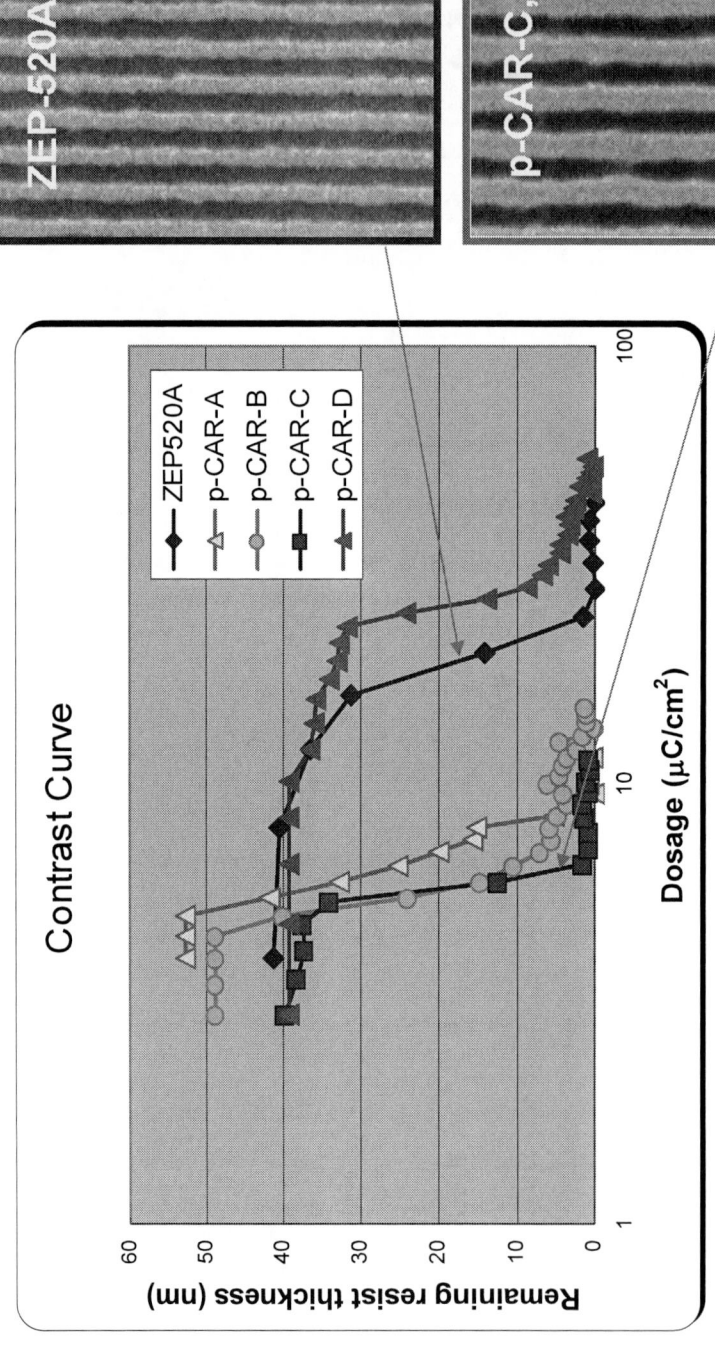

- ZEP-520A proved high resolution of EBL tool.

- P-CAR-C, with polymer bound PAG, showed highest contrast at 5 keV.

- Severe T-topping due to airborne contamination.

23 Oct 2009, 6th International Symposium on Immersion Lithography Extensions

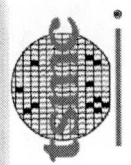

Throughput Challenge – 1: Source

$$WPH = \frac{3600}{t}$$

$$t = t_w + t_m + t_o$$

$$t = \frac{QA}{n \cdot I} + \sum t_m \times r_m + t_o$$

t_w: time for writing all features on a wafer

t_m: time for movements between shots, including over-scans, turnovers of changing scanning direction, and so on

t_o: overhead time between wafers

However, $I = \dfrac{\pi^2}{4} B_r V \alpha^2 d_I^2$

Monte-Carlo simulation by MOSES:

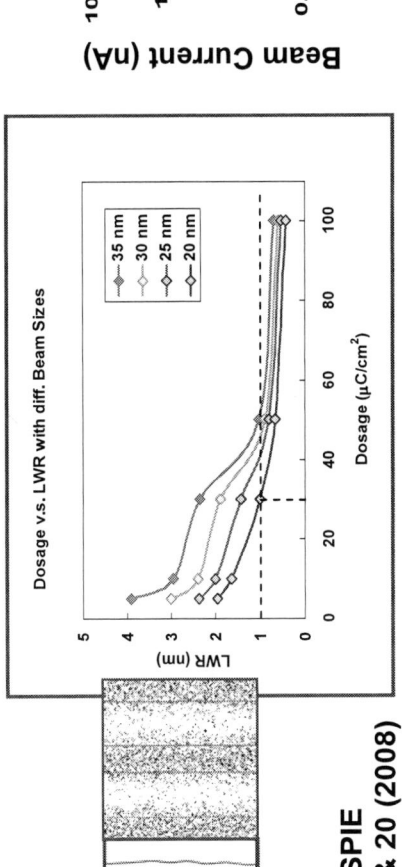

Refer to SPIE
6921-19 & 20 (2008)

➜ Required source Brightness ~ 10^7 A/m^{2}Sr^{2}V!

➜ Or need a solution for ~50x increment on writing area for a normal source brightness!

23 Oct 2009, 6th

© 2009 TSMC, Ltd

The Proven Path to Success SM

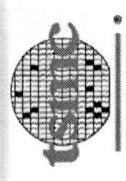

Throughput Challenge – 2: Data Rate

$$WPH = \frac{3600}{t}$$

$$t = t_w + t_m + t_o$$

$$= \frac{TotalData}{n \cdot DataRate} + \sum t_m \times n$$

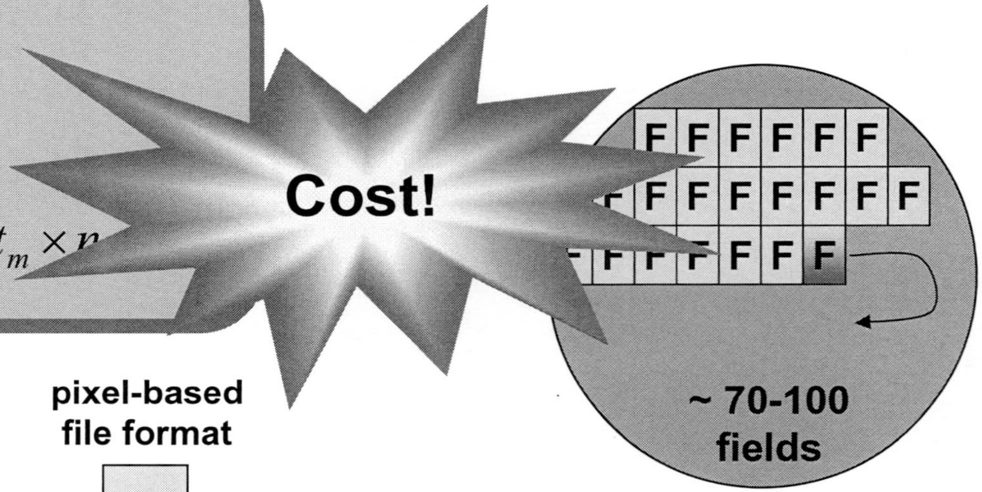

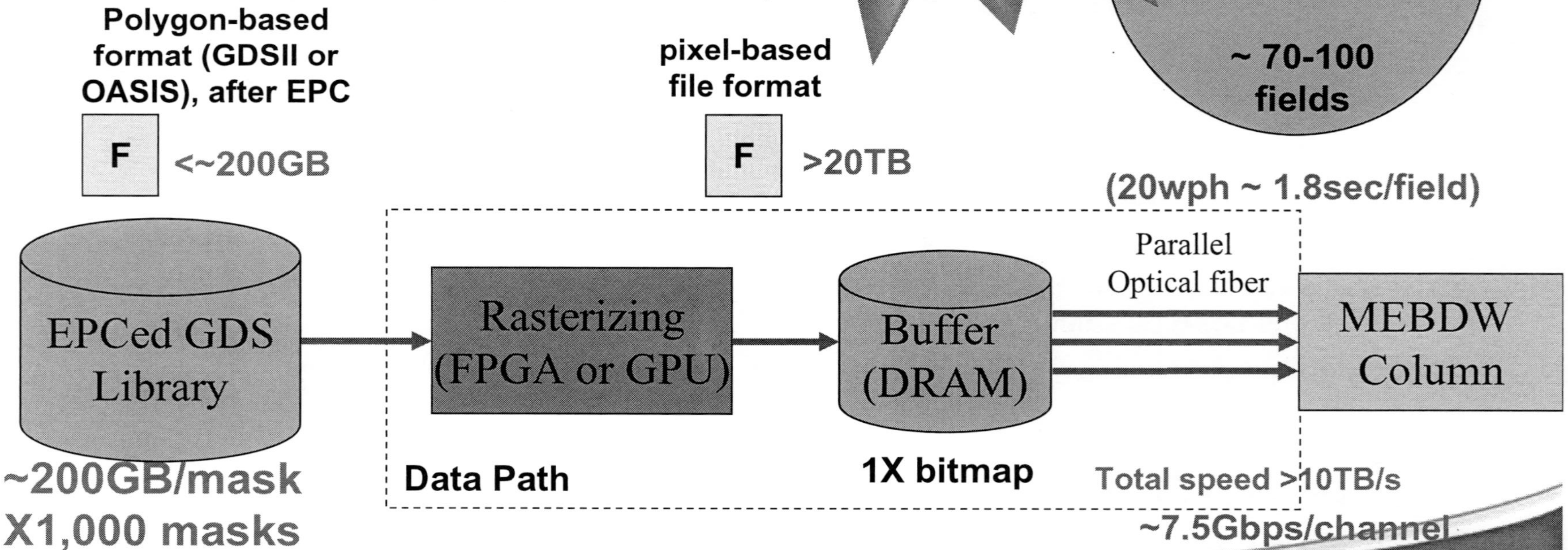

Cost!

~ 70-100 fields

Polygon-based format (GDSII or OASIS), after EPC

pixel-based file format

| F | <~200GB |

| F | >20TB |

EPCed GDS Library

Rasterizing (FPGA or GPU)

Buffer (DRAM)

1X bitmap

Parallel Optical fiber

MEBDW Column

(20wph ~ 1.8sec/field)

Total speed >10TB/s

~7.5Gbps/channel

Data Path

~200GB/mask
X1,000 masks

- 23 Oct 2009, 6th International Symposium on Immersion Lithography Extensions

The Proven Path to Success

© 2009 TSMC, Ltd

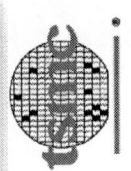

Conclusions

- MAPPER Pre-Alpha tool has been installed in manufacturing environment, and 45-nm HP resolution by 110 beams has been successfully proven.

- High resolution down to 30nm HP at 5keV has been demonstrated, and EPC by shape modulation has been proven.

- Clustered MEB can achieve 100WPH at scanner footprint, and thus in-line to track. CAR is also feasible. So the existing single patterning lithography concept and operation can continue.

- Ebeam maskless lithography is the most desirable NGL if succeeds! Since maskless, as long as the MEB tool is ready, the technology is ready!

23 Oct 2009, 6th International Symposium on Immersion Lithography Extensions

The Proven Path to Success ℠

© 2009 TSMC, Ltd

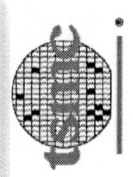

Acknowledgement

- Mr. Hill Liao and Mr. Te-Wei Tsai from TSMC, Hsinchu, for their support on tool installation and resist testing.

- Prof. J.Y. Yen, Prof. K.Y. Tsai, Prof. C.H. Kuan and their group from National Taiwan University for providing EBL tool and lab facility.

- Dr. Yoshio Kawai, from ShinEtsu, Japan, for providing CAR resist sample.

- Mr. Maurits Weeda, Mr. Tijs Teepen, Mr. Abdi Farah, and other colleagues from MAPPER Lithography, Delft, who contributed to the pre-alpha tool.

23 Oct 2009, 6th International Symposium on Immersion Lithography Extensions

The Proven Path to Success

© 2009 TSMC, Ltd

The End

Questions?

23 Oct 2009, 6th International Symposium on Immersion Lithography Extensions

The Proven Path to Success℠

© 2009 TSMC, Ltd

Full-field Liquid Immersion Interference Lithography

James Jacob[1], John Hoffnagle[1], John Burnett[2], Eric Benck[2], Darrell Armstrong[3], Arlee Smith[4]

Acknowledgment: Michael Fritze, DARPA

1. Actinix, Scotts Valley, CA
2. NIST, Gaithersburg, MD
3. Sandia National Labs, Albuquerque, NM
4. A-S Photonics, Albuquerque, NM

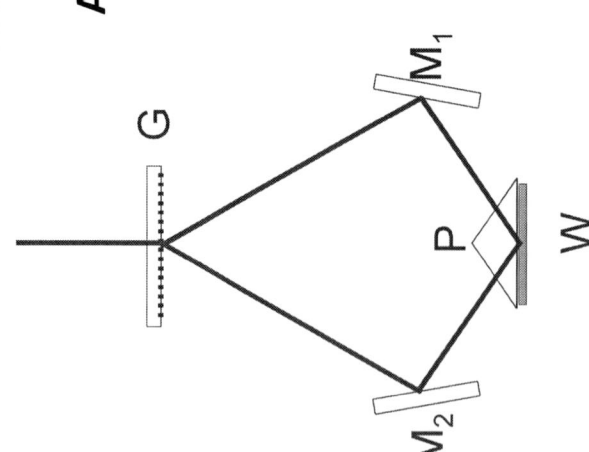

Liquid immersion interference lithography (LIIL)

- Why LIIL? Potentially very low cost, maskless process for 35-25 nm hp using water and gen 2 and 3 fluids

- Main Issues: Full-field exposures and new circuit libraries (2D to 1D)

- Enabling technology: long coherence length, far-UV laser (λ=197 nm) being developed by Actinix/Sandia under a DARPA STTR for low volume nano-fab

- This laser development project is related to upcoming DARPA program, GRATE: Gratings of Regular Arrays and Trim Exposures

- Pending proposal on a full-field LIIL R&D tool for testing new nano-scale metrology methods and protocols essential for next generation semiconductor device production

- Tool will support investigations into fringe doubling and directed self-assembly (DSA) as viable next generation litho processes (< 16 nm hp).

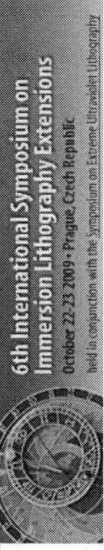

Gratings of Regular Arrays and Trim Exposures (GRATE)

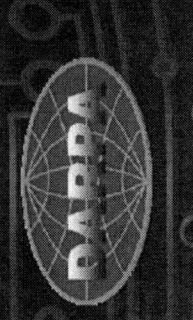

Goals:
- Demonstrate grating-based design paradigm
- Extend the capability of BOTH trailing and leading edge lithography tools using GRATE
- Develop interference lithography tool prototype to enable maskless GRATE

Deliverables:
- Grating-based 1D Design Paradigm
- Design Conversion Software (2D to 1D)
- Interference Lithography Prototype Tool

Today's Typical Design

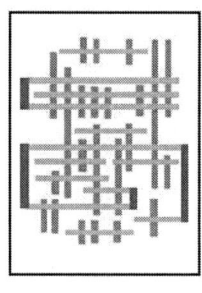

2D layout with complex active geometries

"GRATE" Design

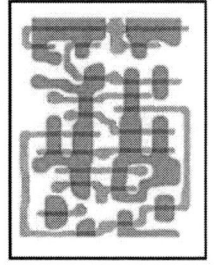

Routing using "stitched" line segments

Much simpler geometries are key to GRATE's benefits

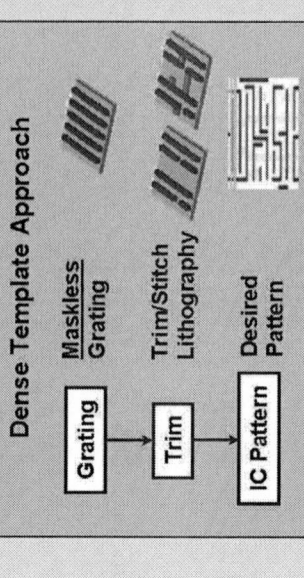

Dense Template Approach

Grating → Maskless Grating

Trim → Trim/Stitch Lithography

IC Pattern → Desired Pattern

GRATE Template Approach
- Fine features all simple line segments
- Cover wafer with simple gratings
  - Mask-based or Maskless interference
- Form IC patterns by "trimming & stitching"
  - Mask-based, Direct write (e-beam, ZPAL, etc)

Actinix

6th International Symposium on Immersion Lithography Extensions

October 22-23 2009 • Prague, Czech Republic
held in conjunction with the Symposium on Extreme Ultraviolet Lithography

Feasibility of "trimming and stitching" gratings demonstrated for circuit-like patterns (Qi, Purdue)

Objective:
- Demonstrate 32-nm Manhattan geometry IC patterns with atomically smooth sidewalls along critical dimensions
- Demonstrate maskless lithography with ≥ 10X throughput compared with current state-of-the-art

Impact:
- Key building block for demonstrating high throughput cost effective maskless lithography

32 nm Node gratings demonstrated

Atomically smooth [111] sidewalls on [110] silicon surface

Cutting with E-beam + RIE demonstrated

Stitching with HSQ resist demonstrated

Initial work with E-beam
Following up with optical interference gratings

Take Advantage of the Manhattan Grating Structure

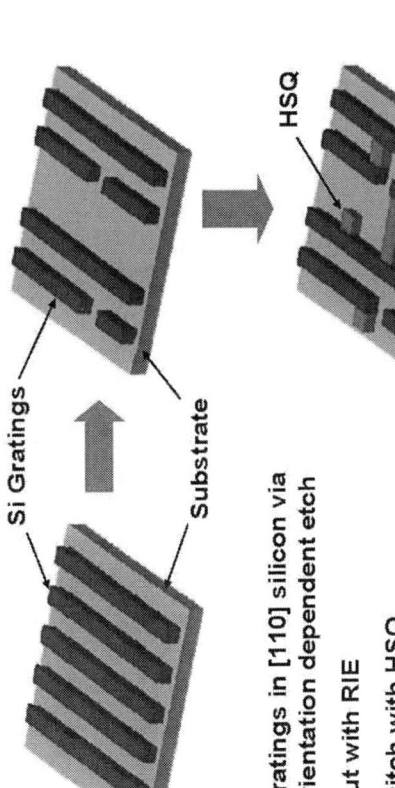

Si Gratings

Substrate

HSQ

- Gratings in [110] silicon via orientation dependent etch
- Cut with RIE
- Stitch with HSQ
- Nanoimprint

Technical Challenges:
- Show effective crystallographic etching for very dense grating patterns ≤ 32 nm
- Show HSQ glass resist can be used for stitching together customized patterns

Technical Approach:
- Employ crystallographic etching to achieve dense 32-nm gratings with atomically smooth sidewalls
- Use E-beam with HSQ glass resist to customize preformed silicon grating templates

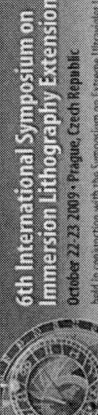

6th International Symposium on
Immersion Lithography Extensions
October 22-23 2009 • Prague, Czech Republic
Held in conjunction with the Symposium on Extreme Ultraviolet Lithography

Interference Lithography Tools - Some Examples

Micro-exposure Tools

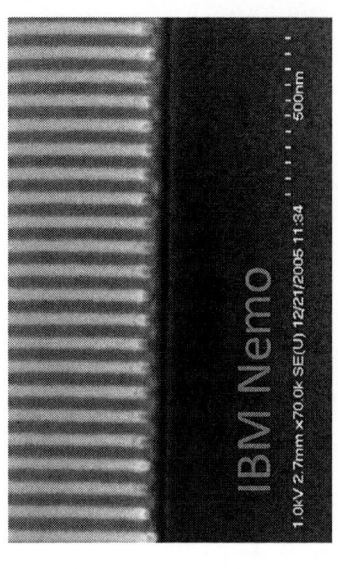

IBM Nemo
1.0kV 2.7mm x70.0k SE(U) 12/21/2005 11:34   500nm

- 257 nm - IBM, UNM
- 248 nm - RIT
- 213 nm - UNM
- 193 nm - RIT (Amphibian), IBM Nemo
- 157 nm - MIT LL
- EUV - Intel, Hanyang Univ., Paul Scherrer Inst., Univ. of Wisconsin

MIT NanoRuler - Scanning beam interference litho (SBIL)

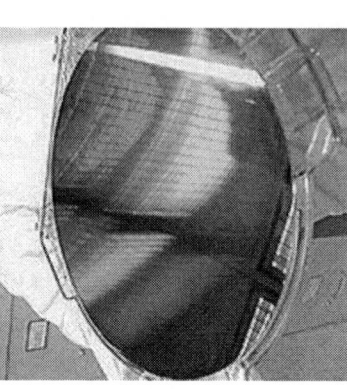

- <u>Meter scale gratings</u>
- 351 nm (Ar ion), Dry
- High precision metrology
- Transparent substrates
- Scanned small field, 1 mm
- Approx. 100 nm hp

http://snl.mit.edu/

Actinix

6th International Symposium on
Immersion Lithography Extensions
October 22-23 2009 • Prague, Czech Republic
held in conjunction with the Symposium on Extreme Ultraviolet Lithography

LIL Tool: IBM Nemo

- Laser: 193 nm, 7 pm, 25 mW, 5 kHz
- High spatial coherence, TEM00
- Liquid Immersion: H_2O, Gen II
- 35 - 30 nm hp
- Millimeter field size
- High contrast > 90%

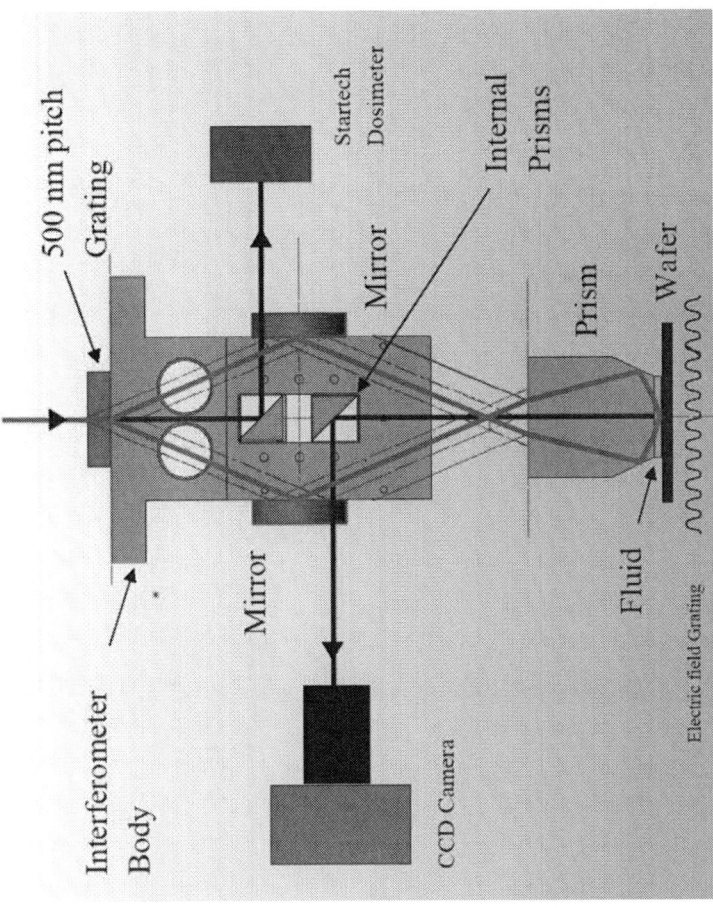

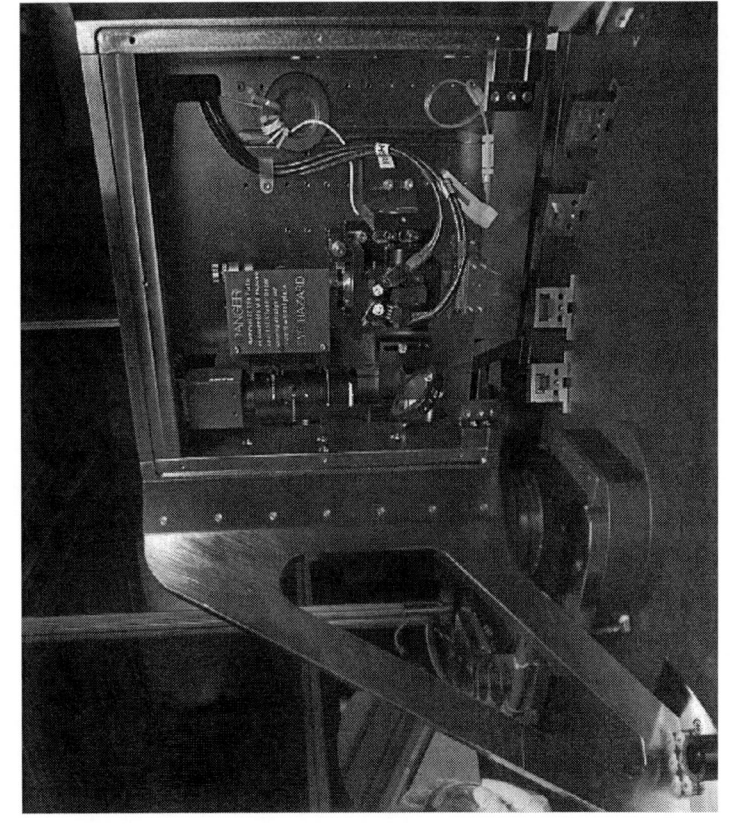

197 nm vs 193 nm : High index fluids have less transmission loss

Lower α → **Less heating** → **Lower Δn** → **Less Distortion**

- Increase source λ by 2%
- Move away from band edge
- Higher index liquids more feasible

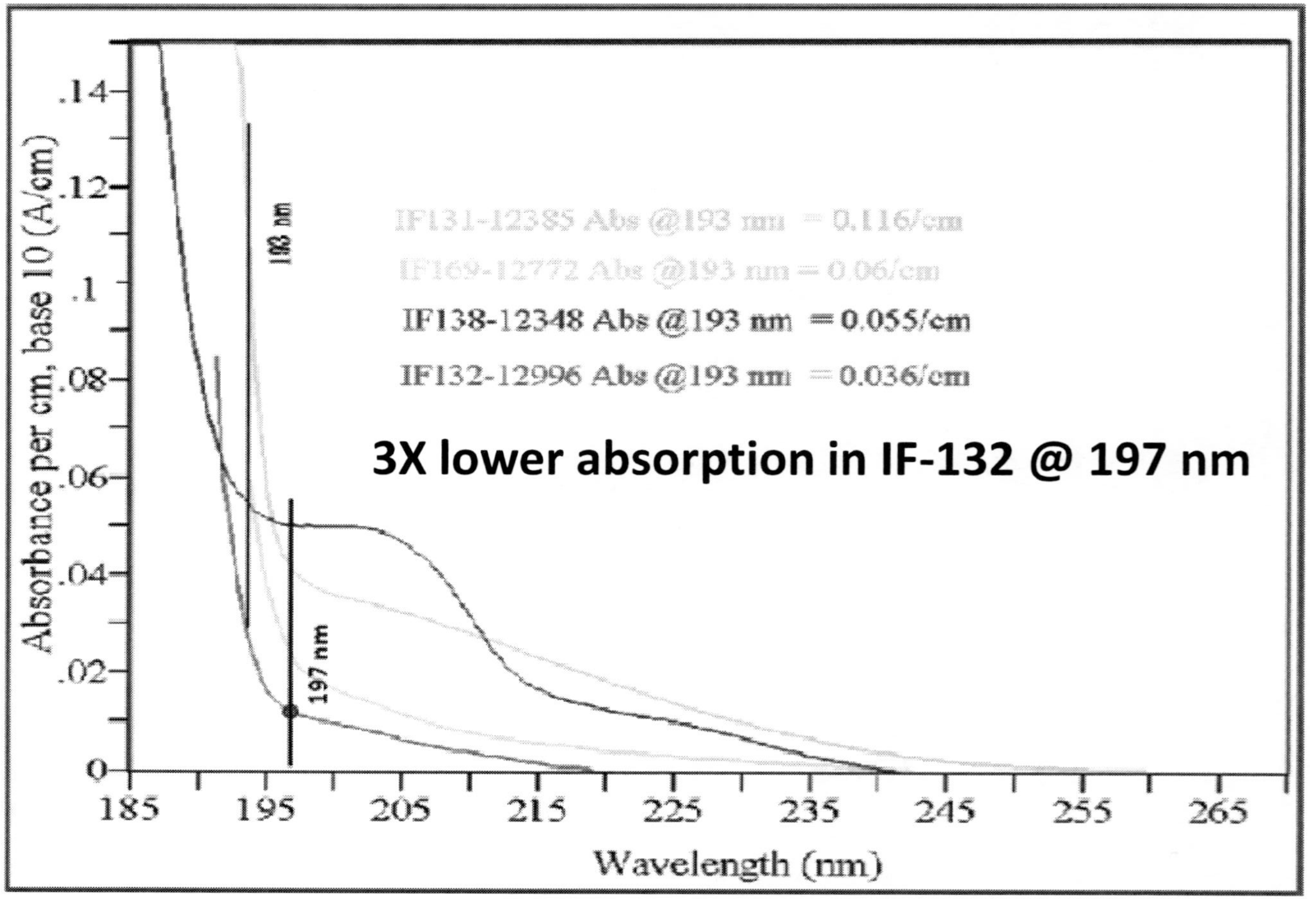

IF131-12385 Abs @193 nm = 0.116/cm
IF169-12772 Abs @193 nm = 0.06/cm
IF138-12348 Abs @193 nm = 0.055/cm
IF132-12996 Abs @193 nm = 0.036/cm

3X lower absorption in IF-132 @ 197 nm

183 nm

197 nm

Absorbance per cm, base 10 (A/cm)

.14 .12 .1 .08 .06 .04 .02 0

Wavelength (nm)

185 195 205 215 225 235 245 255 265

Second Generation Fluids for 193 nm Immersion Lithography, R. French, et al. SPIE Microlithography 2007.

Actinix

6th International Symposium on
Immersion Lithography Extensions
October 22-23 2009 • Prague Czech Republic

Technical Challenges for LIL to Print Full-field Nano-scale Patterns

- Long coherence length - L_c of 5 cm prints 35 nm hp over 1 cm field @ 98% V
- High spatial coherence - pattern fidelity, uniformity
- Power - @ 50 mJ/cm² dose, 1 watt => 49 w/hr, assuming 50% overall optical efficiency
- Wavelength stability - pitch control
- Energy stability - dose control
- Wavefront flatness - suppression of hyperbolic fringes
- Rectangular fields - mask circular/elliptical exposure beams
- Metrology - measure fringe straightness, pitch, registration
- Optical quality - minimize figure errors

Modify the Nemo design for full field patterning:
- *Increase beam dia. to >3 cm, or*
- *Use elliptical pattern, 0.5 x 2.6 cm, and scan*
- *Mask to form rectangular exposures*
- *Employ metrology for registration*

Single full-field

Scanned sub-field

or

26 mm

33 mm

Coherence Properties of Light Source

Temporal Coherence

$$L_c = c/\pi\Delta\nu \quad \text{(assumes Lorentzian lineshape)}$$

@ 197 nm: L_c of 4.7 cm => $\Delta\nu$ of 2 GHz => $\Delta\lambda$ of 0.26 pm

Spatial Coherence

Solid-state 193 nm laser mode quality

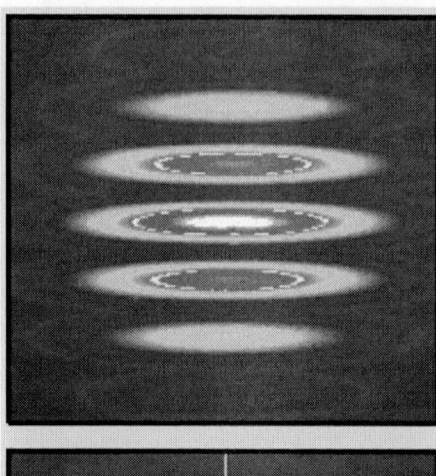
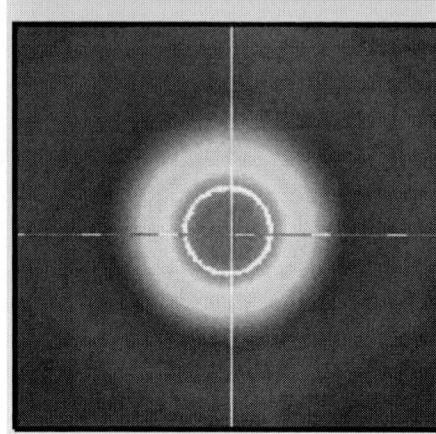

TEM00, $M^2 = 1.2$ Spatial Coherence Test

High spatial coherence is necessary for pattern fidelity and uniformity

Ray A is symmetrical - paths are equal
Ray B lacks symmetry - paths are unequal
Coherence length needs to accommodate OPD

BW	Pitch	Contrast	Field size
0.26 pm	70 nm	98%	10 mm
0.26 pm	70 nm	92%	22 mm
0.26 pm	70 nm	83%	33 mm

Interferometer - symmetrical grating-mirror design

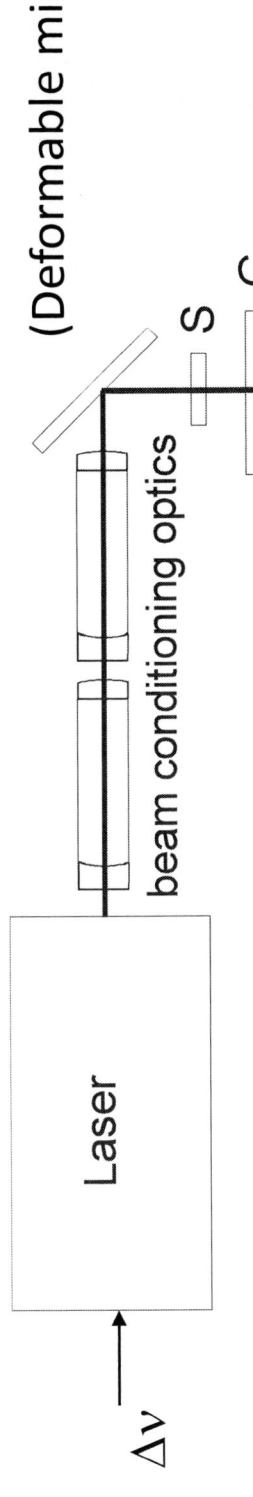

- $\Delta\nu$ - fine freq control
- G - Grating
- Mirrors M1, M2 control pitch
- D - dose monitor
- S - shutter
- L - Inspection lens
- C - CCD camera
- P - coupling prism
- W - wafer
- MG - Metrology grating

Beam shaping

Super-gaussian profiles

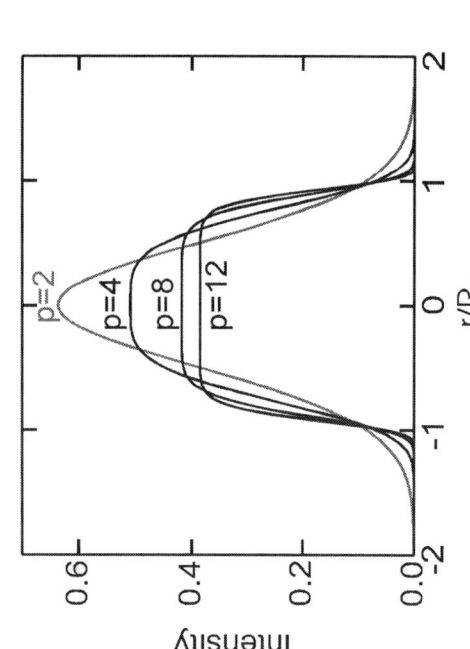

Efficiency vs uniformity

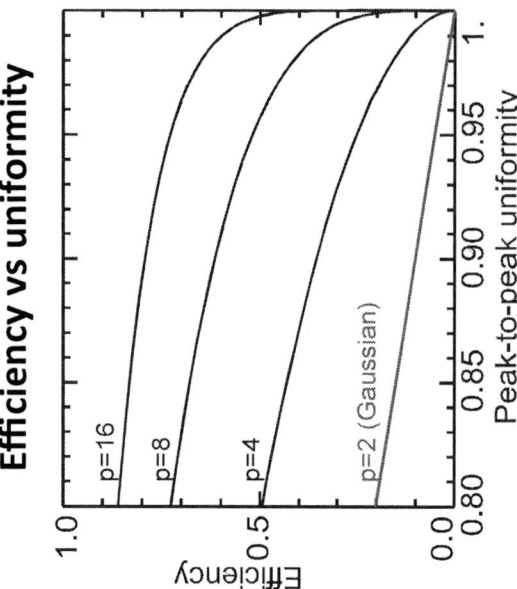

Keplerian-type anamorphic telescope

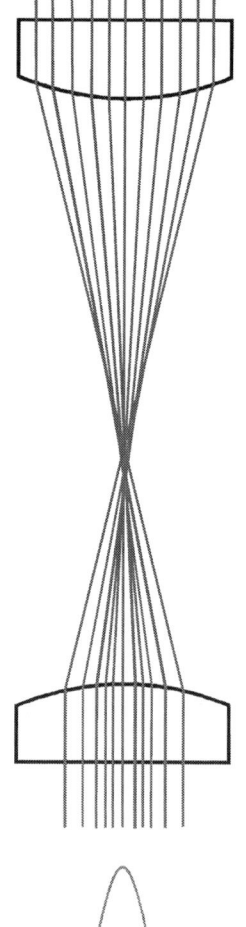

Collimated beam with planar wavefront

Second lens corrects the wavefront distortion due to the first lens,

First lens redistributes rays to transform the intensity profile

Metrology – Pitch Control

- Rectangular interference pattern illumination - full chip field (or scanned over field)

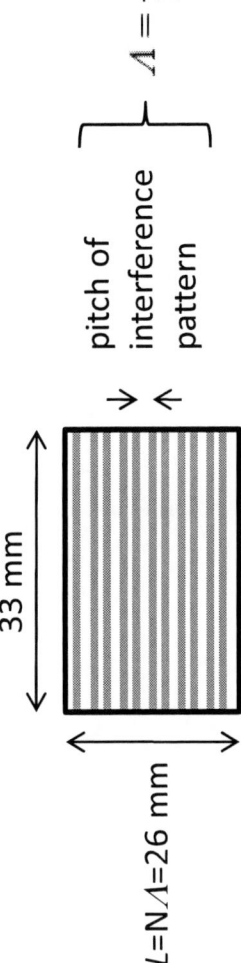

$$\Lambda = \frac{\lambda}{2n\sin\theta}$$

pitch of interference pattern

33 mm

$L = N\Lambda = 26$ mm

- Registration requirements (~10% of pitch) = 0.1 x 70 nm = 7 nm

$\Rightarrow$ distortion $\Delta L \leq 7$ nm over field

$\Rightarrow$ tolerance on pitch Λ

$$\frac{\Delta\Lambda}{\Lambda} = \frac{\Delta L \text{ (distortion)}}{L \text{ (field size)}} = \frac{7 \text{ nm}}{26 \text{ mm}} \approx 0.3 \times 10^{-6}$$

- Tolerance on: $\theta, n, \Delta T, \lambda$

$$\Delta\Lambda = -\Lambda\cot\theta\left(\frac{\Delta\theta}{\theta}\right) - \Lambda\left(\frac{\Delta n}{n}\right) + \Lambda\left(\frac{\Delta\lambda}{\lambda}\right)$$

$$\Delta\theta = -\tan\theta(\Delta\Lambda/\Lambda) \approx -1.0 \times 10^{-6} \text{ rad}$$

$$\Delta n = -n(\Delta\Lambda/\Lambda) \approx -0.4 \times 10^{-6}$$

$$\Rightarrow \Delta T = 0.004 \text{ °C} \quad \text{(for water } dn/dT = 1.0 \times 10^{-4}\text{)}$$

$$\Delta\lambda = \lambda(\Delta\Lambda/\Lambda) \approx 0.06 \times 10^{-12}\text{m} \quad (\Delta\nu \approx 460 \text{ MHz})$$

- Can use precise control of λ to help maintain registration across field

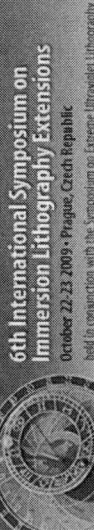

6th International Symposium on
Immersion Lithography Extensions

October 22-23 2009 • Prague, Czech Republic
held in cooperation with the Symposium on Extreme Ultraviolet Lithography

Metrology – Registration with Reference Grating

- Need to control pattern pitch, straightness, position, and orientation
- Overlap laser interference pattern with reference reflection grating having desired pitch
- ⇒ reflected Moiré pattern

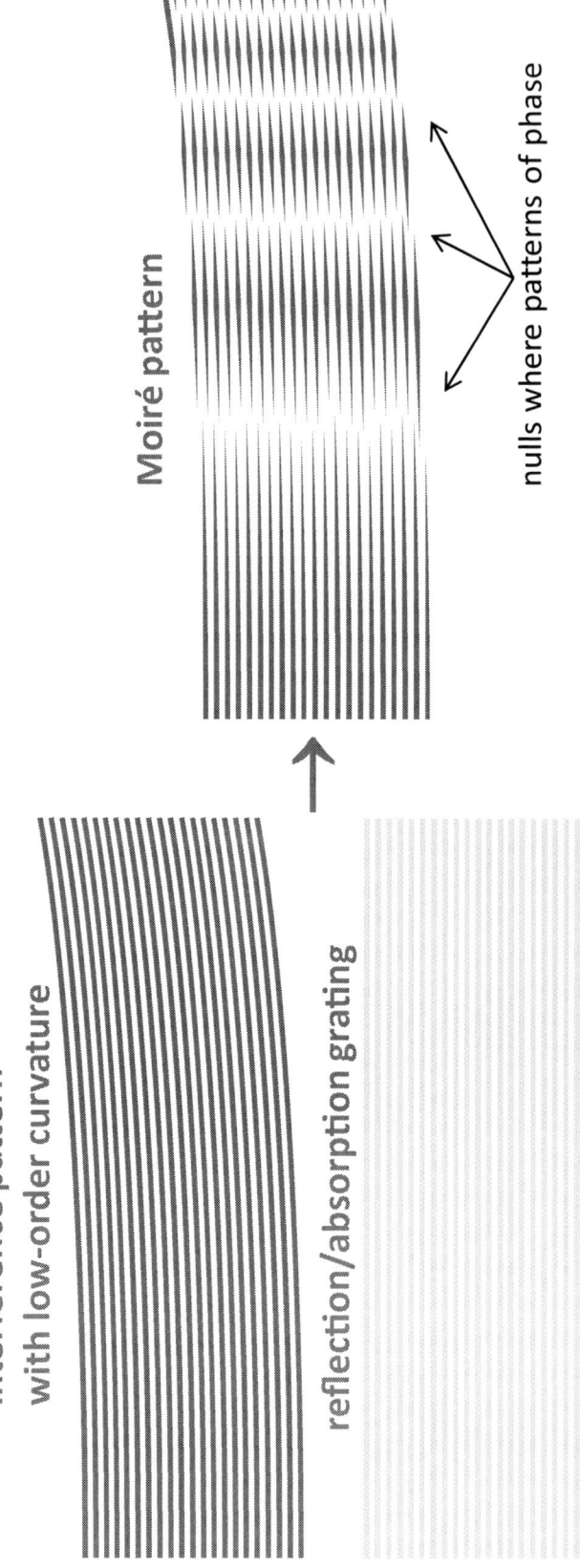

interference pattern
with low-order curvature

reflection/absorption grating

Moiré pattern

nulls where patterns of phase

- View Moiré pattern with CCD (e.g., 100 fringes per pixel) - *unresolved interference*
- Eliminate Moiré nulls with stage rotation, translation, λ control, and adaptive optics (for low-order curvature)
- ⇒ can verify registration with grating across field
- Achieve registration on wafer with help of wafer alignment gratings

Long Coherence Length Light Source - Key Enabler

Initial Specs : 197 nm; Avg. power = 250 mW; BW = 0.26 pm; PRF = 1 MHz

Scaled Specs : 197 nm; Avg. power = 1 W; BW = 0.09 pm; PRF = 4 MHz

- **Fiber laser based light source -**
  - Long storage time in fiber amps, pulsed oscillator appears CW
  - Pulse repetition frequency can therefore be easily increased 4X
  - Requires higher pump diode power - current design only 6 W/amp

- **CLBO mixing crystal used for 197 nm generation**
  - Low absorption
  - Can be non-critically phase-matched (no walkoff)

- **Other fiber laser based sub-200 nm systems:**
  - Nikon - 8th harmonic of Er fiber (193 nm)
  - Megaopto - Er/Yb fibers mixing to 198 nm

6th International Symposium on
Immersion Lithography Extensions
October 22-23 2009 • Prague, Czech Republic
held in conjunction with the Symposium on Extreme Ultraviolet Lithography

Solid-state Far-UV Generation

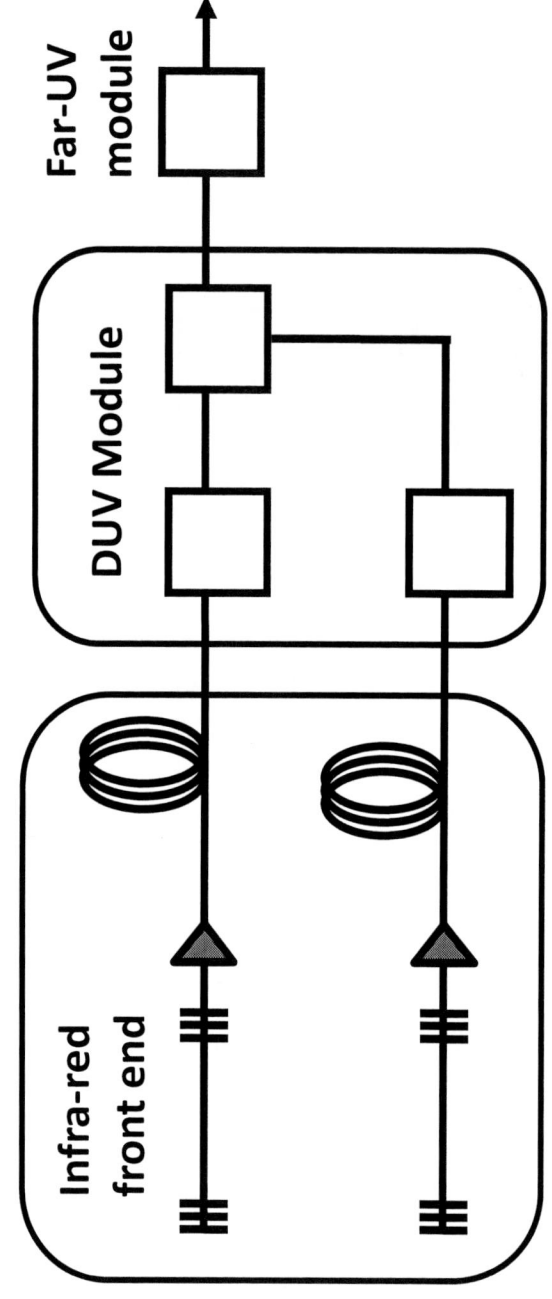

Fiber Amplifiers: Design for Low Self Phase Modulation

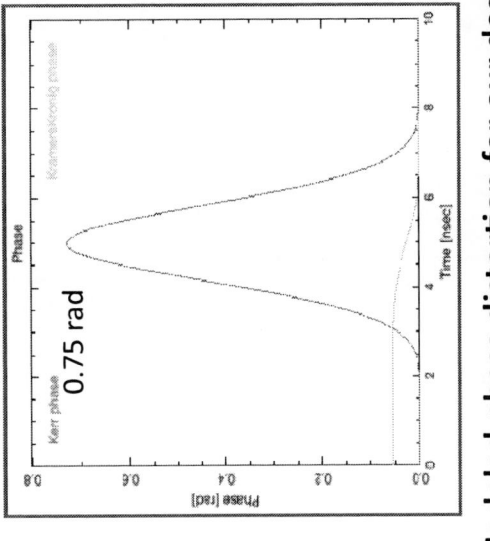

Modeled phase distortion for our design

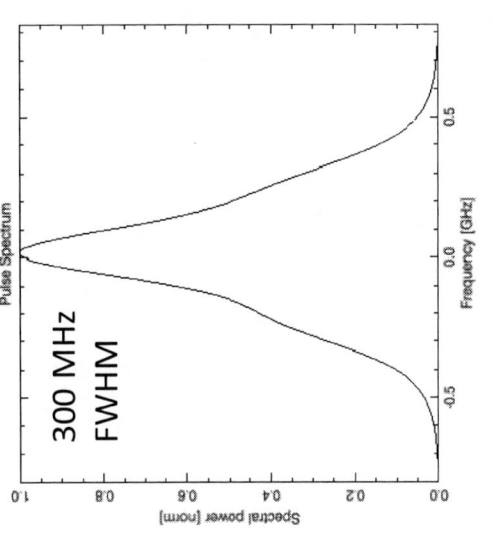

Modeled spectrum assuming 2.0 rad phase

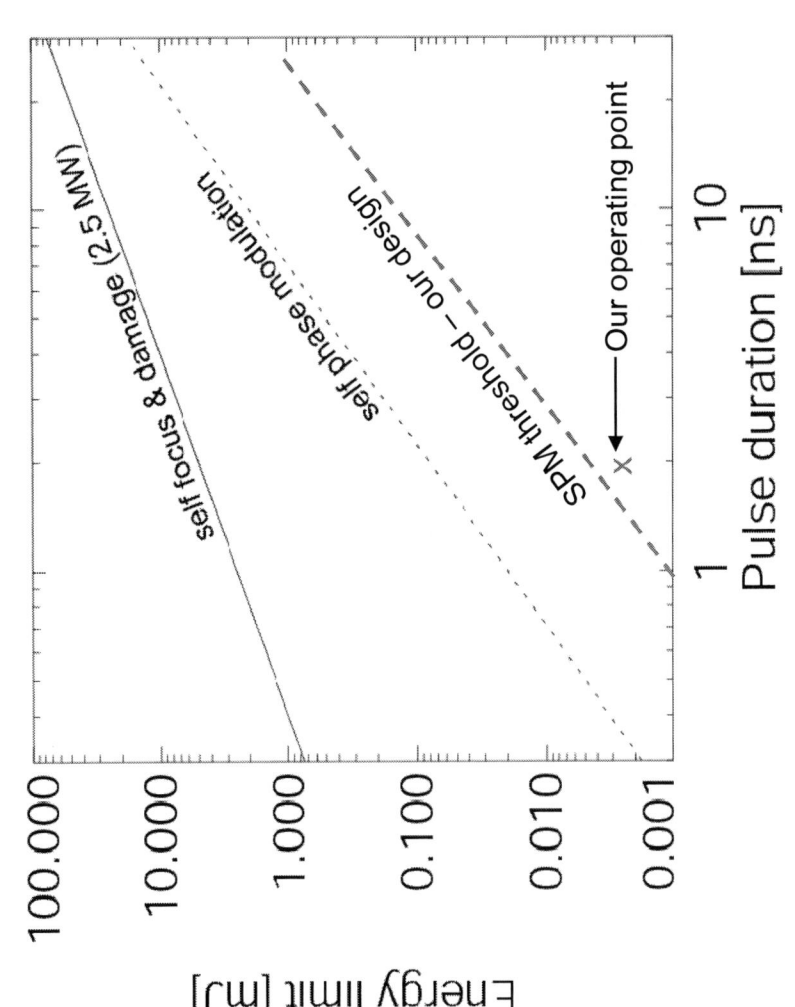

Pulse duration = 2 ns; Energy = 2.6 μJ (P = 2.6 watts)

Power scalable - increase PRF to 4 KHz, P = 10 watts

Efficient NLO ⇨ Lower Power Fiber Amp ⇨ Narrower Bandwidth

In birefringent crystals Poynting vector walk-off reduces effective length

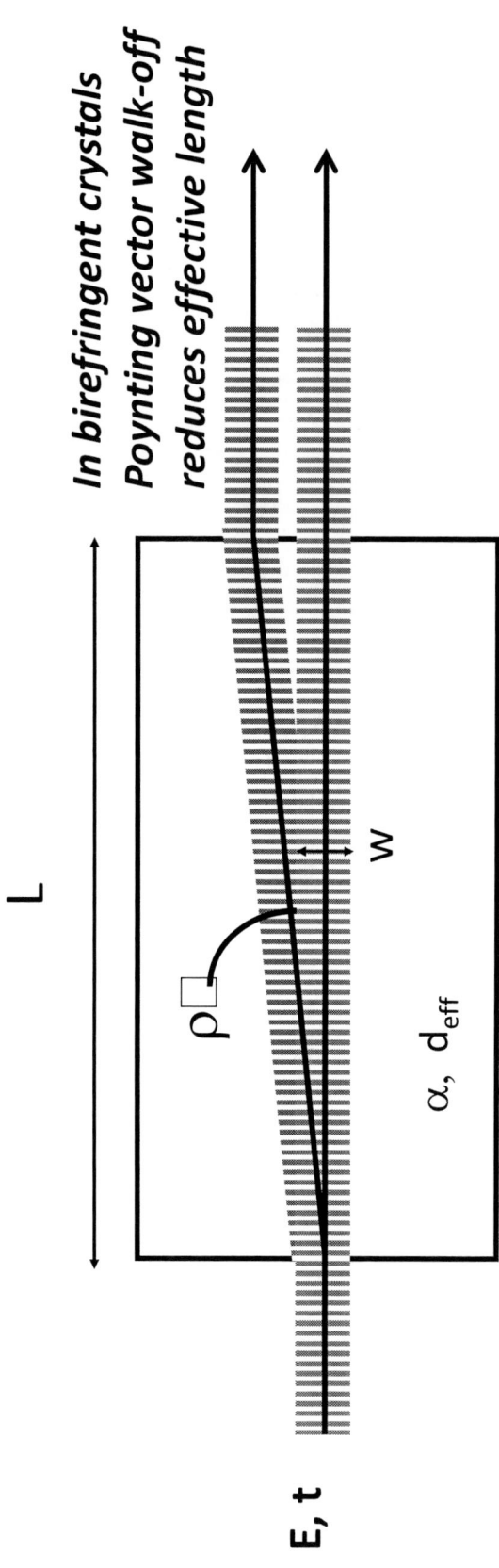

For highest efficiency:

- Low absorption, α
- High non-linear coefficient, d_{eff}
- Long crystal length, L
- No walkoff, non-critical PM, $\rho = 0$
- Small spot size, w

Efficiency relation for NLO interaction

$$\eta \sim E^2 ,\ 1/t^2 ,\ L^2 ,\ d^2 ,\ 1/\alpha ,\ 1/w^2$$

E and 1/t chosen to minimize SPM

Actinix

6th International Symposium on
Immersion Lithography Extensions
October 22-23 2009 · Prague Czech Republic

Fiber Laser System Architecture

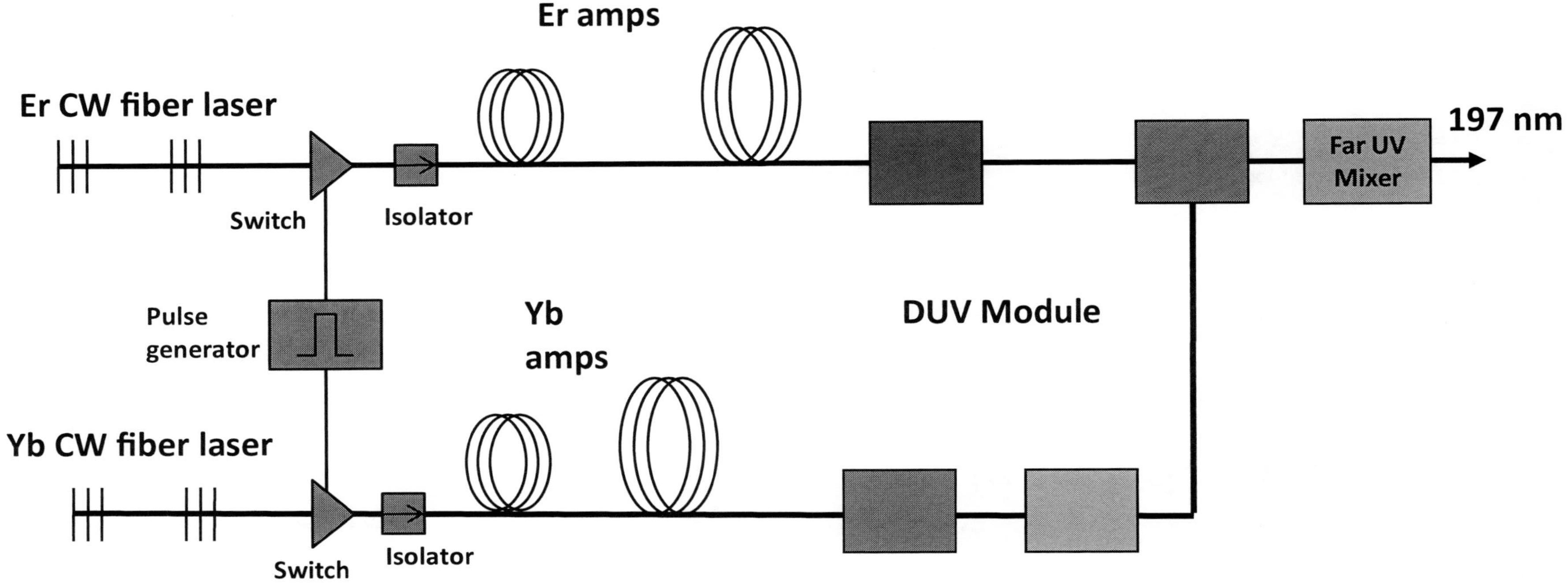

Er CW fiber laser

Switch

Isolator

Er amps

Far UV Mixer

197 nm

DUV Module

Yb amps

Pulse generator

Yb CW fiber laser

Switch

Isolator

- CW sources: 50 MHz frequency stability (7 fm @ 197 nm)
- 50 MHz frequency tuning resolution
- Multiplexed fiber laser architecture
  - ➢ allows power scaling while maintaining bandwidth
- Efficient frequency conversion scheme
  - ➢ All NLO crystals non-critically phase-matched (or quasi-PM)

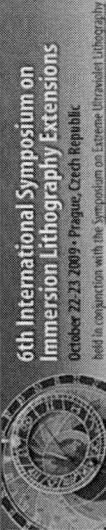

6th International Symposium on
Immersion Lithography Extensions
October 22-23 2009 - Prague, Czech Republic
Held in conjunction with the Symposium on Extreme Ultraviolet Lithography

Actinix

Summary

• Liquid immersion interference lithography is a tractable low cost solution for single exposure patterning of full field grating structures down to 25 nm hp

• 197 nm fiber-based laser will provide coherence length, beam quality and power suitable for printing high contrast gratings at reasonable throughput

• Interferometer design needs to incorporate metrology and possibly adaptive optical techniques to control fringes over full-field

Interference Assisted Lithography (I.A.L.)

A Way to Contain the Lithography Costs for the 32nm and 22nm Half-Pitch Device Generations

Rudi Hendel; David Markle;
John S. Petersen; Andrew Barada;
Periodic Structures, Inc.

Zhilong Rao
Applied Materials, Inc.

Presentation Overview

- **Semiconductor Manufacturing Costs:**
  A quick calibration

- **Cost Model Overview**
  Basics and Assumptions

- **Interference Assisted Lithography**
  An introduction into the Concept

- **Process and Lithography-related cost factors**
  and their impact

- **Results and Discussion**

- **Conclusions**

October 23, 2009

Periodic Structures, Inc.
Immersion Lithography Extensions 2009, Prague

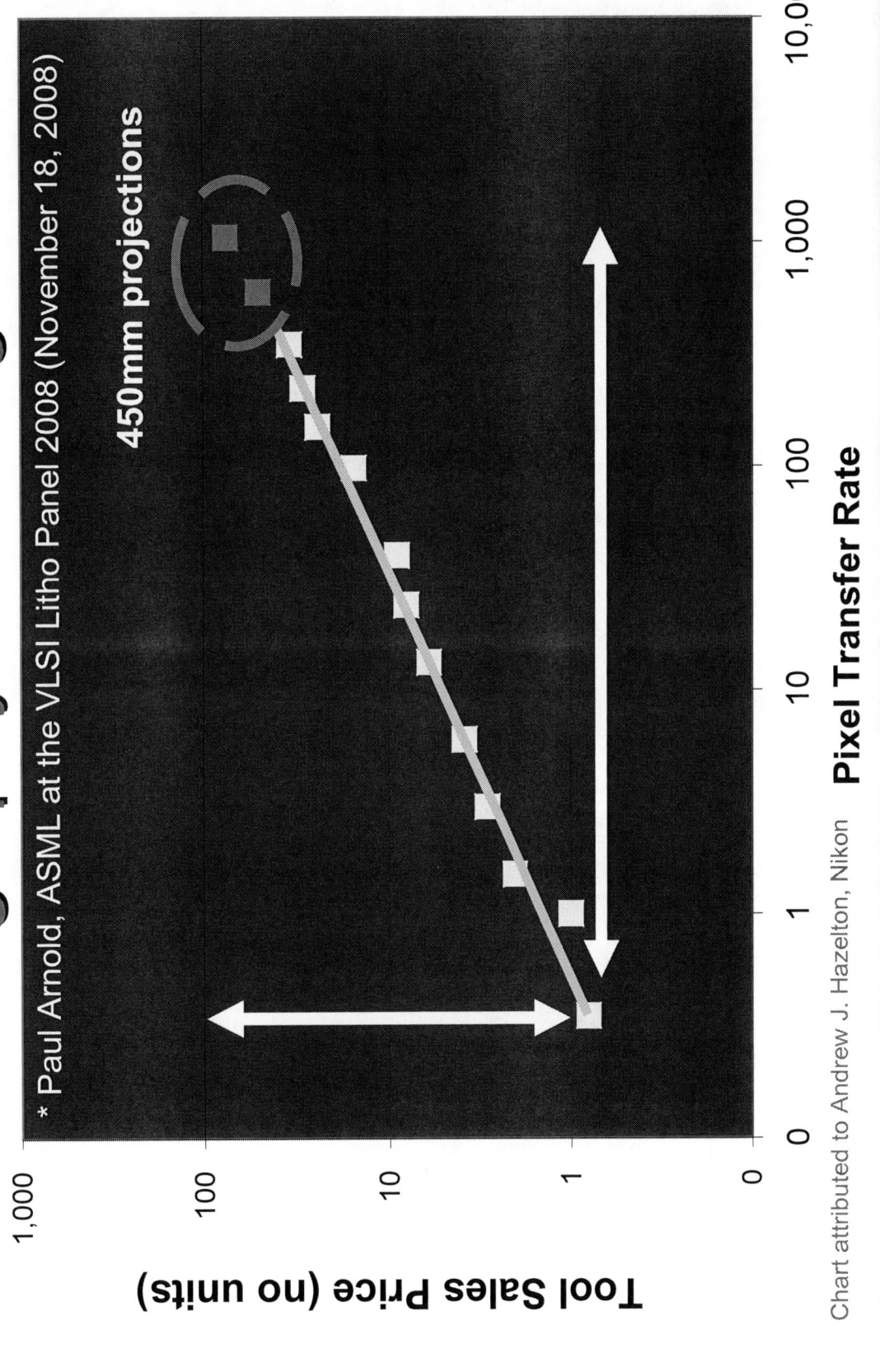

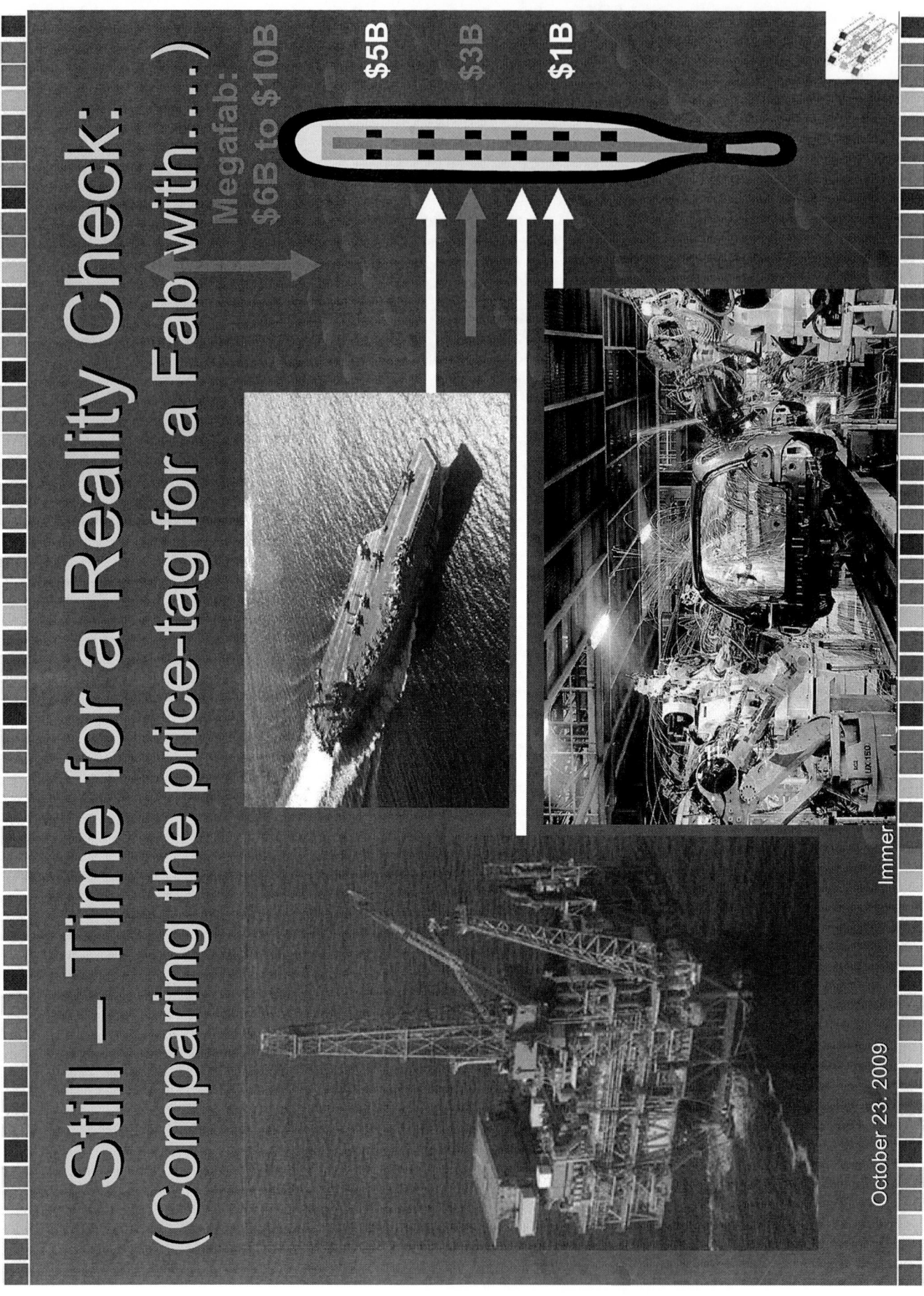

Should the size of Fab Price Tags be a concern?

Net Income % for Semiconductor Manufacturers

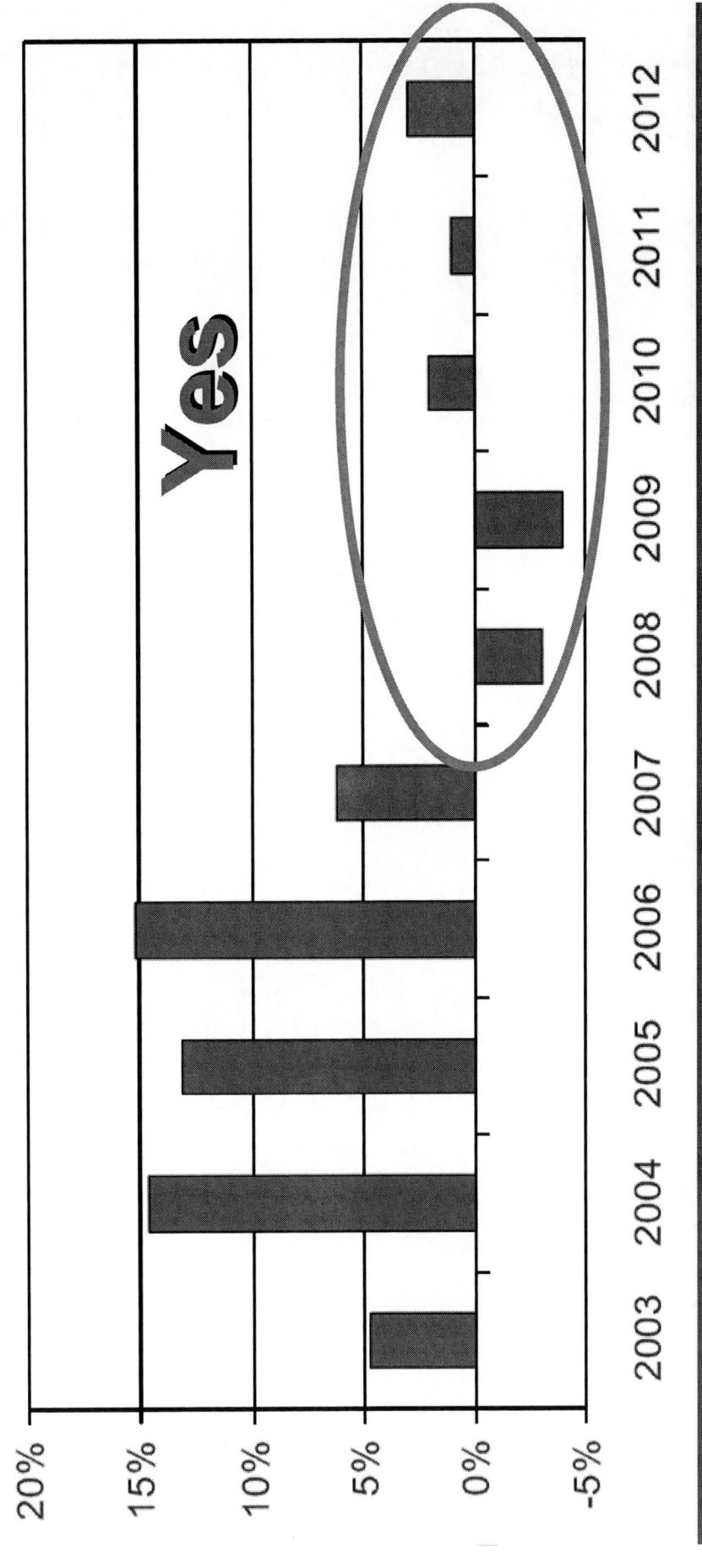

Bob Johnson (Gartner DQ) at ISS 2009
Periodic Structures, Inc.
Immersion Lithography Extensions 2009, Prague
October 23. 2009

The basis of our Cost Model:

$$c_w = p_w + \frac{m_0}{m_e} + \frac{e_0 + d \cdot e_m}{d \cdot t_s \cdot u_s \cdot 365 \cdot 24}$$

Tool component – Depreciation and Tool Maintenance

Allocation of Mask Cost

Process Cost (Resist, Develop, BARC, Metrology, Inspection, etc.

Where,

c_w == Cost to expose one wafer
p_w == Process cost per wafer including track, metrology, inspection
m_0 == Cost of reticle
m_e == Exposures per mask (lifetime)
e_0 == Cost of exposure system
d == Depreciation of exposure system in years
e_m == Annual maintenance cost of exposure tool
t_s == Wafer throughput in wafers/hour
u_s == Utilization of exposure system

Model by Sreenivasan et al – Proc. SPIE V4688, pg. 903 – 909, 2002
Periodic Structures, Inc.

Immersion Lithography Extensions 2009, Prague

October 23, 2009

Cost of Lithography

- **Direct Impact Costs**
  - Cost and Complexity of the Patterning Process
  - Cost of additional Tools required (ex.: S.I.T.)
  - Cost of Mask(s)
  - Depreciation Cost and Maintenance Expense of the Exposure Tool
    - Tool Price to Customer
    - Effective Throughput
    - Scheduled & unscheduled Maintenance

- **Indirect Impact Costs** (Factors of high importance to a customer on which the Choice of Lithography has a critical impact)
  - Cost of Design
  - Time to Design Implementation
  - Risk of First Silicon Fail

October 23, 2009

Periodic Structures, Inc.
Immersion Lithography Extensions 2009, Prague

Replacement Contenders – 2013 HVM

Per Yan Borodovsky; Semicon West; July 15, 2009

- EUVL, Imprint and Maskless - all under significant commercialization efforts to replace 193i ASAP

- Transition from SE to PD Patterning creates window of opportunity for these approaches to replace 193i/PD.

- All parts of Replacement technology must be available, have equal or lower defects levels and significantly better die cost to offset large risks of new technology introduction.

Technology	Tool Ready	'09/'11 Defects Gap Mask	'09/'11 Defects Gap Wafer	Cost
EUV 0.25NA	2011	100X	10X	High
Imprint	2011	10X - 30X	10X - 30X	High
MP EBDW	2012-13	None	Unknown	High

Periodic Structures, Inc.
Immersion Lithography Extensions 2009, Prague

October 23, 2009

How to reduce cost? Start with the Tool

Either:
Simplify the pattern, reducing Pixel Transfer Rate which leads to lower capital cost

Or:
Trade off general purpose capability with acceptable restrictions leading to cost savings

Tool Sales Price (no units)

Pixel Transfer Rate

October 23, 2009   Immersion Lithography Extensions 2009, Prague

Effective Means to segment (and reduce) Information Content contained in a pattern:

- Interference Assisted Lithography (I.A.L.)
  - It separates the high frequency component of the image (dense line-spaces) from the design content (line-ends)

- Simplified Imaging Rules allow lower tool cost
  - without sacrificing minimum resolution

- Several technology approaches are available
  - The final choice of I.A.L. Exposure Technology will accommodate the key requirements for both, logic and memory applications.

October 23, 2009

Periodic Structures, Inc.
Immersion Lithography Extensions 2009, Prague

What is Interference Assisted Lithography?

- I.A.L. is a two-exposure process:
  - Exposure One – Fixed Pitch Pattern (a grating)
  - Exposure Two – a Block-Mask, determining where the lines defined in the first exposure end and start

October 23. 2009

Periodic Structures, Inc.
Immersion Lithography Extensions 2009, Prague

The I.A.L. Technology Principle has been demonstrated

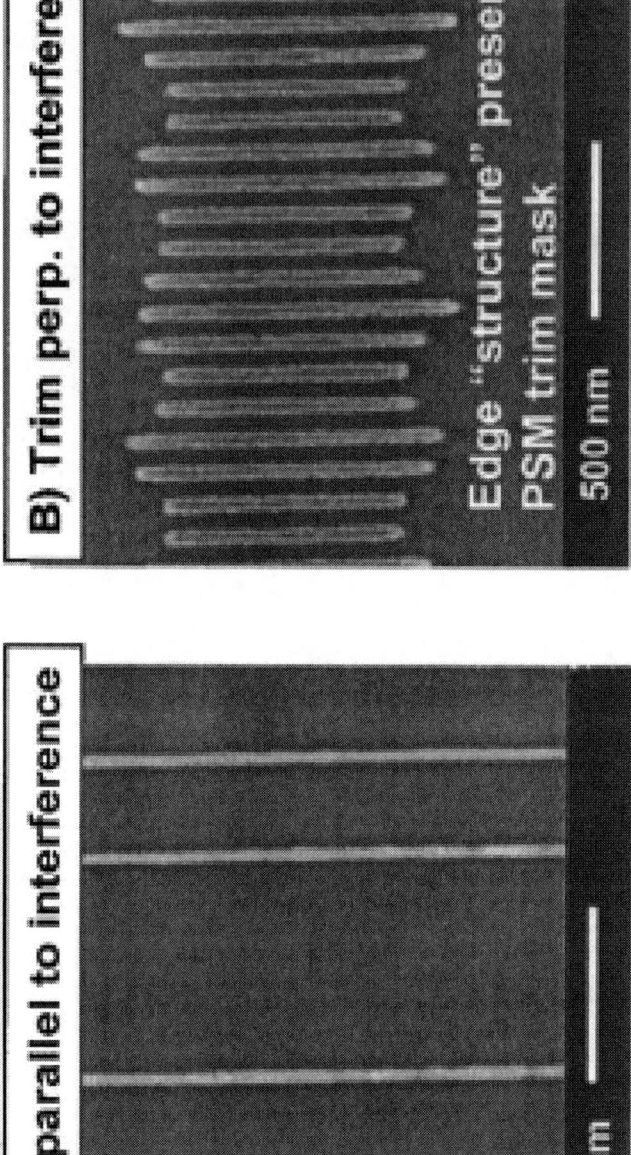

A) Trim parallel to interference

B) Trim perp. to interference

Edge "structure" present in PSM trim mask

500 nm

Source: MIT Lincoln Lab (2005)

October 23. 2009

Periodic Structures, Inc.
Immersion Lithography Extensions 2009, Prague

Benefits of Interference-based Exposure Tools

Cost benefits

- Simplified Optics and thus lower Tool Cost
- OL requirements for I.A.L. are relaxed compared to OL requirements for traditional Double Patterning
- Significant savings in Cost of Masks I.A.L. requires only one critical mask, Double Patterning requires 2 +

Technical Benefits

- Wider process window
- Increased Contrast/DOF

Extendibility to high-index of refraction

October 23, 2009

Periodic Structures, Inc.
Immersion Lithography Extensions 2009, Prague

Cost Model Assumptions

- We have used model input parameters consistent with Information available from the Literature – for example:

  - Phil Seidel – EUVL Cost of Ownership Considerations – 2007 International EUVL Symposium, Sapporo, Japan, October 29 2007

  - Andrew J. Hazelton et al. – Cost of Ownership for Future Lithography Technologies – Proc. SPIE 7140, 7140 Q-1, 2008

  - Andrea Wuest et al. – Estimation of Cost Comparison of Lithography Technologies at the 22nm Half Pitch Node – Proc. SPIE 7271, 7271Y-1, 2009

For details of our specific Cost Model Assumptions, please contact the Author

October 23, 2009

Periodic Structures, Inc.
Immersion Lithography Extensions 2009, Prague

Double Patterning: – Comparing Traditional D.P. vs. I.A.L. D.P.

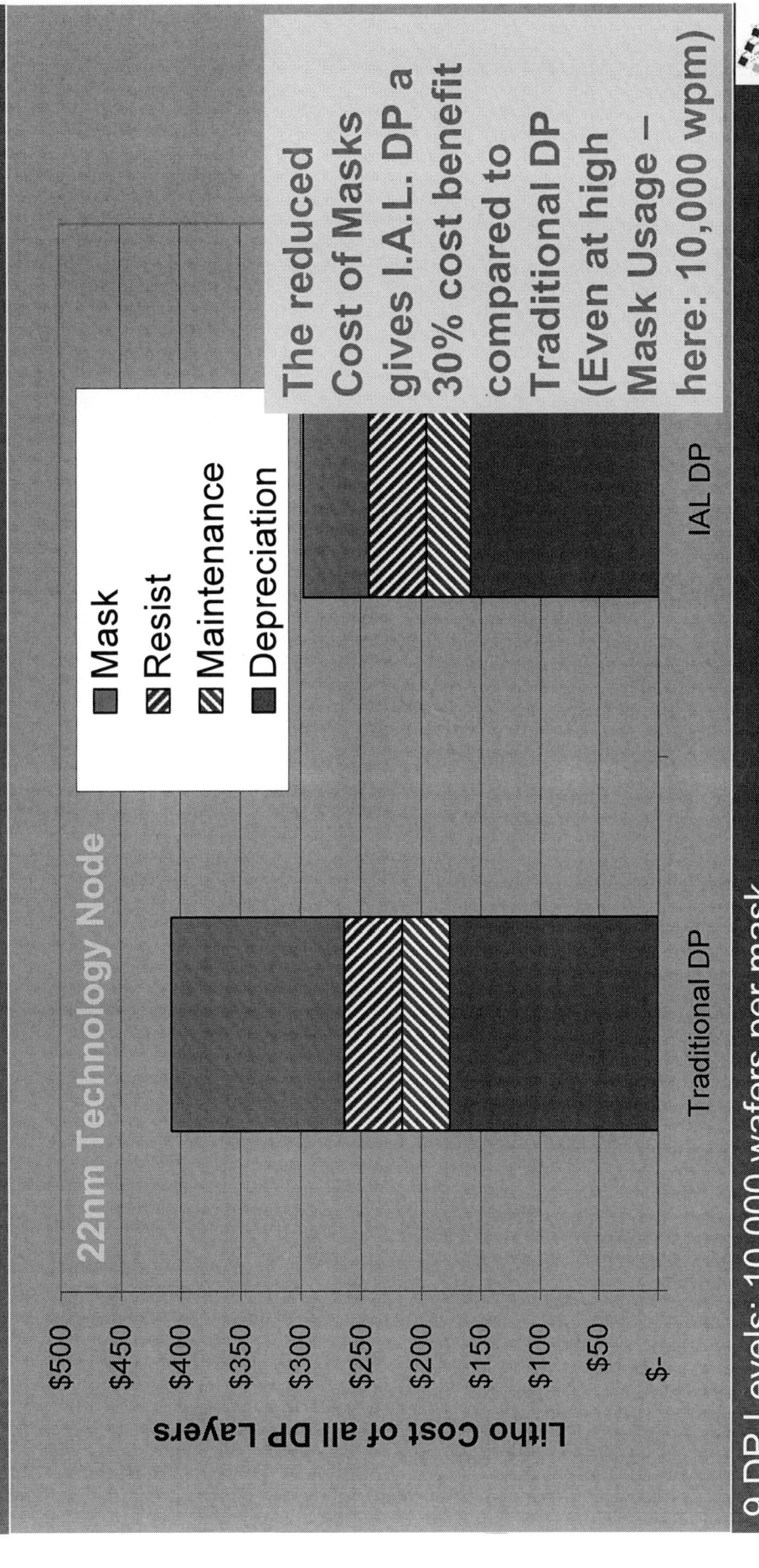

The reduced Cost of Masks gives I.A.L. DP a 30% cost benefit compared to Traditional DP (Even at high Mask Usage – here: 10,000 wpm)

Legend:
- Mask
- Resist
- Maintenance
- Depreciation

22nm Technology Node

Litho Cost of all DP Layers

$500
$450
$400
$350
$300
$250
$200
$150
$100
$50
$-

Traditional DP   IAL DP

9 DP Levels; 10,000 wafers per mask
October 23, 2009

Periodic Structures, Inc.
Immersion Lithography Extensions 2009, Prague

Estimating the Total Cost of Litho by Node: Assignment of Mask Levels

In order to estimate the total Cost of Lithography at a given Node (in this case for Logic), we have assigned levels to Lithography Technologies according to this Table:

Node and Minimum CD	45	32	22	16
Minimum Pitch	130	90	64	45
EUVL Levels	0	0	2	11
193i-DP Levels or 193-IAL Levels	0	10	9	8
193i-SE Levels	7	10	21	21
193d-SE Levels	19	17	8	0
248-SE Levels	7	0	4	4
I-line Levels	7	7	6	6
Total number of Levels in Node	40	44	50	50

October 23, 2009

Periodic Structures, Inc.
Immersion Lithography Extensions 2009, Prague

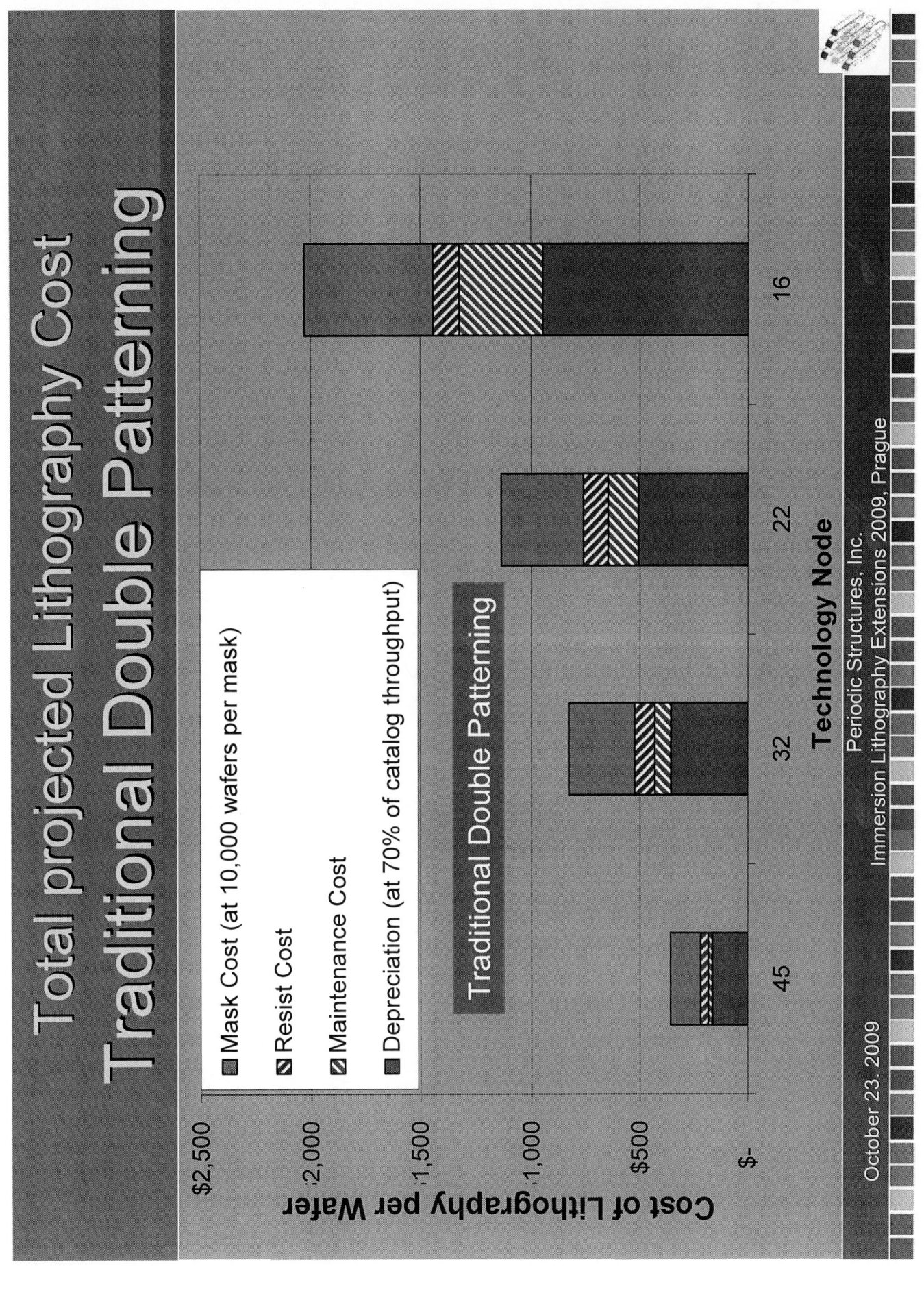

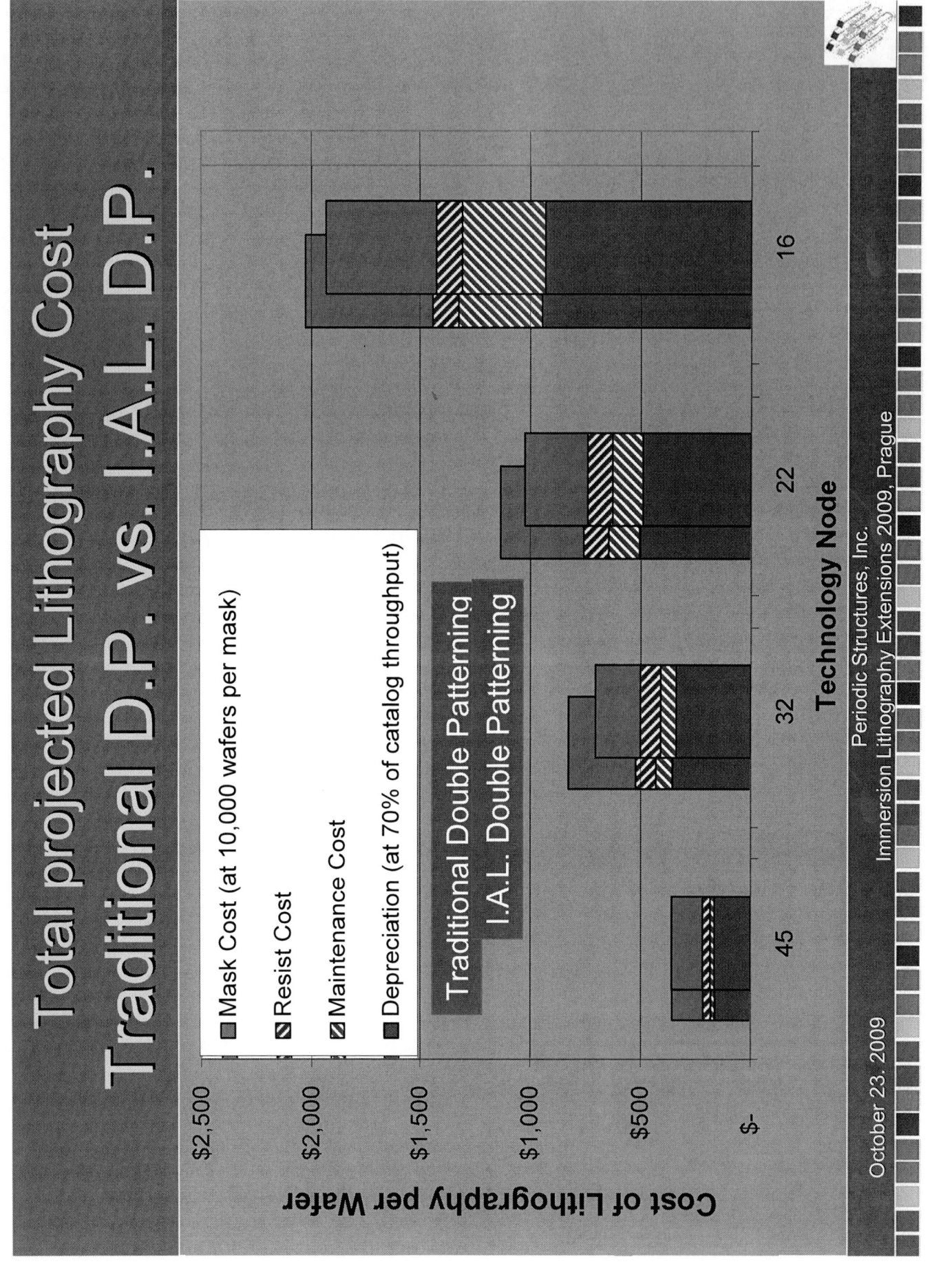

Observations and Discussion:

Even though I.A.L. reduces the cost of Lithography at a given level by 30%, the impact on total Litho Cost is reduced due to the limited number of Levels it was applied to and the rapidly increasing usage and high level cost of Critical Level Litho Choices (EUVL).

We have assumed the use of a SOA "Generation-N" Exposure Tool for the I.A.L. Block Level. This is a worst case assumption. If the Block Level can use Gen(N-1) or even Gen(N-2) Exposure Tools, the cost advantage of I.A.L. over traditional DP Lithography is much more pronounced.

Once the 1-D Design Methodology is adopted and I.A.L. becomes an option for Critical Level Exposure, the projected rapid Increase in the Cost of Litho for future device generations can be contained with I.A.L.. (please contact the author for details)

October 23, 2009

Periodic Structures, Inc.
Immersion Lithography Extensions 2009, Prague

Additional Benefits Possible with I.A.L.
(System Architecture Dependent)

- Reduced Capital Costs and Reduced Reticle Costs discussed earlier apply to all conditions (all Lot Sizes and Exposure Field Sizes)

- In the case of a reduced Exposure Field (Not all applications require large Exposure Fields)
  - Further Mask Cost Reduction is possible
    - Minimizing Mask writing/inspection costs
    - Particularly benefitting smaller lot sizes, which are highly sensitive to Mask Cost
  - Reduced Systems Cost (Stepper architecture (vs. Scanner) allows systems simplifications)
  - Additional efficiency particularly applies for small devices

October 23, 2009

Periodic Structures, Inc.
Immersion Lithography Extensions 2009, Prague

Assumption for
Mask Cost vs. Mask Patterned Area

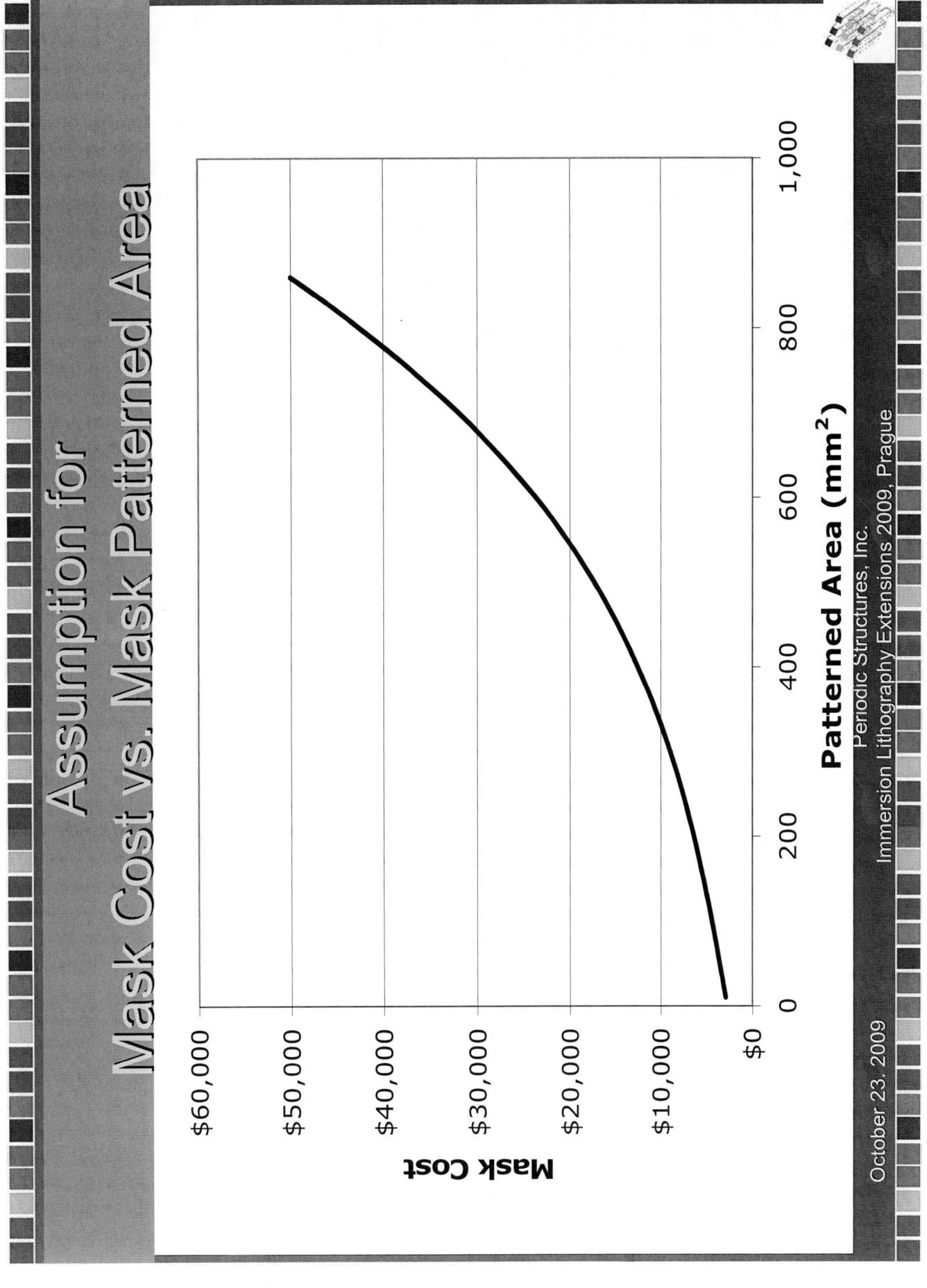

October 23, 2009

Periodic Structures, Inc.
Immersion Lithography Extensions 2009, Prague

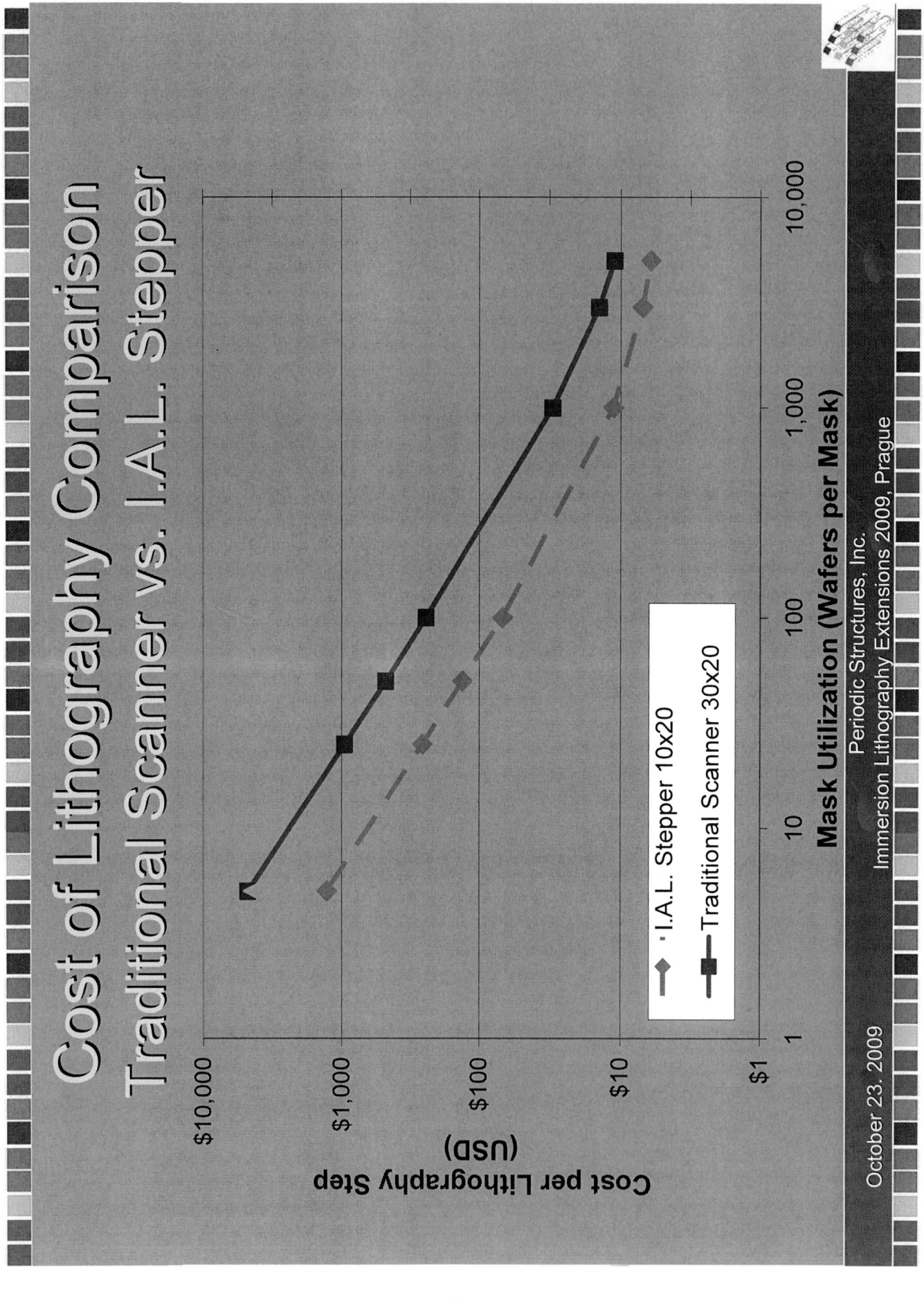

I.A.L. Limitations

- High Pattern Regularity forces Design Restrictions

  - I.A.L. Restrictions are similar to Sidewall Image Transfer (S.I.T.) Restrictions

    - Yet, S.I.T. is seriously considered as one viable option for extending resolution.

    - Designs compatible with S.I.T. are expected to be compatible with I.A.L..

  - Adopting such restrictions early and transitioning to regularity in one single step promises significant cost benefits.

October 23, 2009

Periodic Structures, Inc.
Immersion Lithography Extensions 2009, Prague

Layouts can be modified to become I.A.L. compatible

Example: SRAM Cell:

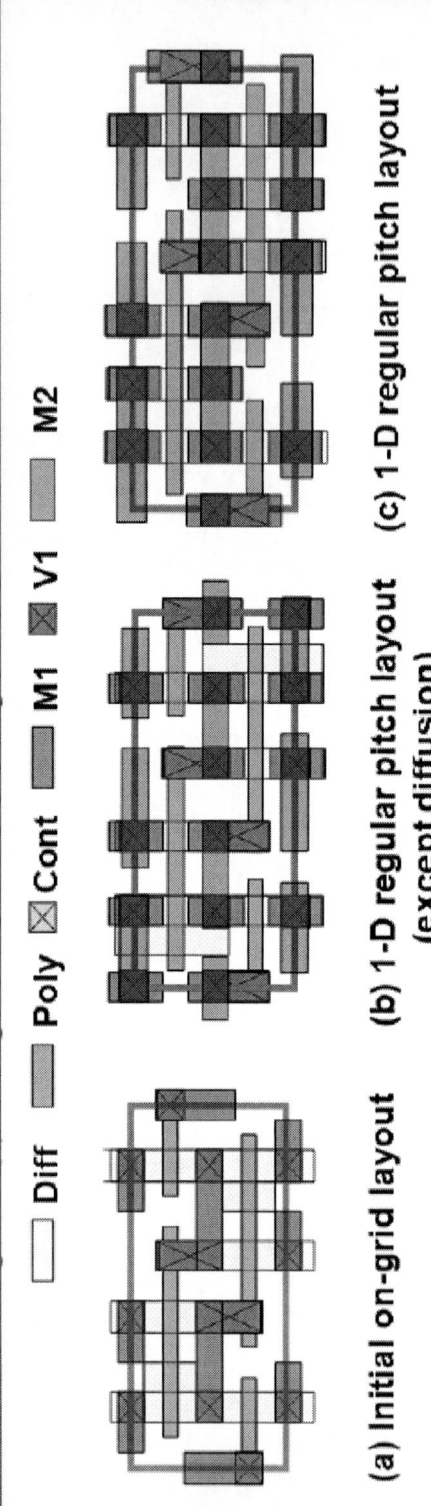

Legend: Diff, Poly, Cont ⊠, M1, V1 ⊠, M2

(a) Initial on-grid layout

(b) 1-D regular pitch layout (except diffusion)

(c) 1-D regular pitch layout

IAL-compatible bitcell layout.

Ref: Interference Assisted Lithography for Patterning of 1D Gridded Design
Robert T. Greenway, Rudolf Hendel, Kwangok Jeong, Andrew B. Kahng,
John S. Petersen, Zhilong Rao, and Michael C. Smayling,
Proc. SPIE 7271, 72712U (2009), DOI:10.1117/12.812033
Periodic Structures, Inc.
Immersion Lithography Extensions 2009, Prague

October 23, 2009

In Summary – Cost Benefits of I.A.L.:

Direct:

Significant reduction of Mask Cost
- Comparable layers (vs. traditional DP)

Containment of Litho Cost Increases
- If applied to all Critical Levels in the process flow

Additional Benefits are possible:
- Reduced Die Size, Small Lot-size, (leading to smaller Exposure Fields)
- Enhanced Maskless Lithography Economics
  - Reduced Image (Pixel) Density for Block Mask
  - Enables competitive MLL Throughput (prototyping)

October 23. 2009

Periodic Structures, Inc.
Immersion Lithography Extensions 2009, Prague

In Summary – Cost Benefits of I.A.L.:

Indirect:

- I.A.L. can become a Technology Enabler for a Market Segment currently underserved:
  - Opening Access to Leading Edge Geometries for low/medium mask usage applications (ASIC/ASSP) by:
    - Reduced Mask Costs (and Capital Cost)
    - Simplified DR complexity
    - Reduced time to Design Completion
    - Reduced risk to first time success

October 23, 2009

Periodic Structures, Inc.
Immersion Lithography Extensions 2009, Prague

Acknowledgements

- Al Bergendahl, Bergendahl Enterprises
- Marc Levenson
- Mark Pinto, Applied Materials
- Hans Stork, Applied Materials
- Bill Tobey, ACT Consulting

October 23, 2009

Periodic Structures, Inc.
Immersion Lithography Extensions 2009, Prague

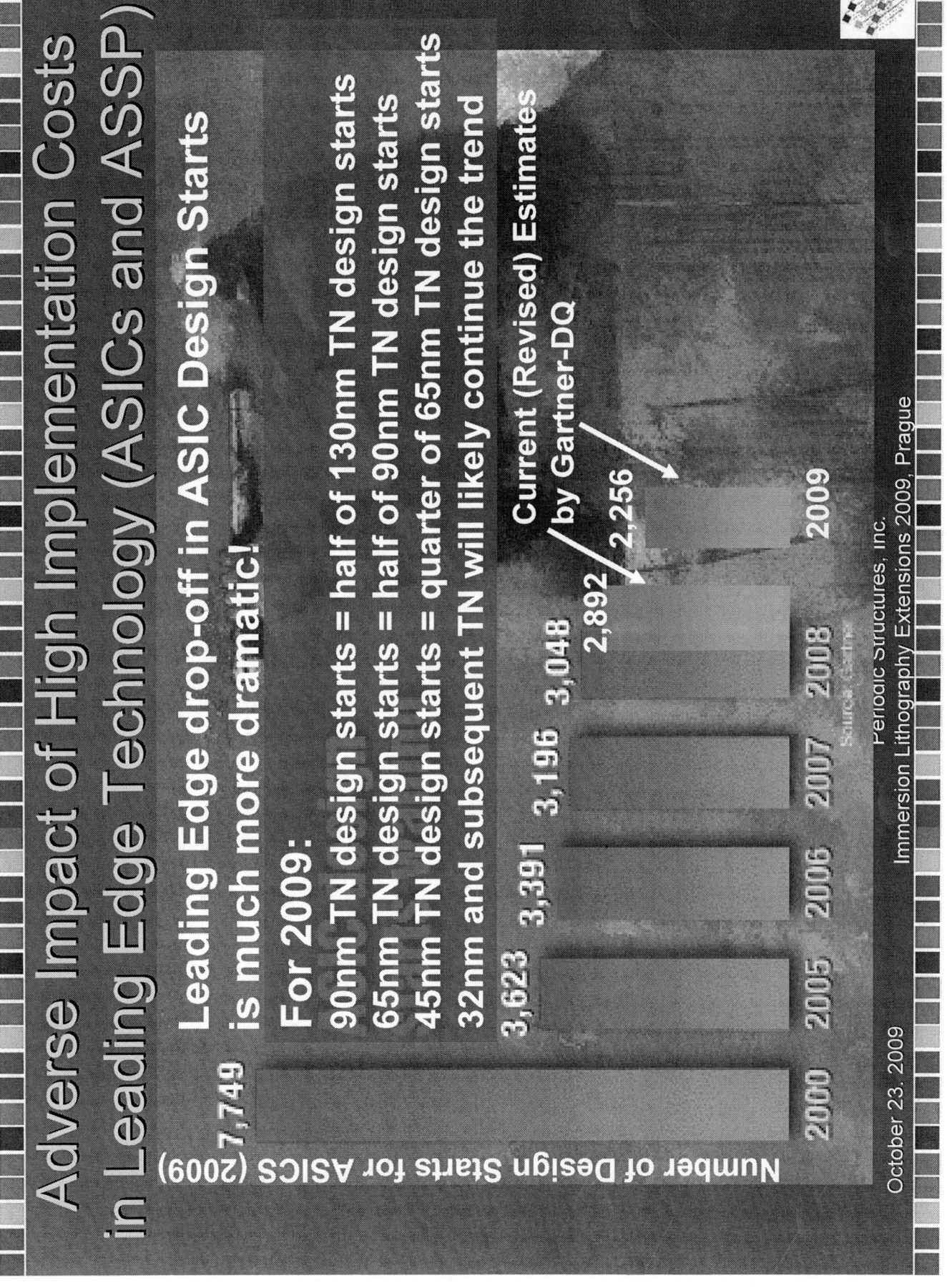

Adverse Impact of High Implementation Costs in Leading Edge Technology (ASICs and ASSP)

Leading Edge drop-off in ASIC Design Starts is much more dramatic!

For 2009:
90nm TN design starts = half of 130nm TN design starts
65nm TN design starts = half of 90nm TN design starts
45nm TN design starts = quarter of 65nm TN design starts
32nm and subsequent TN will likely continue the trend

Current (Revised) Estimates by Gartner-DQ

Number of Design Starts for ASICS (2009)

2000	2005	2006	2007	2008	2009
7,749	3,623	3,391	3,196	3,048	2,892 / 2,256

Source: Gartner

Periodic Structures, Inc.
Immersion Lithography Extensions 2009, Prague

October 23, 2009

FUJIFILM

6th International Symposium on Immersion Lithography

Resist material for
negative tone development process

FUJIFILM Corporation
Electronic Materials Research Laboratories

October, 2009

FUJIFILM

6th International Symposium on Immersion Lithography

Outline

1. Advantages of negative tone imaging for DP

2. Process maturity of negative tone development

3. Resist material progress for negative tone development

4. Summary

5. Acknowledgement

October, 2009

FUJiFILM

6th International Symposium on Immersion Lithography

Overview of double patterning processes

Double Exposure
Resist Coat
First Exposure
Second Exposure
Dev.

Double Development
Resist Coat
Exposure
First Dev.
Second Dev.

Freezing
First Resist Coat
First Exposure
First Dev.
Freezing
Second Resist Coat
Second Exposure
Second Dev.

Spacer Defined
Resist Coat
Exposure
Dev.
Etching
Sidewall form
Etching
Etching

Double Line
Resist Coat
Exposure
Dev.
Etching
Resist Coat
Exposure
Dev.
Etching

Double Trench
Resist Coat
Exposure
Dev.
Etching
Resist Coat
Exposure
Dev.
Etching

October, 2009

FUJIFILM

6th International Symposium on Immersion Lithography

Trench pattern formation with DP

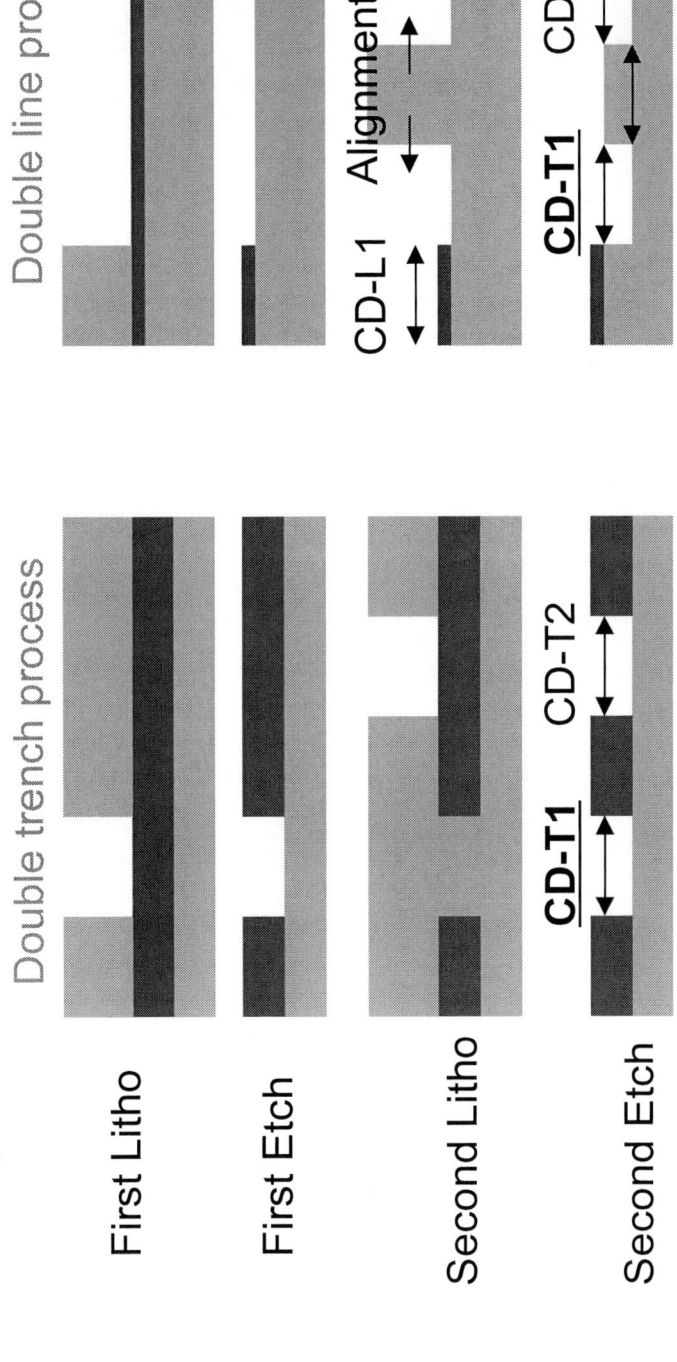

Causes of CD error of CD-T1

Double trench process	Double Line process
CD-T1	CD-L1, CD-L2, Alignment error

CD error of trench pattern:
Double trench process << Double line process (Freezing process)

October, 2009

FUJiFILM

6th International Symposium on Immersion Lithography

Advantage of negative tone imaging in trench pattern printing

NA = 1.2, Immersion (Water), Y Oriented Polarization, Dipole Radius: 0.1
128 nm Pitch 1:3 Pattern Simulation

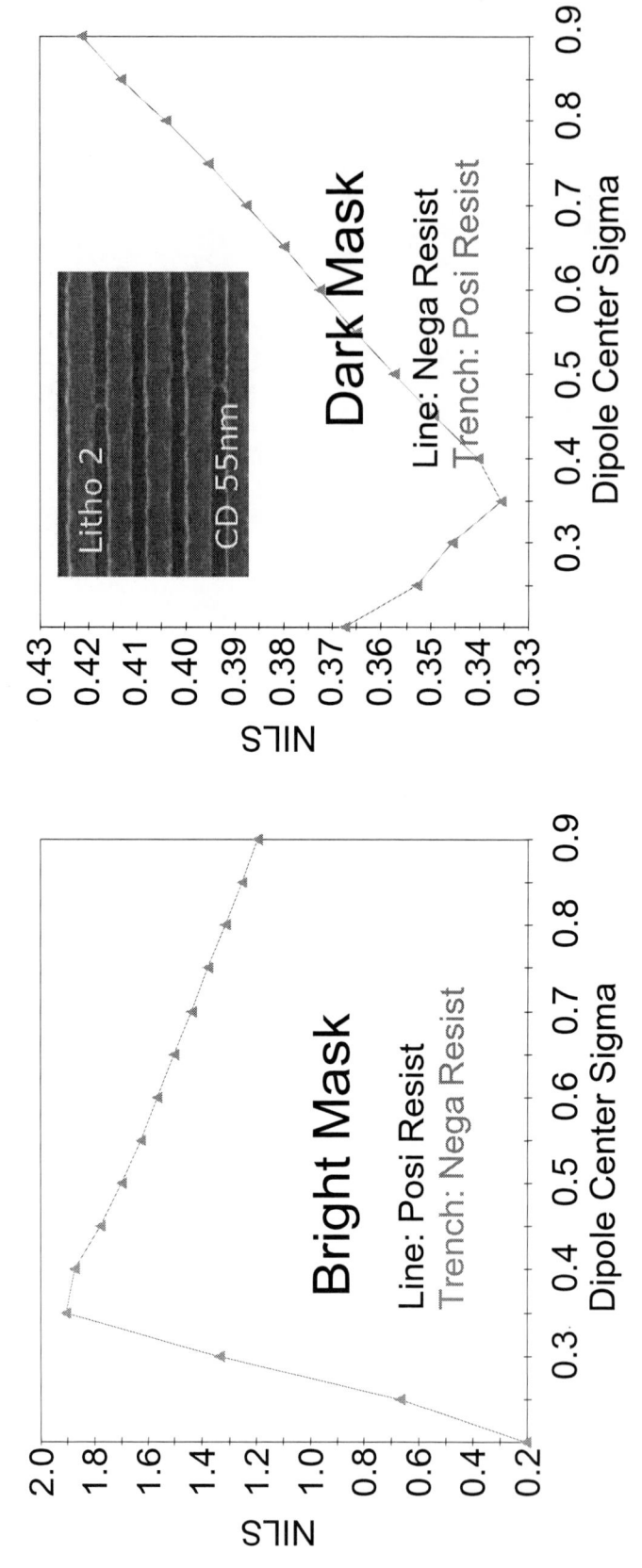

Much higher optical image contrast can be obtained with negative tone imaging

October, 2009

FUJIFILM

6th International Symposium on Immersion Lithography

Negative tone development (NTD), 32 nm trench performance

High frequency LWR (Mask: 64 nm 1:1 B / W)

Nega   Posi

High frequency

4.1 nm

Low frequency
(Rectangle scan)

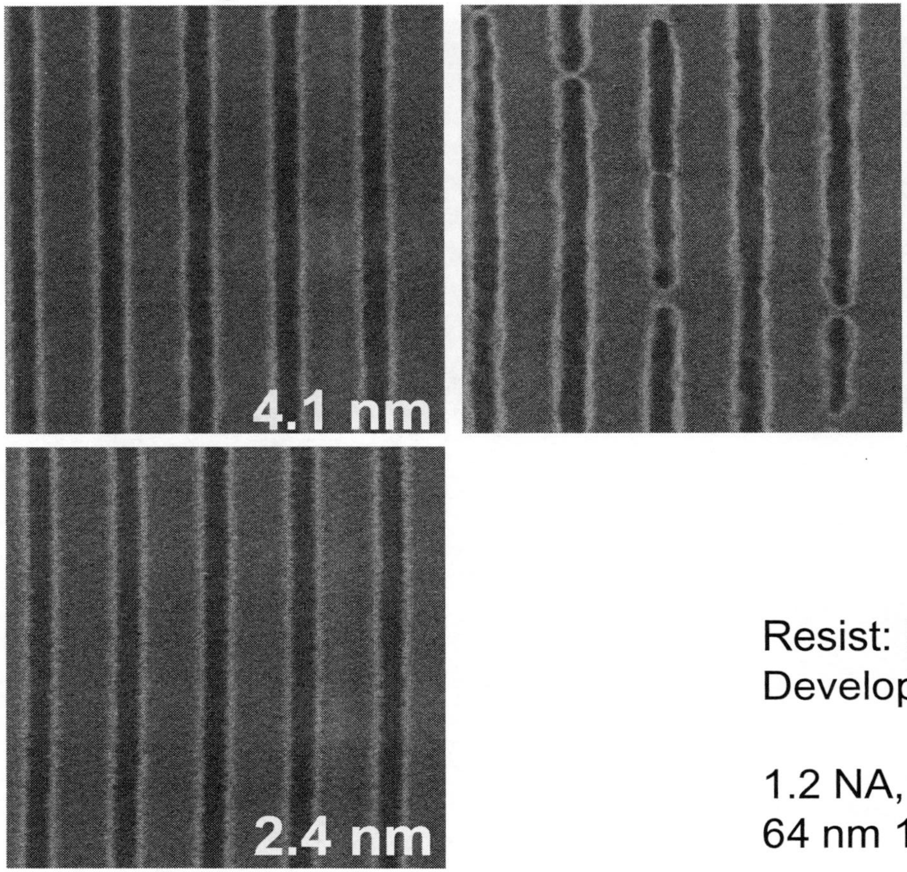

2.4 nm

Resist: FAiRS-9521A02
Developer: FN-DP001

1.2 NA, dipole
64 nm 1:1 binary mask

October, 2009

FUJIFILM & imec

6th International Symposium on Immersion Lithography

Other feasibility, C/H printing by double exposure

Obtained by double line exposures (horizontal and vertical)
1.20NA (ASML XT:1700i)

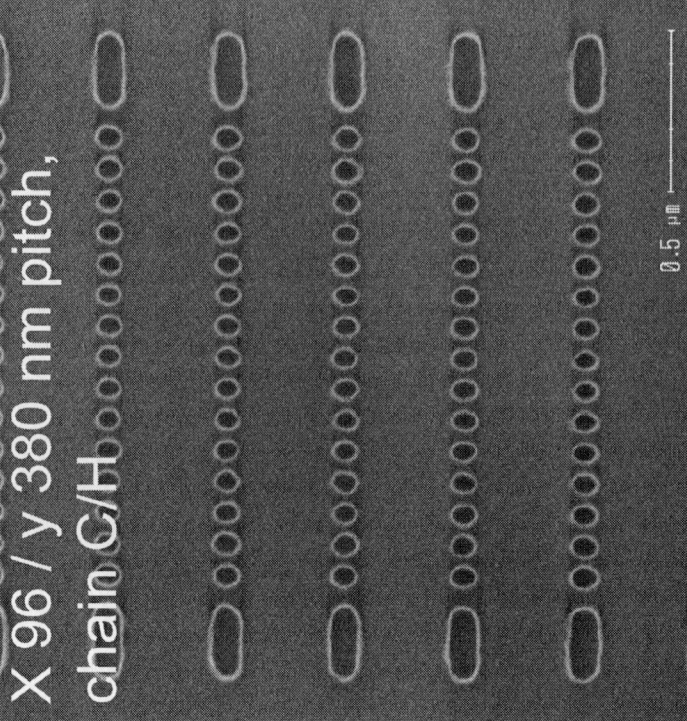

X 96 / y 380 nm pitch, chain C/H

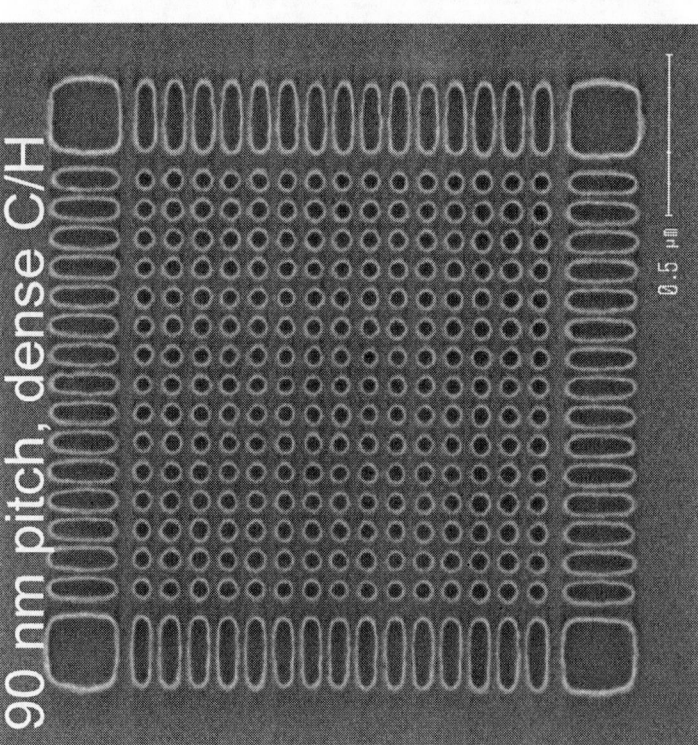

90 nm pitch, dense C/H

Joost Bekaert (IMEC), *et. al.,* See DS-02, September 24, this symposium.

October, 2009

6th International Symposium on Immersion Lithography

Outline

1. Advantages of negative tone imaging for DP

2. Process maturity of negative tone development

3. Resist material progress for negative tone development

4. Summary

5. Acknowledgement

FUJIFILM

October, 2009

FUJiFILM

6th International Symposium on Immersion Lithography

October, 2009

Blanket etching rate comparison to PTD

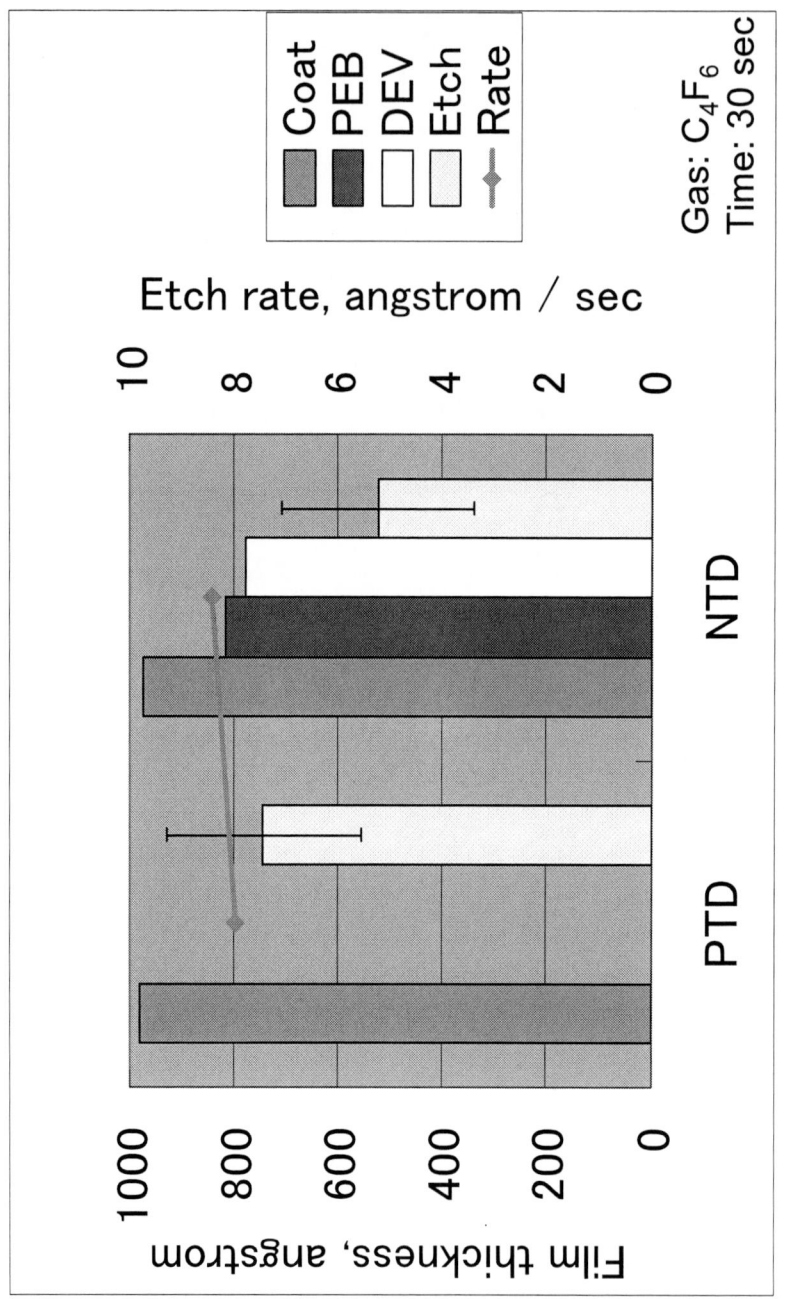

5% faster etch rate than PTD
20% film shrink at PEB due to de-protection reaction

6th International Symposium on Immersion Lithography

Pattern etch result with 90 nm pitch dense C/H

October, 2009

TiN ETCH

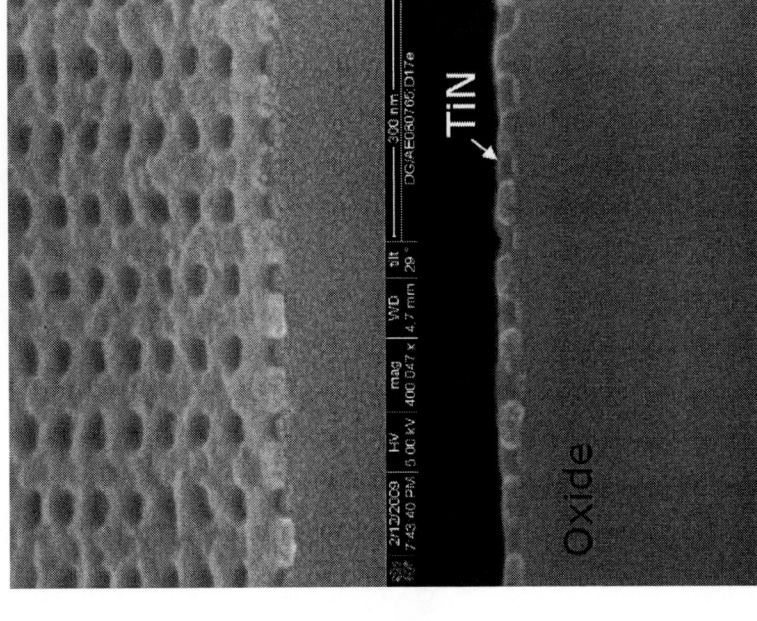

LITHO

CH patterns in PR after Neg. Tone development

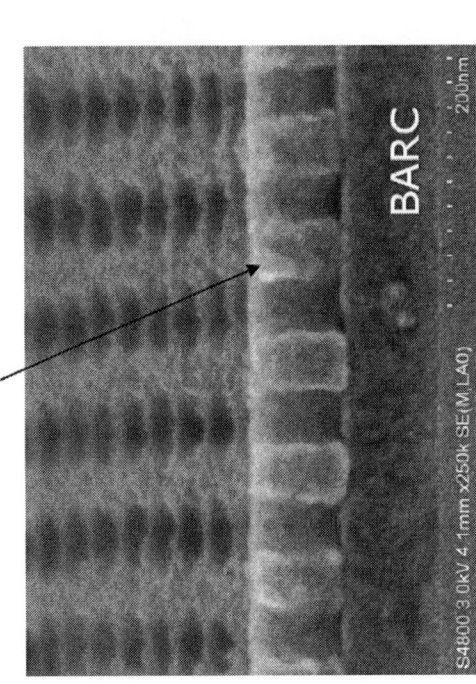

FUJIFILM

October, 2009

6th International Symposium on Immersion Lithography

CD uniformity data, FAiRS-9521A02 NTD and PTD

NTD
Developer: FN-DP001
Dynamic dev.-1

PTD
Developer: OPD5262
Dynamic dev.-2

45nm trench
128nm pitch
NA1.2
dipole illumination

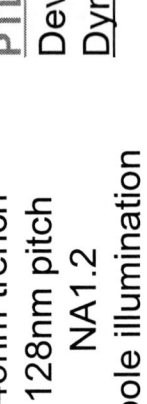

	NTD	**PTD**
Mean	43.8 nm	42.6 nm
3 x STD dev.	3.3 nm	4.0 nm
Mean	42.7 nm	41.2 nm
3 x STD dev.	4.2 nm	5.1 nm

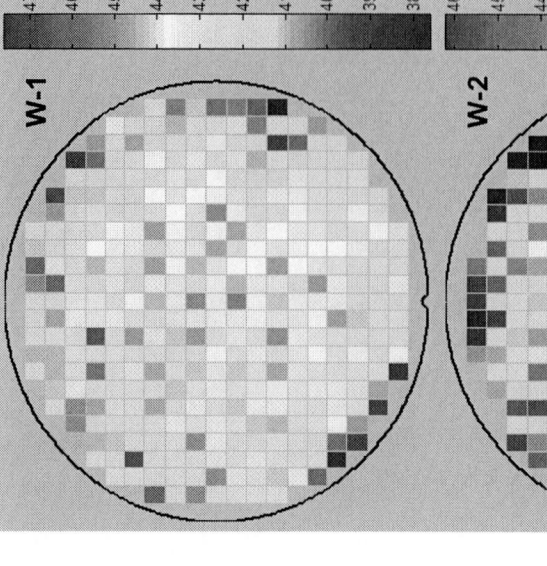

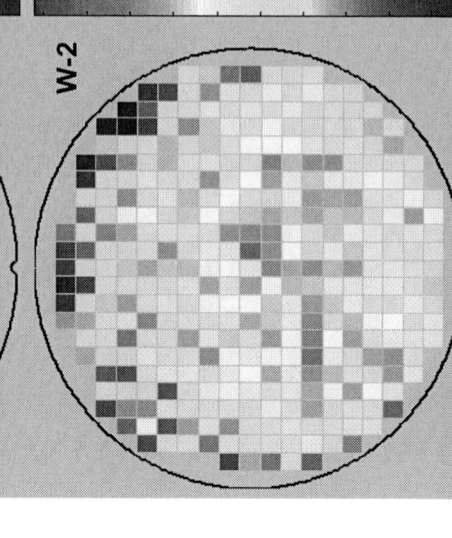

FUJIFILM

6th International Symposium on Immersion Lithography

75 nm L/S defectivity data

Resist   :FAiRS-9521A02
Substrate :Organic BARC
Exposure :Dry tool, 0.75NA
Developer:FN-DP001, resist nozzle dispense
Rinse   :FN-RP001, thinner nozzle dispense

n=2

Total count : 4
D.D. = 0.046 / cm²

n=1

Total count : 2
D.D. = 0.023 / cm²

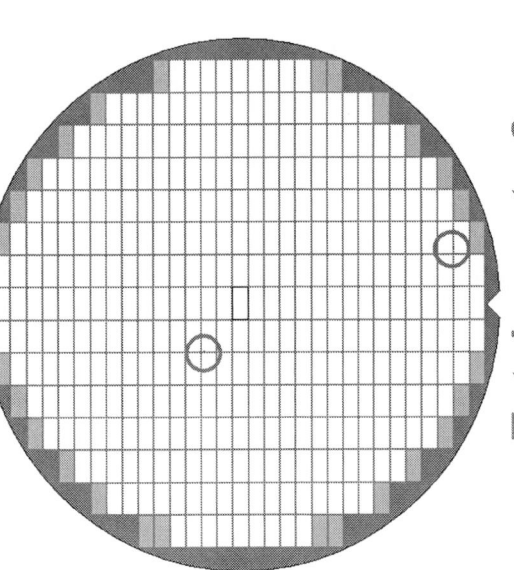

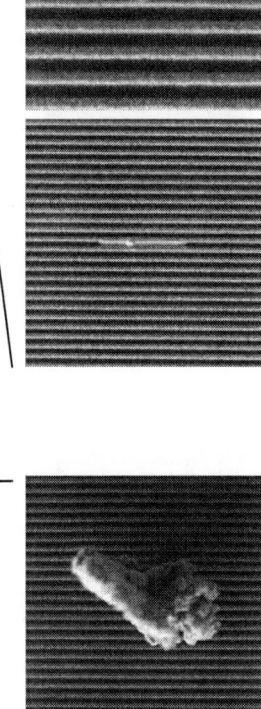

October, 2009

FUJIFILM

6th International Symposium on Immersion Lithography

October, 2009

Outline

1. Advantages of negative tone imaging for DP

2. Process maturity of negative tone development

3. Resist material progress for negative tone development

4. Summary

5. Acknowledgement

FUJIFILM

6th International Symposium on Immersion Lithography

NTD extension ability – dense patterning property for 22 nm HP

October, 2009

FAiRS-9101A19C with 1.35NA immersion exposure

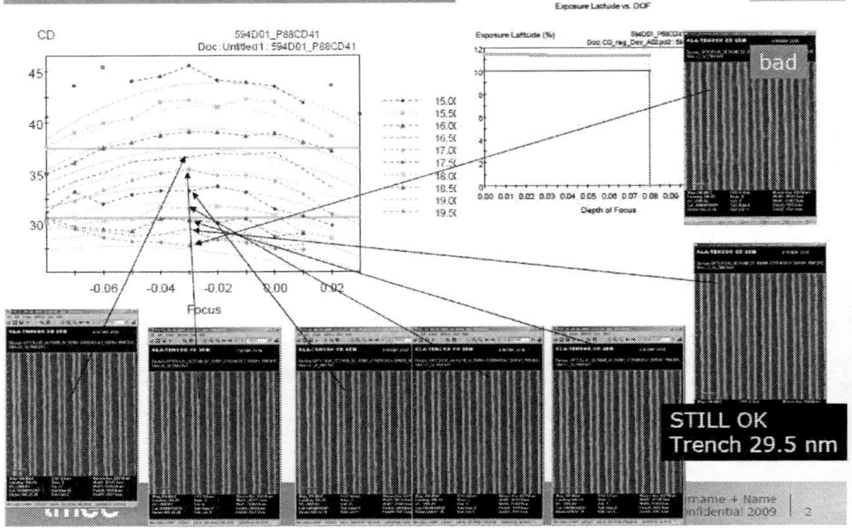

CQuad **PTD** 987D04 **P88 CD 52** - FEM
target 34 nm

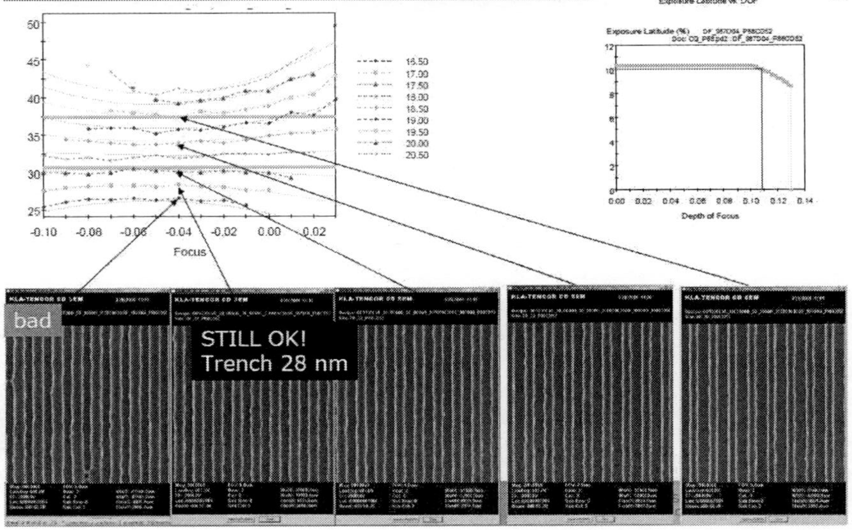

CQ- **NTD** – 594D01 – LF P88 CD 41 FEM
target 34 nm

STILL OK!
Trench 28 nm

bad

STILL OK
Trench 29.5 nm

bad

Wider DOF and smaller trench were obtained with PTD.
Dissolution contrast difference should arise these difference.

Development of resist dedicated to NTD is necessary.

FUJIFILM

6th International Symposium on Immersion Lithography

Dissolution contrast comparison – NTD and PTD

<u>Dissolution Contrast</u>

Nega (Unexposed/Exposed) = 1.3×10^3

Posi (Exposed/Unexposed) = 1.3×10^5

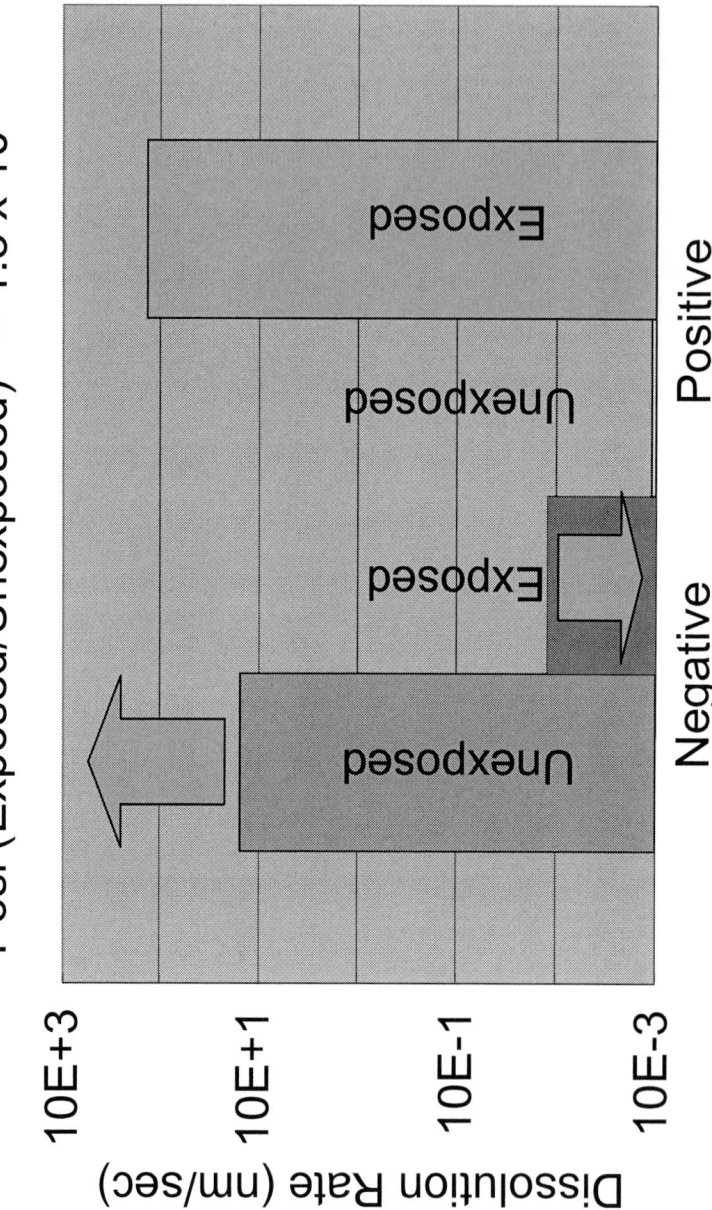

*Positive tone developer: 2.38% TMAH solution

October, 2009

FUJIFILM

6th International Symposium on Immersion Lithography

How to improve dissolution contrast

Method	Concern
Rmax improvement with low Mw.	Trade-off relation to Rmin.
Rmax improvement with SP value control.	No concern of trade-off relation.

$SP = f(\delta_p, \delta_h, \delta_d)$

δ_p : related to polarity
δ_h : related to hydrogen bond
δ_d : related to diffusion

All material has its own SP.
Similar SP value should leads faster dissolution rate.

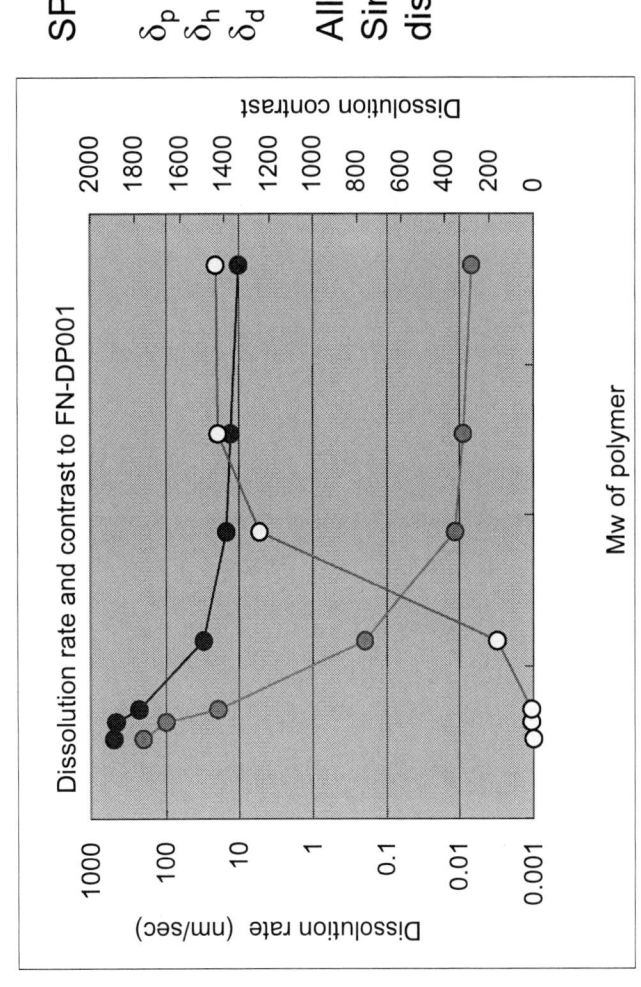

Dissolution rate and contrast to FN-DP001

- ● Unexposed
- ● Exposed(15mJ)
- ○ Dissolution contrast

October, 2009

6th International Symposium on Immersion Lithography

FUJIFILM

October, 2009

Rmax dependence on ΔSP

$$\text{(Mean square of } \Delta SP\text{)} =$$
$$\text{Sqrt}\left((\delta_p^{polymer} - \delta_p^{dev})^2 + (\delta_h^{polymer} - \delta_h^{dev})^2 + (\delta_d^{polymer} - \delta_d^{dev})^2\right)$$

δ_p : related to polarity
δ_h : related to hydrogen bond
δ_d : related to diffusion

Measurement on same Mw, different monomer ratio

SP value control enables faster dissolution rate at un-exposed area.

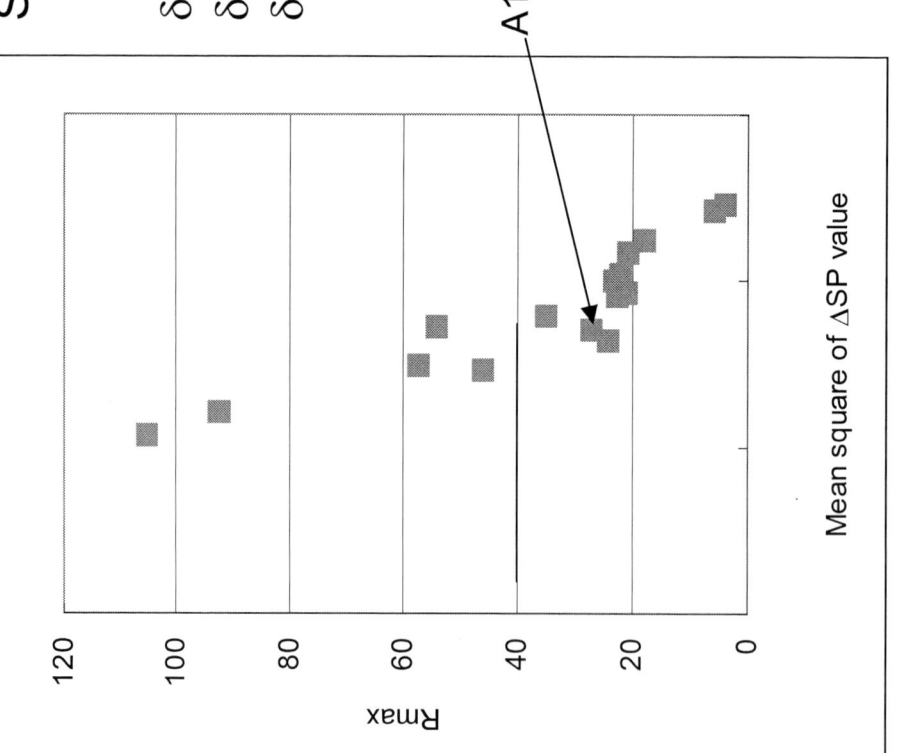

A19C

Mean square of ΔSP value

Rmax

FUJiFILM

6th International Symposium on Immersion Lithography

Dissolution rate of new polymer

October, 2009

FAiRS-9101A19C Contrast curve

FAiRS-9521A08D Contrast curve

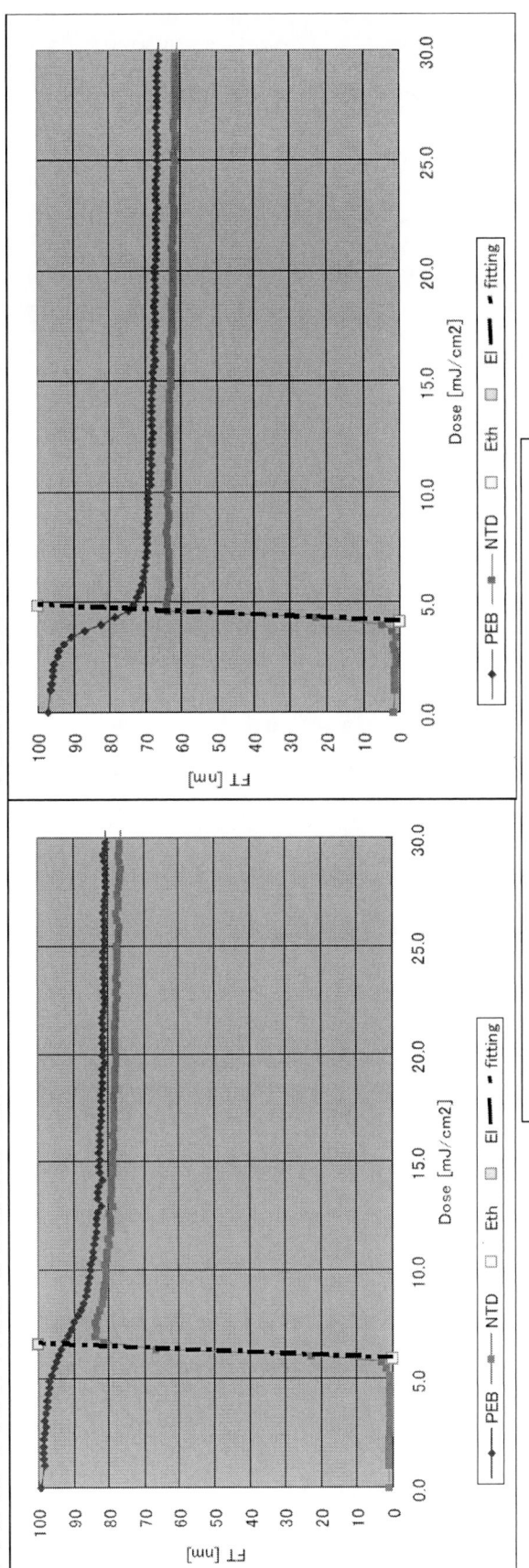

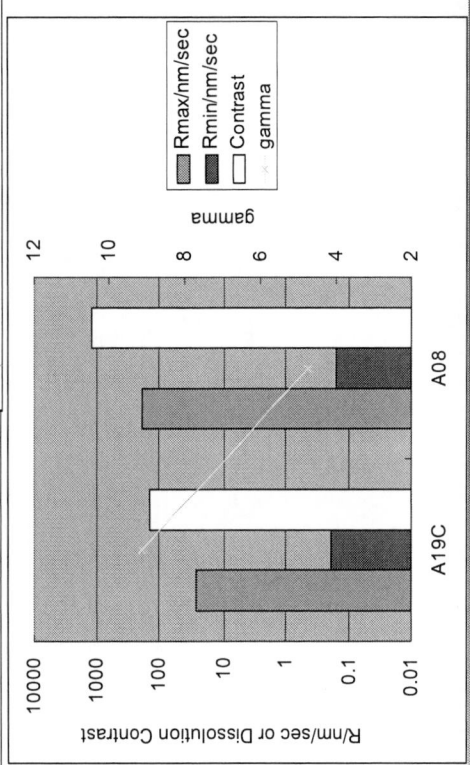

FUJIFILM

6th International Symposium on Immersion Lithography

DOF improvement with ΔSP decrease

NA0.75, Annular, Dry,
90nm dense CH (V/H double exposure)

FAiRS-9521A08D, SB/PEB = 100/95 deg, 60 sec, 23 mJ/cm^2

-0.30 μm -0.25 μm -0.20 μm -0.15 μm   B. F.   +0.15 μm +0.20 μm +0.25 μm +0.30 μm

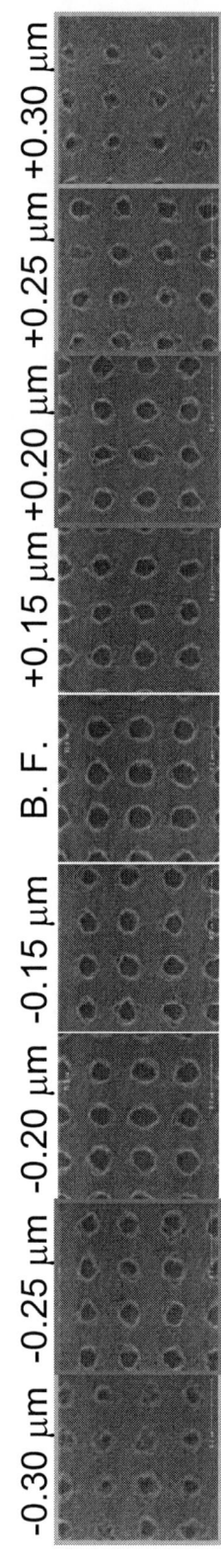

FAiRS-9101A19C, SB/PEB = 100/105 deg, 60 sec, 25 mJ/cm^2

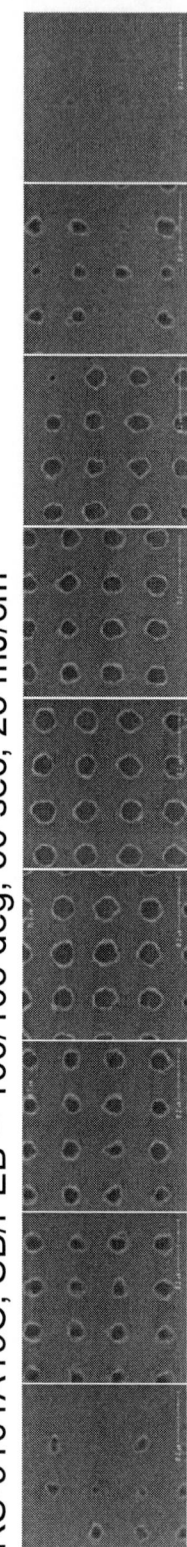

Wider DOF was achieved in A08D with fast Rmax (~200 nm/sec) polymer.

October, 2009

FUJIFILM

6th International Symposium on Immersion Lithography

Another method for DOF – larger γ contrast formulation

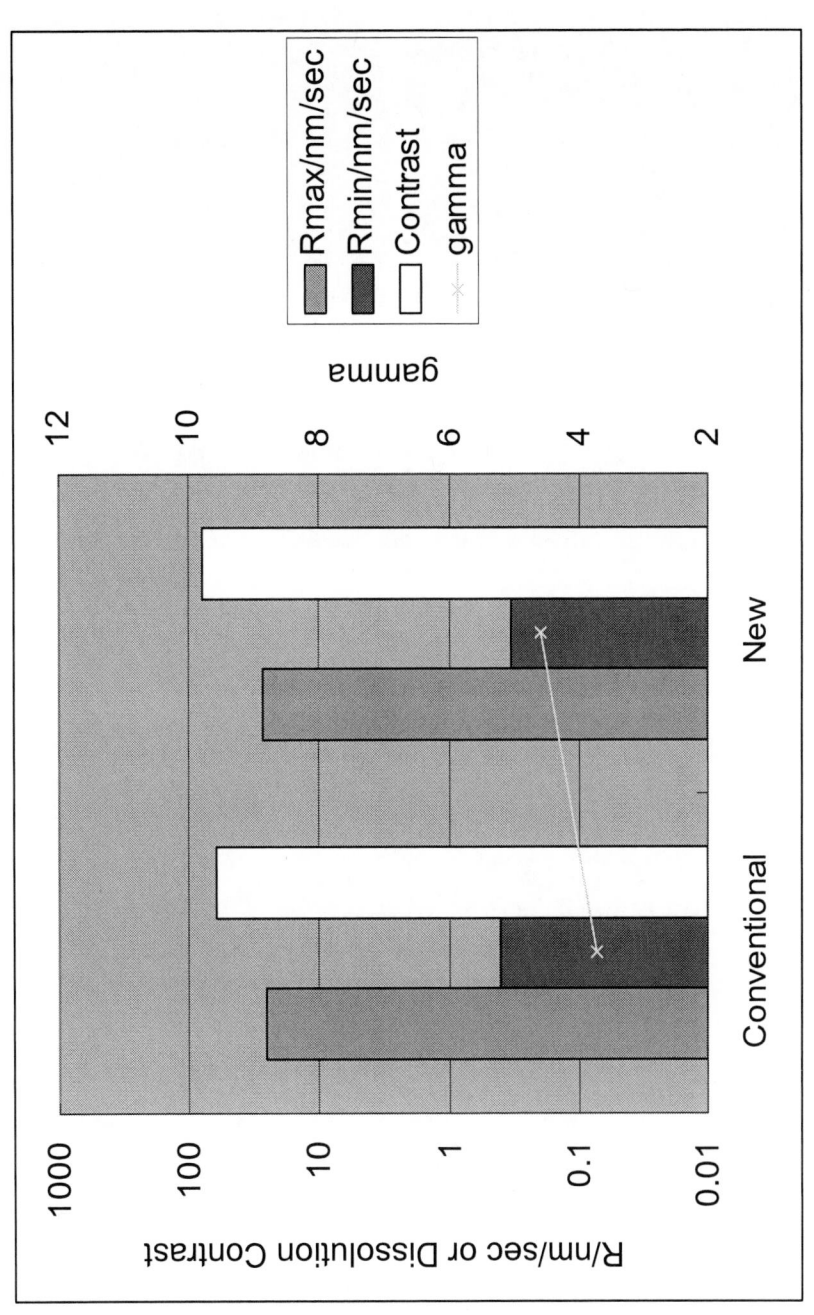

Larger γ was obtained with new formulation.

October, 2009

FUJIFILM

6th International Symposium on Immersion Lithography

Dry exposure lithography performance

NA0.75, Annular, Dry exposure
90nm dense CH (V/H double exposure)

FAiRS-9521A06G
Straight profile

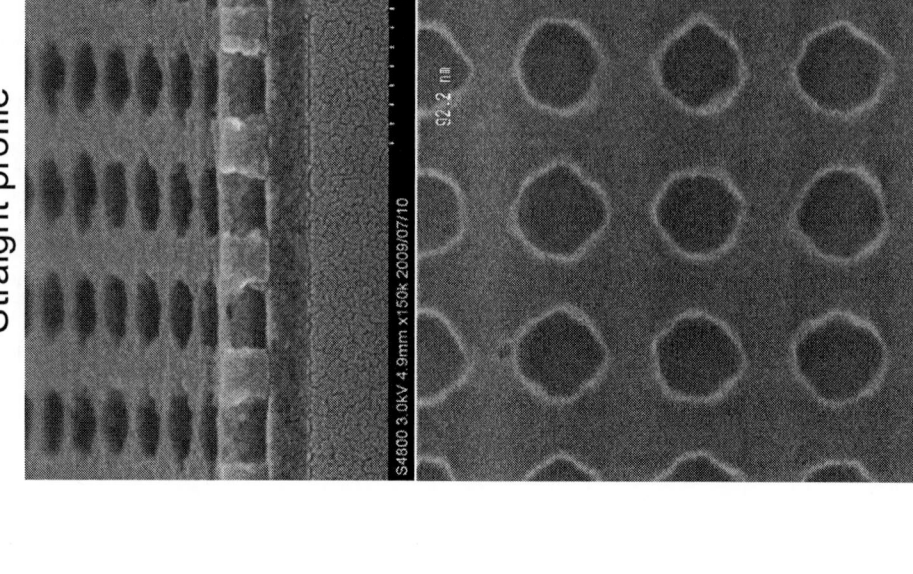

FAiRS-9101A19C
Cuspy profile, underdose

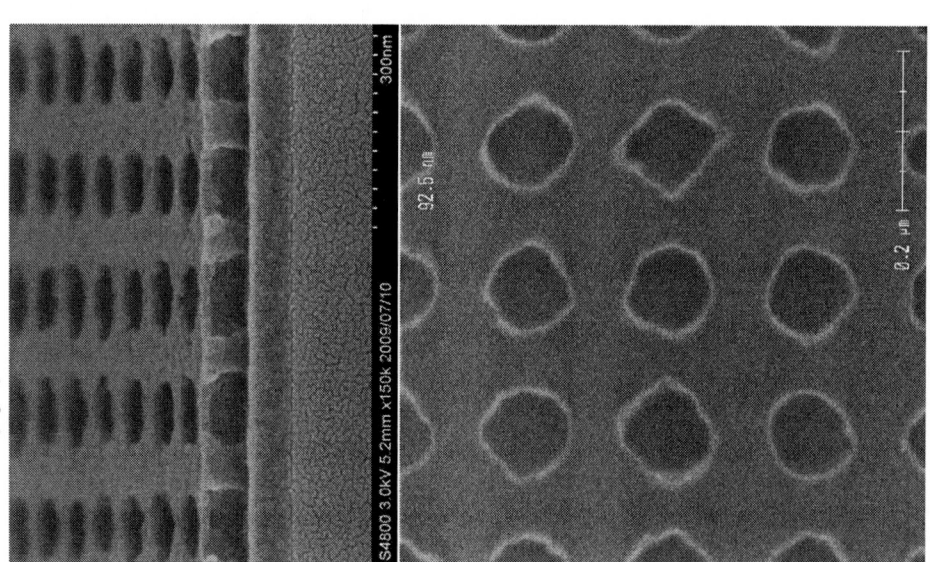

October, 2009

FUJIFILM

6th International Symposium on Immersion Lithography

Dry exposure lithography performance

NA0.75, Annular, Dry exposure
90nm dense CH (V/H double exposure)

October, 2009

A06G

	-0.20um	-0.15um	-0.10um	-0.05um	B.F.	+0.05um	+0.10um	+0.15um	+0.20um
23.7mJ/cm^2	92	89	98	102	100	100	101	102	96
25.0mJ/cm^2	77	64	86	88	92	90	94	87	91
26.3mJ/cm^2	54	73	75	83	80	85	83	81	75

A19C

	-0.20um	-0.15um	-0.10um	-0.05um	B.F.	+0.05um	+0.10um	+0.15um	+0.20um
21.1mJ/cm^2	98	109	110	110	108	112	109	108	112
22.4mJ/cm^2	95	95	97	94	93	97	98	95	95
23.7mJ/cm^2	72	70	71	86	83	83	89	84	73

6th International Symposium on Immersion Lithography

Immersion exposure lithography performance

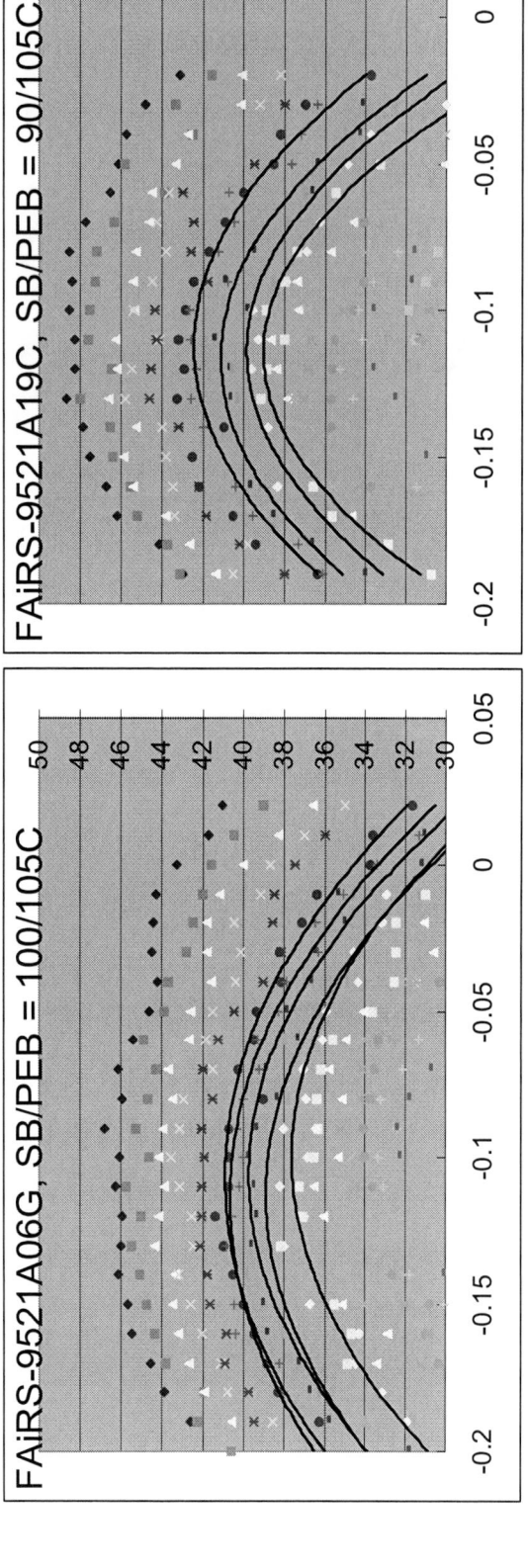

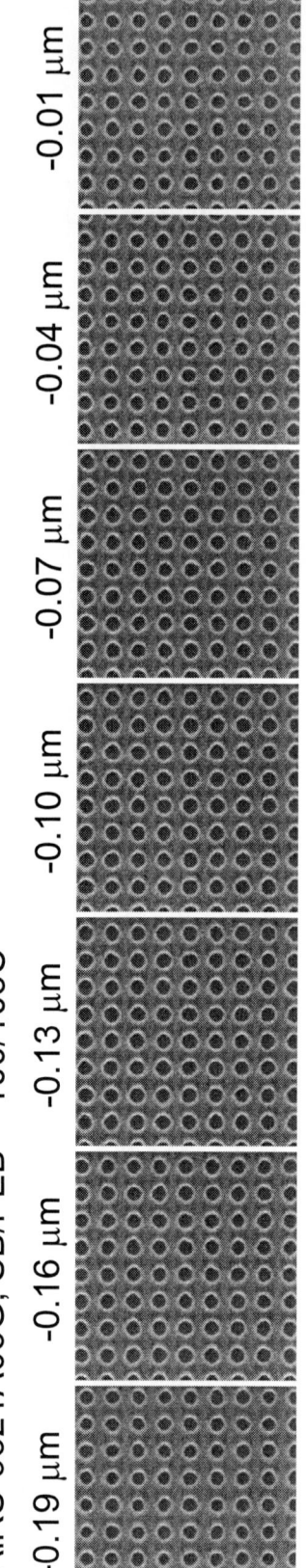

October, 2009

FUJIFILM

6th International Symposium on Immersion Lithography

October, 2009

Outline

1. Advantages of negative tone imaging for DP

2. Process maturity of negative tone development

3. Resist material progress for negative tone development

4. Summary

5. Acknowledgement

FUJIFILM

6th International Symposium on Immersion Lithography

Summary

✓ Negative tone development process is a good candidate for C/H and trench pattern lithography, since good enough optical image can be achieved with bright mask application.

✓ Process maturity was demonstrated in the viewpoint of etch resistance, defectivity, and CD uniformity.

✓ The resist platform of FAiRS-9101A19C has a issue of narrower DOF at dense pattern, compared to its positive tone development. This issue comes from the not large enough dissolution contrast.

✓ Dissolution rate can be controlled by the dissolution parameter, SP value, of polymer. Faster Rmax polymer gave 10 times larger contrast, and wider DOF margin.

✓ Another formulation study gave wider DOF margin.

✓ DOF can be improved with dissolution contrast improvement and formulation.

October, 2009

6th International Symposium on Immersion Lithography

October, 2009

Acknowledgement

- Dr. Roel Gronheid, Dr. Mireille Maenhoud, Dr. Joost Bekaert, Dr. Lieve Van Look at IMEC.

- Grozdan Grozev, Mario Reybrouck, and Veerle Van Driessche at FUJIFILM ELECTRONIC MATERIALS (EUROPE) N.V.

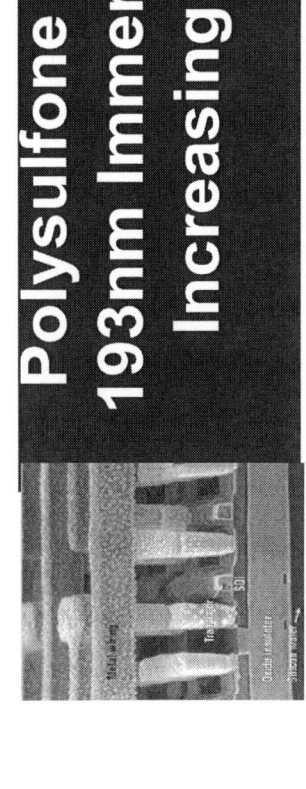

Polysulfone Based Non-CA Resists for 193nm Immersion Lithography: Effect of Increasing Polymer Absorbance on Sensitivity

Idriss Blakey, Lan Chen, Yong-Keng Goh, Kirsten Lawrie, Andrew K. Whittaker
The University of Queensland, Australia

Emil Piscani,
SEMATECH

Paul Zimmerman
Intel assignee to SEMATECH

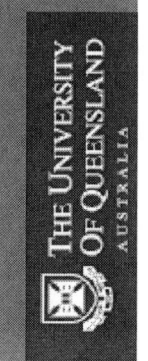

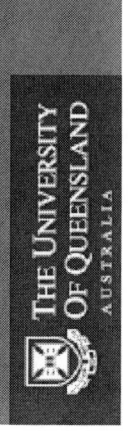

Outline

- Introduction
- Results & Discussion
  - Effect of Irradiation
  - Effect of PEB
  - Effect of Absorbance
  - Demonstration of Patterning
- Conclusions

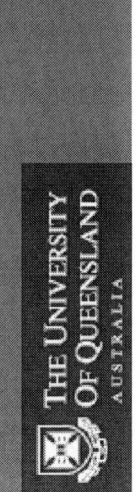

Chemical Amplification vs. Degradation

Chemically Amplified Resist

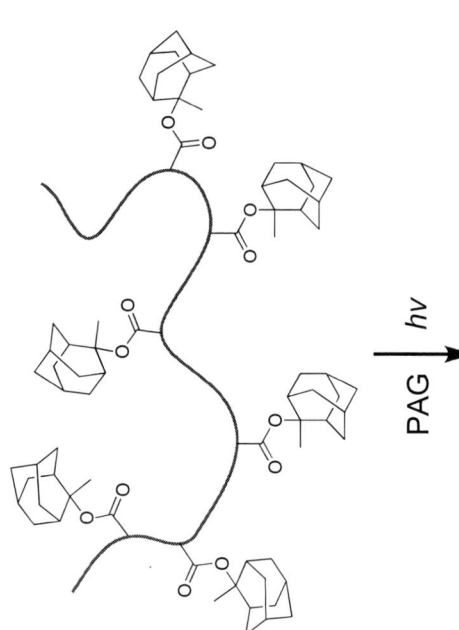

polarity solubility switch

Relies on
Diffusion
Limiting
Resolution
and LER

Non-CAR
Degradation Resist

Relies on
Direct
Photolysis
Limiting
Sensitivity

molecular weight
solubility switch

Effect of Non-CAR on Imaging

Non-CAR PMMA

$\sim 160 - 550$ mJ cm^{-2}

CA Resist

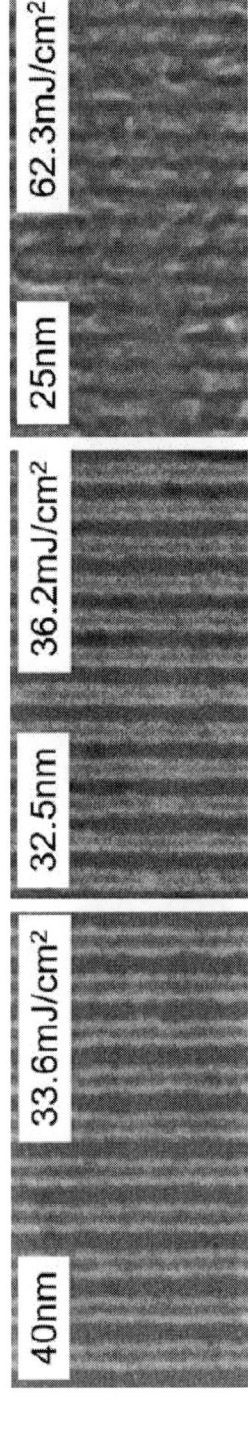

Gronheid, et al., Microelectronic Engineering (2006), 83(4-9), 1103

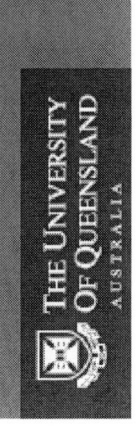

Accelerating the next technology revolution.

Advances in Laser Power

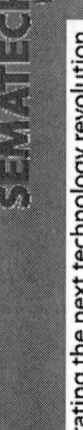

- In 2007, ArF lasers for double patterning were released with powers of 90W, which was double that of lasers in 2005.

  – Prospect for lasers with even higher powers are good

  – There will be surplus photons with current stepper rates

  – Non-CAR/chain scission resists becoming more realistic

  – Still need to significantly improve sensitivity

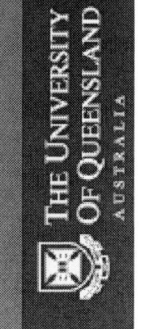

Self-Developing Resists

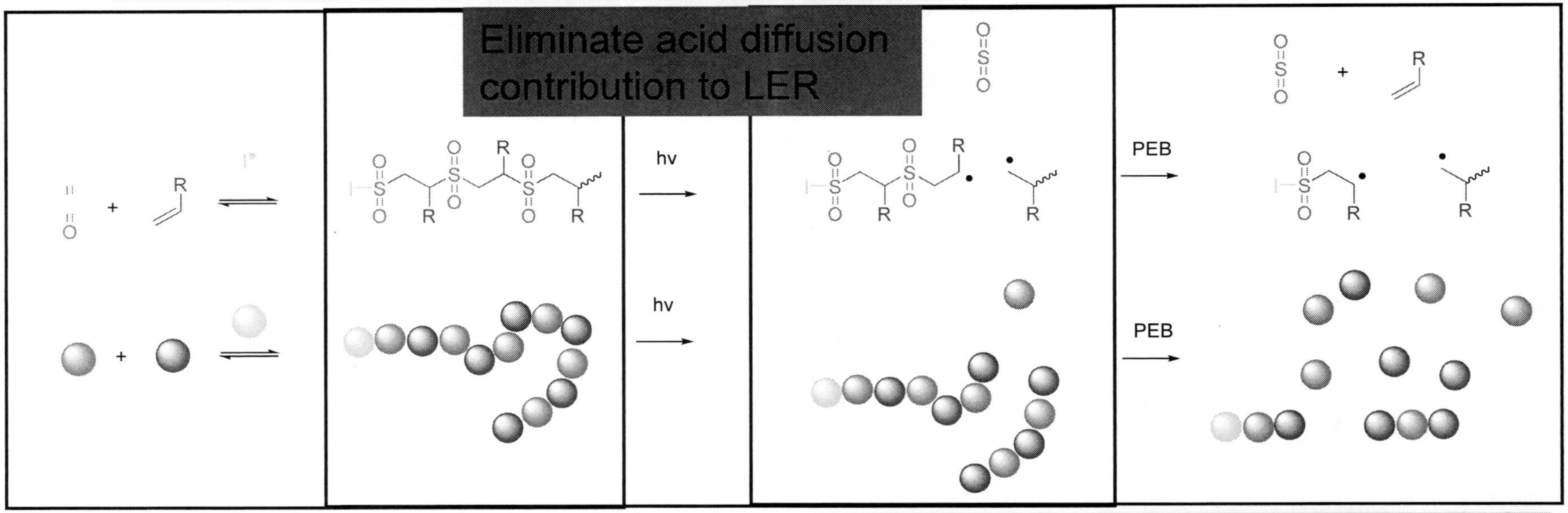

Eliminate acid diffusion contribution to LER

Gain understanding of LER contribution from development

- Synthesis
  - Alternating copolymer of sulfur dioxide and an alkene

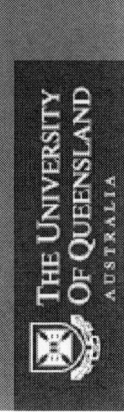

Polysulfones Synthesized

Structure	Molecular Weight[a]	Feed Ratio (x:y:z)	NMR Ratio	T_g (°C)	n_{193nm}	Abs (μm^{-1})
	M_w: 1,216k	100	100	120	1.74	0.17
	M_w: 190k	70:30	82:18	109	1.75	4.49
	M_w: 3,390k	63:37	65:35	113	1.69	6.10
	M_w: 310k	60:20:20	52:16:32	TBD	1.72	3

[a] Molecular weights are reported relative to polystyrene standards.

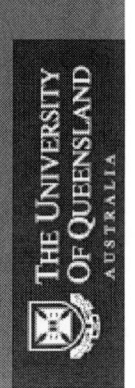

Effect of 193 nm Irradiation on Polysulfone Non-CARs

SEMATECH

Accelerating the next technology revolution.

THE UNIVERSITY OF QUEENSLAND
AUSTRALIA

Chemical Contrast Curves

- Linear decrease in thickness (ellipsometry) and sulfone peak (GATR) with irradiation dose

- ~5 mol% loss after 100 mJ cm^{-2}

  - For a DP of ~12800 → ~640 chain scissions per chain ~6 chain scissions per chain per mJ cm^{-2}

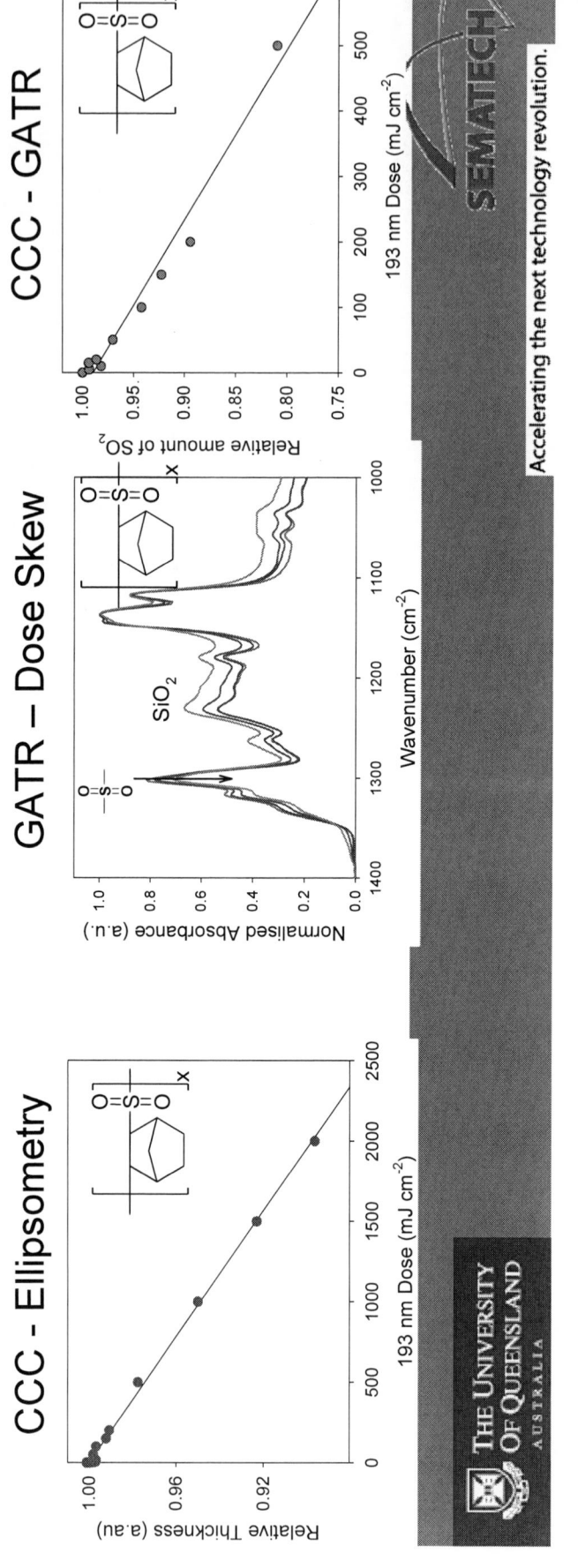

Summary of effect of irradiation

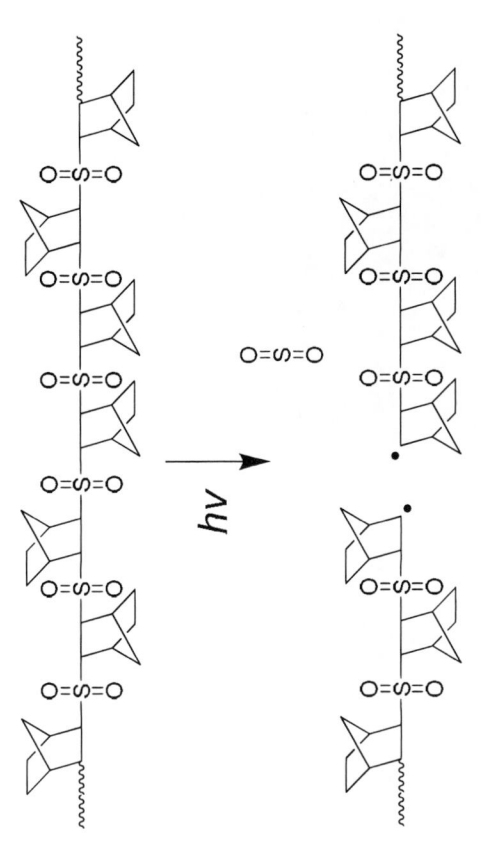

- GATR/ellipsometry indicates that there is film thinning and loss of SO_2 as a result of photon absorption

- Scission is induced by direct absorption of a photon by the polymer and process is not mediated by diffusion

  – Should eliminate blur that results from acid diffusion in CAR resists

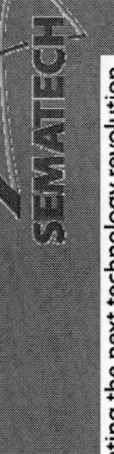

Accelerating the next technology revolution.

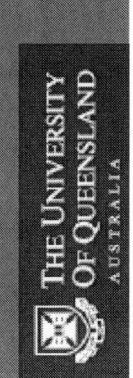

THE UNIVERSITY OF QUEENSLAND
AUSTRALIA

Effect of Post Exposure Bake on Polysulfone Non-CARs

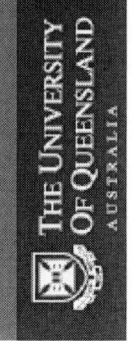

Accelerating the next technology revolution.

THE UNIVERSITY OF QUEENSLAND
AUSTRALIA

Thermal Stability – Dark Loss

- All polymers thermally stable to at least 200 °C

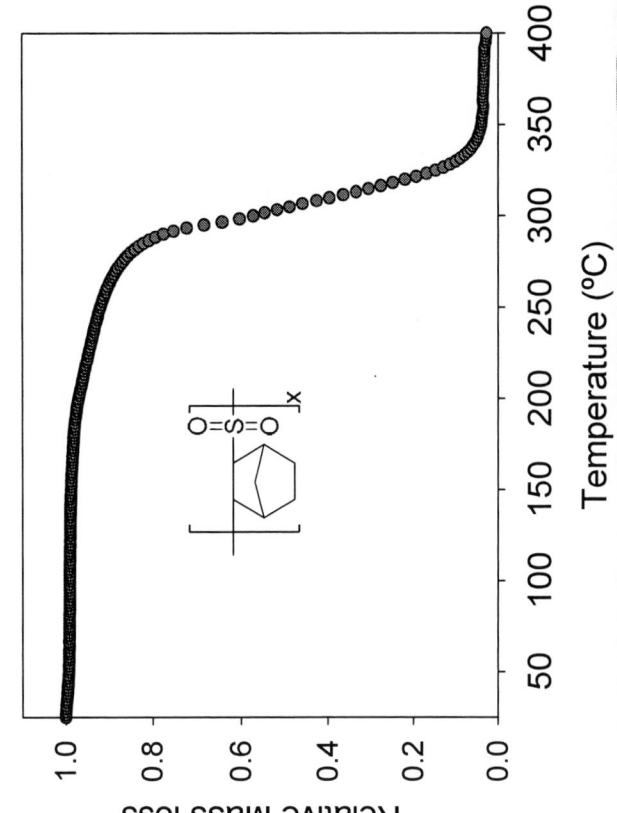

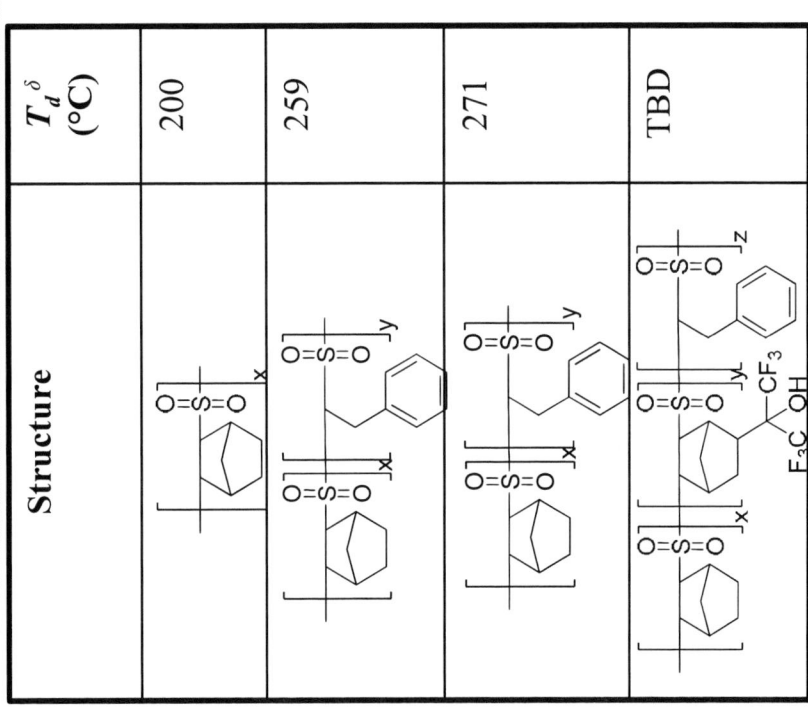

Structure	$T_d^{\,\delta}$ (°C)
	200
	259
	271
	TBD

$^{\delta}T_{\mathrm{d}}$ is measured at the onset of decomposition using TGA.

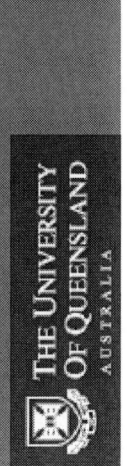

Effect of PEB Temperature on CCC

- Increased thinning is observed for as a function of increasing PEB temperature and dose

  - Indicates that depolymerisation is occurring as a result of PEB

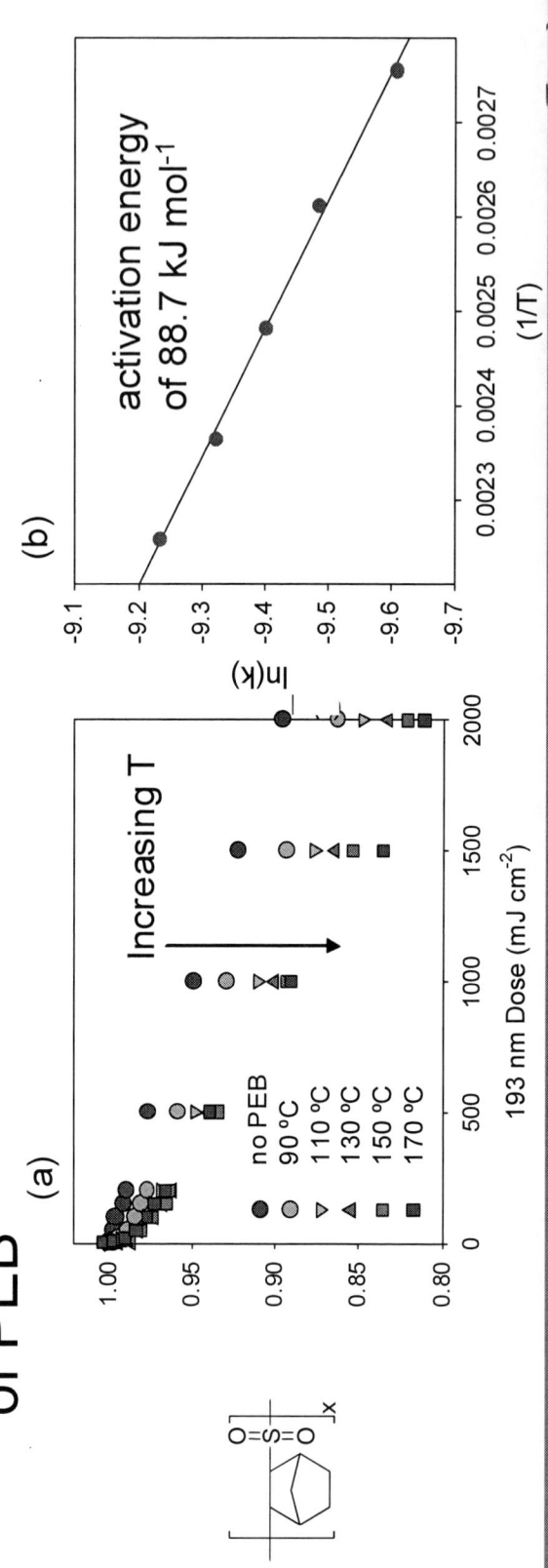

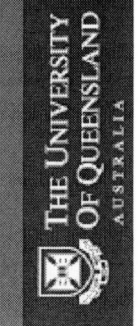

Accelerating the next technology revolution.

GATR- effect of PEB

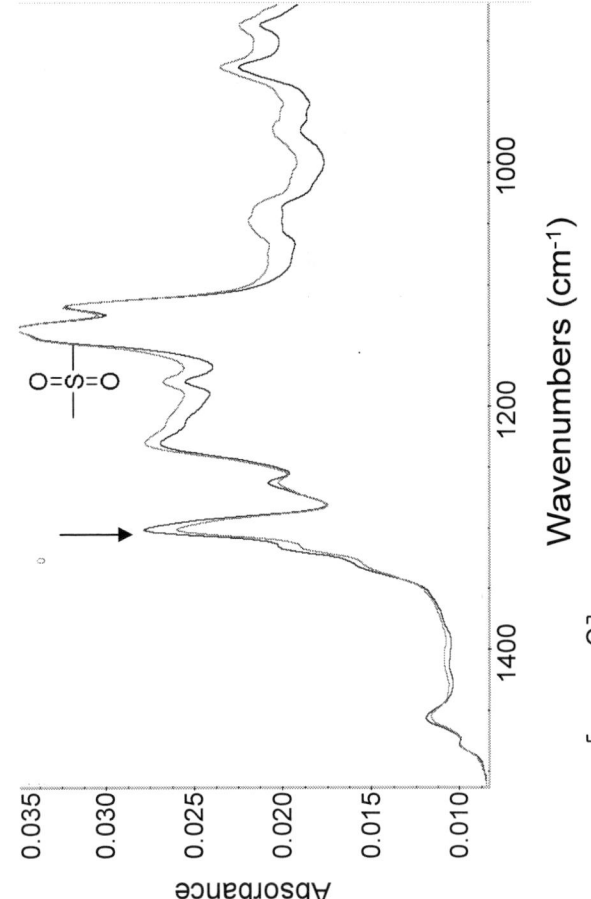

PBHS - 500 mJ cm^{-2} no PEB

PHBS - 500 mJ cm^{-2} and PEB 170 °C

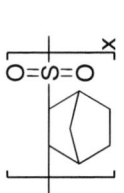

- FTIR shows further loss of sulfone group during PEB

  – indicative of liberation of SO$_2$ during PEB step

  – Most of the SO$_2$ is lost during PEB step not in the 193i tool

Accelerating the next technology revolution.

Summary of effect of PEB

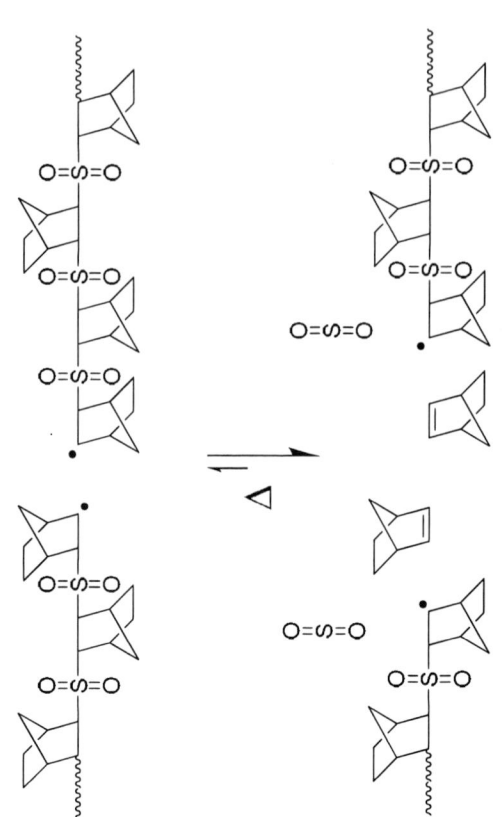

- Ellipsometry/GATR data shows that depolymerisation is occurring during PEB

- Selection of alternative structures may allow a thermal development process
  - Will also require better control over polymer MWt

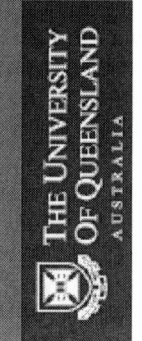

Effect of Polymer Absorbance on Sensitivity

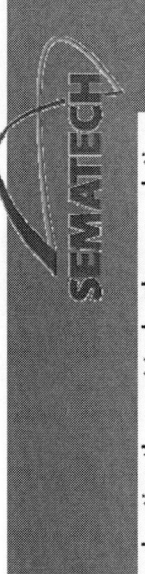

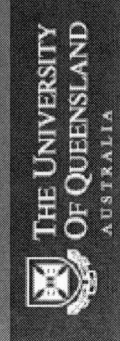

Accelerating the next technology revolution.

THE UNIVERSITY OF QUEENSLAND
AUSTRALIA

Effect of Absorbance on CCC

Chemical contrast curve

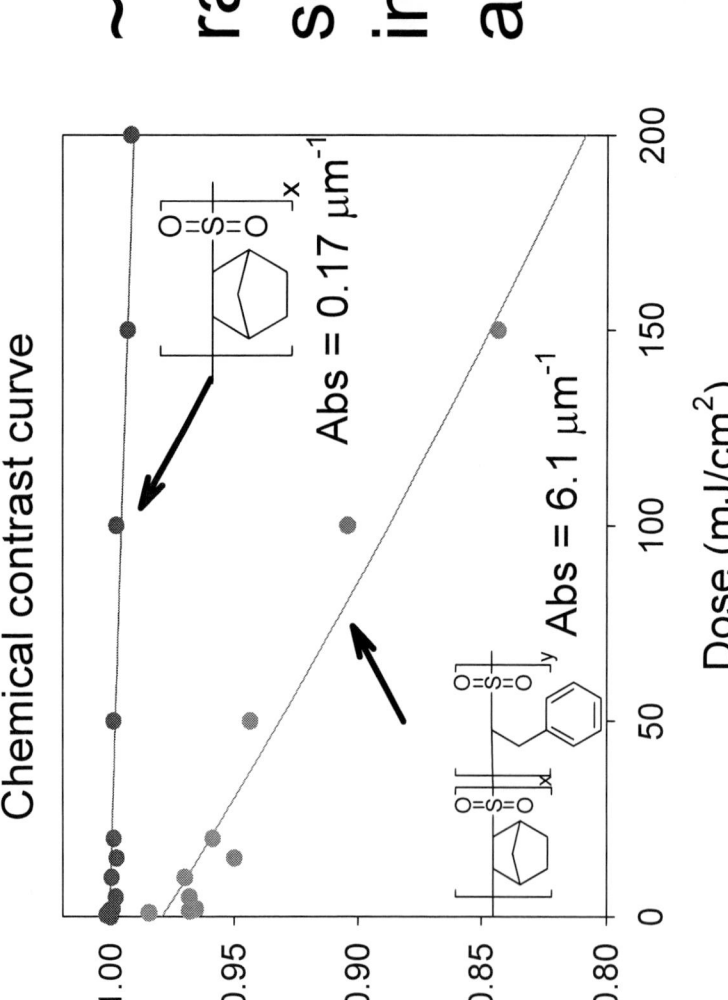

~25 x increase in rate of chain scissions for increased absorbance

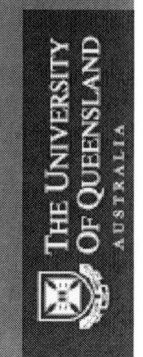

Effect of PEB for Absorbing Polymers

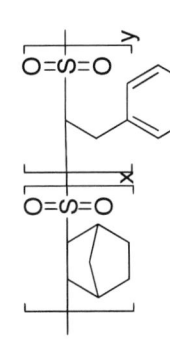

- No PEB effect observed for absorbing polymer

- Allylic radical is more stable – depolymerisation is inhibited

To take advantage of depolymerisation effect, alternative absorbing monomer structures need to be identified

Chemical contrast curve

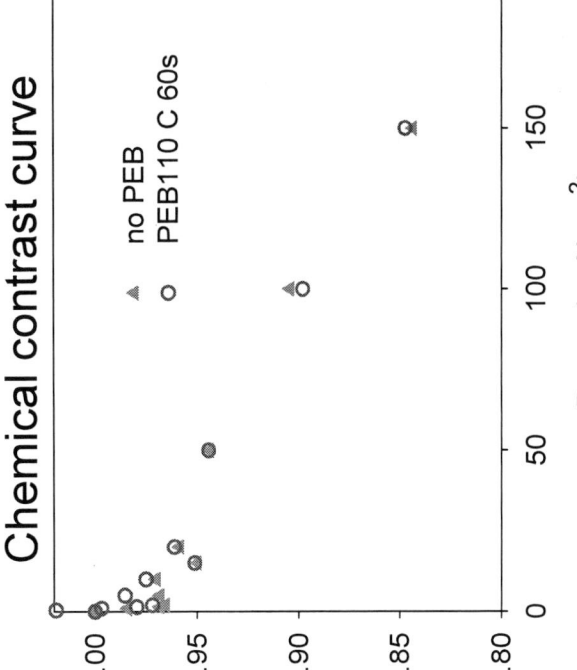

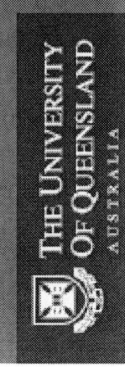

Accelerating the next technology revolution.

Effect of Absorbance on E_0 Values

Items				
Abs, 1/μm	0.17	4.49	6.10	3
PEB, °C	120	110	110	110
Contrast	-	0.71	0.71	0.53
E_0, mJ/cm²	1000	50	50	150
Composition	100	82:18	65:35	TBD

Contrast is poor compared to ArF CA resists

Absorbance significantly decreases E_0

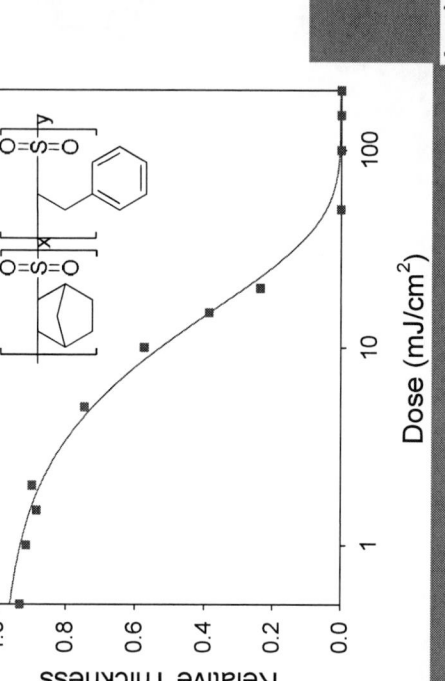

SEMATECH

Accelerating the next technology revolution.

THE UNIVERSITY OF QUEENSLAND
AUSTRALIA

Dose Skew on Amphibian at RIT

40 mJ cm⁻²   50 mJ cm⁻²   60 mJ cm⁻²   70 mJ cm⁻²

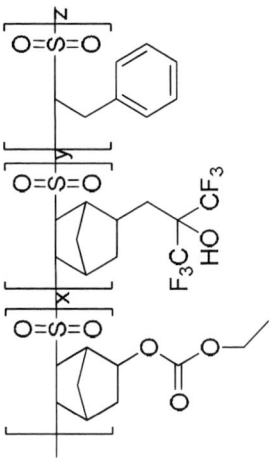

- Best performance observed at 60 mJ cm⁻²

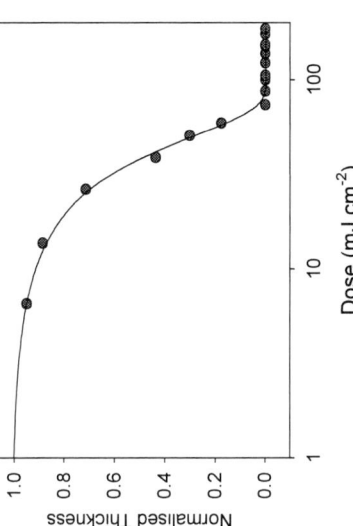

Conditions

➤ BARC coated: 74.6nm
- BARC29
- PAB: 200°C, 60s

➤ Resist Thickness: 73.6nm
- PAB: 110°C, 60s

➤ Exposure
- Prism 0.8 NA in air
- Half-pitch: 60 nm
- PEB 120 °C

➤ Development
- IPA/MIBK (80:20), 60s

SEMATECH

Accelerating the next technology revolution.

THE UNIVERSITY OF QUEENSLAND
AUSTRALIA

Amount of SO₂ Released

- Worst case scenario
  - 100% SO_2 released upon irradiation
    - 0.032 nmol SO_2 cm^{-2}

- More realistic scenario based on FTIR
  - At 100 mJ ~5% released
    - 0.0016 nmol SO_2 cm^{-2}

- In both cases, this is a very low amount

- Solubility of SO_2 in water is ~100 g/L at RT

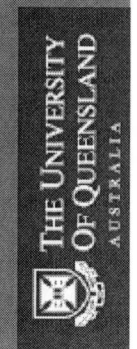

Key Challenges

- Increase contrast
  - Investigating dissolution inhibitors as well as other structures

- Overcome inhibition of depolymerisation
  - Alternative absorbing structures are being investigated
  - To reduce issues such as pattern collapse it would be desirable to identify structures that clear to the wafer during PEB.

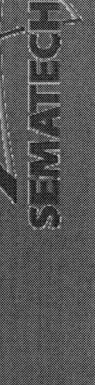

Accelerating the next technology revolution.

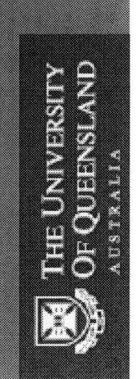

THE UNIVERSITY OF QUEENSLAND
AUSTRALIA

Conclusions

- Irradiation of polysulfones results in
  - Film thinning
  - Decrease in SO_2 content

- Depolymerisation was demonstrated during PEB for certain structures
  - Structure of repeat unit is important

- Absorbance significantly decreases E_0
  - Gains of up to 20x observed
  - Further gains possible if PEB inhibition can be overcome

- Patterning at 60 nm hp has been demonstrated

- Amount of SO_2 liberated during irradiation will be easily dissolved in immersion fluid

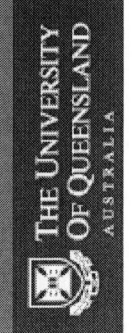

Acknowledgements

- People
  - Prof. Bruce Smith and his group (RIT)
  - Emil Piscani and Paul Zimmerman (SEMATECH)
  - Lauren Butler (ANFF, UQ)

- Funding and Facilities
  - Australian Research Council – Linkage Projects (LP0882551)
  - SEMATECH
  - Australian National Fabrication Facility (ANFF)
  - Australian Institute for Bioengineering and Nanotechnology (AIBN)
  - Centre for Magnetic Resonance (CMR)

Accelerating the next technology revolution.

SEMATECH and the SEMATECH logo, are registered service marks of SEMATECH, Inc. All other service marks and trademarks are the property of their respective owners.

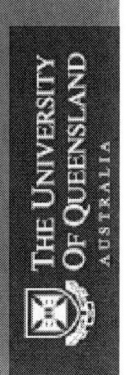

THE UNIVERSITY OF QUEENSLAND
AUSTRALIA

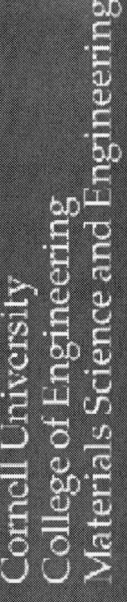

Cornell University
College of Engineering
Materials Science and Engineering

INORGANIC PHOTRESISTS BASED ON HAFNIUM OXIDE

Markos Trikeriotis, Woo-Jin Bae, Evan Schwartz, Marie Krysak, Christopher K. Ober and Emmanuel P. Giannelis
Cornell University

Neal Lafferty, Peng Xie and Bruce Smith
Rochester Institute of Technology

Paul A. Zimmerman
Intel assignee to SEMATECH

22-23 October 2009

6th International Symposium on
Immersion Lithography Extensions

Cornell University
College of Engineering
Materials Science and Engineering

Outline

- Inorganic Resists
  - Why do we need them?
  - How do they work?

- Hafnium oxide nanoparticles
  - Film preparation and characterization

- Inorganic Resist Lithography
  - Photolithography
  - Electron Beam Lithography
  - LER/LWR and Etch Resistance

- Conclusions

22-23 October 2009

6th International Symposium on
Immersion Lithography Extensions

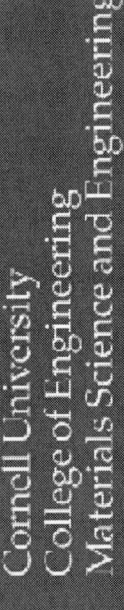

Cornell University
College of Engineering
Materials Science and Engineering

Inorganic Photoresists
Why do we need them?

Aspect ratio → Small features, Thin films

Pattern transfer → Etch Resistance

Inorganic-based instead of polymer-based:

Improved chemical stability

Higher Refractive Index → Increased DOF

22-23 October 2009

6th International Symposium on
Immersion Lithography Extensions

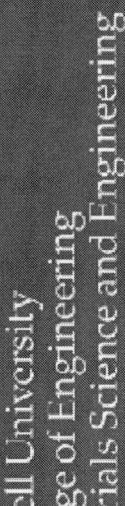

Cornell University
College of Engineering
Materials Science and Engineering

Inorganic Photoresists
How do they work?

Nanoparticle suspension

Spin coating

Sol-gel film

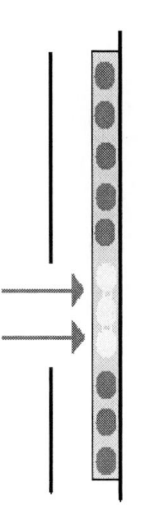

Chemistry or photochemistry of the surface ligands

22-23 October 2009

6th International Symposium on
Immersion Lithography Extensions

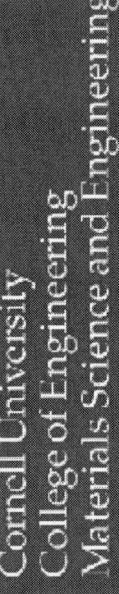

Cornell University
College of Engineering
Materials Science and Engineering

Hafnium Oxide Nanoparticles

- Surface stabilizing ligands (variable inorganic content)
- Water or organic soluble (dual-tone)
- Spin coated sol-gel films
- EHS study shows similar toxicity to bulk HfO_2

- Increased etch resistance (thinner resist films)
- Decreased LER/LWR (consistent device performance)
- Increased refractive index (better DOF)

22-23 October 2009

6th International Symposium on
Immersion Lithography Extensions

Cornell University
College of Engineering
Materials Science and Engineering

Hafnium Oxide Nanoparticles

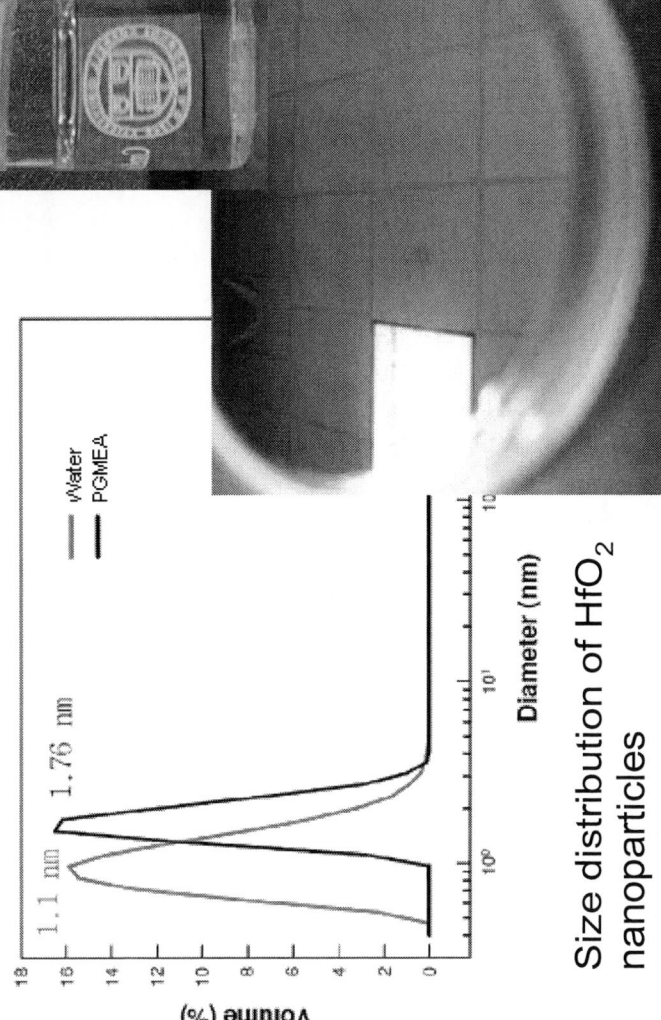

Size distribution of HfO_2 nanoparticles

- ◆ Small particles are consistently obtained (< 4.0 nm)

- ◆ Transparent, stable, suspensions in water or PGMEA

- ◆ High RI films

- ◆ Acceptable film absorbance achieved

22-23 October 2009

6th International Symposium on Immersion Lithography Extensions

Cornell University
College of Engineering
Materials Science and Engineering

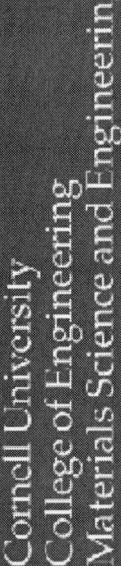

Hafnium Oxide Nanoparticles

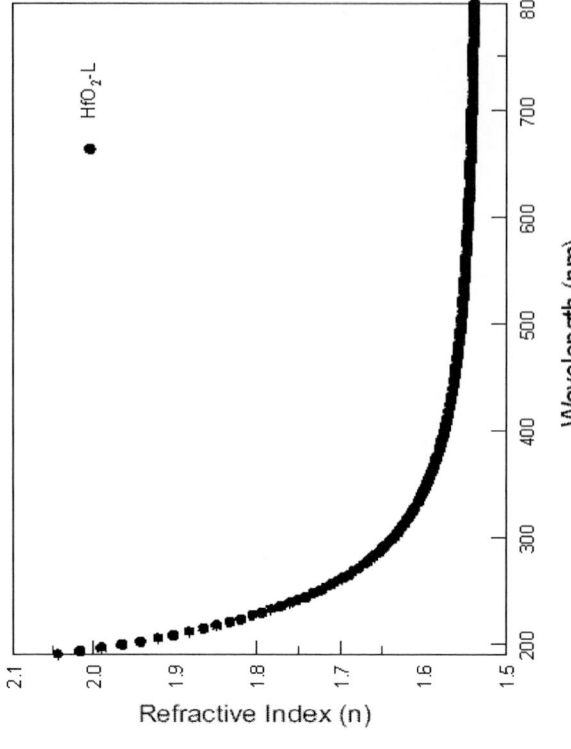

Film refractive index
(~2.0 @ 193 nm)

Film absorbance per micron
(Typically ~2.5/µm)

6th International Symposium on
Immersion Lithography Extensions

22-23 October 2009

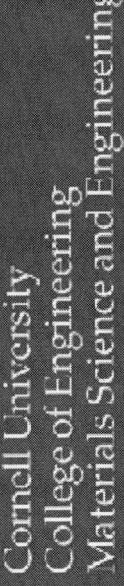

Cornell University
College of Engineering
Materials Science and Engineering

Hafnium Oxide Nanoparticles

High inorganic content: 60% - 75% w/w

The absolute inorganic content depends on the stabilizing ligand used

Ultimate target is 80-85%

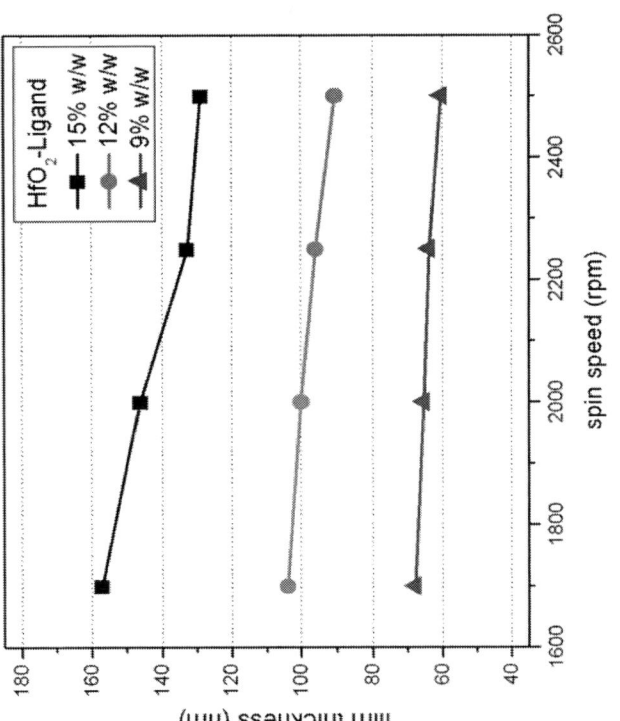

Film thickness versus spin speed and nanoparticle loading

22-23 October 2009

6th International Symposium on Immersion Lithography Extensions

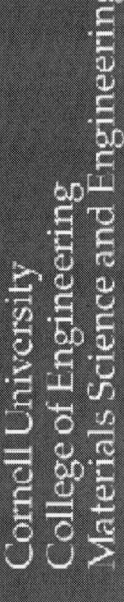

Cornell University
College of Engineering
Materials Science and Engineering

Inorganic Resist Lithography

Photolithography

- 254nm and 193nm patterning:

    - Both positive- and negative-tone images achieved independent of the stabilizing ligand

    - Dark Loss measurement

Electron Beam Lithography

- Positive- and negative-tone images

22-23 October 2009

6th International Symposium on
Immersion Lithography Extensions

Cornell University
College of Engineering
Materials Science and Engineering

254 nm Photolithography

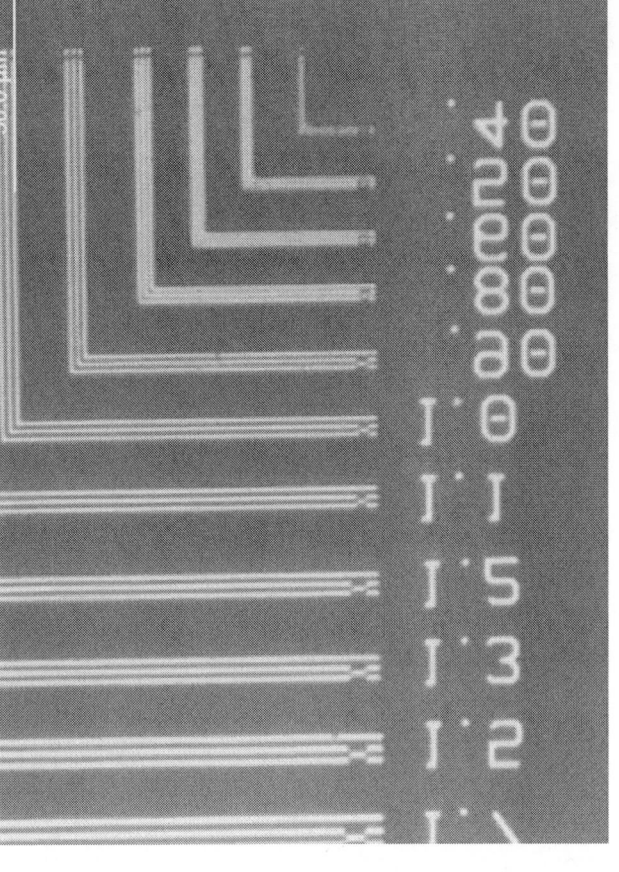

Positive-tone 25mJ/cm²

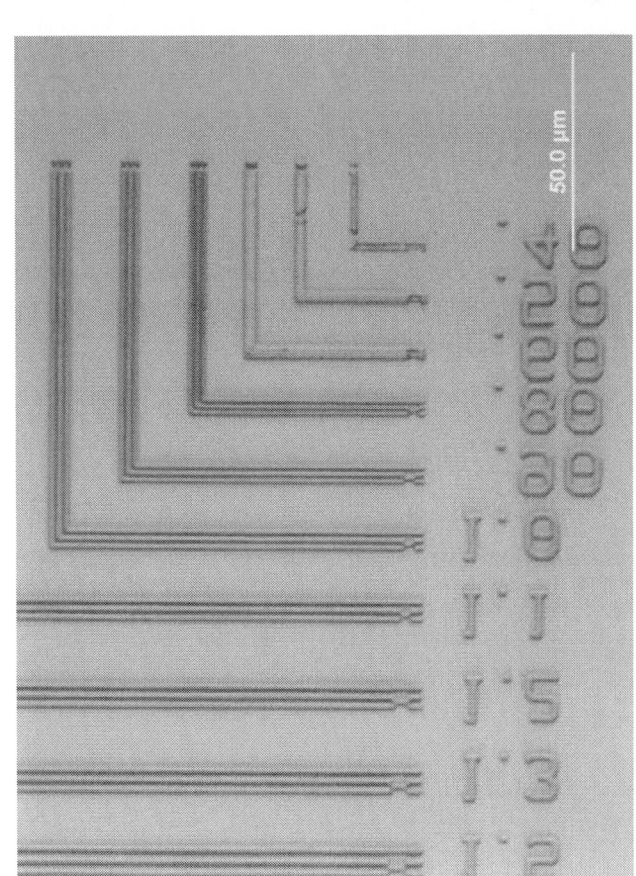

Negative-tone 520mJ/cm²

6th International Symposium on
Immersion Lithography Extensions

22-23 October 2009

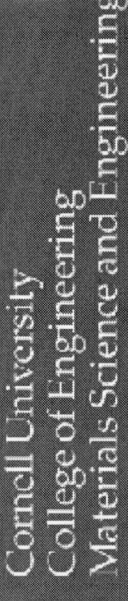

Cornell University
College of Engineering
Materials Science and Engineering

193nm Photolithography

0.32 NA, 150nm half-pitch

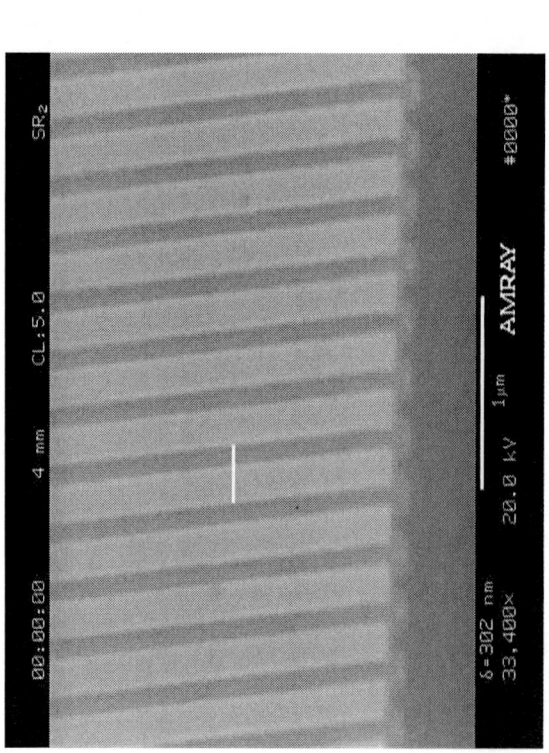

Negative-tone, 76 mJ/cm^2

Positive-tone, 34 mJ/cm^2

0.80 NA, 60nm half-pitch currently underway

6th International Symposium on
Immersion Lithography Extensions

22-23 October 2009

Cornell University
College of Engineering
Materials Science and Engineering

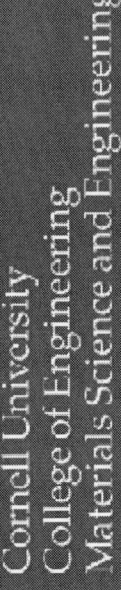

Dark Loss measurement

Positive-tone

Poor performance:
More than 50% of the film is lost

Exploring the use of a dissolution
Inhibitor to reduce loss

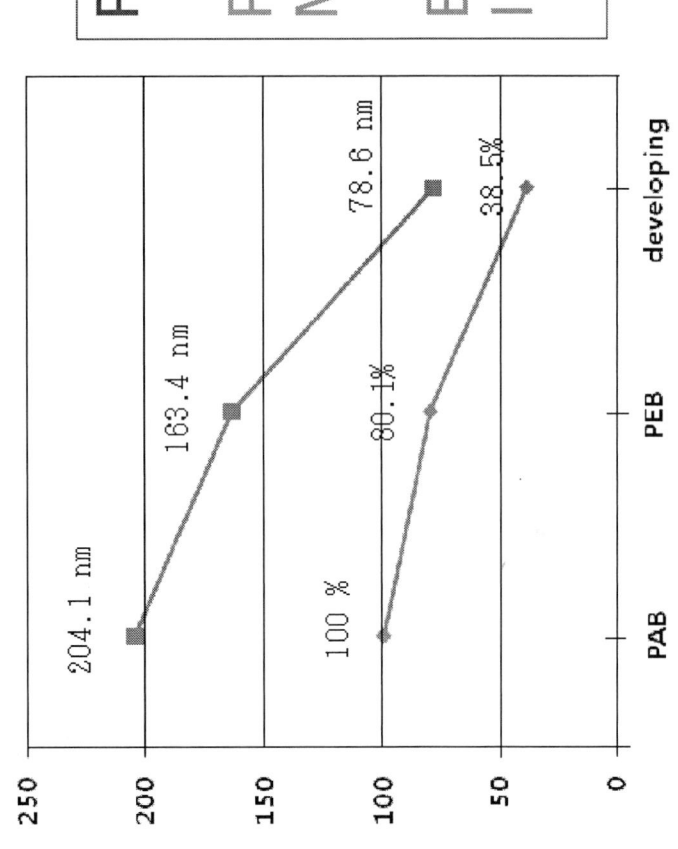

22-23 October 2009

6th International Symposium on
Immersion Lithography Extensions

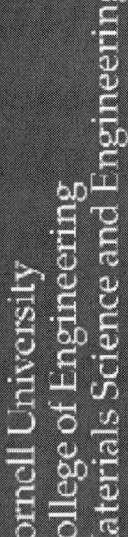

Cornell University
College of Engineering
Materials Science and Engineering

Electron Beam Lithography

negative tone image

Dose: 213 µC/cm^2
Developing in IPA 2min

Positive tone image

Dose: 240 µC/cm^2
PEB: 130 °C / 5min
Developing in water 5min

22-23 October 2009

6th International Symposium on
Immersion Lithography Extensions

Cornell University
College of Engineering
Materials Science and Engineering

Electron Beam Lithography

Negative-tone images

Dose = 102.7 μC/cm²

50 nm dense lines and isolated lines

Higher resolution is also possible: **30nm lines**

200 nm

Aperture Size = 30.00 μm   Date :13 Oct 2009
EHT = 1.50 kV   Pixel Size = 2.6 nm   Time :16:39:11

CNF

22-23 October 2009

6th International Symposium on
Immersion Lithography Extensions

LWR and LER measurement

Cornell University
College of Engineering
Materials Science and Engineering

E-beam lithography
Dose= 199 μC/cm²

LWR and LER Data
from Summit software

35 nm isolated lines

LER = 4.7 nm

LWR = 6.4 nm

CD=35.7 nm
LER=4.7 nm

30 nm

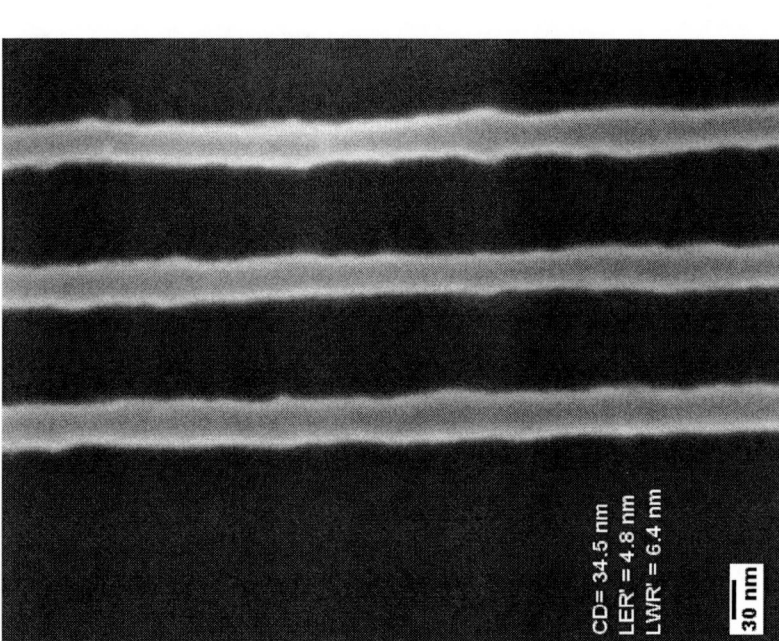

CD= 34.5 nm
LER' = 4.8 nm
LWR' = 6.4 nm

30 nm

22-23 October 2009

6th International Symposium on
Immersion Lithography Extensions

Cornell University
College of Engineering
Materials Science and Engineering

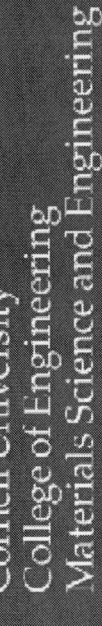

Etch Resistance

SF_6/O_2 plasma etching for 10s

Negative tone

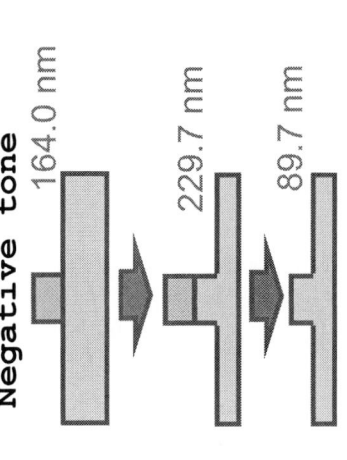

164.0 nm

229.7 nm

89.7 nm

Positive tone

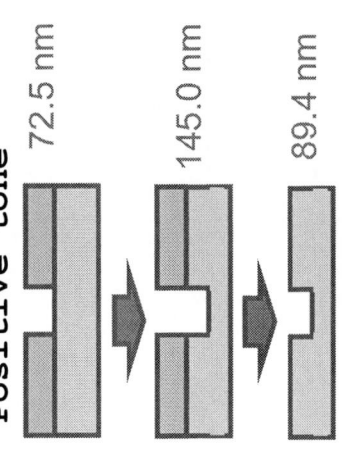

72.5 nm

145.0 nm

89.4 nm

Goal is 10 times slower rate than PHOST:

Increase inorganic content

	Resist etching rate	Si etching rate
Positive tone	1.7 nm/sec	8.97 nm/sec
Negative tone	2.4 nm/sec	8.94 nm/sec
PHOST	4.83 nm/sec	8.99 nm/sec

6th International Symposium on
Immersion Lithography Extensions

22-23 October 2009

Cornell University
College of Engineering
Materials Science and Engineering

Conclusions

◆ The use of HfO_2 nanoparticles allows the formulation of a non-CAR capable of optical and e-beam lithography.

◆ Images can be obtained on both positive- and negative-tone with reasonable sensitivity.

◆ Material shows both improved etch resistance (thinner film) and higher refractive index (DOF).

◆ Preliminary LER/LWR measurement shows promising trends that can be improved with modified formulations.

22-23 October 2009

6th International Symposium on Immersion Lithography Extensions

Cornell University
College of Engineering
Materials Science and Engineering

Future Work

- Optimize formulations to improve overall resist performance

  - Resolution
  - LER/LWR
  - Etch resistance

- Test for EUV imaging

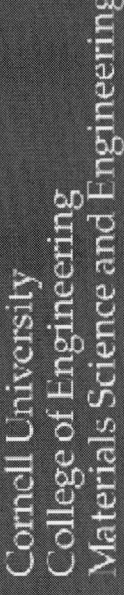

22-23 October 2009

6th International Symposium on
Immersion Lithography Extensions

SEMATECH

Cornell University
College of Engineering
Materials Science and Engineering

Acknowledgments and Disclaimers

CORNELL:

Emmanuel Giannelis
Markos Trikeriotis
Robert Rodriguez
Michael Zettel
Aris Bakandritsos

Christopher Ober
Woo-Jin Bae
Evan Schwarz
Marie Krysak
Jing Sha

RIT:

Bruce Smith
Neal Lafferty
Peng Xie

CNF
Cornell NanoScale
Science and Technology Facility

Disclaimer:

Copyright ©2008 SEMATECH, Inc. SEMATECH, and the SEMATECH logo are registered servicemarks of SEMATECH, Inc. International SEMATECH Manufacturing Initiative, ISMI, Advanced Materials Research Center and AMRC are servicemarks of SEMATECH, Inc. All other servicemarks and trademarks are the property of their respective owners.

22-23 October 2009

6th International Symposium on
Immersion Lithography Extensions

A PECVD Bi-layer DARC Solution for Immersion Lithography

Michael Lin, Betty Tang, Martin Seamons, Ran Ding, Liyan Miao, Huixiong Dai

think it. apply it.

APPLIED MATERIALS.

Outline

- Introduction
  - Optical Lithography Limitations
  - Immersion Lithography Challenges
  - APF/DARC Stack for Patterning
- Bi-Layer DARC Development
  - Reflectivity Control
  - Film Properties Optimization
- Immersion Lithography Performance
- Etch Performance
- Process Manufacturability
- Summary

APPLIED MATERIALS.

Limitations of Optical Lithography

Numerical Aperture

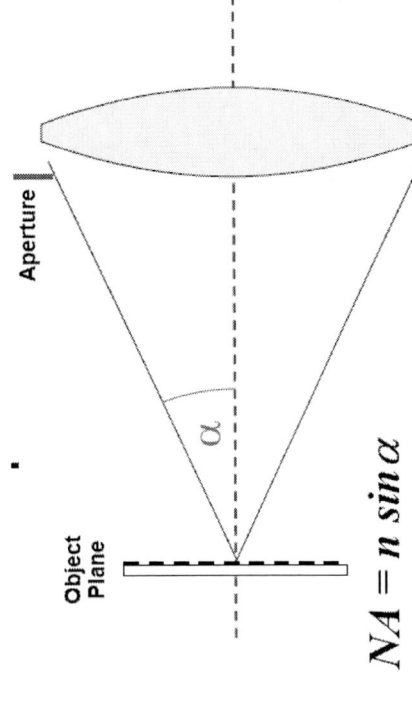

Aperture

Object Plane

$$NA = n\,sin\,\alpha$$

n = index of refraction of media
α = maximum half-angle of light making it through the lens

Resolution

$$R = k_1 * \left(\frac{\lambda}{NA}\right)$$

$k_1 \sim 0.29$ (best case for dry litho)

$n_{air} = 1.0$

$NA_{max} = 1.0$, (Reality) $NA_{best} = 0.95$

$$R_{best} \sim \frac{0.27 * 193nm}{0.95} \sim 60nm$$

Dry lithography hits limit at 60 nm. Immersion lithography extends optical lithography to sub-45nm feature size

APPLIED MATERIALS.

Challenges of Immersion Lithography

- **Reflectivity Control**
  - Wider range of incident angles
  - Conventional organic bottom antireflective coating (BARC) may not be sufficient to control reflectivity below 1% at all incident angles

- **Pattern Transfer**
  - Smaller depth of focus (DOF) resulting from increased NA
  - Thinner photoresist is used which may not provide enough etch resistance for pattern transfer to under-layers
  - BARC has limited selectivity to PR and under-layers

- **Pattern Integrity**
  - Resist collapse with increased aspect ratio
  - Adhesion between resist and under-layer is critical with reduced line width

APPLIED MATERIALS.

APF™/DARC® Stack for Patterning

Patterning Film Stack

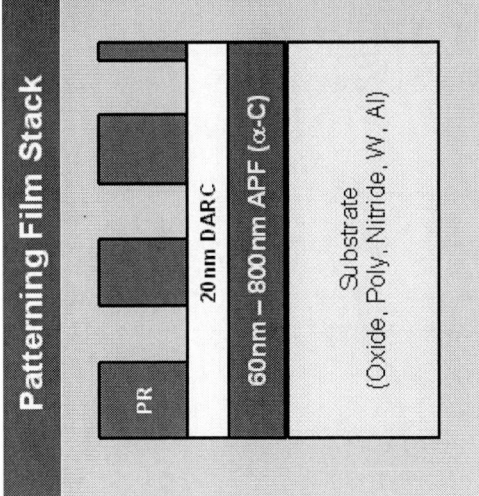

Product	APF™	DARC®
Name	Advanced Patterning Film	Dielectric Anti-reflective Coating
Description	PECVD amorphous carbon hardmask	PECVD silicon-rich oxide or oxynitride
Key Properties	• High etch selectivity to oxide, poly-Si and nitride • Easily strippable with plasma oxygen ash • UV-absorbing for low reflectivity	• Transfer PR pattern into APF with high selectivity (>10:1) • Tunable thickness and optical properties to control reflectivity • Good adhesion to PR with no poisoning

APPLIED MATERIALS.

Optical Properties Correlation of DARC Film

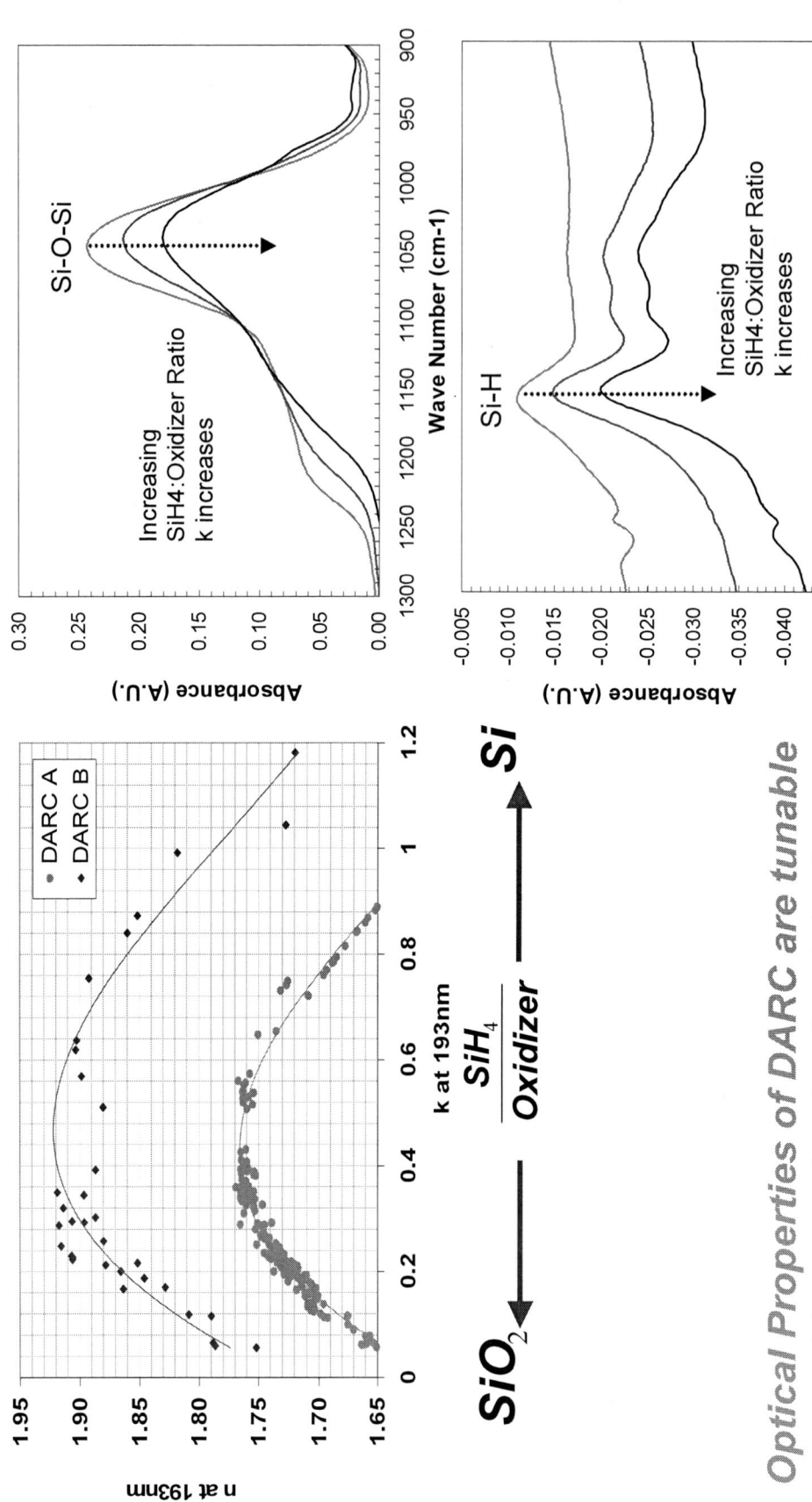

Optical Properties of DARC are tunable to achieve optimal reflectivity for a given film stack

APPLIED MATERIALS.

Multi-Layer DARC Reflectivity Control

- Bi-Layer DARC
  - Low k DARC: Phase shift cancellation through destructive interference
  - High k DARC: Absorbing layer
- Graded DARC:
  - Match n and k with top layer resist
  - Continuously increase k of DARC to absorb UV light

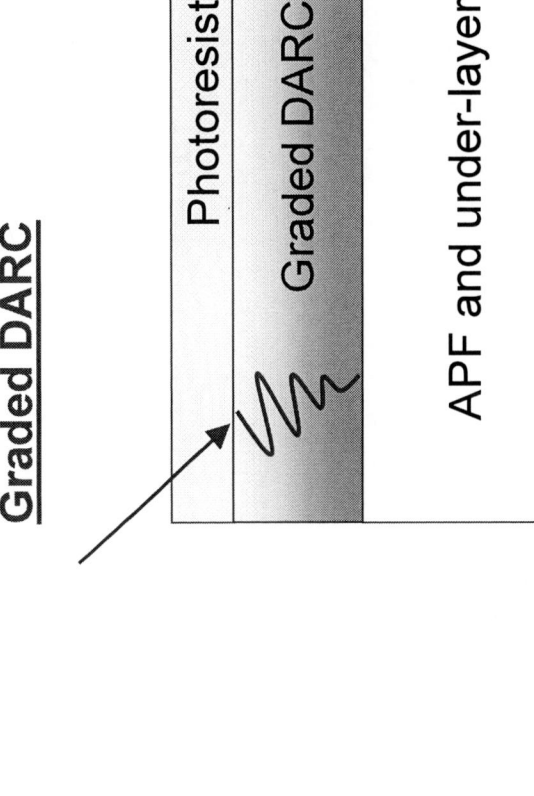

Bi-Layer DARC

Graded DARC

APPLIED MATERIALS

Immersion Lithography Simulation Results

- Bi-layer DARC
  - 10nm high k DARC + 30nm low k DARC
  - 40nm gives <0.3% reflectivity

- Graded DARC
  - Extreme thick DARC is needed to achieve <0.5% reflectivity

*Lithography simulation done with "Synopsys Sentaurus Lithography" software

Bi-Layer DARC

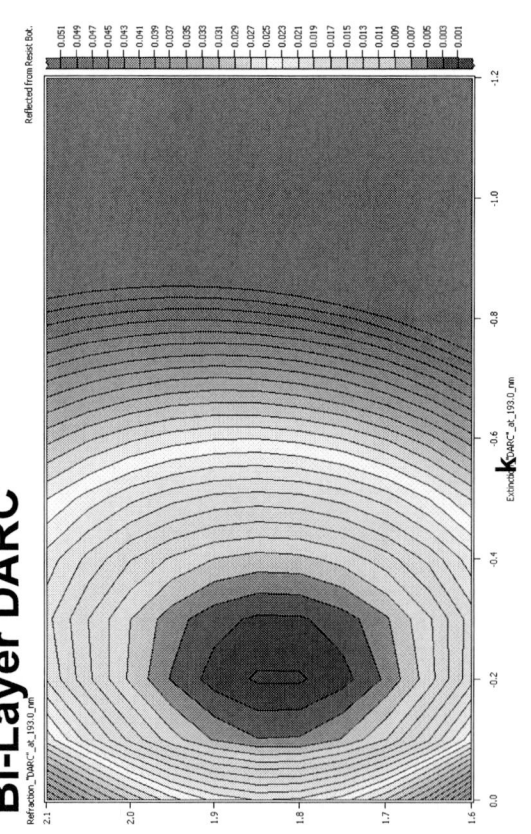

Graded DARC (40nm)

Bi-layer DARC allows better reflectivity control with lower film thickness which improves etch margin for pattern transfer

APPLIED MATERIALS.

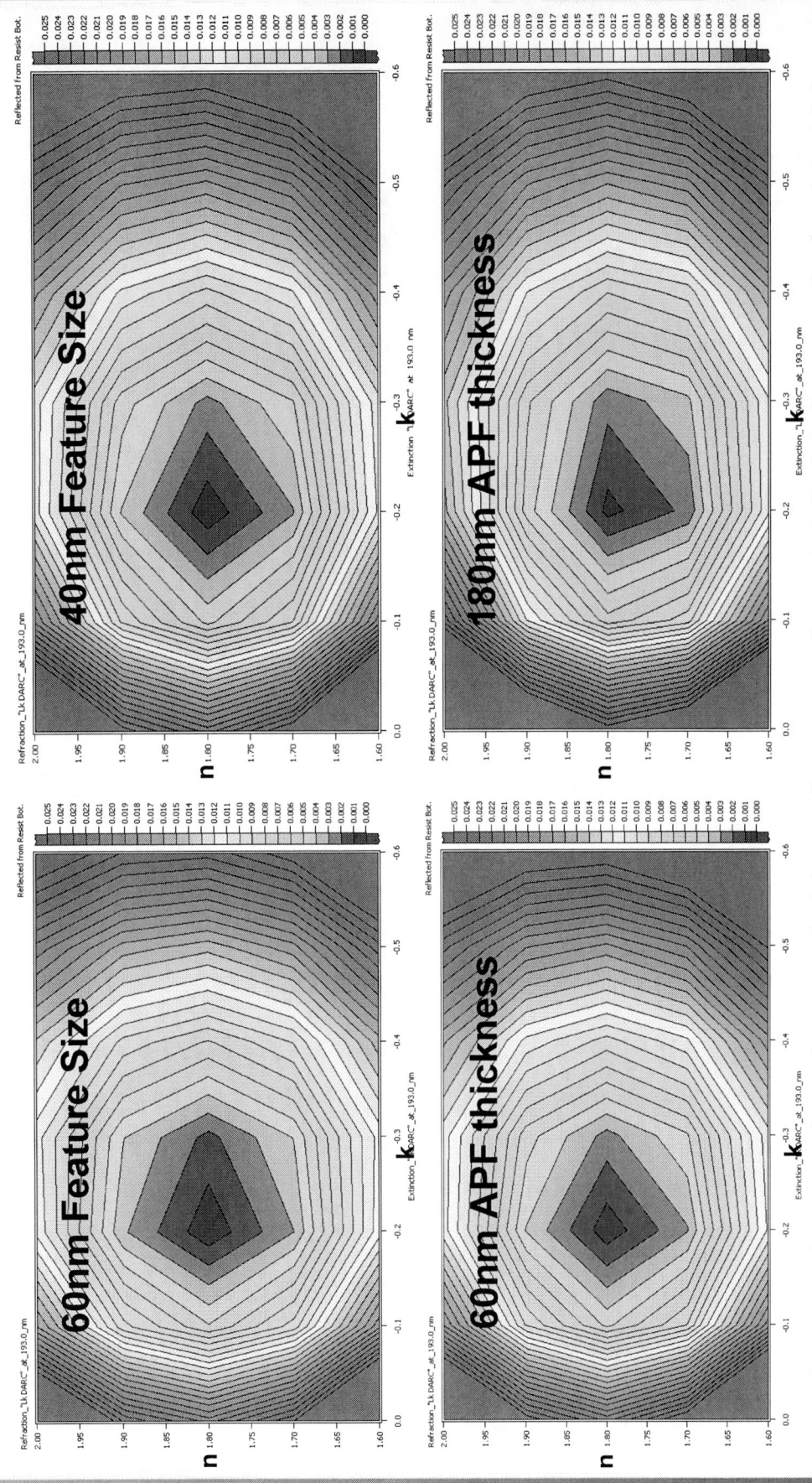

Reflectivity Control: Bi-Layer DARC vs. Single DARC+BARC

- Bi-Layer DARC can control reflectivity <0.3% with wide process window

- Single layer DARC+BARC scheme can achieve <0.5% reflectivity

- Single layer DARC can achieve <1.5% reflectivity in best case

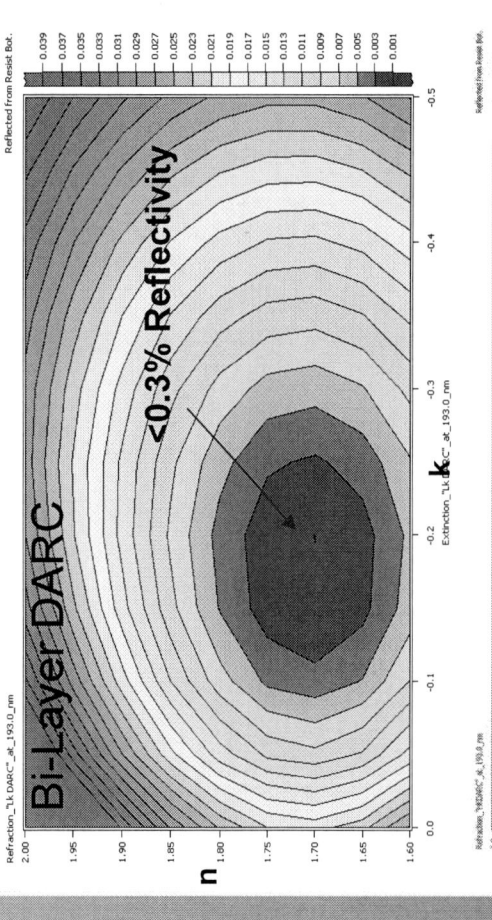

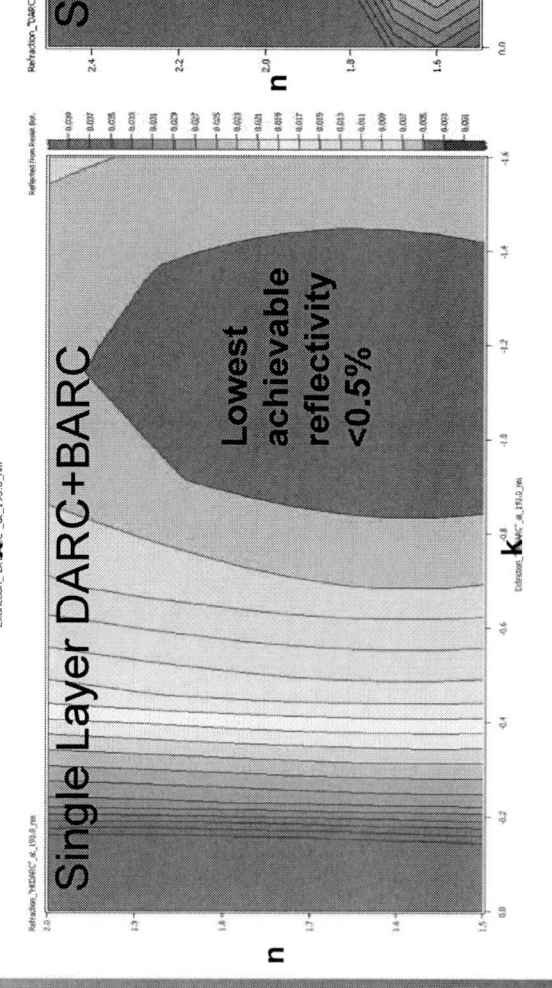

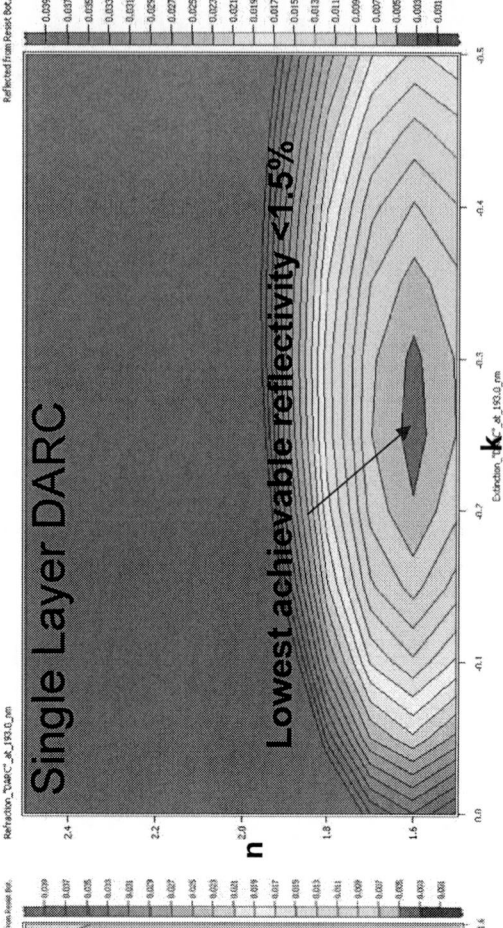

APPLIED MATERIALS.

Immersion Lithograph Process Window: NA=1.35 with 90 nm Pitch

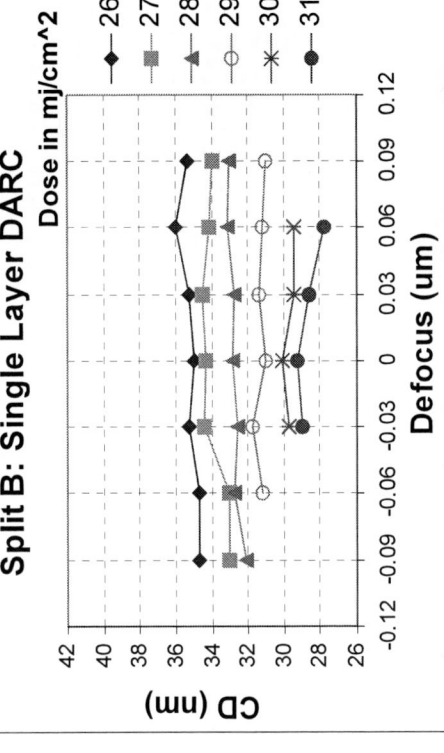

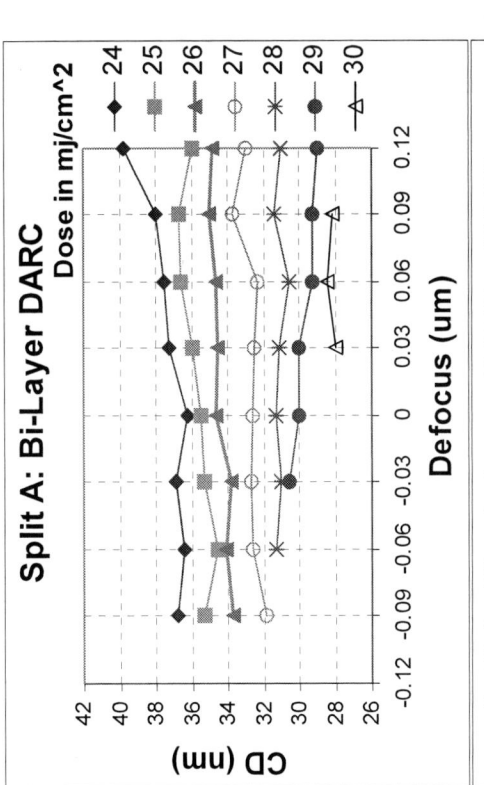

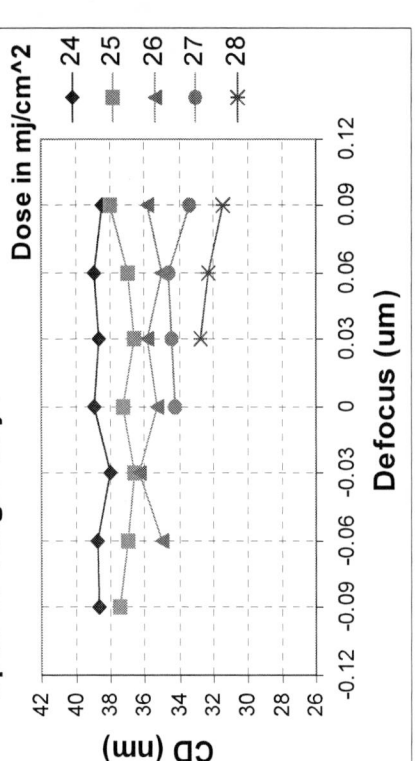

Split	Min CD (nm) With ≥150nm DOF
40 nm Bi-Layer DARC	30
20 nm Single Layer DARC	32
25 nm DARC+ 25 nm BARC	36

Bi-Layer DARC enables wider depth of focus window and smaller line width compared to single layer DARC or conventional DARC+BARC schemes

APPLIED MATERIALS.

Lithography/Etch Performance: Bi-Layer DARC

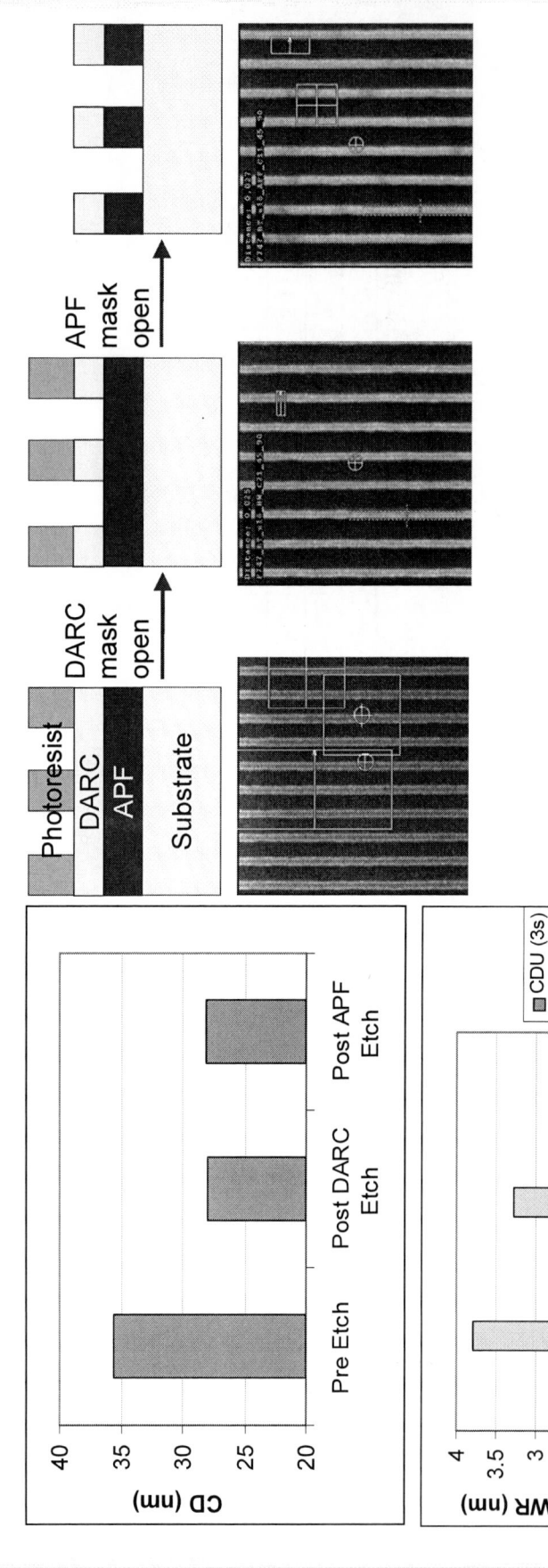

ITRS: Lithography Requirement for 2012

	CD: 1/2 Pitch (nm)	CD control (3 sigma) (nm)
DRAM	36	3.7
MPU	36	2.3
Bi-Layer DARC	36	**2.1**

Bi-layer DARC CD uniformity, LER, LWR are maintained or improved after DARC and APF etch. This performance is capable of achieving ITRS lithography requirement for CD control to 2012 and beyond

APPLIED MATERIALS

Rework Capability of Bi-Layer DARC

CD Performance after Multiple Rework Cycles

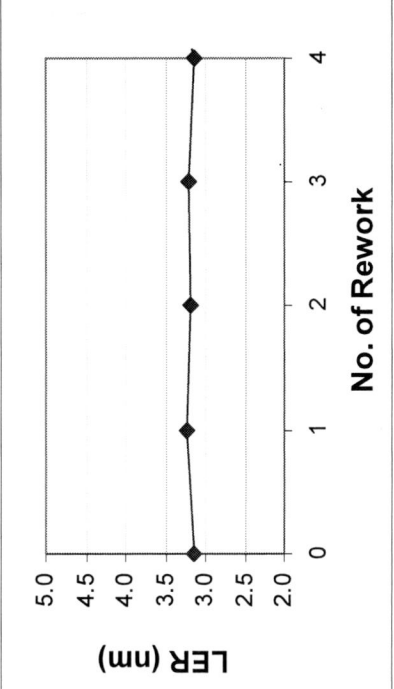

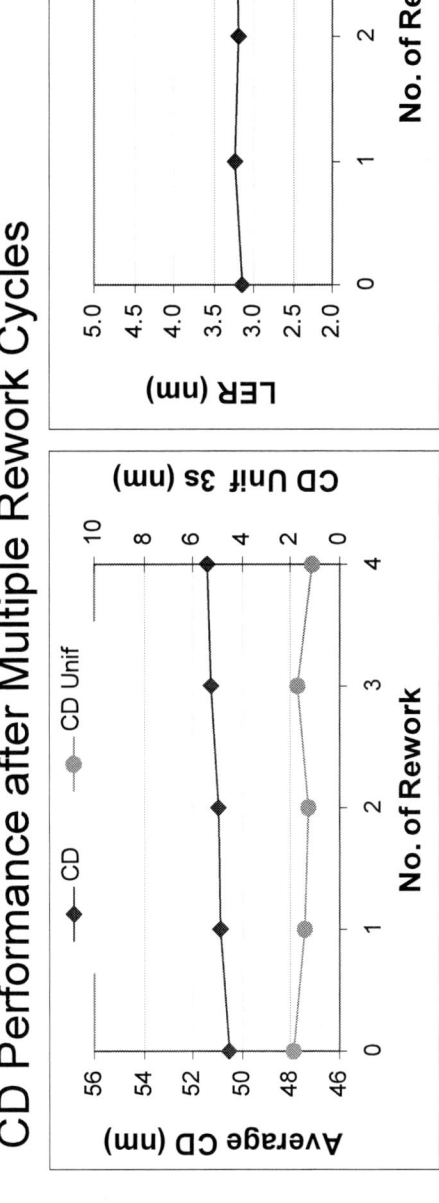

Optical Properties after Multiple Rework Cycles

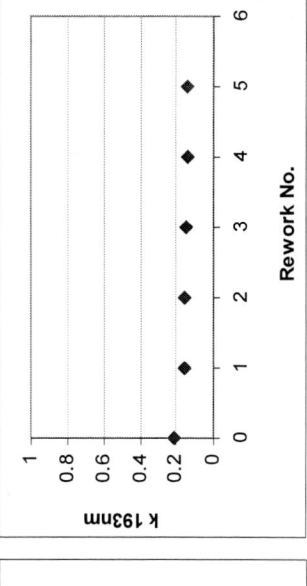

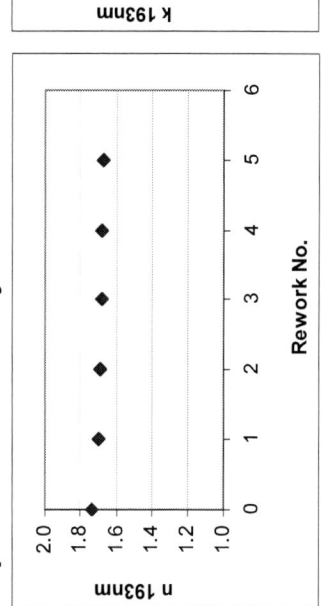

Multiple rework cycles have no impact on CD, uniformity and LER performance of Bi-Layer DARC

APPLIED MATERIALS.

Bi-Layer DARC Manufacturability

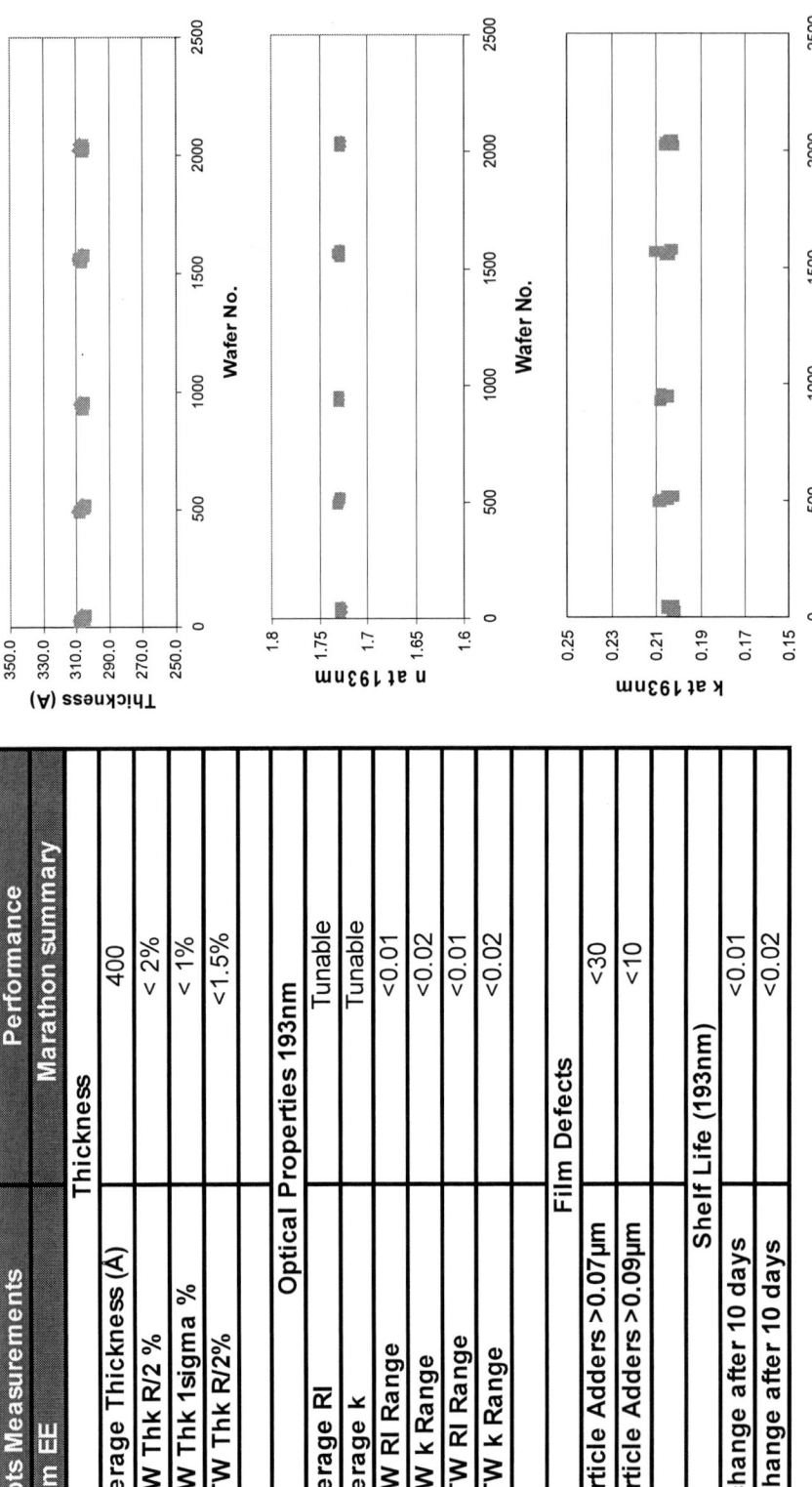

49pts Measurements	Performance
2mm EE	Marathon summary
Thickness	
Average Thickness (Å)	400
WIW Thk R/2 %	< 2%
WIW Thk 1sigma %	< 1%
WTW Thk R/2%	<1.5%
Optical Properties 193nm	
Average RI	Tunable
Average k	Tunable
WIW RI Range	<0.01
WIW k Range	<0.02
WTW RI Range	<0.01
WTW k Range	<0.02
Film Defects	
Particle Adders >0.07μm	<30
Particle Adders >0.09μm	<10
Shelf Life (193nm)	
n change after 10 days	<0.01
k change after 10 days	<0.02

Film properties, defectivity and shelf life of Bi-Layer DARC are stable.

APPLIED MATERIALS.

Summary

- Bi-layer DARC film has been developed for high NA immersion lithography, qualified at a logic manufacturer for 45nm device.

- Reflectivity <0.3% was achieved with wide process window.

- Depth of focus window is improved as a result of low reflectivity compared to single layer DARC, and DARC + BARC schemes.

- Bi-Layer DARC is capable of patterning 30 nm line width with good adhesion to photoresist. Its performance is capable of achieving ITRS 2012 lithography requirements for both memory and MPU.

- Bi-Layer DARC demonstrated stable shelf life and multiple rework capability.

- Manufacturability of bi-layer DARC process has been demonstrated. Film properties and defectivity are stable.

- Etch margin is improved by eliminating the need for a BARC layer and BARC etch steps.

APPLIED MATERIALS.

Acknowledgements

- Dena Blanco for Providing SEM Support
- Liyan Miao, Yongmei Chen, Jaklyn Jin for Etch Development

APPLIED MATERIALS.

Nissan Chemical Industries, LTD.

6th International Symposium on Immersion Lithography Extensions

Development of reverse materials and BARCs for Double patterning process

Yasushi Sakaida, Hiroaki Yaguchi, Rikimaru Sakamoto, Bang-Ching Ho

Nissan Chemical Industries, LTD.

Electronic Materials Research Laboratories.

Semiconductor Materials Research Department

 日産化学工業株式会社

Nissan Chemical Industries, LTD.

Introduction

Materials and processes for double patterning using 193nm immersion lithography has been developed for the 32/22 nm node device generations. LELE and LFLE process have been investigating but there are still many problems remaining on each process.

	LELE	LFLE
Issue	Throughput CoO	Film Loss, Top Rounding Profile (collapse, footing, scum)

Target

Proposal and development new process & materials for lithography on 32/22 nm node device ⇒ Reverse process

日産化学工業株式会社

Nissan Chemical Industries, LTD.

Application of Reverse materials

One of issue for next generation lithography : How to obtain the small hole pattern ?

Line mask   Line mask   Pattern image
Y-direction   X-direction   (C/H or pillar)

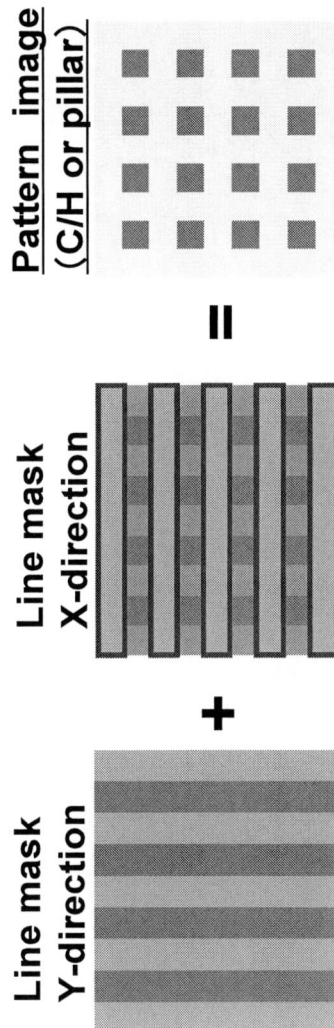

Reverse material can be used for the C/H pattern making.

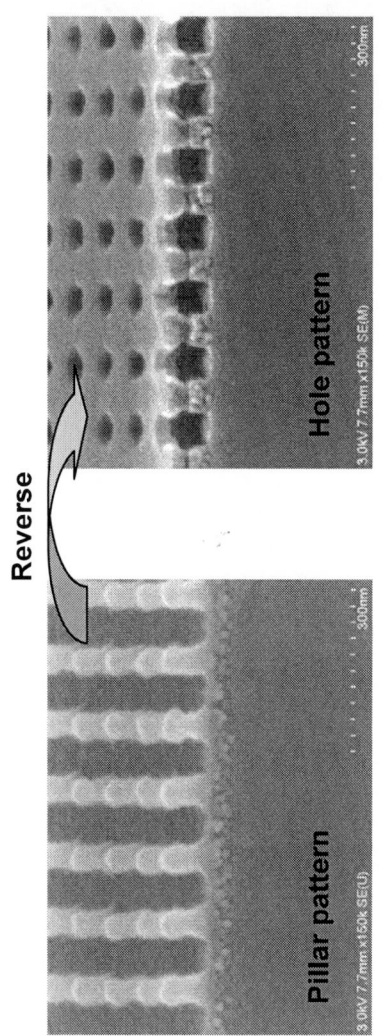

日産化学工業株式会社

Nissan Chemical Industries, LTD.

Process comparison

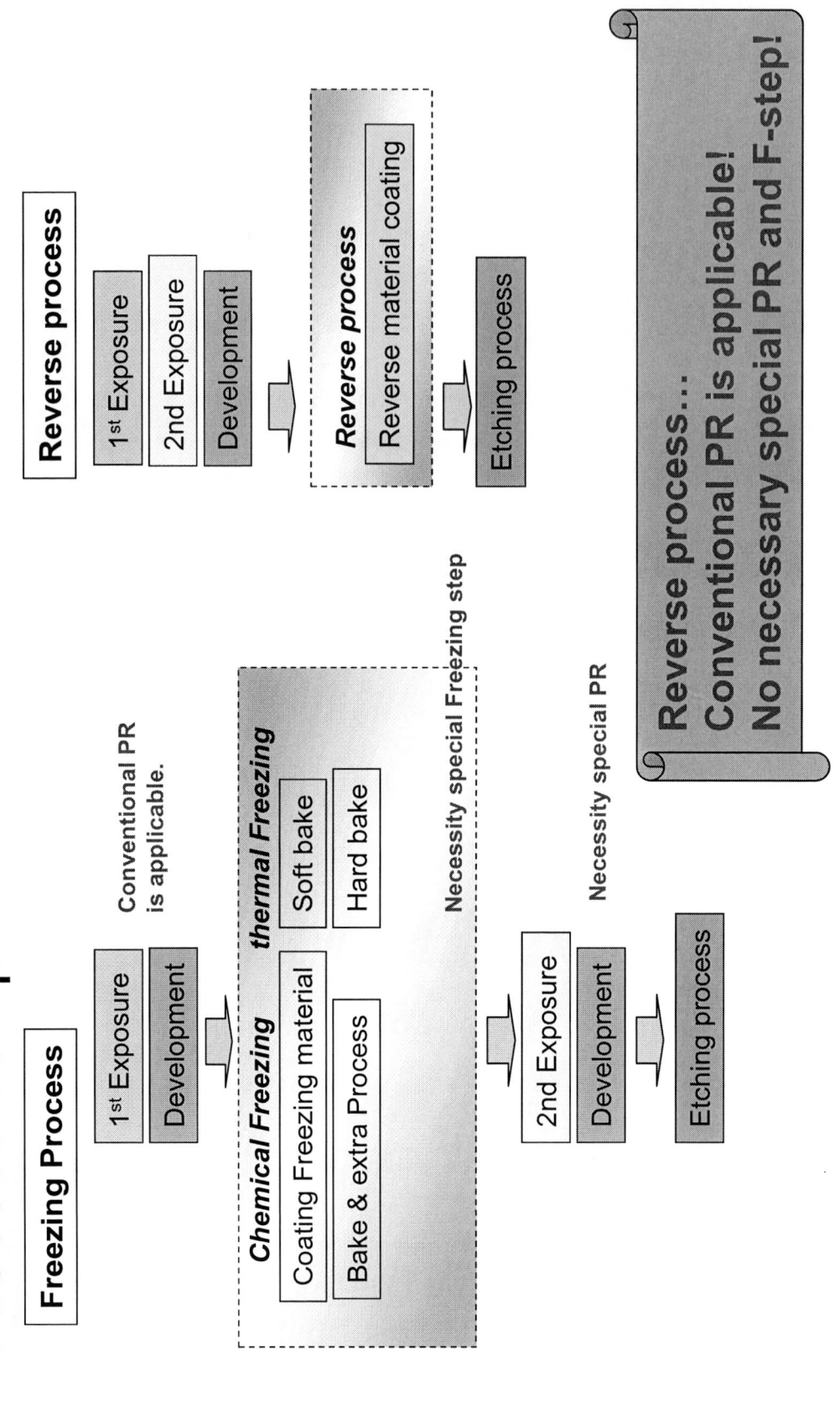

Nissan Chemical Industries, LTD.

Reverse Process

➤ Process scheme

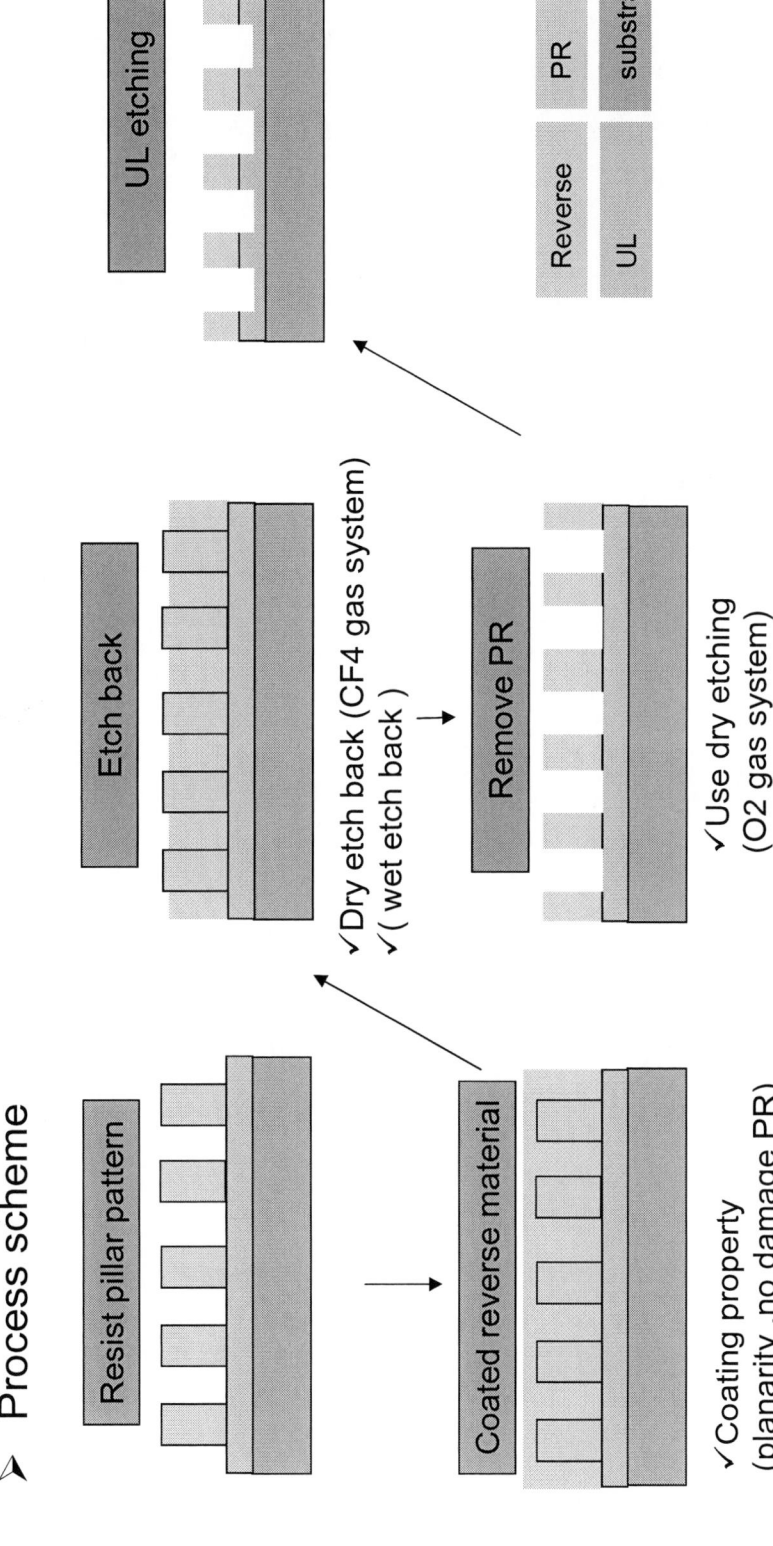

Resist pillar pattern	Etch back	UL etching

Coated reverse material

Remove PR

✓Dry etch back (CF4 gas system)
✓(wet etch back)

✓Use dry etching
(O2 gas system)

✓Coating property
(planarity ,no damage PR)

Reverse	PR
UL	substrate

➤ Reverse process is not so complicate compared with LELE and LFLE methods.

⭐ 日産化学工業株式会社

Nissan Chemical Industries, LTD.

Sample property 1

Sample	Si content	Solvent	n value @633nm	n/k value @193nm
NCR2	32%	MIBC	1.45	1.55/0.08

110degC bake condition

Polymer

Solvent : 4-methyl-2-pentanol (MIBC)

Compatibility test (Mixing Test)

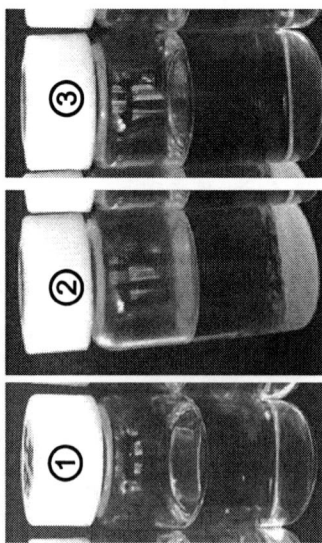

① NCR2 / PR thinner = 1/1 、→ clear
② NCR2 / PR= 1/1 、→ precipitation
③ NCR2 / TopCoat = 1/1 、→ clear

*PR thinner : PGME, PGMEA, EL, Cy, IPA

Reverse material
⇒ ○applicable TC coater cup

★ 日産化学工業株式会社

Nissan Chemical Industries, LTD.

Sample property 2

Wafer edge shape (EBR compatibility)

EBR thinner : PGME/PGMEA=70/30

8 inch wafer

Film THK of NCR2 : 100nm

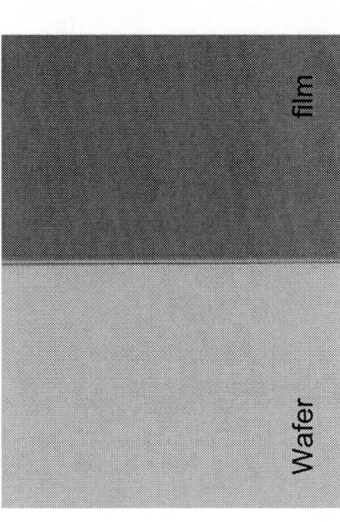

film

Wafer

Tool is difference.

Time delay exist between coat and etch back process

⇒Necessity for no changing of film property

Thickness Change

Coated reverse material

Etch back

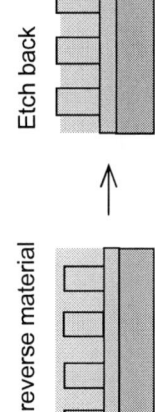

Coater & track

Dry etcher

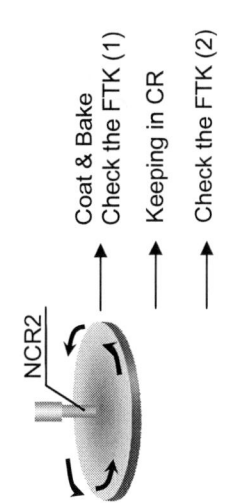

NCR2

Coat & Bake
Check the FTK (1)

Keeping in CR

Check the FTK (2)

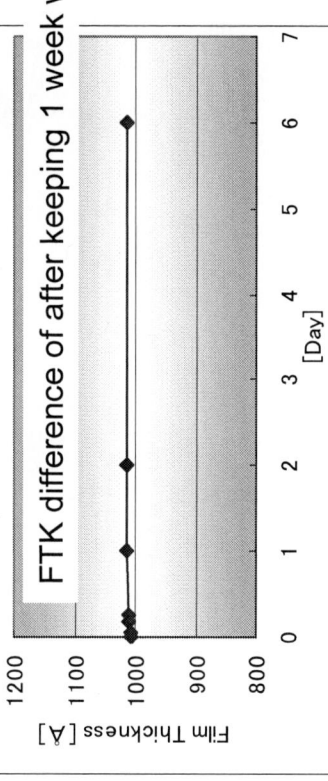

FTK difference of after keeping 1 week was 0.5nm.

Fig. Film thickness vs keeping day

⊙ 日産化学工業株式会社

Nissan Chemical Industries, LTD.

Sample property 3

Etching rate

Samco RIE-10NR
CF4/Ar : 50/200 sccm, 15Pa,200W O2/N2 : 10/20 sccm ,1Pa ,300W

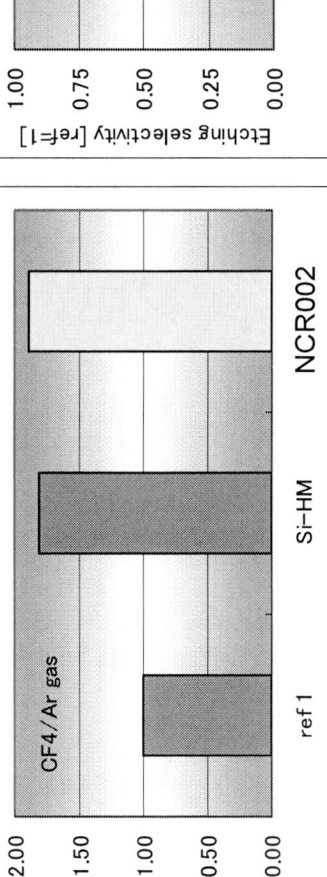

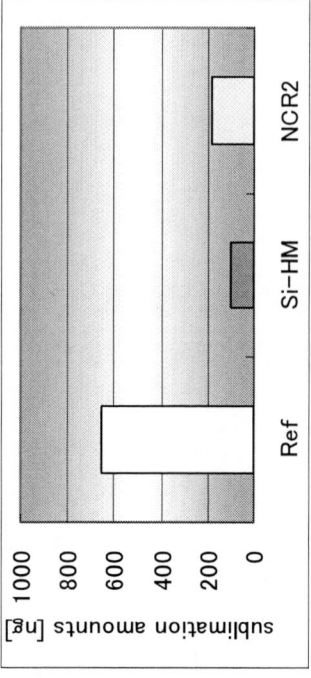

Have a sufficient etching selectivity

Sublimation

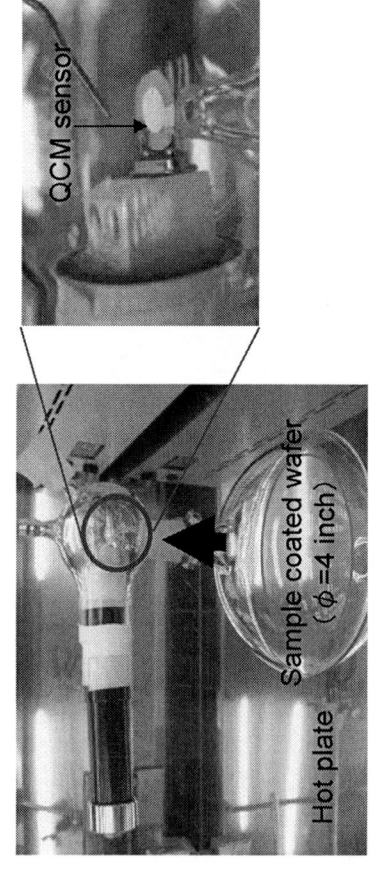

Sublimation amounts : low revel

⊛ 日産化学工業株式会社

Nissan Chemical Industries, LTD.

Patterning data 1

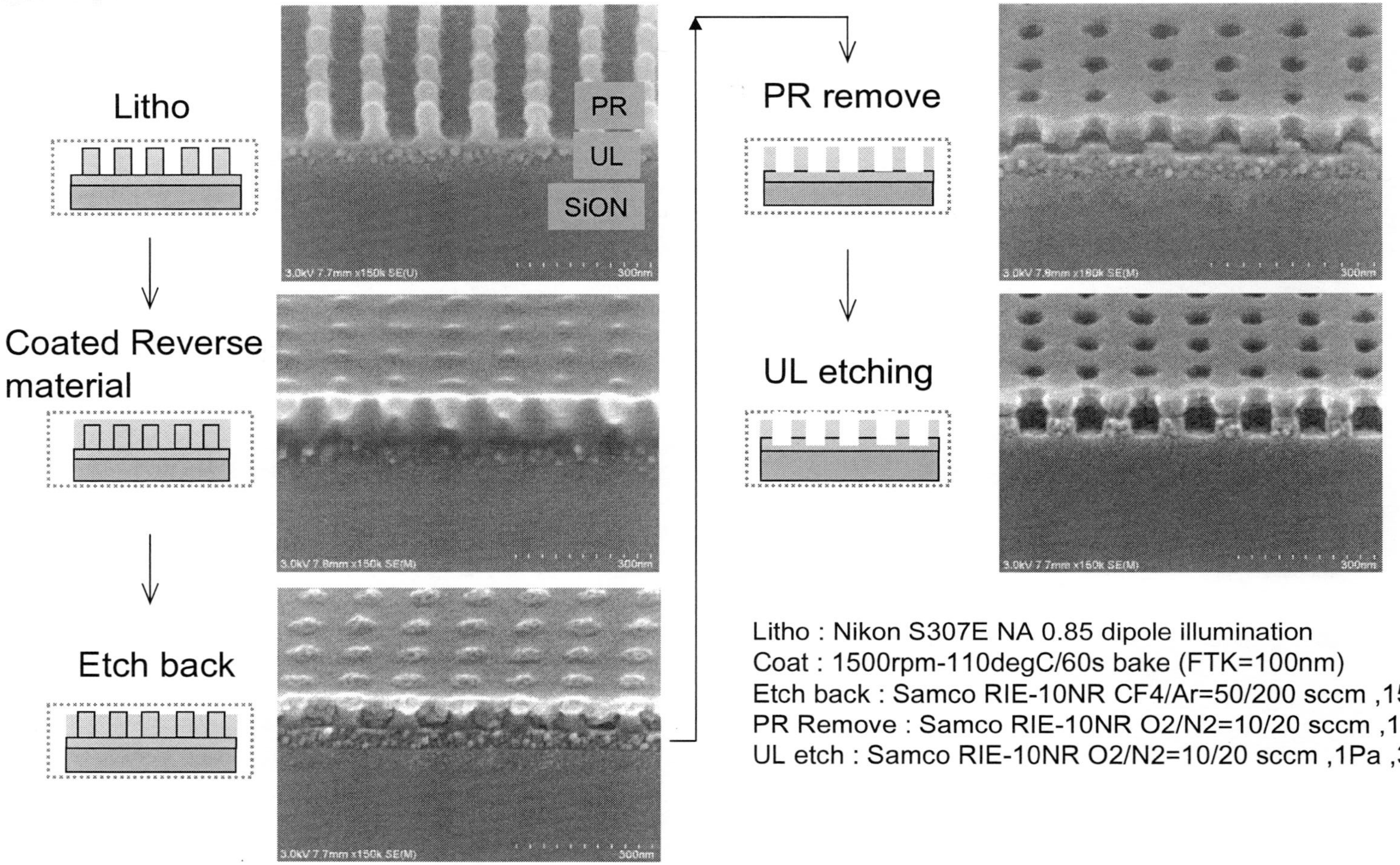

Litho

Coated Reverse material

Etch back

PR remove

UL etching

PR
UL
SiON

Litho : Nikon S307E NA 0.85 dipole illumination
Coat : 1500rpm-110degC/60s bake (FTK=100nm)
Etch back : Samco RIE-10NR CF4/Ar=50/200 sccm ,15Pa ,200W
PR Remove : Samco RIE-10NR O2/N2=10/20 sccm ,1Pa ,300W
UL etch : Samco RIE-10NR O2/N2=10/20 sccm ,1Pa ,300W

日産化学工業株式会社

Nissan Chemical Industries, LTD.

Patterning data 2

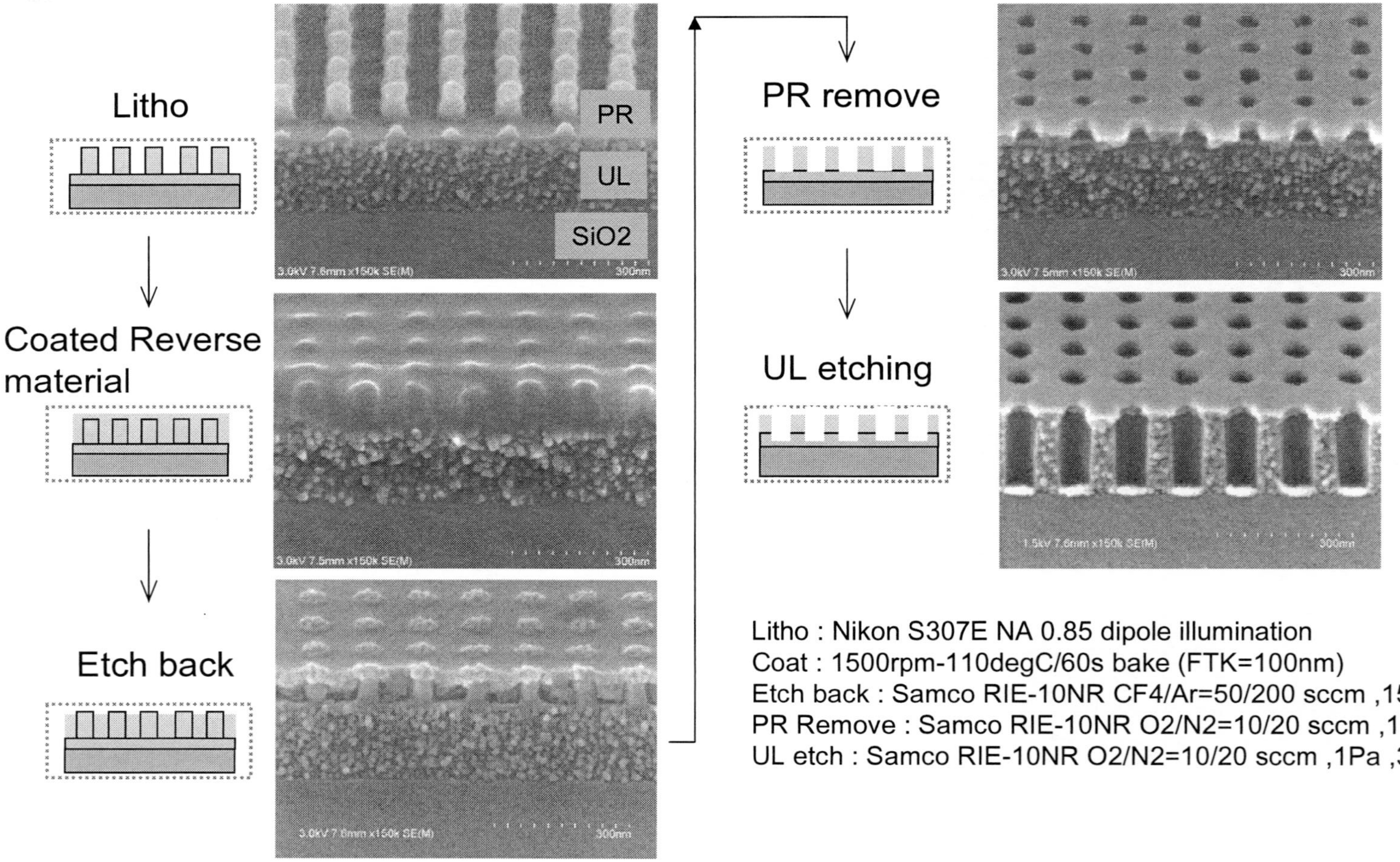

Litho

Coated Reverse material

Etch back

PR remove

UL etching

PR
UL
SiO2

Litho : Nikon S307E NA 0.85 dipole illumination
Coat : 1500rpm-110degC/60s bake (FTK=100nm)
Etch back : Samco RIE-10NR CF4/Ar=50/200 sccm ,15Pa ,200W
PR Remove : Samco RIE-10NR O2/N2=10/20 sccm ,1Pa ,300W
UL etch : Samco RIE-10NR O2/N2=10/20 sccm ,1Pa ,300W

➤ Hole pattern was obtained by using NCR2

日産化学工業株式会社

Nissan Chemical Industries, LTD.

Patterning data 3

Stack : bareSi / SOC-BARC / PR
Resist : FTK 105nm
BARC : NCA9189 FTK71nm (205degC/60s)
Exposure : Tool:ASML XT:1900Gi
1st exposre & 2nd exposre
NA:1.35 Illumination:Dipole σ:outer/inner : 0.98 / 0.81

Measurement Point	209	
Maximum	47.12	nm
Minimum	44.27	nm
Mean	45.57	nm
Range	2.85	nm
3 Sigma	1.63	

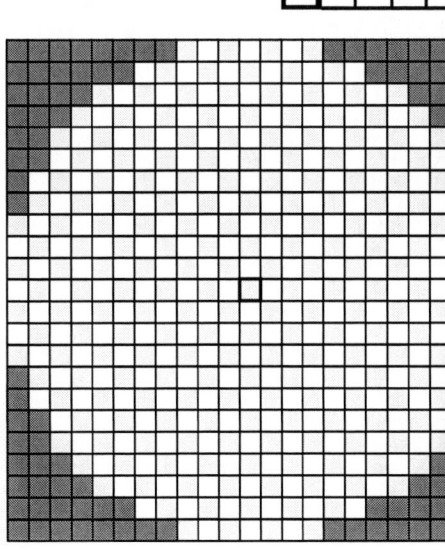

☐ total : 418 shot
☐ CD measurement : 209 shot

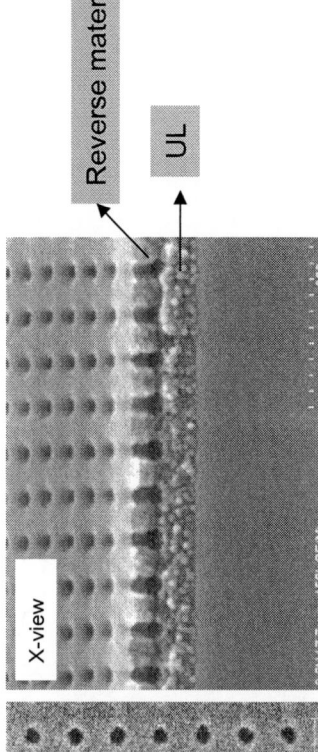

Reverse material

UL

X-view

Top-view

➢ CDU (after litho) was no problem, and obtained hole pattern

 日産化学工業株式会社

Nissan Chemical Industries, LTD.

Summary

⋀ Reverse process is one candidate for making small hole pattern.
⋀ NCI developed NCR2 of reverse material as Si contained type.
⋀ EBR ,Sublimation ,thickness change of NCR2 have been optimized.
⋀ Pillar pattern were obtained by dry and immersion scanner.
⋀ Small hole pattern was obtained by using reverse material of NCR2

日産化学工業株式会社

6th International Symposium on Immersion Lithography Extensions

Bottom-Anti-Reflective coatings (BARC) for LFLE Double patterning process

Rikimaru Sakamoto, Takafumi Endo, Bang-Ching Ho
Shigeo Kimura, Tomohisa Ishida, Masakazu Kato, Noriaki Fujitani,
Ryuji Onishi, Yoshiomi Hiroi, Daisuke Maruyama

Nissan Chemical Industries, LTD.
Electronic Materials Research Laboratory
Semiconductor Materials Research Department

Anti-Reflective Coatings for DPT. Nissan Chemical Industries, LTD

1. Introduction

Several kind of Freezing process were proposed and investigated.
Same ARC is used at 1st and 2nd Litho
At each LFLE process.

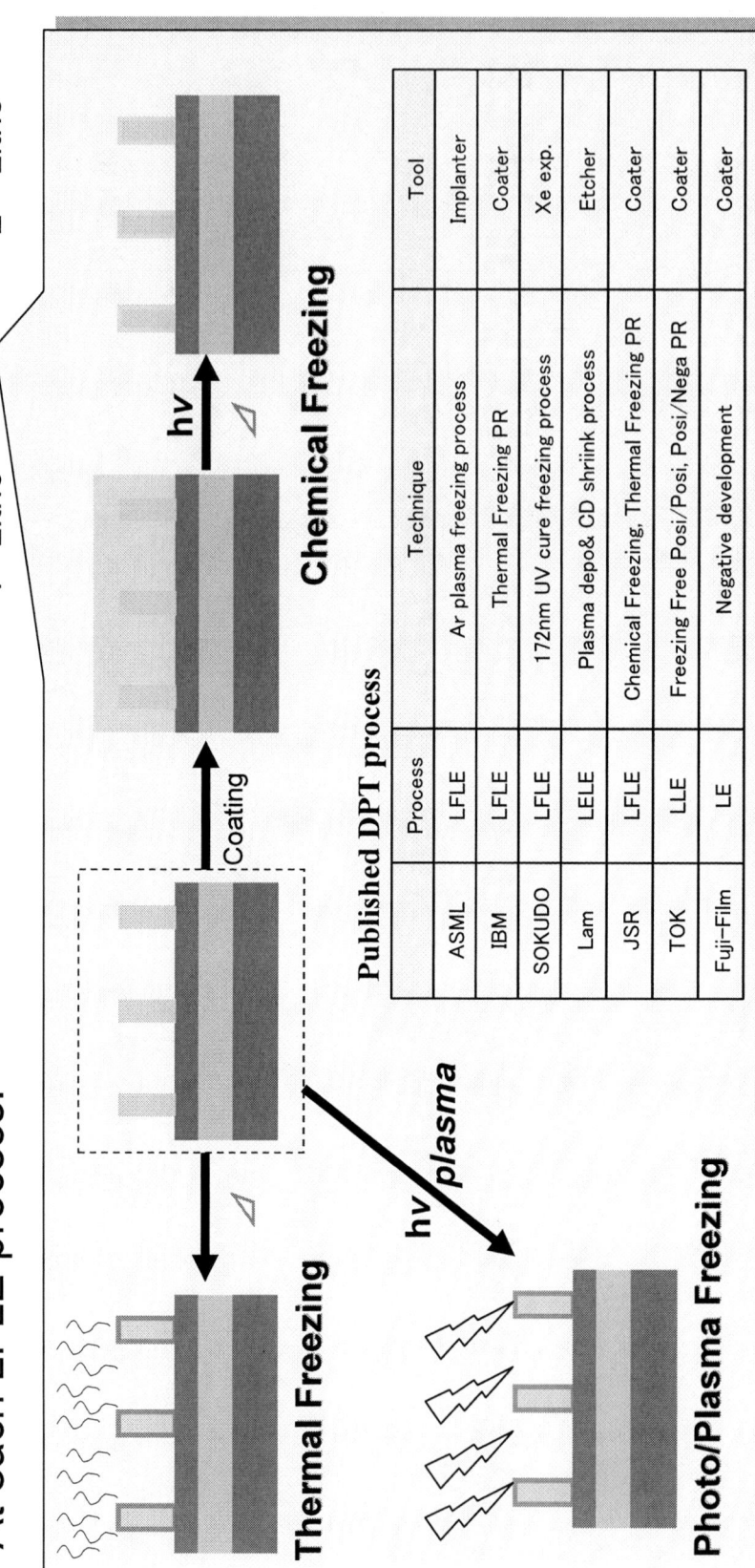

	Process	Technique	Tool
ASML	LFLE	Ar plasma freezing process	Implanter
IBM	LFLE	Thermal Freezing PR	Coater
SOKUDO	LFLE	172nm UV cure freezing process	Xe exp.
Lam	LELE	Plasma depo& CD shrink process	Etcher
JSR	LFLE	Chemical Freezing, Thermal Freezing PR	Coater
TOK	LLE	Freezing Free Posi/Posi, Posi/Nega PR	Coater
Fuji-Film	LE	Negative development	Coater

Anti-Reflective Coatings for DPT, Nissan Chemical Industries, LTD

2. Possible issue @ LFLE process

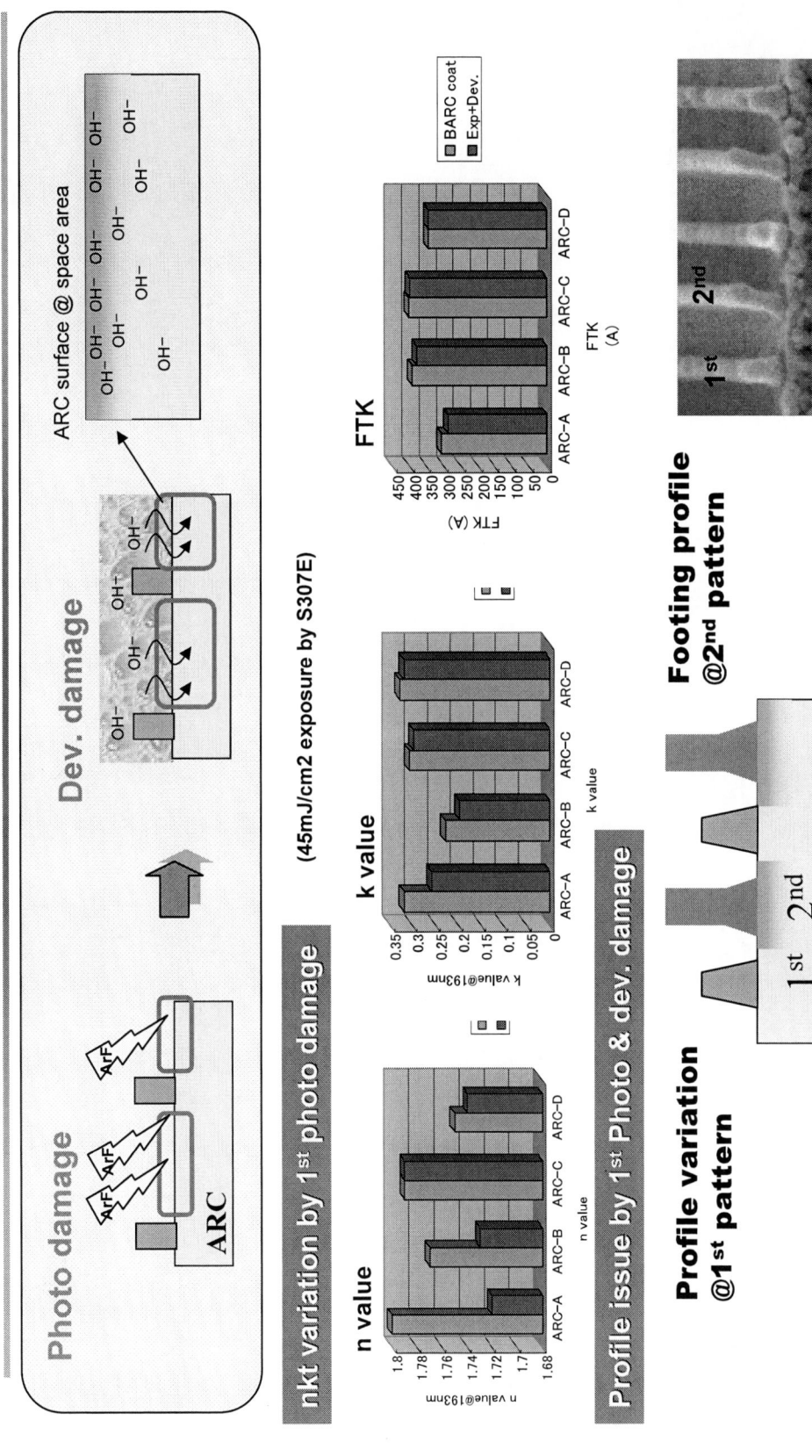

3. ARC base polymer effect for LFLE performance

Process condition

Exp.
S307E (Nikon)
NA=0.85
Dipole (0.92/0.85)
L/S=1/3 (65nm)
Sub. Bare-Si

Dev.
ACT-8 (TEL)
LD nozzle 60s

PR & Freezing chemical (JSR)
1st Litho: ArF-Imm. PR-A
2nd Litho: ArF-Imm. PR-A
Freezing: Freezing Chemical-B

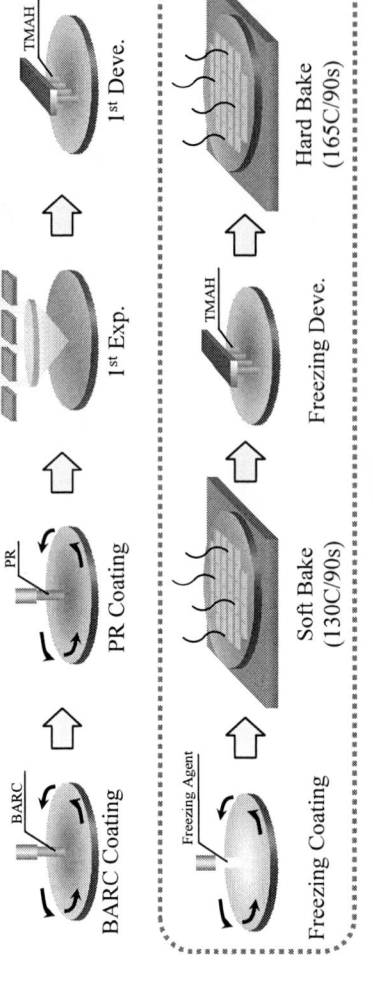

BARC Coating

PR Coating

1st Exp.

1st Deve.

Freezing Coating

Soft Bake (130C/90s)

Freezing Deve.

Hard Bake (165C/90s)

PR Coating

2nd Exp.

2nd Deve.

Polyester platform

Hetero cyclic structure

Varied structure to change polymer property

Material (ARC)

Sample	Structure	n/k @ 193nm	FTK(nm)
ARC-A	Hetero cyclic	1.82/0.32	85
ARC-B	Flexible Chain	1.87/0.24	80
ARC-C	Terminal acid-A	1.86/0.25	80
ARC-D	Terminal acid-B	1.80/0.32	85
ARC-E	Aliphatic OH	1.85/0.25	80
ARC-F	Rigid structure	1.79/0.32	85

Anti-Reflective Coatings for DPT. Nissan Chemical Industries. LTD

3. ARC base polymer effect for LFLE performance

Anti-Reflective Coatings for DPT, Nissan Chemical Industries, LTD

3. ARC base polymer effect for LFLE performance

Anti-Reflective Coatings for DPT. Nissan Chemical Industries. LTD

3. ARC base polymer effect for LFLE performance

Result

ARC-B

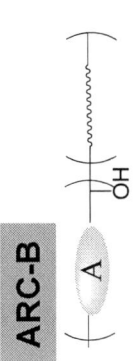

Flexible chain

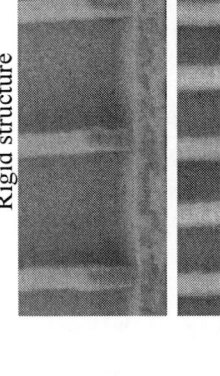

ARC-F

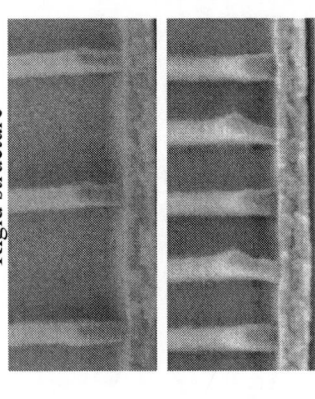

Rigid structure

Issue

1st pattern

1. PR thickness loss
2. CD variation (CD increasing~10nm)
3. Pattern profile variation (Tapering).
4. CD variation and pattern profile were varied with different ARC.

2nd pattern

1. Easy to be footing or scumming.
2. 2nd Litho shape was varied with different ARC.

Anti-Reflective Coatings for DPT, Nissan Chemical Industries, LTD

4. Investigation of profile issue

1st Pattern Variation

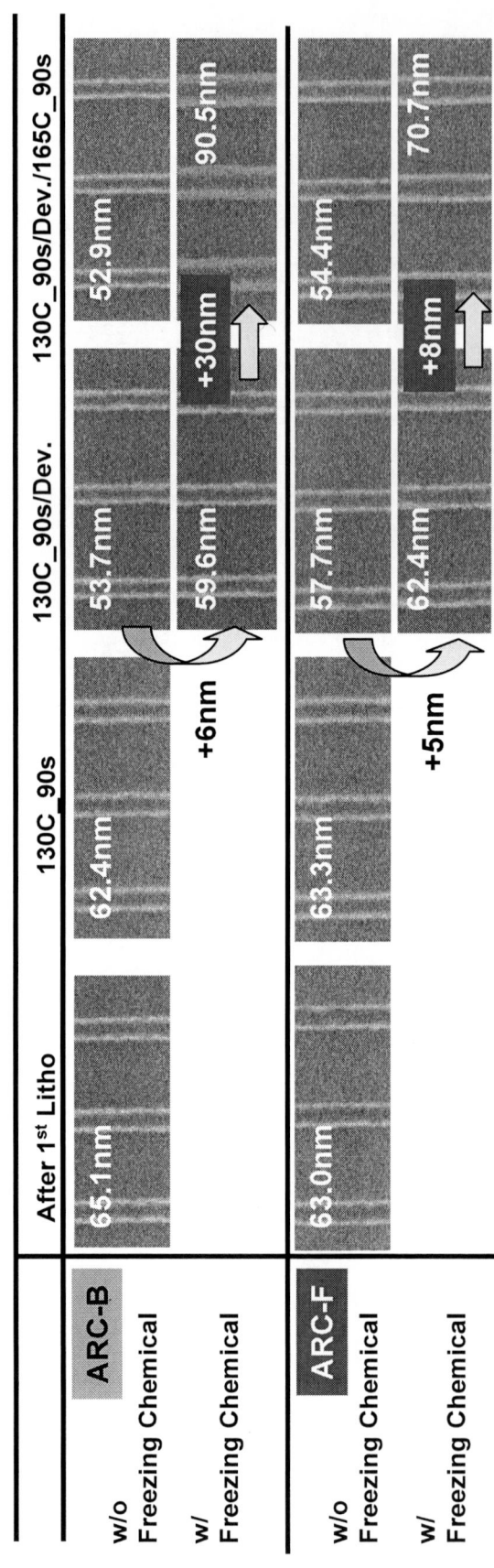

PR pattern wasn't reflowed by only baking up to 165C.

Pattern variation was related to Freezing Chemical Coating step.

What is happening during Freezing Step?
Why BARC kinds can impact to Freezing property?

Anti-Reflective Coatings for DPT. Nissan Chemical Industries, LTD

4. Investigation of profile issue

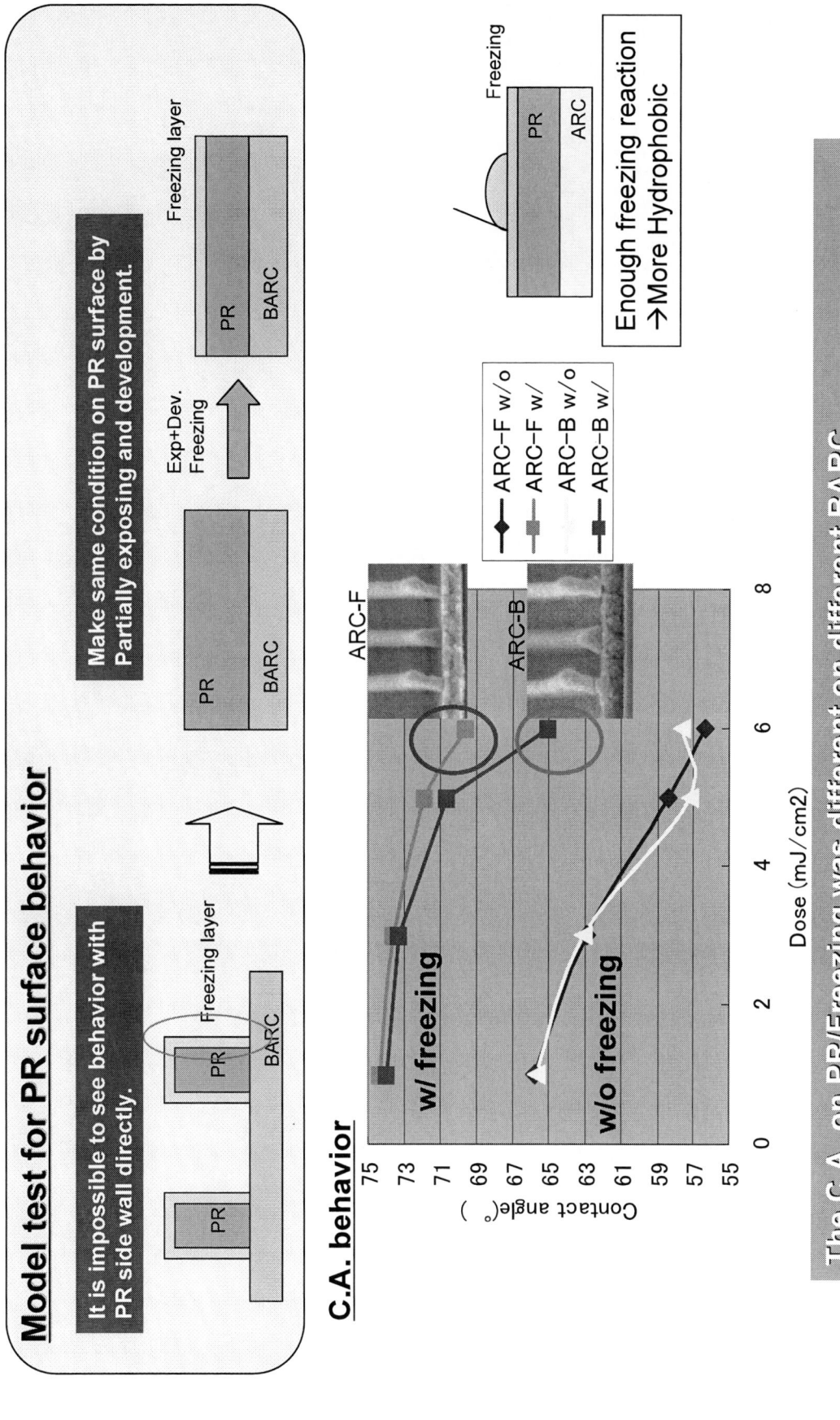

4. Investigation of profile issue

1st Pattern Variation

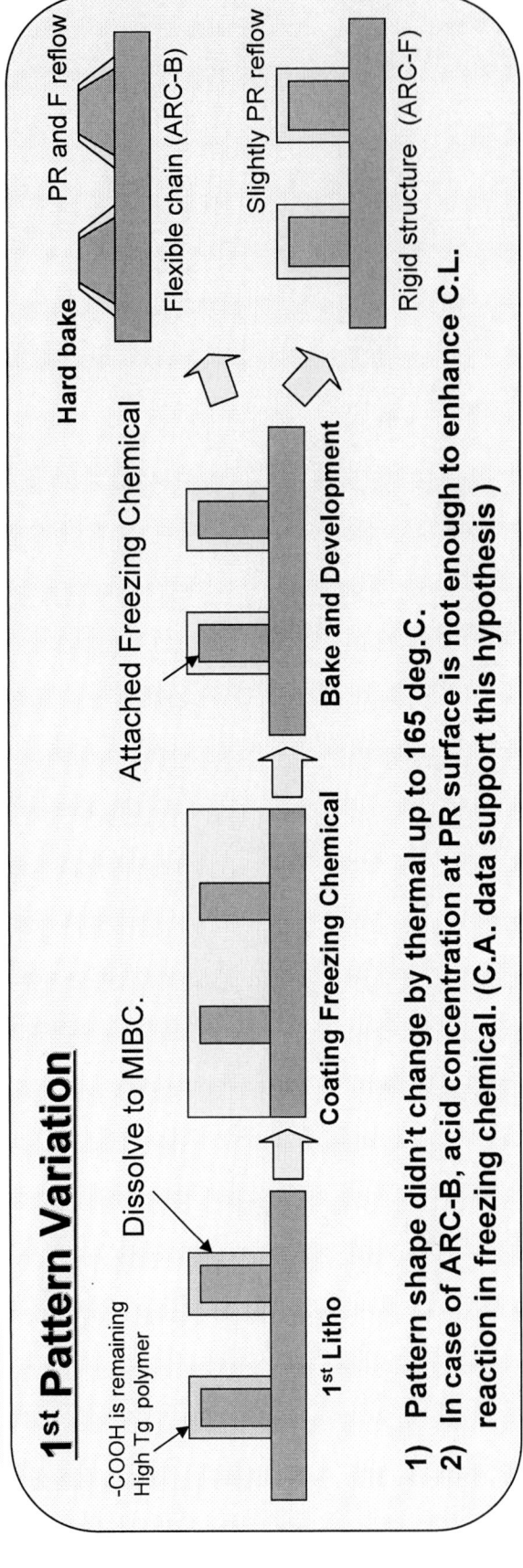

1) Pattern shape didn't change by thermal up to 165 deg.C.
2) In case of ARC-B, acid concentration at PR surface is not enough to enhance C.L. reaction in freezing chemical. (C.A. data support this hypothesis)

2nd Pattern Footing

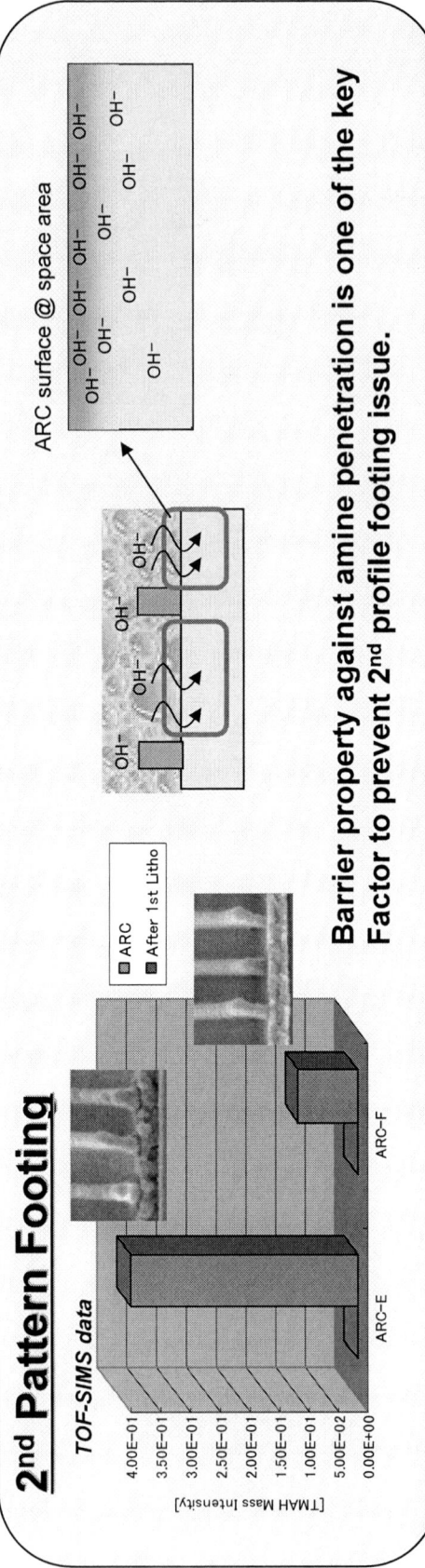

Barrier property against amine penetration is one of the key
Factor to prevent 2nd profile footing issue.

5. LFLE Litho performance on HM stack

Exposure condition

Tool: ASML XT:1900Gi

NA: 1.00 (water immersion), Sigma: outer/inner: 0.85/0.65

Illumination: Dipole40X(Blade angle: 40deg.), Y-polarized

PR&F-chemical: ArF-Imm.PR-C(1st) / Freezing Chem.-B / ArF-Imm.PR-D(2nd) (JSR)

imec

	44nm DHP	32nm DHP
Ref. (29nm)	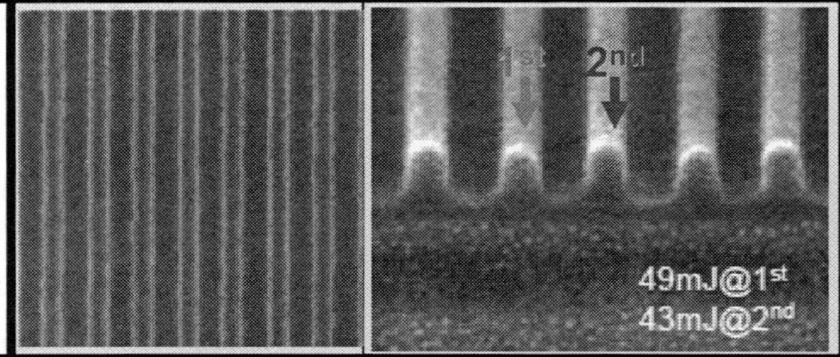 49mJ@1st 43mJ@2nd	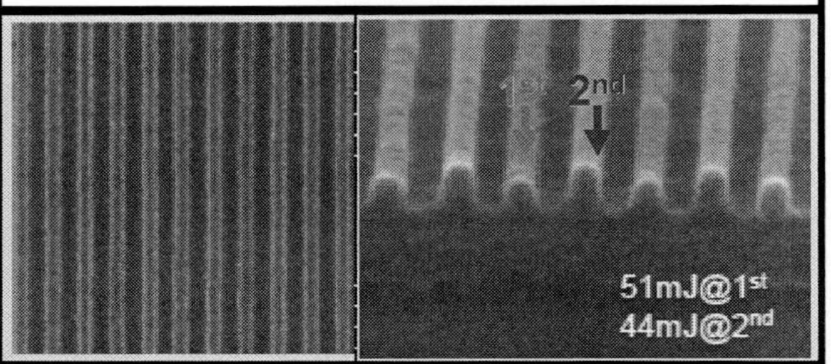 51mJ@1st 44mJ@2nd
LFLE-ARC A (n/k=1.79/0.32) (32nm)	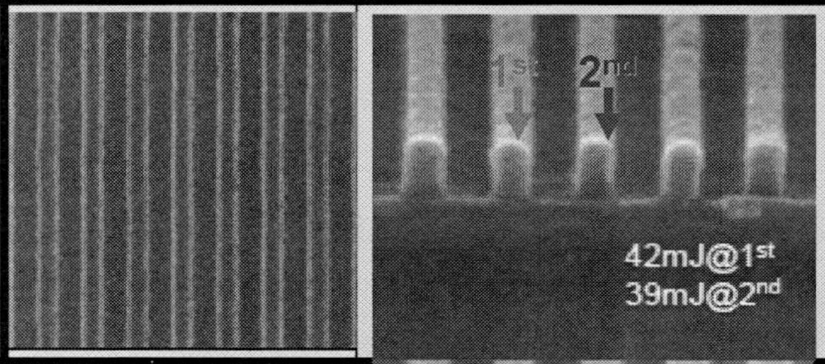 42mJ@1st 39mJ@2nd	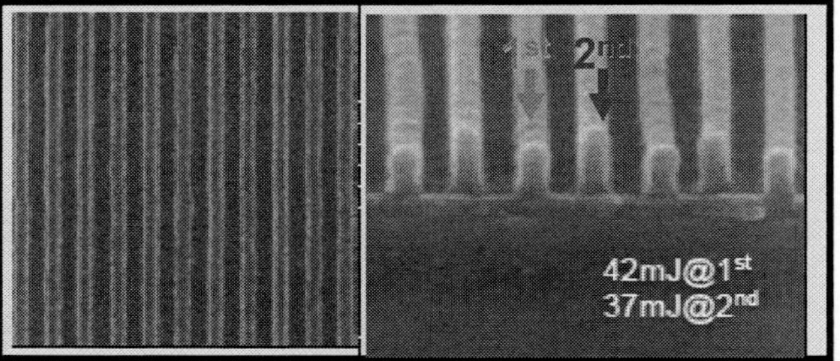 42mJ@1st 37mJ@2nd
LFLE-ARC B (n/k=1.81/0.25) (32nm)	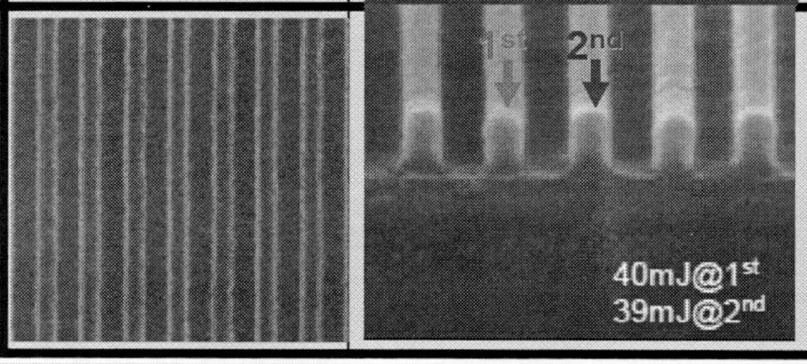 40mJ@1st 39mJ@2nd	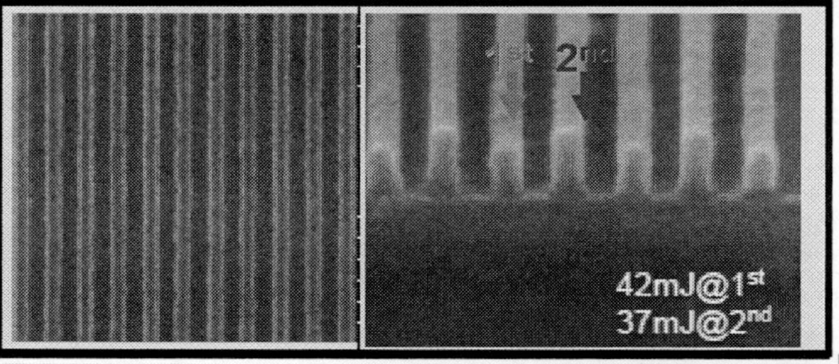 42mJ@1st 37mJ@2nd

Anti-Reflective Coatings for DPT, Nissan Chemical Industries, LTD

6. Conclusion

1. In LFLE process, ARC is used 2 times with same coated film.

2. BARC surface can be basic condition by 1st exposure and Develoment.

3. Several kind of ARC were tested with chemical Freezing LFLE process, and showed different behavior with different polymer backbone.

4. 1st pattern profile variation and 2nd pattern footing issue are identified as the key issue.

5. BARC design impact freezing reaction efficiency.
   Rigid base polymer structure showed better performance for 1st pattern shape compare to soft structure.

6. TMAH penetration seems to be the factor for 2nd pattern footing issue.

7. Optimized ARC (LFLE-ARC A, B) showed good profile at DHP44 and 32nm compare to conventional ARC system.

Acknowledgement

JSR Corporation: Providing resist and Freezing chemical.

IMEC: Providing HM stack substrate.

Anti-Reflective Coatings for DPT. Nissan Chemical Industries, LTD

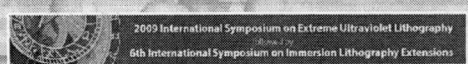

Understanding the Relationship between the Evaporation Behavior of Water Droplets and the Formation of Stains on Polymer Surfaces

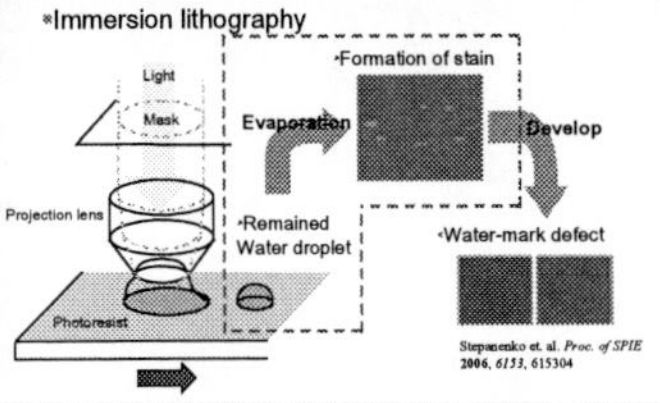

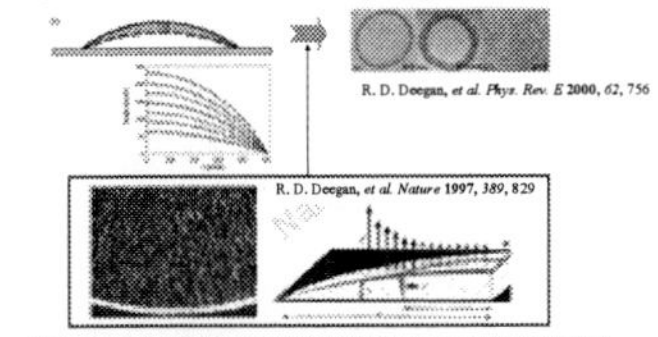

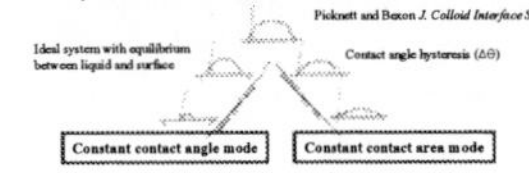

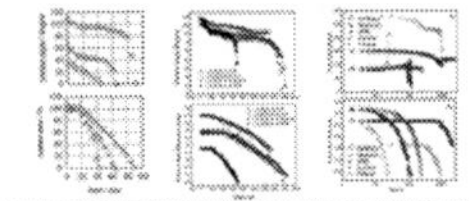

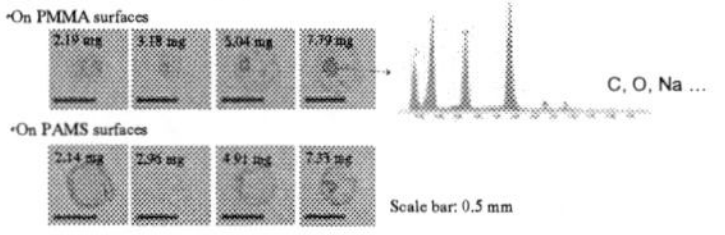

Jung-Hoon Kim[1], Jae Hyun Kim[2], Wang-Cheol Zin[1]*

[1]Department of Materials Science and Engineering, Pohang University of Science and Technology, Pohang, Korea
[2]Material Engineering Group, Device Solution Network Division, Samsung Electronics Co. Ltd., Korea
*Corresponding Author: wczin@postech.ac.kr

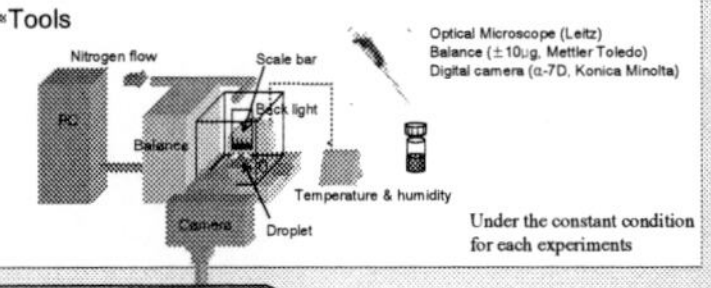

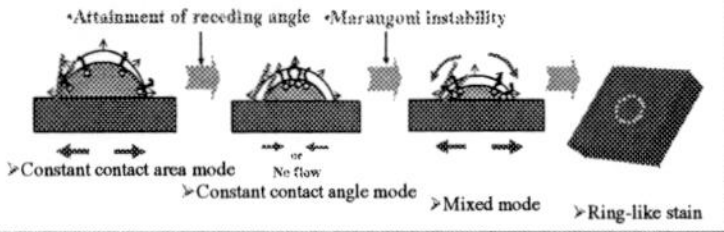

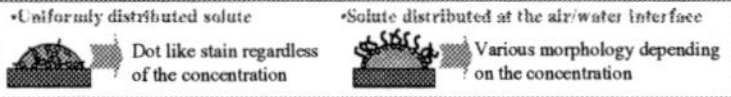

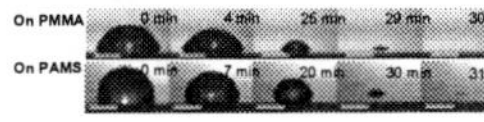

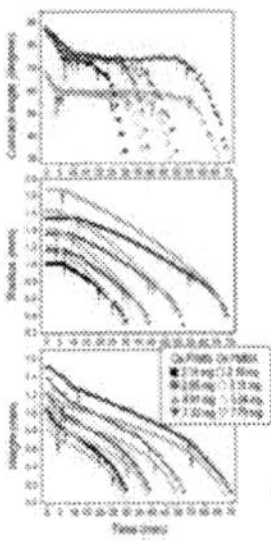

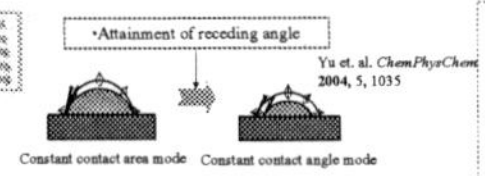

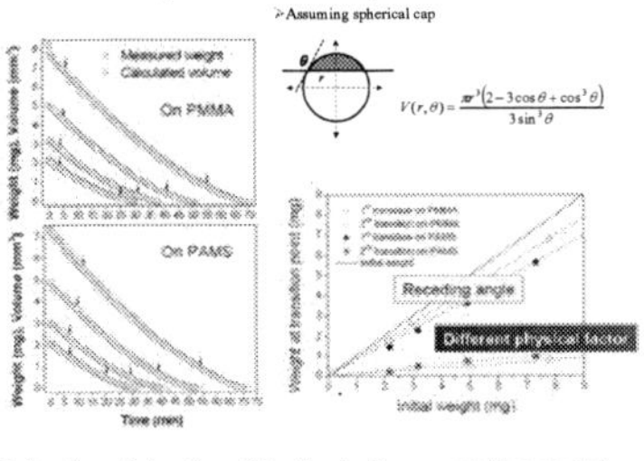

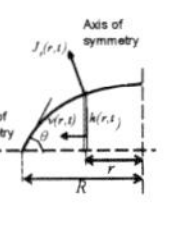

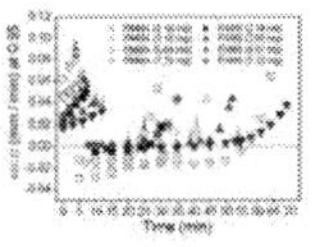

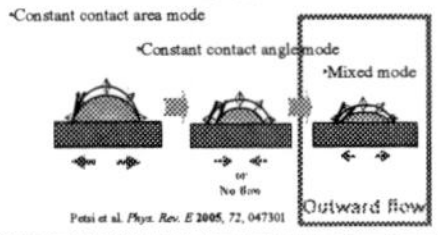

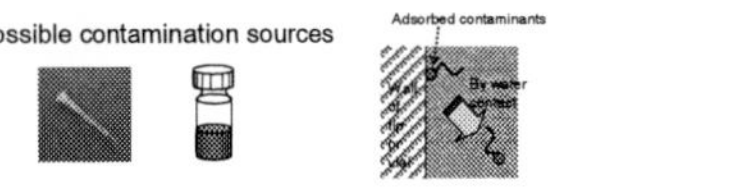

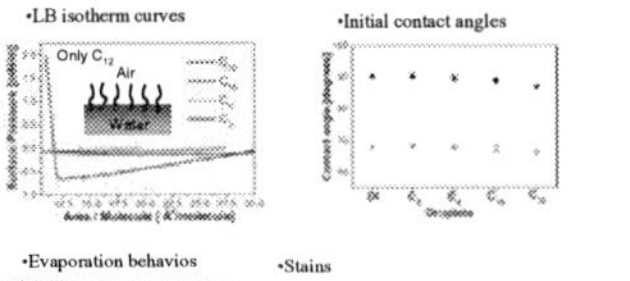

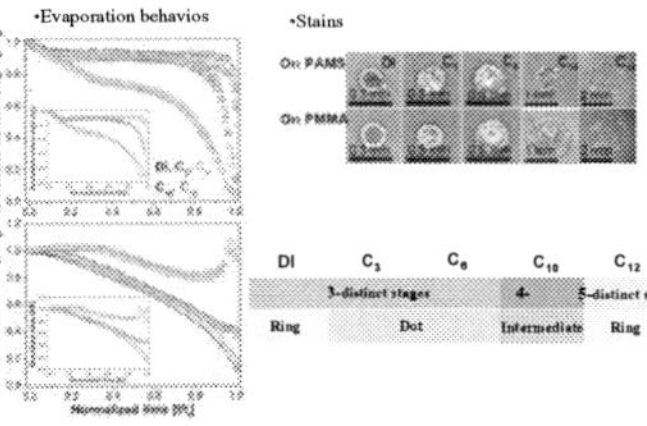

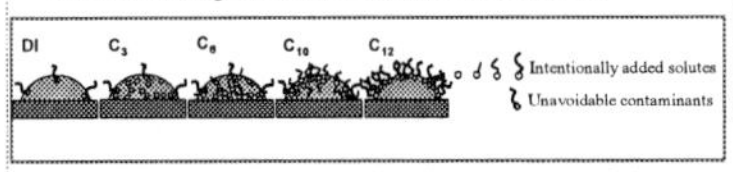

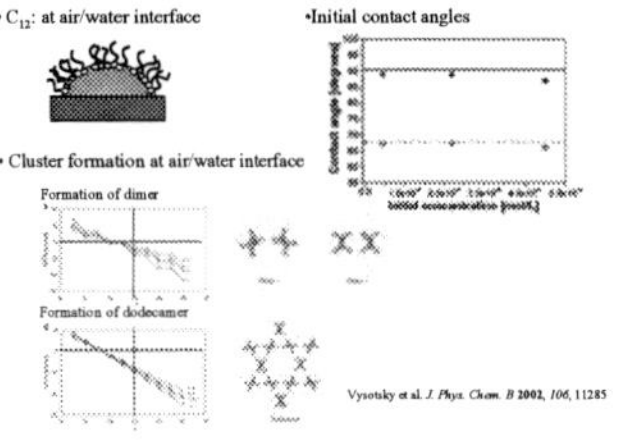

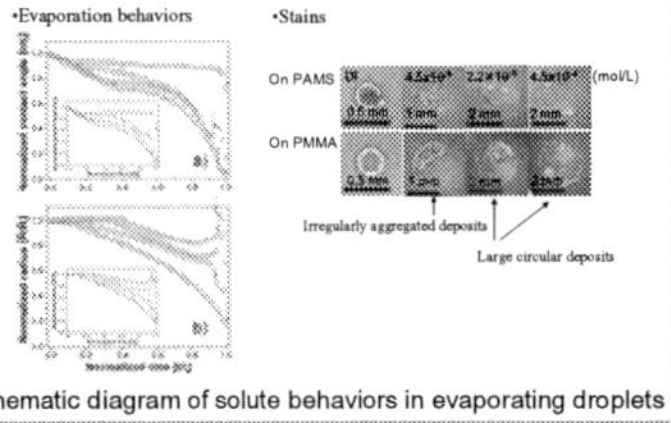

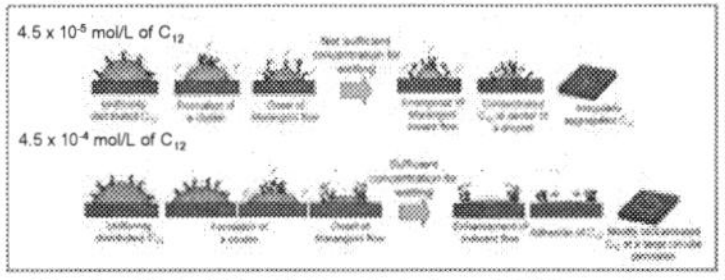

FUJIFILM

6th International Symposium on Immersion Lithography

October, 2009

Process parameter influence to negative tone development process for double patterning

Shinji Tarutani, Sou Kamimura, Jiro Yokoyama
FUJIFILM Corporation
Electronic Materials Research Laboratories

FUJIFILM

6th International Symposium on Immersion Lithography

Introduction

Negative tone development process is a good candidate for C/H and trench pattern lithography, since good enough optical image can be achieved with bright mask application.

Process maturity was demonstrated in the viewpoint of etch resistance, defectivity, and CD uniformity (see the presentation on Friday afternoon).

However, no one demonstrated the process parameter impacts on CD and defectivity. These are the important points for designing total process.

1. Small trenches

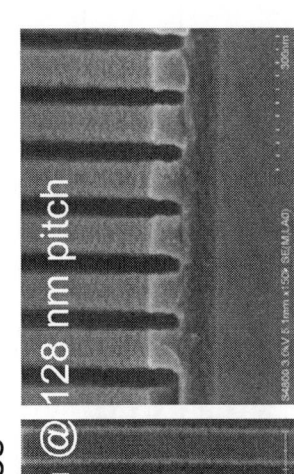

32 nm @ 128 nm pitch

2. Variety of C/H with Single exposure

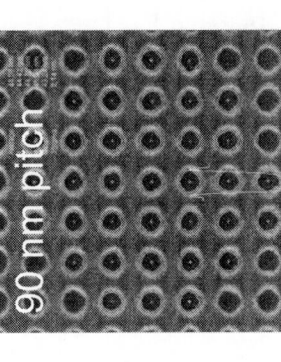

90 nm pitch

90 nm pitch

3. Variety of C/H with V/H line double exposure

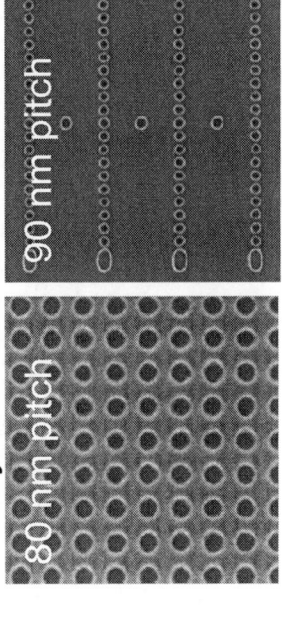

80 nm pitch

90 nm pitch

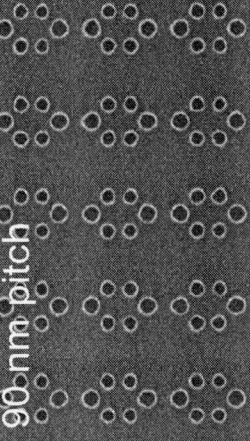

90 nm pitch

October, 2009

FUJiFILM

6th International Symposium on Immersion Lithography

Objective

Extract the risks at the development process, and additionally, the proper risks in double exposure process.
Evaluate these risks and estimate process parameter influence on CD and defectivity performance.

Extracted and evaluated risks

Process	Detail process	Risk	Assessment
Exposure	VV.../HH...	Exp.-exp. delay	CD measurement
	VHVH(one reticle)	Throughput	Throughput calculation
	VHVH(two reticle)	Throughput,	Throughput calculation
	VHHVVH	Throughput	Throughput calculation
	Single expsosure	-	
Development	Development time	CDU, defectivity	CD measurement and defectivity evaluation
	Developer temp.	CDU, defectivity	Dissolution rate measurement
	Developer volume	CDU, defectivity	CD measurement and defectivity evaluation
	Rinse time	CDU, defectivity	CD measurement and defectivity evaluation
	Dry-up time	CDU, defectivity	CD measurement and defectivity evaluation
	Inline-system	Vapor contamination	Vapor concentration analysis in track
	Offline-system	Exp.-PEB delay	CD measurement
		PEB-DEV delay	CD measurement

Developer: FN-DP001
Rinse: FN-RP002

October, 2009

FUJIFILM

6th International Symposium on Immersion Lithography

Result (1) Exposure-exposure delay

Wafer process scheme

CD-y (2nd exp.)

CD-x (1st exp.)

1. Resist coat (FAiRS-9101A19C)
2. (TC coat, TCX-041)
3. Immersion exposure 1 (with Mask 1, V-line)
4. (Delay in FOUP)
5. Immersion exposure 2 (with Mask 2, H-line)
6. PEB
7. DEV

417 points measurement across the wafer
Two wafers were exposed for each delay time
90 nm pitch
Delay time: 0 (minimum), 15, 30, 60 min.

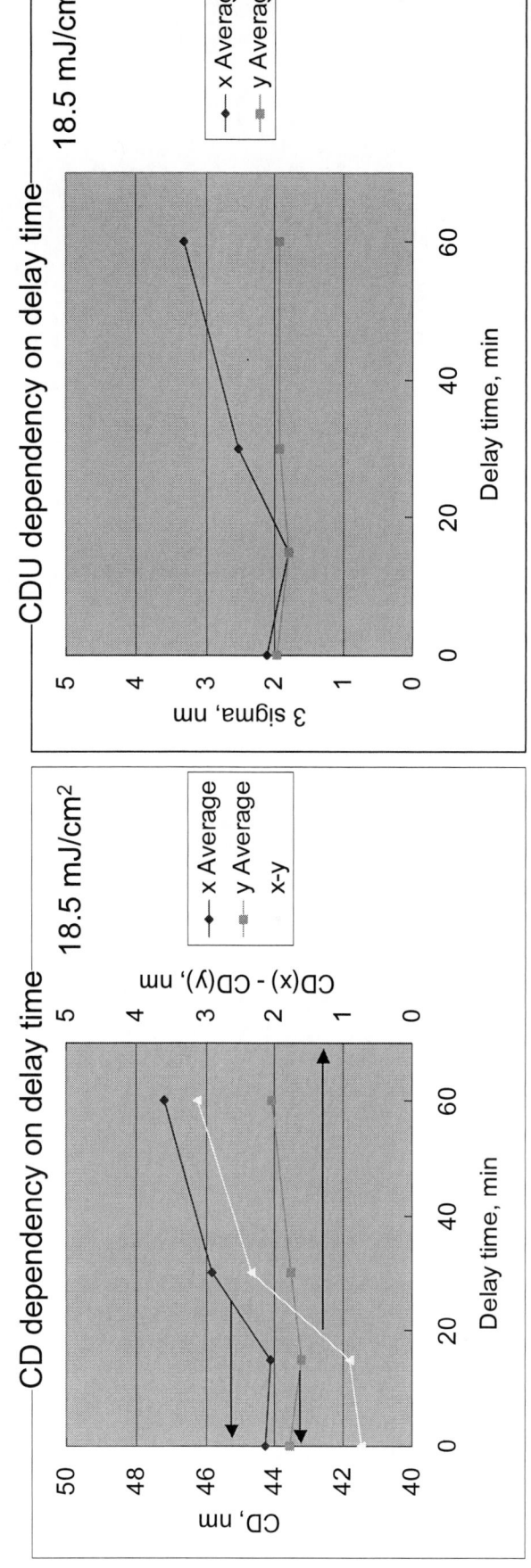

Delay time will be about 20 minutes for 25 wafers per lot with 100 wph exposure machine.
15 minutes of delay time gives no change in CD and CDU.
30 minutes of delay time gives around 2 nm increase of CD and 0.5 nm increase of CDU.

October, 2009

FUJIFILM

6th International Symposium on Immersion Lithography

Result (1) Exposure-exposure delay

TC-less: 18.5 mJ/cm^2
TC: 16.5 mJ/cm^2

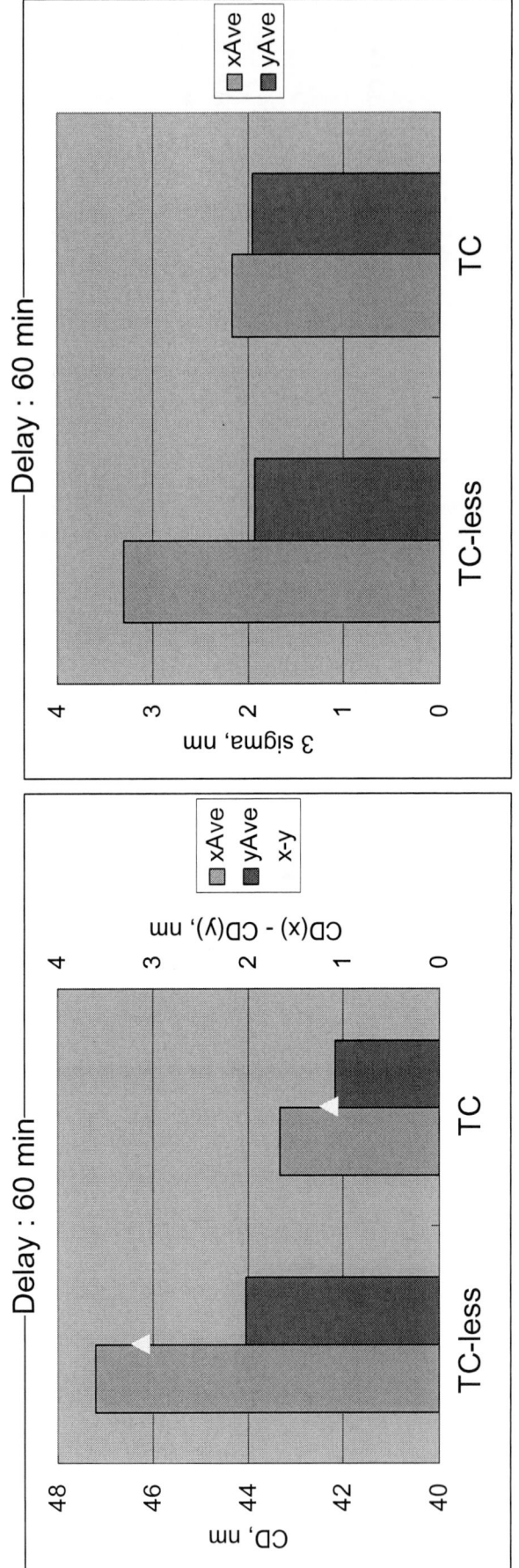

TC process reduced CD change during delay.

October, 2009

FUJiFILM

6th International Symposium on Immersion Lithography

October, 2009

Result (2) Exposure-PEB delay

Possibility of off-line development

Wafer process scheme

1. Resist coat (FAiRS-9101A19C)
2. Immersion exposure 1 (with Mask 1)
3. (Delay in FOUP)
4. PEB
5. DEV

417 points measurement across the wafer
60nm trench measurement

Delay impact:
~ 0.5 nm / 30 min

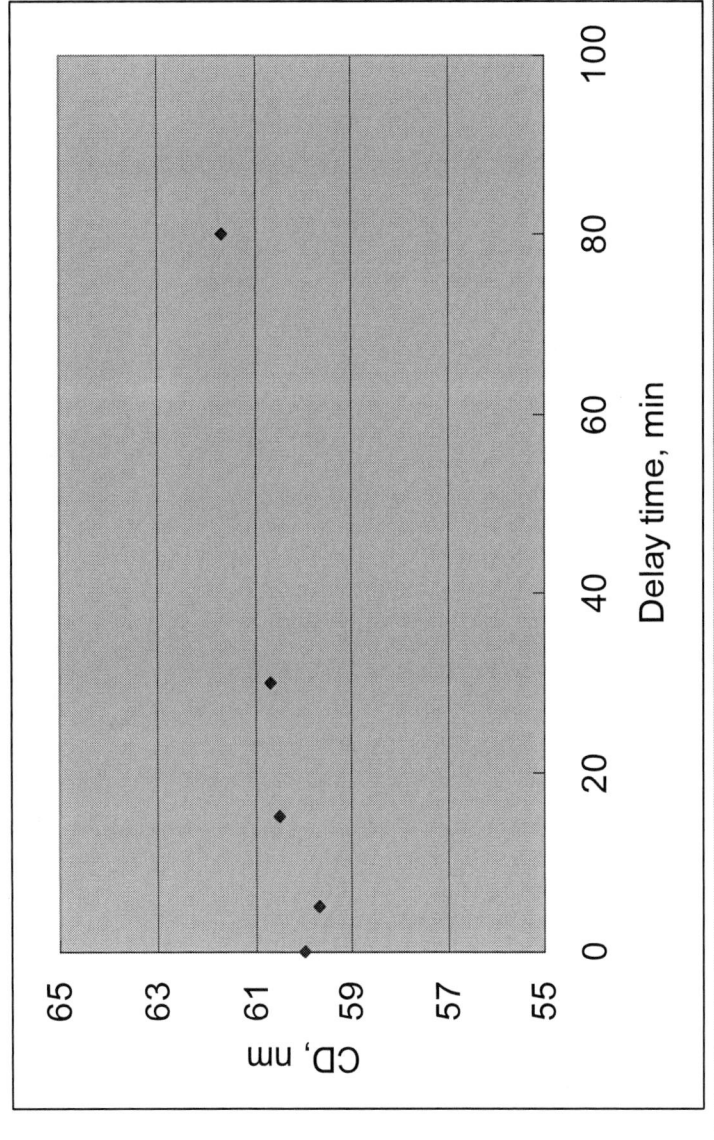

FUJIFILM

6th International Symposium on Immersion Lithography

October, 2009

Result (3) PEB-DEV delay

Possibility of off-line development

Wafer process scheme

1. Resist coat (FAiRS-9101A19C)
2. Dry exposure
4. PEB
5. (Delay in FOUP)
6. DEV

Dry exposure,
NA 0.75, Dipole
6% HTM
75 nm 1:1 L/S, trench measurement

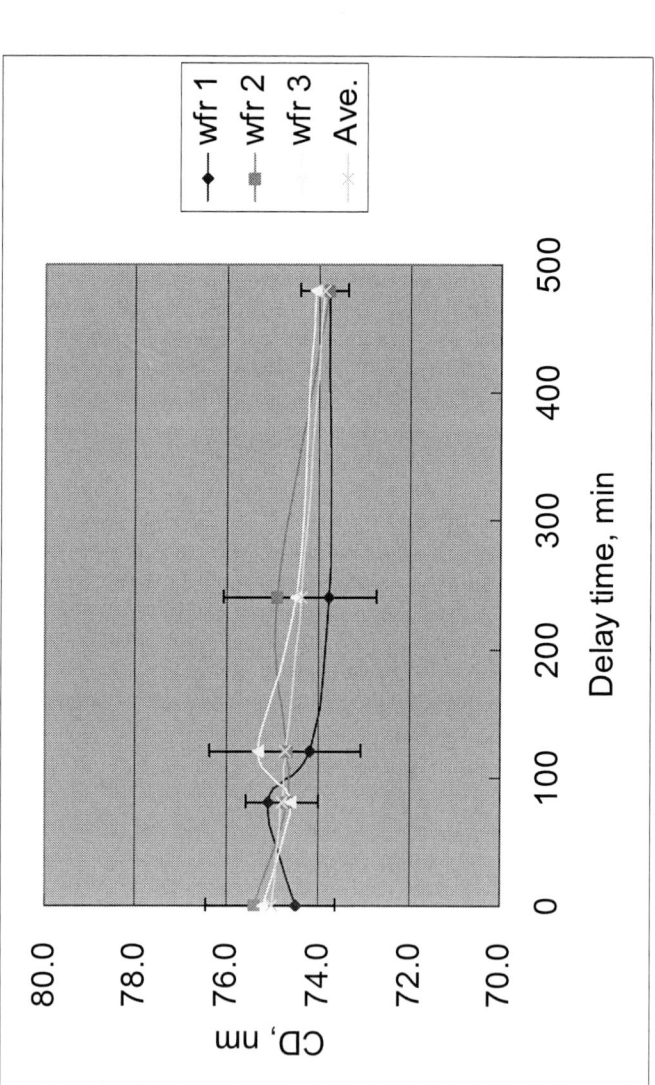

Delay impact:
~ 1.0 nm / 8h

FUJIFILM

6th International Symposium on Immersion Lithography

October, 2009

Result (4) Development time influence on CD

Dry exposure,
NA 0.75, Dipole
6% HTM
75 or 72 nm 1:1 L/S, trench measurement

Resist: FAiRS-9521A02, 100nm
75 nm L/S NILS: 1.53
72 nm L/S NILS: 1.02

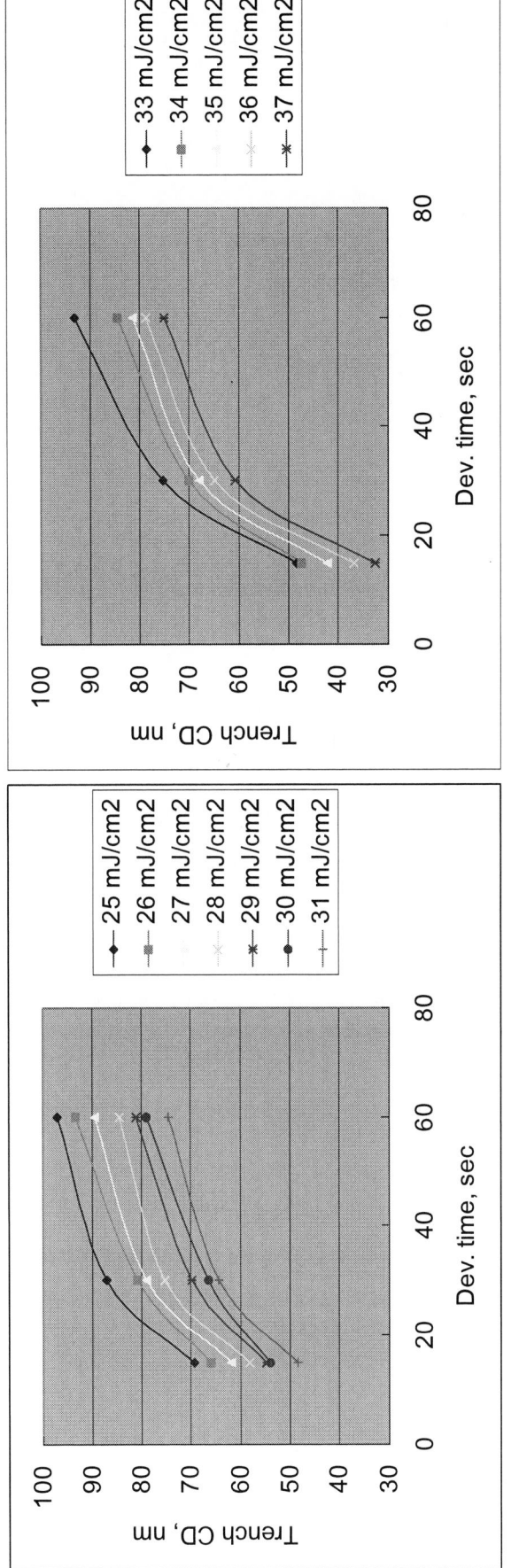

CD depends on development time strongly.
Additionally, lower NILS gave larger development time dependency.
This behavior should be related to the slow dissolution rate in this process.
Increase of Rmax and γ-contrast should relax this behavior.

6th International Symposium on Immersion Lithography

Result (5) Developer temperature influence on dissolution rate

Due to the difficulty in change of developer temperature in track, dissolution rates at different temperatures are measured by QCM method with un-exposed resist film.

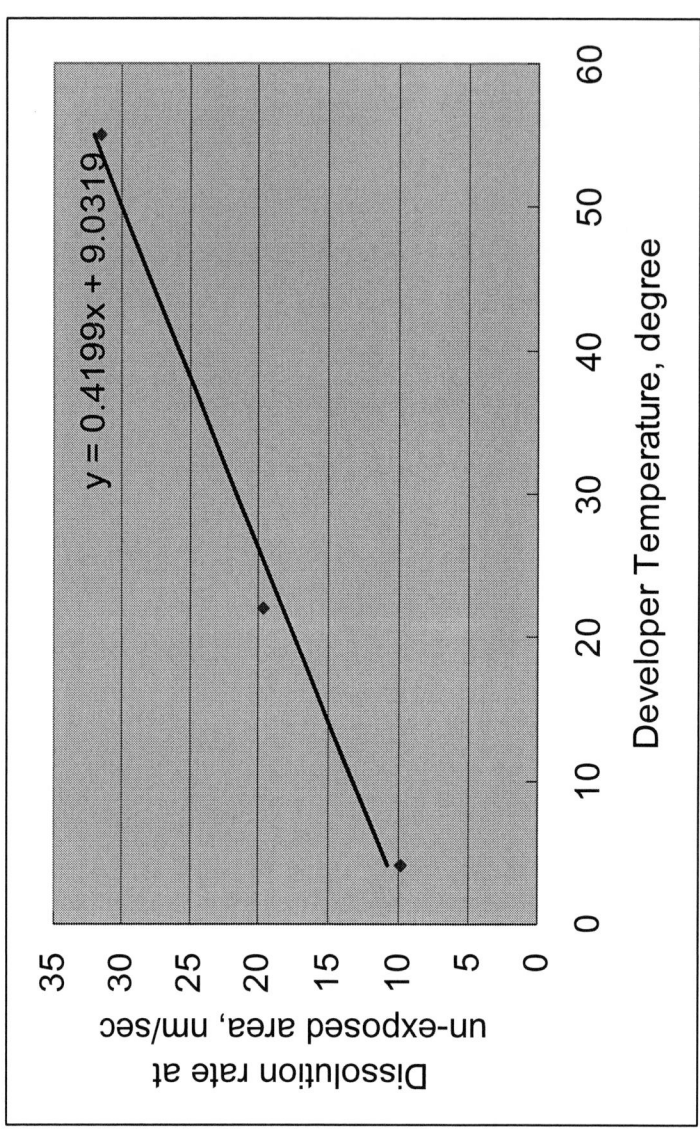

$y = 0.4199x + 9.0319$

Not so much impact.
Dissolution rate change 0.4-0.6 nm / sec / degree.
This is 2-3% of dissolution rate per degree.

October, 2009

FUJiFILM

6th International Symposium on Immersion Lithography

October, 2009

Result (6) Developer volume and rinse time influence on CD

1.3 cc/sec for DP001 and RP002

Dry exposure, NA 0.75, Dipole
6% HTM
75 nm 1:1 L/S, line measurement

Resist: FAiRS-9521A02, 100nm
75 nm L/S

Developer 2 sec dispense

Developer 10 sec dispense

Larger dispense volume of developer gave smaller line CD.
Dispense volume of developer did not give obvious influence to CDU.

Rinsing time does not relates to CD (no rinse results small CD??).
Slightly worse CDU with longer rinsing time.

Result (7) Developer volume influence on defectivity

1.3 cc/sec for DP001 and RP002

Dry exposure, NA 0.75, Dipole
6% HTM
75 nm 1:1 L/S

10 sec dispense (std)

1 sec dispense

Reviewed Result

	10 sec		1 sec	
	n1	n2	n1	n2
Fall on	0.03	0.04	0.19	0.13
Embedded	0.11	0.01	0.03	0.06
Slimming	0.01	0	0	0
Residue	0.00	0.01	0	0.08
Bridge	0.00	0.01	0	0.04
Collapse	0.00	0.01	0	0
Peeling	0.00	0.01	0	0.01
False	0.96	1.06	1.66	1.27
Raw D.D.	1.11	1.19	1.89	1.60
Final D.D.	0.16	0.10	0.22	0.33

Counts / cm^2

Small dispense volume gave much false type, and larger number of final defect density.

6th International Symposium on Immersion Lithography

Result (8) Rinse time influence on defectivity

1.3 cc/sec for DP001 and RP002

Dry exposure, NA 0.75, Dipole
6% HTM
75 nm 1:1 L/S

Reviewed Result

	16 sec		0 sec	
	n1	n2	n1	n2
Fall on	0.03	0.04	0.02	0.02
Embedded	0.11	0.01	0	0
Slimming	0.01	0	0	0
Residue	0.00	0.01	0	0
Bridge	0.00	0.01	0	0.01
Collapse	0.00	0.01	0	0
Peeling	0.00	0.01	0	0
False	0.96	106	5.96	3.17
Raw D.D.	1.11	1.19	5.98	3.21
Final D.D.	0.16	0.10	0.02	0.03

Counts / cm^2

Short rinse time increased only false type defect.

Rinse 0 sec

Rinse 16 sec (std)

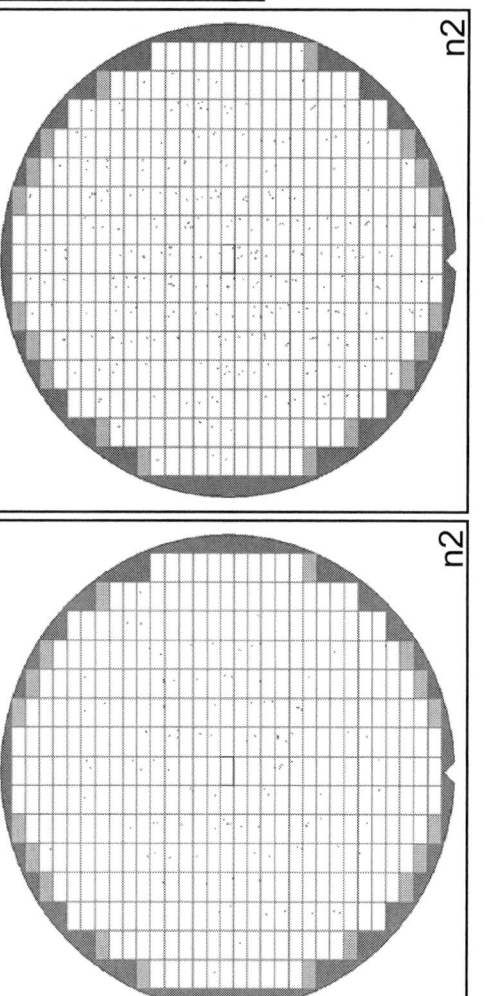

October, 2009

6th International Symposium on Immersion Lithography

Summary

Exposure to exposure delay gave CD and CDU change. In the largest case, 2 nm CD shift and 0.5 nm worse CDU are predicted in the HVM. It is important to minimize the delay, otherwise, TC process reduced the CD change.

Exposure to PEB gave slightly large CD-shift, and PEB to development delay gave minor impact. Even if off-line developer system is needed, exposure to PEB should be in-lined.

Pattern size strongly depends on development time. This behavior should relate to the slow dissolution rate in this process. Further dissolution rate should be improved by material development.

Developer temperature was one of big concerns, but from the dissolution rate study, little influence to CD is supposed.

Small volume of developer and short time rinse both resulted large number of defects, but most of all are non-visible with SEM. After etch inspection should be needed to make clear if these defects are killer type.

From these data, there are no big issue in designing total process, but further detail study should be needed to optimize for each fab condition.

October, 2009

 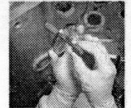     

DEFECTIVITY INVESTIGATION WITH POINT-OF-USE FILTRATION PARAMETER CHANGES

W. Schollaert, X. Buch, K. Hoshiko – JSR Micro N.V., Technologielaan 8, B-3001 Leuven, Belgium

J. Braggin – Entegris, Inc. Assignee to IMEC, Kapeldreef 75, 3000 Leuven, Belgium

Abstract

Many of the major defect challenges of immersion lithography have been tackled over the years, including water marks and stain defects. However, a few familiar defects from dry lithography, such as microbridging and residues, are still present in immersion lithography processing with topcoatless resists. Although these defects are not new, their existence on ever-shrinking patterns is a good reason for alarm when driving towards more challenging yield targets.

In this study, the effect of point-of-use filtration and filtration parameters has been studied to understand their effect on immersion defectivity levels with topcoatless resists. The effect of filter choice, including membrane design and material, is one important factor in reducing defectivity. What may be more important, however, is the effect of a proper filtration setup to capture defects before they appear on the wafer.

Experimental

1. Illumination Conditions:
Sokudo® RF3Si, XT: 1900 Gi

- Mask: IMEC® DEF 45 (45 nm line/space, 90 nm pitch)
- Stack: 95 nm BARC, 105 nm resist
- Resist A illumination settings: NA = 1.35; Dose = 21.5 mJ; dipole 40X, σ_i = 0.69; σ_o = 0.93; y-polarization
- Resist B illumination settings: NA = 1.35; Dose = 26.0 mJ; dipole 40X, σ_i = 0.69; σ_o = 0.93; y-polarization

2. Design of Experiments

Two different JSR resist samples were provided to see the effect of filtration parameters on different resist systems.

- Resist chemistry: Resist A and B
- Filter material and design: UPE, Duo (polyamide/UPE), polyamide
- Filtration rate: Low, medium, high

Irregular Bridging   Bridging   Residue

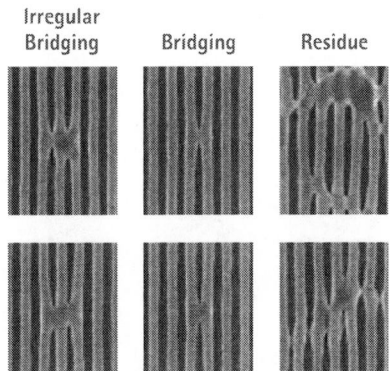

Overall Defectivity

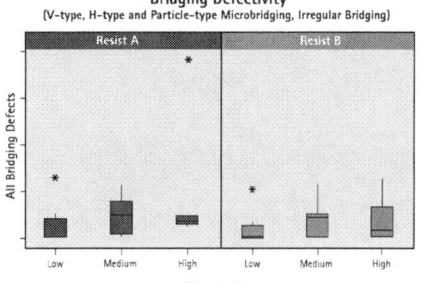

Effect of Filtration Rate:
Resist A shows a clear trend with the lowest filtration rate being preferred. Due to the high level of defectivity from Resist B, the effect is not as clear.

Effect of Filter Membrane:
The two resists show different performance with the different filters. Resist B has the best performance with the duo (polyamide/UPE) filter.

Bridging Defects

Bridging Defectivity
(V-type, H-type and Particle-type Microbridging, Irregular Bridging)

Bridging Defectivity
(V-type, H-type and Particle-type Microbridging, Irregular Bridging)

Effect of Filtration Rate:
Both resists show a trend that the lowest filtration rate has the best performance for microbridging defects.

Effect of Filter Membrane:
Both resists show the best performance for microbridging defects with the use of the duo (polyamide/UPE) membrane.

Residue Defects

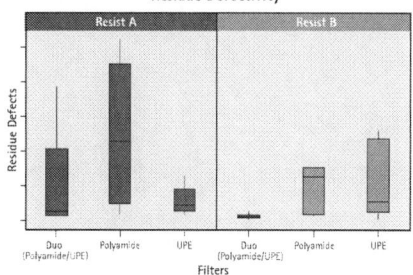

Residue Defectivity

Residue Defectivity

Effect of Filtration Rate:
Although filtration rate did not have a significant effect on residue defects, the low filtration rate has the best repeatability.

Effect of Filter Membrane:
Resist A had the best performance with a UPE membrane, whereas Resist B had the best performance with the duo (polyamide/UPE) membrane.

Conclusions

- Point-of-use filtration affects microbridging and residue defectivity.
- Lower filtration rate shows a lower level of overall defectivity.
- Careful filter membrane choice is important in overall defect reduction.

6th International Symposium on
Immersion Lithography Extensions

Robust material and processes for Double Patterning

Tokyo Electron Limited

Leading-edge Process Development Center

Hidetami Yaegashi

TOKYO ELECTRON

OUTLINE

- Introduction
- Photolithography trend
- Double patterning application
- Introduction of New material for Double Patterning
- Property of SiO2 film
- Demonstration results
- Next plan
- Summary

Tokyo Electron limited   6th International Symposium on Immersion Lithography Extensions
H. Yaeasghi / LPDC / LPDC-091022-rev.-001   October 22 – Prague, Czech Republic

Introduction

The Numerical aperture (NA) has been significantly improved to 1.35 by the introduction of water base immersion 193nm exposure tool, however, realistic minimum feature size is still limited to 40nm, even with the help of robust RETs.

Double patterning processes are technique that may be adopted for fabricating etching mask patterns for the 32nm node, and possible also for the 22nm node. Although several double patterning processes have been introduced such as LELE, LLE and self-aligned spacer process, newly developed film materials should be needed to optimize DP process finely.

In this paper, robust material and innovative demonstration result to fabricate 22nm node pattern would be introduced.

Tokyo Electron limited
H. Yaeasghi / LPDC / LPDC-091022-rev.-001

6th International Symposium on Immersion Lithography Extensions
October 22 – Prague, Czech Republic

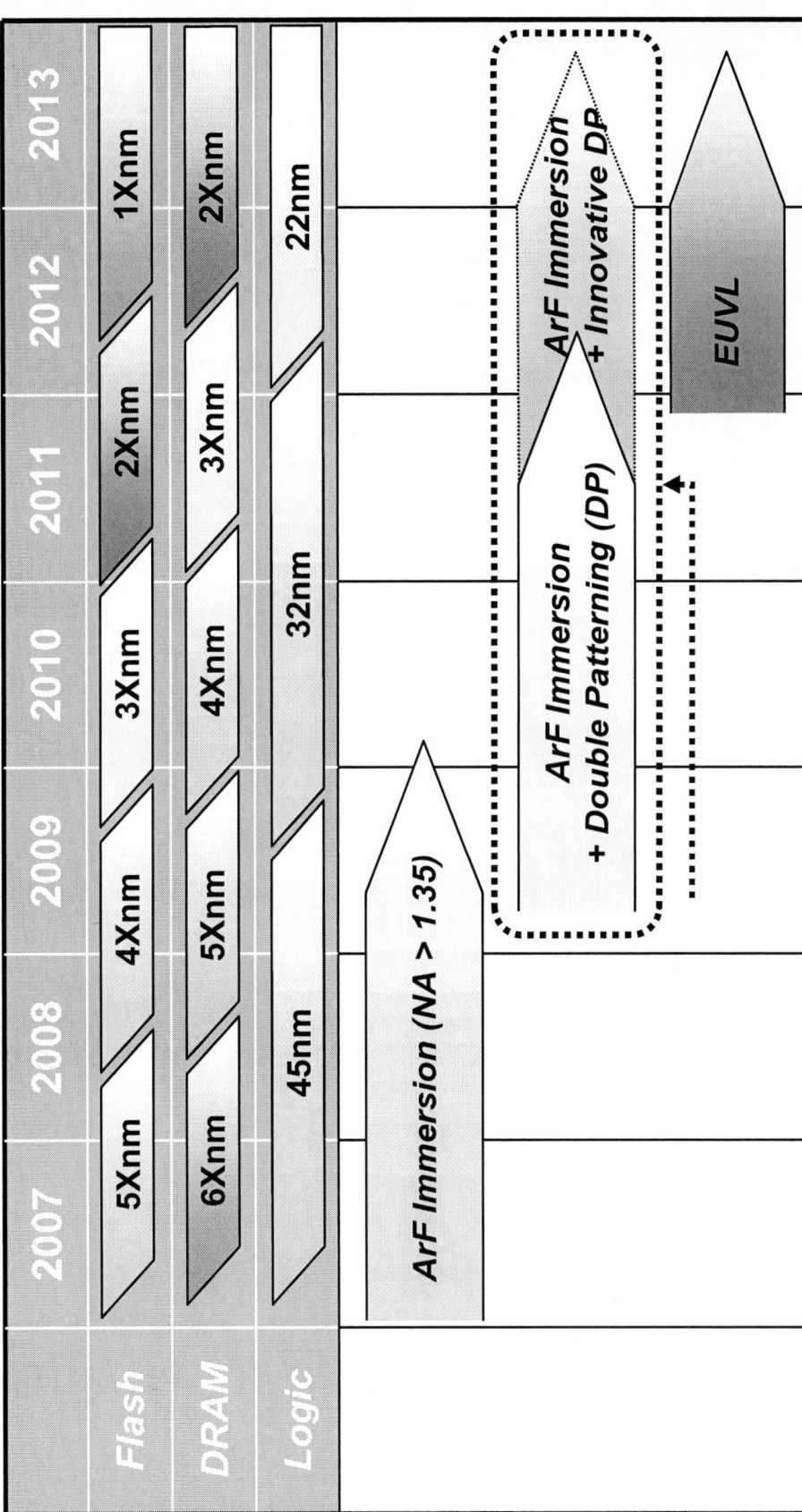

Key issues in DP process(1)

Pattern splitting method

Process step	Single-Expo.	LELE(1) hole/trench	LELE(2) line	LLE w/Frz	Cross-LLE
PR material	O	O	O	☆	☆
PR Coat	O	O	O	O	O
Exposure	O	O	O	O	O
PR deve.	O	O	O	O	O
PR CD control	→	☆	→	☆	→
Etching	O	O	O	→	→
Freezing	→	→	→	☆	→
PR material	→	O	O	O	☆
PR coat	→	☆	☆	O	O
Exposure	→	☆	☆	☆	☆
PR deve.	→	O	O	O	O
PR CD control	→	☆	→	☆	→
H/M Etch	O	O	☆	O	O

Key issues in DP process(2)

Self-aligned method

Process	Single-Expo.	Spacer DP — Positive	Spacer DP — Negative	DTD
S/F depo.	○	→	○	→
PR coat	○	○	○	○
Exposure	○	○	○	○
PR deve.	○	○	○	○
2nd PR deve.	→	→	→	☆
Core etch	→	○	○	→
Spacer depo.①	→	○	☆	→
Spacer depo.②	→	○	○	→
Etch-back	→	→	○	→
Core strip	→	○	→	→
Spacer Etch	→	☆	☆	→
H/M Etch	○	○	○	→

Tokyo Electron limited
H.Yaeasghi / LPDC / LPDC-091022-rev.-001

6th International Symposium on Immersion Lithography Extensions
October 22 – Prague, Czech Republic

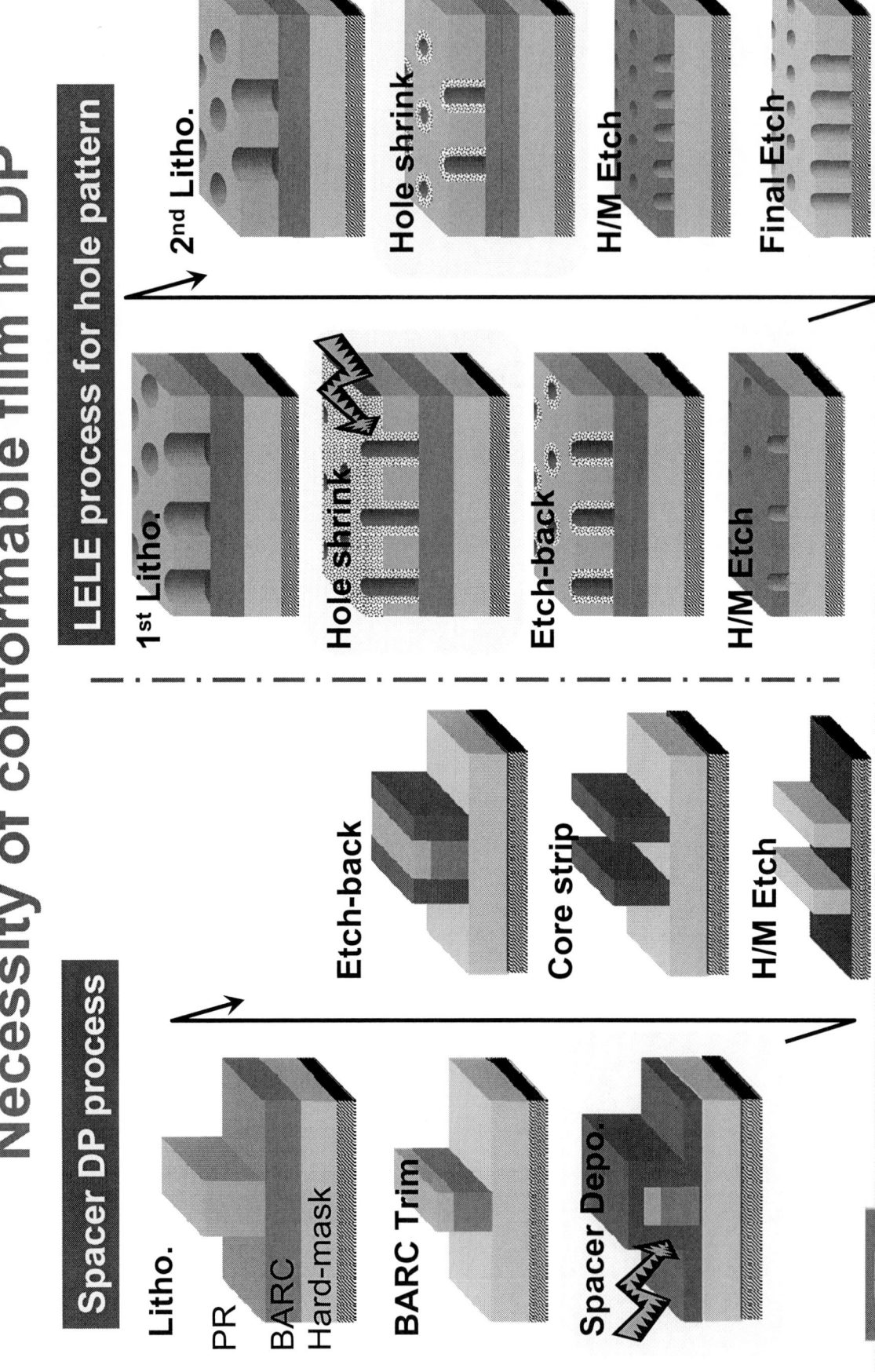

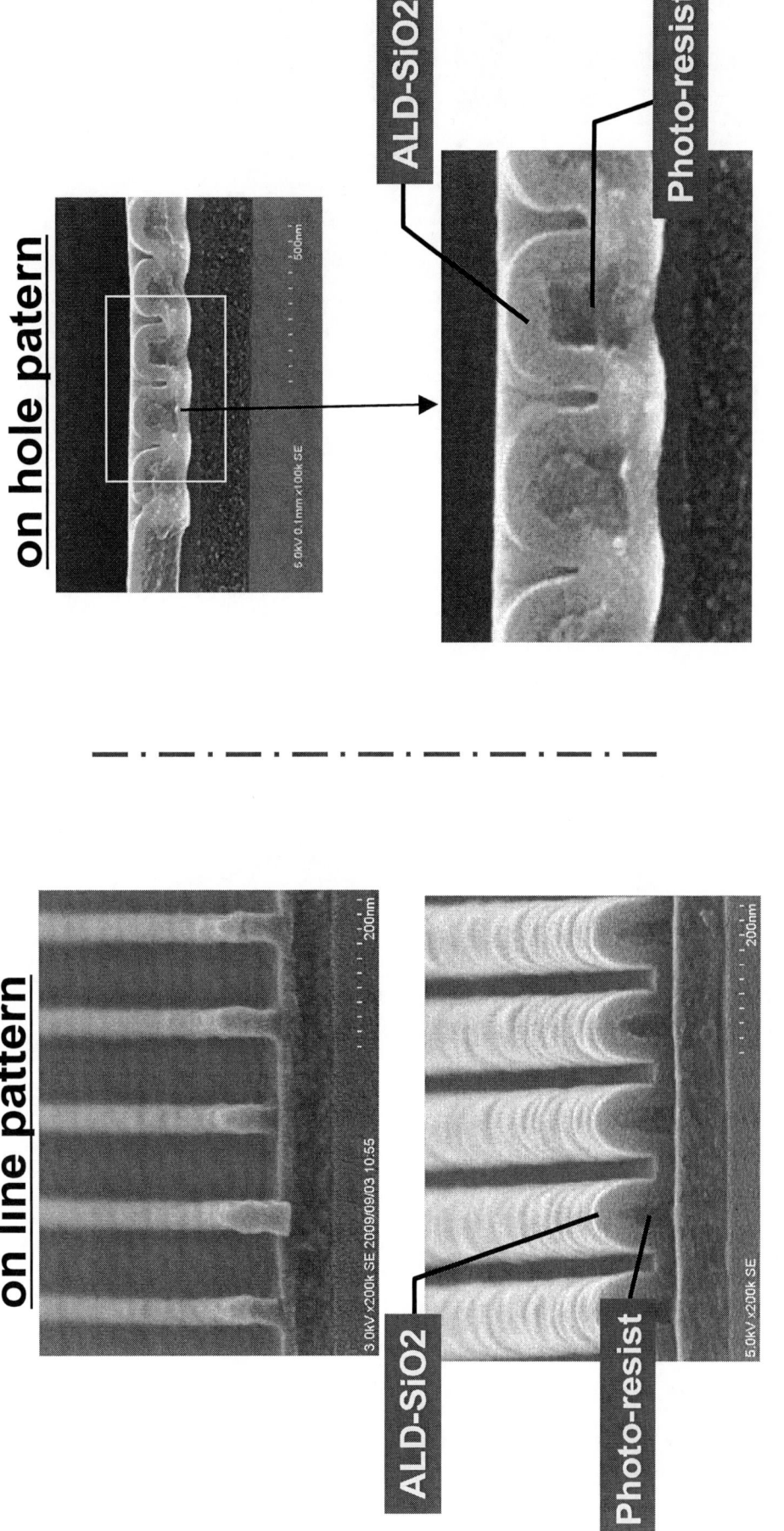

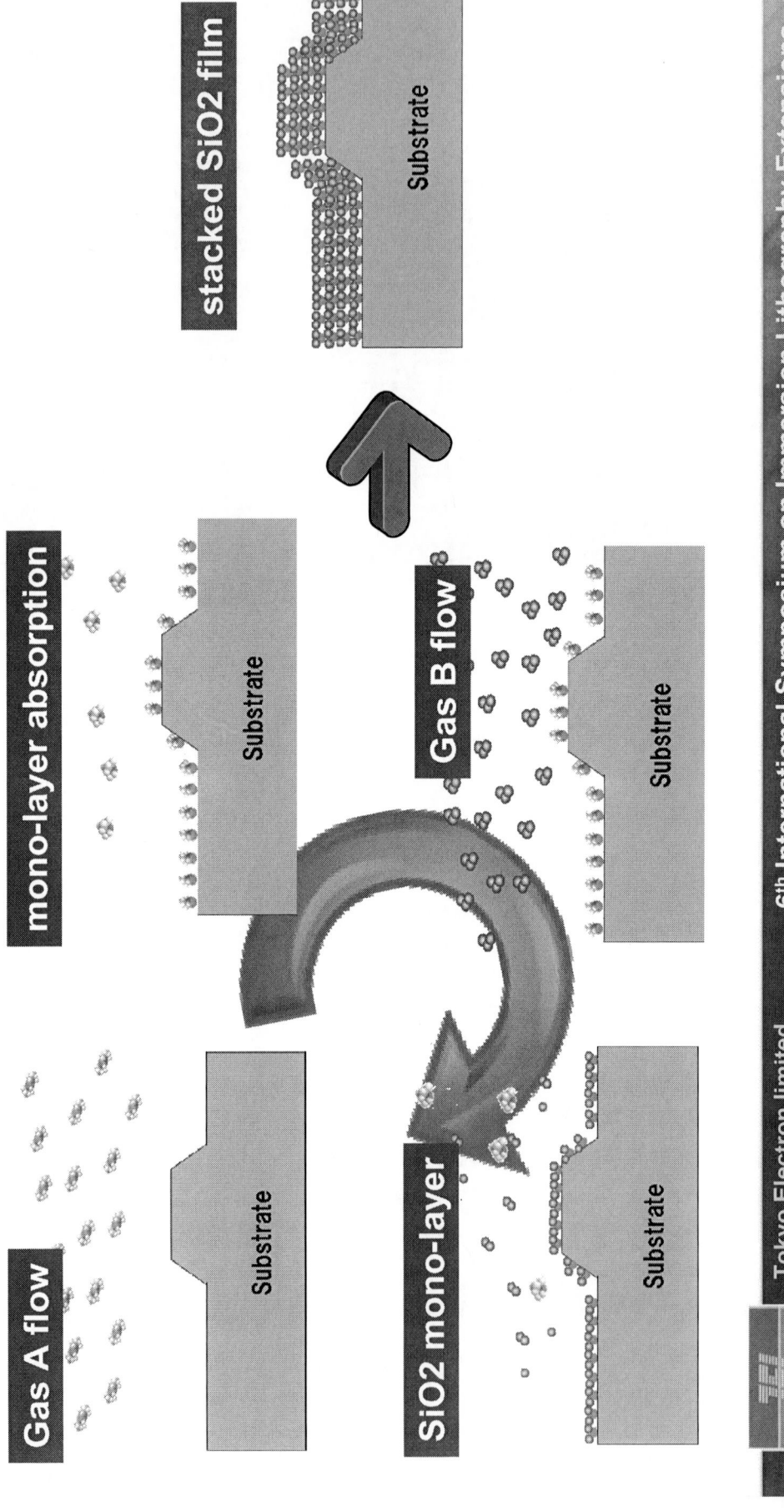

Thickness controllability & Conformability

Ultra-low-temperature SiO2 film

control

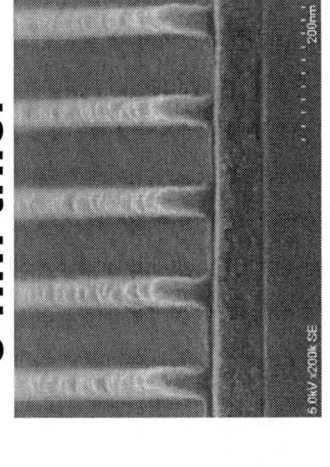

3 nm thic

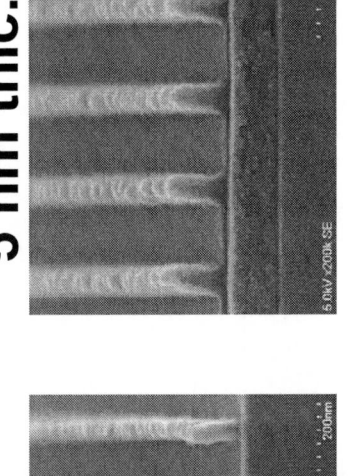

5 nm thic.

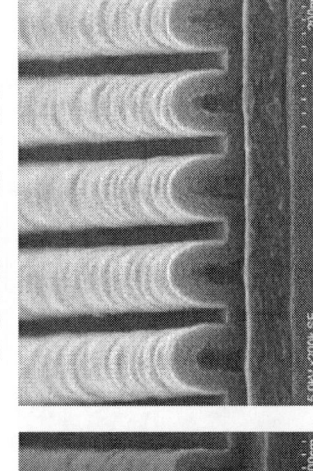

10 nm thic.

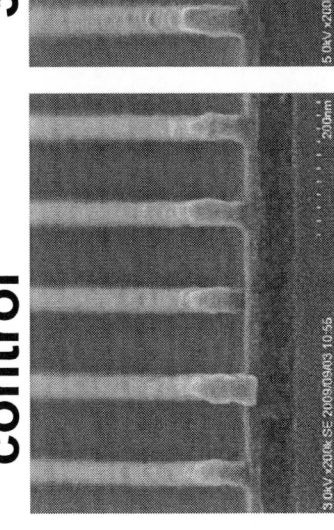

20nm thic.

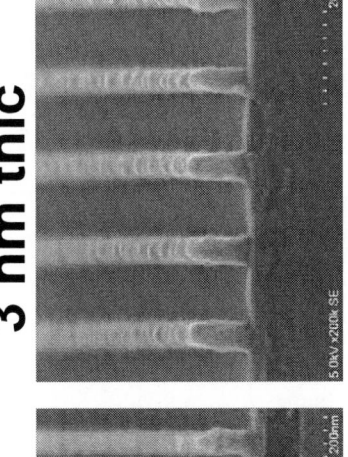

30nm thic.

Various thick film can be deposited on resist directly without any profile change

Tokyo Electron limited
H. Yaeasghi / LPDC / LPDC-091022-rev.-001

6th International Symposium on Immersion Lithography Extensions
October 22 – Prague, Czech Republic

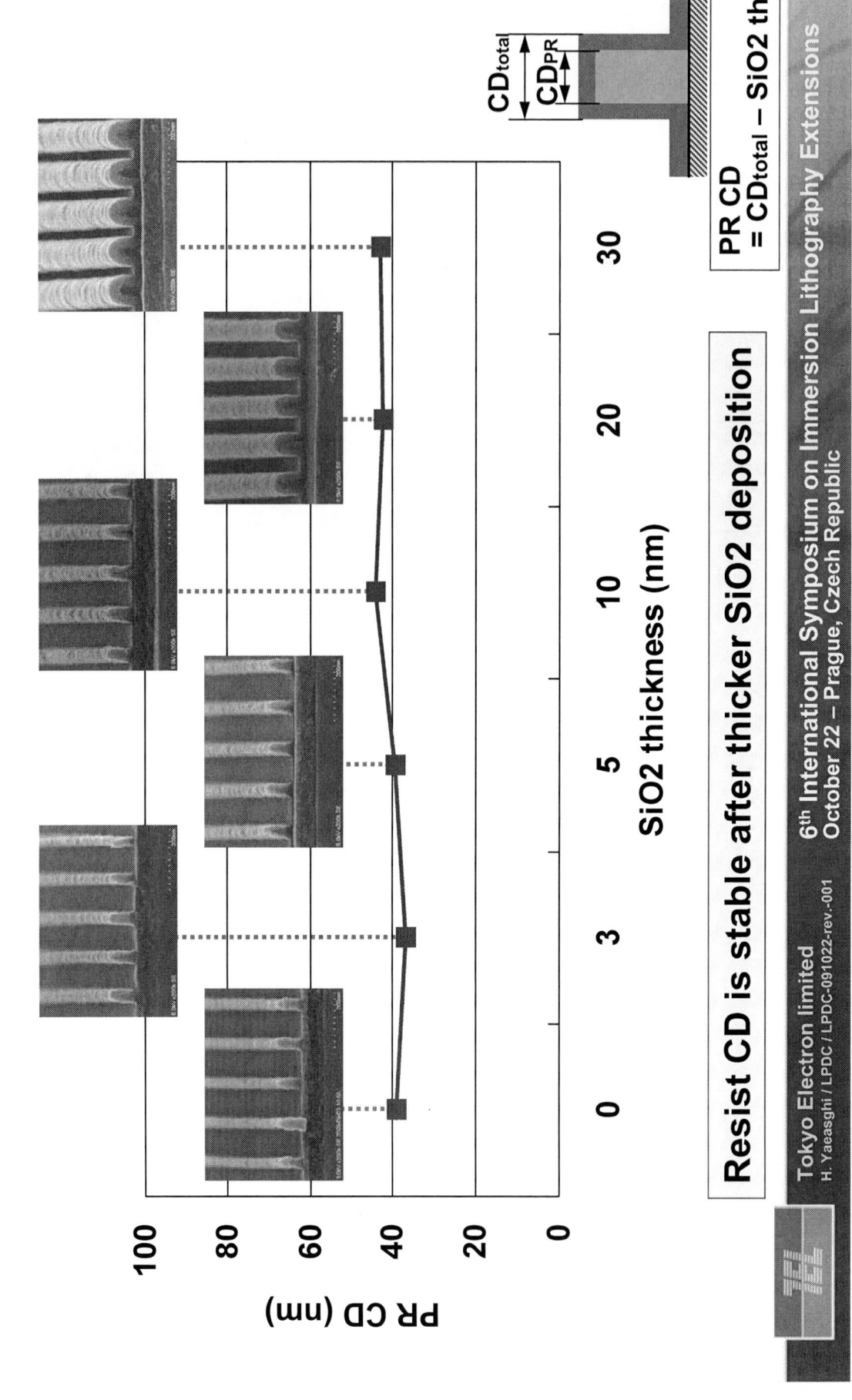

Stable Photo-resist profile after deposition

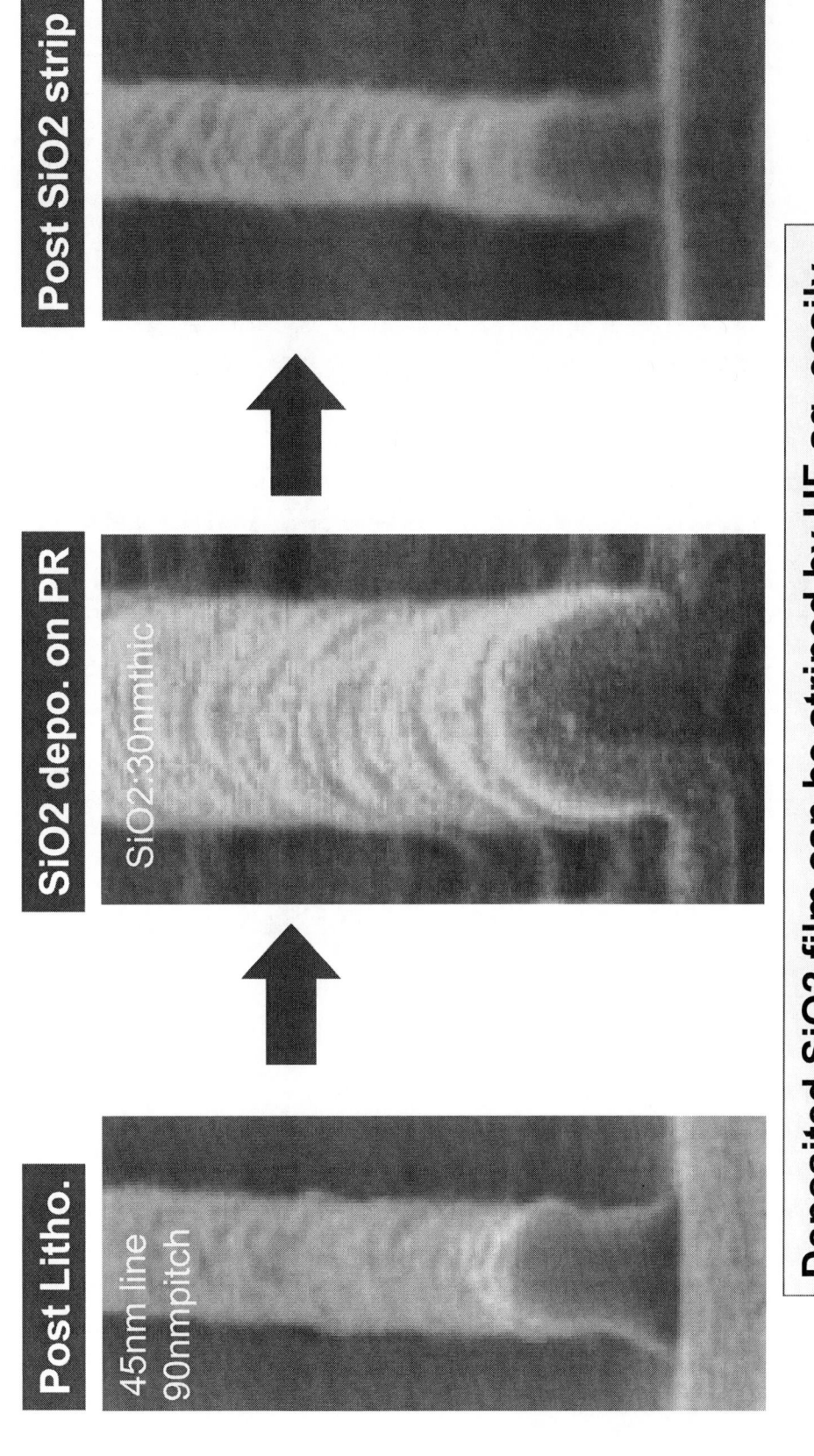

Post Litho.

45nm line
90nmpitch

SiO2 depo. on PR

SiO2:30nmthic

Post SiO2 strip

Deposited SiO2 film can be striped by HF aq. easily, and same feature resist pattern apear.

Tokyo Electron limited
H. Yaeasghi / LPDC / LPDC-091022-rev.-001

6th International Symposium on Immersion Lithography Extensions
October 22 – Prague, Czech Republic

Analytical result on SiO2/PR
TEM-EELS atomic mapping

Control

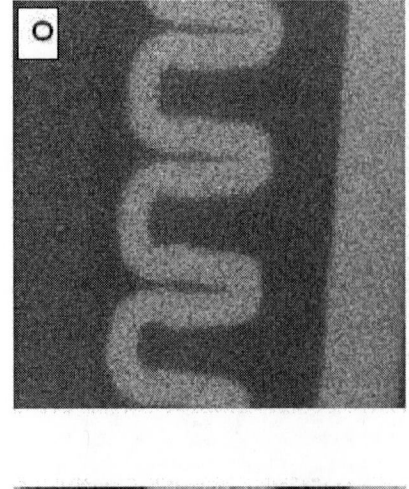

Si

O

Bright field

C

N

Any interaction between Si and Resist cannot be detected.

Tokyo Electron limited
H. Yaeasghi / LPDC / LPDC-091022-rev.-001

6th International Symposium on Immersion Lithography Extensions
October 22 – Prague, Czech Republic

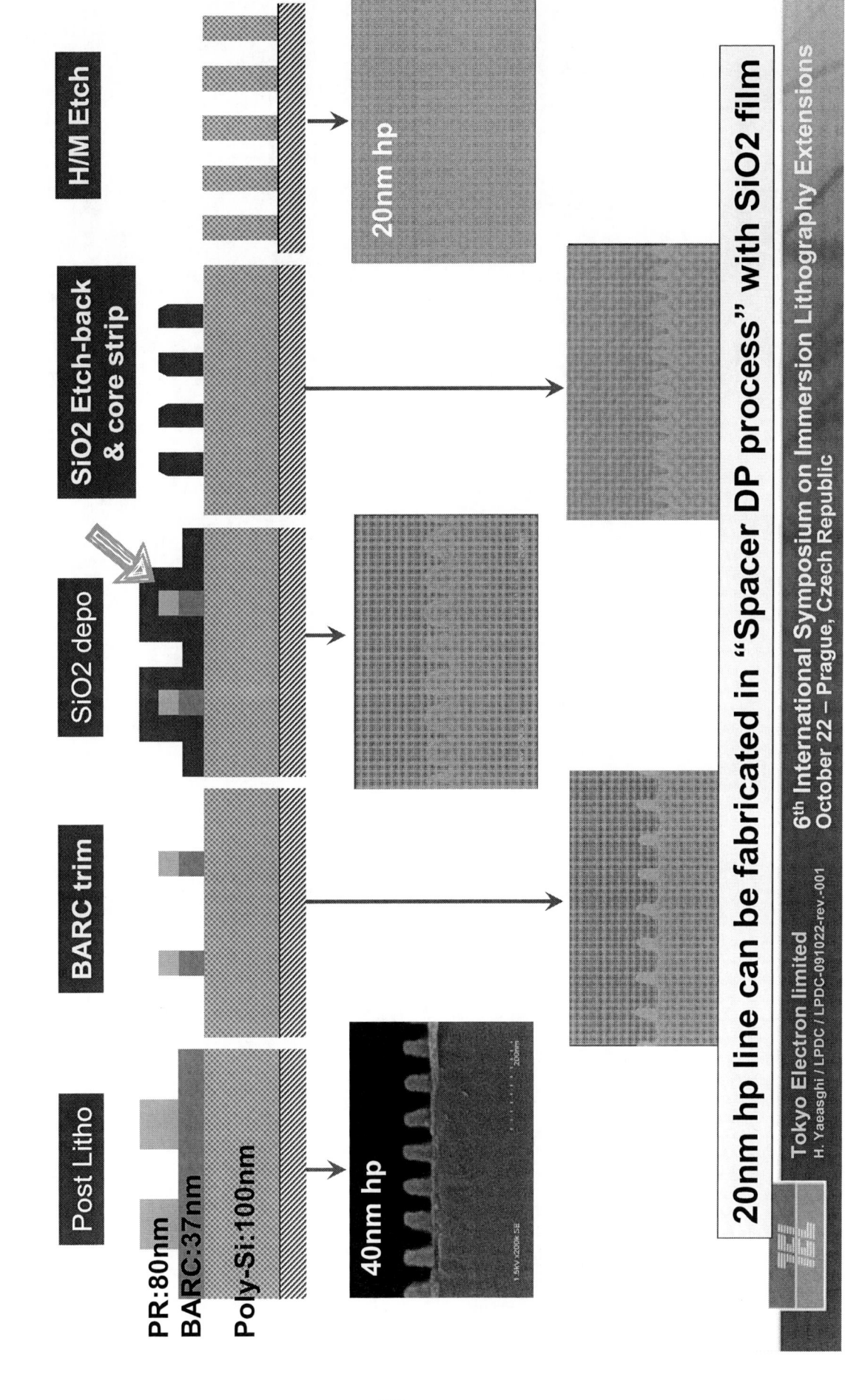

Controllable hole CD

ALD-SiO2

Photo-resist

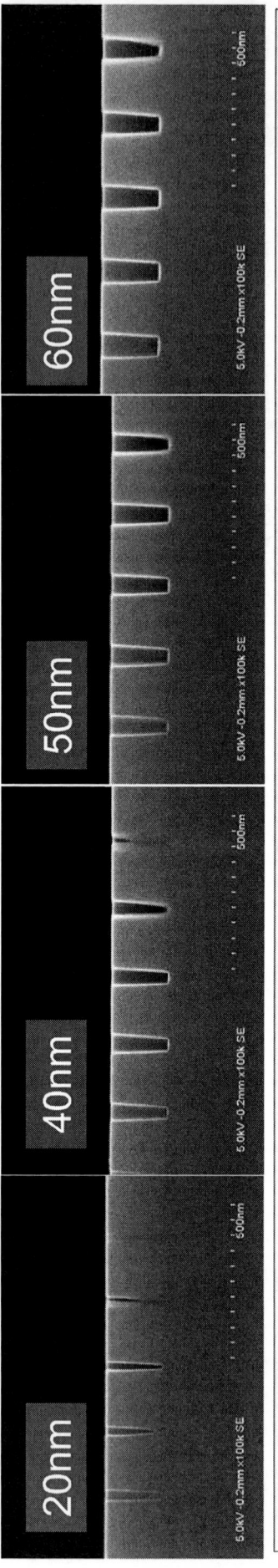

Various CD control on TEOS

20nm   40nm   50nm   60nm

Precise SiO2 film thickness controllability can correct hole CD for any hole size.

Tokyo Electron limited
H. Yaeasghi / LPDC / LPDC-091022-rev.-001

6th International Symposium on Immersion Lithography Extensions
October 22 – Prague, Czech Republic

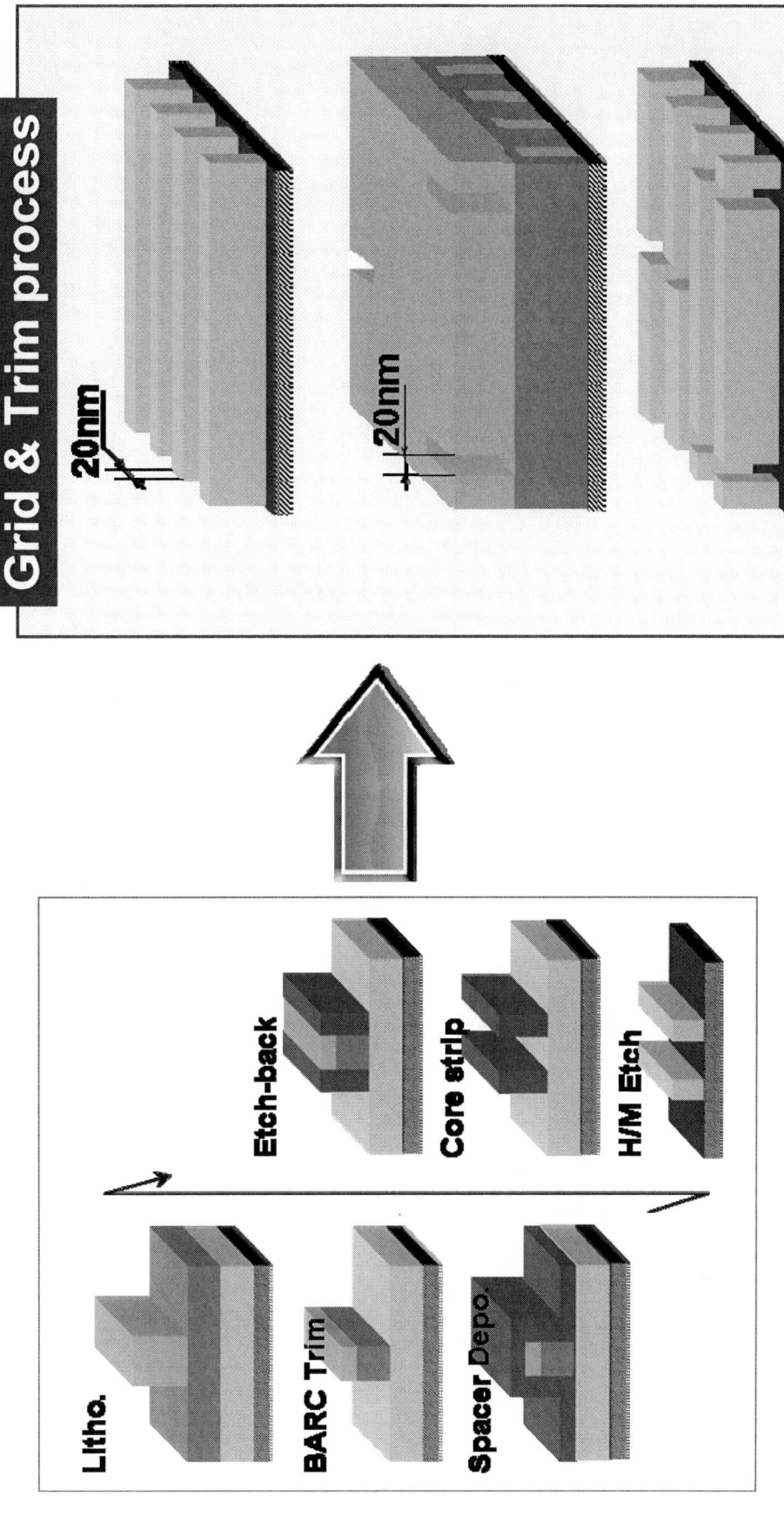

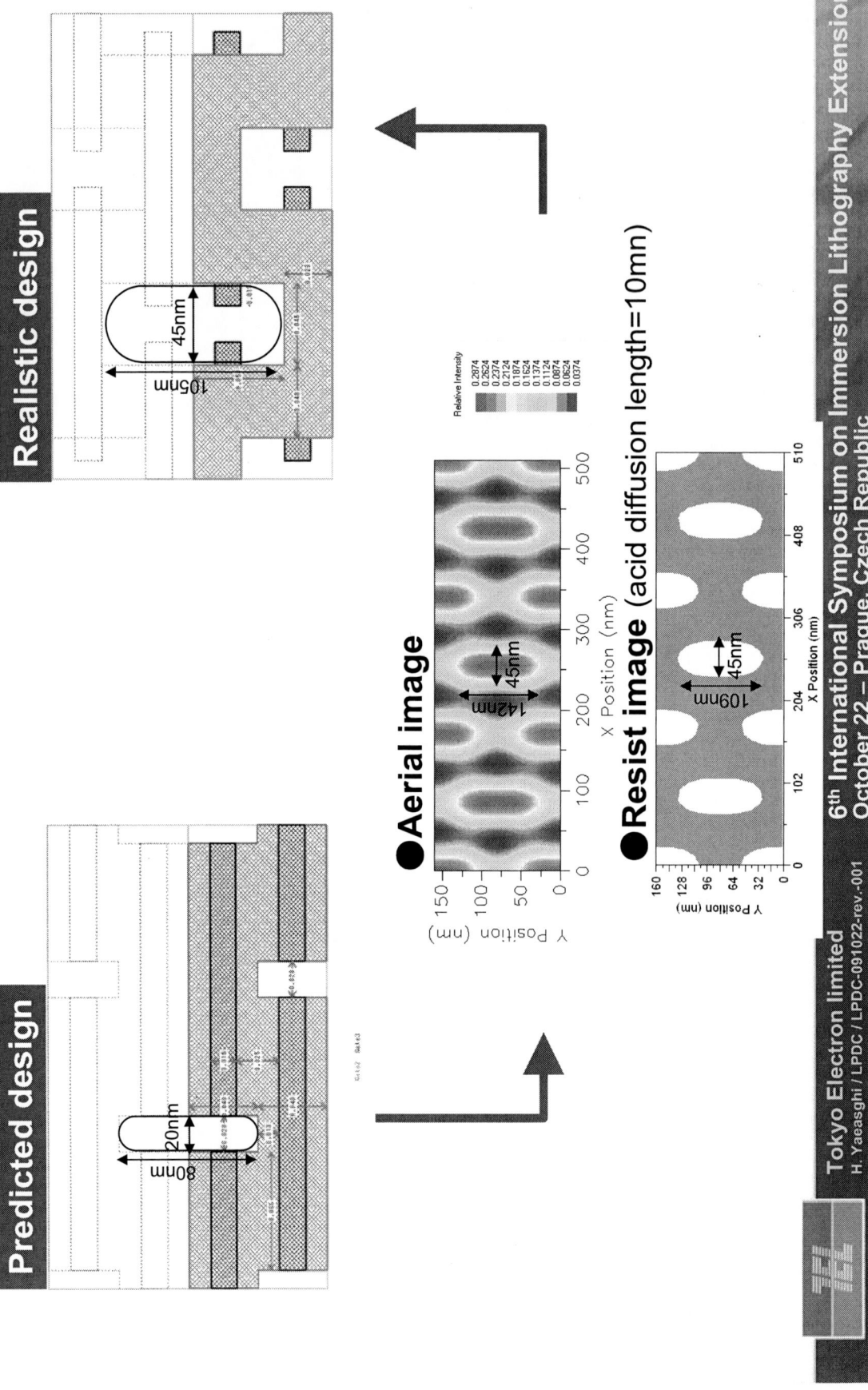

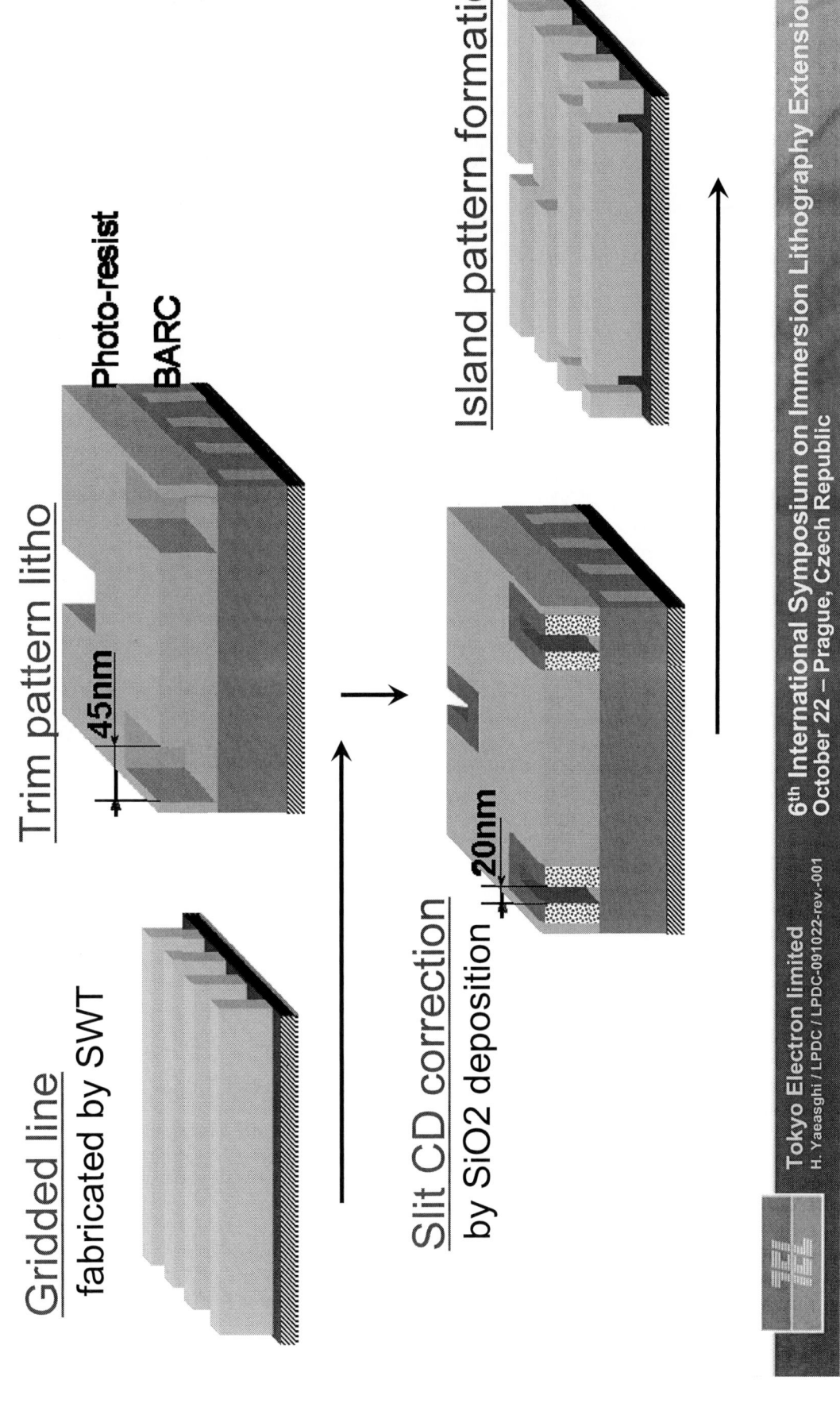

Summary

☐ Several Double patterning process scheme have been demonstrated to enable the fine pattern fabrication, however, newly developed film material will be needed for precise process control and saving the process cost.

☐ Although SiO_2 film might not be familiar in photolithography area, it has pretty good property to match up double patterning process.

☐ Positive feature of ALD-SiO_2 film

   -Conformable deposition property

   -Variable thickness controllability, even if in narrow gap

   -Damage free to photo-resist pattern

☐ Applicable DP process / Demonstration results

   -Spacer DP process: 20nm hp line can be described

   -Hole shrink process in LELE: 30nm dens hole

☐ Next plan

   "Grid and trim process" will be demonstrated on spacer DP process combined with trimming process and hole shrink process.

Tokyo Electron limited
H. Yaeasghi / LPDC / LPDC-091022-rev.-001

6th International Symposium on Immersion Lithography Extensions
October 22 – Prague, Czech Republic

COLUMBIA UNIVERSITY
IN THE CITY OF NEW YORK

Accelerating the next technology revolution

Tethered Naphthalene Derivatives as Sensitizers for Sequential Two-Photon Photoacid Generators for Double Exposure Photolithography

Arun Kumar Sundaresan, Yongjun Li, Naphthali O'Connor, Steffen Jockusch, and Nicholas J. Turro
Chemistry Department,
Columbia University, New York

Tomoki Nagai, Toshiyuki Ogata, Younjin Cho, Saul Lee, Adam Berro, Xinyu Gu, and C. Grant Willson
Department of Chemical Engineering,
The University of Texas at Austin, Texas

Paul A. Zimmerman, SEMATECH

Copyright ©2008
SEMATECH, Inc. SEMATECH, and the SEMATECH logo are registered servicemarks of SEMATECH, Inc. International SEMATECH Manufacturing Initiative, ISMI, Advanced Materials Research Center and AMRC are servicemarks of SEMATECH, Inc. All other servicemarks and trademarks are the property of their respective owners.

Limitations of Conventional Lithography

SEMATECH

- κ_1 A processing parameter
- λ Wavelength of light used
- N A   Numerical Aperture.

Resolution $= \kappa_1 \lambda / N A$

Resolution: "the width of a patterned line or the distance between two lines, monitored to maintain device performance consistency."

With $\lambda = 193$ nm; $\kappa_1 = 0.25$; NA (water) $= 1.35$, maximum resolution ≈ 38 nm

Alternative resolution enhancement technologies that use existing tools include

Double Patterning: Two-exposure process with an intermediate chemical development step.

Double Exposure: Two-exposure pass process that does not require the removal of the wafer between exposures.

Double Patterning (DP)

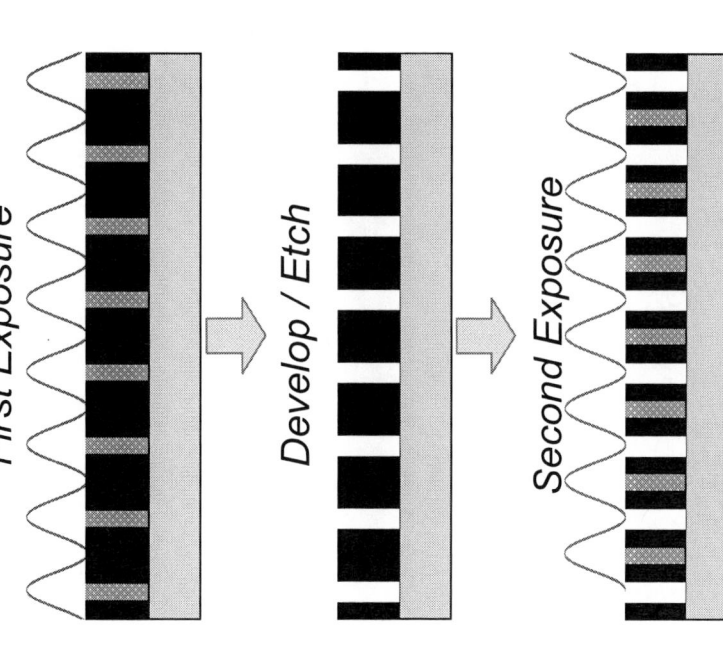

First Exposure

Develop / Etch

Second Exposure

Double Exposure (DE)

SEMATECH

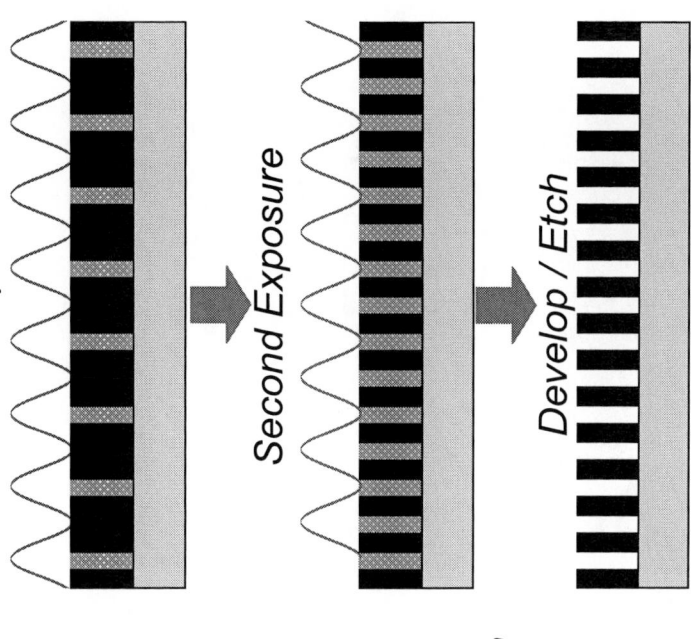

First Exposure

Second Exposure

Develop / Etch

Develop / Etch

Advantages of DE Approach

- Faster - Wafer not removed between exposures
- Cost-effective - One additional exposure step
  (compared to conventional lithography)

Need for Newer Materials for DE Lithography

SEMATECH

Effect of Two Exposures

A - Amplitude Coefficient
B - Minimum Image Intensity

$$I_{\text{exposure}1} = A\cos^2\left(\frac{\pi x}{\text{pitch}}\right) + B$$

$$I_{\text{exposure}2} = A\cos^2\left(\frac{\pi x}{\text{pitch}} + \frac{\pi}{2}\right) + B = A\sin^2\left(\frac{\pi x}{\text{pitch}}\right) + B$$

$$I_{\text{sum}} = I_{\text{exposure}1} + I_{\text{exposure}2} = A\cos^2\left(\frac{\pi x}{\text{pitch}}\right) + A\sin^2\left(\frac{\pi x}{\text{pitch}}\right) + 2B = \boxed{A + 2B}$$

A Constant

Needed: Materials with Non-Linear Dose - Response Behavior

$$\boxed{I_{\text{sum}} \neq I_{\text{exposure}1} + I_{\text{exposure}2}}$$

1. *Two-Photon Materials*

Reactant + $2\,h\nu$ $\longrightarrow$ Product

2. *Intermediate State Two-Photon Materials (ISTP)*

Reactant $\underset{h\nu}{\overset{h\nu}{\rightleftarrows}}$ Intermediate $\overset{h\nu}{\longrightarrow}$ Product

Materials with Nonlinear Dose - Response Ratio

Two major requirements must be met to successfully demonstrate double exposure lithography

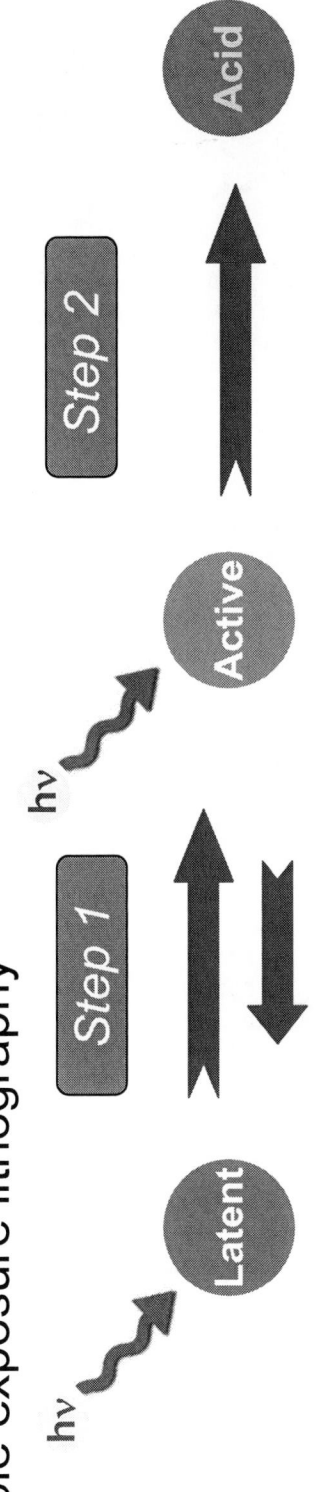

Latent

hν

Step 1

hν

Active

Step 2

Acid

Requirement 1:

Photons absorbed sequentially, only generate acid upon absorption of 2nd photon

Requirement 2:

Intermediate species must reverse or decay to maintain nonlinearity

SEMATECH

Simplification of System Requirements

Difficult to incorporate both requirements into one molecule.
Solution: Decouple functions.
Use intermediate step like electron transfer to sensitize PAG

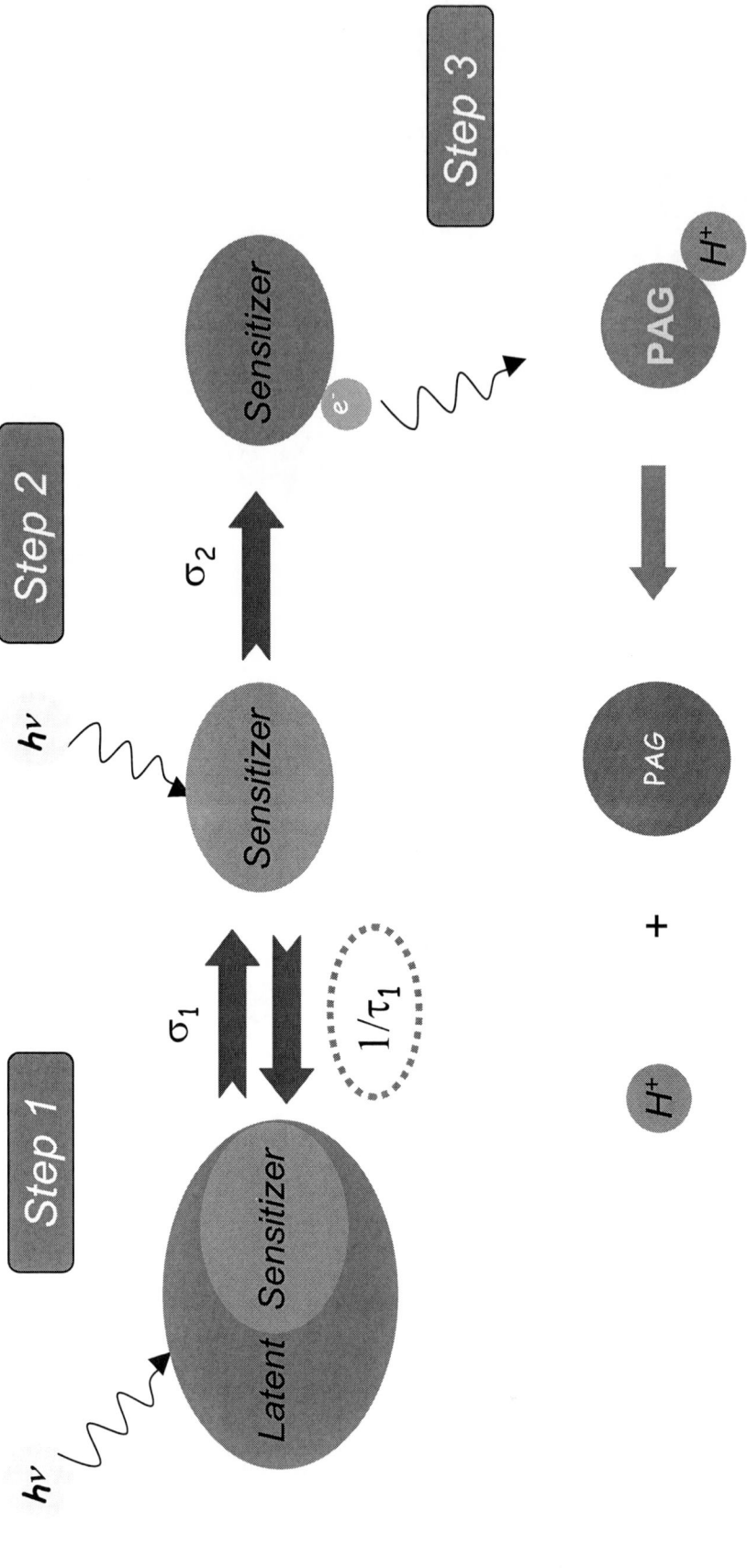

Naphthalene Cycloadduct as Model Compound

- Photoproduct from 4 + 4 cycloaddition of naphthalene can be used as the latent sensitizer.

- 2-Methoxynaphthalene chromophore has high enough reduction potential to react with acid generators.

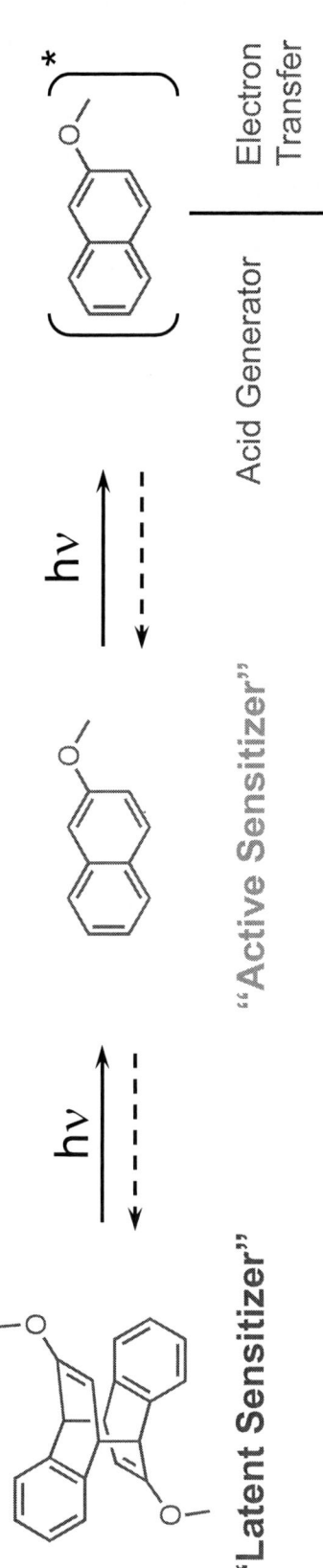

"Latent Sensitizer"

$h\nu$

"Active Sensitizer"

$h\nu$

Electron Transfer

Acid Generator

Acid

Acid Generators

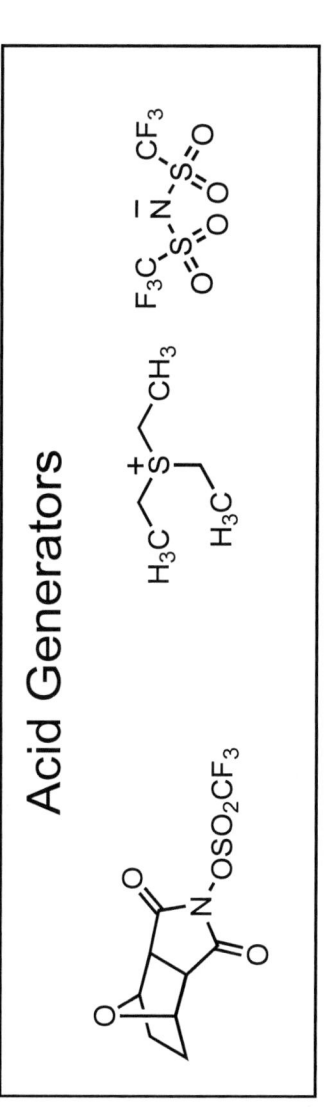

Tethering Increases Rate of Cycloaddition

Cycloaddition occurs efficiently in naphthalenes with a three-member tether.

DNP

MeO-DNP

DNME

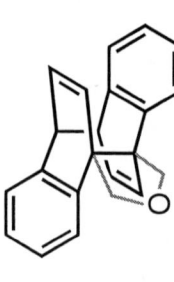

exo-cDNME

endo-cDNME

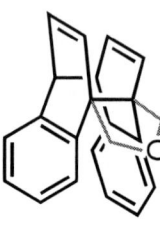

cMeO-DNP

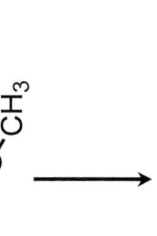

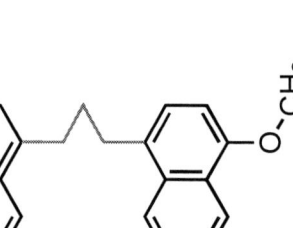

cDNP

Reversible Cycloaddition / Ring Opening

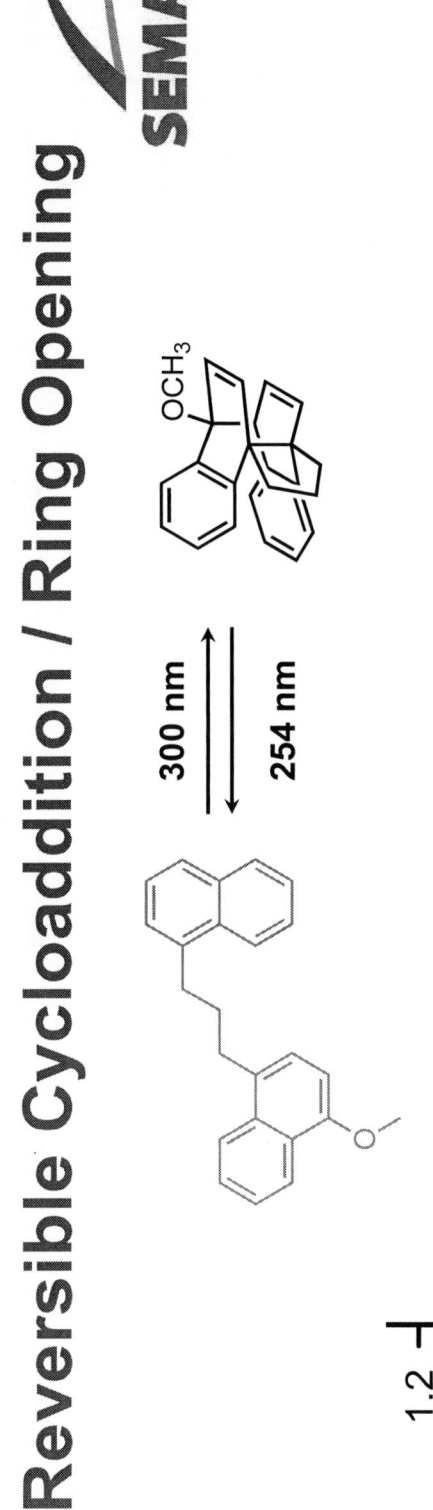

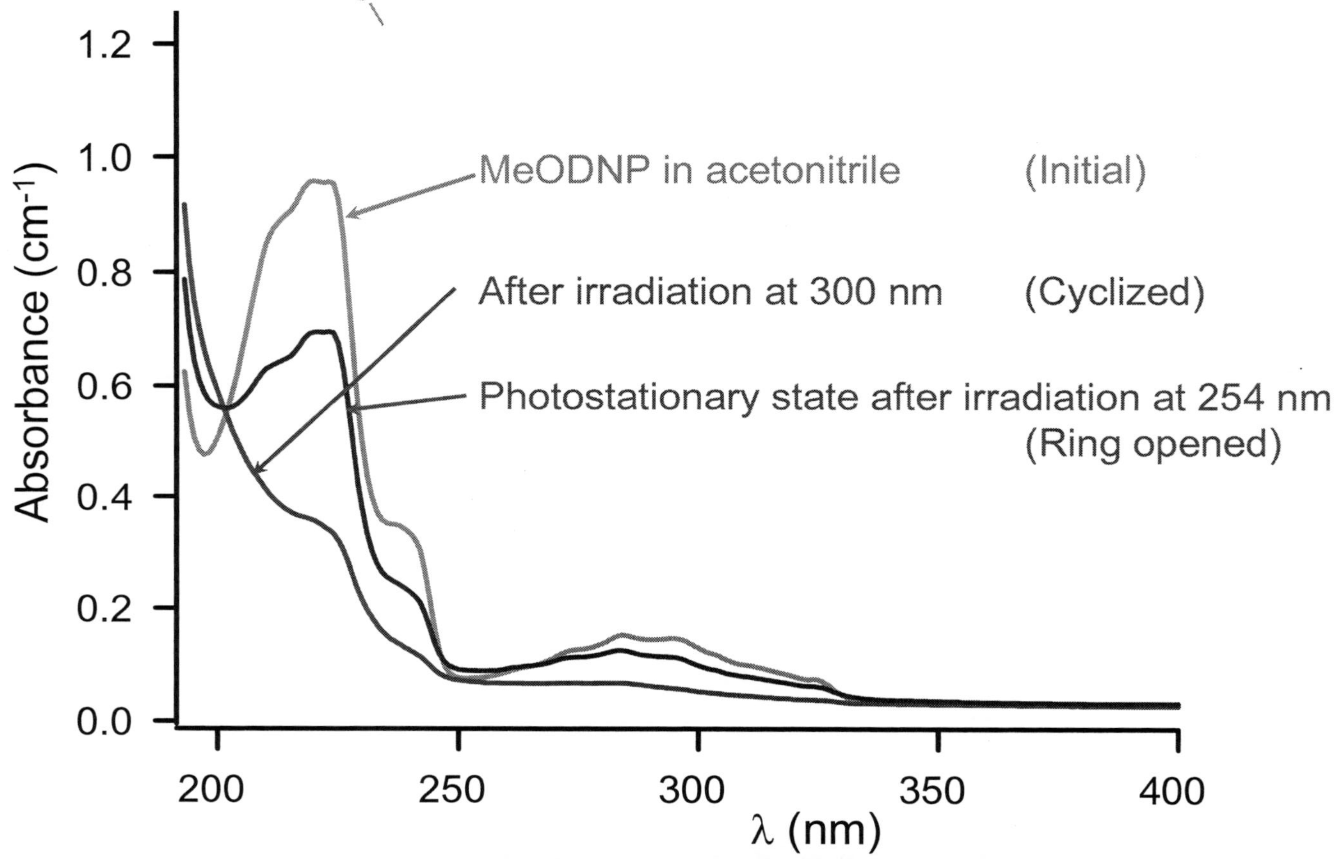

SEMATECH

300 nm
254 nm

MeODNP in acetonitrile   (Initial)

After irradiation at 300 nm   (Cyclized)

Photostationary state after irradiation at 254 nm
(Ring opened)

Sensitization of Acid Generator

R = H or OMe

$h\nu$
> 300 nm

Acid Released

Fluorescein as acid sensor

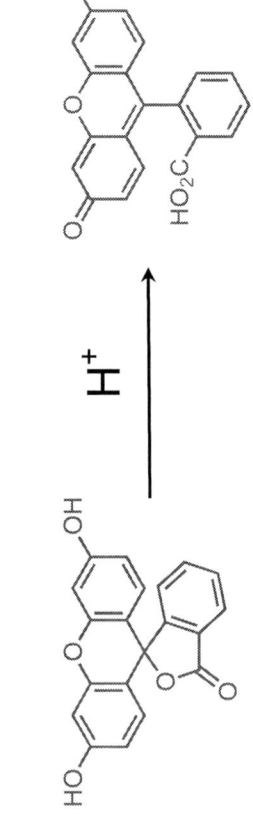

H^+

Absorption @ 450 nm

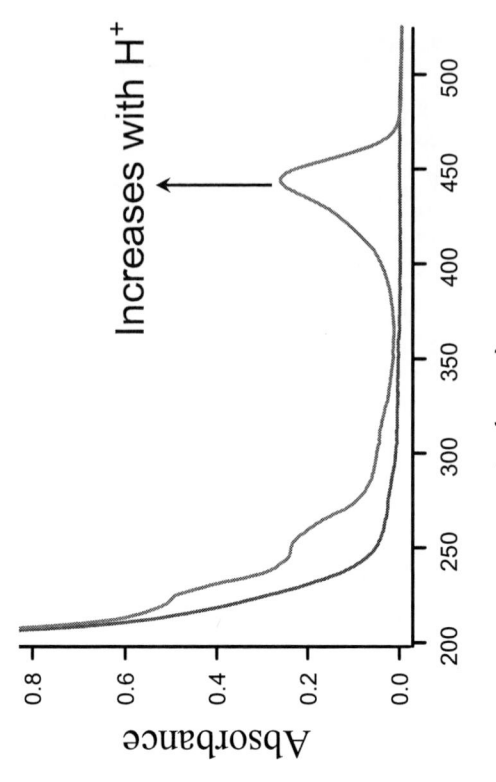

Increases with H^+

Representation of ISTP as Energy Diagram

Ring opening (*Step 1*), excitation (*Step 2*), and sensitization by electron transfer (*Step 3*) reactions to generate acid

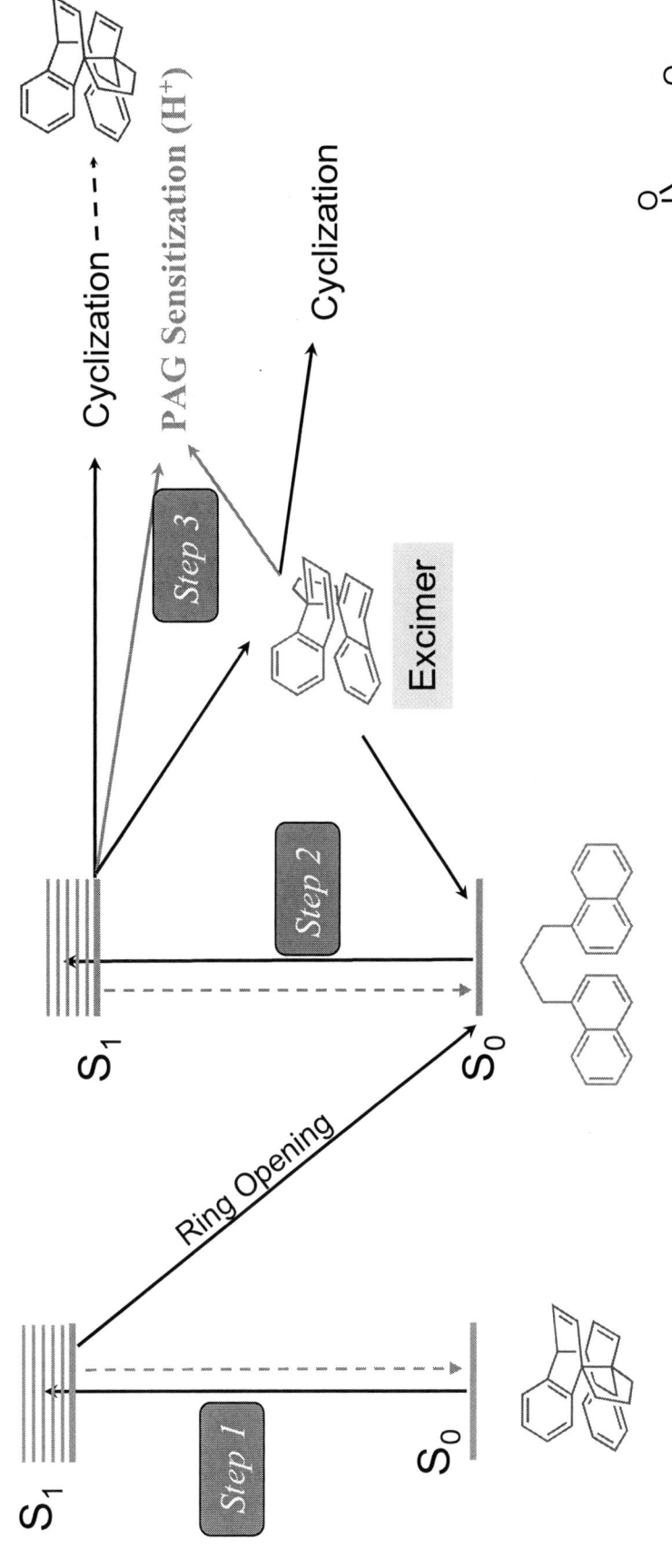

Quencher:

Stern Volmer Constant: $k_Q \sim 2 \times 10^9 \ M^{-1}s^{-1}$

Coumarin-6 as Acid Sensor in Polymer Films

Absorption and emission of protonated form are red-shifted

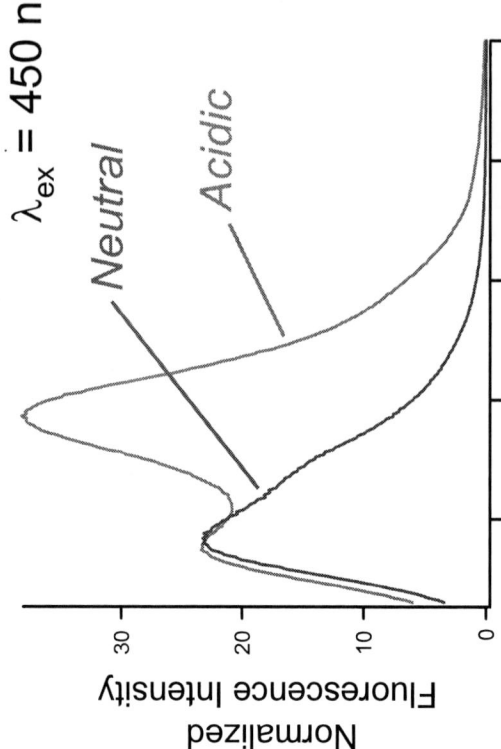

Protonated (C6-H^+)

H^+

Coumarin-6 (C6)

Emission in PMMA Film

λ_{ex} = 450 nm

Neutral

Acidic

Normalized Fluorescence Intensity

λ (nm)

Absorption (in CD_3CN)

Acidic

Neutral

Absorbance

λ (nm)

Mason, et al., J. Phys. Chem. B, 2003, 107, 4219–14224

Sensitization of TES-6 by DNP in Polymer Films

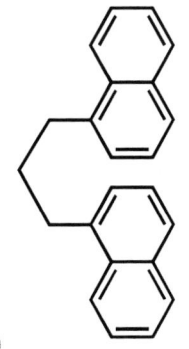

+ $Et_3S^+ (Tf)_2^-$

$$\xrightarrow[300 \text{ nm}]{\text{Polymer, C6}}$$

Acid Released

- Irradiation of DNP - TES-6 mixture in polymer film led to protonation of C6.

- However, open and cycloadducts generate acid (Adiabatic photochemistry?)

Acid detected by protonation of C6

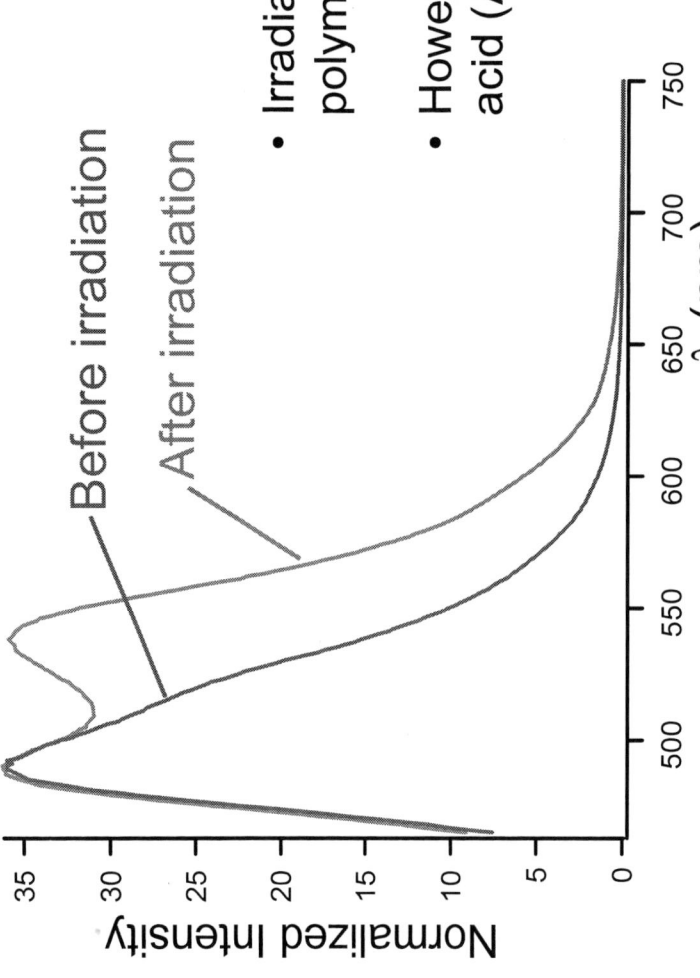

Before irradiation

After irradiation

λ (nm)

Normalized Intensity

Diabatic vs. Adiabatic Photoreactions

SEMATECH

<u>Diabatic Reaction</u>

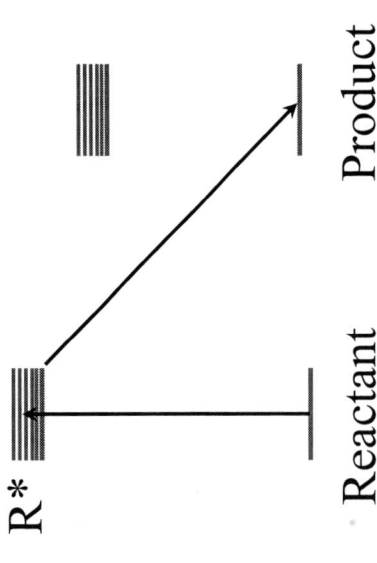

R*

Reactant   Product

- Primary photoproduct formed in ground state.

- Common with most photochemical reactions.

<u>Adiabatic Reaction</u>

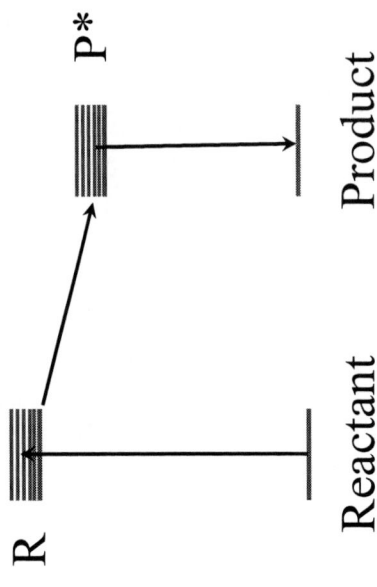

R   P*

Reactant   Product

- Primary photoproduct formed in excited state.

- Rarely observed. However, some cycloreversions occur adiabatically.

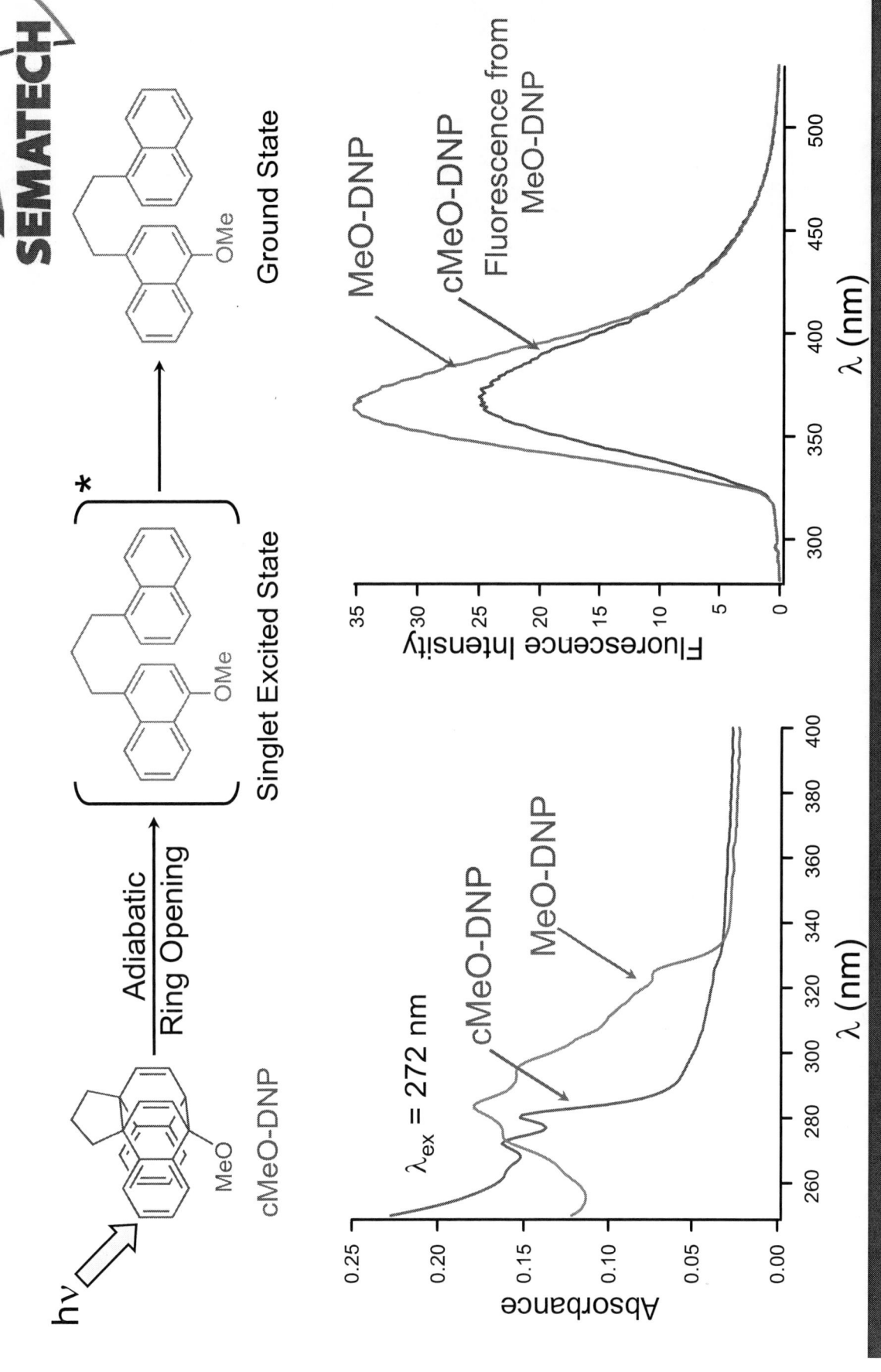

Effect of Adiabatic Reaction on Sensitization

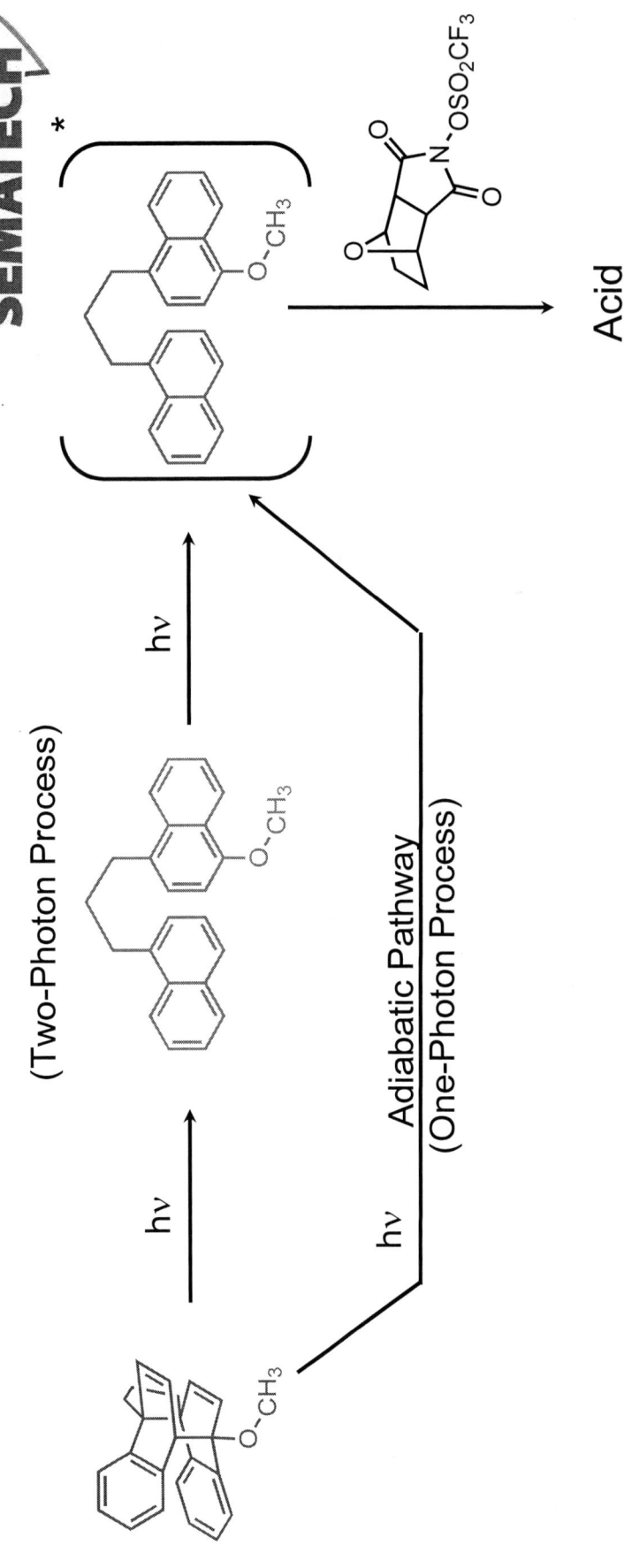

- Adiabatic ring opening translates sensitization into one photon process (loss of nonlinear dose – response).

- Extent of adiabatic ring opening must be reduced to use tethered naphthalenes for double exposure lithography.

Observed Sensitization by Adiabatic Pathway

Adiabatic ring opening followed by PAG sensitization (red arrows) occurs efficiently after absorption of the first photon.

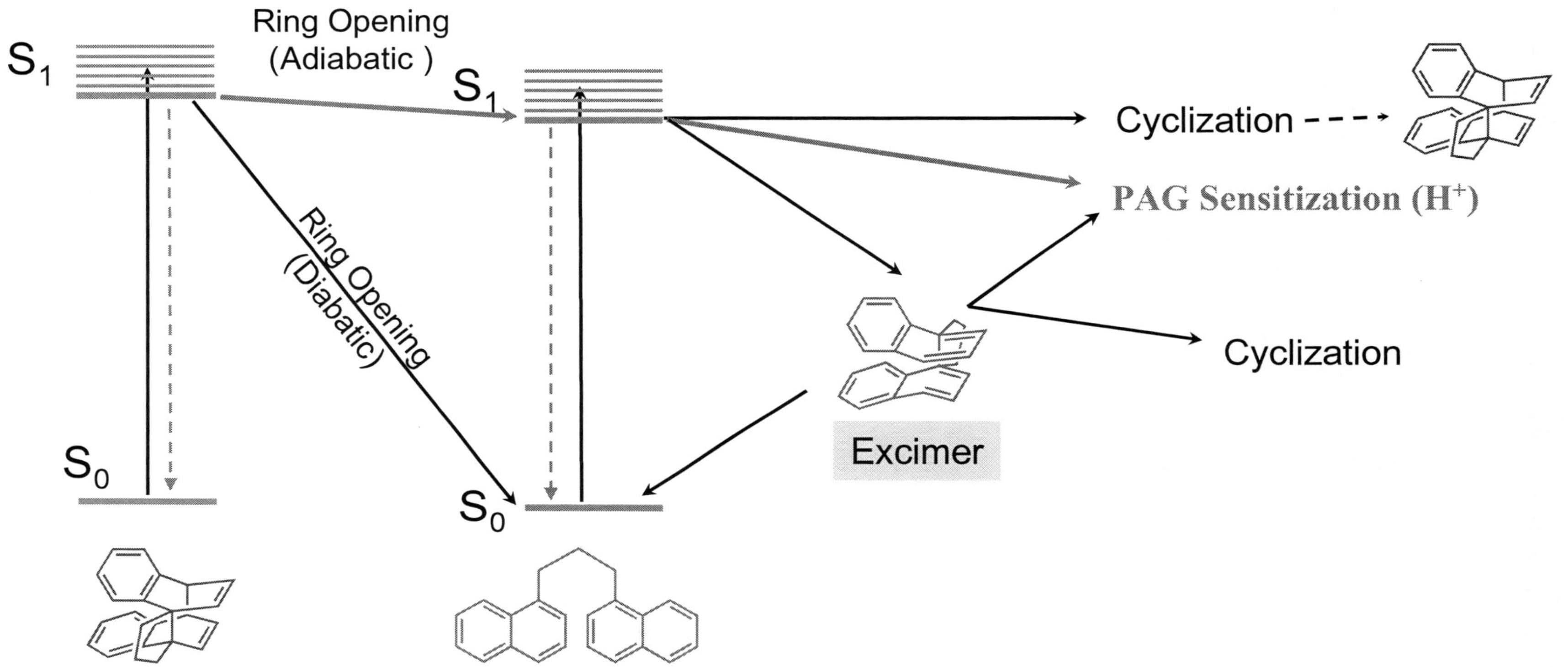

Efficiency of Adiabatic Cycloreversion

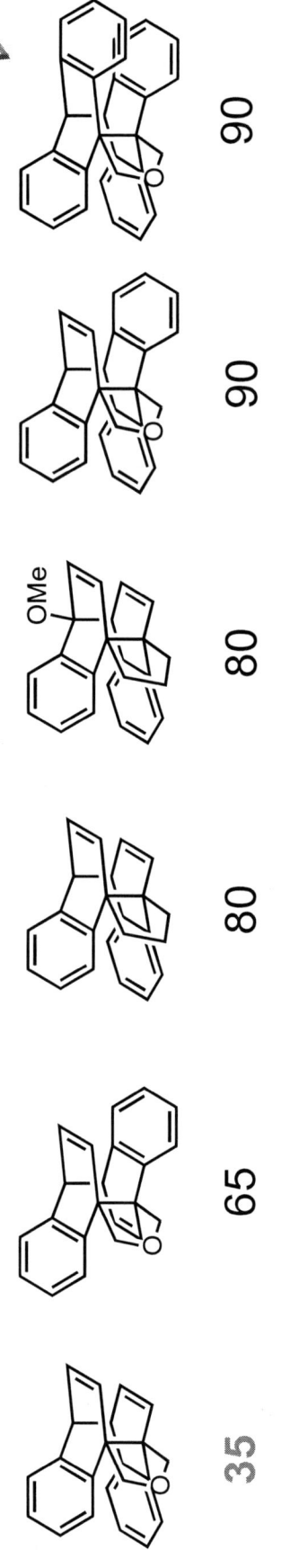

- Among naphthalene cycloadducts studied, all but one were found to revert efficiently to naphthalene by an adiabatic pathway.

- Similar adiabatic cycloreversion was also observed with naphthalene - anthracene and anthracene - anthracene cycloadducts.

- Only with *endo*-cDNME (and its Cope-rearrangement product) was the diabatic contribution to cycloreversion > 60%.

- Need to identify cycloadducts wherein cycloreversion is primarily diabatic.

Summary

❖ Tethered naphthalene derivatives are promising candidates for DE Lithography.

❖ Efficient photocycloaddition and cycloreversion occur in the two substrates.

❖ Both DNP and MeO-DNP successfully sensitize the acid generator to produce acid.

❖ Preliminary studies show adiabatic ring opening of the cycloadducts can be a potential limitation towards the application of tethered naphthalenes for double exposure lithography.

Acknowledgements

☐ The authors thank SEMATECH for financial support and Katherine Esswein, Ke Min Robert Bristol and James Blackwell at Intel Corporation for technical help.

COLUMBIA UNIVERSITY
IN THE CITY OF NEW YORK

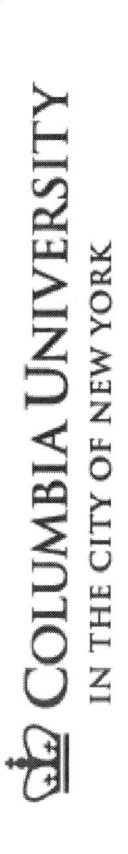

Alternatives to Chemical Amplification for Sub-32nm Photoresists

Bruce Smith, Thomas Smith, Burak Baylav, Meng Zhao, Ran Yin, Peng Xie, Christopher Scholz, and Paul Zimmerman*

Rochester Institute of Technology
*Intel assignee to SEMATECH

Copyright (c)2009 SEMATECH, Inc. SEMATECH, Inc. SEMATECH, and the SEMATECH logo are registered servicemarks of SEMATECH, Inc. International SEMATECH Manufacturing Initiative, ISMI, Advanced Materials Research Center and AMRC are servicemarks of SEMATECH, Inc. All other servicemarks and trademarks are the property of their respective owners.

Non-CAR and Low-CAR Goals

- A program is underway to explore alternative thin film (<60nm) resists for low LER and high resolution with some sacrifice to sensitivity
- Lower resist sensitivity may be tolerable for sub-32nm
  - Higher power excimers
  - More CaF_2 in projections lens near pupil and image planes
- With more photons available, resist design opportunities open
- Various new and old platforms become feasible

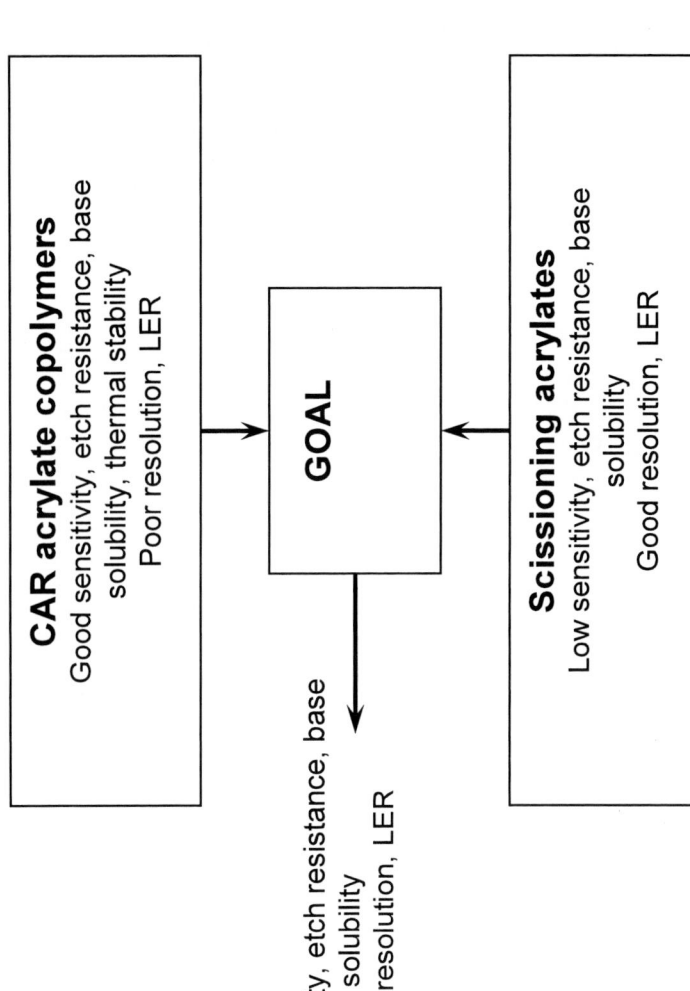

CAR acrylate copolymers
Good sensitivity, etch resistance, base solubility, thermal stability
Poor resolution, LER

GOAL

Scissioning acrylates
Low sensitivity, etch resistance, base solubility
Good resolution, LER

Good sensitivity, etch resistance, base solubility
Good resolution, LER

Pathways Considered

Dissolution inhibitors (DI)

1. DNQs
2. o-nitrobenzyl chemistry
3. Poly sulfones
4. Others

Scissioning enhancement

1. α-CH_3
2. α-halides
3. Others

Polymeric systems

1. Acrylates and derivatives (MMA, MAA, TBMA, TBCA, MAN, others)
2. Polystyrenes and derivatives (AMS, TBMS, Photo-Fries, PMIPK, others)
3. Poly sulfones and derivatives (UQ)

- Scissioning with improved sensitivity through co-polymerization
- Improvements in scissioning yield through substitution
- CAR combined with scissioning through co-polymerization
- Low- or no-chemical amplification deprotection (Φ<1)
- Dissolution inhibition systems
- Direct crosslinking or hybrid CAR negative resists

Dissolution Inhibitor Approach

- **Bile acid esters:** o-nitrobenzyl cholate (Reichmanis *et al.*)

  (R' is cholic acid derivative)

- **Diazonapthoquinone (DNQ):**

- **Poly olefin sulfones [Poly (1-pentene sulfone)]**

Cholate Ester Dissolution Inhibitors

Structure of o-nitrobenzyl cholate

Functionalization:

o-nitrobenzyl cholate has 3 hydroxyl groups that can be substituted with trimethylsiyl, pivaloyl, or acetyl groups. Higher contrast is possible with o,o,o tripivaloyl-o-nitrobenzyl cholate. Esterfication of the three hydroxyl groups results in less dark loss (less than 5%).

- No gaseous byproducts
- High alicyclic carbon content (RIE resistance)
- Bulky group (large fraction of resist matrix undergoes photochemical change upon radiation)
- Functionality due to the 3 hydroxyl groups on the cholate (can add substituents to improve performance)

DI Contrast, Sensitivity, and Dark Loss

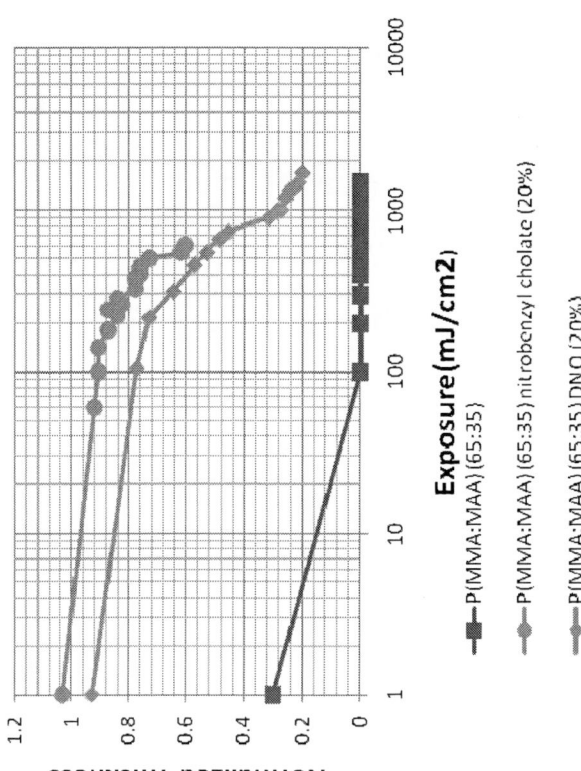

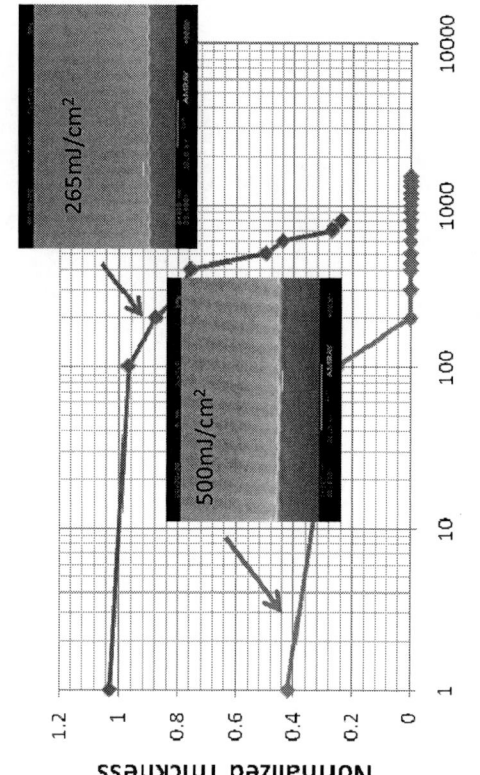

Both nitrobenzyl cholate and DNQ inhibit dissolution of P(MMA:MAA). The loading of each inhibitor can be increased with absorbance less than 10/µm.

DI – Polymer Absorbance Data

Absorbance of nitrobenzyl cholate and DNQ loaded at 20% and 30% in copolymer of MMA and MAA and poly(norbornene/maleic anhydride/acrylic acid) matrices.

Maximum tolerable absorbance is ~8/μm. Loading of DNQ in the P(NB/MA/AA) system must be less than 30%. 193nm absorbance of nitrobenzyl cholate at 30% loading is less than 5/μm.

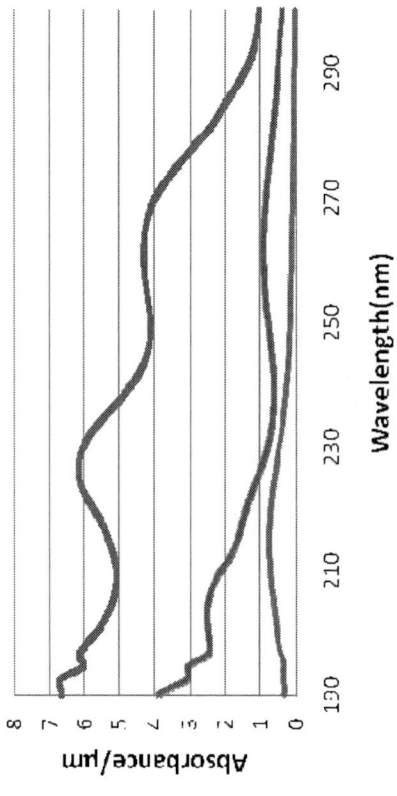

Absorbance/μm in P(MMA/MAA) Matrix

—— No inhibitor —— nitrobenzyl cholate (20%) —— diazonapthoquinone (20%)

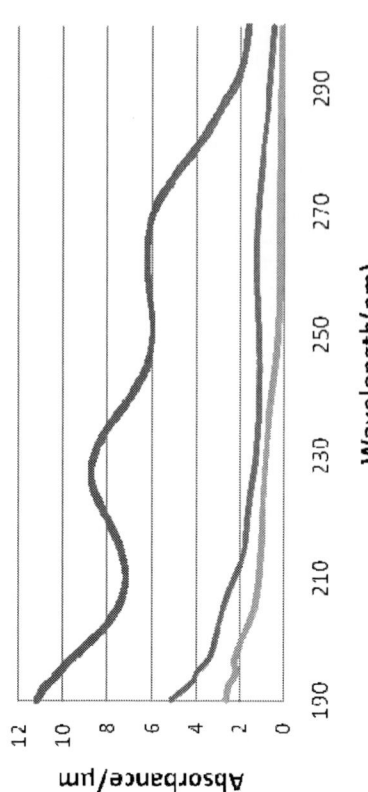

Absorbance/μm in P(NB/MA/AA) Matrix

—— No inhibitor —— nitrobenzyl cholate (30%) —— diazonapthoquinone (30%)

Optimization of DI Systems

1) Optimize Base Resin: for better resolution and higher RIE resistance

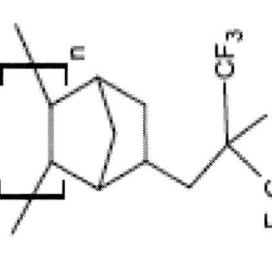

Poly(NB/MA/AA) with varying ratios of acrylic acid

Norbornene hexafluoro alcohol and copolymer derivatives

2) Optimize Inhibitor - variations on o-nbc:
- Tripivaloyl nitrobenzyl cholate (less dark loss, higher contrast)
- Dinitrobenzyl cholate (3X quantum yield at 254nm)

For this, nitrobenzyl cholate-type inhibitors are investigated in transparent, base developable resin matrices to give good resolution and etch resistance charecteristics.

Chain-Scission Pathway
α-MEST chain scissioning

- Main-chain scission has long been explored as a solution for LER problem.

- Components into the polymer backbone create an additional photo energy-absorbing point.

- Absorption data for PMMA and P(MMA-αMEST) (with αMEST 1% and 3% by weight) – photo-induced chain breakage.

PMMA-αMEST Absorbance

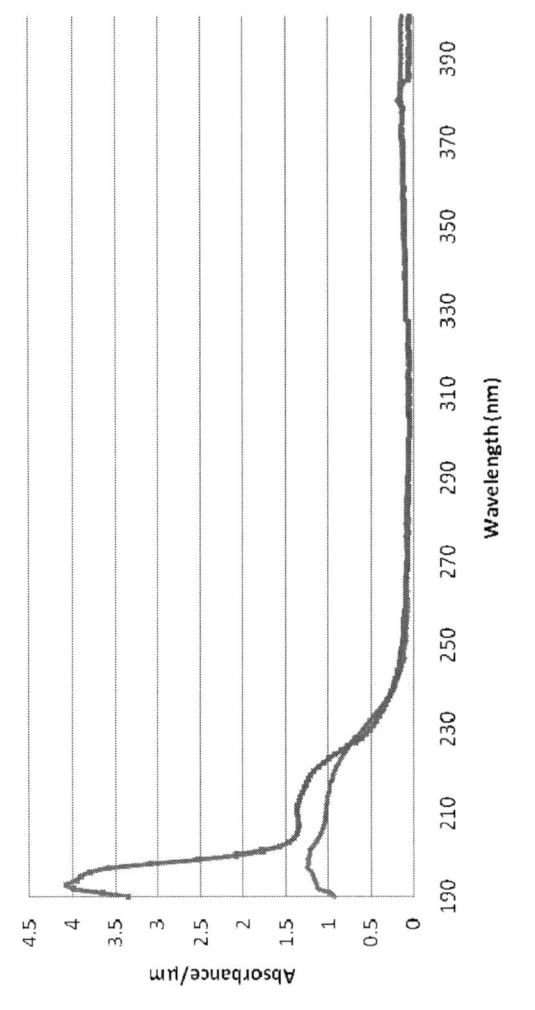

Chain-Scission Pathway
Improving imaging performance

- Thin film (50nm) sensitivity is increased 3X with 3% αMEST

Polymer	Mw	E_0(mJ/cm2)
PMMA	79K	1400
P(MMA- 3% aMEST)	77K	420

- Higher αMEST content may be feasible (7%) and within the acceptable absorption range.

- Copolymerization is required to address base-solubility (MAA, TBMA), etch resistance (norbornene and adamantane), dark loss, and image contrast.

- P(MMA-MAA) with 30%, 25%, 15% MAA were made and evaluated. Copolymer with 15% MAA was chosen considering to minimize dark loss.

Polymer	Mw (before exposure)	Mw after	Efficiency
P(MMA-MAA) 85:15	103k	1364	98.7%
P(MMA-TBMA-MAA) 42.5:42.5:15	132k	1300	99.02%

Chain-Scission Pathway
Contrast curve

- A four-component polymer of P(MMA-TBMA-MAA- αMEST) (39:39:15:7) was made and lithographically evaluated at 45nm.

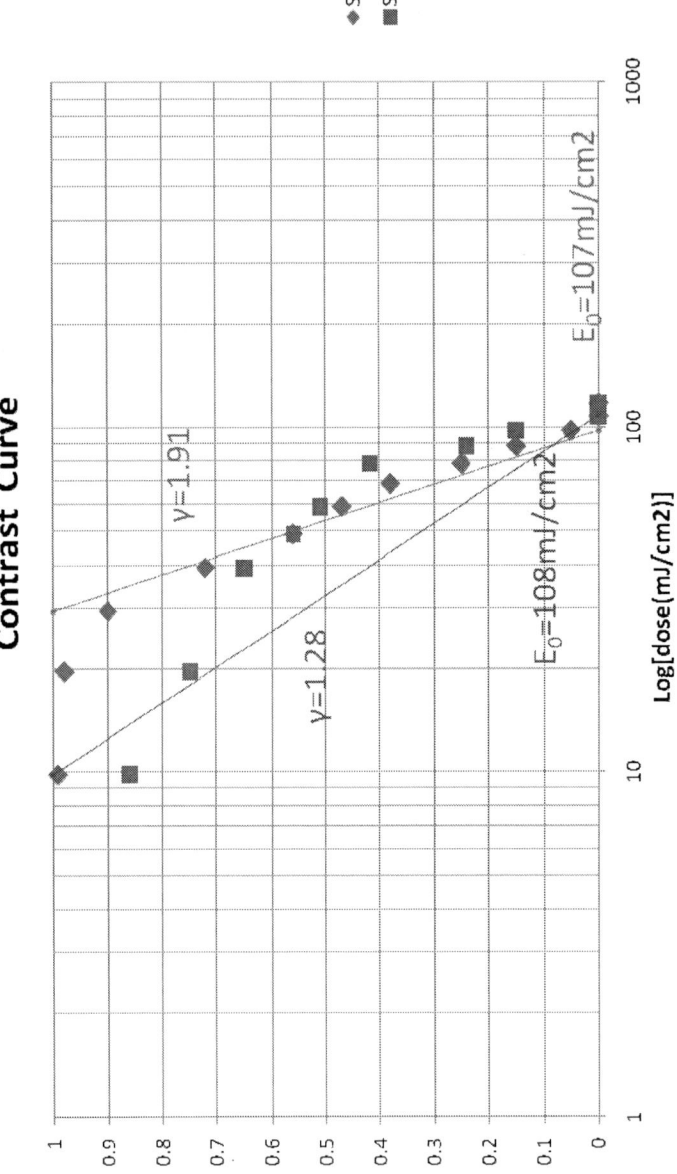

Chain-Scission Pathway
Contrast improvement

- A base-developable scissioning system using P(MMA-TBMA-MAA- αMEST) with 15% MAA and 7% αMEST shows sensitivity at 108mJ/cm^2 with contrast near 1.9 at 45nm film thickness.

- Methods to introduce etch resistance utilize adamantyl and norbornyl derivatives.

- Contrast of the current approach is limited by MMA. Alternative polymer system is investigated - poly(AM-TBMA-MAA-αMEST).

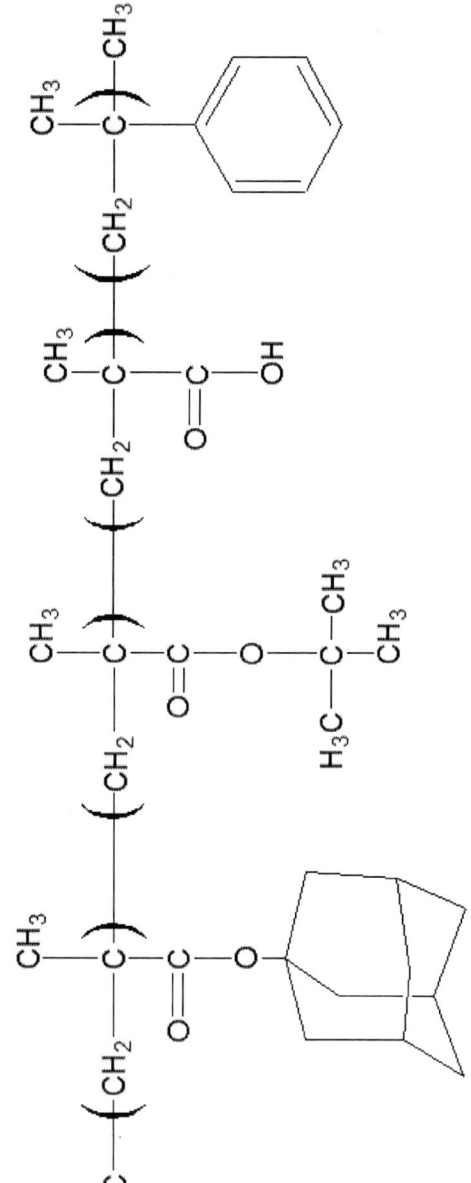

Photo-Fries Pathway

Transformation of FOxS to HO-Sty

The absorbance of the poly(formyloxystyrene), PFOxS, homopolmyer is significant at 193nm but optimization of design parameters for copolymeric resist systems offers potential.

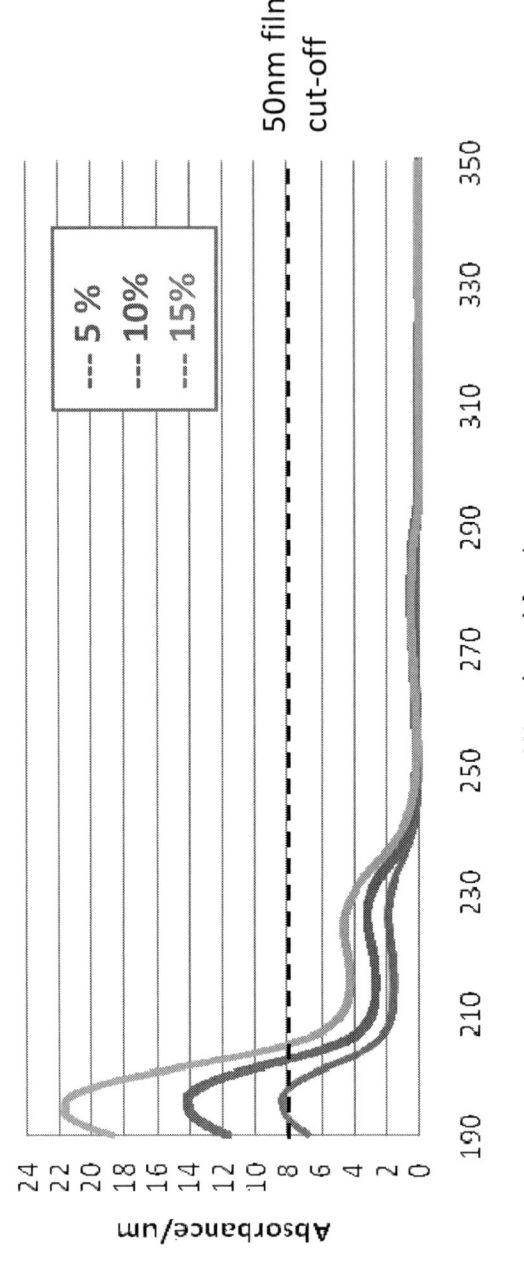

Methyl methacrylate (MMA) and methylacrylic acid (MAA) polymers containing 5, 10, and 15 mol% of FOxS were synthesized at MAA levels to set a threshold for base solubility (10-20 mol%). Sensitivity of 10 mol% MAA and 5 mol% FOxS copolymer is <250 mJ/ cm^2.

Photo-Fries Pathway

Sensitivity, base solubility, and etch resistance

- Three-component copolymers of norbornene, hexafluoroisopropylmethyl norbornene, and norbornenes having phenylformate and phenylacetate substituents are being explored.

  - The threshold for base solubility will be set by the level of hexafluoroisopropylmethyl norbornene.

  - The level of the phenyl formate or phenylacetate substituted norbornene will provide the differential solubility for image development.

4-(bicyclo[2.2.1]hept-5-en-2-yl) phenylformate and 4-(bicyclo[2.2.1]hept-5-en-2-yl) phenylacetate have been synthesized by a Diels-Alder reaction between dicyclopentadiene and formyloxystyrene and acetoxystyrene.

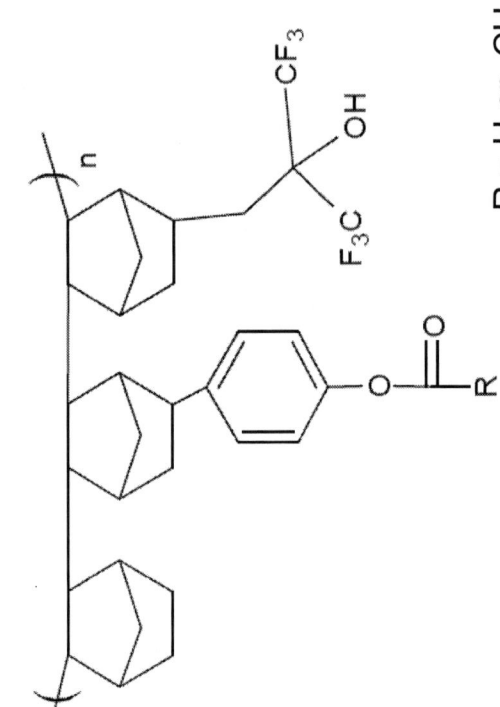

$R = H$ or $-CH_3$

Conclusions

Thin film non-CAR or low-CAR resists may offer alternatives to current 193nm systems. Several avenues are being pursued:

- <u>Scissioning Pathway</u> - Enhancement of chain-scissioning acrylates is possible but contrast remains limited. Etch resistance and image improvement is addressed with adamantly and norbornene groups.

- <u>Dissolution Inhibition Pathway</u> – Moderate concentrations of DIs (<30%) allow for adequate transmission ($\alpha < 8\ \mu m^{-1}$) and the enhancement of sensitivity and base solubility of acrylic polymers. Nitrobenzyl cholates are promising candidates.

- <u>Photo-Fried Pathway</u> – Small levels of FOxS incorporated into resist polymers allows for adequate transmission and photoconversion to base solubility. Sensitivity <250 mJ/cm^2 is possible. Norbornene copolymerization is being explored.

High Fluence Testing of Optical Materials for 193-nm Lithography Extensions Applications

P. A. Zimmerman

Intel/SEMATECH, Austin, TX 78741

V. Liberman, S. Palmacci, G. P. Geurtsen, M. Rothschild

Lincoln Laboratory, Massachusetts Institute of Technology

Lexington, MA 02420

October 22, 2009

The Lincoln Laboratory portion of this work was sponsored by a Cooperative Research and Development Agreement between Lincoln Laboratory and SEMATECH, Inc. and the Research Foundation of State University of New York. Opinions, interpretations, conclusions, and recommendations are those of the authors and are not necessarily endorsed by the United States Government.

Copyright ©2008
SEMATECH, Inc. SEMATECH, and the SEMATECH logo are registered servicemarks of SEMATECH, Inc. International SEMATECH Manufacturing Initiative, ISMI, Advanced Materials Research Center and AMRC are servicemarks of SEMATECH, Inc. All other servicemarks and trademarks are the property of their respective owners.

HiFluence 10/29/2009

MIT Lincoln Laboratory

Outline

- **Study goals and logistics**
- **Experimental approach**
- **Materials test results**
  - Bulk SiO_2
  - Bulk CaF_2
  - Thin film coatings
- **Summary and outlook**

MIT Lincoln Laboratory

HiFluence 10/29/2009

Motivation

- **Optical materials in an immersion lithography tool are subjected to ever-increasing laser doses and pulse fluence**
  - Illuminator
    - Incident laser power >300 W at 6 kHz expected
  - Lens
    - Move to non-CA resists contemplated
    - ⇒ Fluence per pulse may increase up to 10x

- **Durability of *all* components of optical train should be verified for high fluence *and* large pulse counts**
  - Bulk materials and thin films
  - Fluence ≥ 50 mJ/cm^2/pulse and pulse counts of ≈10^9 shots

MIT Lincoln Laboratory

HiFluence 10/29/2009

Study Logistics

- **Optical materials from leading suppliers of lithographic equipment are being tested**
  - 2 suppliers of bulk SiO_2
  - 2 suppliers of bulk CaF_2
  - 3 suppliers of thin film coatings
    - On both SiO_2 and CaF_2 substrates

- **Results are fully shared with individual suppliers**
  - Feedback for assessment of different grades
  - Drive process improvement

- **Coded results presented in public forum**
  - Technology assessment for the industry

MIT Lincoln Laboratory

HiFluence 10/29/2009

Experimental Techniques

- **In-situ laser irradiation**
  - 4000 Hz laser, 11 ns FWHM / 22 ns TIS
  - Focused down to 2 mm x 2 mm spot, 30 – 50 mJ/cm^2/pulse
  - Up to 700 Mp accumulated per sample

- **In-situ diagnostics**
  - 193-nm laser-based ratiometric transmission
  - Laser-induced fluorescence (LIF)
    Sensitive to material composition and impurities/defects

- **Ex-situ diagnostics**
  - UV-Vis Spectrophotometry
  - Spatio-spectral ellipsometry maps
    For thin films
  - Spatial birefringence maps
    For bulk materials
  - Atomic Force Microscopy (AFM)

MIT Lincoln Laboratory

HiFluence 10/29/2009

Irradiation Geometry

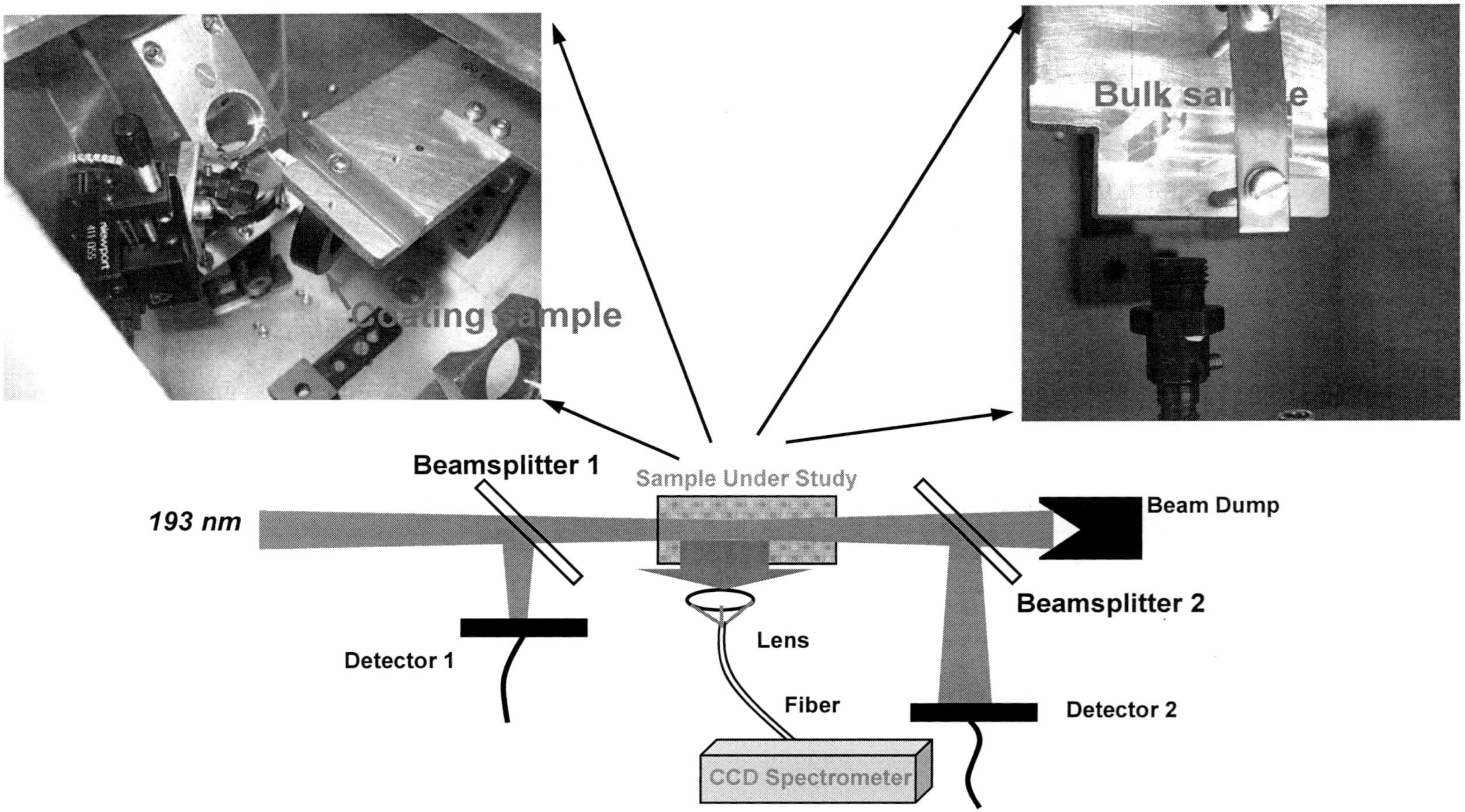

Bulk sample

Beam Dump

Beamsplitter 2

Detector 2

Sample Under Study

Lens

Fiber

CCD Spectrometer

Beamsplitter 1

Detector 1

193 nm

Experiment enclosed in a N$_2$-purged chamber

MIT Lincoln Laboratory

HiFluence 10/29/2009

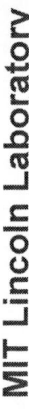

Bulk SiO$_2$
4 different grades exposed

Pre- and Post-measurement For Two Samples

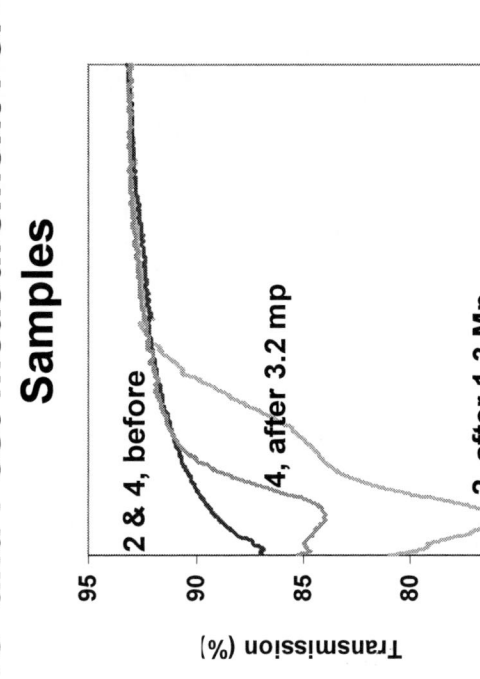

In-situ, 193 nm

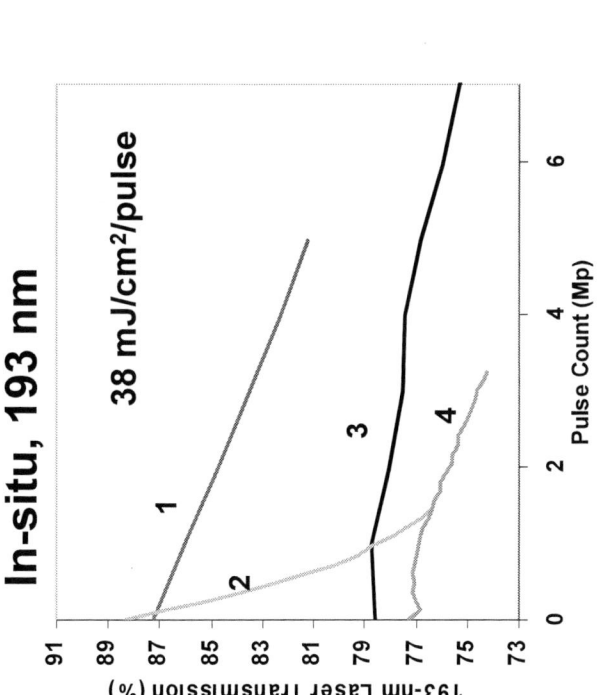

- **From in-situ data**
  - Anomalously low starting transmission in samples 3 and 4
- **Combining in-situ data with pre- and post-measurements**
  - Sample 4 recovers after irradiation
  - May exhibit rapid *transient* absorption with 193-nm laser
    - Even before/during 1st transmission measurement

MIT Lincoln Laboratory

HiFluence 10/29/2009

Laser Induced Fluorescence of SiO$_2$

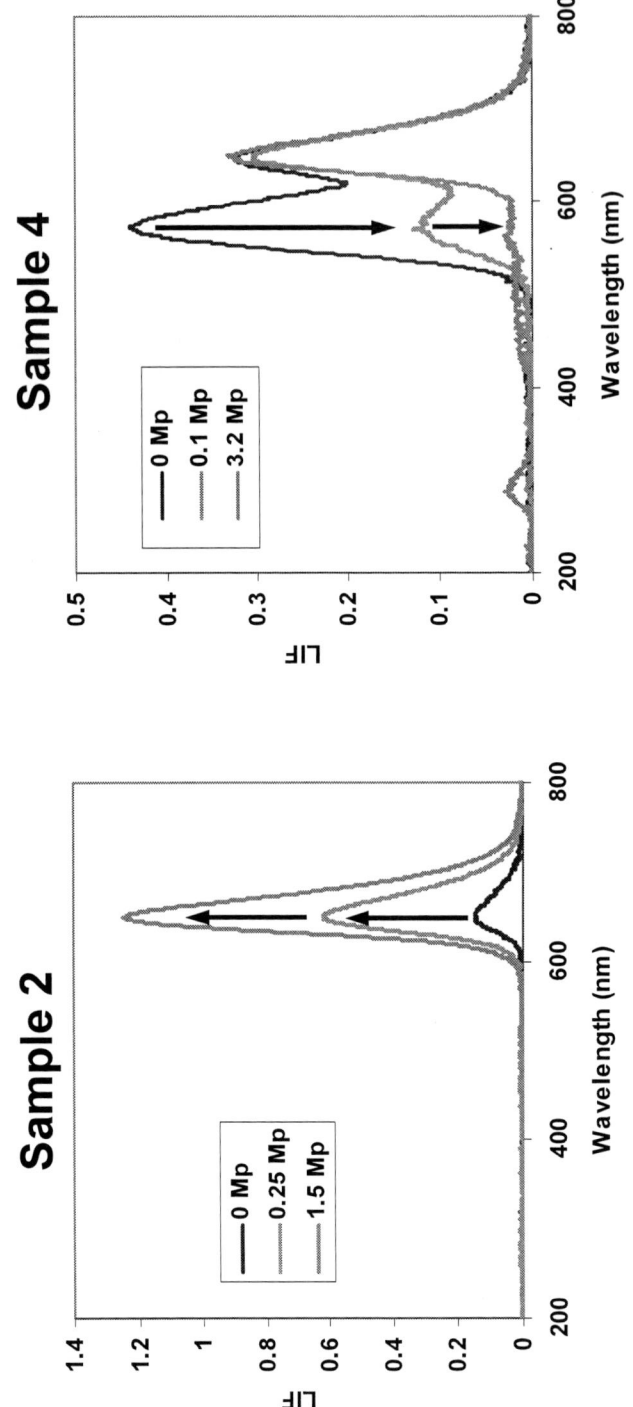

- **Complicated fluorescence behavior with increasing dose**
- **Evolution of peaks related to creation and destruction of various defect centers in the material**

MIT Lincoln Laboratory

HiFluence 10/29/2009

Laser-Induced Birefringence in SiO$_2$

Sample 2
Induced Birefringence < 1 nm

Sample 4
Induced Birefringence > 4 nm

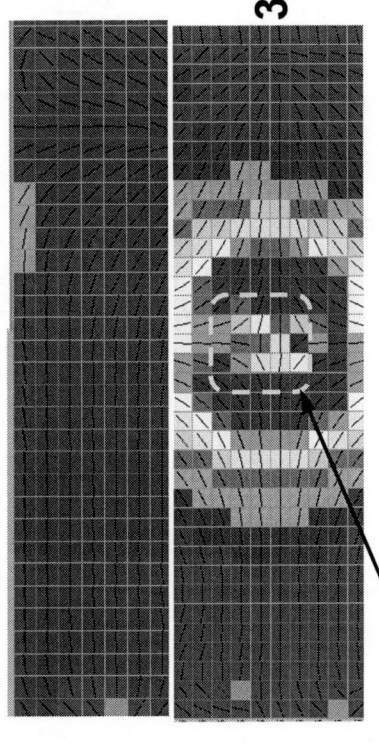

0 Mp

3.3 Mp

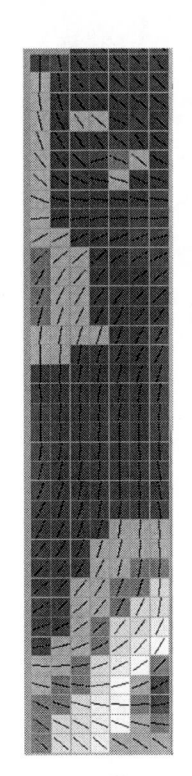

0 Mp

1.5 Mp

Laser Spot Location

- Induced birefringence may be grade-dependent
  - *Sample 4* received 2.2 x the dose of *Sample 2*, but developed >4x birefringence
- Not related to transmission degradation
  - *Sample 2* shows >2x induced absorbance over *Sample 4* but no detectable induced birefringence

MIT Lincoln Laboratory

HiFluence 10/29/2009

Laser-Induced Birefringence in SiO$_2$

SEMATECH

Sample 2
Induced Birefringence < 1 nm

0 Mp

Legend [nm]

- 0.000 - 0.799
- 0.800 - 1.374
- 1.375 - 1.949
- 1.950 - 2.524
- 2.525 - 3.099
- 3.100 - 3.674
- 3.675 - 4.249
- 4.250 - 4.824
- 4.825 - 5.399

1.5 Mp

Legend [nm]

- 0.000 - 0.799
- 0.800 - 1.487
- 1.488 - 2.174
- 2.175 - 2.862
- 2.863 - 3.549
- 3.550 - 4.237
- 4.238 - 4.924
- 4.925 - 5.612
- 5.613 - 6.299
- 6.300 - 100.000

Sample 4
Induced Birefringence > 4 nm

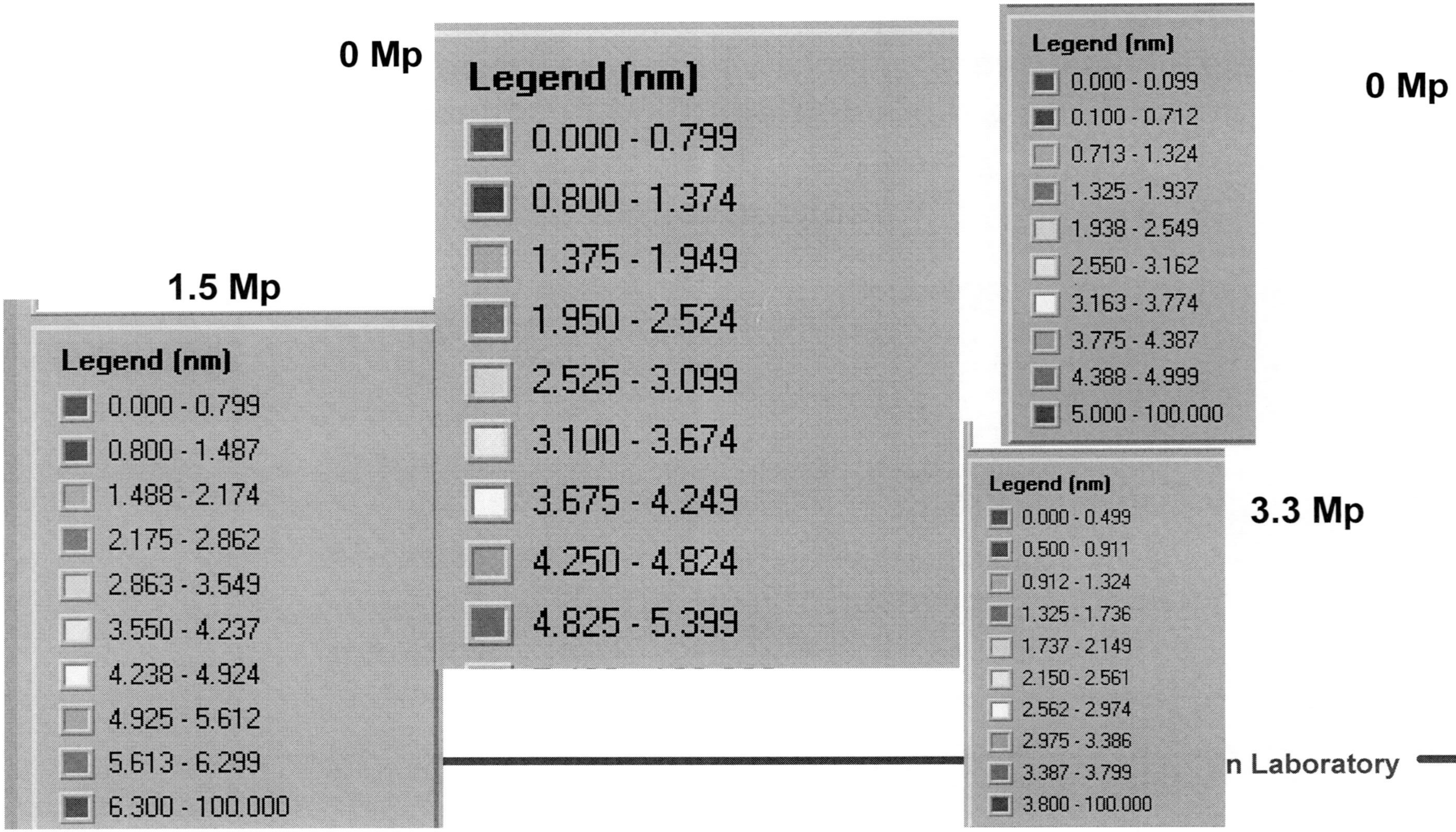

0 Mp

Legend (nm)

- 0.000 - 0.099
- 0.100 - 0.712
- 0.713 - 1.324
- 1.325 - 1.937
- 1.938 - 2.549
- 2.550 - 3.162
- 3.163 - 3.774
- 3.775 - 4.387
- 4.388 - 4.999
- 5.000 - 100.000

3.3 Mp

Legend (nm)

- 0.000 - 0.499
- 0.500 - 0.911
- 0.912 - 1.324
- 1.325 - 1.736
- 1.737 - 2.149
- 2.150 - 2.561
- 2.562 - 2.974
- 2.975 - 3.386
- 3.387 - 3.799
- 3.800 - 100.000

n Laboratory

Bulk CaF₂

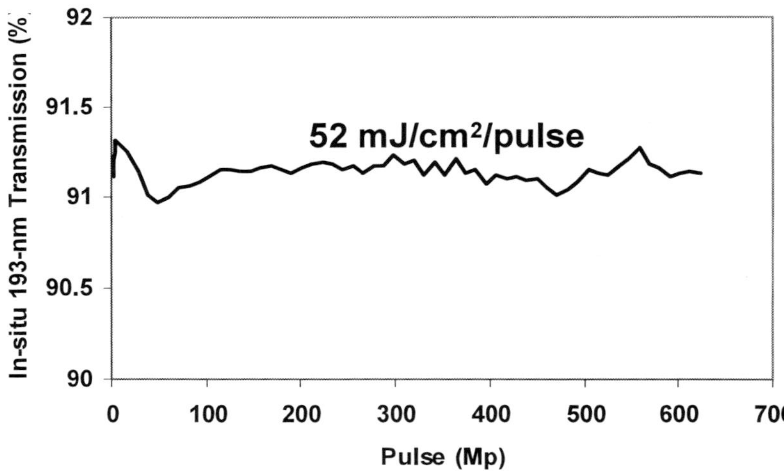

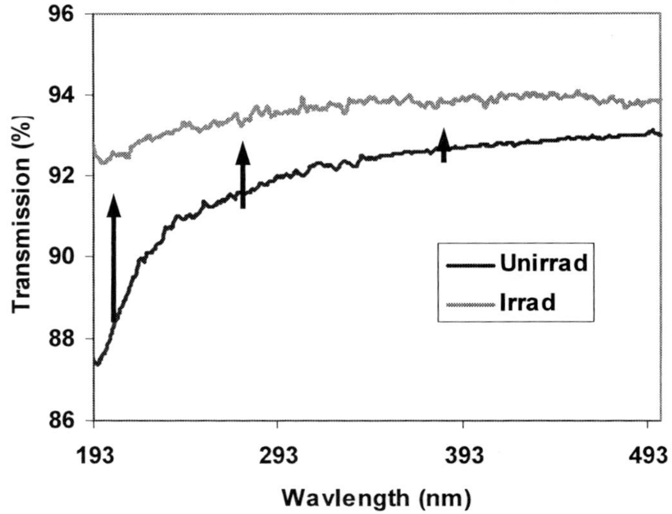

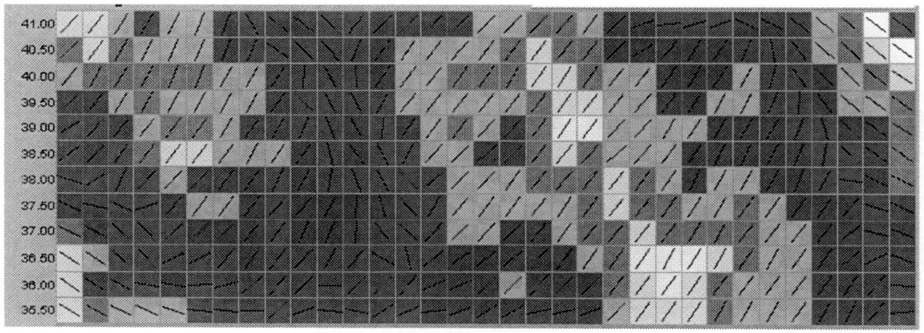

Legend [nm]
- 0.000 - 0.399
- 0.400 - 0.737
- 0.738 - 1.074
- 1.075 - 1.412
- 1.413 - 1.749
- 1.750 - 2.087
- 2.087 - 2.424
- 2.425 - 2.762
- 2.763 - 3.099
- 3.100 - 100.000

650 Mp

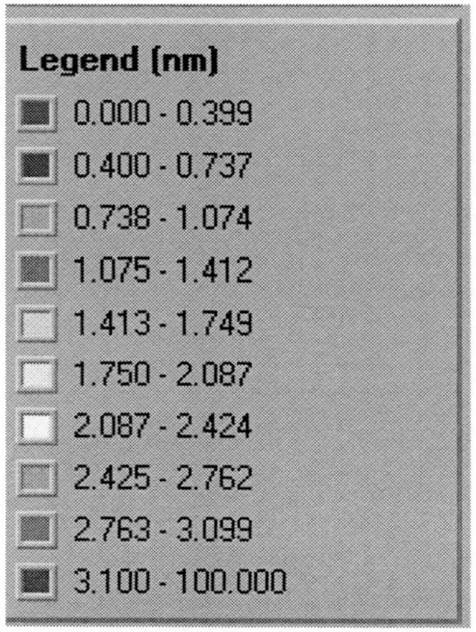

- No transmission degradation
  - Laser-induced cleaning
- No induced birefringence

MIT Lincoln Laboratory

HiFluence 10/29/2009

Bulk CaF$_2$
Laser Induced Fluorescence

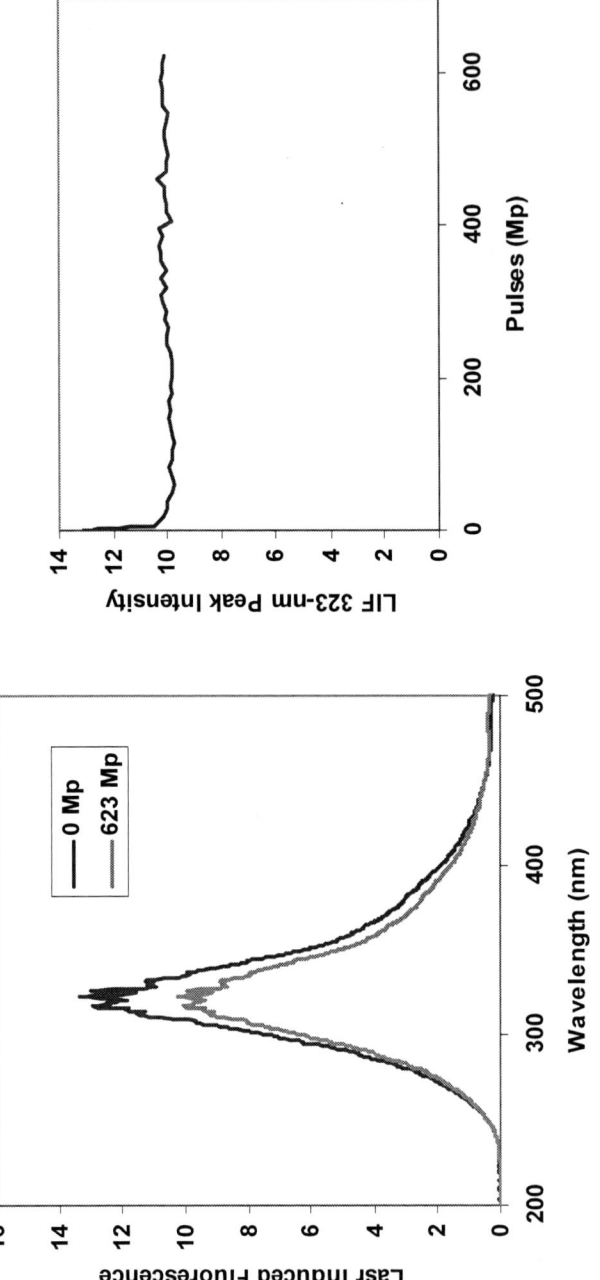

- ## LIF changes may indicate subtle annealing effect, but no major spectral changes

MIT Lincoln Laboratory

HiFluence 10/29/2009

Thin Film Coatings– Examples
(CaF$_2$ substrate)

In-situ 193 nm data
4 different coating stacks

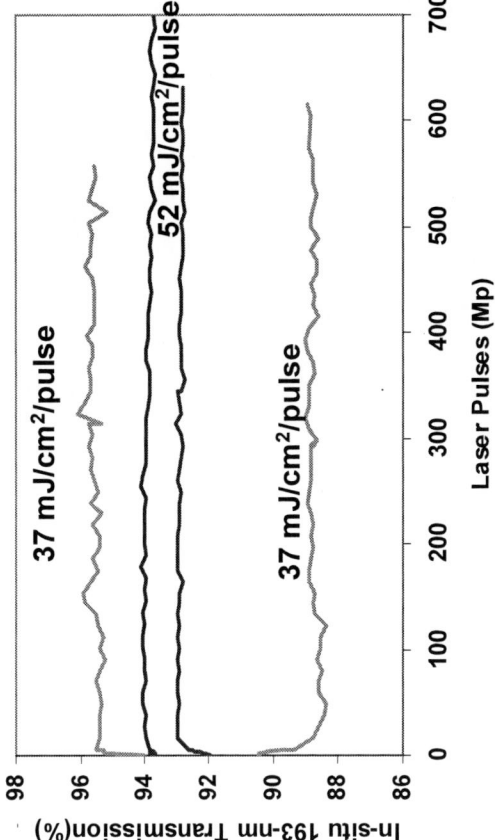

UV-Vis
Typical Laser-Induced Recovery

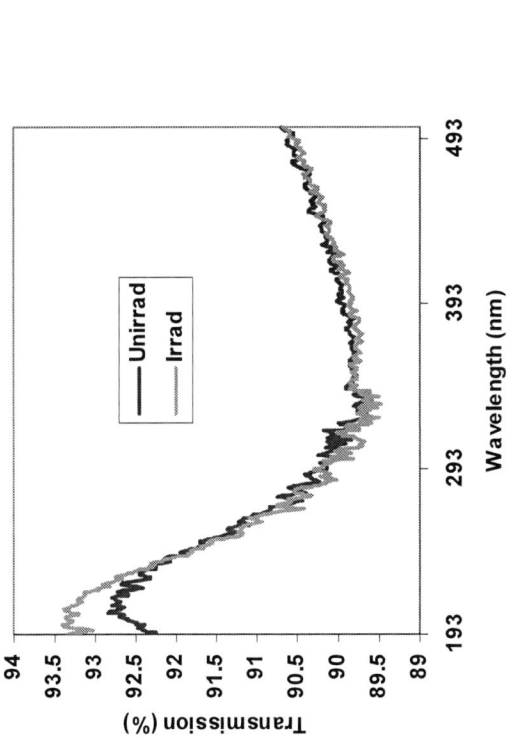

- **Various coating designs are being tested**
  - Both single layers and multiple-layer stacks
- **Subtle transmission changes are observed**
  - No catastrophic failure detected

MIT Lincoln Laboratory

HiFluence 10/29/2009

Thin Film Coatings
Hydrocarbon Removal Monitored by LIF

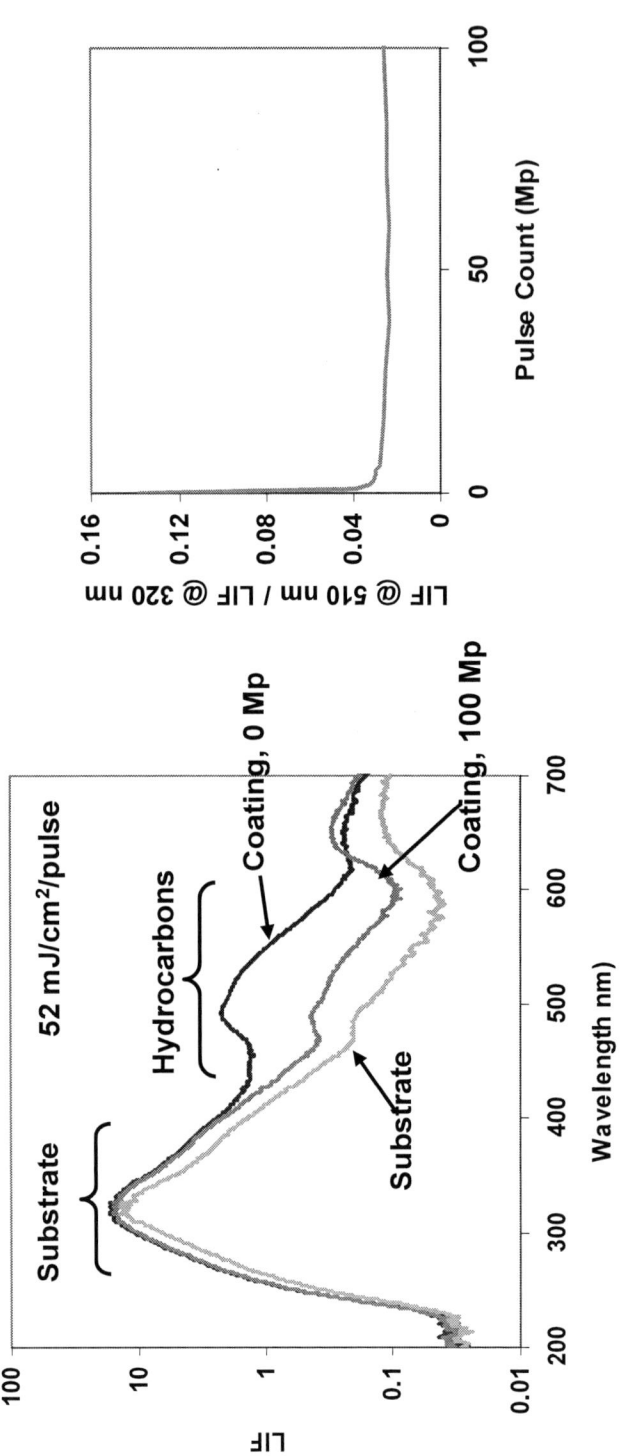

- **Evolution of fluorescence peak at ≈ 500 nm tracks removal of hydrocarbons incorporated into the thin film coating**

  – See, e.g., Heber et al., Appl. Phys. A 76 (2003) 123.

MIT Lincoln Laboratory

HiFluence 10/29/2009

Thin Film Coatings
Post Irradiation Ellipsometry Mapping

- Interpret changes of the coating top layer due to irradiation

300 Mp
30 mJ/cm²/pulse

Spectroscopic Ψ, Δ

350 nm
Δ in degrees
14.0
13.0
12.0
11.0
10.0
9.0

400 nm
Δ in degrees
5.0
4.0
3.0
2.0
1.0

450 nm
Δ in degrees
0.0
-1.0
-2.0

500 nm
Δ in degrees
-1.0
-2.0
-3.0
-4.0

Data Reduction

$$\rho = \tan \Psi e^{i\Delta}$$

Optical Constants

$$n_{top\_layer} = A + \frac{B}{\lambda^2}$$

Index increase of ≈0.01 in irradiated area

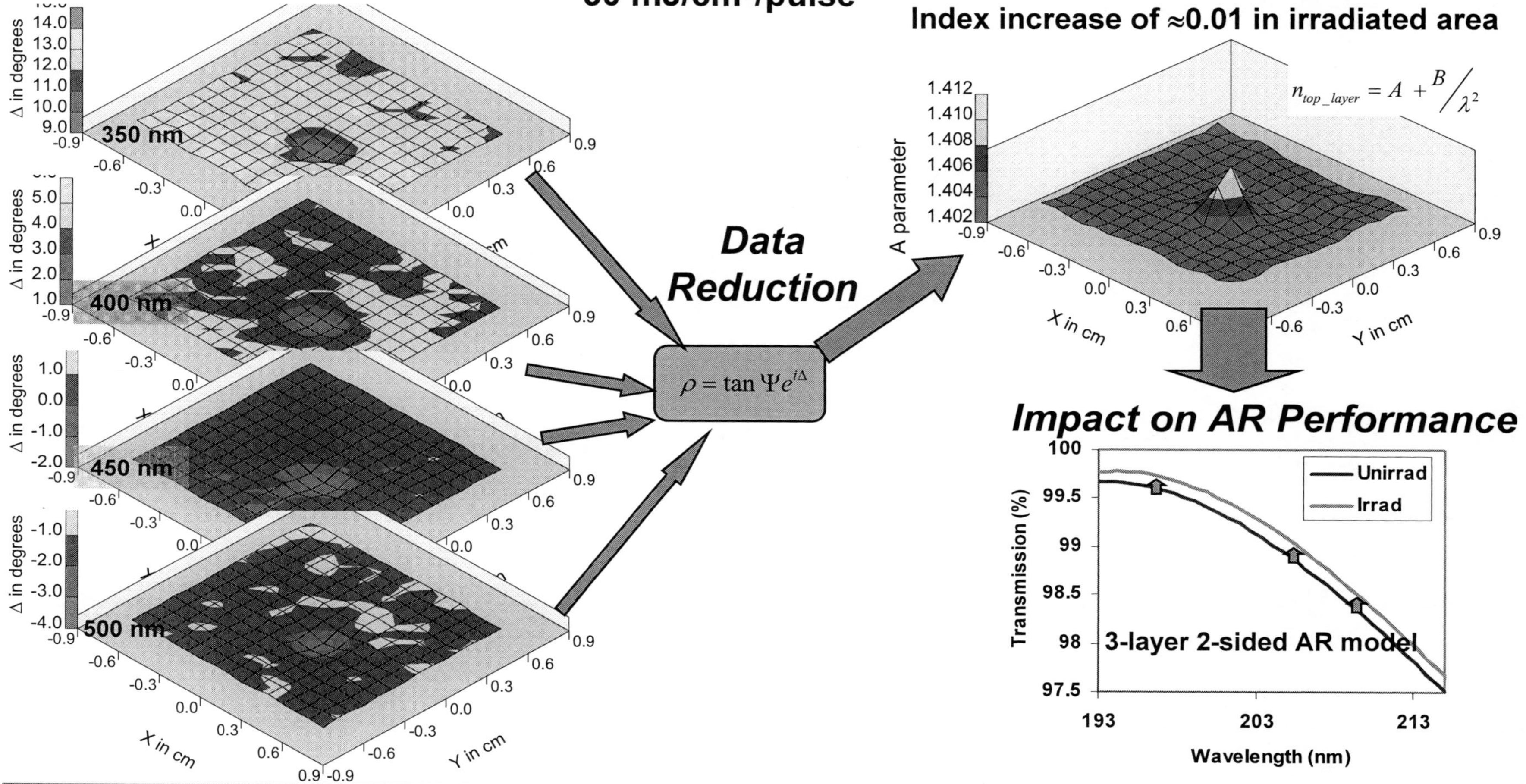

A parameter
1.412
1.410
1.408
1.406
1.404
1.402

Impact on AR Performance

Transmission (%)

Unirrad
Irrad

3-layer 2-sided AR model

Wavelength (nm)

MIT Lincoln Laboratory

HiFluence 10/29/2009

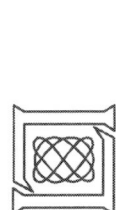

Outlook – Bulk Materials

- **Bulk SiO_2 shows degradation under moderately high-fluence 193-nm irradiation**
  - 38 mJ/cm²/pulse and < 10 Mp
  - 193-nm transmission degradation accompanied by color center formation
  - Rapid transient absorption in some grades
    - Further study needed for better understanding
  - Laser-induced birefringence observed
    - Material-grade dependent

- **Bulk CaF_2 shows no degradation under intense 193-nm irradiation**
  - 52 mJ/cm²/pulse up to 650 Mp
  - No laser-induced birefringence observed

MIT Lincoln Laboratory

HiFluence 10/29/2009

Outlook – Thin Film Coatings

- Thin-film coated CaF_2 substrates only show subtle changes under long term irradiation

  - 37 to 52 mJ/cm²/pulse, up to 700 Mp

  - No catastrophic failure is observed

  - Removal of incorporated hydrocarbons can be tracked by laser-induced fluorescence

  - Top layer densification occurs for some films

    This densification may be beneficial to an antireflective stack transmission performance

MIT Lincoln Laboratory

HiFluence 10/29/2009

Future Work

- **Complete testing of samples in-house**
  - One more supplier of bulk SiO_2
  - One more supplier of bulk CaF_2

- **Further study of the "transient damage" behavior of SiO_2**
  - Laser-based transmission as a function of incident fluence

- **Further data analysis and reduction**
  - Understand fluorescence behavior of bulk SiO_2 samples
  - Understand laser-induced ellipsometric coating changes
    - Closed loop with individual suppliers

MIT Lincoln Laboratory

HiFluence 10/29/2009

The Lincoln Laboratory portion of this work was sponsored by a Cooperative Research and Development Agreement between Lincoln Laboratory and SEMATECH, Inc. and the Research Foundation of State University of New York. Opinions, interpretations, conclusions, and recommendations are those of the authors and are not necessarily endorsed by the United States Government.

MIT Lincoln Laboratory

HiFluence 10/29/2009

AUTHOR INDEX

Altamirano-Sanchez, E.................331
Armstrong, D..............................684
Bae, W-J...................................782
Bae, Y......................................526
Baerts, C...................................331
Barada, A..................................703
Barnes, L..................................234
Barnola, S..................................526
Baron, S....................................351
Bates, C....................................294
Baylav, B...................................894
Bekaert, J............................203, 351
Benck, E....................................684
Berro, A....................................875
Blakey, I....................................758
Bouma, A...................................351
Braggin, J..................................855
Bristol, R..................................294
Brus, S......................................331
Buch, X.....................................855
Burnett, J...................................684
Cardolaccia, T............................526
Carpaij, R..................................351
Carriere, J..................................388
Chang, S....................................665
Chen, J......................................665
Chen, L......................................758
Cheng, S....................................618
Cho, Y.................................294, 875
Colburn, M................................112
Cork, C......................................234
Costner, E..................................294
Crowell, R..................................310
Dai, H.......................................801
De Backer, J...............................331
De Boer, G..................................665
De Jong, F..................................573
Delvaux, C..................................331
Demand, M.................................331
Ding, R......................................801
Doytcheva, M..............................618
Dusa, M.....................................351
Endo, T......................................829
Engelen, A..................................573
Ercken, M...................................331
Fang, T......................................665
Fenger, G.............................89, 274
Fonseca, C..................................159
Fujimoto, J..................................596
Fujitani, N...................................829
Gao, W......................................234
Gaugiran, S.................................526
Geurtsen, G.................................909
Giannelis, E.................................782
Goh, Y-K....................................758
Golotsvan, A................................112

Graupner, P.................................351
Gronheid, R.................................159
Gu, X..................................294, 875
Guerin, I....................................526
Hamatani, M................................547
Hendel, R...................................703
Hennerkes, C...............................351
Himel, M....................................388
Hirano, K...................................547
Hiroi, Y.....................................829
Ho, B-C..............................817, 829
Hoffnagle, J.................................684
Holmes, S...................................112
Horiguchi, N................................331
Hoshiko, K..................................855
Hsu, S.......................................351
Iriuchijima, Y...............................547
Ishida, T....................................829
Ishikawa, J..................................547
Izikson, P...................................112
Jacob, J.....................................684
Jockusch, S.................................875
Kamimura, S................................842
Kampherbeek, B-J..........................665
Kathman, A.................................388
Kato, M.....................................829
Kawasuji, Y.................................596
Kim, Jae Hyun..............................841
Kim, Jung-Hoon.............................841
Kimura, S...................................829
Koay, C-S...................................112
Komirenko, S...............................274
Krecinic, F...................................665
Krysak, M...................................782
Kumazaki, T.................................596
Kurosu, A...................................596
Kuwahara, Y................................159
Lacour, P....................................274
Laenens, B..................................351
Lafferty, N..................................782
Laidler, D...................................618
Laske, F.....................................112
Lawrie, K...................................758
Lazzarino, F.................................351
Lee, S.......................................875
Leray, P.....................................618
Li, X..234
Li, Y...875
Liberman, V.................................909
Lin, B.......................................665
Lin, M.......................................801
Lin, S..665
Litt, L.......................................639
Liu, Y.......................................526
Locorotondo, S..............................331
Lucas, K.....................................234

AUTHOR INDEX

Luk-Pat, G. 234
Mack, C. 64
Maenhoudt, M. 203, 500
Markle, D. 703
Maruyama, D. 829
Matsumoto, S. 596
Matsunaga, T. 596
Medeiros, D. 468
Menguelti, K. 526
Miao, L. 801
Miloslavsky, A. 234
Mizoguchi, H. 596
Mulder, M. 351, 573
Mulkens, J. 573
Murdoch, G. 500
Nafus, K. 159
Nagai, T. 294, 875
Nagaswami, V. 112
Nakamura, T. 139
Nakano, M. 310
Ning, K. 351
Ober, C. 782
O'Connor, N. 875
Ogata, T. 294, 875
Ohmori, K. 139
Onishi, R. 829
Owa, S. 547
Palmacci, S. 909
Pargon, E. 526
Petersen, J. 703
Petrillo, K. 112
Pieczulewski, C. 181
Pikon, A. 526
Piscani, E. 758
Rao, Z. 703
Ratin, C. 526
Reybrouck, M. 203
Robinson, J. 112
Rosslee, C. 181
Rothschild, M. 909
Saito, H. 139
Sakaida, Y. 817
Sakamoto, R. 817, 829
Scheer, S. 159, 310
Scheffers, P. 665
Schipper, B. 665
Schollaert, W. 855
Scholz, C. 894
Schreel, K. 351
Schwartz, E. 782
Seamons, M. 801
Shibazaki, Y. 547
Sinha, J. 112
Sivakumar, S. 426
Smith, A. 684
Smith, B. 782, 894

Smith, T. 894
Socha, R. 351
Somervell, M. 159
Stack, J. 388
Sundaresan, A. 294, 875
Suzuki, T. 596
Takasu, R. 139
Takeshita, M. 139
Tanaka, H. 596
Tanaka, S. 596
Tang, B. 801
Tarutani, S. 159, 203, 732, 842
Tenner, M. 618
Tien, D. 112
Trefonas, P. 526
Trikeriotis, M. 782
Tritchkov, A. 274
Trivkovic, D. 351
Truffert, V. 203
Tsai, M.-C. 351
Tsushima, H. 596
Turo, N. 875
Turro, N. 294
Umeda, H. 596
Van Adrichem, P. 351
Van Den Berg, C. 665
Van Der Heijden, E. 351
Van Haren, R. 618
Van Look, L. 203, 351
Vandenberghe, G. 203, 351
Vandeweyer, T. 331
Vangoidsenhoven, D. 500
Veeraraghavan, S. 112
Veloso, A. 331
Verhaegen, S. 89, 331, 351, 500
Viswanathan, A. 310
Vleeming, B. 573
Wang, W. 665
Watanabe, H. 596
Whittaker, A. 758
Wiaux, V. 89, 203, 234, 274, 500
Wieland, M. 665
Willson, G. 1, 294, 875
Wong, P. 89, 500
Xie, P. 782, 894
Yaegashi, H. 856
Yaguchi, H. 817
Yin, R. 894
Yokoya, J. 139
Yokoyama, J. 842
Yoshii, Y. 139
Yoshino, M. 596
Zhao, M. 894
Zimmerman, P. 294, 758, 782, 875, 894, 909
Zimmermann, J. 351
Zin, Wang-Cheol 841

SEMATECH
2706 Montopolis Drive
Austin, Texas 78741

ISBN 978-1-61782-107-3